COMPUTATIONAL
SOLID
MECHANICS

COMPUTATIONAL
SOLID
MECHANICS

Variational Formulation and High Order Approximation

MARCO L. BITTENCOURT

Department of Integrated Systems, Faculty of Mechanical Engineering
University of Campinas, Brazil

CRC Press
Taylor & Francis Group
Boca Raton   London   New York

CRC Press is an imprint of the
Taylor & Francis Group, an **informa** business

CRC Press
Taylor & Francis Group
6000 Broken Sound Parkway NW, Suite 300
Boca Raton, FL 33487-2742

© 2015 by Taylor & Francis Group, LLC
CRC Press is an imprint of Taylor & Francis Group, an Informa business

No claim to original U.S. Government works

Printed on acid-free paper
Version Date: 20140724

International Standard Book Number-13: 978-1-4398-6001-4 (Hardback)

Visit the Taylor & Francis Web site at
http://www.taylorandfrancis.com

and the CRC Press Web site at
http://www.crcpress.com

Dedication

I dedicate this book to my wife Maristela, my son Felipe, and my daughter Ana Luísa. I would like to express my gratitude to my parents, Fábio and Terezinha, who have provided me with very good educational opportunities.

Contents

Preface

Nowadays computational simulation is a basic tool in the development of engineering projects. Due to the increase of computing resources, it has become possible to consider together many aspects of the mechanical models.

Simulation software, such as ANSYS, ABAQUS, and NASTRAN, implement a discrete version of continuous mathematical models. As the numerical models become more complex, engineers require a strong and fundamental background to confidently use the software features. Learning the fundamental concepts of mechanics and approximation in a general way should be the starting point for the application of computers in solving real engineering applications. The impossibility of understanding clearly the fundamental assumptions and limitations of the mechanical and numerical models makes it highly likely that engineers will obtain computational solutions which do not represent the actual behavior of mechanical components.

There have been many books presenting the formulation and finite element approximation of solid mechanical models. Relative to the books on formulation of mechanical models, particular cases in nonpedagogical and/or very formal approaches are presented. Books on finite elements do not make clear the boundary between the mechanical models and their approximations. These aspects make the learning process of engineering students difficult. Another aspect is that in the computer era, it is crucial to organize the way that the information is supplied to students, as they have access to many information sources from the Internet. Using standard procedures to formulate and approximate models and at the same time illustrating their application by software are very important aspects. This book intends to address these points.

In terms of formulation of mechanical models, the basic tool considered here is the variational formulation based on the principle of virtual work (PVW). All models are presented following the same sequence of steps, which includes kinematics, strain measure, rigid body deformation, internal loads, external loads, equilibrium, constitutive equations, and structural design. This sequence allows the reader to establish a logical reasoning for the treatment of any mechanical model. In addition, all aspects of a mechanical model are presented in each chapter and not spread out in many chapters, as is common in many books on solid mechanics. Mechanical models for plates and solids models are also considered using the same approach.

In terms of finite element approximation, the book starts with simple applications of low-order approximation to bars and shafts elements. The main concepts are introduced gradually in the others chapters, including high-order approximations. As in the formulation, all approximations are presented following the same sequence of steps which includes the definition of strong form, weak form, global and local approximations, finite elements, and applications.

In terms of software, MATLAB scripts and an object-oriented high-order program are supplied with the book to run examples.

Taking into consideration these three main aspects, readers should learn the limitations and strengths of the considered mechanical models, their approximations, and how they are implemented in computer software.

The book is intended for the use by undergraduate and beginning graduate students in engineering. Most of the chapters include many examples, problems, and software applications. This edition will be limited to models with small deformation and linear material behavior. The book is organized in 10 chapters. Chapter 1 provides a general introduction to variational formulation and an overview of the mechanical models to be presented in the other chapters. Chapter 2 presents a review about the Newton and variational formulations, the principle of virtual work, and the equilibrium of particles and rigid bodies using the PVW. The main idea is to use the concepts on equilibrium that readers should already have to introduce basic notions on kinematics, virtual work, and the PVW.

Chapters 3 to 10 present mechanical models, approximations, and applications to bars, shafts, beams, beams with shear, general two- and three-dimensional beams, solids, plane models, and general torsion, and plates. In particular, Chapter 8 presents the most general case of solids using two approaches. The first one follows the basic idea of the other chapters. In the second approach, the concept of second-order tensor is introduced using a Taylor expansion, and the solid model is reformulated using again the same previously formulation steps. In this case, small and large deformations are considered. After the presentation of elastic solids, the kinematical hypotheses of the previously considered problems are introduced in this model. It is then possible to observe where the simplifications are introduced in the solids to formulate the previous cases. Chapter 9 presents a more formal introduction to variational formulation based on the general steps applied to the other chapters.

I believe that the main features of the book are: the systematic and pedagogical approaches to formulate and approximate solid mechanical models, starting from simple cases and going to more complex models; a clear separation of formulation and finite element approximation; and the user-friendly MATLAB software.

I have used this material at the University of Campinas in Brazil for about 15 years. We have two one-semester courses on solid mechanics and another one-semester course on numerical methods in engineering for undergraduate students. I have also used the material included in this book for a one-semester course for beginner graduate students. I used also part of this material for a graduate course at the Division of Applied Mathematics at Brown University in 2010.

I would like to express my invaluable gratitude to my PhD advisor, Professor Raúl A. Feijóo of the National Laboratory of Scientific Computation at Petropólis in Brazil, who introduced me the fundamentals of mechanics and variational formulation. I would like also to thank Professor George Em Karniadakis of the Division of Applied Mathematics at Brown University for his careful review of the book proposal and suggestions. I would like to thank my students for their collaboration during the time I have written this material. In particular, I would like to thank Cláudio A. C. Silva, Rodrigo A. Augusto, Jorge L. Suzuki, and Allan P. C. Dias for their invaluable help in the preparation of the many versions that led to this book. Finally, I would like to thank CRC Press for the publication of the book.

The MATLAB scripts used in this book are available online at `www.facebook.com/ComputationalSolidMechanics`.

Marco L. Bittencourt
Campinas, SP, Brazil
2014

List of Figures

List of Tables

1 INTRODUCTION

Several famous scientists, including Cauchy, Lagrange, Leibnitz, and Newton, studied motion and deformation phenomena. Their main objective was to describe the behavior of bodies subjected to general loading, determining the internal forces and stress and strain states.

Considering geometric simplifications on the general case of solid bodies, the discipline of strength of materials studies one-dimensional linear elastic models. Generally, it studies the internal forces and deformation of such one-dimensional structural elements as bars, beams, and shafts. To this purpose, it is assumed that any applied load deforms the body only, without any rigid body motion. In this context, many expressions are determined to calculate the stresses and strains in structures which are composed of these basic structural elements. Therefore, a one-dimensional description of general solid mechanics concepts is considered.

Nowadays, engineering problems have multidisciplinary features. This aspect can be partially justified by the availability of effective computational resources to simulate models. In this way, engineers should know the fundamental concepts of mechanics to deal with different real engineering problems. From a teaching point of view, this fact requires an approach that emphasizes the basic and fundamental concepts of mechanics. This approach should provide a broad view for formulation and approximation of mechanical models, being able, for instance, to handle problems of solids and fluids using the same conceptual backgorund.

Such fact is the starting point for the application of computers to solve real engineering problems. The misunderstanding of the mechanical model formulation makes impossible the clear comprehension of its fundamental hypothesis and limitations. Because of that, the engineer will very likely obtain computational solutions that do not represent the real behavior of the problem under consideration. Therefore, knowledge of the mechanical model is the fundamental starting point to the confident use of computational simulation techniques.

This broader approach of studying the formulation of mechanical problems should use the concepts of continuum mechanics, which is based on the notion of continuum media, and consequently on the concept of the infinitesimal. This is the main reason why engineering curricula include courses of differential calculus.

However, the usual approach of teaching basic engineering courses, such as Strength of Materials, Dynamics, and Fluid Dynamics, does not usually link appropriately differential calculus and continuum mechanics to the contents of these courses. Particular concepts are in general presented which should be obtained from the application of the fundamentals principles of continuum mechanics to simple models. This creates a gap in the engineer's knowledge when facing complex real problems: engineers will not be able to identify where the hypotheses that resulted in the simplified theories of the traditional teaching approaches should be changed to deal with more complex models. The traditional teaching approach may be totally disconnected from the contemporary engineering multidisplinary problems and the use of computational simulation.

Another possible solution could be a standard course on continuum mechanics. In general, it is required to first teach tensorial analysis, deformation, stress, and the fundamentals principles before applying them to mechanical models. This path may seem very formal to undergraduate students.

This book aims to present another path to the teaching of solid mechanics from the conceptual bases. As the classical strength of materials theory, we aim to introduce the basics of mechanics from one-dimensional models, with emphasis on the simplification aspects required in the formulation of these models. Later, other mechanical models will be introduced, such as solids, plane stress/strain, and plates. From these models, readers should be capable of dealing with different types of problems not considered here.

Another fundamental aspect is to present the model formulations following the same method-

1

ology. In this book, we consider analytical mechanics and their variational principles, mainly the principle of virtual work (PVW). The same sequence of steps will be employed to formulate all the models considered here. In the meantime, model approximations by the finite element method (FEM) will be presented. The approximation concepts will be gradually introduced in the chapters. Particularly, high-order approximations are considered that avoid common problems in structural analysis such as the numerical solution locking. MATLAB scripts are extensively used to calculate analytic and approximated solutions of the considered mechanical models. In this way, each chapter will present the mechanical model formulation, the associated approximation, and the use of programs to obtain the analytic and approximated solutions for the given examples.

This introduction aims to motivate the approach here considered by presenting mechanical models, their main hypothesis, and some applications. The examples are considered in ascending order of complexity, with the objective of stimulating the applicability of the mechanical models. The basic theoretical tools to the problem formulation and approximation are differential calculus and vector algebra, complemented by other mathematical concepts which will be introduced as they are required. The concepts are gradually presented, exploring the reader's intuition. Before considering the mechanical models, some initial concepts are presented. The models considered here are detailed in the following chapters.

1.1   INITIAL ASPECTS

1.1.1   OBJECTIVES OF CONTINUUM MECHANICS

It is well known that matter is not continuous but is formed by molecules, which are formed by particles. However, several physical phenomena can be analyzed without considering the molecular structure of materials. For that purpose, the continuum mechanics theory is used, which describes these phenomena by neglecting the material behavior at the microscale.

The continuum theory starts from the basic hypothesis of indefinite divisibility of matter. The idea of an infinitesimal volume of material called particle is considered. Thus, there is always matter in the neighborhood of an infinitesimal portion of material. The validity of this statement depends on the case under consideration. However, the applicability of continuum mechanics concepts can be justified for the problems considered in this book.

Basically, continuum mechanics studies the behavior of materials for different loading conditions. This theory is divided in two main branches [30, 36]:

- Common general principles for several media
- Constitutive equations for idealized materials

The general principles are axioms obtained from observations of physical phenomena and include the conservation of mass and energy as well as the principles of conservation of linear and angular momentum. The second part of the theory concerns the constitutive equations employed in the definition of idealized material behavior, like the linear elastic solid and Newtonian fluid.

1.1.2   DEFINITION OF BODIES

The continuum mechanics workspace is the pointwise three-dimensional Euclidean space, denoted by $\mathscr{E}$, and their elements are denominated points.

A body has the physical characteristic of occupying regions of the pointwise Euclidean space $\mathscr{E}$ [30]. A body can occupy different regions of $\mathscr{E}$ along different time instants. Any of these regions may be associated with the boby, but it is convenient to select one of them, denominated as reference configuration and denoted by $\mathscr{B}$, which identifies body points with their positions in $\mathscr{B}$. Based on that, a body $\mathscr{B}$ becomes a regular region of $\mathscr{E}$, with the points $\mathbf{x} \in \mathscr{B}$ denominated material points.

Any limited regular subregion of $\mathscr{B}$ is called a part and indicated by $\mathscr{P}$. The boundaries of $\mathscr{B}$ and $\mathscr{P}$ are denoted, respectively, by $\partial\mathscr{B}$ and $\partial\mathscr{P}$. These concepts are illustrated in Figure 1.1.

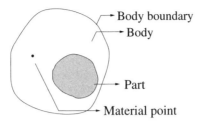

Figure 1.1   Reference configuration, part, and material point.

As the body can occupy different regions along the deformation process, it is necessary to introduce the parameter $t \in [t_0, t_f]$, designating a given configuration $\mathscr{B}_t$ of the body. When the body is under deformation, t represents time. In other problems, t does not necessarily represent time but denotes different configurations which are occupied by the body.

1.1.3   ANALYTIC AND NEWTONIAN FORMULATIONS

This section presents general aspects of Newtonian and analytic formulations for the modeling of mechanical systems according to [39, 37, 22, 28].

One of the main difficulties in the history of mechanics was finding a satisfactory physical and mathematical representation to the concept of the action of a body over the configuration of another body.

From the postulates of motion established by Newton, mechanics developed along two main approaches. The first, called vector mechanics, starts directly from Newton's laws and represents the action between two bodies through forces represented by vectors according to a certain reference system. Therefore, the concept of force is an abstract entity, which is unrelated to the adopted kinematics of the considered problem. This approach is applied in the development of Newtonian physics, where the the main concern is the analysis and synthesis of forces and moments.

Leibniz, Newton's contemporary, introduced a second approach called analytic mechanics, which is based on the study of equilibrium and motion using two basic scalar quantities, the kinetic and potential energies. The main elements adopted to characterize the action between bodies are the movement action and the power (work) required to perform it. From this point, the concept of force comes naturally, not as an *apriori* abstract definition, but as a connecting element between the body movement action and the power required to perform this action. This approach represents the mathematical statement of physical experiments in our daily life:

· When we want to estimate the weight of an object, we slightly raise it, evaluating the weight by the power (or work) required to perform such movement. In other words, we introduce a movement action in the vertical direction that displaces the object from its natural resting state (see Figure 1.2).
· In a similar manner, in order to know the tension in a belt, we displace it from its natural configuration. Thus, we perform a movement action and evaluate the tension according to the consumed power (see Figure 1.3).

The latter methodology differs considerably from vector mechanics. The object weight and belt tension are determined by introducing an appropriate movement action for each case. Based on the power or work required to perform the movement action, it is possible to evaluate the weight of the object or the tension on the belt. This action is called virtual because it is not a natural

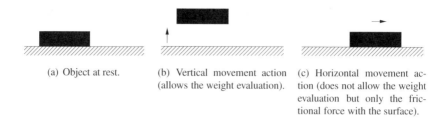

(a) Object at rest.   (b) Vertical movement action (allows the weight evaluation).   (c) Horizontal movement action (does not allow the weight evaluation but only the frictional force with the surface).

Figure 1.2   Weight evaluation of an object.

movement action, because both the object and the belt are at rest. An appropriate virtual action is introduced only to assess the states of the external or internal forces acting on the object and the belt, respectively. As illustrated in Figure 1.2(b), the appropriate movement action to evaluate the weight of the object must be in the vertical direction. A horizontal movement does not determine the weight of the object but the frictional force between the surfaces [see Figure 1.2(c)]. In the belt case, a movement action which is normal to the belt evaluates the tension, as shown in Figure 1.3(b). A tangential action estimates not the tension but the frictional force between the belt and the pulleys, as illustrated in Figure 1.3(c).

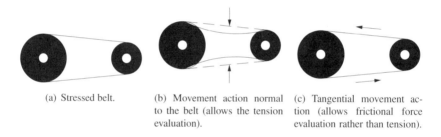

(a) Stressed belt.   (b) Movement action normal to the belt (allows the tension evaluation).   (c) Tangential movement action (allows frictional force evaluation rather than tension).

Figure 1.3   Tension evaluation on the belt.

The fundamental law of motion established by Newton is valid only for a single particle. This law was deduced for the motion of a particle in the Earth's gravitational field and applied to the study of the motion of the planets around the sun. In these two problems, the moving bodies can be idealized as particles, which are points with a certain mass and negligible dimensions. Newton's law provides a differential equation of motion which solves the problem of dynamic equilibrium.

However, in the case of solids or fluids, particles are linked to each other. To apply Newton's law, every particle must be isolated from all others, and the interacting forces of neighboring particles must be determined. Thus, each particle is an independent unit under the law of motion. This analysis is cumbersome in terms of force equilibrium, because the general nature of the interaction forces is unknown. To address this limitation, Newton introduced the principle of action and reaction as the third law of motion. However, not all problems can be solved using this postulate and will require further assumptions. For example, in the case of rigid bodies, the distance between two arbitrary points must remain constant. It is also verified that the Newtonian approach cannot provide a single answer to more complex problems.

Analytic mechanics treats problems in a different way. The particle is no longer isolated, but is part of a whole system. A mechanical system is an assembly of particles that interact with each other. In this way, a single particle is not important, but is a part the whole system. Unlike the vector approach, in which each particle should be individually considered and the associated acting force determined independently from the other particles, the analytic approach has a single function describing the forces acting on the particles in the system.

Another fundamental difference relates to the treatment of auxiliary conditions, such as unknown kinematic relations of the system under consideration. For example, the particles of a solid can move as if the body were rigid, that is, the distance between any two points should remain fixed. In the case of Newtonian mechanics, forces are required to maintain this condition. In the analytic approach, there is no need to know these forces, and only the established kinematic condition should be taken into account. Similarly, in the case of fluids, it is not necessary to know what types of forces exist among the particles; it only takes into account the empirical fact that a fluid resists any attempt to change significantly its volume, while it has less resistance to actions that change its shape. Thus, the nature of forces between particles is neglected. The kinematic conditions that preserve the volume of any part of the fluid during a movement action must be established.

The main difference between the two approaches lies in a single principle on which analytic mechanics is based. For a complex system, the number of equations of motion can be quite large. The variational principles of analytic mechanics allow a single basis, from which all equations are derived. Given the fundamental concept of action, the stationarity principle of this action results in the set of differential equations of the system. In addition, this formulation is invariant with respect to any coordinate transformation.

The four main differences between vector and analytical mechanics can be summarized as [37]:

1. Vector mechanics isolates the particle and treats it individually; analytical mechanics considers the entire system.
2. Vector mechanics considers the resultant of forces for each particle; analytical mechanics considers a single work function or potential energy relating all forces of interest.
3. Vector mechanics considers a set of forces that are necessary to maintain any established relation between the system coordinates; in analytical mechanics, any kinematic condition represents another known parameter about the system.
4. In the analytical approach, every set of equations can be developed from unique principle, which minimizes a certain action. This principle is independent of the system of coordinates, allowing the choice of the most natural one, according to the considered problem.

1.1.4 FORMULATION METHODOLOGY

As mentioned previously, the variational principles of analytical mechanics are employed here for the formulation of the considered mechanical models. The principle of virtual power (PVP) is used to determine the dynamic equilibrium of bodies. The principle of virtual work (PVW) is a particular case of the PVP for static equilibrium. A detailed presentation of variational principles can be found in [22] and in the references cited in this work.

The goal here is to use a standard procedure employing the PVW for the formulation of mechanical models. In this context, the initial step is to determine the kinematics of the model and the corresponding strain measure. By employing the concept of internal strain work, the compatible internal loads with the defined kinematics are obtained. As will be seen, contrary to what is commonly assumed, external loads are not always represented by vectors. For models with nonzero strain measures, loads are represented by continuous functions over the domain of the body. As continuum mechanics assumes that matter is continuous, it is natural that all involved quantities, including kinematics and internal and external loads, are represented by continuous functions. The PVW can also be applied to obtain the solution to specific examples.

In the most general case, the standard procedure for the formulation of the mechanical models considered here is given by the following steps:

1. Definition of the geometrical hypothesis of the model under consideration
2. Definition of the model kinematics
3. Determination of the strain measure compatible with the defined kinematics

4. Determination of rigid actions
5. Determination of the kinematically compatible internal loads by the application of the concept of internal strain work
6. Determination of the external loads compatible with the internal loads
7. Application of the PVW to obtain the equilibirum boundary value problem (BVP) and the characterization of the rigid body equilibrium conditions
8. Definition of the material behavior and application of the constitutive equation to the equilibrium BVP to obtain the PVW in terms of kinematics, as well as the stress distributions
9. Design of the body to determine the minimal characteristic dimensions to avoid permanent strains

Steps 1 through 6 determine the main characteristics of the model, such as kinematics, strain measure, rigid actions, and internal and external loads. By applying the PVW, we obtain the equilibrium BVP, defined by one or more differential equations and boundary conditions. In step 7, we introduce the material behavior to obtain the BVP in terms of kinematics and stress distributions. For cases in which the analytical solution of BVPs is possible, another standard solution procedure is employed. The solution steps will be illustrated by examples.

In the following sections, we consider an overview of the main features of some models which are considered in the following chapters of this book.

1.2 BARS

Figure 1.4 illustrates a bar of length L and circular cross-section of diameter d and area A. The used Cartesian coordinate system is also shown in Figure 1.4, with the origin at the geometrical center (GC) of the cross-section of the left end. Axis x is located along the length of the bar and is denominated the longitudinal or axial axis.

The main geometric feature of the bar model is the fact that its length is much larger than the cross-section dimensions. Based on that, the bar is modeled as a one-dimensional mathematical model. Its behavior is analyzed along the parallel direction to the longitudinal dimension, i.e., the x axis of the reference system shown in Figure 1.4.

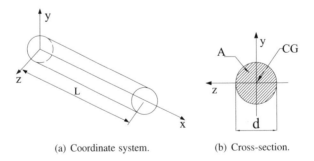

(a) Coordinate system. (b) Cross-section.

Figure 1.4 Bar of length L and the Cartesian coordinate system.

The kinematics of the bar model consists of axial displacement actions. The cross-sections remain perpendicular to the longitudinal axis of the bar, as illustrated in Figure 1.5. In this way, a bar stretches or shortens only and the final length is L'. The axial displacement is given by the continuous function $u_x(x)$. The rigid body actions are translations in the direction of the x axis. The internal and external loads compatible with the defined kinematics are axial forces (see Chapter 3).

The equilibrium differential equation of the bar is

$$\frac{dN_x(x)}{dx} = -q_x(x), \tag{1.1}$$

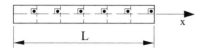

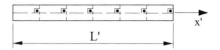

(a) Initial configuration (cross-sections are perpendicular to the bar axis).

(b) Deformed configuration (the cross-sections remain perpendicular to the bar axis).

Figure 1.5 Kinematics of the bar model.

with $N_x(x)$ the function that describes the axial internal force, called normal force, for each cross-section of the bar. The function $q_x(x)$ represents the external axial force distributed along the bar length. It should be noted that the internal and external forces are represented by continuous functions, and not vectors.

The indefinite integral of the previous expression results in

$$N_x(x) = -\int q_x(x)dx + C_1, \tag{1.2}$$

where C_1 is an arbitrary integration constant, determined from the boundary conditions. Therefore, after the integration of the differential equation (1.1), we obtain the function $N_x(x)$ which describes the behavior of the normal force along the bar.

The normal strain component $\varepsilon_{xx}(x)$ in each cross-section of the bar is given by the derivative of the axial displacement $u_x(x)$. Therefore,

$$\varepsilon_{xx}(x) = \frac{du_x(x)}{dx}. \tag{1.3}$$

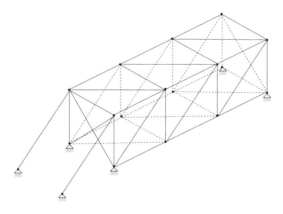

Figure 1.6 Truss structure.

Employing the material behavior based on Hooke's law, we obtain the differential equation of the bar in terms of the axial displacement, that is,

$$EA\frac{d^2u_x(x)}{dx^2} = -q_x(x), \tag{1.4}$$

and E is a material property denominated longitudinal elastic or Young's modulus.

Truss structures are the main example of the application of bar elements. They consist of bars which are connected by perfect joints, as illustrated in Figure 1.6. The transmission of external forces to the structure is done exclusively by the strength of the bars with respect to traction and compression.

1.3   SHAFTS

Shafts are elements of circular cross-section which, analogously to bars, have a predominant longitudinal dimension. The kinematics of the model consists only of rotation of the cross-sections about the x axis. Axial rotation is characterized by the torsion angle, given by the continuous functions $\theta_x(x)$, as illustrated in Figure 1.7.

The respective equilibrium differential equation is

$$\frac{dM_x(x)}{dx} = -m_x(x), \tag{1.5}$$

where $M_x(x)$ is the internal torsional moment and $m_x(x)$ the external applied torque, both along the shaft length.

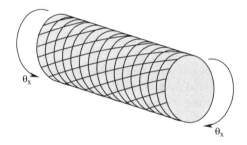

Figure 1.7   Kinematics of circular torsion.

By applying Hooke's law to the circular torsion model, we obtain the following differential equation in terms of the torsion angle $\theta_x(x)$:

$$GI_p\frac{d^2\theta_x(x)}{dx^2} = -m_x(x), \tag{1.6}$$

with I_p the polar moment of area of the cross-section and G a material property called transversal elastic modulus.

While the bar kinematics is expressed in terms of change of its length, there are axial rotations of the cross-sections for the circular torsion (see Chapter 4).

The main applications of the circular torsion elements are the shafts used in most mechanical systems.

1.4   BEAMS

Consider the rack illustrated in Figure 1.8 with a deflection due to the weight of the books. Taking an arbitrary section x, it is observed in Figure 1.8 that the deflection is characterized by a vertical displacement u_y in the transversal direction y and a rotation θ_z about the z axis called bending angle. Elements subjected to bending are called beams here.

The model, which represents beams in pure bending, assumes that the possible kinematic actions are such that the cross-sections remain flat, undeformed, and orthogonal to the tangent of the beam axis (see Chapter 5).

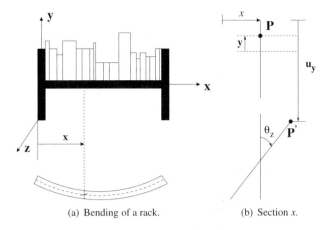

(a) Bending of a rack.

(b) Section x.

Figure 1.8 Book rack under pure bending.

The internal loads in the beam cross-sections for pure bending are given by the continuous functions $V_y(x)$ and $M_z(x)$, which define, respectively, the shear force in the y direction and the bending moment in the z direction.

As in the cases of bars and shafts, it is possible to obtain the equilibrium of forces and moments in beams, which results in the following differential equations in terms of the shear force and bending moment, respectively,

$$\frac{dV_y(x)}{dx} = q_y(x),$$ (1.7)

$$\frac{dM_z(x)}{dx} = V_y(x),$$ (1.8)

where $q_y(x)$ is the external transversal distributed load over the beam. Substituting (1.7) in (1.8) yields the following second-order differential equation in terms of the bending moment:

$$\frac{d^2M_z(x)}{dx^2} = q_y(x),$$ (1.9)

Again, by employing the Hookean material constitutive equation, the BVP in terms of the kinematic is obtained, i.e.,

$$EI_z \frac{d^4u_y(x)}{dx^4} = q_y(x),$$ (1.10)

and $u_y(x)$ is the deflection function along the beam subjected to the transverse distributed load function $q_y(x)$; I_z is the second moment of area of the cross-section relative to the z axis of the adopted reference system.

The pure bending beam model does not include the shear effect in the transversal and longitudinal sections of the beam. This effect will be considered in Chapter 6.

The bending effect is of fundamental importance and occurs in the majority of engineering components and structures.

The effects of traction, torsion, and bending occur simultaneously in shafts of machinery and equipment. They are studied separately for pedagogical aspects only. Subsequently, these effects are combined to study other cases as illustrated in Figure 1.9 (see Chapter 7).

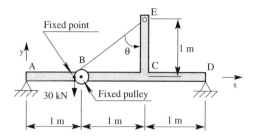

Figure 1.9 Structure loaded with traction and bending loads.

1.5 TWO-DIMENSIONAL MODELS

There are several three-dimensional components which can be analyzed using two-dimensional models. Figure 1.10 illustrates a hook modeled by a plane stress model (see Chapter 9).

The equilibrium differential equations to the plane stress model with a Hookean material are given by

$$\mu \left(\frac{\partial^2 u_x}{\partial x^2} + \frac{\partial^2 u_x}{\partial y^2} \right) + (\mu + \lambda) \left(\frac{\partial^2 u_x}{\partial x^2} + \frac{\partial^2 u_y}{\partial x \partial y} \right) + b_x = 0 \tag{1.11}$$

$$\mu \left(\frac{\partial^2 u_y}{\partial x^2} + \frac{\partial^2 u_y}{\partial y^2} \right) + (\mu + \lambda) \left(\frac{\partial^2 u_x}{\partial x \partial y} + \frac{\partial^2 u_y}{\partial y^2} \right) + b_y = 0, \tag{1.12}$$

with $u_x(x, y)$ and $u_y(x, y)$ the displacement functions in the x and y directions, respectively; $b_x(x, y)$ and $b_y(x, y)$ are the body force components; μ and λ are the Lam coefficients of the material.

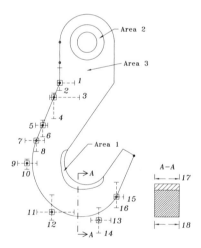

Figure 1.10 Example of a plane stress problem [47].

1.6 PLATES

Plates are structural components subjected to bending loads with the thickness t much smaller than the other dimensions. This geometrical feature allows the use of two-dimensional models for the medium surface, reducing the original three-dimensional problem to a two-dimensional one.

The classical models for plates are the formulations by Kirchhoff and Reissner-Mindlin (see Chapter 10).

The Kirchhoff's equilibrium equation of a rectangular plate with small deflections and Hooke's material is

$$\frac{\partial^4 u_z}{\partial x^4} + 2\frac{\partial^4 u_z}{\partial x^2 \partial y^2} + \frac{\partial^4 u_z}{\partial y^4} = \frac{q_z}{D}, \tag{1.13}$$

with $u_z(x,y)$ the normal deflection function on each point of coordinates (x,y) of the medium surface of the plate; $q_z(x,y)$ is the normal distributed load on the medium surface; and D is given by the following equation:

$$D = \frac{Et^3}{12(1-v^2)}, \tag{1.14}$$

where v is the Poisson's ratio of the material.

1.7 LINEAR ELASTIC SOLIDS

Bodies will deform when subjected to external loads. If the material behavior is such that the strains become zero when the loads are removed, this material is called elastic. In the same way, when the material properties are independent of the direction in which they are considered, the material is denominated isotropic. In this section, the equilibrium equations for three-dimensional isotropic linear elastic problems are presented [30, 36].

In the case of solids, there is no simplifying geometric hypothesis. Consequently, the possible kinematic actions for a three-dimensional elastic solid is given by the following smooth and continuous vectorial field **u** (see Chapter 8) denoted by:

$$\mathbf{u}(x,y,z) = \left\{ \begin{array}{c} u_x(x,y,z) \\ u_y(x,y,z) \\ u_z(x,y,z) \end{array} \right\}, \tag{1.15}$$

where u_x, u_y, and u_z are the scalar functions representing the displacement components in the x, y, and z directions, respectively.

The general equations which describe the equilibrium for small deformation of a three-dimensional elastic solid and the linear elastic isotropic material are given by [30, 36]

$$\begin{aligned} \operatorname{div} \mathbf{T} + \mathbf{b} &= \mathbf{0}, \\ \mathbf{T} &= 2\mu\mathbf{E} + \lambda(\operatorname{tr}\mathbf{E})\mathbf{I}, \\ \mathbf{E} &= \tfrac{1}{2}(\nabla\mathbf{u} + \nabla\mathbf{u}^T), \end{aligned} \tag{1.16}$$

with **T** the tensorial field of Cauchy stresses; **b** the vectorial field of body forces; μ and λ the Lamé material coefficients; and **E** the small strain tensorial field. The first equation in (1.16) formally describes the equilibrium for any continuous medium. The second equation, called the constitutive equation, characterizes the particular behavior of a solid for the Hookean material. The third equation is the definition of the infinitesimal strain tensorial field in terms of the displacement vectorial field.

The analytical solution of the system of equations (1.16) can be obtained in particular cases only. In most of the cases, numerical techniques are employed, such as the finite element method (FEM). Figure 1.11 illustrates a finite element mesh for a three-dimensional pump.

Although only solid mechanics models are covered in this book, it is interesting to explore the equilibrium equation of a continuous medium given in (1.16) for the case of a Newtonian fluid. In this way, the reader may observe that while the equilibrium equation is valid for any continuous three-dimensional medium, the response of a particular material depends on the considered constitutive equation.

The main characteristic of a fluid is to have continuous deformation when subjected to shear stresses. It is usual to define a fluid as a class of idealized materials that do not have any shear stress

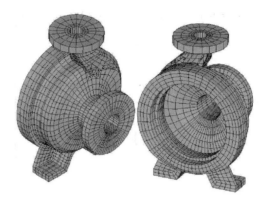

Figure 1.11   Finite element mesh for a three-dimensional elastic body.

resistance. When the density remains constant, regardless of the state of stress, the fluid is called incompressible. The basic assumptions of Newtonian fluids (linear or viscous fluids) are isotropy at any configuration and stress response that depends linearly on the deformation rates only.

The equations describing the behavior of these fluids are similar, in terms of mathematical structure, to those that describe the behavior of linear elastic solids. The equilibrium of a fluid is given by [30, 36]

$$\begin{aligned}
\mathrm{div}\mathbf{T} + \mathbf{b} &= \rho\mathbf{a}, \\
\mathbf{T} &= -p\mathbf{I} + 2\mu\mathbf{D}, \\
\mathbf{D} &= \tfrac{1}{2}(\nabla\mathbf{v} + \nabla\mathbf{v}^T),
\end{aligned} \tag{1.17}$$

with $\mathbf{T}$ and $\mathbf{b}$ the Cauchy stress tensorial field and the body force vectorial field, respectively; $\mathbf{a}$ is the acceleration vectorial field; ρ is the fluid density; p is the hydrostatic pressure scalar field; $\mathbf{D}$ is the deformation rate tensorial field; μ is the fluid viscosity; and $\mathbf{v}$ is the velocity vectorial field.

Figure 1.12 illustrates two types of pipe flow which were solved by the FEM. The first one considers laminar flow [Figure 1.12(a)] and the second a turbulent one [Figure 1.12(b)]. We observe that the first case can be dealt entirely by the equations described in (1.17). However, the second flow needs a model which includes turbulent effects.

(a) Laminar flow.   (b) Turbulent flow.

Figure 1.12   Pipe flow.

We can observe that the mechanical models become more complex as fewer simplifying hypotheses are considered. In the context of multidisciplinary engineering problems, it is important to know the limitations and the capabilities of these models for the construction of robust and reliable approximated solutions.

Since it is pratically impossible to obtain analytical solutions for real engineering cases, approximated solutions using computer simulation programs are the reality for the design of complex systems. To address these problems effectively, engineers should have a strong conceptual basis in the fundamentals of mechanics. From there, new concepts may be added to this conceptual basis in an organized manner, looking to link these new concepts to the previously acquired ones. This is the adopted methodology of this book.

Initially, we consider equilibrium of particles and rigid bodies to introduce some concepts of the PVP and the PVW. Subsequently, deformable bodies are presented, including the study of bars, shafts, beams, plates, and solids.

2 EQUILIBRIUM OF PARTICLES AND RIGID BODIES

2.1 INTRODUCTION

In this chapter, we review the equilibrium conditions of particles and rigid bodies subjected to external loads. The main objective is to use the already known equilibrium concepts to introduce some useful notions for later chapters, such as virtual actions and the principle of virtual work.

It is assumed that the particles and rigid bodies are static and do not move. To satisfy this assumption, a sufficient number of kinematic constraints represented by supports should be used to avoid any rigid motion.

Initially, we present diagrammatic conventions for supports and loadings. Subsequently, the classical Newtonian conditions for equilibrium of particles are considered, which are based on the concepts of forces and force resultant on the particle. The analytical approach is also introduced, using the concept of virtual motion (or displacement) and the principle of virtual power (or work). These concepts are extended to rigid body equilibrium. At the end of the chapter, several examples for equilibrium of particles and rigid bodies are presented.

For notation purposes, all vectorial quantities are represented by lowercase bold letters.

2.2 DIAGRAMMATIC CONVENTIONS

As mentioned before, the kinematic constraints are represented by supports. They are needed in a sufficient number to avoid any rigid body motion. The kinematic constraints must be satisfied for any motion of the considered mechanical system constituted by particles and rigid bodies. The following sections present the most used diagrams to represent loads and supports.

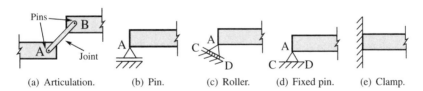

(a) Articulation.　(b) Pin.　(c) Roller.　(d) Fixed pin.　(e) Clamp.

Figure 2.1 Supports.

2.2.1 SUPPORTS

It is important to define symbols to represent the supports, which are responsible for maintaining a mechanical system at rest when subjected to external loads. Basically, supports are identified by the type of the kinematic constraint, or equivalently, by the reaction loads they generate when external loads are applied.

Figures 2.1(a), 2.1(b), and 2.1(c) present an articulated joint, a pin, and a roller, respectively. In these cases, the displacement is zero and a reaction force on a specific line of action exists. In Figure 2.1(a), any motion must satisfy the kinematic constraint of zero displacement along the AB line. From a vectorial mechanics standpoint, when we analyze each body separately, the articulation gives reaction forces in the AB direction only. The pin in Figure 2.1(b) has zero displacement in the vertical direction, leading to a vertical reaction force acting on point A. However, this point is free to move in the horizontal direction. In the roller of Figure 2.1(c), we have zero displacement and a reaction force both perpendicular to the plane *CD*. For all the above-mentioned cases, there is a single reaction or unknown force in the equilibrium equations.

The pin illustrated in Figure 2.1(d) is another type of support. We observe that displacements are zero, both in the horizontal and vertical directions, giving rise to two reaction forces. Figure 2.1(e) illustrates a clamp, for which the displacements in both directions and the rotation about the perpendicular axis to the mechanical system plane are zero. Therefore, this type of support resists forces applied to any direction and also moments in the perpendicular direction. Figure 2.2 summarizes the differences among the discussed supporting conditions, emphasizing the kinematic constraints, as well as the respective reaction loads.

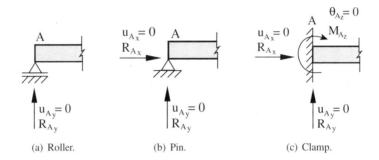

(a) Roller.   (b) Pin.   (c) Clamp.

Figure 2.2 Kinematic constraints and respective reactions for different supporting conditions.

We observe that the kinematic constraints for the supports given in this section are usual for the study of equilibrium. In the case of deformable bodies, the kinematic restriction supplied by the support will depend on the body kinematics, as will be seen in the following chapters.

2.2.2 LOADINGS

The loads applied on a mechanical component can be idealized as concentrated (pointwise), distributed, or volume loads. The concentrated loads arise, for instance, by a pillar, a lever, or a riveted component, as shown in Figure 2.3(a). It should be observed that these elements generate loads at a limited portion of the body and are idealized as concentrated forces, as shown in the free body diagram of Figure 2.3(b).

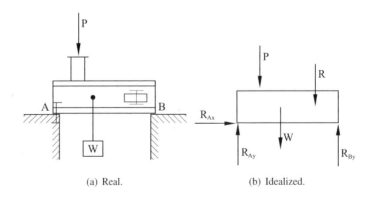

(a) Real.   (b) Idealized.

Figure 2.3 Concentrated loads in a body.

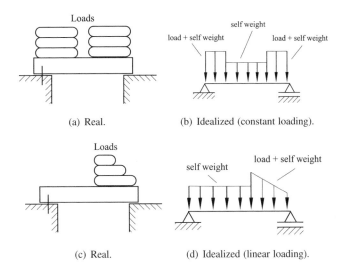

(a) Real.

(b) Idealized (constant loading).

(c) Real.

(d) Idealized (linear loading).

Figure 2.4 Constant and linear distributed loadings.

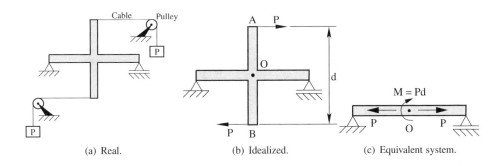

(a) Real.

(b) Idealized.

(c) Equivalent system.

Figure 2.5 Couple and equivalent concentrated moment.

In other cases, the forces are applied over a larger portion of the body and are called distributed loads. Figure 2.4 illustrates constant and linearly varying distributed forces, with the corresponding idealizations. Finally, a concentrated moment can be applied to a mechanical system, as shown in Figure 2.5. In this case, the concentrated forces of intensity P acting on the wires are transferred to points A and B of the body [Figure 2.5(b)]. This pair of forces with same intensity and opposite directions is called couple. It results in a pure moment of intensity $M = Pd$ applied to point O, as shown in Figure 2.5(c). It should be observed that when the couple applied at A and B is transferred to point O, we obtain a zero resultant force in the horizontal direction. The volume forces are uniformly distributed over the body as for the gravitational and electromagnetic forces.

2.3 EQUILIBRIUM OF PARTICLES

A particle is a material point with a certain mass m and whose dimensions are irrelevant. In the Newtonian formulation, a particle is in equilibrium if the resultant $\mathbf{f}$ of external forces acting on the particle is zero, that is

$$\mathbf{f} = \mathbf{0}. \tag{2.1}$$

When analyzing a system of particles, we isolate each particle and substitute the kinematic constraints by the respective reaction forces. Thus, force is an *apriori* concept for Newton mechanics.

Adopting a Cartesian reference system xyz, the previous equilibrium condition implies that the resultant of external forces in directions x, y, and z is zero. Therefore,

$$\sum f_x = 0,$$
$$\sum f_y = 0, \tag{2.2}$$
$$\sum f_z = 0.$$

Figure 2.6 illustrates a truss subjected to external forces P_1 and P_2 applied on node A. F_1, F_2, and F_3 are the internal forces on the elements that share node A. Considering the free body diagram (FBD) shown in Figure 2.6(b), the equilibrium conditions of particle A are

1. $\sum f_x = 0 : P_1 + F_2 \cos \theta + F_3 = 0$;
2. $\sum f_y = 0 : P_2 + F_1 + F_2 \sin \theta = 0$.

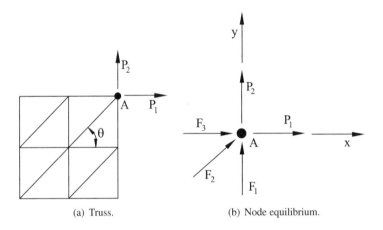

| (a) Truss. | (b) Node equilibrium. |

Figure 2.6   Nodal forces in a truss.

In the analytical approach, we consider firstly the kinematical action of the particle. From the definition of work (or power) for the given kinematics, we recover the classical definition that the resultant of forces in a particle is given by a vector.

Consider particle P, which is free of any kinematical restriction, on the three-dimensional Cartesian system xyz, denoted by $\mathfrak{R}^3$ and illustrated in Figure 2.7. The movement of P is described by the velocity vector $\mathbf{v}$ of $\mathfrak{R}^3$. The external power P_e associated with $\mathbf{v}$ is generally denoted by

$$f: \quad \mathfrak{R}^3 \quad \rightarrow \quad \mathfrak{R}$$
$$\mathbf{v} \quad \rightarrow \quad f(\mathbf{v}) = P_e$$

Therefore, the external power P_e is defined by the function f which operates on vectors $\mathbf{v} \in \mathfrak{R}^3$ and results in a scalar.

As the kinematics of P is given by the vector $\mathbf{v}$, the operation over this vector, which results in a scalar representing the external power P_e, is the dot or scalar product. Therefore, there is the vector $\mathbf{f} \in \mathfrak{R}^3$ associated to the power P_e and the kinematical vector $\mathbf{v}$, such that the power can be written as

$$P_e = f(\mathbf{v}) = \langle \mathbf{f}, \mathbf{v} \rangle = \mathbf{f} \cdot \mathbf{v}. \tag{2.3}$$

Consider a dimensional analysis of the above expression in the International System (S.I.), with P_e and $\mathbf{v}$ expressed in Watts=Nm/s and m/s, respectively. Thus, vector $\mathbf{f}$ must have units of force N.

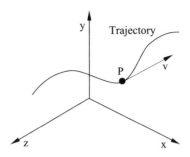

Figure 2.7   Kinematics of a particle given by the velocity vector **v**.

Hence, starting from the particle kinematics (which is expressed by the velocity vector **v**), and the required power to perform this action, we obtain the classical concept that the resultant of forces **f** on a particle is described by a vector. Therefore, the compatible external forces with the defined kinematics are also given by vectors **f**.

2.4   EQUILIBRIUM OF RIGID BODIES

According to the definition in Section 1.1.2, rigid bodies are constituted of an infinite number of particles. The main feature of a rigid body is that the distance between any two particles remains constant for any movement of the body.

From Newtonian mechanics, the equilibrium conditions for three-dimensional rigid bodies are that the resultants of external forces **f** and moments **m** are zero, i.e.,

$$\mathbf{f} = \mathbf{0}, \tag{2.4}$$

$$\mathbf{m} = \mathbf{0}. \tag{2.5}$$

In terms of the Cartesian system, the equilibrium conditions are obtained by imposing that the force and moment resultants are zero in the x, y, and z directions. Thus, analogous to the case of particle kinematics, the equilibrium equations for the forces are

$$\begin{aligned}
\sum f_x &= 0, \\
\sum f_y &= 0, \\
\sum f_z &= 0,
\end{aligned} \tag{2.6}$$

and for the moments

$$\begin{aligned}
\sum m_x &= 0, \\
\sum m_y &= 0, \\
\sum m_z &= 0.
\end{aligned} \tag{2.7}$$

Figure 2.8 shows the possible rigid body planar motions, which are translations in x and y and the rotation about the z axis. In this case, the equilibrium conditions are

$$\begin{aligned}
\sum f_x &= 0, \\
\sum f_y &= 0, \\
\sum m_z &= 0.
\end{aligned} \tag{2.8}$$

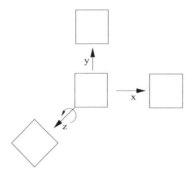

Figure 2.8   Rigid body motions in the *xy* plane.

We consider now the rigid body $\mathscr{B}$ of Figure 2.9, which is free of any kinematic constraints. The admissible kinematics should maintain the hypothesis that the body is rigid, i.e., any displacements or rotations do not change any dimension of the body.

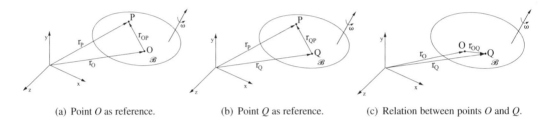

(a) Point *O* as reference.   (b) Point *Q* as reference.   (c) Relation between points *O* and *Q*.

Figure 2.9   Rigid body motion.

Figure 2.9(a) illustrates the Cartesian reference system *xyz* and the reference point *O*, adopted to describe the rigid body motion. Then, the position vector of any point *P* is given by

$$\mathbf{r}_P = \mathbf{r}_O + \mathbf{r}_{OP}, \tag{2.9}$$

where $\mathbf{r}_O$ and $\mathbf{r}_P$ are the position vectors of points *O* and *P* from the origin, respectively, and $\mathbf{r}_{OP}$ is the position vector of point *P* from point *O*.

The position vectors $\mathbf{r}_P$, $\mathbf{r}_O$, and $\mathbf{r}_{OP}$ change over time for any motion of the body. The rate of change of these vectors represents the instantaneous velocity of the respective points. Taking the time derivative of the above expression, we get the absolute linear velocity $\mathbf{v}_P$ of point *P* as [36]

$$\mathbf{v}_P = \mathbf{v}_O + \boldsymbol{\omega} \times \mathbf{r}_{OP}, \tag{2.10}$$

where $\mathbf{v}_O$ is the velocity vector of point *O*, representing the translation of the body $\mathscr{B}$; $\boldsymbol{\omega}$ is the angular velocity vector, which describes the instantaneous rotation of the body; $\times$ is the cross-product operator of two vectors.

If we decompose the vectors $\mathbf{v}_O$ and $\boldsymbol{\omega}$, along the axes of the Cartesian system illustrated in Figure 2.9, there are six components to describe the rigid body motion. Three of them are the translational components v_{O_x}, v_{O_y}, and v_{O_z} of $\mathbf{v}_O$; the other three are related to $\boldsymbol{\omega}$ and given by ω_x, ω_y, and ω_z, and describe the rigid rotations of the body relative to *x*, *y*, and *z* axes, respectively. Figure 2.11 illustrates the three translational and rotational components of a rigid body, according to the Cartesian system *xyz*.

Example 2.1 *Consider the cylindrical body given in Figure 2.10(a). It rotates with constant angular velocity ω_x about the x axis. For the right end section and the considered coordinate system, the velocity of the center O is zero, that is, $\mathbf{v}_O = \mathbf{0}$. The velocity of any point P with coordinates (y, z) at the boundary of this section is given from (2.10) by*

$$\mathbf{v}_P = \omega \times \mathbf{r}_{OP} = \begin{vmatrix} \mathbf{e}_x & \mathbf{e}_y & \mathbf{e}_z \\ \omega_x & 0 & 0 \\ 0 & y & z \end{vmatrix} = (-z\mathbf{e}_y + y\mathbf{e}_z)\omega_x.$$

The velocity of point P is along the tangent direction as shown in Figure 2.10(b).

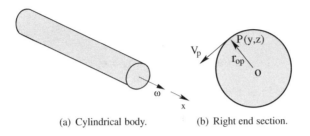

(a) Cylindrical body.   (b) Right end section.

Figure 2.10 Example 2.1: cylindrical body rotating about the *x* axis.

□

The expression for the associated power to the rigid body motion (2.10) is given by (2.3) as

$$P_e = f(\mathbf{v}_P) = f(\mathbf{v}_O + \omega \times \mathbf{r}_{OP}). \tag{2.11}$$

The velocity of any point P of the rigid body is given by the vector $\mathbf{v}_P$ and the power P_e related to $\mathbf{v}_P$ is a scalar. Analogously to the case of particles, there is a vector $\mathbf{f}_O$ associated to the kinematics and the power, in such way that the power P_e is given by the following dot product of the vectors $\mathbf{v}_P$ and $\mathbf{f}_O$:

$$P_e = f(\mathbf{v}_P) = \langle \mathbf{f}_O, \mathbf{v}_P \rangle = \mathbf{f}_O \cdot \mathbf{v}_P = \mathbf{f}_O \cdot [\mathbf{v}_O + \omega \times \mathbf{r}_{OP}]. \tag{2.12}$$

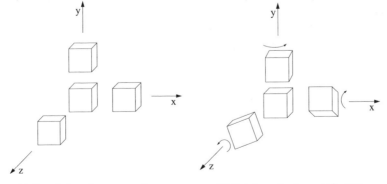

(a) Translations along *x*, *y*, and *z* axes.   (b) Rotations about *x*, *y*, and *z* axes.

Figure 2.11 Three-dimensional rigid body motions.

Applying the distributive property of the dot product, the above expression can be rewritten as

$$P_e = \mathbf{f}_O \cdot \mathbf{v}_O + \mathbf{f}_O \cdot (\omega \times \mathbf{r}_{OP}). \tag{2.13}$$

We can commute the dot and cross-product order in the second term of the right-hand side of the above expression. Generally, given three vectors **a**, **b**, and **c**, the following relation is valid:

$$\mathbf{a} \cdot (\mathbf{b} \times \mathbf{c}) = (\mathbf{c} \times \mathbf{a}) \cdot \mathbf{b} = (\mathbf{b} \times \mathbf{c}) \cdot \mathbf{a}. \tag{2.14}$$

Example 2.2 *Verify the above relation to the vectors $\{\mathbf{a}\} = \{1\ 1\ 1\}^T$, $\{\mathbf{b}\} = \{2\ 1\ 1\}^T$, and $\{\mathbf{c}\} = \{1\ 2\ 1\}^T$.*

The cross-products in equation (2.14) for the given vectors are obtained as

$$\mathbf{b} \times \mathbf{c} = \begin{vmatrix} \mathbf{e}_x & \mathbf{e}_y & \mathbf{e}_z \\ 2 & 1 & 1 \\ 1 & 2 & 1 \end{vmatrix} = -\mathbf{e}_x - \mathbf{e}_y + 3\mathbf{e}_z,$$

$$\mathbf{c} \times \mathbf{a} = \begin{vmatrix} \mathbf{e}_x & \mathbf{e}_y & \mathbf{e}_z \\ 1 & 2 & 1 \\ 1 & 1 & 1 \end{vmatrix} = \mathbf{e}_x + \mathbf{e}_z,$$

$$\mathbf{b} \times \mathbf{c} = \begin{vmatrix} \mathbf{e}_x & \mathbf{e}_y & \mathbf{e}_z \\ 2 & 1 & 1 \\ 1 & 2 & 1 \end{vmatrix} = -\mathbf{e}_x - \mathbf{e}_y + 3\mathbf{e}_z.$$

The dot products in (2.14) are

$$\begin{aligned} \mathbf{a} \cdot (\mathbf{b} \times \mathbf{c}) &= -1 - 1 + 3 = 1, \\ (\mathbf{c} \times \mathbf{a}) \cdot \mathbf{b} &= 2 + 0 - 1 = 1, \\ (\mathbf{b} \times \mathbf{c}) \cdot \mathbf{a} &= -1 - 1 + 3 = 1. \end{aligned}$$

Then, equation (2.14) is verified for the given vectors.
□

Applying the previous property to the second term on the right-hand side of equation (2.13), we obtain

$$P_e = \mathbf{f}_O \cdot \mathbf{v}_O + (\mathbf{r}_{OP} \times \mathbf{f}_O) \cdot \boldsymbol{\omega} = \mathbf{f}_O \cdot \mathbf{v}_O + \mathbf{m}_O \cdot \boldsymbol{\omega}. \tag{2.15}$$

The vectors $\mathbf{f}_O$ and $\mathbf{m}_O$ represent, respectively, the resultant of forces and moments generated by $\mathbf{f}_O$ relative to point O.

We shall observe that the choice of point O to represent the motion of $\mathscr{B}$ is arbitrary. Taking another distinct point Q, we obtain the following relation to the position vector of P [see Figure 2.9(b)]

$$\mathbf{r}_P = \mathbf{r}_Q + \mathbf{r}_{QP}.$$

Consequently, the instantaneous velocity of P is now given by

$$\mathbf{v}_P = \mathbf{v}_Q + \boldsymbol{\omega} \times \mathbf{r}_{QP}.$$

Following the previous procedure, but now taking the new reference point Q, the power P_e associated to the motion of P is

$$P_e = \mathbf{f}_Q \cdot \mathbf{v}_Q + \mathbf{m}_Q \cdot \boldsymbol{\omega} = \mathbf{f}_Q \cdot \mathbf{v}_Q + (\mathbf{r}_{QP} \times \mathbf{f}_Q) \cdot \boldsymbol{\omega}. \tag{2.16}$$

The velocity of point Q can be rewritten, taking point O as the reference [see Figure 2.9(c)]. The following relations are now valid, respectively, for the position and velocity vectors of point Q:

$$\mathbf{r}_Q = \mathbf{r}_O + \mathbf{r}_{OQ},$$

$$\mathbf{v}_Q = \mathbf{v}_O + \boldsymbol{\omega} \times \mathbf{r}_{OQ}. \tag{2.17}$$

As the motion is the same independently of changing its representation, the resulting power is also the same when we take points O and Q as references. Hence, equating (2.15) and (2.16), we obtain

$$\mathbf{f}_O \cdot \mathbf{v}_O + \mathbf{m}_O \cdot \boldsymbol{\omega} = \mathbf{f}_Q \cdot \mathbf{v}_Q + \mathbf{m}_Q \cdot \boldsymbol{\omega},$$

or

$$(\mathbf{f}_O \cdot \mathbf{v}_O - \mathbf{f}_Q \cdot \mathbf{v}_Q) + (\mathbf{m}_O - \mathbf{m}_Q) \cdot \boldsymbol{\omega} = 0.$$

Substituting (2.17) and simplifying the expressions, we have

$$0 = [\mathbf{f}_O - \mathbf{f}_Q] \cdot \mathbf{v}_O + [\mathbf{m}_O - \mathbf{m}_Q - (\mathbf{r}_{OQ} \times \mathbf{f}_Q)] \cdot \boldsymbol{\omega}.$$

To satisfy the previous relation, we must enforce that for any rigid motion described by vectors $\mathbf{v}_O$ and $\boldsymbol{\omega}$, the terms in the brackets must be simultaneously zero, which results in

$$\mathbf{f}_O = \mathbf{f}_Q = \mathbf{f} \qquad \text{and} \qquad \mathbf{m}_O - \mathbf{m}_Q = \mathbf{r}_{OQ} \times \mathbf{f}_Q.$$

Hence, the classical results of rigid body mechanics are obtained. The forces are given by a resultant force vector $\mathbf{f}$, which is independent of the chosen point to describe the kinematics. Moreover, there is a resulting moment vector $\mathbf{m}_O$, generated by the external force resultant, which is dependent of the chosen point.

An important aspect in the study of an engineering problem is the transition from the physical model to a convenient mathematical representation. This aspect is called modeling. From the equilibrium point of view, treating a body as a particle or rigid body depends on aspects such as the involved dimensions and purpose of the analysis. For example, when studying the orbits of the planets, Newton considered the planets and the sun as particles. Figure 2.12 illustrates the piston-rod-crank mechanism of an internal combustion engine. The piston is modeled as a particle and the rod and crank as rigid bodies. In this case, the piston is modeled as a particle due to the fact that the interest is only in the translational movement of the piston.

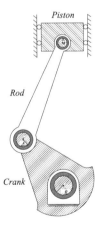

Figure 2.12 Piston-rod-crank mechanism (adapted from [26]).

2.5   PRINCIPLE OF VIRTUAL POWER (PVP)

Consider two spheres resting on the convex and concave surfaces of Figures 2.13(a) and 2.13(b), respectively. In order to analyze the stability of these equilibrium positions, we intuitively induce a motion around the highlighted positions. In the first case, the sphere oscillates around the original position, and then returns to equilibrium. That means that the total power spent by this motion is zero. In the second case, the sphere no longer returns to the initial position, and we conclude that the original equilibrium position is unstable. The important point in this example is that in order to evaluate the equilibrium state of the sphere, a perturbation or variation of the initial configuration is considered. As the total power is zero for the first case, we conclude that the corresponding original position is stable.

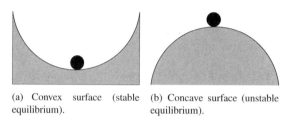

(a) Convex surface (stable equilibrium).   (b) Concave surface (unstable equilibrium).

Figure 2.13   Equilibrium position analysis of a sphere.

The imposed perturbations from the equilibrium position are called virtual actions. They are called virtual because they are not physically performed by the particle or body, but are supposedly applied to evaluate the original equilibrium state.

Virtual actions are used in the principle of virtual power (PVP) to evaluate the general equilibrium state. The principle states that if the power is zero for any virtual kinematic admissible motion from the equilibrium position, the considered mechanical system stands in equilibrium. We should observe that kinematically admissible motions do respect the constraint conditions of the mechanical system under consderation. We assume the case of idealized systems, where any dissipative effects, such as friction, are not considered.

From the preceding sections, we know that the equilibrium conditions for particles and rigid bodies are obtained making zero the external power for any kinematically admissible virtual motion from the equilibrium position. In this way, it is possible to recover the equilibrium conditions of the Newtonian mechanics.

Consider point A of the body subjected to several forces as shown in Figure 2.14. Suppose the body is displaced, in a compatible way with the established kinematics, and point A moves to position A'. The forces may be in static equilibrium, and the body stays at rest or moves to the AA' direction. The considered motion is then imaginary. It is called a virtual motion and designated by $\delta\mathbf{v}$. If point A is in equilibrium, $\delta\mathbf{v}$ represents a variation of the point's position related to the equilibrium state. Hence, the PVP results in the zero power δP_e associated to $\delta\mathbf{v}$, i.e.,

$$\delta P_e = \mathbf{f} \cdot \delta\mathbf{v} = 0. \tag{2.18}$$

Consequently, the resultant of forces $\mathbf{f}$ acting on the particle must be zero, because the action $\delta\mathbf{v}$ is arbitrary. Hence, we recover the Newtonian mechanics equilibrium condition, which states that the resultant of external forces $\mathbf{f}$ must be zero at the particle equilibrium. The external power is denoted by δP_e, because it is associated to a virtual action $\delta\mathbf{v}$.

We should emphasize that the preceding condition is sufficient and necessary for the equilibrium. If the power is zero to any virtual action $\delta\mathbf{v}$, then the scalar product $\mathbf{f} \cdot \delta\mathbf{v}$ is also zero, which implies that the resultant of forces $\mathbf{f}$ must be zero, because the action $\delta\mathbf{v}$ is arbitrary. Analogously, if particle A is in equilibrium, the resultant of forces is zero, and thus the virtual power is also zero.

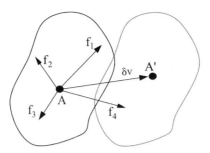

Figure 2.14 Forces acting on a particle.

By taking a rigid body virtual action $\delta\mathbf{v} = \delta\mathbf{v}_O + \delta\boldsymbol{\omega} \times \mathbf{r}$, with point O as reference, we have zero power at the equilibrium position. Hence, from the preceding section, the power is given as

$$\delta P_e = \mathbf{f}_O \cdot \delta\mathbf{v}_O + \mathbf{m}_O \cdot \delta\boldsymbol{\omega} = 0, \tag{2.19}$$

implying that the resultant in terms of forces and moments must be zero to any arbitrary virtual motion described by the vectors $\delta\mathbf{v}_O$ and $\delta\boldsymbol{\omega}$. Therefore, $\mathbf{f}_O = \mathbf{0}$ and $\mathbf{m}_O = \mathbf{0}$.

These equilibrium conditions are the same as those of Newtonian mechanics. The basic difference is that in analytical mechanics we start from the motion, and using the definition of power, we find the compatible external loads with the kinematics of particles and rigid bodies. In this sense, the concepts of kinematics and power are more natural (remember the examples of the weight of an object and tension in a belt) and may be physically observed. In the Newtonian mechanics approach, force is defined as the fundamental interacting concept among bodies and their environment. However, we emphasize that for a moving vehicle, what is indeed observed is the kinematics, and not the acting forces and moments.

Taking the example illustrated in Figure 1.2 to evaluate the weight of the object, it is not necessary to impose a large displacement to measure its weight. In the attempt to displace the object, we can estimate its weight. This implies that the imposed virtual actions can be arbitrarily small, and may be conveniently described by differentials to evaluate the state of equilibrium of a particle or rigid body. The use of a differential for virtual actions is convenient as will be seen in the following examples. In addition, remember that infinitesimal rotations are commutative and can be represented by vectors [32, 7].

The power generated by internal forces in a rigid body is zero, as illustrated in Figure 2.15. Taking points A and B, the interaction forces between them are $\mathbf{f}_{AB}$ and $\mathbf{f}_{BA} = -\mathbf{f}_{AB}$. Even if we consider two distinct virtual actions $\delta\mathbf{v}_A$ and $\delta\mathbf{v}_B$ to points A and B, the components of these vectors along line AB must be equal, because the distance between two points must be constant for rigid bodies. Designating these components as $\delta\mathbf{v}$, the power associated to the motion of points A and B are given by

$$\delta P_e = \mathbf{f}_{AB} \cdot \delta\mathbf{v} + \mathbf{f}_{BA} \cdot \delta\mathbf{v} = (\mathbf{f}_{AB} - \mathbf{f}_{AB}) \cdot \delta\mathbf{v} = 0.$$

Hence, the power associated to internal forces between any two points of a rigid body is zero, independently of the considered virtual motion used to evaluate the equilibrium. As the virtual motion and the kinematics of the body are compatible, the supports do not perform any work.

In the analysis of bodies in static equilibrium, there are no associated velocities; the principle of virtual power is applied in terms of virtual displacements and is called the principle of virtual work (PVW). The PVW states that, if an idealized mechanical system is in equilibrium, the total virtual work performed by the external forces is zero for any virtual displacement compatible with the kinematics of the body.

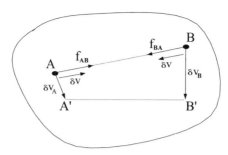

Figure 2.15 Power of internal forces in a rigid body.

This postulate is not restricted to static cases. It is also applicable to dynamic equilibrium. In this case, the PVP is generalized by D'Alembert's principle, taking the inertial forces as external forces applied to the body. As all other variational principles (Euler, Lagrange, Jacobi, Hamilton) of analytical mechanics are alternative mathematical formulations of D'Alembert's principle, the PVP may be considered as the main postulate of analytical mechanics and is of fundamental importance.

The physical interpretation of the principle of virtual work becomes interesting when we take the equilibrium of a particle. According to the Newtonian mechanics, the resultant of forces in the state of equilibrium, expressed as the sum of external and reaction forces acting over any particle of the system, is zero. Since in the equilibrium, the principle requires that the work of these forces is zero, the virtual work caused by the external forces can be replaced by the one of the reaction forces. Hence, the PVW can be reformulated by the following postulate: The virtual work of the reaction forces is always zero for any virtual displacement compatible with the kinematic constraints.

As the work of the reaction force R is zero, we have $W_e = Ru = R(0) = 0$, with u being the displacement in the direction of R. This means that R can be zero, implying that the body or particle under consideration is not being loaded; or not zero, which is the case of any particle or body subjected to external forces.

Following, an example regarding the calculation of reaction forces in a rigid body is presented using the PVW. We adopt the following procedure:

1. Draw the free body diagram (FBD) to each of the considered bodies, indicating the applied external forces and the respective reaction forces.
2. Apply a convenient displacement or rotation, so that the desired reaction force or moment performs work. We should apply a virtual displacement or rotation in a manner that the respective reaction will be the only unknown force or moment of the body to perform work.
3. Calculate the work generated by all forces and moments considered in the FBD to the virtual action and apply the PVW to solve the obtained equation.
4. Repeat items 2 and 3 for each reaction force and moment.

Some observations can be stated regarding the previous procedure. The points subjected to kinematic constraints do not rotate and displace, depending on the supporting condition (see Figure 2.1). When drawing the FBD, we replace the supports by the respective reaction forces. Thus, when the body is loaded, reaction forces are necessary to satisfy the supporting conditions of zero rotations and displacements. In the FBD, the body under consideration is free of any kinematic restriction compatible with the support.

Once the kinematical supports are removed from the FBD, we can apply convenient virtual displacements/rotations from the equilibrium position, and determine the necessary reaction forces to maintain the body in equilibrium. In this procedure, we must establish a relationship between the virtual displacements/rotations of the points with applied forces/moments and the reactions. For this purpose, trigonometric relations, similarity of triangles, or Taylor series expansions are used, as illustrated in the following example. This procedure is valid for systems with one principal degree of freedom and the virtual displacements/rotations described in terms of it. Dissipative effects, like friction, are not considered.

Example 2.3 *Consider the articulated lever of Figure 2.16(a). We want to determine the force exerted by the lever on the block using the PVW, when a force F is applied on C. The friction between the lever and the contact surface is negligible.*

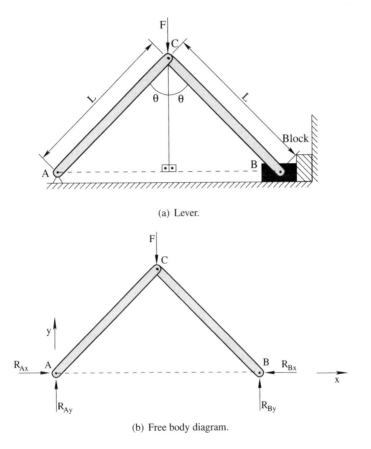

(a) Lever.

(b) Free body diagram.

Figure 2.16 Example 2.3: articulated lever with a force F.

Next, we apply the aforementioned steps to solve this example.

- *Construction of the free body diagram*
  The diagram for the lever illustrated in Figure 2.16(b) is constructed with the external and reaction forces. The origin of the adopted Cartesian coordinate system xyz is at A.
- *Definition of a convenient virtual displacement*
  The coordinates x_B and y_C of points B and C of the lever can be expressed in terms of the angle θ as

$$x_B(\theta) = 2L\sin\theta, \tag{2.20}$$
$$y_C(\theta) = L\cos\theta. \tag{2.21}$$

Thus, θ is the only independent degree of freedom of the lever system.

Initially, we consider a positive angle increment $\delta\theta$ for θ and the lever applies a horizontal force on the block (see Figure 2.17). There are variations δx_B and δy_C, respectively, in the dimensions x_B and y_C. Reactions R_{Ax}, R_{Ay}, and R_{By} will not perform any work for the considered virtual action. Therefore, it is necessary to calculate only the work performed by F and R_{Bx}.

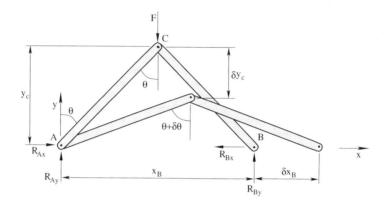

Figure 2.17   Example 2.3: virtual displacement to calculate R_{Bx}.

Imposing a virtual variation $\delta\theta$ from the initial equilibrium position given by θ, the final angle is $\theta + \delta\theta$, as indicated in Figure 2.17. Hence, the new positions B' and C' of points B and C will be, respectively (see Figure 2.17),

$$x_{B'} = x_B(\theta + \delta\theta) = 2L\sin(\theta + \delta\theta), \tag{2.22}$$
$$y_{C'} = y_C(\theta + \delta\theta) = L\cos(\theta + \delta\theta). \tag{2.23}$$

Expanding $\sin(\theta + \delta\theta)$ and $\cos(\theta + \delta\theta)$, we have

$$x_{B'} = 2L(\sin\theta\cos\delta\theta + \cos\theta\sin\delta\theta),$$
$$y_{C'} = L(\cos\theta\cos\delta\theta - \sin\theta\sin\delta\theta).$$

By considering $\delta\theta$ small, we obtain $\cos\delta\theta \approx 1$ and $\sin\delta\theta \approx \delta\theta$. Thus,

$$x_{B'} = 2L\sin\theta + 2L\cos\theta(\delta\theta),$$
$$y_{C'} = L\cos\theta - L\sin\theta(\delta\theta).$$

Consequently, the virtual increments δx_B and δy_C are given by

$$\delta x_B = x_{B'} - x_B = 2L\cos\theta(\delta\theta),$$
$$\delta y_C = y_{C'} - y_C = -L\sin\theta(\delta\theta).$$

Another method to obtain the expressions for δx_B and δy_C is using the Taylor series expansion. As indicated in (2.20) and (2.21), positions x_B and y_C of points B and C are functions of angle θ.

Given a single variable function f, remember that the Taylor series expansion around x is given by

$$f(y) = f(x) + f'(x)(y - x) + \frac{1}{2}f''(x)(y - x) + \cdots.$$

Expanding $\sin(\theta + \delta\theta)$ and $\cos(\theta + \delta\theta)$ in Taylor series and neglecting the terms from the second-order derivative (which means that $\delta\theta$ is small), we obtain

$$
\begin{aligned}
\sin(\theta + \delta\theta) &= \sin\theta + \cos\theta(\theta + \delta\theta - \theta) = \sin\theta + \cos\theta(\delta\theta), \\
\cos(\theta + \delta\theta) &= \cos\theta - \sin(\theta + \delta\theta - \theta) = \cos\theta - \sin\theta(\delta\theta).
\end{aligned}
$$

Substituting these equations in (2.22) and (2.23) results in

$$
\begin{aligned}
x_{B'} &= 2L\sin\theta + 2L\cos\theta(\delta\theta) = x_B + 2L\cos\theta(\delta\theta), \\
y_{C'} &= L\cos\theta - L\sin\theta(\delta\theta) = y_C - L\sin\theta(\delta\theta).
\end{aligned}
$$

Thus,

$$
\begin{aligned}
\delta x_B &= x_{B'} - x_B = 2L\cos\theta(\delta\theta), \\
\delta y_C &= y_{C'} - y_C = -L\sin\theta(\delta\theta).
\end{aligned}
$$

A third method of calculating δx_B and δy_C is by taking the differentials of x_B and y_C, i.e.,

$$
\begin{aligned}
\delta x_B &= \frac{dx_B}{d\theta}\delta\theta = 2L\cos\theta(\delta\theta), \\
\delta y_C &= \frac{dy_C}{d\theta}\delta\theta = -L\sin\theta(\delta\theta).
\end{aligned}
$$

Hence, we observe that calculating the virtual variations δx_B and δy_C is analogous to calculating the differential of x_B and y_C.

- *Write the expression of the virtual work for the mechanical system*
  As R_{Bx} and δx_B have opposite signs, the respective virtual work is negative, that is, $\delta W_{R_{Bx}} = -R_{Bx}\delta x_B$. As F and the increment δy_C have the same sign, the corresponding virtual work is positive, and $\delta W_F = F\delta y_C$. The other supporting reactions do not perform any work for the considered virtual motion. Hence, the total virtual work is obtained as

$$
\delta W_e = \delta W_{R_{Bx}} + \delta W_F = -R_{Bx}\delta x_B + F\delta y_C = -2R_{Bx}L\cos\theta(\delta\theta) + FL\sin\theta(\delta\theta).
$$

- *Solve the equations*
  From the PVW, the lever system is in equilibrium if $\delta W_e = 0$. Thus,

$$
2R_{Bx}L\cos\theta = FL\sin\theta \rightarrow R_{Bx} = \frac{F}{2}\tan\theta.
$$

To calculate the reaction force R_{By}, a virtual displacement δy_B in the direction of R_{By} is considered and point A is fixed. This kinematical action is shown in Figure 2.18. Point C is displaced by δy_C. Only the forces F and R_{By} perform work, and the virtual work equation for equilibrium is

$$
\delta W_e = R_{By}\delta y_B - F\delta y_C = 0.
$$

Assuming that δy_B and δy_C are small and using similar triangles as illustrated in Figure 2.18, $\frac{\delta y_C}{\delta y_B} = \frac{1}{2}$. Hence,

$$
R_{By} = \frac{\delta y_C}{\delta y_B}F = \frac{F}{2}.
$$

Therefore, to calculate the reaction forces, we impose a virtual displacement in the same direction of the force. The same procedure can be applied to determine R_{Ax} and R_{Ay}.

□

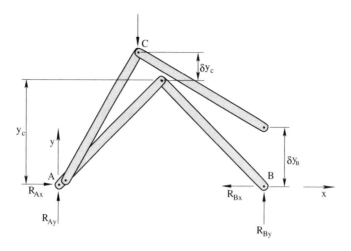

Figure 2.18   Example 2.3: virtual displacement to calculate the reaction force R_{By}.

From the preceding example, the reader should observe that it is necessary to think about the applied virtual actions for the body. The use of virtual displacements given by differentials simplifies the solution of the equilibrium problem.

Example 2.4 *Solve Example 2.3 employing the Newtonian approach.*

- *Construction of the free body diagram*
  The free body diagram was constructed in the last example and is illustrated in Figure 2.16(b).
- *Determination of the system of equilibrium equations*
  From the free body diagram, we have the following equilibrium conditions for the lever:

$$
\begin{aligned}
&i) && \sum f_x = 0: && R_{Ax} - R_{Bx} = 0, \\
&ii) && \sum f_y = 0: && R_{Ay} + R_{By} - F = 0, \\
&iii) && \sum m_{z_A} = 0: && 2L\sin\theta R_{By} - L\sin\theta F = 0.
\end{aligned}
$$

In this case, the sum of moments relative to point A of the lever was considered.
We observe that the lever is a hyperstatic system, since the number of unknowns $(R_{Ax}, R_{Bx}, R_{Ay}, R_{By})$ is greater than the number of equilibrium conditions. We must find an additional condition (generally in terms of the geometry or deformation of the component) to solve this example. Considering that the lever is constituted of two elements AC and BC, there are only resulting axial forces along AC and BC. From the geometry of the problem, the following expression is valid:

$$
iv) \quad \tan\theta = \frac{R_{Bx}}{R_{By}}.
$$

- *Solution of equations for the unknowns*
  From Equations iii) and iv), we obtain R_{Bx} and R_{By} as

$$
R_{By} = \frac{F}{2} \quad and \quad R_{Bx} = \frac{F}{2}\tan\theta.
$$

The other reactions can be found using R_{Bx} and R_{By} and employing Equations i) and ii). Thus,

$$
R_{Ax} = \frac{F}{2}\tan\theta \quad and \quad R_{Ay} = \frac{F}{2}.
$$

☐

Example 2.5 *Calculate the reaction forces of the rigid body of Figure 2.19(a) using the principle of virtual work and Newton equilibrium conditions. The weight of the body is negligible.*

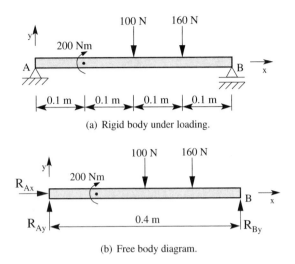

(a) Rigid body under loading.

(b) Free body diagram.

Figure 2.19 Example 2.5: rigid body equilibrium.

- *Construction of the free body diagram*
  The free body diagram is illustrated in Figure 2.19(b), along with the adopted coordinate system. There are two unknown reaction components in A, because there is a fixed support. There is only a vertical reaction in B because it is roller support.
- *Determination of R_{By}*
  Considering the virtual rotation $\delta\theta$ about point A from the original position of the body, as shown in Figure 2.20(a), we obtain a virtual displacement of point B in the direction of the unknown R_{By}. In this way, we have the following expression for the work performed by external and reaction forces and moments:

$$\delta W_e = 200\delta\theta + 100(0.2)\sin\delta\theta + 160(0.3)\sin\delta\theta - R_{By}(0.4)\sin\delta\theta.$$

As the body is in equilibrium, the PVW establishes that the work performed by external and reaction forces and moments is zero, that is, $\delta W_e = 0$. Hence,

$$200\delta\theta + 100(0.2)\sin\delta\theta + 160(0.3)\sin\delta\theta - R_{By}(0.4)\sin\delta\theta = 0.$$

For a small variation $\delta\theta$, we can assume the approximation $\sin\delta\theta \approx \delta\theta$ to be valid and

$$200\delta\theta + 100(0.2)\delta\theta + 160(0.3)\delta\theta - R_{By}(0.4)\delta\theta = 0.$$

From this equation, $R_{By} = 670$ N.
- *Determination of R_{Ay}*
  Analogously, considering a virtual rotation $\delta\theta$ about B, there is the displacement of point A in the direction of the unknown reaction R_{Ay}, according to Figure 2.20(b). Writing the expression of the virtual work of the external and reaction loads, we obtain

$$\delta W_e = -R_{Ay}(0.4)\sin\delta\theta - 200\delta\theta + 100(0.2)\sin\delta\theta + 160(0.1)\sin\delta\theta.$$

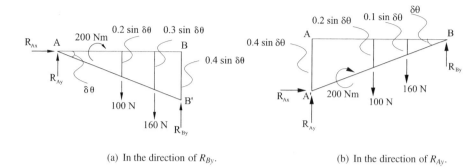

(a) In the direction of R_{By}.   (b) In the direction of R_{Ay}.

Figure 2.20   Example 2.5: virtual displacements for the calculation of R_{By} and R_{Ay}.

Again, for a small angle $\delta\theta$, $\sin\delta\theta \approx \delta\theta$ is valid, and consequently,

$$-R_{Ay}(0.4)\delta\theta - 200\delta\theta + 100(0.2)\delta\theta + 160(0.1)\delta\theta = 0.$$

Hence, $R_{Ay} = -410$ N. As the value is negative, the direction is opposite from the one indicated in the free body diagram of Figure 2.19(b).

· *Determination of R_{Ax}*

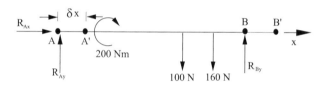

Figure 2.21   Example 2.5: virtual displacement in the direction of R_{Ax}.

In this case, we assume a horizontal displacement δx in the x direction, as shown in Figure 2.21. Thus, the expression for the virtual work is

$$\delta W_e = R_{Ax}\delta x = 0 \rightarrow R_{Ax} = 0.$$

Because there are no external forces in the axial direction, the reaction force R_{Ax} is zero.

From the free body diagram of Figure 2.19(b), it is possible to write the following equilibrium conditions:

$i)\ \sum f_x = 0:\qquad R_{Ax} = 0.$

$ii)\ \sum m_{z_A} = 0:\qquad -200 - 100(0.2) - 160(0.3) + R_{By}(0.4) = 0 \rightarrow R_{By} = 670$ N.

$iii)\ \sum m_{z_B} = 0:\qquad -R_{Ay}(0.4) - 200 + 100(0.2) + 160(0.1) = 0 \rightarrow R_{Ay} = -410$ N.

In this case, two equilibrium conditions in terms of the resulting moments related to points A and B were employed to determine directly the reactions R_{Ay} and R_{By}.

We can check the obtained results by taking the force summation in the y direction, that is,

$$\sum f_y = 670 - 410 - 100 - 160 = 0.$$

The body is in equilibrium, because the resultant of forces in the y direction is zero.

□

Example 2.6 *Determine the reaction forces in the rigid body shown in Figure 2.22(a) using the principle of virtual work and Newton equilibrium conditions.*

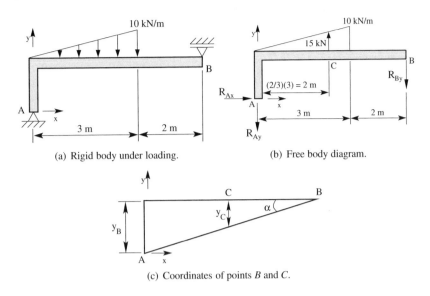

(a) Rigid body under loading.   (b) Free body diagram.

(c) Coordinates of points B and C.

Figure 2.22   Example 2.6: rigid body equilibrium.

- *Construction of the free body diagram*
  The free body diagram is shown in Figure 2.22(b) with the adopted Cartesian coordinate system. The linear distributed load was converted into a concentrated force of intensity 15 kN. It is applied at one-third of the triangle height, which is formed by the distributed load intensity and the portion of the body where it is applied.
- *Determination of R_{By}*
  As illustrated in Figure 2.22(c), the coordinates y_B and y_C of points B and C are expressed in terms of angle α, respectively, by

$$y_B(\alpha) = 5\tan\alpha,$$
$$y_C(\alpha) = 2\tan\alpha.$$

The virtual displacements of points B and C can be obtained by taking the differentials of the previous expressions, i.e.,

$$\delta y_B(\alpha) = 5\sec^2\alpha(\delta\alpha),$$
$$\delta y_C(\alpha) = 2\sec^2\alpha(\delta\alpha).$$

The ratio between the previous virtual displacements is

$$\frac{\delta y_C}{\delta y_B} = \frac{2}{5}.$$

The same virtual displacements can be determined from Figure 2.23(a). Considering Figures 2.23(b) and 2.23(c), the coordinates of points B' and C' are

$$y_{B'} = 5\tan(\alpha + \delta\alpha) = 5\frac{\tan\alpha + \tan\delta\alpha}{1 - \tan\alpha\tan\delta\alpha},$$
$$y_{C'} = 2\tan(\alpha + \delta\alpha) = 2\frac{\tan\alpha + \tan\delta\alpha}{1 - \tan\alpha\tan\delta\alpha}.$$

The corresponding virtual displacements for these points are obtained as

$$\delta y_B \;=\; y_{B'} - y_B = 5\frac{\tan\alpha + \tan\delta\alpha}{1 - \tan\alpha\tan\delta\alpha} - 5\tan\alpha,$$

$$\delta y_C \;=\; y_{C'} - y_C = 2\frac{\tan\alpha + \tan\delta\alpha}{1 - \tan\alpha\tan\delta\alpha} - 2\tan\alpha.$$

Assuming $\delta\alpha$ small, $\tan\delta\alpha \approx \delta\alpha$ and consequently $1 - \tan\alpha(\delta\alpha) \approx 1$. Therefore,

$$\frac{\delta y_C}{\delta y_B} = \frac{2}{5}.$$

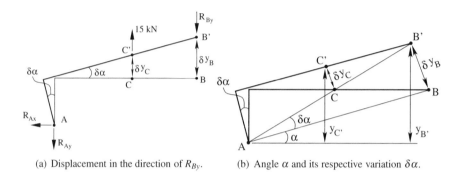

(a) Displacement in the direction of R_{By}.   (b) Angle α and its respective variation $\delta\alpha$.

(c) Relationship between the displacements of points B and C.

Figure 2.23   Example 2.6: virtual displacements of points B and C.

The same relation between the virtual displacements δy_B and δy_C can be obtained using the similar triangles given in Figure 2.23(c). Therefore,

$$\frac{\delta y_C}{\delta y_B} = \frac{2}{5}.$$

The work done by the external and reaction forces for the assumed virtual displacement is given by

$$\delta W_e = 15\delta y_C - R_{By}\delta y_B.$$

The PVW states that δW_e is zero if the body is in equilibrium. Hence,

$$\delta W_e = 15\delta y_C - R_{By}\delta y_B = 0 \rightarrow R_{By} = 15\frac{\delta y_C}{\delta y_B} = 6\ \text{kN}.$$

- *Determination of R_{Ay}*

We adopt the virtual displacement shown in Figure 2.24(a) to find R_{Ay}. Using the similar triangles indicated in Figure 2.24(b), we obtain the following ratio between the virtual displacements of points A and C:

$$\frac{\delta y_C}{\delta y_A} = \frac{3}{5}.$$

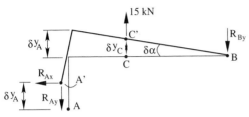

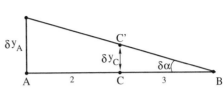

(a) Displacement in the R_{Ay} direction obtained from the rotation about B.

(b) Relation between displacements of points A and C.

Figure 2.24   Example 2.6: virtual displacements of points A and C.

The expression of the work of the external and reaction forces to the given virtual displacement is

$$\delta W_e = 15\delta y_C - R_{Ay}\delta y_A = 0 \rightarrow R_{Ay} = 9 \text{ kN}.$$

- *Determination of R_{Ax}*
  We have the following virtual work expression for the δx_A virtual displacement in the x direction shown in Figure 2.25:

$$\delta W_e = R_{Ax}\delta x_A = 0 \rightarrow R_{Ax} = 0.$$

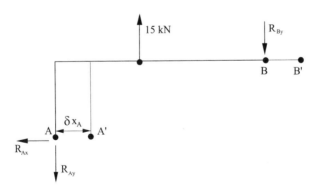

Figure 2.25   Example 2.6: virtual displacement in the R_{Ax} direction.

The following Newton equilibrium equations are valid:

$$i)\sum f_x = 0: \qquad R_{Ax} = 0.$$
$$ii)\sum m_{z_A} = 0: \qquad 15(2) - R_{By}(5) = 0 \rightarrow R_{By} = 6 \text{ kN}.$$
$$iii)\sum m_{z_B} = 0: \qquad -R_{Ay}(5) + 15(3) = 0 \rightarrow R_{Ay} = 9 \text{ kN}.$$

Two equilibrium conditions in terms of moments were used in order to directly calculate the reactions R_{Ay} and R_{By}. In the case of using the condition $\sum f_y = 0$, the following equation in terms *of R_{Ay} and R_{By} would be obtained:*

$$-R_{Ay} - R_{By} + 15 = 0,$$

being impossible to determine directly the unknowns with the above equilibrium condition.

   To check the values of the calculated reactions, it is sufficient to take the summation of forces in the y direction, that is,

$$\sum f_y = -6 - 9 + 15 = 0.$$

As the previous equation is zero, the body is in equilibrium and the reaction forces are calculated correctly.

   $\square$

Example 2.7 *Determine the reaction forces for the rigid body shown in Figure 2.26(a) using the principle of virtual work and Newton equilibrium conditions.*

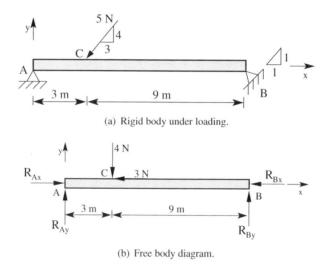

(a) Rigid body under loading.

(b) Free body diagram.

Figure 2.26 Example 2.7: rigid body equilibrium.

- *Construction of the free body diagram.*
  Figure 2.26(b) illustrates the free body diagram to the considered problem. There are two unknown reactions at A, namely R_{Ax} and R_{Ay}. The reaction force R_B is in the normal direction of the support. It is convenient to replace this force by two components, R_{Bx} and R_{By}, which are numerically equal in this problem. Analogously, the 5 N inclined force is decomposed in two components shown in Figure 2.26(b).
- *Determination of R_{Bx} and R_{By}*
  Considering the small virtual rotation $\delta\theta$ illustrated in Figure 2.27(a), the displacements of points B and C are, respectively,

$$\delta y_B = 12\sin\delta\theta \approx 12\delta\theta,$$
$$\delta y_C = 3\sin\delta\theta \approx 3\delta\theta.$$

The virtual work expression is given by

$$\delta W_e = -4(3)(\delta\theta) + R_{By}(12)(\delta\theta) = 0 \rightarrow R_{By} = 1 \text{ N}.$$

We know that $|R_{Bx}| = |R_{By}|$, because the support at B has an inclination of 45°. Thus, $R_{Bx} = 1$ N.

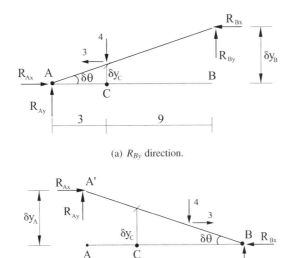

(a) R_{By} direction.

(b) R_{Ay} direction.

Figure 2.27 Example 2.7: virtual displacements.

- *Determination of R_{Ay}*
  According to Figure 2.27(b), we have the following virtual displacements for points A and C:
  $$\delta y_A = 12 \sin \delta\theta \approx 12\delta\theta,$$
  $$\delta y_C = 9 \sin \delta\theta \approx 9\delta\theta.$$

  The expression of the virtual work of the external and reaction forces is
  $$\delta W_e = -4(9\delta\theta) + R_{Ay}(12\delta\theta) = 0 \rightarrow R_{Ay} = 3 \text{ N}.$$

- *Determination of R_{Ax}*
  According to the virtual displacement shown in Figure 2.28, we have the following expression for the virtual work:
  $$\delta W_e = R_{Ax}(\delta x) - 3(\delta x) - R_{Bx}(\delta x) = 0 \rightarrow R_{Ax} = 4 \text{ N}.$$

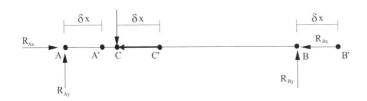

Figure 2.28 Example 2.7: virtual displacement in the R_{Ax} direction.

The equilibrium conditions of the Newtonian mechanics are:

$i) \sum m_{z_A} = 0:$ $\quad 4(3) - R_{By}(12) = 0 \rightarrow R_{By} = 1 \text{ N} \rightarrow R_{Bx} = 1 \text{ N}.$

$ii) \sum m_{z_B} = 0:$ $\quad -R_{Ay}(12) + 4(9) = 0 \rightarrow R_{Ay} = 3 \text{ N}.$

$iii) \sum f_x = 0:$ $\quad R_{Ax} - R_{By} - 3 = 0 \rightarrow R_{Ax} = 4 \text{ N}.$

The absolute values for reactions R_A and R_B are given by

$$R_A = \sqrt{4^2 + 3^2} = 5 \text{ N} \quad and \quad R_B = \sqrt{1^2 + 1^2} = \sqrt{2} \text{ N}.$$

To verify the calculated reactions, we sum the forces in the y direction, that is,

$$\sum f_y = 3 - 4 + 1 = 0.$$

As the resultant of forces in the y direction is zero, the body is in equilibrium with the calculated reaction forces.

☐

Example 2.8 *Consider the truss illustrated in Figure 2.29(a) subjected to forces $P_1 = 5$ kN and $P_2 = 3$ kN and made of steel bars with cross-section areas of $A_1 = 1$ cm², $A_2 = 2$ cm² and $A_3 = 3$ cm². Determine the forces on each bar using the principle of virtual work.*

The free body diagram of the truss is illustrated in Figure 2.29(b). There are six reaction forces, but only three equilibrium conditions, making the truss hyperstatic.

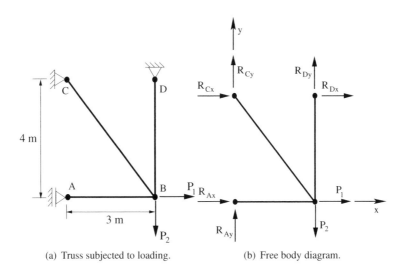

(a) Truss subjected to loading.   (b) Free body diagram.

Figure 2.29   Example 2.8: truss equilibrium.

Node B can be displaced in the horizontal and vertical directions because of the application of external forces P_1 and P_2. Even if we apply force P_1 only, node B would still displace in both directions, because the cross-section areas are distinct. Figures 2.30(a) and 2.30(b) illustrate the displacements Δ_1 and Δ_2 at node B in the horizontal and vertical directions, respectively. Assuming small deformation, the axial displacement of bar 2 is given by the linear combination of the projections of the horizontal and vertical displacements along the direction of bar 2. Thus, the axial displacements of each bar are expressed by

$$u_1 = \Delta_1, \quad u_2 = \frac{3}{5}\Delta_1 - \frac{4}{5}\Delta_2 \quad e \quad u_3 = -\Delta_2. \tag{2.24}$$

The minus sign of the projection of vertical displacement Δ_2 for bar 2 arises because Δ_2 is directed downwards, in the negative direction of y.

We should notice that this truss has two degrees of freedom in terms of displacements Δ_1 and Δ_2. Hence, the PVW is used for each degree of freedom. Taking a virtual displacement $\delta\Delta_1$ in the

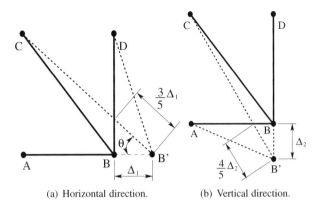

(a) Horizontal direction.   (b) Vertical direction.

Figure 2.30   Example 2.8: real displacements of node B.

horizontal direction, the virtual work expression is given by

$$F_1 \delta \Delta_1 + F_2 \left(\frac{3}{5} \delta \Delta_1 \right) = P_1 \delta \Delta_1.$$

Now, considering a virtual displacement $\delta \Delta_2$ pointing downwards in the vertical direction, the virtual work expression at equilibrium is

$$F_3 \delta \Delta_2 + F_2 \left(\frac{4}{5} \delta \Delta_2 \right) = P_2 \delta \Delta_2,$$

with F_1, F_2, and F_3 the internal forces in the bars. In this case, the PVW was applied considering that, for an arbitrary virtual displacement, the work done by internal and external forces, δW_i and δW_e, respectively, must be equal, that is,

$$\delta W_e = \delta W_i.$$

This general form of the PVW will be used in the following chapters for the formulation of deformable mechanical models.

   Simplifying the above expressions, we have

$$\begin{cases} F_1 + \frac{3}{5} F_2 = P_1 \\[2mm] F_3 + \frac{4}{5} F_2 = P_2 \end{cases} . \tag{2.25}$$

This system of equations is indeterminate, because we have three unknown forces and only two equilibrium equations.

   To solve this indeterminate system, a constitutive relation must be established between the internal forces and axial displacements for each bar. It will be seen in Chapter 3 that for a bar subjected to a concentrated load at one end, the constitutive relation is given by

$$F_i = k_i u_i = \frac{E_i A_i}{L_i} u_i. \tag{2.26}$$

where k_i is the stiffness coefficient of the bar, given in terms of the material Young's modulus (E_i), the cross-section area (A_i), and the length (L_i); u_i is the axial displacement of the bar at $x = L$.

By substituting the above constitutive relation in the system of equilibrium equations (2.25) and the displacements given by (2.24), we obtain the following system in terms of Δ_1 and Δ_2:

$$\begin{cases} \left(k_1 + \frac{9}{25}k_2\right)\Delta_1 - \frac{12}{25}k_2\Delta_2 = P_1 \\[2mm] -\frac{12}{25}k_2\Delta_1 + \left(k_3 + \frac{16}{25}k_2\right)\Delta_2 = -P_2 \end{cases} \tag{2.27}$$

Considering the given values for the cross-section areas and lengths of the bars, and adopting $E = 240\,\text{GPa} = 240 \times 10^9\,\text{N/m}^2$, the stiffness coefficients are

$$k_1 = \frac{EA_1}{L_1} = \frac{(240 \times 10^9)(1 \times 10^{-4})}{3} = 8000\,\text{kN/m},$$

$$k_2 = \frac{EA_2}{L_2} = \frac{(240 \times 10^9)(2 \times 10^{-4})}{5} = 9600\,\text{kN/m},$$

$$k_3 = \frac{EA_3}{L_3} = \frac{(240 \times 10^9)(3 \times 10^{-4})}{4} = 18000\,\text{kN/m}.$$

Substituting these values in (2.27), we can determine the following system of equations:

$$\begin{cases} 11456\Delta_1 - 4608\Delta_2 = 5 \\ -4608\Delta_1 + 24144\Delta_2 = -3 \end{cases},$$

which reults in $\Delta_1 = 0.4186$ mm and $\Delta_2 = -0.0444$ mm. After they are substituted in (2.24), we have the following axial displacements: $u_1 = 0.4186$ mm, $u_2 = 0.2867$ mm, and $u_3 = 0.0444$ mm. Finally, the forces acting on the bars are calculated from (2.26) and $F_1 = 3.35$ kN, $F_2 = 2.75$ kN, and $F_3 = 0.80$ kN.

□

The PVW can be used to determine the equilibrium of a body, as well as if the equilibrium position is stable, unstable, or neutral, as illustrated in the following example.

Example 2.9 *Consider the ladder with length $L = 5$ m and weight $P = 160$ N illustrated in Figure 2.31(b). A horizontal force $F = 100$ N is applied at the lower end. We want to determine the minimum distances a and b, to maintain the ladder in equilibrium from the vertical position shown in Figure 2.31(a). Verify if the equilibrium refers to a stable condition. The friction between the ladder and the contact surfaces is negligible.*

The equilibrium position of the ladder can be indicated by dimensions a and b or by the angle θ, from the vertical position as illustrated in Figures 2.31(b) and 2.31(c). These variables are related as

$$a(\theta) = -L\cos\alpha = L\cos\theta \quad and \quad b(\theta) = L\sin\alpha = L\sin\theta.$$

The angles α and θ are complementary, and α is measured from the x axis to the ladder. Once θ is determined, dimensions a and b are calculated using the above expressions. Hence, the angle θ is the only degree of freedom and main unknown of the problem.

For mechanical systems without dissipative effects and with a single degree of freedom, it is simpler to write the function of external work for a displacement from an initial to a specific position. The final position of the ladder is defined by the angle θ and the work function is given by

$$\begin{aligned} W_e(\theta) &= -100a(\theta) + 160\frac{L - b(\theta)}{2} \\ &= -(100)(5)\cos\theta + (80)(5)(1 - \sin\theta) \\ &= -500\cos\theta + 400(1 - \sin\theta). \end{aligned}$$

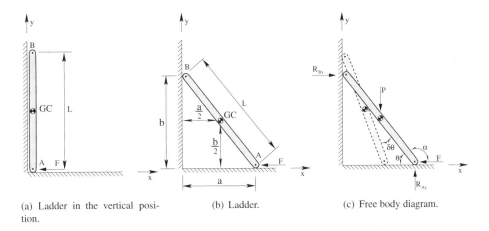

(a) Ladder in the vertical posi-   (b) Ladder.   (c) Free body diagram.
tion.

Figure 2.31   Example 2.9: ladder stability.

We observe that the reaction forces between the ladder and the wall do not add any work, accord-ing to the FBD in Figure 2.31(c), for a virtual increment $\delta\theta$ from the ladder equilibrium position.

Assuming that $\delta\theta$ is expressed by a differential, the virtual work can be determined by the deriva-tive of the above expression as

$$\delta W_e = \frac{dW_e(\theta)}{d\theta}\delta\theta = (500\sin\theta - 400\cos\theta)\delta\theta.$$

The equilibrium condition using the PVW requires the above expression to be zero, that is,

$$\delta W_e = (500\sin\theta - 400\cos\theta)\delta\theta = 0,$$

resulting in

$$\tan\theta = \frac{4}{5} \rightarrow \theta = 38.7^o.$$

Hence, the equilibrium position is obtained by imposing a zero derivative of the work function. It means also that the function has a maximum or minimum value. For $\theta = 38.7^o$, we determine $a = 3.9$ m and $b = 3.1$ m.

For a body in a stable equilibrium condition, the forces perform a negative work for a given virtual motion from the equilibrium position. Consequently, the work function reaches a maximum value. Thus, while the first derivative of the work function must be zero for equilibrium, its second derivative must be evaluated to verify the state of equilibrium. If negative, the work function assumes a maximum value, and the equilibrium position is stable. In the case of unstable equilibrium, the forces perform a positive work for the given virtual motion from the unstable equilibrium position. The work function assumes a minimum value and its second derivative has a positive value. If the second derivative is zero, the equilibrium position is called neutral.

To determine the equilibrium state for the ladder, we substitute the obtained value for θ to the equilibrium position into the second derivative of the work function, that is,

$$\frac{dW_e^2(\theta)}{d\theta^2} = 500\cos\theta + 400\sin\theta|_{\theta=38.7^o} = 614.3.$$

The obtained value is positive, and thus the equilibrium position is unstable.

□

As mentioned in Section 1.1.3, analytical mechanics considers a single work function from which the equilibrium condition, position, and state can be determined.

2.6   SOME ASPECTS ABOUT THE DEFINITION OF POWER

A function, such as the external power P_e, which associates a scalar value with each element of a vector space is called a functional. The power function is linear in $\mathbf{v}$, that is, the greater the magnitude of the velocity vector $\mathbf{v}$, the greater will be the value of power P_e. In the same way, if the magnitude of $\mathbf{v}$ decreases, power P_e will decrease proportionally. Generally, power P_e is a linear functional of the kinematics described by $\mathbf{v}$.

Thus, given a real number α and two velocity vectors $\mathbf{v}_1$ and $\mathbf{v}_2$, the following expressions are valid:

$$P_e = f(\alpha\mathbf{v}) = \mathbf{f} \cdot (\alpha\mathbf{v}) = \alpha(\mathbf{f} \cdot \mathbf{v}) = \alpha P_e,$$
$$P_e = f(\mathbf{v}_1 + \mathbf{v}_2) = \mathbf{f} \cdot (\mathbf{v}_1 + \mathbf{v}_2) = (\mathbf{f} \cdot \mathbf{v}_1) + (\mathbf{f} \cdot \mathbf{v}_2) = P_{e_1} + P_{e_2}, \qquad (2.28)$$

where $P_{e_1} = \mathbf{f} \cdot \mathbf{v}_1$ and $P_{e_2} = \mathbf{f} \cdot \mathbf{v}_2$ are the powers associated to the movement actions $\mathbf{v}_1$ and $\mathbf{v}_2$, respectively. The two previous properties show the linearity of the power functional, because when we multiply $\mathbf{v}$ by a scalar α, the power is also multiplied by α. Besides that, the power P_e associated with the sum of two actions $\mathbf{v}_1$ and $\mathbf{v}_2$ is equal to the sum of the individual powers P_{e_1} and P_{e_2}. The above properties can be rewritten by a single expression for scalars α and β and velocity vectors $\mathbf{v}_1$ and $\mathbf{v}_2$ as

$$
\begin{aligned}
P_e &= f(\alpha\mathbf{v}_1 + \beta\mathbf{v}_2) \\
&= \mathbf{f} \cdot (\alpha\mathbf{v}_1 + \beta\mathbf{v}_2) \\
&= \mathbf{f} \cdot (\alpha\mathbf{v}_1) + \mathbf{f} \cdot (\beta\mathbf{v}_2) \\
&= \alpha(\mathbf{f} \cdot \mathbf{v}_1) + \beta(\mathbf{f} \cdot \mathbf{v}_2) \\
&= \alpha P_{e_1} + \beta P_{e_2}.
\end{aligned}
$$

We observe that the linearity property comes from the fact that the scalar product of two vectors is also a linear operation.

The Cartesian space $\Re^3$ is an example of the general concept of vector spaces. Figure 2.32 illustrates the relationship between the sets of kinematical actions $\mathbf{v}$ and the force vectors $\mathbf{f}$ for particles in $\Re^3$, using the power associated to the kinematics. This relationship is denominated duality relation between the kinematical action $\mathbf{v}$ and the force resultant $\mathbf{f}$ in the particle. The duality concept will be constantly employed in this book.

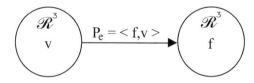

Figure 2.32   Duality relationship between the spaces of motion and forces for a particle.

2.7   FINAL COMMENTS

In this chapter, we reviewed the concept of particles and rigid bodies, along with the static equilibrium conditions using the Newtonian and analytical approaches.

Initially, diagrammatic conventions for supports and loadings were considered. After that, the equilibrium conditions for particles and rigid bodies were presented, as well as the PVP and the PVW.

In reference to the analytical approach, the main concepts that must be assimilated are motion, work, virtual displacement, and the PVW. Such concepts will be employed in the next chapters to model deformable bodies. To do this, we will consider the strain internal work to find the internal forces, which are compatible with the model kinematics.

There is no doubt that the analytical approach induces the reader to think, and more effort is required to model the physics of the problem. But when this is done once, the fundamental concepts are assimilated and will never be forgotten.

2.8 PROBLEMS

1. Determine the reactions of the rigid bodies illustrated in Figure 2.33, employing Newton equilibrium conditions and the principle of virtual work.

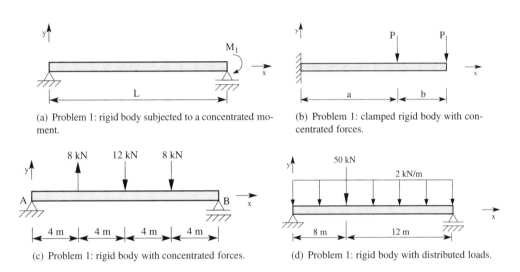

(a) Problem 1: rigid body subjected to a concentrated moment.

(b) Problem 1: clamped rigid body with concentrated forces.

(c) Problem 1: rigid body with concentrated forces.

(d) Problem 1: rigid body with distributed loads.

Figure 2.33 Problem 1.

2. Determine the reaction forces acting on the truss shown in Figure 2.34. Use the Newton method and the principle of virtual work. The weight of the truss is negligible.

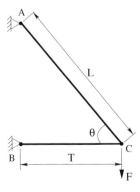

Figure 2.34 Problem 2.

3. Using the principle of virtual work, find the reactions at point A of the body in Figure 2.35. Check the obtained results by applying the Newton method.

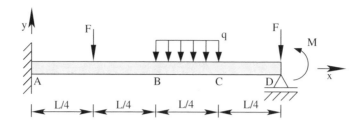

Figure 2.35 Problem 3.

4. Using the principle of virtual work, find the force P that maintains the equilibrium of the structure illustrated in Figure 2.36.

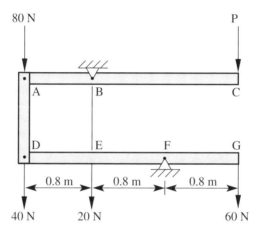

Figure 2.36 Problem 4.

5. Consider the ladder with length $L = 5$ m and weight $P = 100$ N illustrated in Figure 2.37, subjected to forces $F_1 = 200$ N and $F_2 = 150$ N. Calculate the minimum required dimension a such that it stands in equilibrium.

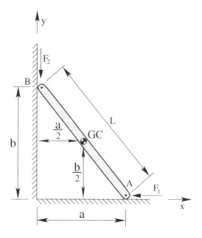

Figure 2.37 Problem 5.

3 FORMULATION AND APPROXIMATION OF BARS

3.1 INTRODUCTION

In this chapter, we present the mechanical model of bars. They are subjected to stretching and shortening, giving rise to traction and compression loads. This is the first case of deformable bodies considered in this book.

We employ the PVW to obtain the bar model. In the case of deformable bodies, the PVW states that the work done by the internal and external loads must be equal for any compatible virtual kinematic action from the deformed equilibrium position. The notion of virtual displacements, work, and PVW presented in the previous chapter will be important in the formulation of the bar model. We assume small displacements and strains and linear elastic material behavior whose constitutive equation is given by Hooke's law. The variational formulation of the bar model follows the steps presented in Section 1.1.4.

Subsequently, we consider the variational numerical methods applied to the bar model, based on the weighted residual methods, such as collocation, least squares and Galerkin methods. The low-order finite element method (FEM) is applied to bars and MATLAB programs for the analysis and design of trusses are presented.

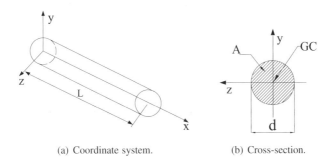

(a) Coordinate system. (b) Cross-section.

Figure 3.1 Bar of length L with the adopted coordinate system.

3.2 KINEMATICS

A bar is a structural element whose main geometric feature is to have the length much larger than the cross-section dimensions. This allows us to model it with a one-dimensional mathematical model, by analyzing its behavior along the longitudinal axis x shown in Figure 3.1. For a bar of length L and circular cross-section of diameter d, the geometrical hypothesis is equivalent to assuming that $L \gg d$. The origin of the adopted coordinate system is at the geometric center (GC) of the left end cross-section. The x axis is called longitudinal, while y and z are the transversal axes.

Figure 3.2(a) illustrates a bar in the initial undeformed configuration. The cross-sections of the bar are indicated by the vertical lines which are perpendicular to the longitudinal axis x. The kinematics of the bar model is such that all points of cross-section x have the same axial displacement $u_x(x)$ in the longitudinal direction x, as illustrated in Figure 3.2. This implies that the cross-sections remain perpendicular to the deformed axis x', as shown in Figure 3.2(b).

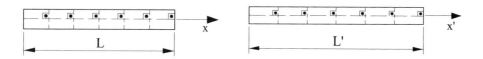

(a) Initial configuration (cross-sections are perpendicular to the bar axis).

(b) Deformed configuration (the cross-sections remain perpendicular to the longitudinal axis).

Figure 3.2   Bar model kinematics.

As all points of any cross-section have the same axial displacement, it is enough to consider only one point of each section in order to derive the bar model. In this case, we take the point located at the geometric center of the cross-section. Thus, in spite of the bar being a physically three-dimensional body, we can employ a one-dimensional mathematical model due to the adopted geometrical and kinematic hypotheses.

It is also verified that the cross-sections remain plane and normal to the x axis, as shown in Figure 3.3. We consider in this case the bar fixed at the left end $x = 0$ only to make simpler the representation of the kinematics. For any bar cross-section x, the axial displacement is given by the difference between the final x' and initial x positions of the section, that is,

$$u_x(x) = x' - x. \tag{3.1}$$

If $u_x(x) > 0$, there is a stretching of the x cross-section. If $u_x(x) < 0$, there is a shortening of the x cross-section. Hence, the one-dimensional kinematics of the bar model consists of axial displacements $u_x(x)$, causing only stretching or shortening of the bar cross-sections. The axial denomination indicates that the displacement is along the bar length, that is, in the x axis of the adopted reference system. The function $u_x(x)$ is assumed to be continuous.

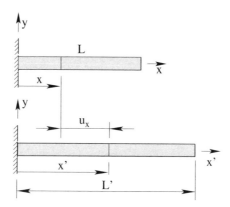

Figure 3.3   Axial displacement $u_x(x)$ of the cross-section x after the bar stretching.

The kinematic constraints are represented by supports. For instance, the bar is fixed at $x = 0$ in Figure 3.3. This implies that the axial displacement is zero in this section, i.e., $u_x(x = 0) = 0$.

Despite the use of a clamp symbol to represent the support in Figure 3.3, the only valid restriction is $u_x(x=0)=0$, because the bar kinematics is being considered in this case.

The following example illustrates an axial displacement function to a bar subjected to displacement at the free end.

Example 3.1 *Consider the bar with length L, supported at $x = 0$ and subjected to an axial displacement δ at the free end $x = L$, as illustrated in Figure 3.4(a).*

The function $u_x(x)$, which gives the axial displacement of the points of any bar section, in this example is

$$u_x(x) = \frac{\delta}{L}x.$$

This function will be determined later by the solution of the equations obtained at the end of the bar model formulation (see Example 3.7). Therefore, from the formulation point of view, the starting point is the kinematics. However, the determination of the axial displacement function for a bar subjected to certain loads is only possible by solving the differential equation, which will be obtained at the end of the formulation.

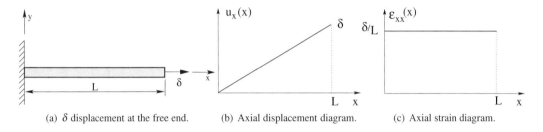

(a) δ displacement at the free end.   (b) Axial displacement diagram.   (c) Axial strain diagram.

Figure 3.4   Example 3.1: bar with displacement at the free end.

The axial displacement diagram of $u_x(x)$ is a plot of the function which provides the value of axial displacement of the points for any cross-section x. The displacement diagram of this example is illustrated in Figure 3.4(b). It is observed that $u_x(x)$ satisfies the kinematic constraint of zero displacement at $x = 0$, that is, $u_x(x=0)=0$. Additionally, $u_x(x)$ is a continuous function.
    □

Figures 3.5(a) and 3.5(d) illustrate the constant axial displacement distribution in a bar cross-section x to the stretching and shortening cases.

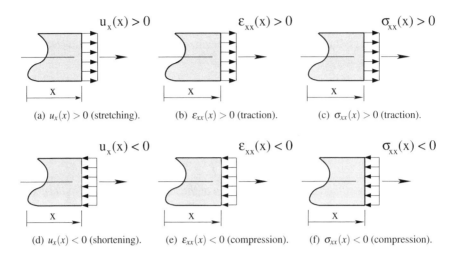

(a) $u_x(x) > 0$ (stretching).   (b) $\varepsilon_{xx}(x) > 0$ (traction).   (c) $\sigma_{xx}(x) > 0$ (traction).

(d) $u_x(x) < 0$ (shortening).   (e) $\varepsilon_{xx}(x) < 0$ (compression).   (f) $\sigma_{xx}(x) < 0$ (compression).

Figure 3.5 Axial displacement, normal strains, and normal stress distributions in the bar x cross-section for stretching and shortening cases.

3.3 STRAIN MEASURE

Suppose that the maximum axial displacement in a bar is 1 mm. This absolute value cannot exactly represent the physical bar deformation process. For a bar with length 100 mm, the given axial displacement represents 1% of the bar length, allowing us to conclude that it is in the range of small deformation. However, if the length is 10 mm, the displacement represents 10% of the initial length, representing a regime of large deformation. Thus, a specific measure of relative displacements, also referred as strain measure, carries information about the actual deformation process of the body. Generally, the strain measure represents the relative specific kinematics between any two points of a body, as will be seen below.

Consider the bar of length L illustrated in Figure 3.6. To obtain the strain measure at the points of each section x of this bar, we compare the relative displacements of two sections, $x + \Delta x$ and x, as illustrated in Figure 3.6.

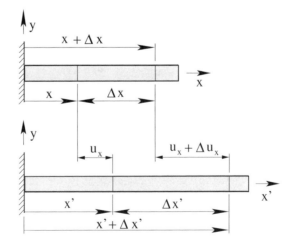

Figure 3.6 Relative kinematics between sections x and $x + \Delta x$ of the bar.

The axial displacements of sections x and $x + \Delta x$ are given from the kinematics defined in (3.1), respectively, by

$$
\begin{aligned}
u_x(x) &= x' - x, \\
u_x(x + \Delta x) &= (x' + \Delta x') - (x + \Delta x) = (x' - x) + (\Delta x' - \Delta x).
\end{aligned}
$$

From the previous expressions, we obtain the following relationship for the displacements of sections x and $x + \Delta x$:

$$
u_x(x + \Delta x) = u_x(x) + \Delta x' - \Delta x. \tag{3.2}
$$

The relative displacement between the points of the considered sections is defined as

$$
\begin{aligned}
\Delta u_x &= u_x(x + \Delta x) - u_x(x) \\
&= (x' - x) + (\Delta x' - \Delta x) - (x' - x) \\
&= \Delta x' - \Delta x. \tag{3.3}
\end{aligned}
$$

We obtain the specific relative displacement by taking the ratio of the previous expression and the initial distance between the sections, in other words,

$$
\frac{\Delta u_x}{\Delta x} = \frac{\Delta x' - \Delta x}{\Delta x} = \frac{u_x(x + \Delta x) - u_x(x)}{\Delta x}.
$$

As we work with the continuous media hypothesis, we can take the distance Δx as small as desired. Thus, considering the two sections x and $x + \Delta x$ infinitesimally close to each other, we define the longitudinal or normal specific strain measure $\varepsilon_{xx}(x)$ for the x cross-section by taking the limit of the previous equation for Δx going to zero, i.e.,

$$
\varepsilon_{xx}(x) = \lim_{\Delta x \to 0} \frac{\Delta u_x}{\Delta x}.
$$

Using the derivative definition of a single variable function, the general expression of the specific normal strain measure ε_{xx} is

$$
\varepsilon_{xx}(x) = \lim_{\Delta x \to 0} \frac{\Delta u_x}{\Delta x} = \frac{du_x(x)}{dx}. \tag{3.4}
$$

The first subscript x in ε_{xx} indicates the plane in which the strain measure is present, and the second subscript indicates the direction. Therefore, the specific normal strain ε_{xx} is at the x normal plane to the cross-section and in the direction of the x axis. It represents the rate at which the axial displacement varies in each section of the bar.

Substituting (3.3) in (3.2), we have

$$
u_x(x + \Delta x) = u_x(x) + \Delta u_x.
$$

Taking the limit for $\Delta x \to 0$ in the previous equation, and employing definition (3.4), we obtain

$$
u_x(x + dx) = u_x(x) + du_x = u_x(x) + \varepsilon_{xx}(x)dx. \tag{3.5}
$$

A question that could be raised in the definition of the specific strain measure is why divide the relative axial displacement by the initial distance Δx and not by the final distance $\Delta x'$ between the cross-sections. The answer comes from the fact that we assume a small deformation regime. In this case, both distances are almost equal, i.e., $\Delta x \approx \Delta x'$ and thus the difference is negligible when using any of the two distances.

Example 3.2 *Consider the bar of Example 3.1. The specific normal strain ε_{xx} can be obtained with the derivation of the displacement equation $u_x(x) = \frac{\delta}{L}x$, namely,*

$$\varepsilon_{xx}(x) = \frac{\delta}{L}.$$

The strain measure is constant in all sections of the bar under consideration, as illustrated in Figure 3.4(c).

$\square$

The strain measure $\varepsilon_{xx}(x)$ is called specific, since it is a dimensionless quantity indicating the percentage of stretching or shortening in each section of the bar. The term longitudinal indicates that strain occurs along the bar length. This is consistent with the defined kinematics. Note that if $u_x(x)$ is positive, we have a tractive strain measure in section x of the bar. Likewise, if $u_x(x)$ is negative, there is a compressive strain measure in section x. Figures 3.5(b) and 3.5(e) illustrate these two cases for a generic x cross-section.

3.4 RIGID ACTIONS

A rigid body action is that for which there is no relative displacement between any two points of the body. Thus, the strain measure components are zero.

For the bar case, this implies that the strain measure $\varepsilon_{xx}(x)$ is zero for any cross-section. Hence,

$$\varepsilon_{xx}(x) = \frac{du_x(x)}{dx} = 0, \quad x \in (0, L). \tag{3.6}$$

To satisfy the previous condition, the axial displacement function must be constant (cte) for every x cross-section, i.e., $u_x(x) = u_x =$ cte. Physically, this implies that the rigid action is a translation of the bar along the x axis, as illustrated in Figure 3.7.

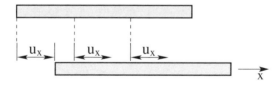

Figure 3.7 Rigid body displacement for the bar: translation u_x along x axis.

Example 3.3 *For the bar of Example 3.1, we have $u_x(x = 0) = 0$, because the bar is fixed at the end $x = 0$. As the rigid displacement is constant for all sections of the bar, and the kinematic constraint must be satisfied, the rigid displacement in this case is $u_x = 0$.*

$\square$

3.5 DETERMINATION OF INTERNAL LOADS

Consider the bar of length L and the differential volume element $dV = dA \, dx$ illustrated in Figure 3.8(a). After stretching, the bar will have a final length L' and the volume element will elongate du_x in the axial direction, as shown in Figure 3.8(b). Intuitively, it is known that the larger the bar stretches, the larger the work or energy required to perform it. As this work is associated with the straining of the bar, it is named strain internal work and denoted by W_i. Thus, the internal work is a function of the strain measure, namely,

$$W_i = f(\varepsilon_{xx}). \tag{3.7}$$

In order to quantify it, it is necessary to describe the internal loads on the points of the bar under the deformation process.

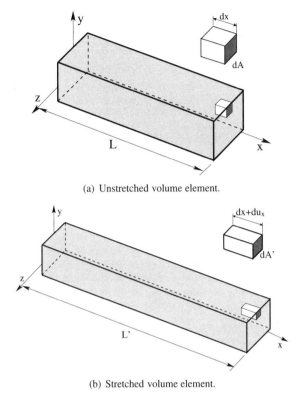

(a) Unstretched volume element.

(b) Stretched volume element.

Figure 3.8 Differential volume elements before and after the bar stretching.

In the case of the volume element dV, we consider the internal work density dw_i, given by the ratio of the strain internal work dW_i and the volume dV, i.e.,

$$dw_i = \frac{dW_i}{dV} = g(\varepsilon_{xx}). \tag{3.8}$$

Due to the proportionality between the internal work and the strain measure, the relationship between these quantities is linear and the previous equation can be written as

$$dw_i = \frac{dW_i}{dV} = \sigma_{xx}\varepsilon_{xx}. \tag{3.9}$$

Considering a dimensional analysis of the previous expression in the International System of Units (SI), we have

$$\frac{\text{Nm}}{\text{m}^3} = \frac{\text{N}}{\text{m}^2}\frac{\text{m}}{\text{m}}.$$

Therefore, for the left side to represent the work density, the internal forces in the bar must necessarily be expressed by a force density per area, called stress. The term σ_{xx} is designated as the normal stress, because it is present in the perpendicular direction of each cross-section, as illustrated in Figures 3.5(c) and 3.5(f). Positive values indicate tractive normal stresses [see Figure 3.5(c)]. Negative values indicate compressive normal stresses [see Figure 3.5(f)]. Thus, the internal loads for a deformed bar are given by the normal stress component σ_{xx}. This result is a consequence of the concepts of strain and internal work.

From (3.9) and (3.4), the strain internal work in the volume element dV is

$$dW_i = \sigma_{xx}\frac{du_x}{dx}dV. \tag{3.10}$$

The strain internal work of the bar is obtained by taking the sum of the work for each differential volume element. Due to the continuity assumption, there are infinite differential volume elements, and an infinite sum is represented by the following Riemann integral

$$W_i = \int_V dW_i = \int_V \sigma_{xx}(x)\varepsilon_{xx}(x)dV = \int_V \sigma_{xx}(x)\frac{du_x(x)}{dx}dV. \tag{3.11}$$

Due to the assumed one-dimensional kinematics for the bar, the stress and strain components vary only from one section to another, namely, only with the x coordinate. Thus, we can express the previous volume integral as the product of an area integral and integral over the bar length. Hence,

$$W_i = \int_0^L \left(\int_A \sigma_{xx}(x)dA \right) \frac{du_x(x)}{dx}dx. \tag{3.12}$$

The product

$$dN_x = \sigma_{xx}(x)dA$$

has units of force and represents a differential normal force dN_x acting on the area element dA, as shown in Figure 3.9. Note that dN_x is in the same x direction of the normal stress σ_{xx}.

The normal force $N_x(x)$ in the cross-section x is given by the sum of the normal force differentials dN_x acting on each dA area element. Again, as there are infinite area elements in the x section, the normal force is given by the following area integral:

$$N_x(x) = \int_A dN_x = \int_A \sigma_{xx}(x)dA. \tag{3.13}$$

As the stress is constant to all points of a cross-section, the previous expression assumes the following form:

$$N_x(x) = \sigma_{xx}(x)\int_A dA = \sigma_{xx}(x)A(x). \tag{3.14}$$

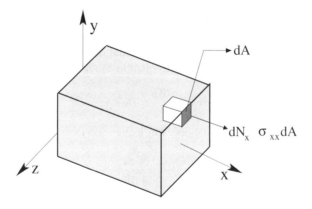

Figure 3.9   Normal force differential dN_x in the differential area element dA.

Based on that, the internal work expression (3.12) summarizes to

$$W_i = \int_0^L N_x(x)\frac{du_x(x)}{dx}dx. \tag{3.15}$$

In order to write the previous expression in the most usual form of work, given by a product of force and displacement, we apply the integration by parts, resulting in

$$W_i = -\int_0^L \frac{dN_x(x)}{dx} u_x(x)dx + N_x(L)u_x(L) - N_x(0)u_x(0). \tag{3.16}$$

Thus, the compatible internal loads with the adopted kinematics for a bar are illustrated in Figure 3.10(a) and are denominated as

- $\frac{dN_x(x)}{dx}$: distributed normal force along the bar length, with SI units of N/m, and represented by a continuous function
- $N_x(0)$ and $N_x(L)$: concentrated normal forces at the ends $x = 0$ and $x = L$

It should be observed that the direction of forces in the FBD of Figure 3.10(a) is indicated by the signs of expression (3.16).

3.6 DETERMINATION OF EXTERNAL LOADS

The internal forces which are present in a strained bar are indicated in equation (3.16). The external forces that may be applied to a bar are only those ones that can be balanced by the kinematically compatible internal forces. Thus, we have the following external forces acting on a bar, as shown in the FBD of Figure 3.10(b):

- $q_x(x)$: axial distributed force along the bar length, with SI units of N/m, and represented by a continuous function
- P_0 and P_L: axial concentrated forces at the $x = 0$ and $x = L$ ends

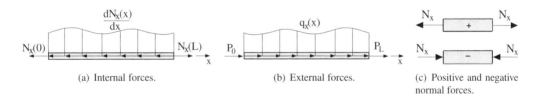

(a) Internal forces.   (b) External forces.   (c) Positive and negative normal forces.

Figure 3.10 Free body diagrams of internal and external forces of a bar and positive and negative normal forces.

Analogous to (3.16), the work W_e of the external loads for any axial displacement $u_x(x)$ is

$$W_e = P_0 u_x(0) + P_L u_x(L) + \int_0^L q_x(x)u_x(x)dx. \tag{3.17}$$

Note that external loads are represented in the positive direction of x in the FBD of Figure 3.10, analogously to the equilibrium cases of the previous chapter. If any of these loads are found to be negative at the end of the equilibrium process, it will be necessary to change the direction of that load.

It is verified that the external and internal concentrated forces, which are compatible with the bar kinematics, are present only at the bar ends.

3.7 EQUILIBRIUM

The balance between the internal and external loads will be determined using the PVW. For deformable bodies, this principle states that if the body is in static equilibrium, the external (δW_e) and internal (δW_i) works are equal for any virtual displacement from the deformed equilibrium position of the body, namely,

$$\delta W_e = \delta W_i. \tag{3.18}$$

Substituting (3.16) and (3.17) in the previous PVW statement, for any virtual displacement δu_x, we obtain

$$P_0 \delta u_x(0) + P_L \delta u_x(L) + \int_0^L q_x(x) \delta u_x(x) dx = N_x(L) \delta u_x(L) - N_x(0) \delta u_x(0)$$

$$- \int_0^L \frac{dN_x(x)}{dx} \delta u_x(x) dx.$$

Rearranging the previous expression, we have

$$- [N_x(0) + P_0] \delta u_x(0) + [N_x(L) - P_L] \delta u_x(L) - \int_0^L \left[\frac{dN_x(x)}{dx} + q_x(x) \right] \delta u_x(x) dx = 0. \tag{3.19}$$

As the virtual displacement $\delta u_x(x)$ is arbitrary, all terms inside the brackets must be simultaneously zero resulting in

$$\begin{cases} \dfrac{dN_x(x)}{dx} + q_x(x) = 0 & \text{in } x \in (0, L) \\ N_x(L) = P_L & \text{in } x = L \\ N_x(0) = -P_0 & \text{in } x = 0 \end{cases}. \tag{3.20}$$

The above expression defines the local form of the bar equilibrium problem free of any kinematic constraints. It is given by the differential equation in terms of the normal force and two boundary conditions. This set (differential equation + boundary conditions) defines the boundary value problem (BVP). In this case, it is an equilibrium BVP, since both the differential equation and the boundary conditions express the equilibrium of internal and external forces. The solution of the previous BVP results in the function for the normal force $N_x(x)$ along the x axis of the bar. A positive value of N_x indicates that section x is under tensile forces. On the other hand, a negative value of $N_x(x)$ represents a compressive normal force in the x section. The graph of the function $N_x(x)$ is commonly known as the normal force diagram.

The boundary conditions in (3.20) indicate that the positive normal force $N_x(L)$ at the bar's right end has the same direction as the external axial force P_L. Similarly, the positive normal force $N_x(0)$ at the bar left end has the opposite direction of the external axial force P_0. In this case, both internal $N_x(0)$ and $N_x(L)$ or external P_0 and P_L forces physically indicate a traction of the bar, in the absence of distributed loads. If the signals are changed, the bar is under compression. Figure 3.10(c) illustrates tractive and compressive normal forces, according to the signs of the boundary conditions in (3.20).

If $\delta u_x(x)$ is a rigid virtual displacement, the strain measure is zero and hence the internal work is also zero. In this case, the PVW given in (3.18) states that for any virtual rigid action $\delta u_x(x)$ for the bar in equilibrium, the external work, given by (3.17) is zero, namely,

$$P_0 \delta u_x(0) + P_L \delta u_x(L) + \int_0^L q_x(x) \delta u_x(x) dx = 0. \tag{3.21}$$

The bar rigid actions are constant axial displacements for all cross-sections, i.e., translations in the x direction. Therefore, we have $\delta u_x(x) = \delta u_x = \text{cte}$ and substituting in the previous expression becomes

$$\left(P_0 + P_L + \int_0^L q_x(x) dx \right) \delta u_x = 0.$$

From this point, the rigid body equilibrium condition for the bar is obtained, stating that the resultant of external forces in the axial direction x must be zero

$$\sum f_x = 0: \ P_0 + P_L + \int_0^L q_x(x)dx = 0. \tag{3.22}$$

The term resulting from the integral along the bar length represents the equivalent concentrated force of the external axial distributed load $q_x(x)$. In terms of equilibrium point of view, the integral may be replaced by an axial concentrated force. However, for deformed bars, this substitution may result in large differences in terms of the distribution of internal forces in the bar.

Before defining the material behavior, we present some examples to obtain the normal force function for the bar by integrating the BVP (3.20). To solve this type of problem and to determine the normal force diagram, we apply the following general steps which will also be used in other chapters:

1. Write the equation of external distributed loads.
2. Write the boundary conditions and additional constraints (supports along the length of the structural element).
3. Integrate the differential equation of the considered problem.
4. Determine the integration constants by applying the boundary conditions.
5. Write the final equations.
6. Plot the diagrams.
7. Determine the support reaction forces.

The important point to be observed is that the support reactions are automatically obtained from the BVP solution, because the differential equation indicates explicitly the balance of internal and external loads. Thus, the same procedure can be applied without modifications for hyperstatic problems, but using the equations in terms of kinematics to be obtained after the consideration of the material behavior.

Example 3.4 *Plot the normal force diagram to the bar illustrated in Figure 3.11, subjected to a concentrated force P at the free end, by integrating the differential equation of equilibrium in terms of the normal force.*

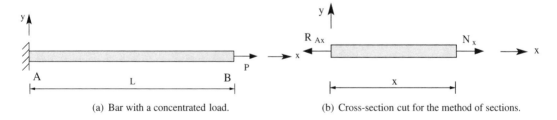

(a) Bar with a concentrated load.   (b) Cross-section cut for the method of sections.

Figure 3.11 Example 3.4: bar subjected to a concentrated force P at the free end.

We apply the previous procedure in order to solve this problem.

- *Distributed load equation*
  As there are no axial distributed loads applied on the bar, we have $q_x(x) = 0$.
- *Boundary conditions*
  There is a concentrated force P at the bar free end ($x = L$), which will be handled as a boundary condition. As its effect is to traction the bar, we have $N_x(x = L) = P$. The bar is fixed at the left end $x = 0$, and thus the axial displacement is zero, $u_x(x = 0) = 0$. However,

in this example we integrate the differential equation in terms of the normal force, and this boundary condition is of no use to solve the present problem. Subsequently, the use of the boundary conditions in terms of the axial displacement will be presented. In the $x = 0$ cross-section, we have also $N_x(x = 0) = -R_{Ax}$ as a boundary condition, where R_{Ax} is the support reaction force at $x = 0$. This reaction force is still unknown and therefore the above boundary condition will not be used. Obviously, this reaction force could be calculated by using the $\sum f_x = 0$ equilibrium condition. However, this is not necessary because the support reaction will naturally be obtained from the solution of the BVP given in (3.20).

· *Integration of the differential equation*
  The integration of the equilibrium differential equation in terms of the normal force gives

$$\frac{dN_x(x)}{dx} = -q_x(x) = 0 \rightarrow N_x(x) = C_1,$$

where C_1 is the integration constant.

· *Determination of the integration constant*
  In order to determine C_1, we apply the boundary condition in terms of the normal force. Thus,

$$N_x(x = L) = C_1 = P \rightarrow C_1 = P.$$

· *Final equation*
  By substituting C_1, the final normal force equation is obtained, namely,

$$N_x(x) = P.$$

· *Normal force diagram*
  The normal force diagram is shown in Figure 3.12 for $P = 100$ N and $L = 1$ m. It is observed that the normal force is constant for all cross-sections of the bar.

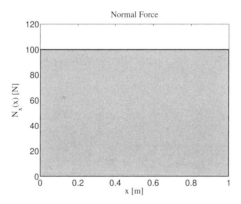

Figure 3.12   Example 3.4: normal force diagram.

· *Support reaction*
  In this case, it is initially assumed that the support reaction R_{Ax} is in the positive direction of the x axis. There are two ways to determine the reaction. The simplest one is to use the boundary condition at $x = 0$ of the equilibrium BVP in terms of the normal force given in (3.20). In this case, the R_{Ax} reaction is P_0 and

$$R_{Ax} = -N_x(x = 0) = P = -100 \text{ N}.$$

The value of R_{Ax} is negative, meaning that the positive reaction is to the left and therefore opposite to the positive direction of axis x.

The second way is to calculate the reaction employing the equation for the rigid body equilibrium given in (3.22). Hence,

$$\sum f_x = 0: \ R_{Ax} + P + \int_0^L 0 dx = 0 \rightarrow R_{Ax} = -P = -100 \text{ N}.$$

This value can also be obtained directly from the normal force diagram.
The interpretation of the normal force diagram is shown in Figure 3.11(b). There is a cut on the generic section x of the bar and the left portion is considered. It is observed that there must be a normal force $N_x(x)$ in this section which balances the support reaction. By applying the balance of forces in the x direction, we have

$$\sum f_x = 0: \ -R_{Ax} + N_x(x) = 0 \rightarrow N_x(x) = 100 \text{ N}.$$

This procedure of cutting through different sections and taking the balance of isolated portions of the bar is called the method of sections and is usually employed in strength of materials.

☐

Example 3.5 *Plot the normal force diagram for the bar illustrated in Figure 3.13, subjected to a distributed axial force of constant intensity p_0, by integrating the differential equation.*

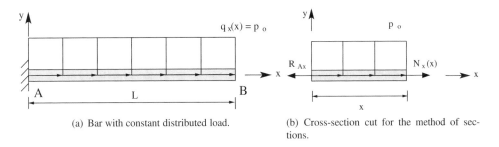

(a) Bar with constant distributed load.　　(b) Cross-section cut for the method of sections.

Figure 3.13　Example 3.5: bar subjected to a constant distributed axial force.

- *Distributed load equation*
  The bar is subjected to a distributed axial of constant intensity p_0 in the positive direction of x, and thus, $q_x(x) = p_0$.
- *Boundary conditions*
  The axial displacement at $x = 0$ is zero, because the bar is fixed at this end. Thus, $u_x(x = 0) = 0$. As there are no concentrated forces applied at the free end, the normal force is zero, in other words, $N_x(x = L) = 0$.
- *Integration of the differential equation*
  The integration of the differential equation in terms of the normal force gives

$$\frac{dN_x(x)}{dx} = -q_x(x) = -p_0 \rightarrow N_x(x) = -p_0 x + C_1,$$

  where C_1 is the integration constant.
- *Determination of the integration constant*
  The boundary condition in terms of the normal force is used to determine C_1, i.e.,

$$N_x(x = L) = -p_0(L) + C_1 = 0 \rightarrow C_1 = p_0 L.$$

- *Final equation*
  The final expression for the normal force is obtained by substituting the integration constant, i.e.,

$$N_x(x) = -p_0(x - L).$$

- *Normal force diagram*
  The normal force diagram is shown in Figure 3.14 for $L = 2$ m and $p_0 = 10$ N/m. Due to the constant distributed load applied on the bar, the normal force varies linearly.

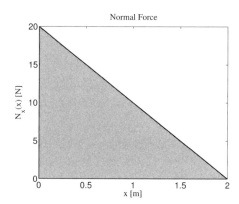

Figure 3.14   Example 3.5: normal force diagram.

- *Support reaction*
  The support reaction R_{Ax} at the $x = 0$ bar end may be obtained by the boundary condition of the BVP given in (3.20), namely, $R_{Ax} = -N_x(x = 0) = -p_0 L = -20$ N. Therefore, the positive direction of the reaction is leftwards, and thus opposite to the positive x axis. The intensity of the reaction force can also be obtained directly from the normal force diagram. Another way of calculating the reaction force is using the rigid body equilibrium equation given in (3.22), i.e.,

$$\sum f_x = 0: \ R_{Ax} + \int_0^L p_0 dx = 0 \rightarrow R_{Ax} = -p_0 L = -20 \text{ N}.$$

The interpretation of the normal force diagram is shown in Figure 3.13(b), by using the method of sections. A cut is made in a generic section x and the left portion of the bar is isolated. It is observed that a normal force $N_x(x)$ must exist in section x to balance the reaction force and the distributed load. Taking the balance of forces in the x direction, we have

$$\sum f_x = 0: \ -R_{Ax} + p_0 x + N_x(x) = 0 \rightarrow N_x(x) = 20 - 10x = 10(2 - x).$$

Thus, we recover the same expression for the normal force obtained by integrating the differential equilibrium equation.

☐

Example 3.6 *Plot the normal force diagram for the bar illustrated in Figure 3.15, subjected to a linear distributed load, by integration of the differential equation.*

- *Distributed load equation*
  The expression for the linear distributed force is $q_x(x) = p_0 \dfrac{x}{L}$. Observe that $q_x(x = 0) = 0$ and $q_x(x = L) = p_0 = 2000$ N/m.

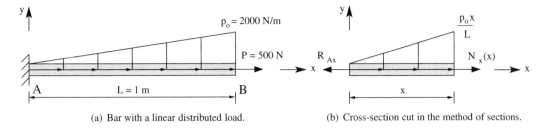

(a) Bar with a linear distributed load.

(b) Cross-section cut in the method of sections.

Figure 3.15 Example 3.6: bar subjected to a linear distributed axial force.

- *Boundary conditions*
  The axial displacement at $x = 0$ is zero, because the bar is supported at this end. Thus, $u_x(x = 0) = 0$. Because of the concentrated force applied at $x = L$, we have $N_x(x = L) = P = 500$ N.
- *Integration of the differential equation*
  The integration of the differential equation of equilibrium in terms of the normal force gives

$$\frac{dN_x(x)}{dx} = -q_x(x) = -p_0\frac{x}{L} \rightarrow N_x(x) = -\frac{p_0}{2L}x^2 + C_1.$$

- *Determination of the integration constant*
  By applying the boundary condition in terms of the normal force, we have

$$N_x(x = L) = -\frac{p_0}{2L}(L)^2 + C_1 = P \rightarrow C_1 = P + \frac{p_0 L}{2}.$$

- *Final equation*
  The final equation for the normal force is obtained by substituting the integration constant, i.e.,

$$N_x(x) = -\frac{p_0}{2L}x^2 + \left(P + \frac{p_0 L}{2}\right).$$

- *Normal force diagram*
  The normal force expression for $L = 1$ m, $p_0 = 2000$ N/m and $P = 500$ N is given by

$$N_x(x) = -1000x^2 + 1500.$$

Note that due to the linear distributed load, the normal force varies as a parabola. The normal force diagram is illustrated in Figure 3.16.
- *Support reaction*
  It is initially assumed that the support reaction R_{Ax} is oriented in the positive direction of the x axis. In this case, the reaction R_{Ax} at bar end $x = 0$ is given by the boundary condition $R_{Ax} = -N_x(x = 0) = -1500$ N. Thus, the positive direction of the reaction is leftwards and therefore contrary to the positive x axis direction. The reaction force intensity can also be obtained from the normal force diagram.
  Another way of calculating the reaction force is via the rigid body equilibrium equation given in (3.22), i.e.,

$$\sum f_x = 0: R_{Ax} + P + \int_0^L \frac{p_0}{L}x\,dx = 0 \rightarrow R_{Ax} = -P - \frac{p_0 L}{2} = -1500 \text{ N}.$$

The interpretation of the normal force diagram is shown in Figure 3.15(b). Again, the method of sections is applied. A cut is made in a generic x section of the bar and the left

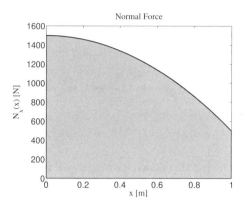

Figure 3.16   Example 3.6: normal force diagram.

portion is isolated. It is observed that this section must have a normal force $N_x(x)$ which balances the support reaction and the distributed load. Making the balance of forces in the x direction, we have

$$\sum f_x = 0: \; -R_{Ax} + \int_0^x \frac{p_0}{L} x dx + N_x(x) = 0 \rightarrow N_x(x) = -1000x^2 + 1500.$$

And thus recovering the same expression of the normal force as the one obtained by integrating the differential equation.

□

3.8 MATERIAL BEHAVIOR

Until this point of the bar formulation, the only assumed hypothesis concerned small displacements and strains, without considering the material behavior. In this section, a class of materials is presented, characterized by the constitutive equation called Hooke's law. Generally, a constitutive equation defines a relationship between the stress and strain measures. In the case of a bar, these measures are given, respectively, by the normal stress σ_{xx} and the normal strain ε_{xx}. Hooke's law will be introduced from the tensile test diagram.

3.8.1 EXPERIMENTAL TRACTION AND COMPRESSION DIAGRAMS

Consider the bar of length L and cross-section area A, subjected to a tensile axial force P at the its free end [see Figure 3.17(a)]. Due to the P force, the normal force at each cross-section x is constant with intensity P, namely, $N_x(x) = P$, as shown in Example 3.1. Consequently, the normal stress $\sigma_{xx}(x) = \sigma_{xx}$ will also be constant in each cross- section and given from expression (3.14) by

$$\sigma_{xx} = \frac{P}{A}. \tag{3.23}$$

The bar will elongate by δ, given by the difference of the final L' and initial L lengths of the bar, i.e.,

$$\delta = L' - L. \tag{3.24}$$

As the normal force and stress are constant in all the bar cross-sections, the specific normal strain $\varepsilon_{xx}(x) = \varepsilon_{xx}$ is also constant and is given by the quotient of the stretching δ and the initial length L

(see Example 3.2). Hence,

$$\varepsilon_{xx} = \frac{\delta}{L}.$$ (3.25)

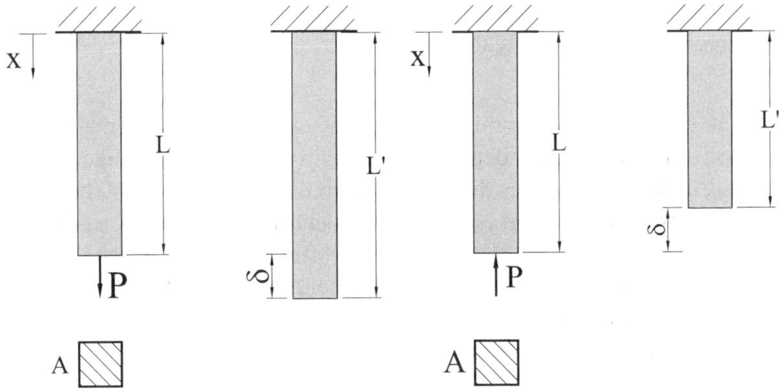

(a) Axial traction force.   (b) Axial compressive force.

Figure 3.17 Bar subjected to a P axial force at the end.

Figure 3.17(b) shows the bar subjected to an axial compressive force P. The normal force N_x, the normal compressive stress σ_{xx}, shortening δ, and the longitudinal strain ε_{xx} are given, respectively, by

$$N_x(x) = -P, \qquad \sigma_{xx}(x) = -\frac{P}{A}, \qquad \delta = L' - L, \quad \text{and} \quad \varepsilon_{xx}(x) = \frac{\delta}{L}.$$ (3.26)

The compressive stresses are considered negative. It is observed that the shortening δ is also negative, because the final length L' is smaller than the initial bar length L. Therefore, the strain measure ε_{xx} and normal stress σ_{xx} are also negative and therefore compressive.

For the bar in Figure 3.17, we intuitively know that if it is made of steel, the strain values will be smaller compared to a bar made of aluminum. Thus, each material has a distinct mechanical stiffness. It is given in terms of material properties, which are determined by experiments. The type of test to be performed is indicated by the constitutive equation of the material. This equation represents the behavior of an idealized material class and generally establishes a relation between the stress and strain measures.

An example of a constitutive equation is Hooke's law, which is valid for homogeneous isotropic linear elastic materials. The denomination elastic means that the body deforms under loading, but returns to its initial shape when the loads are removed. Linear means there is a linear or proportional relation between the stress and strain measures. The term isotropic indicates that the material properties are the same for any direction under consideration. Homogeneous means that the constitution of the material is the same for any considered region. Metallic materials at certain temperature limits fit satisfactorily with the expected behavior predicted by Hooke's law.

The tensile and compression tests are performed to characterize the material properties. For this purpose, a cylindrical specimen with standardized dimensions is used [see Figure 3.18]. The specimen is assembled in a testing machine, which gradually applies an axial concentrated load P at the specimen end. For each value of P, the distance L' between two points, which is initially at distance L, is measured, determining the elongation or shortening $\delta = L' - L$. The normal stress $\sigma_{xx} = \frac{P}{A}$ and the strain $\varepsilon_{xx} = \frac{\delta}{L}$ are then calculated and plotted in the $\sigma_{xx} \times \varepsilon_{xx}$ stress-strain diagram. It is also possible to measure the diameter variation $\Delta d = d' - d$ of the specimen, allowing the characterization of strains at the cross-section, as will be seen below.

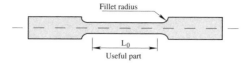

Figure 3.18 Specimen for traction and compression testing.

The shape of the stress-strain diagram varies significantly for different materials. For a given material, different results are obtained depending on environmental factors like the specimen temperature and loading conditions, such as the rate at which the load is applied. The common features found in traction and compression diagrams allow the identification of brittle and ductile classes of materials.

The main characteristic of ductile materials is to present a very pronounced yield phase at room temperature. Materials such as steel, aluminum, and copper are classified as ductile. Figure 3.19 illustrates typical diagrams for steel and aluminum.

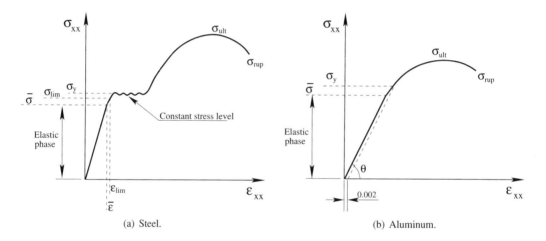

(a) Steel.   (b) Aluminum.

Figure 3.19 Tensile and compressive test diagrams for a ductile material specimen.

There is a region of the diagram in which the stress-strain relation is linear. This means that when an increasing load P is applied, the bar length L' increases proportionally to the applied load. Thus, the initial part of the diagram is well approximated by a straight line with a large angular coefficient. This linearity continues until reaching the the proportionality limit normal stress, denoted as $\sigma_{\lim}$. Stress values within the $0 \leq \sigma_{xx} \leq \sigma_{\lim}$ range characterize the elastic phase of the material; in other words, the material behaves elastically. This means that when force P is removed, the specimen recovers its original form. Generally, we design mechanical components in order to remain in the elastic phase, i.e., the maximum normal stress $\sigma_{xx}^{\max}$ should be within the interval $0 \leq \sigma_{xx}^{\max} \leq \sigma_{\lim}$.

For values above $\sigma_{\lim}$, the specimen begins to yield; in other words, it has large strains with a small load increment. In ductile materials, this strain is caused by the relative sliding of material layers along the oblique surfaces. At this stage of the diagram, the test specimen presents a permanent deformation, in such way that, when the P force is removed, the specimen returns to an unstressed state across the line BB', parallel to AA', as shown in Figure 3.20.

From the diagrams of Figure 3.19, it is observed that the stretching of the material during the yield phase is considerably larger than that observed in the linear elastic range. This greater elongation during yielding identifies, for some materials, a practically constant stress level in the diagram [see Figure 3.19(a)]. This value is called yield normal stress of the material and denoted by σ_y.

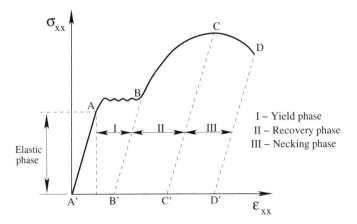

Figure 3.20 Phases of a traction/compression test diagram.

There is a recovery phase of the specimen after the yield stress. This phase is characterized by a stiffening of the material caused by a rearrangement of its particles, increasing the stiffness and, consequently, requiring larger loading values to deform the specimen. The maximum reached stress in this phase is called ultimate or maximum normal stress, and denoted by σ_{ult}.

After reaching this maximum value, a narrowing occurs in the specimen and the diameter decreases. Thus, a smaller force increment is sufficient to maintain the deformation process in the specimen until attaining rupture. The stress at which rupture occurs is referred to as rupture normal stress and is indicated as σ_{rup}. This fact proves that the rupture of ductile materials is due to the effect of friction between the cross-sections, which characterizes the shear stresses. Therefore, the maximum shear stresses due to the axial load occur in planes that form a 45° angle with the axial direction.

Referring to Figure 3.19(b), it is difficult to exactly determine the stress $\sigma_{\lim}$ which characterizes the linear elastic range. Thus, for safety reasons, we define the admissible normal stress $\bar{\sigma}$ of the material, given by the ratio of the yield stress σ_y and the safety factor k_s as

$$\bar{\sigma} = \frac{\sigma_y}{k_s}. \tag{3.27}$$

Therefore, the elastic range is characterized by normal stress values in the $0 \leq \sigma_{xx} \leq \bar{\sigma}$ interval.

Materials such as cast iron, glass, and rock are classified as brittle and characterized by a rupture of the specimen without any visible change in the mode of deformation. Thus, there is no difference between the ultimate and rupture stresses. Furthermore, no narrowing is observed in the specimen during the tensile test. Rupture occurs on a surface which is perpendicular to the axial loading line. Thus, it appears that the rupture in a brittle material is mainly due to the action of normal stresses. Figure 3.21 shows a typical tensile test diagram for a brittle material. It is also observed that the strain magnitude is much smaller in brittle materials when compared with ductile ones.

The angular coefficient $\tan \theta$ of the straight line of the linear elastic phase for brittle or ductile material test diagrams defines a material property called longitudinal elasticity modulus or Young modulus, which is denoted by the E. Thus, the straight-line equation which defines the elastic range is given by

$$\sigma_{xx} = E\varepsilon_{xx}. \tag{3.28}$$

This equation is known as Hooke's law for the case of simple traction and compression.

The longitudinal elasticity modulus provides an estimate of the stiffness of the material in the linear elastic range, i.e., the resistance to deform when subjected to loads. As ε_{xx} is a dimensionless

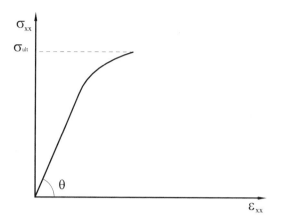

Figure 3.21   Typical tensile test diagram for a brittle material.

quantity, units of E are the same as the normal stress σ_{xx}, such as N/m^2 and Kgf/cm^2. The units N/m^2 are defined as Pascal (Pa). The reader should also remember the following relation for unit conversion

$$1\,\frac{Kgf}{cm^2} \approx \frac{10}{10^{-4}}\,\frac{N}{m^2} = 10^5\,\frac{N}{m^2} = 10^5\,Pa = 0.1\,MPa. \tag{3.29}$$

Typical values of the elasticity modulus for some materials are: steel ($E = 21 \times 10^5\,Kgf/cm^2 = 21 \times 10^{10}\,N/m^2 = 210\,GPa$), aluminum ($E = 7 \times 10^5\,Kgf/cm^2 = 7 \times 10^{10}\,N/m^2 = 70\,GPa$ and copper ($E = 11 \times 10^5\,Kgf/cm^2 = 11 \times 10^{10}\,N/m^2 = 110\,GPa$).

In general, the same behavior is observed in the test diagram for a ductile material subjected to an axial tensile or compressive force. Thus, the admissible tensile ($\bar{\sigma}_t$) and compressive ($\bar{\sigma}_c$) stresses are equal. For brittle materials, the compression stiffness is greater than the traction stiffness, implying $\bar{\sigma}_t < \bar{\sigma}_c$. The values of admissible ($\bar{\sigma}_t, \bar{\sigma}_c$), yield σ_y, rupture σ_{rup}, and ultimate σ_{ult} stresses are tabulated for several materials. These stress values can be changed by adding metallic alloys, thermal treatment, and manufacturing processes as illustrated in Figure 3.22. However, the elasticity modulus remains the same.

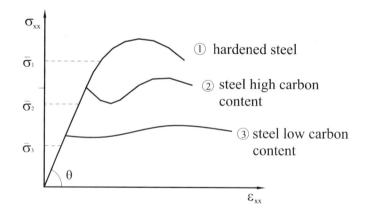

Figure 3.22   Behavior of the tensile test for steels of different chemical compositions.

By substituting (3.23) and (3.25) in (3.28), we have

$$P = \frac{EA}{L}\delta. \tag{3.30}$$

The term

$$k_b = \frac{EA}{L} \tag{3.31}$$

is denominated stiffness of the bar subjected to an axial force P at the free end. The bar of linear elastic material behaves as a spring. Remember that for a spring with elastic constant k_e, the force F in the spring and the elongation x are related by $F = k_e x$. In the case of a bar, we have $F = P$, $k_e = k_b$, and $x = \delta$. The analogy between the bar and spring elements is illustrated in Figure 3.23.

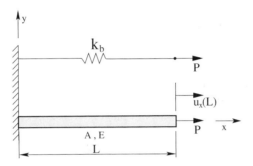

Figure 3.23 Bar-spring analogy.

3.8.2 POISSON RATIO

Due to the δ stretching of the specimen under the tensile test, there is a decrease of cross-section area. Similarly, in a compression test, there is a shortening δ, and a consequent increase of the cross-section dimensions. Therefore, it is necessary to characterize the cross-section deformation of the specimen, modeled here as a bar. The elastic range of the material is in the regime of small strain and the geometric shape of the section does not change, but only its dimensions.

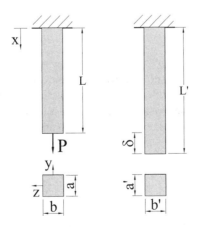

Figure 3.24 Transversal deformation in a bar.

Consider a bar of rectangular cross-section shown in Figure 3.24, subjected to a tensile force P. The force produces a specific longitudinal strain ε_{xx} given by (3.25). The cross-section also has two components of specific strain ε_{yy} and ε_{zz} in the y and z directions, due to variations of the a and b dimensions to a' and b', respectively. These transversal strain components are obtained analogously to ε_{xx} in (3.25). We take the variations $\Delta a = a' - a$ and $\Delta b = b' - b$ and divide them by the initial a and b dimensions, that is,

$$\varepsilon_{yy} = \frac{\Delta a}{a} = \frac{a' - a}{a} \quad \text{and} \quad \varepsilon_{zz} = \frac{\Delta b}{b} = \frac{b' - b}{b}. \tag{3.32}$$

It may be verified experimentally that all dimensions of the cross-section have the same specific transversal strain ε_t, that is,

$$\varepsilon_{yy} = \varepsilon_{zz} = \varepsilon_t. \tag{3.33}$$

Furthermore, the strain component ε_t is proportional to the specific longitudinal strain ε_{xx}, that is,

$$\varepsilon_t = \varepsilon_{yy} = \varepsilon_{zz} = -v\varepsilon_{xx}, \tag{3.34}$$

where v is a characteristic property of the material, called Poisson ratio. The negative sign is introduced only to represent the observed physical phenomenon. When the bar is under traction, the longitudinal strain ε_{xx} is positive, while ε_t is negative, because the cross-section dimensions decrease. In the case of compression, ε_{xx} is negative and ε_t is positive.

Assuming that the bar has a circular cross-section with initial d and deformed d' diameters, the strain components ε_{yy} and ε_{zz} are equal. With $\Delta d = d' - d$ the diameter variation after deformation and employing (3.34), we have

$$\varepsilon_{yy} = \varepsilon_{zz} = \frac{\Delta d}{d} = -v\varepsilon_{xx}. \tag{3.35}$$

By substituting (3.25) and the δ displacement given in (3.30) in the previous relation, we obtain

$$\Delta d = -v\frac{Pd}{AE}.$$

The Poisson ratio is determined from the tensile or compression test, by measuring the cross-section dimensions. A typical behavior is illustrated in Figure 3.25. This coefficient remains constant during the elastic range. Subsequently, during the yield phase of the specimen, the Poisson ratio increases, until attaining an asymptotic value. Generally, $v = 0.3$ for steel and $v = 0.27$ for aluminum.

3.8.3   HOOKE'S LAW

As mentioned before, Hooke's law is a constitutive equation which is valid for homogeneous linear elastic materials. In the case of a bar with a Hooke material, there is a normal stress component σ_{xx} and three specific normal strain components (ε_{xx}, ε_{yy}, and ε_{zz}), which are related by the elasticity modulus E and the Poisson ratio v from expressions (3.28) and (3.34), that is,

$$\begin{aligned}
\sigma_{xx}(x) &= E(x)\varepsilon_{xx}(x), & (3.36) \\
\varepsilon_{yy}(x) &= -v(x)\varepsilon_{xx}(x), & (3.37) \\
\varepsilon_{zz}(x) &= -v(x)\varepsilon_{xx}(x). & (3.38)
\end{aligned}$$

There is an uniaxial stress state in the bar, described by the normal stress component $\sigma_{xx}(x)$. This state is independent of the material behavior. When introducing the hypothesis of an elastic material according to Hooke's law, a strain state with three components ε_{xx}, ε_{yy}, and ε_{zz} is obtained.

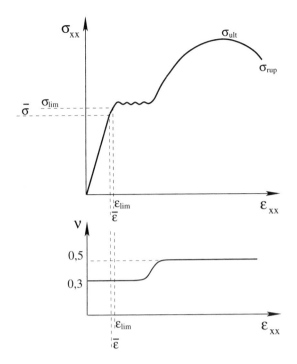

Figure 3.25 Poisson ratio behavior in a tensile test.

The strain components can be expressed in terms of the normal stress components σ_{xx} as

$$\varepsilon_{xx}(x) = \frac{1}{E(x)}\sigma_{xx}(x), \tag{3.39}$$

$$\varepsilon_{yy}(x) = -\frac{v(x)}{E(x)}\sigma_{xx}(x), \tag{3.40}$$

$$\varepsilon_{zz}(x) = -\frac{v(x)}{E(x)}\sigma_{xx}(x). \tag{3.41}$$

3.9 APPLICATION OF THE CONSTITUTIVE EQUATION

In this section, we perform the last step of the formulation presented in Section 1.1.4, applying Hooke's law to the bar model.

The normal stress $\sigma_{xx}(x)$, given by (3.36), can be replaced in the normal force expression (3.14) and

$$N_x(x) = E(x)A(x)\varepsilon_{xx}(x). \tag{3.42}$$

By using equation (3.4) for $\varepsilon_{xx}(x)$, we have

$$N_x(x) = E(x)A(x)\frac{du_x(x)}{dx}. \tag{3.43}$$

Aiming to a generalized formulation, it is assumed that the elasticity modulus may vary as a function of x, that is, $E = E(x)$, as in the case of a bar made of different materials along its length. Similarly, the cross-sectional area may also vary along the length of the bar, denoting $A = A(x)$.

By substituting equation (3.43) in the equilibrium differential equation given in (3.20), a differential equation in terms of the axial displacement $u_x(x)$ for $x \in (0, L)$ is obtained

$$\frac{d}{dx}\left(E(x)A(x)\frac{du_x(x)}{dx}\right) + q_x(x) = 0. \tag{3.44}$$

For the case of the constant elasticity modulus and cross-section area, the previous expression simplifies to

$$EA\frac{d^2u_x(x)}{dx^2} + q_x(x) = 0. \tag{3.45}$$

Thus, a second-order differential equation is obtained for a bar with a homogeneous isotropic linear elastic material. This equation must be integrated twice to determine the equation of bar axial displacement $u_x(x)$. The first integration provides the normal force $N_x(x) = E(x)A(x)\frac{du_x(x)}{dx}$. The boundary conditions can now be stated in terms of axial displacements, such as the supports present at the bar ends, as illustrated in Figure 3.26.

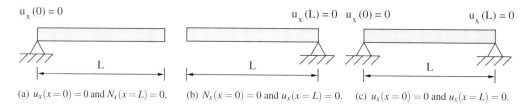

(a) $u_x(x = 0) = 0$ and $N_x(x = L) = 0$.   (b) $N_x(x = 0) = 0$ and $u_x(x = L) = 0$.   (c) $u_x(x = 0) = 0$ and $u_x(x = L) = 0$.

Figure 3.26   Boundary conditions in terms of the axial displacement and normal force for bars.

Example 3.7 *Derive the expression of the bar axial displacement in Example 3.1, with an axial displacement δ at the free end.*

The same procedure given in Section 3.7 is applied for the differential equation (3.45).

- *Distributed load equation*
  There is no distributed load applied on the bar and $q_x(x) = 0$.
- *Boundary conditions*
  The bar is fixed at the $x = 0$ end and thus the axial displacement is zero, that is, $u_x(x = 0) = 0$. Due to the axial displacement δ at the bar free end $(x = L)$, the boundary condition is $u_x(x = L) = \delta$.
- *Integration of the differential equation*
  The differential equation in terms of the axial displacement is considered. Thus,

$$EA\frac{du_x^2(x)}{dx^2} = -q_x(x) = 0.$$

The first indefinite integration provides the normal force expression

$$N_x(x) = EA\frac{du_x(x)}{dx} = C_1.$$

The second integration results in the axial displacement

$$u_x(x) = \frac{1}{EA}(C_1x + C_2),$$

with C_1 and C_2 the integration constants.

- *Determination of the integration constants*
  The boundary conditions in terms of the axial displacement are applied to determine the integration constants C_1 and C_2. Hence,

$$u_x(x = 0) = \frac{1}{EA}\left[C_1(0) + C_2\right] = 0 \rightarrow C_2 = 0,$$

$$u_x(x = L) = \frac{C_1(L)}{EA} = \delta \rightarrow C_1 = \frac{EA}{L}\delta.$$

- *Final equations*
  By substituting the constants C_1 and C_2, we obtain the following final equations for the normal force and axial displacement:

$$N_x(x) = \frac{EA}{L}\delta,$$

$$u_x(x) = \frac{\delta}{L}x.$$

Therefore, we recover the same expression for the axial displacement given in Example 3.1.

☐

Example 3.8 *Plot the normal force and axial displacement diagrams, by integrating the differential equation (3.45) for the double-supported bar illustrated in Figure 3.27, subjected to a distributed axial load of p_0 intensity.*

There are two support reactions R_{A_x} and R_{B_x} at the A and B bar ends, and only one equilibrium condition $(\sum f_x = 0)$. Hence, the bar is hyperstatic and the differential equation in terms of the kinematics should necessarily be considered.

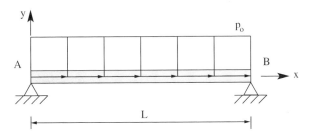

Figure 3.27 Example 3.8: double-supported bar subjected to a constant distributed load.

- *Distributed load equation*
  As the bar is subjected to an axial distributed load of constant intensity p_0, in the positive direction of x, we have $q_x(x) = p_0$.
- *Boundary conditions*
  The axial displacements at $x = 0$ and $x = L$ are zero, because the bar is fixed in both ends. Thus, $u_x(x = 0) = 0$ and $u_x(x = L) = 0$.
- *Integration of the differential equation*
  By integrating the differential equation in terms of the axial displacement,

$$EA\frac{du_x^2(x)}{dx^2} = -q_x(x) = -p_0,$$

we obtain, respectively, the expressions for the normal force and axial displacement

$$N_x(x) = -p_0 x + C_1,$$

$$u_x(x) = \frac{1}{EA}\left(-\frac{p_0}{2}x^2 + C_1 x + C_2\right),$$

with C_1 and C_2 the integration constants.

- *Determination of the integration constants*
  The boundary conditions in terms of the axial displacement are used to determine the integration constants C_1 and C_2, in other words,

$$u_x(x=0) = \frac{1}{EA}\left(-\frac{p_0}{2}(0)^2 + C_1(0) + C_2\right) = 0 \rightarrow C_2 = 0,$$

$$u_x(x=L) = \frac{1}{EA}\left(-\frac{p_0}{2}(L)^2 + C_1(L)\right) = 0 \rightarrow C_1 = \frac{p_0}{2}L.$$

- *Final equations*
  The final expressions for the normal force and axial displacement are obtained with the substitution of the integration constants C_1 and C_2. Thus,

$$N_x(x) = -\frac{p_0}{2}(2x - L),$$

$$u_x(x) = -\frac{p_0}{2EA}\left(x^2 - Lx\right).$$

- *Normal force and axial displacement diagrams*
  The normal force and the axial displacement vary linearly and parabolically with the x coordinate along the bar length. As the sign of the term at x^2 is negative, the concavity of the parabola points downwards. The diagrams for the normal force and axial displacement are shown in Figure 3.28 for $L = 2$ m, $p_0 = 10$ N/m, $E = 200$ GPa, and $A = 5 \times 10^{-4}$ m^2.

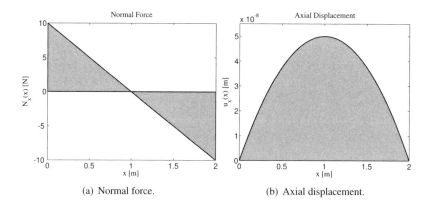

(a) Normal force.   (b) Axial displacement.

Figure 3.28   Example 3.8: diagrams for the normal force and axial displacement.

- *Support reactions*
  In this case, the support reactions R_{Ax} and R_{Bx} at $x = 0$ and $x = L$ bar ends are given by the boundary conditions of the equilibrium BVP (3.20), that is, $R_{Ax} = -N_x(x=0) = -10$ N and $R_{Bx} = N_x(x=L) = -10$ N. Thus, the reactions are directed leftwards, and therefore contrary to the positive direction of the x axis. These values can also be obtained directly from the normal force diagram.

In this case, the rigid body equilibrium equation given in (3.22) cannot be applied, since the problem is statically indeterminate, i.e.,

$$\sum f_x = 0: \ R_{Ax} + R_{Bx} + \int_0^L p_0 dx = 0.$$

As there are two unknowns and one rigid body equilibrium equation, the problem is statically indeterminate. Such fact did not restrict the application of the same solution procedure.

☐

3.10 DESIGN AND VERIFICATION

In general, it is desired to determine the minimum dimensions of the mechanical components in such a way that they remain in the linear elastic range. This implies that the component will deform under the action of external loads, but will return to its original shape when the load is removed.

The bar length is determined by constructive restrictions. Thus, designing the bar means calculating the minimum dimensions of the cross-section, so that the bar remains in the linear elastic range. The design here considered is based on the maximum normal stress value in the bar. For bars subjected to axial compressive forces, it is necessary also to verify the effect of buckling, as discussed in Section 5.12.

The following steps are considered in the design of bars based on the maximum normal stress:

- Determine the normal force function $N_x(x)$ and plot its respective diagram. If the problem is statically determined, the equilibrium differential equation (3.20) in terms of the normal force should be integrated. If the problem is statically indeterminate, the differential equation (3.45) in terms of the axial displacement should be integrate.
- Find the critical section of the bar using the normal force diagram. This is the section in which there is the largest absolute value of the normal force. This value is denoted by $N_x^{\max}$.
- The maximum normal stress $\sigma_{xx}^{\max}$ is given by expression (3.14), i.e.,

$$\sigma_{xx}^{\max} = \frac{N_x^{\max}}{A}. \tag{3.46}$$

In order to maintain the bar in the linear elastic range, the maximum normal stress must be less than the tensile $\bar{\sigma}_t$ and compressive $\bar{\sigma}_c$ admissible normal stresses, in cases where the critical section is in tension and compression, respectively. Hence,

$$\sigma_{xx}^{\max} = \frac{N_x^{\max}}{A} \leq \bar{\sigma}_t \qquad \text{or} \qquad \sigma_{xx}^{\max} = \frac{N_x^{\max}}{A} \leq \bar{\sigma}_c. \tag{3.47}$$

- The minimum area $A^{\min}$ required to maintain the bar in the linear elastic range is obtained by taking the equality of the previous expressions, that is,

$$A^{\min} = \frac{N_x^{\max}}{\bar{\sigma}_t} \quad \text{or } A^{\min} = \frac{N_x^{\max}}{\bar{\sigma}_c}. \tag{3.48}$$

The largest value of the area determined by the previous equations should be considered. It is common to have more than one section with the same maximum absolute normal force value, but some in traction and others in compression (see Example 3.11). The area is determined such that the minimum area keeps the bar in the linear elastic range.

- With the determination of the cross-section area, we can calculate its dimensions. For example, for a circular rod of diameter d, the area is given by $A = \frac{\pi d^2}{4}$. From the expressions indicated in (3.48), d is determined by

$$d = \sqrt{\frac{4N_x^{\max}}{\pi \bar{\sigma}_t}} \text{ or } d = \sqrt{\frac{4N_x^{\max}}{\pi \bar{\sigma}_c}}. \tag{3.49}$$

Generally, cross-sections with standard dimensions are used for economical reasons. The geometric characteristics of these sections are given in tables provided by manufacturers. By using these tables, it is possible to select the cross-section with an area greater than or equal to the area calculated above, that is, the final area is

$$A > A^{\min}.$$

In the case of bar verification, the cross-section dimensions are known and it is desired to verify if it remains in the linear elastic range when subjected to a certain load. To accomplish this, the maximum normal stress $\sigma_{xx}^{\max}$ should be calculated using (3.46). With this maximum stress, one should just check if the value is less than $\bar{\sigma}_t$ for bars under tension. For bars under compression, the value is compared with the admissible normal stress $\bar{\sigma}_c$. To summarize, one should verify that

$$\sigma_{xx}^{\max} \leq \bar{\sigma}_t \text{ or } \sigma_{xx}^{\max} \leq \bar{\sigma}_c. \tag{3.50}$$

In this case, the bar remains in the linear elastic range. If either of the two previous conditions is invalid, the bar should be redesigned using the previous procedure.

Example 3.9 *A steel bar of length* 100 cm *and with a hollow square cross-section is subjected to a* 100 kN *axial traction force. It elongated of* 2×10^{-3} cm. *Calculate the specific longitudinal normal strain and the tensile normal stress in the bar. For an admissible stress of* $\bar{\sigma} = 100$ MPa, *design the bar given* $\dfrac{a_1}{a_2} = 0.7$, *where* a_1 *and* a_2 *the sizes of the internal and external cross-section edges, respectively.*

The specific longitudinal strain is calculated as

$$\varepsilon_{xx} = \frac{\delta}{L} = \frac{2 \times 10^{-3}}{100} = 2.0 \times 10^{-5}.$$

The normal stress is obtained from Hooke's law, that is,

$$\sigma_{xx} = E\varepsilon_{xx} = (210 \times 10^9)(2 \times 10^{-5}) = 4.2 \text{ MPa}.$$

The cross-section area of the bar is calculated as

$$A = \frac{P}{\bar{\sigma}} = \frac{100000}{100 \times 10^6} = 10.0 \text{ cm}^2.$$

For a hollow section with larger a_2 *and smaller* a_1 *sizes, we have* $A = a_2^2 - a_1^2$. *By using* $a_1 = 0.7a_2$, *we have*

$$a_2^2 - (0,7a_2)^2 = 10 \rightarrow a_2 = 4.4 \text{ cm}.$$

Consequently, $a_1 = 0.7a_2 = 3.1$ cm.

□

Example 3.10 *Design the diameter of the bar in Example 3.6, with* $E = 210$ GPa *and* $\bar{\sigma} = 60$ MPa. *Assume* $p_0 = 200$ kN/m *and* $P = 50$ kN.

Based on the normal force diagram shown in Figure 3.16, the critical bar section is determined. This is the section with the largest absolute value of the normal force. In this case, section $x = 0$ is the critical and for $p_0 = 200$ kN/m and $P = 50$ kN, we have $N_x^{max} = 150$ kN.

The next step is to calculate the maximum normal stress corresponding to N_x^{max}, that is,

$$\sigma_{xx}^{max} = \frac{N_x^{max}}{A} = \frac{150000}{A}.$$

This stress cannot exceed the admissible normal stress $\bar{\sigma}$, in order to keep the bar in the linear elastic range, i.e.,

$$\bar{\sigma} = \sigma_{xx}^{max} = \frac{N_x^{max}}{A^{min}} = \frac{150000}{A^{min}}.$$

Thus,

$$A^{min} = \frac{N_x^{max}}{\bar{\sigma}} = \frac{150000}{60 \times 10^6} = 2.50 \times 10^{-3} \text{ m}^2.$$

As the cross-section area is $A = \frac{\pi d^2}{4}$, the bar diameter becomes

$$A^{min} = \frac{\pi d^2}{4} = 2.50 \times 10^{-3} \text{ m}^2 \rightarrow d = \sqrt{\frac{(4)(2.50 \times 10^{-3})}{\pi}} = 5.64 \text{ cm}.$$

This is the minimum diameter of the bar to remain in the linear elastic range. With a table of standard sections, we should select a bar with a diameter greater than or equal to 5.64 cm.

□

Example 3.11 *Design the diameter of the steel bar of Example 3.8 for $\bar{\sigma}_c = 60$ MPa and $\bar{\sigma}_t = 80$ MPa. Assume that $p_0 = 100$ kN/m.*

Based on the normal force diagram of Figure 3.28(b), the critical section of the bar is determined. In this case, sections $x = 0$ and $x = 2$ m are the critical ones and for $p_0 = 100$ kN/m, we have $N_x^{max} = 100$ kN. As the admissible tensile and compressive stresses are different, two sections are designed and we select the one with the largest diameter. Thus, as section $x = 0$ is under traction

$$d = \sqrt{\frac{4N_x^{max}}{\pi \bar{\sigma}_t}} = \sqrt{\frac{(4)(100000)}{\pi(80 \times 10^6)}} = 3.99 \text{ cm}.$$

For section $x = 2$ m under compression, we have

$$d = \sqrt{\frac{4N_x^{max}}{\pi \bar{\sigma}_c}} = \sqrt{\frac{(4)(100000)}{\pi(60 \times 10^6)}} = 4.61 \text{ cm}.$$

Hence, the minimum bar diameter to remain in the linear elastic range is $d = 4.61$ cm. In the cases with more than one critical section, it is only necessary to design the bar for the smallest normal admissible stress for traction or compression.

□

3.11 BARS SUBJECTED TO TEMPERATURE CHANGES

Consider a bar of length L subjected to a positive temperature variation ΔT, as illustrated in Figure 3.29. There is an elongation at the bar end and the final length L' is given by

$$L' = L(1 + \alpha \Delta T), \tag{3.51}$$

with α the thermal dilatation coefficient of the material.

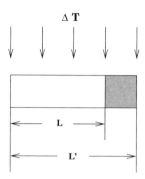

Figure 3.29 Bar subjected to temperature variation.

The previous expression can be rewritten as

$$\frac{L' - L}{L} = \alpha \Delta T.$$

The term on the left side represents the bar elongation $\delta = L' - L$ divided by the initial length L. It is a specific longitudinal thermal strain measure, denoted as ε_T. Hence,

$$\varepsilon_T = \alpha \Delta T. \tag{3.52}$$

The total strain measure ε is the sum of the strain measures due to the axial forces and temperature variation, that is,

$$\varepsilon = \varepsilon_{xx} + \varepsilon_T. \tag{3.53}$$

The thermal normal stress σ_T is given by Hooke's law as

$$\sigma_T = E \varepsilon_T. \tag{3.54}$$

Example 3.12 *A steel bar of length L is mounted between two fixed bulkheads [see Figure 3.30(a)]. When raising the temperature 50° C, what is the normal stress in the bar?*

In the initial condition, as there is no axial load applied, the normal stress and strain are zero. When the temperature has a 50° C increase, the total elongation of the bar will be zero due to the constraints at both ends. For a homogeneous bar with uniform section, the specific thermal strain at any point is $\varepsilon_T = \delta_T / L$, with δ_T being the elongation due to the temperature variation. The thermal strain measure ε_T is also zero.

In order for the total elongation to be zero, the bulkheads will apply axial reaction forces P over the bar, as shown in Figure 3.30(b). Such reactions will cause normal stresses in the bar, with zero specific strains.

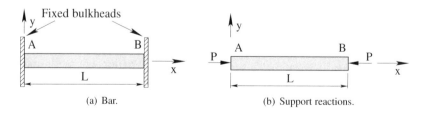

(a) Bar.   (b) Support reactions.

Figure 3.30 Example 3.12: bar subjected to temperature variation.

To calculate the normal stress σ_{xx}, we first determine the bar elongation δ_T due to temperature variation without considering one of the bulkheads (B, for instance), as shown in Figure 3.31. The corresponding elongation due to temperature variation is given by

$$\delta_T = \alpha\left(\Delta T\right)L.$$

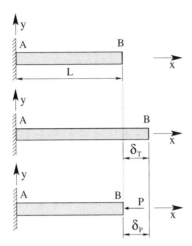

Figure 3.31   Example 3.12: bar without the support in *B*, elongation due to temperature variation and due to the *P* force.

We should sum the elongations due to the reaction P in B, that is,

$$\delta_P = -\frac{PL}{AE}.$$

As we know that the total elongation should be zero, we have

$$\delta = \delta_T + \delta_P = \alpha\left(\Delta T\right)L - \frac{PL}{AE} = 0.$$

Therefore,

$$P = AE\alpha\left(\Delta T\right).$$

The normal stress acting of the bar due to a ΔT temperature change is

$$\sigma_{xx} = \frac{P}{A} = E\alpha\left(\Delta T\right).$$

This solution is only applied to the case of bars with a uniform cross-section and homogeneous material.

□

3.12   VOLUME AND AREA STRAIN MEASURES

The mechanical models considered here assume the conservation of mass. This does not mean that the volume of the considered component cannot change. But the density of the material should also vary consistently, to ensure the conservation of mass. For small strains, there are also small changes in volume and hence in the density. In this section, expressions for area and volume strain measures in a bar will be deduced.

Consider the bar of initial length L and initial rectangular cross-section dimensions a and b illustrated in Figure 3.24. It is known that the bar has a specific longitudinal strain ε_{xx} and material Poisson ratio v.

The deformed dimensions a' and b' of the cross-section are obtained using (3.32) and (3.34). Hence,

$$a' = a(1 - v\varepsilon_{xx}) \quad \text{and} \quad b' = b(1 - v\varepsilon_{xx}).$$

Based on that, the new cross-section area A' is

$$A' = a'b' = ab(1 - v\varepsilon_{xx})^2 = A(1 - 2v\varepsilon_{xx} + v^2\varepsilon_{xx}^2).$$

For small strains, ε_{xx}^2 is much smaller than ε_{xx}. Consequently, the previous expression simplifies to

$$A' = A(1 - 2v\varepsilon_{xx}). \tag{3.55}$$

The area specific strain measure can be defined as the ratio of area variation and the initial area, that is,

$$\varepsilon_A = \frac{A' - A}{A} = \frac{\Delta A}{A} = -2v\varepsilon_{xx}. \tag{3.56}$$

The deformed length L' is obtained by

$$L' = L(1 + \varepsilon_{xx}). \tag{3.57}$$

Therefore, the deformed bar volume is given by

$$V' = a'b'L' = V(1 + \varepsilon_{xx} - 2v\varepsilon_{xx} - 2v^2\varepsilon_{xx}^2).$$

For $\varepsilon_{xx}^2 << \varepsilon_{xx}$, we have

$$V' = V(1 + \varepsilon_{xx} - 2v\varepsilon_{xx}). \tag{3.58}$$

The specific volume strain is defined as

$$\varepsilon_V = \frac{V' - V}{V} = \frac{\Delta V}{V} = (1 - 2v)\varepsilon_{xx}. \tag{3.59}$$

Example 3.13 *A steel bar of length* 1 m *was subjected to an axial normal force of* 80 kN *and had an elongation of* 0.05 cm. *Calculate the area and volume strain measures.*

From (3.57), we get

$$\varepsilon_{xx} = \frac{\delta}{L} = \frac{5 \times 10^{-4}}{1} = 5 \times 10^{-4}.$$

The area and volume strain measures are determined using (3.56) and (3.59), respectively. Thus,

$$\varepsilon_A = -2v\varepsilon_{xx} = -2(0.3)(5 \times 10^{-4}) = -3.0 \times 10^{-4},$$

$$\varepsilon_V = (1 - 2v)\varepsilon_{xx} = (1 - 2(0.3))(5 \times 10^{-4}) = 2.0 \times 10^{-4}.$$

The negative value in ε_A *is consistent, because the bar is under traction and the cross-section becomes smaller after the deformation process.*

□

3.13 SINGULARITY FUNCTIONS FOR EXTERNAL LOADING REPRESENTATION

The external loads which are compatible with the bar kinematics are given by a continuous function $q_x(x)$ along the length, representing the distributed axial force, and the concentrated forces P_0 and P_L at bar ends. Thus, concentrated forces along the bar length are not compatible with the presented model. However, the use of concentrated forces is very common in modeling engineering problems. To allow its use in bars, we make use of singularity functions to include concentrated forces in the expression for $q_x(x)$.

The following single variable function $f(x)$ is singular at $x = 2$:

$$f(x) = \frac{1}{x-2} = (x-2)^{-1}.$$

Notice that $f(x)$ has an infinite value or singularity at $x = 2$.

Suppose that an axial concentrated force P is applied to the bar section $x = a$, with $0 < a < L$. This force will be represented in the loading function as

$$q_x(x) = P < x - a >^{-1}. \tag{3.60}$$

The following notation is employed:

$$< x - a >^{-1} = \begin{cases} 0 & \text{if } x < a \\ (x-a)^{-1} & \text{if } x \geq a \end{cases}. \tag{3.61}$$

The integration of the previous term for $x \geq a$ is given by

$$\int < x - a >^{-1} dx = < x - a >^{0}. \tag{3.62}$$

The justification for the use of the notation given in (3.61) and integration (3.62) will be presented in Chapter 5. It should be noted that for the example shown in (3.60), the concentrated force P is active only to the right of section $x = a$.

Consider the following examples that employ singularity functions to represent the bar distributed loading function.

Example 3.14 *Consider the bar illustrated in Figure 3.32, subjected to concentrated forces along the length. Plot the normal force and axial displacement diagrams. Use $A = 10^{-4}$ m^2, $E = 100$ GPa, $L = 1$ m, $P = 10$ kN.*

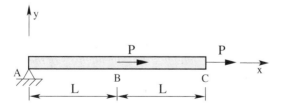

Figure 3.32 Example 3.14: bar subjected to concentrated forces.

- *Distributed load equation*
  The concentrated axial force P at $x = L$ is represented using the singularity function notation. Thus, the distributed load equation is given by

$$q_x(x) = P < x - L >^{-1} = \begin{cases} 0 & \text{if } x < L \\ P(x-L)^{-1} & \text{if } x \geq L \end{cases}.$$

- *Boundary conditions*
  The bar is fixed at $x = 0$ and consequently $u_x(x = 0) = 0$. The normal force at the right end is such that $N_x(x = 2L) = P$.
- *Integration of the differential equation*
  The axial displacement differential equation is integrated, that is,

$$EA\frac{d^2 u_x(x)}{dx^2} = -q_x(x) = -P <x-L>^{-1}.$$

The normal force and axial displacement are given, respectively, by

$$N_x(x) = EA\frac{du_x(x)}{dx} = -P <x-L>^0 +C_1,$$

$$u_x(x) = \frac{1}{EA}\left(-P <x-L>^1 +C_1 x + C_2\right),$$

with C_1 and C_2 the integration constants.
- *Determination of the integration constants*
  We use the normal force and axial displacement boundary conditions to determine C_1 and C_2, namely,

$$N_x(x = 2L) = -P(2L-L)^0 + C_1 = P \rightarrow C_1 = 2P,$$
$$u_x(x = 0) = \frac{1}{EA}(0 + C_1(0) + C_2) = 0 \rightarrow C_2 = 0.$$

- *Final equations*
  The final expressions for the normal force and axial displacement are obtained by substituting the integration constants. Thus,

$$N_x(x) = -P <x-L>^0 +2P,$$
$$u_x(x) = \frac{1}{EA}\left(-P <x-L>^1 +2Px\right).$$

By using the given values, the previous expressions reduce to

$$N_x(x) = -10000 <x-1>^0 +20000,$$

$$u_x(x) = (-10000 <x-1>^1 +20000x) \times 10^{-7}.$$

- *Normal force and axial displacement diagrams*
  The normal force diagram shows discontinuities in the sections where the concentrated forces are applied with the same magnitude as the intensity of the forces.
  For this purpose, we must split the expressions of the normal force and axial displacements for each set of sections between the concentrated forces. Thus,
  - *0 m $< x < 1$ m interval:*

$$N_x(x) = 20000 = \begin{cases} N_x(x \rightarrow 0^+) = 20000 \text{ N} \\ N_x(x \rightarrow 1^-) = 20000 \text{ N} \end{cases},$$

$$u_x(x) = 2.0 \times 10^{-3}x = \begin{cases} u_x(x \rightarrow 0^+) = 0 \text{ m} \\ u_x(x \rightarrow 1^-) = 2.0 \times 10^{-3} \text{ m} = 2 \text{ mm} \end{cases}.$$

- $1 \text{ m} < x < 2 \text{ m}$ *interval:*

$$N_x(x) = 10000 = \begin{cases} N_x(x \to 1^+) = 10000 \text{ N} \\ N_x(x \to 2^-) = 10000 \text{ N} \end{cases},$$

$$\begin{aligned} u_x(x) &= \left(-10000(x-1)^1 + 20000x\right) \times 10^{-7} \\ &= (1.0x + 1.0) \times 10^{-3} = \begin{cases} u_x(x \to 1^+) = 2.0 \times 10^{-3} \text{ m} = 2.0 \text{ mm} \\ u_x(x \to 2^-) = -3.0 \times 10^{-3} \text{ m} = 3.0 \text{ mm} \end{cases}. \end{aligned}$$

The $x \to a^-$ and $x \to a^+$ notations indicate, respectively, the neighboring sections to the left and to the right of $x = a$.

Figure 3.33 illustrates the normal force and axial displacement diagrams. While the axial displacement diagram is continuous, the normal force diagram is discontinuous at section $x = 1$ m where the concentrated force is applied. The jump magnitude at discontinuity is 10 kN which is the intensity of the concentrated force.

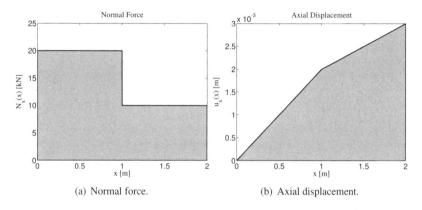

(a) Normal force.   (b) Axial displacement.

Figure 3.33 Example 3.14: normal force and axial displacement diagrams.

- *Support reaction*
  The support reaction at the left bar end is given by the boundary condition. Thus,

$$R_{A_x} = -N_x(x = 0) = -2P = -20000 \text{ N}.$$

The support reaction is in the negative direction of the x axis, because the sign is negative.

□

Example 3.15 *Consider the bar subjected to the loading illustrated in Figure 3.34. Plot the normal force and axial displacement diagrams. Use $A = 10^{-4}$ m^2, $E = 100$ GPa, $L = 1$ m and $p_0 = 10000$ N/m.*

- *Distributed load equation*
  In this case, the singularity function notation is used to represent the distributed load in the bar's right half span. Moreover, as there is the support at $x = L$, the respective reaction force R_{Bx} is included in the loading equation. It is still unknown and the condition of zero displacement at $x = L$ is also used. Therefore, the distributed load function is given by

$$\begin{aligned} q_x(x) &= p_0 <x-L>^0 + R_{Bx} <x-L>^{-1} \\ &= \begin{cases} 0 & 0 < x < L \\ p_0 + R_{Bx}(x-L)^{-1} & L \leq x < 2L \end{cases}. \end{aligned}$$

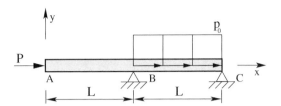

Figure 3.34 Example 3.15: hyperstatic bar with a support in the middle and subjected to a constant distributed load in the right half span.

- *Boundary conditions and additional constraint (overhang)*
  The boundary conditions of this example are $N_x(x = 0) = -P$ and $u_x(x = 2L) = 0$. Moreover, the bar is supported at $x = L$, and thus there is an additional constraint of zero displacement in this section. Thus, $u_x(x = L) = 0$.
- *Integration of the differential equation*
  The axial displacement differential equation is integrated, that is,

$$EA\frac{d^2u_x(x)}{dx^2} = -q_x(x) = -p_0 <x-L>^0 -R_{Bx} <x-L>^{-1}.$$

The normal force and axial displacement are given, respectively, by

$$N_x(x) = EA\frac{du_x(x)}{dx} = -p_0 <x-L>^1 -R_{Bx} <x-L>^0 +C_1,$$

$$u_x(x) = \frac{1}{EA}\left(-\frac{p_0}{2} <x-L>^2 -R_{Bx} <x-L>^1 +C_1x+C_2\right).$$

with C_1 and C_2 the integration constants.
- *Determination of integration constants*
  To find C_1 and C_2, we use the boundary conditions in terms of the normal force and axial displacement and also the additional constraint, i.e.,

$$\begin{aligned}
N_x(x = 0) &= -p_0(0) - R_{Bx}(0) + C_1 = -P \rightarrow C_1 = -P,\\
u_x(x = 2L) &= \frac{1}{EA}\left(-\frac{p_0}{2}(2L-L)^2 - R_{Bx}(2L-L)^1 - P(2L) + C_2\right) = 0\\
&\rightarrow -R_{Bx}L + C_2 = 2PL + \frac{p_0}{2}L^2,\\
u_x(x = L) &= \frac{1}{EA}\left(-\frac{p_0}{2}(L-L)^2 - R_{Bx}(L-L)^1 - P(L) + C_2\right) = 0\\
&\rightarrow C_2 = PL.
\end{aligned}$$

By substituting $C_2 = PL$ in the second of the previous expressions, we obtain $R_{Bx} = -P - \frac{p_0}{2}L$.
- *Final equations*
  The final expressions of the normal force and axial displacement are obtained by replacing the integration constants and the reaction R_{Bx}, namely,

$$\begin{aligned}
N_x(x) &= -p_0 <x-L>^1 +\left(P+\frac{p_0}{2}L\right) <x-L>^0 -P,\\
u_x(x) &= \frac{1}{EA}\left(-\frac{p_0}{2} <x-L>^2 +\left(P+\frac{p_0}{2}L\right) <x-L>^1 -Px+PL\right).
\end{aligned}$$

Considering the given values, the previous expressions reduce to

$$N_x(x) = -10000 <x-1>^1 +6000 <x-1>^0 -1000,$$

$$u_x(x) = (-5000 < x - 1 >^2 + 6000 < x - 1 >^1 - 1000x + 1000) \times 10^{-7}.$$

- *Normal force and axial displacement diagrams*
  As in the previous example, the expressions for the normal force and axial displacement should be split for any set of sections between the concentrated loads, in this case the reaction force R_{Bx}. Thus,
  - 0 m $< x < 1$ m *interval:*

$$N_x(x) = -1000 = \begin{cases} N_x(x \to 0^+) = -1000 \text{ N} \\ N_x(x \to 1^-) = -1000 \text{ N} \end{cases},$$

$$u_x(x) = (-1.0x + 1.0) \times 10^{-3} = \begin{cases} u_x(x \to 0^+) = 1.0 \times 10^{-4} \text{ m} = 0.1 \text{ mm} \\ u_x(x \to 1^-) = 0.0 \text{ m} \end{cases}.$$

  - 1 m $< x < 2$ m *interval:*

$$\begin{aligned} N_x(x) &= -10000(x-1)^1 + 6000(x-1)^0 - 1000 \\ &= -10000x + 15000 = \begin{cases} N_x(x \to 1^+) = 5000 \text{ N} \\ N_x(x \to 2^-) = -5000 \text{ N} \end{cases}, \end{aligned}$$

$$\begin{aligned} u_x(x) &= (-5000 < x - 1 >^2 + 6000 < x - 1 >^1 - 1000x + 1000) \times 10^{-7} \\ &= (-5x^2 + 15x - 10) \times 10^{-4} = \begin{cases} u_x(x \to 1^+) = 0.0 \text{ m} \\ u_x(x \to 2^-) = 0.0 \text{ m} \end{cases}. \end{aligned}$$

Figure 3.35 illustrates the diagrams for the normal force and axial displacement. It should be observed that at $x = l$ m, the normal force diagram has a discontinuity with the same magnitude as the support reaction $R_{Bx} = -6000$ N.

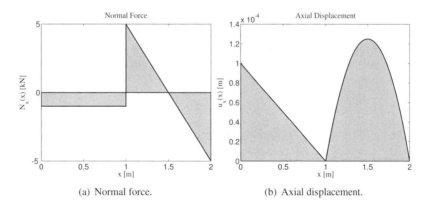

(a) Normal force.
(b) Axial displacement.

Figure 3.35 Example 3.15: normal force and axial displacement diagrams.

- *Support reaction*
  The reaction for the support at the right bar end is given by the boundary condition of the BVP. Thus,

$$R_{Cx} = N_x(x = 2L) = -\frac{p_0}{2}L = -5000 \text{ N}.$$

Note that the sign is negative and thus the support reaction is in the negative direction of x axis.

□

Example 3.16 *Consider the bar subjected to the loading illustrated in Figure 3.36. Plot the normal force and axial displacement diagrams. Consider $P_1 = 100\,\text{N}$, $P_2 = 200\,\text{N}$, $P_3 = 300\,\text{N}$, $A = 10^{-4}\,\text{m}^2$, $E = 100\,\text{GPa}$.*

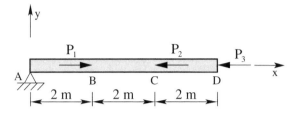

Figure 3.36   Example 3.16: bar subjected to concentrated forces.

- *Distributed load equation*
  The axial concentrated forces P_1 and P_2 are represented using the singularity function notation for the distributed load equation. Thus,

$$q_x(x) = P_1 <x-2>^{-1} -P_2 <x-4>^{-1}.$$

  with $x = 2$ m and $x = 4$ m the sections where P_1 and P_2 are applied, respectively.
- *Boundary conditions*
  The bar is fixed at $x = 0$ and thus $u_x(x = 0) = 0$. The normal force at the right bar end is $N_x(x = 6) = -P_3 = -300\,\text{N}$.
  Note that the concentrated force P_3 can be included in the load equation as

$$q_x(x) = P_1 <x-2>^{-1} -P_2 <x-4>^{-1} -P_3 <x-6>^{-1}.$$

  In this case, the boundary condition at $x = 6$ m is $N_x(x = 6) = 0$.
- *Integration of the differential equation*
  The differential equation in terms of the bar axial displacement is integrated and

$$EA\frac{d^2 u_x(x)}{dx^2} = -q_x(x) = -P_1 <x-2>^{-1} +P_2 <x-4>^{-1}.$$

  The normal force and axial displacements are given, respectively, by

$$N_x(x) = EA\frac{du_x(x)}{dx} = -P_1 <x-2>^0 +P_2 <x-4>^0 +C_1,$$

$$u_x(x) = \frac{1}{EA}\left(-P_1 <x-2>^1 +P_2 <x-4>^1 +C_1 x + C_2\right),$$

  with C_1 and C_2 the integration constants.
- *Calculation of the integration constants*
  The boundary conditions in terms of the normal force and axial displacement are used in order to determine C_1 and C_2, that is,

$$N_x(x = 6) = -P_1 + P_2 + C_1 = -P_3 \rightarrow C_1 = P_1 - P_2 - P_3,$$
$$u_x(x = 0) = \frac{1}{EA}(0 + 0 + C_1(0) + C_2) = 0 \rightarrow C_2 = 0.$$

- *Final equations*
  The final expressions for the normal force and axial displacement are obtained by substituting the integration constants, namely,

$$N_x(x) \quad = \quad -P_1 <x-2>^0 +P_2 <x-4>^0 +P_1 - P_2 - P_3,$$

$$u_x(x) \quad = \quad \frac{1}{EA} \left(-P_1 <x-2>^1 +P_2 <x-4>^1 +(P_1 - P_2 - P_3)x \right).$$

Using the given values, the previous expressions become

$$N_x(x) = -100 <x-2>^0 -200 <x-4>^0 -400,$$

$$u_x(x) = (-100 <x-2>^1 -200 <x-4>^1 -400x) \times 10^{-7}.$$

- *Diagrams of normal force and axial displacement*
  The diagram of the normal force will have discontinuities at the sections where the concentrated forces are applied, whose magnitudes are equal to the values of the axial forces. The expressions of the normal force and axial displacement must be considered for each set of sections between the concentrated forces. Thus,
  - $0 \text{ m} < x < 2 \text{ m}$ *interval:*

$$N_x(x) = -400 = \begin{cases} N_x(x \to 0^+) = -400 \text{ N} \\ N_x(x \to 2^-) = -400 \text{ N} \end{cases},$$

$$u_x(x) = -400 \times 10^{-7} x = \begin{cases} u_x(x \to 0^+) = 0 \text{ m} \\ u_x(x \to 2^-) = -8.0 \times 10^{-5} \text{ m} \end{cases}.$$

  - $2 \text{ m} < x < 4 \text{ m}$ *interval:*

$$N_x(x) = -500 = \begin{cases} N_x(x \to 2^+) = 500 \text{ N} \\ N_x(x \to 4^-) = 500 \text{ N} \end{cases},$$

$$u_x(x) = (-500x + 200) \times 10^{-7} = \begin{cases} u_x(x \to 2^+) = -8.0 \times 10^{-5} \text{ m} \\ u_x(x \to 4^-) = -1.8 \times 10^{-4} \text{ m} \end{cases}.$$

  - $4 \text{ m} < x < 6 \text{ m}$ *interval:*

$$N_x(x) = -300 = \begin{cases} N_x(x \to 4^+) = -300 \text{ N} \\ N_x(x \to 6^-) = -300 \text{ N} \end{cases},$$

$$u_x(x) = (-300x - 600) \times 10^{-7} = \begin{cases} u_x(x \to 4^+) = -1.8 \times 10^{-4} \text{ m} \\ u_x(x \to 6^-) = -2.4 \times 10^{-4} \text{ m} \end{cases}.$$

Figure 3.37 illustrates the normal force and axial displacement diagrams.
- *Support reaction*
  The support reaction force at the left bar end is given by the boundary condition of the BVP. Thus,

$$R_{A_x} = -N_x(x=0) = -400 \text{ N}.$$

Because of the negative sign, the support reaction is in the negative direction of the x axis.

□

Example 3.17 *Figure 3.38(a) illustrates a bar with one fixed end, and the other supported on a spring of elastic constant k. Due to the axial force P, the bar applies a traction to the spring, which applies a force $F = ku_{x_0}$ to the bar, with u_{x_0} the displacement of the right end, as illustrated in Figure 3.38(b). We wish to determine the expressions for the normal force and axial displacement of the bar.*

To solve the problem, the previous integration procedure is applied, indicating the concentrated load by using the singularity function notation.

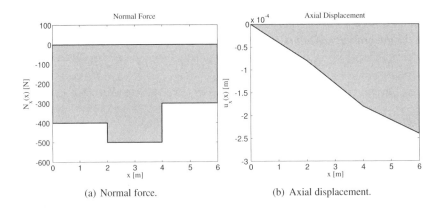

(a) Normal force. (b) Axial displacement.

Figure 3.37 Example 3.16: normal force and axial displacement diagrams.

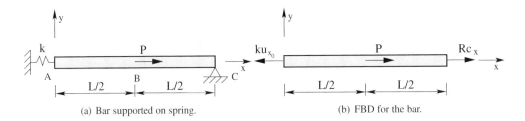

(a) Bar supported on spring. (b) FBD for the bar.

Figure 3.38 Example 3.17: bar supported on spring.

- *Distributed load equation:* $q_x(x) = P < x - \frac{L}{2} >^{-1}$.
- *Boundary conditions*
  The bar is supported on the spring at $x = 0$ and the normal force is such that $N_x(x = 0) = ku_{x_0}$. The bar is fixed in the other end, and consequently $u_x(x = L) = 0$.
- *Integration of the differential equation*
  The differential equation in terms of the axial displacement should be integrated, that is,

$$EA \frac{d^2 u_x(x)}{dx} = -q_x(x) = -P < x - \frac{L}{2} >^{-1}.$$

The first integration results in the normal force expression

$$N_x(x) = EA \frac{du_x(x)}{dx} = -P < x - \frac{L}{2} >^0 + C_1.$$

The second integration gives the axial displacement expression

$$u_x(x) = \frac{1}{EA} \left(-P < x - \frac{L}{2} >^1 + C_1 x + C_2 \right).$$

- *Calculation of the integration constants*
  The boundary conditions are applied to determine the integration constants C_1 and C_2. Thus,

$$N_x(x = 0) = 0 + C_1 = ku_{x_0} \rightarrow C_1 = ku_{x_0},$$

$$u_x(x = L) = \frac{1}{EA} \left[-P \left(L - \frac{L}{2} \right) + ku_{x_0} L + C_2 \right] = 0 \rightarrow C_2 = \frac{PL}{2} - ku_{x_0} L.$$

- *Final equations*
  By substituting the integration constants, we can determine the expressions for the normal force and axial displacement, that is,

$$N_x(x) = -P < x - \frac{L}{2} >^0 + ku_{x_0},$$

$$u_x(x) = \frac{1}{EA} \left(-P < x - \frac{L}{2} >^1 + ku_{x_0}(x-L) + \frac{PL}{2} \right).$$

- *Diagrams of normal force and axial displacement*
  Because there is a concentrated force along the bar length, the expressions are written for the following bar intervals:
  - $0 < x < \frac{L}{2}$ *interval*

$$N_x(x) = ku_{x_0},$$

$$u_x(x) = \frac{1}{EA} \left(ku_{x_0}(x-L) + \frac{PL}{2} \right).$$

  - $\frac{L}{2} \leq x < L$ *interval*

$$N_x(x) = -P + ku_{x_0},$$

$$u_x(x) = \frac{1}{EA} \left(-P + ku_{x_0} \right)(x-L).$$

Figure 3.39 illustrates the normal force and axial displacement diagrams for $L = 1$ m, $E = 100$ GPa, $A = 10$ cm^2, $k = 10000$ N/m, $u_{x_0} = 0.005$ m, and $P = 500$ N.

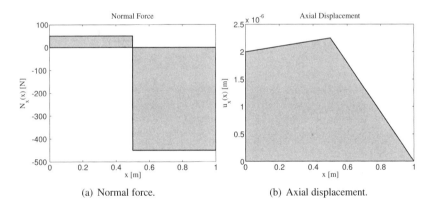

(a) Normal force.   (b) Axial displacement.

Figure 3.39   Example 3.17: normal force and axial displacement diagrams.

- *Support reaction*
  From the FBD of Figure 3.38(b), the support reaction at C is given in terms of the normal force by

$$R_{C_x} = N_x(x=L) = -P + ku_{x_0} = -450 \text{ N}.$$

□

The use of structural elements with different cross-sections and materials is quite common in engineering applications. The following example shows how to solve this problem using the bar differential equation. We must consider the equilibrium and compatibility conditions in sections for area discontinuity and material change. The integration of the differential equation is taken to the sections in which the cross-section area and the material properties are constant.

Example 3.18 *Plot the diagrams for the normal force, axial displacement, longitudinal strain, and normal stress for the bar shown in Figure 3.40. The bar has two segments of lengths L_1 and L_2 with cross-section areas A_1 and A_2 and Young's modulus E_1 and E_2, respectively. Use $L_1 = 4$ m, $L_2 = 3$ m, $p_0 = 25$ kN/m, $P = 125$ kN, $A_1 = 6$ cm^2, $A_2 = 12$ cm^2, $E_1 = 210$ GPa, and $E_2 = 70$ GPa.*

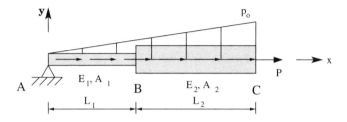

Figure 3.40   Example 3.18: bar with varying cross-sections and materials.

In this example, the cross-section area and Young's modulus vary along the bar length. For its solution, we should consider the sections AB and BC separately, as illustrated in Figure 3.41. In this case, N_1 and u_1 represent respectively, the normal force and the axial displacement at the interface between the two bar segments. They will be determined in the solution process. The differential equation in terms of the axial displacement for each bar segment is integrated. It is necessary also to consider the compatibility condition of axial displacement at the $x = L_1$ interface between the two bar segments.

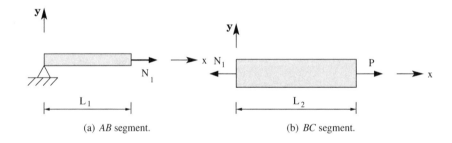

(a) *AB* segment.	(b) *BC* segment.

Figure 3.41   Example 3.18: forces at the interface of the bar segments.

Segment AB is initially considered and $0 < x < L_1$. The unknowns are the integration constants C_1 and C_2 and the normal force N_1 at the interface of the two segments. The same previous procedure is applied to integrate the differential equation for the axial displacement of the AB segment.

- *Distributed load equation*

$$q_{x_1}(x) = \frac{p_0}{L}x = \begin{cases} 0 & x = 0 \\ p_0\frac{L_1}{L} & x = L_1 \end{cases},$$

  with $L = L_1 + L_2$.
- *Boundary conditions*

$$u_{x_1}(x = 0) = 0 \quad and \quad N_{x_1}(x = L_1) = N_1.$$

- *Integration of the differential equation*

$$E_1 A_1 \frac{d^2 u_{x_1}(x)}{dx^2} = -q_{x_1}(x) = -\frac{p_0}{L}x.$$

The first integration gives the normal force expression

$$N_{x_1}(x) = -\frac{p_0}{2L}x^2 + C_1.$$

The second one results in the axial displacement expression

$$u_{x_1}(x) = \frac{1}{E_1 A_1}\left(-\frac{p_0}{6L}x^3 + C_1 x + C_2\right).$$

- *Determination of C_1 and C_2*
  The boundary conditions are used to determine the integration constants C_1 and C_2. Thus,

$$N_{x_1}(x = L_1) = -\frac{p_0}{2L}L_1^2 + C_1 = N_1 \rightarrow C_1 = N_1 + \frac{p_0}{2L}L_1^2,$$

$$u_{x_1}(x = 0) = \frac{1}{E_1 A_1}\left(-\frac{p_0}{6L}(0)^3 + C_1(0) + C_2\right) \rightarrow C_2 = 0.$$

- *Final equations*
  By substituting the integration constants, the expressions for the normal force and axial displacement are given in terms of N_1, respectively, by

$$N_{x_1}(x) = \frac{p_0}{2L}\left(L_1^2 - x^2\right) + N_1,$$

$$u_{x_1}(x) = \frac{1}{E_1 A_1}\left[\frac{p_0}{6L}(3L_1^2 - x^2)x + N_1 x\right].$$

We now consider the bar segment BC with $L_1 < x < L_2$. In this case, the unknowns are the integration constants C_3 and C_4 and the axial displacement u_1 at the interface of the two segments. The same integration procedure is applied to the differential equation in terms of the axial displacement.

- *Distributed load equation*

$$q_{x_2}(x) = \frac{p_0}{L}x = \begin{cases} p_0\frac{L_1}{L} & x = L_1 \\ p_0 & x = L \end{cases}.$$

- *Boundary conditions*

$$u_{x_2}(x = L_1) = u_1 \text{ and } N_{x_2}(x = L) = P.$$

- *Integration of the differential equation*

$$E_2 A_2 \frac{d^2 u_{x_2}(x)}{dx^2} = -q_{x_2}(x) = -\frac{p_0}{L}x.$$

The first integration results in the normal force expression

$$N_{x_2}(x) = -\frac{p_0}{2L}x^2 + C_3.$$

The second integration gives the axial displacement equation

$$u_{x_2}(x) = \frac{1}{E_2 A_2}\left(-\frac{p_0}{6L}x^3 + C_3 x + C_4\right).$$

- Determination of C_3 and C_4
  The boundary conditions are used to find the integration constants C_3 and C_4, that is,

$$N_{x_2}(x = L) = -\frac{p_0}{2L}L^2 + C_3 = P \rightarrow C_3 = P + \frac{p_0 L}{2},$$

$$u_{x_2}(x = L_1) = \frac{1}{E_2 A_2}\left[-\frac{p_0}{6L}L_1^3 + \left(P + \frac{p_0 L}{2}\right)L_1 + C_4\right] = u_1$$

Thus,

$$C_4 = E_2 A_2 u_1 + \frac{p_0}{6L}(L_1^3 - 3L_1 L^2) - P L_1.$$

- Final equations
  By substituting the integration constants C_3 and C_4, the final expressions for the normal force and axial displacement are given, respectively, by

$$N_{x_2}(x) = \frac{p_0}{2L}(L^2 - x^2) + P,$$

$$u_{x_2}(x) = \frac{1}{E_2 A_2}\left[\frac{p_0}{6L}\left((3L^2 - x^2)x + L_1(L_1^2 - 3L^2)\right) + P(x - L_1)\right] + u_1.$$

By making $x = L_1$ in the previous normal force expression, the following expression for N_1 is obtained:

$$N_1 = N_{x_2}(x = L_1) = \frac{p_0}{2L}(L^2 - L_1^2) + P.$$

By replacing N_1 in the normal force and axial displacement equations of the AB segment, we obtain the following final expressions:

$$N_{x_1}(x) = \frac{p_0}{2L}\left(L^2 - x^2\right) + P,$$

$$u_{x_1}(x) = \frac{1}{E_1 A_1}\left[\frac{p_0}{6L}\left(L^2 - x^2\right)x + Px\right].$$

By calculating the last expressions for $x = L_1$, the displacement u_1 is obtained

$$u_1 = u_{x_1}(x = L_1) = \frac{1}{E_1 A_1}\left[\frac{p_0}{6L}\left(L^2 - L_1^2\right)L_1 + P L_1\right].$$

Now, substituting u_1 in the axial displacement equation of the BC segment, we have

$$
\begin{aligned}
u_{x_2}(x) ={}& \frac{1}{E_2 A_2}\left[\frac{p_0}{6L}\left((3L^2 - x^2)x + L_1(L_1^2 - 3L^2)\right) + P(x - L_1)\right] \\
&+ \frac{1}{E_1 A_1}\left[\frac{p_0}{6L}\left(L^2 - L_1^2\right)L_1 + P L_1\right].
\end{aligned}
$$

The normal strain measures for both bar segments are calculated by using the expressions:

$$\varepsilon_{xx_1}(x) = \frac{du_{x_1}(x)}{dx} \quad and \quad \varepsilon_{xx_2}(x) = \frac{du_{x_2}(x)}{dx}.$$

The normal stresses for each segment are calculated by using Hooke's law

$$\sigma_{xx_1}(x) = E_1 \varepsilon_{xx_1}(x) \quad and \quad \sigma_{xx_2}(x) = E_2 \varepsilon_{xx_2}(x).$$

Figure 3.42 illustrates the diagrams of the normal force, axial displacement, longitudinal strain, and normal stress after substituting the given data. It is observed that the normal force and axial

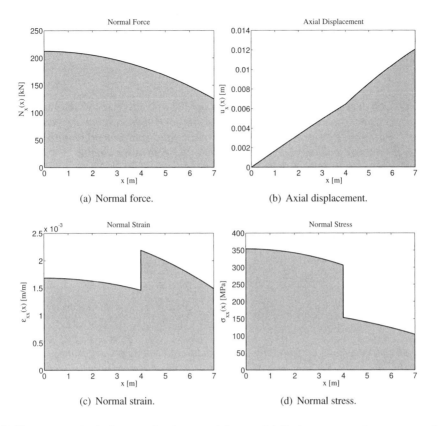

(a) Normal force. (b) Axial displacement.

(c) Normal strain. (d) Normal stress.

Figure 3.42 Example 3.18: diagrams for the normal force, axial displacement, strain measure, and normal stress.

displacement are continuous along the bar length. But the normal strains and stresses are discontinuous at the $x = L_1$ interface section.

The file barexemp18.m implements the solution of this example using the symbolic manipulation toolbox available in MATLAB.

□

Example 3.19 *Consider the bar of the previous example, but subjected to a constant distributed load of intensity p_0 and a concentrated force P at the interface of the two bar segments, as illustrated in Figure 3.43. Plot the diagrams of the normal force, axial displacement, longitudinal strain, and normal stress for the bar using the same parameters of the previous example.*

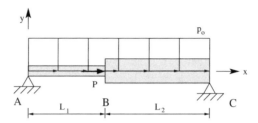

Figure 3.43 Example 3.19: bar with varying material and cross-sections and a concentrated force at the interface of two segments.

To solve this problem, we consider the AB and BC bar segments separately as illustrated in Figure 3.44. In this case, N_1 and N_2 are, respectively, the normal forces in sections $x = L_1^-$ of the AB segment and $x = L_1^+$ of the BC segment. These forces will be determined by the solution procedure. To this purpose, we need consider explicitly the equilibrium of the interface between the two segments, as illustrated in Figure 3.44(b). In addition, we need the compatibility condition of axial displacement at the $x = L_1$ interface between the two bar segments.

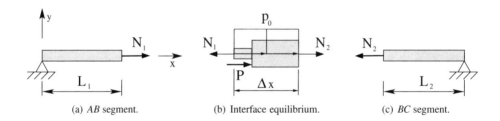

(a) *AB segment.*   (b) Interface equilibrium.   (c) *BC segment.*

Figure 3.44 Example 3.19: forces in the segments and interface.

We initially consider the AB bar segment with $0 < x < L_1$. In this case, the unknowns are the C_1 and C_2 integration constants and the normal force N_1. The same previous procedure of integrating the differential equation is applied, for the axial displacement of the AB segment.

• *Distributed load equation*

$$q_{x_1}(x) = p_0.$$

• *Boundary conditions*

$$u_{x_1}(x = 0) = 0 \quad and \quad N_{x_1}(x = L_1) = N_1.$$

• *Integration of the differential equation*

$$E_1 A_1 \frac{d^2 u_{x_1}(x)}{dx^2} = -q_{x_1}(x) = -p_0.$$

The first integration yields the normal force expression

$$N_{x_1}(x) = -p_0 x + C_1.$$

The second one yields the axial displacement equation

$$u_{x_1}(x) = \frac{1}{E_1 A_1}\left(-\frac{p_0}{2}x^2 + C_1 x + C_2\right).$$

- *Determination of C_1 and C_2*

  The boundary conditions are applied to determine the C_1 and C_2 integration constants, that is,

$$N_{x_1}(x = L_1) = -p_0 L_1 + C_1 = N_1 \rightarrow C_1 = N_1 + p_0 L_1,$$

$$u_{x_1}(x = 0) = \frac{1}{E_1 A_1}\left(-\frac{p_0}{2}(0)^2 + C_1(0) + C_2\right) \rightarrow C_2 = 0.$$

- *Final equations*

  By substituting the C_1 and C_2 integration constants, the normal force and axial displacement expressions in terms of N_1 can be determined, respectively, as

$$N_{x_1}(x) = p_0(L_1 - x) + N_1,$$

$$u_{x_1}(x) = \frac{1}{E_1 A_1}\left[\frac{p_0}{2}(L_1 - x)x + N_1 x\right].$$

We now consider the BC bar segment with $L_1 < x < L_2$. In this case, the unknowns are the integration constants C_3 and C_4 and the normal force N_2. The same procedure of integrating the differential equation of the axial displacement is applied.

- *Distributed load equation*

$$q_{x_2}(x) = p_0.$$

- *Boundary conditions*

$$u_{x_2}(x = L_1) = u_1 \quad and \quad N_{x_2}(x = L) = 0.$$

- *Integration of the differential equation*

$$E_2 A_2 \frac{d^2 u_{x_2}(x)}{dx^2} = -q_{x_2}(x) = -p_0.$$

The first integration yields the normal force expression

$$N_{x_2}(x) = -p_0 x + C_3.$$

The second one results in the axial displacement equation

$$u_{x_2}(x) = \frac{1}{E_2 A_2}\left(-\frac{p_0}{2}x^2 + C_3 x + C_4\right).$$

- *Determination of C_3 and C_4*

  The boundary conditions are applied to find C_3 and C_4. Thus,

$$N_{x_2}(x = L) = -p_0 L + C_3 = 0 \rightarrow C_3 = p_0 L,$$

$$u_{x_2}(x = L_1) = \frac{1}{E_2 A_2}\left(-\frac{p_0}{2}L_1^2 + p_0 L L_1 + C_4\right) = u_1$$

$$\rightarrow C_4 = E_2 A_2 u_1 - \frac{p_0}{2}L_1(2L - L_1).$$

- *Final equations*

  By substituting the C_3 and C_4 integration constants, we can determine, respectively, the final expressions for the normal force and axial displacement

  $$N_{x_2}(x) = p_0(L - x),$$

  $$u_{x_2}(x) = \frac{p_0}{2E_2 A_2}\left[(2L - x)x - L_1(2L - L_1)\right] + u_1.$$

  By taking $x = L_1$ in the previous normal force expression, we calculate N_2 as

  $$N_2 = N_{x_2}(x = L_1) = p_0(L - L_1) = p_0 L_2.$$

For this problem, the equilibrium of forces in the x direction should be considered at the interface of segments AB and BC, as illustrated in Figure 3.44(b). Thus,

$$\sum f_x = 0: \quad -N_1 + P + N_2 + p_0 \Delta x = 0 \rightarrow N_1 - N_2 = P - p_0 \Delta x.$$

Taking the limit for $\Delta x \rightarrow 0$, we have

$$N_1 - N_2 = P.$$

Substituting the expression for N_2, the normal force N_1 is determined as

$$N_1 = P + p_0 L_2.$$

The final expressions for the normal force and axial displacement of the AB segment are, respectively,

$$N_{x_1}(x) = p_0(L_1 - x) + P + p_0 L_2 = p_0(L - x) + P,$$

$$u_{x_1}(x) = \frac{1}{E_1 A_1}\left[\frac{p_0}{2}(L + L_2 - x)x + Px\right].$$

Using the kinematic compatibility, the axial displacements u_{x_1} and u_{x_2} of the AB and BC segments must be equal at the $x = L_1$ interface. Thus,

$$u_{x_1}(x = L_1) = u_{x_2}(x = L_1) = u_1.$$

Taking $x = L_1$ in the $u_{x_1}(x)$ expression, we have

$$u_1 = u_{x_1}(x = L_1) = \frac{p_0 L_1 L_2}{E_1 A_1} + \frac{P L_1}{E_1 A_1}.$$

By substituting u_1 in the u_{x_2} equation, we obtain the final expression for $u_{x_2}(x)$, that is,

$$u_{x_2}(x) = \frac{p_0}{2E_2 A_2}\left[(2L - x)x - L_1(2L - L_1)\right] + \frac{L_1}{E_1 A_1}(p_0 L_2 + P).$$

Figure 3.45 illustrates the diagrams of the normal force, axial displacement, longitudinal normal strain, and normal stress, which were obtained after substituting the given values. In this case, it is observed that the normal force is also discontinuous due to the concentrated force P. The discontinuity value of the normal force at $x = L_1$ is numerically equal to the value of P. The remaining diagrams preserve the same behavior of the previous example.

The MATLAB file barexemp19.m presents the solution of this example using symbolic manipulation.

□

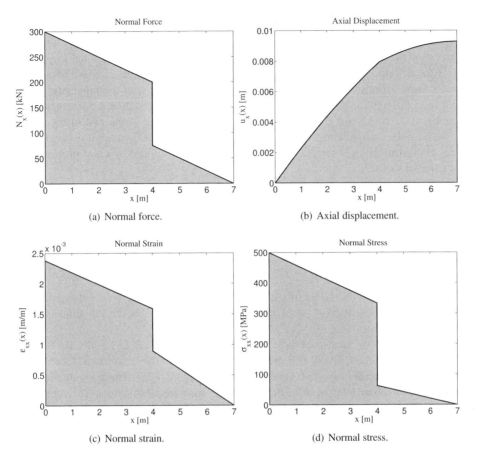

(a) Normal force.

(b) Axial displacement.

(c) Normal strain.

(d) Normal stress.

Figure 3.45 Example 3.19: normal force, axial displacement, strain measure, and normal stress diagrams.

3.14 SUMMARY OF THE VARIATIONAL FORMULATION OF BARS

In this section, a summary of the variational formulation of the bar mechanical model is presented.

To determine the mechanical model under consideration using the variational formulation, the first step is the definition of model kinematics. From the adopted kinematics, we obtain the strain components, rigid actions, internal and external compatible loads with the kinematics, as well as the equilibrium differential equations. Applying the material constitutive model, it is possible to determine the differential equations in terms of kinematics. Thus, at the end of the formulation process, we have the complete mechanical model of the problem. After that, it is possible to apply this model to determine the kinematics and other quantities of interest to a real problem under the action of external loads.

Thus, while the variational formulation of the model starts from the kinematics, when solving a real problem with the formulated model, the kinematics is the unknown to be determined.

A more formal way of presenting the bar model employs a notation that covers all the main elements of the formulation. Thus, with the adopted kinematics, we can define the set $\mathscr{V}$ of the possible displacement actions for the bar as

$$\mathscr{V} = \{u_x(x),\ x \in (0,L)\,,\ u_x(x) \text{ is continuous}\}. \tag{3.63}$$

This notation means that $\mathscr{V}$ is the set of continuous functions of the variable x along the bar length L. This set can also be denoted as $C(0,L)$, where letter C indicates that the functions of this set are continuous.

The boundary of the bar consists of the end sections $x = 0$ and $x = L$. Generally, the bar is fixed in at least one of these sections, giving rise to the displacement kinematic constraints. In this case, the values of $u_x(x)$ are prescribed at $x = 0$ and/or $x = L$ (see Figure 3.26). These kinematic constraints are indicated in the subset Kin_v of $\mathscr{V}$ of the admissible displacement actions. If the bar is fixed at $x = 0$ or at both $x = 0$ and $x = L$ ends, the respective subsets of kinematically admissible actions are given by

$$
\begin{aligned}
Kin_v &= \{u_x(x), u_x(0) = 0,\ x \in (0,L)\,, u_x(x) \text{ is continuous}\}\,, \\
Kin_v &= \{u_x(x), u_x(0) = 0, u_x(L) = 0, x \in (0,L)\,, u_x(x) \text{ is contnuous}\}\,.
\end{aligned}
\tag{3.64}
$$

Therefore, the kinematically admissible displacement actions for bars with physical supports are given by functions that satisfy these restrictions, constituting the subset Kin_v of $\mathscr{V}$. In the case of free bar, that is, without any kinematic constraints, all functions $v \in \mathscr{V}$ are also admissible actions, because in these situations there are no supports and consequently no kinematic constraints.

The normal specific strain measure in section x of the bar is given by the derivative of the kinematics, i.e.,

$$\varepsilon_{xx}(x) = \frac{du_x(x)}{dx}. \tag{3.65}$$

We represent this strain measure by the $\mathscr{D}$ operator. In this case, the $\mathscr{D}$ operator is simply the derivative with respect to x, i.e., $\mathscr{D} = \dfrac{d}{dx}$. Likewise, $\mathscr{W}$ is the set of all scalar functions $\varepsilon_{xx}(x)$, called normal strain measures, and obtained by the derivatives of the displacement actions $u_x(x)$ of $\mathscr{V}$. It is observed that the operator $\mathscr{D} : \mathscr{V} \to \mathscr{W}$ relates the kinematics to the strain measure, i.e.,

$$
\begin{aligned}
\mathscr{D} : \quad & \mathscr{V} \longrightarrow \mathscr{W} \\
& u_x(x) \longrightarrow \varepsilon_{xx}(x) = \mathscr{D}u_x(x) = \frac{du_x(x)}{dx}
\end{aligned}
\tag{3.66}
$$

As shown previously, the work of the internal forces is zero for rigid body displacement actions. This is equivalent to affirming that the normal strain measure is zero for every section x along the bar length. Hence,

$$\varepsilon_{xx}(x) = \frac{du_x(x)}{dx} = 0 \qquad x \in (0,L). \tag{3.67}$$

For this purpose, $u_x(x) = u_x$ must be constant, which physically represents a translation of the bar in the x axis. The set of rigid actions in $\mathcal{V}$, that is, the actions $u_x \in \mathcal{V}$ such that $\varepsilon_{xx}(x) = \mathcal{D}u_x = \dfrac{du_x}{dx} = 0$, defines the subset $\mathcal{N}(\mathcal{D})$ of the bar rigid actions. This subset is formally defined as

$$\mathcal{N}(\mathcal{D}) = \left\{ u_x(x) \in \mathcal{V} \mid u_x(x) = u_x = \text{cte}, \mathcal{D}u_x = \frac{du_x}{dx} = 0 \right\}, \tag{3.68}$$

that is, $\mathcal{N}(\mathcal{D})$ is the subset of all $u_x(x)$ actions of $\mathcal{V}$, such that $u_x(x)$ is constant and consequently the strain measure $\varepsilon_{xx}(x)$ is zero.

In section 3.7, a set of expressions (differential equation + boundary conditions) was developed, which represents the equilibrium problem of the deformed bar, that is,

$$\begin{cases} \dfrac{dN_x(x)}{dx} + q_x(x) = 0 & \text{in } x \in (0, L) \\ N_x(L) = P_L & \text{in } x = L \\ N_x(0) = -P_0 & \text{in } x = 0 \end{cases} \tag{3.69}$$

From these expressions, the equilibrium operator $\mathcal{D}^*$ is defined between the internal and external loads present in the bar. This operator can be described as

$$\mathcal{D}^* N_x(x) = \begin{cases} -\dfrac{d}{dx} N_x(x) & \text{in } x \in (0, L) \\ -\left. N_x(x) \right|_{x=0} & \text{in } x = 0 \\ \left. N_x(x) \right|_{x=L} & \text{in } x = L \end{cases} \tag{3.70}$$

The $\mathcal{D}^*$ operator maps the internal $\mathcal{W}'$ and external $\mathcal{V}'$ vector spaces. In this case, the vector space of the external forces $\mathcal{V}'$ is characterized by a continuous scalar function $q_x(x)$, which indicates the distributed axial load along the bar, and the concentrated forces P_0 and P_L at the bar ends, used as boundary conditions for the problem. Hence, $\mathcal{D}^*$ is denoted as

$$\mathcal{D}^* : \quad \mathcal{W}' \to \mathcal{V}'$$

$$N_x(x) \to \mathcal{D}^* N_x(x) = \begin{cases} -\dfrac{d}{dx} N_x(x) = q_x(x) & \text{in } x \in (0, L) \\ -\left. N_x(x) \right|_{x=0} = P_0 & \text{in } x = 0 \\ \left. N_x(x) \right|_{x=L} = P_L & \text{in } x = L \end{cases} \tag{3.71}$$

A diagram can be constructed with the relations among the $\mathcal{V}$, $\mathcal{W}$, $\mathcal{V}'$, and $\mathcal{W}'$ vector spaces and the $\mathcal{D}$ and $\mathcal{D}^*$ operators, summarizing the bar formulation, as illustrated in Figure 3.46.

The relations between the pair of dual spaces $(\mathcal{V}, \mathcal{V}')$, and $(\mathcal{W}, \mathcal{W}')$ are given, respectively, by the linear functionals W_e and W_i, which represent the work of the internal and external forces. The equilibrium is obtained with the application of the PVW.

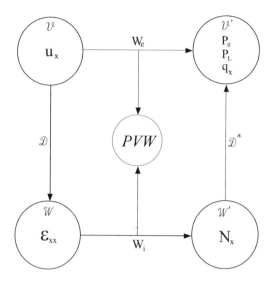

Figure 3.46   Variational formulation of the bar model.

3.15   APPROXIMATED SOLUTION

So far, all the considered examples involved a single bar element and were solved analytically from the equilibrium differential equations (3.20) and (3.45). In many cases, it may be impossible to obtain the analytical solution even for a single bar, due to variations of the cross-section area and the elasticity modulus along the bar length, different boundary conditions, and discontinuous loads.

In addition, in problems with several bar elements, such as the truss illustrated in Figure 1.6, the solution of the equilibrium BVP for each bar and the coupling of these solutions with the other elements constitutes a troublesome task. Accordingly, the use of computational methods based on constructing approximated solutions is an important alternative and extensively applied in engineering nowadays as presented in [33, 53, 35].

The approximated solution $u_{x_N}(x)$ for the bar axial displacement $u_x(x)$ is given by the following interpolation:

$$u_{x_N}(x) = \sum_{i=1}^{N} a_i \Phi_i(x), \tag{3.72}$$

where a_i are the coefficients to be determined and $\Phi_i(x)$ the basis, shape, or interpolation functions.

The a_i coefficients are determined by applying an approximation criterion. That allows the definition of various approximation methods such as collocation, Galerkin, and least squares. These methods belong to the class of weighted residual methods. The selection of interpolation functions defines, among others, the spectral, meshless, and finite element methods.

In this section, we present an introduction to the approximation methods based on weighted residuals, considering the bar mechanical model.

3.15.1   ANALOGY OF THE APPROXIMATED SOLUTION WITH VECTORS

Consider the vector $\mathbf{v}$, illustrated in Figure 3.47(a), with components (v_x, v_y, v_z), according to the adopted Cartesian system xyz. The vector $\mathbf{v}$ can be represented as the linear combination of base vectors $(\mathbf{e}_x, \mathbf{e}_y, \mathbf{e}_z)$, that is,

$$\mathbf{v} = v_x \mathbf{e}_x + v_y \mathbf{e}_y + v_z \mathbf{e}_z. \tag{3.73}$$

The projection of vector $\mathbf{v}$ in the xy plane is given by

$$\mathbf{v}_{xy} = v_x\mathbf{e}_x + v_y\mathbf{e}_y. \tag{3.74}$$

Consequently, vector $\mathbf{v}$ can be rewritten as

$$\mathbf{v} = \mathbf{v}_{xy} + v_z\mathbf{e}_z. \tag{3.75}$$

Considering $\mathbf{v}_{xy}$ as an approximation to the vector $\mathbf{v}$, the associated error vector $\mathbf{e}$ is

$$\mathbf{e} = \mathbf{v} - \mathbf{v}_{xy} = v_z\mathbf{e}_z. \tag{3.76}$$

Thus, the error vector $v_z\mathbf{e}_z$ is orthogonal to the xy plane. Such condition is expressed by the following scalar product of vectors:

$$(\mathbf{e}, \mathbf{v}_{xy}) = v_z\mathbf{e}_z \cdot \mathbf{v}_{xy} = 0, \quad \forall \mathbf{v}_{xy} \text{ in the } xy \text{ plane.} \tag{3.77}$$

In particular, as the vectors $\mathbf{e}_x$ and $\mathbf{e}_y$ are in the xy plane, we have

$$(\mathbf{e}, \mathbf{e}_x) = (\mathbf{e}, \mathbf{e}_y) = 0. \tag{3.78}$$

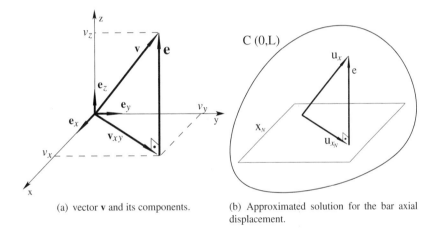

(a) vector $\mathbf{v}$ and its components.

(b) Approximated solution for the bar axial displacement.

Figure 3.47 Approximated solution analogy with vectors.

A similar interpretation can be used to justify the approximated solution representation given in (3.72). For this purpose, consider a continuous function $u_x(x)$ for the bar axial displacement, represented as a vector in the set of continuous functions $C(0,L)$ as illustrated in Figure 3.47(b). The basis functions $\{\Phi_i(x)\}_{i=1}^{N}$ defines a subset X_N of $C(0,L)$. Similarly to the vector $\mathbf{v}_{xy}$, written as a linear combination of the subset of basis vectors $(\mathbf{e}_x, \mathbf{e}_y)$, the approximated solution $u_{x_N}(x)$ is given by the linear combination of basis functions in X_N. Moreover, as $(\mathbf{e}_x, \mathbf{e}_y)$ is a set of linearly independent vectors, X_N constitutes the set of linearly independent continuous functions. Thus, any function $v_N(x)$ of X_N can be written by the following linear combination of the interpolation functions $\{\Phi_i(x)\}_{i=1}^{N}$:

$$v_N(x) = \sum_{j=1}^{N} b_j\Phi_j(x). \tag{3.79}$$

The error between the analytical $u_x(x)$ and approximated $u_{x_N}(x)$ functions is now given by an error function $e(x)$ in such way that

$$e(x) = u_x(x) - u_{x_N}(x). \tag{3.80}$$

One should observe that the error function is orthogonal to any function $v(x)$ given by (3.79). Analogously to (3.77), we have

$$(e(x), v_N(x)) = \int_0^L v_N(x)e(x)dx = 0. \tag{3.81}$$

Consider the continuous functions f and g defined in the closed interval $[a,b]$, as illustrated in Figure 3.48. Also consider the $N+1$ points x_i $(i = 0,\ldots,N)$ and the respective $f_i = f(x_i)$ and $g_i = g(x_i)$ function values. Analogously to the scalar product of vectors, the inner product of f and g is given by

$$(f,g)_N = \sum_{i=0}^N f(x_i)g(x_i) = \sum_{i=0}^N f_i g_i.$$

By taking infinite points x_i, the previous summation represents a Riemann integral, that is,

$$(f,g) = \lim_{N\to\infty} \sum_{i=1}^N f(x_i)g(x_i) = \int_a^b f(x)g(x)dx. \tag{3.82}$$

This equation defines the inner product of functions f and g in the interval $[a,b]$.

The L_2 norm of the f function, defined in the interval (a,b), is

$$\|f\|_{L_2} = \sqrt{(f,f)} = \sqrt{\int_a^b f^2(x)dx}. \tag{3.83}$$

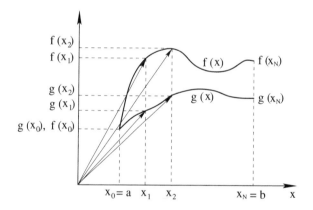

Figure 3.48 Internal product of functions.

Therefore, equation (3.81) represents the orthogonality condition between the error function and any $v_N(x)$ function in X_N. This means, analogously to (3.78), that the error function must be orthogonal to any basis function, that is,

$$(e(x), \Phi_i(x)) = 0 \quad \forall \Phi_i(x) \in X_N. \tag{3.84}$$

Using (3.83), we calculate the L_2 norm of the error function. The relative error is given by

$$\|e_r\|_{L_2} = \frac{\|e\|_{L_2}}{\|u_x\|_{L_2}}, \tag{3.85}$$

with $\|u_x\|_{L_2}$ the norm of the exact solution.

The concepts introduced in this section will be widely used in this book.

3.15.2 COLLOCATION METHOD

Substituting the approximated solution (3.72) in the differential equation (3.45), the distributed internal axial force $EA\dfrac{d^2 u_{x_N}(x)}{dx^2}$ will not balance the distributed external axial force $q_x(x)$, because in the general case, $u_{x_N}(x)$ does not coincide with $u_x(x)$. The residual function $r_N(x)$, associated to the $u_{x_N}(x)$ approximation, is defined by the difference between the approximate internal and external axial forces, i.e.,

$$r_N(x) = EA\frac{d^2 u_{x_N}(x)}{dx^2} + q_x(x), \quad x \in (0,L). \tag{3.86}$$

The collocation method computes the coefficients a_i of the approximation, by imposing a zero residue on some collocation points, with coordinates x_i, in the interval $(0,L)$. Consequently, the approximated solution coincides with the exact solution at the collocation points.

Therefore, given N collocation points with coordinates x_i, the calculated residue in these points should be zero, that is,

$$r_N(x_i) = EA\frac{d^2 u_{x_N}(x_i)}{dx^2} + q_x(x_i) = 0, \quad i = 1,\ldots,N. \tag{3.87}$$

Substituting (3.72) in the previous equation, we have

$$\sum_{j=1}^{N} EA\Phi''_j(x_i)a_j + q_x(x_i) = 0, \quad i = 1,\ldots,N. \tag{3.88}$$

In matrix notation,

$$EA \begin{bmatrix} \Phi''_1(x_1) & \Phi''_2(x_1) & \cdots & \Phi''_N(x_1) \\ \Phi''_1(x_2) & \Phi''_2(x_2) & \cdots & \Phi''_N(x_2) \\ \vdots & \vdots & \cdots & \vdots \\ \Phi''_1(x_N) & \Phi''_2(x_N) & \cdots & \Phi''_N(x_N) \end{bmatrix} \begin{Bmatrix} a_1 \\ a_2 \\ \vdots \\ a_N \end{Bmatrix} = \begin{Bmatrix} q_x(x_1) \\ q_x(x_2) \\ \vdots \\ q_x(x_N) \end{Bmatrix}. \tag{3.89}$$

The following examples illustrate the application of the collocation method.

Example 3.20 *Consider the bar of Example 3.8. For the* $\Phi_1(x) = x(x-L)$ *shape function, the approximated solution is*

$$u_{x_1}(x) = a_1\Phi_1(x) = x(x-L)a_1.$$

Notice that the $\Phi_1(x)$ *function satisfies the boundary conditions of the problem.*

By replacing the previous approximation in (3.87) and using one collocation point with coordinate $x_1 = 0.5$, *we have*

$$EA\Phi''_1(x)a_1 + p_0\big|_{x_1=0.5} = 0 \to a_1 = -\frac{p_0}{2EA}.$$

Hence, the approximated solution is given by

$$u_x(x) = -\frac{p_0}{2EA}x(x-L),$$

which coincides with the exact solution. This happened because the analytical solution and the shape function are both of second order.

□

Example 3.21 *Consider again the bar of Example 3.8, but with an external distributed load function given by* $q_x(x) = \dfrac{\pi}{L}\sin\left(\dfrac{\pi}{L}x\right)$. *Solving the equilibrium differential equation (3.45), the exact solution is given by*

$$u_x(x) = \frac{L}{\pi EA}\sin\left(\frac{\pi}{L}x\right).$$

The approximated solution using the interpolation function $\Phi_1(x) = x(L-x)$ *is*

$$u_{x_1}(x) = a_1\Phi_1(x) = x(L-x)a_1.$$

Substituting the previous approximation in (3.87) and using one collocation point with coordinate $x_1 = 0.5$, *we have*

$$EA\Phi_1''(x)a_1 + \frac{\pi}{L}\sin\left(\frac{\pi}{L}x\right)\Big|_{x_1=0.5} = 0 \rightarrow a_1 = \frac{\pi}{2EAL}\sin\left(\frac{\pi}{2L}\right).$$

For $L = 1$ m *and* $a_1 = \frac{\pi}{2EA}$, *the approximated solution is given by*

$$u_{x_1}(x) = \frac{\pi}{2EA}x(1-x).$$

The L_2 *norm of the relative error between the exact and approximated solutions is given by*

$$\|e_r\|_{L_2} = \sqrt{\frac{\int_0^1 (u_x(x)-u_{x_1}(x))^2\,dx}{\int_0^1 u_x^2(x)dx}} = 27.75\%.$$

Consider now the interpolation functions $\Phi_1(x) = x(1-x)$ *and* $\Phi_2(x) = x^2(1-x)$. *The approximated solution is given by the following linear combination:*

$$u_{x_2}(x) = a_1\Phi_1(x) + a_2\Phi_2(x) = x(1-x)a_1 + x^2(1-x)a_2.$$

Substituting the previous approximation in equation (3.87) results in

$$EA(\Phi_1''(x)a_1 + \Phi_2''(x)a_2) + \frac{\pi}{L}\sin\left(\frac{\pi}{L}x\right)\Big|_{x_i} = 0.$$

Because $\Phi_1''(x) = -2$ *and* $\Phi_2''(x) = 2 - 6x$ *and taking two collocation points with coordinates* $x_1 = 0.25$ *and* $x_2 = 0.75$, *we have the second-order system of equations given by*

$$EA\begin{bmatrix} 2.0 & -0.5 \\ 2.0 & 2.5 \end{bmatrix}\begin{Bmatrix} a_1 \\ a_2 \end{Bmatrix} = \begin{Bmatrix} 2.22 \\ 2.22 \end{Bmatrix},$$

with solution $a_1 = 1.111/EA$ *and* $a_2 = 0.0$. *Thus, the approximated solution is*

$$u_{x_2}(x) = \frac{1.111}{EA}x(1-x).$$

The L_2 *error norm between the exact and approximated solutions is given by*

$$\|e_r\|_{L_2} = \sqrt{\frac{\int_0^1 (u_x(x)-u_{x_2}(x))^2\,dx}{\int_0^1 u_x^2(x)dx}} = 10.54\%.$$

Note that due to the chosen coordinate for the second collocation point, the a_2 *coefficient is zero and the second function did not contribute directly on improving the approximation. However, the use of the collocation coordinate* $x_i = 0.25$ *allowed a substantial improvement on the approximation, because the relative error reduced almost three times. This shows that the choice of the coordinates is important in the collocation method.*

By including another basis function $\Phi_3(x) = x^3(1-x)$ and repeating the process for collocation coordinates $x_1 = 0.25$, $x_2 = 0.50$, and $x_3 = 0.75$, we obtain the following system of equations:

$$EA \begin{bmatrix} 2.00 & -0.50 & -0.75 \\ 2.00 & 1.00 & 0.00 \\ 2.00 & 2.50 & 2.25 \end{bmatrix} \begin{Bmatrix} a_1 \\ a_2 \\ a_3 \end{Bmatrix} = \begin{Bmatrix} 2.22 \\ 3.14 \\ 2.22 \end{Bmatrix},$$

resulting in $a_1 = 0.096/EA$ and $a_2 = a_3 = 0.123/EA$. Notice that the matrix of the system of equations is not symmetric.

The approximated solution and the relative error are given, respectively, by

$$u_{x_3}(x) = \frac{1}{EA}\left(0.096x(1-x) + 0.123x^2(1-x) - 0.123x^3(1-x)\right),$$

$$\|e_r\| = \sqrt{\frac{\int_0^1 \left(u_x(x) - u_{x_3}(x)\right)^2 dx}{\int_0^1 u_x^2(x)dx}} = 0.99\%.$$

Functions $\Phi_i(x)$ satisfy the boundary conditions of the problem. It is observed that when approximating the residue function, the solution of a differential equation is substituted by the solution of a system of equations for the coefficients of the approximated solution.

Figure 3.49 illustrates the analytical solution and the three approximated solutions with $EA = 10^4$ N. It can be observed that the solution with three functions allowed a more pronounced reduction of the error. File barsolapex2.m implements the solution of this example to an arbitrary number of equally spaced collocation points.

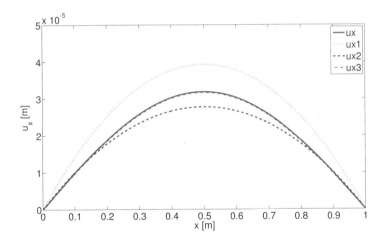

Figure 3.49 Example 3.21: approximated solutions for the collocation method.

☐

A more detailed treatment about collocation methods can be found in [14, 46].

3.15.3 WEIGHTED RESIDUALS METHOD

The collocation method is a particular case of the method of weighted residuals. This method imposes the restriction that the inner product of the residue of the approximation $r_N(x)$ with the test or weight functions $v_j(x)$ $(j = 1,\ldots,N)$ is zero. Therefore,

$$\int_0^L r_N(x)v_j(x)dx = 0, \quad j = 1,\ldots,N. \tag{3.90}$$

The choice of the test functions allows us to define different approximation methods. The set of test functions is denoted by Y_N.

Consider the test functions $v_j(x) = \delta(x - x_j)$, with $\delta(x - x_j)$ the Dirac's delta distribution, illustrated in Figure 3.50. One of the properties of the Dirac's delta for any function $f(x)$ is

$$\int_{-\infty}^{+\infty} f(x)\delta(x - x_j)dx = f(x_j). \tag{3.91}$$

Due to this property, equation (3.90) recovers the zero residue condition at the collocation points, as given in (3.87). Thus, substituting v_j for δ_j in (3.90) we have

$$\int_0^L r_N(x)\delta(x - x_j)dx = r_N(x_j), \quad j = 1,\ldots,N. \tag{3.92}$$

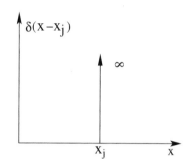

Figure 3.50 Dirac's delta.

3.15.4   LEAST SQUARES METHOD

The least squares method minimizes the square of the L_2 norm of the approximation residue, that is,

$$\min ||r_N||^2 = \min(r_N, r_N) = \min \int_0^L r_N^2(x)dx. \tag{3.93}$$

The previous condition is equivalent to making the derivative of the inner product to the approximation coefficients equal to zero. Thus,

$$\min(r_N, r_N) \Rightarrow \frac{\partial}{\partial a_j}\int_0^L r_N^2(x)dx = 0. \tag{3.94}$$

The following expression is obtained:

$$\int_0^L r_N(x)\frac{\partial r_N(x)}{\partial a_j}dx = 0, \quad j = 1,\ldots,N. \tag{3.95}$$

Comparing the previous expression with (3.90), we verify that the least squares method is a weighted residuals method with the test functions $v_j(x) = \frac{\partial r_N(x)}{\partial a_j}$.

The partial derivatives of the residue function for the bar case relative to the coefficients a_j are given by

$$\frac{\partial r_N(x)}{\partial a_j} = EA\Phi''_j(x).$$

Thus, the following system of equations for the a_j unknowns results, when expanding expression (3.95):

$$EA \int_0^L \begin{bmatrix} \Phi''_1(x)\Phi''_1(x) & \cdots & \Phi_1''(x)\Phi''_j(x) & \cdots & \Phi''_1(x)\Phi''_N(x) \\ \vdots & \cdots & \vdots & \cdots & \vdots \\ \Phi''_i(x)\Phi''_1(x) & \cdots & \Phi''_i(x)\Phi''_j(x) & \cdots & \Phi''_i(x)\Phi''_N(x) \\ \vdots & & \vdots & & \vdots \\ \Phi''_N(x)\Phi''_1(x) & \cdots & \Phi''_N(x)\Phi''_j(x) & \cdots & \Phi''_N(x)\Phi''_N(x) \end{bmatrix} dx \begin{Bmatrix} a_1 \\ \vdots \\ a_i \\ \vdots \\ a_N \end{Bmatrix} =$$

$$- \int_0^L \begin{Bmatrix} q_x(x)\Phi''_1(x) \\ \vdots \\ q_x(x)\Phi''_i(x) \\ \vdots \\ q_x(x)\Phi''_N(x) \end{Bmatrix} dx. \tag{3.96}$$

The matrix of the system of equations is now symmetric.

In compact notation, we have

$$k_{ij}a_j = f_j, \quad i,j = 1,\ldots,N, \tag{3.97}$$

or in matrix notation,

$$[K]\{a\} = \{f\}. \tag{3.98}$$

Example 3.22 *The least squares method is applied to Example 3.21, solving the system of equations (3.96) for one, two, and three interpolation functions $\Phi_i(x) = x^i(L-x)$, $i = 1,2,3$.*

For one interpolation function $\Phi_1(x) = x(L-x)$, we obtain $k_{11} = 4.0$, $f_1 = 4.0$, and $a_1 = 1$. Consequently, the approximated solution and relative error are, respectively,

$$u_{x_1}(x) = \frac{1}{EA}x(1-x).$$

$$||e_r||_{L_2} = \sqrt{\frac{\int_0^1 (u_x(x) - u_{x_1}(x))^2 \, dx}{\int_0^1 u_x^2(x)dx}} = 19.19\%.$$

The following system of equations is obtained with two interpolation functions:

$$EA \begin{bmatrix} 4.0 & 2.0 \\ 2.0 & 4.0 \end{bmatrix} \begin{Bmatrix} a_1 \\ a_2 \end{Bmatrix} = \begin{Bmatrix} 4.0 \\ 2.0 \end{Bmatrix},$$

with solution of $a_1 = 1.0/EA$ and $a_2 = 0.0$. Thus, the approximated solution is

$$u_{x_2}(x) = \frac{1}{EA}x(1-x).$$

As the a_2 coefficient is zero, the same previous approximated solution with one function is obtained. The system of equations for three functions is

$$EA \begin{bmatrix} 4.0 & 2.0 & 2.0 \\ 2.0 & 4.0 & 4.0 \\ 2.0 & 4.0 & 4.8 \end{bmatrix} \begin{Bmatrix} a_1 \\ a_2 \\ a_3 \end{Bmatrix} = \begin{Bmatrix} 4.00 \\ 2.00 \\ 2.14 \end{Bmatrix},$$

with solution $a_1 = 1.0/EA$ and $a_2 = a_3 = 1.08/EA$.

The approximated solution and the relative error are, respectively,

$$u_{x_3}(x) = \frac{1}{EA}x(1-x)\left(1 + 1.08x - 1.08x^2\right),$$

$$||e_r||_{L_2} = \sqrt{\frac{\int_0^1 \left(u_x(x) - u_{x_3}(x)\right)^2 dx}{\int_0^1 u_x^2(x)dx}} = 0.21\%.$$

A great reduction in the relative error is observed when employing three interpolation functions.

Figure 3.51 illustrates the relative error for the solved problem with the collocation and least squares methods, for 1 to 10 interpolation functions given by $\Phi_i(x) = x^i(L-x)$. The collocation points are equally spaced in the open interval $(0,L)$, which excludes the coordinates $x = 0$ and $x = L$. Until $N = 9$, the relative error with the least squares method is smaller than the values for the collocation method. For $N = 10$, the error given by the collocation method becomes smaller.

File barsolapex3.m implements the solution of this example with least squares for an arbitrary number of interpolation functions.

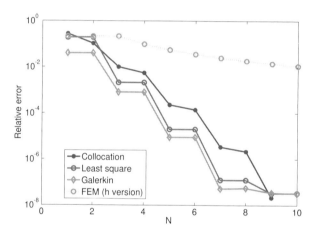

Figure 3.51 Example 3.22: relative errors of the approximated solutions with the methods of collocation, least squares, Galerkin, and FEM.

□

3.15.5 GALERKIN METHOD

In this case, the test functions $v_j(x)$ are the same as the basis functions $\Phi_j(x)$. Thus, the condition (3.90) is written as

$$\int_0^L r_N(x)\Phi_j(x)dx = 0, \quad j = 1,\dots,N. \tag{3.99}$$

This is equivalent to making the residue of approximation $r_N(x)$ orthogonal to the set of interpolation functions X_N, as illustrated in Figure 3.52. Note that the set of test functions in the Galerkin method is also X_N, that is, $Y_N = X_N$.

Substituting the residue in (3.99), we have the following system of equations for the a_i unknowns:

$$\sum_{i=1}^N \left(\int_0^L EA\Phi_i''(x)\Phi_j(x)dx\right)a_i = -\int_0^L q_x(x)\Phi_j(x)dx, \quad j = 1,\dots,N. \tag{3.100}$$

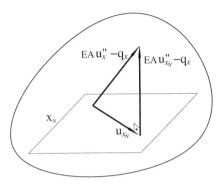

Figure 3.52 Projected residue in the X_N subset for the Galerkin method.

In an expanded form,

$$EA \int_0^L \begin{bmatrix} \Phi_1''(x)\Phi_1(x) & \cdots & \Phi_1''(x)\Phi_j(x) & \cdots & \Phi_1''(x)\Phi_N(x) \\ \vdots & \cdots & \vdots & \cdots & \vdots \\ \Phi_i''(x)\Phi_1(x) & \cdots & \Phi_i''(x)\Phi_j(x) & \cdots & \Phi_i''(x)\Phi_N(x) \\ \vdots & \vdots & \vdots & \vdots & \vdots \\ \Phi_N''(x)\Phi_1(x) & \cdots & \Phi_N''(x)\Phi_j(x) & \cdots & \Phi_N''(x)\Phi_N(x) \end{bmatrix} dx \begin{Bmatrix} a_1 \\ \vdots \\ a_i \\ \vdots \\ a_N \end{Bmatrix} =$$

$$-\int_0^L \begin{Bmatrix} q_x(x)\Phi_1(x) \\ \vdots \\ q_x(x)\Phi_i(x) \\ \vdots \\ q_x(x)\Phi_N(x) \end{Bmatrix} dx. \tag{3.101}$$

Example 3.23 *The Galerkin method is applied to Example 3.21, solving the system of equations (3.101) with the interpolation functions $\Phi_i(x) = x^i(L-x)$, $i = 1, \ldots, 10$.*

For three functions, we have the following system of equations:

$$EA \begin{bmatrix} 0.333 & 0.167 & 0.100 \\ 0.167 & 0.133 & 0.100 \\ 0.100 & 0.100 & 0.086 \end{bmatrix} \begin{Bmatrix} a_1 \\ a_2 \\ a_3 \end{Bmatrix} = \begin{Bmatrix} 0.405 \\ 0.203 \\ 0.115 \end{Bmatrix},$$

with solution $a_1 = 0.991/EA$ and $a_2 = a_3 = 1.125/EA$. Observe that the coefficient matrix is symmetric.

The approximated solution and the relative error are, respectively,

$$u_{x_3}(x) = \frac{1}{EA} x(1-x) \left(0.991 + 1.125x - 1.125x^2 \right),$$

$$\|e_r\|_{L_2} = \sqrt{\frac{\int_0^1 \left(u_x(x) - u_{x_3}(x) \right)^2 dx}{\int_0^1 u_x^2(x) dx}} = 0.081\%.$$

Figure 3.51 also illustrates the relative error obtained with the Galerkin method. It may be observed that until $N = 9$, the decreasing rate of the error is greater with the Galerkin method, when compared to collocation and least squares methods.

File barsolapex4.m implements the solution of this example using the Galerkin method for an arbitrary number of interpolation functions.

□

Example 3.24 *The Galerkin method is applied to Example 3.21, but taking the interpolation functions* $\Phi_i(x) = \sin\dfrac{i\pi}{L}x$. *These functions are used, for instance, in spectral methods [29, 46].*

It is also assumed that the distributed load has a parabolic behavior $q_x(x) = x(L - x)$. *The corresponding analytical solution is*

$$u_x(x) = \frac{1}{12EA}(x^3(x - 2L) + L^3x).$$

The following orthogonality equation is valid for the considered interpolation functions:

$$\int_0^L \sin\frac{i\pi}{L}x\sin\frac{j\pi}{L}x\,dx = \begin{cases} 0 & i \neq j \\ \frac{L}{2} & i = j \end{cases}. \tag{3.102}$$

Hence, the matrix of the system of equations (3.101) is diagonal with the following coefficients:

$$k_{ii} = -EA\left(\frac{i\pi}{L}\right)^2\int_0^L \sin^2\left(\frac{i\pi}{L}x\right)dx = \frac{EA}{2L}\pi^2 i^2.$$

The calculated relative errors for $i = 1, 3, 5, 7, 9$ *are, respectively,* 0.4128%, 0.0326%, 0.00623%, 0.00183%, *and* 0.00070%.

File barsolapex5.m implements the solution of this example using the Galerkin method for an arbitrary number of interpolation functions.

□

All the interpolation functions used until now satisfy the boundary conditions and have sufficient regularity so that the second derivative can be calculated. It is possible to reduce the minimum required regularity of the interpolation functions, which makes the construction of approximated solutions easier.

For this purpose, consider the bar differential equation rewritten as the following residue function:

$$r(x) = EA\frac{d^2u_x(x)}{dx^2} + q_x(x) = 0. \tag{3.103}$$

Multiplying the previous expression by the test function $v(x)$ and integrating along the bar length L, the following integral equilibrium equation is obtained:

$$\int_0^L r(x)v(x)dx = \int_0^L \left(EA\frac{d^2u_x(x)}{dx^2} + q_x(x)\right)v(x)dx = 0. \tag{3.104}$$

This expression represents the continuum equivalent to the main equation (3.90) of the weighted residue method.

The previous expression can be rewritten as

$$\int_0^L EA\frac{d^2u_x(x)}{dx^2}v(x)dx = -\int_0^L q_x(x)v(x)dx. \tag{3.105}$$

Applying integration by parts in the left side of the previous expression, we have

$$\int_0^L EA\frac{du_x(x)}{dx}\frac{dv(x)}{dx}dx = \int_0^L q_x(x)v(x)dx + \underbrace{EA\frac{du_x(x)}{dx}}_{N_x(x)}v(x)|_0^L. \tag{3.106}$$

Using the definition of normal force and the boundary conditions in terms of the normal force given in (3.20), we obtain

$$\int_0^L EA\frac{du_x(x)}{dx}\frac{dv(x)}{dx}dx = \int_0^L q_x(x)v(x)dx + P_Lv(L) + P_0v(0). \tag{3.107}$$

Thus, the previous equation has only the first derivatives of functions $u_x(x)$ and $v(x)$, while the differential equation (3.45) involves the second derivative of $u_x(x)$. The expression (3.107) is called weak form, while (3.45) is the strong form.

Considering the test function $v(x)$ as a virtual displacement $\delta u_x(x)$, equation (3.107) represents the PVW for the bar model, and is also called equilibrium integral equation. The left and right sides of equation (3.107) represent, respectively, the work of internal and external loads for $v(x) = \delta u_x(x)$.

In the Galerkin method, the approximation and test function spaces are the same, that is, $X_N = Y_N$. By substituting equations (3.72) and (3.79) in (3.107), we obtain the following expression for $j = 1, \ldots, N$:

$$\sum_{i,j=1}^{N} \left[\left(\int_0^L EA\Phi_i'(x)\Phi_j'(x)dx \right) a_i - \int_0^L q_x(x)\Phi_j(x)dx - P_L\phi_j(L) - P_0\phi_j(0) \right] b_j = 0.$$

As b_j is an arbitrary coefficient, because the test functions $v(x)$ are also arbitrary, the term inside brackets must be zero for $i = 1, \ldots, N$ resulting in the following system of equations:

$$\sum_{i,j=1}^{N} \left(\int_0^L E(x)A(x)\Phi_i'(x)\Phi_j'(x)dx \right) a_j = \int_0^L q_x(x)\Phi_i(x)dx + P_L\Phi_i(L) + P_0\Phi_i(0). \tag{3.108}$$

In expanded form and $EA = $ cte,

$$EA \int_0^L \begin{bmatrix} \Phi_1'(x)\Phi_1'(x) & \cdots & \Phi_1'(x)\Phi_j'(x) & \cdots & \Phi_1'(x)\Phi_N'(x) \\ \vdots & \cdots & \vdots & \cdots & \vdots \\ \Phi_i'(x)\Phi_1'(x) & \cdots & \Phi_i'(x)\Phi_j'(x) & \cdots & \Phi_i'(x)\Phi_N'(x) \\ \vdots & \vdots & \vdots & \vdots & \vdots \\ \Phi_N'(x)\Phi_1'(x) & \cdots & \Phi_N'(x)\Phi_j'(x) & \cdots & \Phi_N'(x)\Phi_N'(x) \end{bmatrix} dx \begin{Bmatrix} a_1 \\ \vdots \\ a_i \\ \vdots \\ a_N \end{Bmatrix} =$$

$$\int_0^L \begin{Bmatrix} q_x(x)\Phi_1(x) \\ \vdots \\ q_x(x)\Phi_i(x) \\ \vdots \\ q_x(x)\Phi_N(x) \end{Bmatrix} dx + \begin{Bmatrix} P_L\Phi_1(L) + P_0\Phi_1(0) \\ \vdots \\ P_L\Phi_i(L) + P_0\Phi_i(0) \\ \vdots \\ P_L\Phi_N(L) + P_0\Phi_N(0) \end{Bmatrix}, \tag{3.109}$$

or

$$[K]\{a\} = \{f\},$$

where $[K]$ is the symmetric stiffness matrix, $\{a\}$ the vector of unknowns, and $\{f\}$ the load vector.

The axial displacement approximation for $N = 2$ is

$$u_{x_2}(x) = a_1\Phi_1(x) + a_2\Phi_2(x) = \begin{bmatrix} \Phi_1(x) & \Phi_2(x) \end{bmatrix} \begin{Bmatrix} a_1 \\ a_2 \end{Bmatrix} = [N]\{a\}, \tag{3.110}$$

and $[N]$ is the matrix of interpolation functions given by

$$[N] = \begin{bmatrix} \Phi_1(x) & \Phi_2(x) \end{bmatrix}. \tag{3.111}$$

Consider the following basis functions and their respective derivatives:

$$\Phi_1(x) = 1 - \frac{x}{L} \quad \rightarrow \quad \Phi_1'(x) = -\frac{1}{L},$$

$$\Phi_2(x) = \frac{x}{L} \quad \rightarrow \quad \Phi_2'(x) = \frac{1}{L}.$$

These functions and their derivatives are shown in Figure 3.53 for $L = 1$. Note that the maximum values are 1 at the bar ends. Thus, we have, in this case, the following physical interpretation of the approximation coefficients:

$$u_{x_2}(0) = a_1\Phi_1(0) + a_2\Phi_2(0) = a_1 = u_1,$$
$$u_{x_2}(L) = a_1\Phi_1(L) + a_2\Phi_2(L) = a_2 = u_2, \tag{3.112}$$

that is, they represent the axial displacement at the bar ends. This feature is called the collocation property of the interpolation functions.

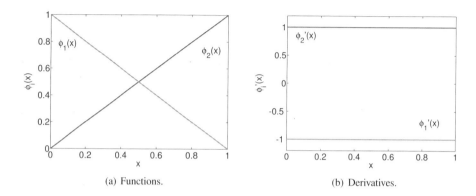

(a) Functions.   (b) Derivatives.

Figure 3.53   Global linear interpolation functions and their respective first derivatives for $L = 1$.

The approximation of the normal strain $\varepsilon_{xx_2}(x)$ is

$$\begin{aligned}
\varepsilon_{xx_2}(x) &= \frac{du_{x_2}(x)}{dx} = a_1\Phi_1'(x) + a_2\Phi_2'(x) \\
&= \begin{bmatrix} \Phi_1'(x) & \Phi_2'(x) \end{bmatrix} \begin{Bmatrix} a_1 \\ a_2 \end{Bmatrix} = [B]\{a\},
\end{aligned} \tag{3.113}$$

with $[B]$ the strain-displacement matrix given by

$$[B] = \begin{bmatrix} \Phi_1'(x) & \Phi_2'(x) \end{bmatrix} = \frac{1}{L} \begin{bmatrix} -1 & 1 \end{bmatrix} . \tag{3.114}$$

The stiffness matrix for $N = 2$ is given from (3.109) by

$$[K] = EA \int_0^L \begin{bmatrix} \Phi_1'(x)\Phi_1'(x) & \Phi_1'(x)\Phi_2'(x) \\ \Phi_1'(x)\Phi_2'(x) & \Phi_2'(x)\Phi_2'(x) \end{bmatrix} dx. \tag{3.115}$$

Substituting the expressions of the derivatives and performing the indicated operations, we obtain

$$[K] = \frac{EA}{L} \begin{bmatrix} 1 & -1 \\ -1 & 1 \end{bmatrix} . \tag{3.116}$$

The stiffness matrix for the bar can be denoted in a general form as

$$[K] = \int_V [B]^T[D][B]dV = \int_0^L [B]^T[D][B]Adx, \tag{3.117}$$

being $[D]$ the elasticity matrix with $[D] = E$ for the bar model.

For a constant distributed load of intensity $q_x(x) = p_0$, the load vector is given by

$$\{f\} = \left\{ \begin{array}{c} \int_0^L p_0 \Phi_1(x)dx + P_0 \Phi_1(0) + P_L \Phi_1(L) \\ \int_0^L p_0 \Phi_2(x)dx + P_0 \Phi_2(0) + P_L \Phi_2(L) \end{array} \right\} = \left\{ \begin{array}{c} \dfrac{p_0 L}{2} + P_0 \\ \dfrac{p_0 L}{2} + P_L \end{array} \right\}. \tag{3.118}$$

The distributed load is transformed in two concentrated equivalent forces of intensity $\dfrac{p_0 L}{2}$. The general expression for the equivalent force is denoted as

$$\{f\} = \int_0^L q_x(x)[N]^T dx. \tag{3.119}$$

The final system of equations is given by

$$\frac{EA}{L} \begin{bmatrix} 1 & -1 \\ -1 & 1 \end{bmatrix} \left\{ \begin{array}{c} u_1 \\ u_2 \end{array} \right\} = \left\{ \begin{array}{c} \dfrac{p_0 L}{2} + P_0 \\ \dfrac{p_0 L}{2} + P_L \end{array} \right\}. \tag{3.120}$$

Example 3.25 *Consider the bar of Example 3.5 with a force P applied at $x = L$, in such way that the boundary condition is $N_x(x = L) = P$. The analytical solution for the axial displacement is given by*

$$u_x(x) = \frac{1}{EA} \left(Px + p_0 Lx - \frac{p_0}{2} x^2 \right).$$

From the system of equations (3.120), we have

$$\frac{EA}{L} \begin{bmatrix} 1 & -1 \\ -1 & 1 \end{bmatrix} \left\{ \begin{array}{c} 0 \\ u_2 \end{array} \right\} = \left\{ \begin{array}{c} \dfrac{p_0 L}{2} + R \\ \dfrac{p_0 L}{2} + P \end{array} \right\},$$

with R the support reaction at $x = 0$ and $u_1 = u_x(x = 0) = 0$. The previous system of equations reduces to

$$\frac{EA}{L} u_2 = -\frac{p_0 L}{2} - R,$$

$$\frac{EA}{L} u_2 = \frac{p_0 L}{2} + P.$$

The first equation has two unknowns, namely u_2 and R. Solving the second equation, we have

$$u_2 = \frac{L}{EA} \left(\frac{p_0 L}{2} + P \right).$$

Substituting u_2 in the first equation, we obtain the reaction R

$$R = -(p_0 L + P).$$

The approximated solution is

$$u_{x_2}(x) = \frac{1}{EA} \left(P + \frac{p_0 L}{2} \right) x.$$

It is observed that the previous function is linear, while the analytical solution is of second order.
The relative L_2-error between the exact and approximated solutions for $L = 1$ m, $p_0 = 100$ N/m and $P = 200$ N is

$$\|e_r\|_{L_2} = 6.0\%.$$

□

The Galerkin method will be adopted throughout this book with the interpolation functions, which will be defined in the following section and in Chapter 4.

3.15.6   FINITE ELEMENT METHOD (FEM)

Until this point, the interpolation functions were defined as nonzero along the entire bar length. Therefore, the matrices of the systems of equations, derived from the approximation procedure, were dense. For problems with a large number of unknowns, dense matrices mean greater demand on memory and numerical processing, due to factors such as a larger condition number of the matrix and a large number of coefficients.

To avoid such problems, we can work with interpolation functions in terms of polynomials by parts, which are nonzero only in an interval of the problem domain.

Consider the bar of Figure 3.54 in which three internal nodes were included and labeled 1, 2 and 3. The boundary points are indicated by 0 and 4. The interpolation functions are associated to each node. The general expression of the interpolation functions is

$$
\phi_i(x) = \begin{cases} 0 & x \notin [x_{i-1}, x_{i+1}] \\ \dfrac{x - x_{i-1}}{x_i - x_{i-1}} & x \in [x_{i-1}, x_i] \\ -\dfrac{x - x_{i+1}}{x_{i+1} - x_i} & x \in [x_i, x_{i+1}] \end{cases} \tag{3.121}
$$

As can be seen in Figure 3.54(a), the functions are nonzero only in parts of the bar. These intervals are denominated support of the functions and such functions are called functions with compact support [45, 40]. The feature of compact support of these functions allows the definition of the finite element concept. The FEM provides a systematic way of constructing polynomial shape functions with compact support [45].

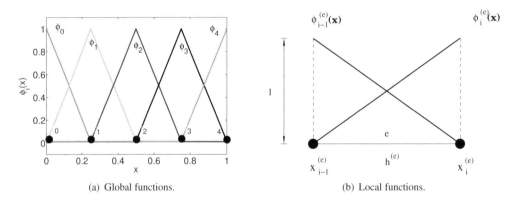

(a) Global functions.   (b) Local functions.

Figure 3.54   Global one-dimensional linear interpolation functions used in the FEM.

For this purpose, a line segment is taken from each of the functions $\phi_{i-1}(x)$ and $\phi_i(x)$, $i = 0, \ldots, 4$ to define the finite element. In this way, the local interpolation functions for element e are

$$
\phi_{i-1}^{(e)}(x) = -\frac{x - x_i^{(e)}}{h^{(e)}}, \tag{3.122}
$$

$$
\phi_i^{(e)}(x) = \frac{x - x_{i-1}^{(e)}}{h^{(e)}}, \tag{3.123}
$$

with $h^{(e)}$ the length of element e. These functions are illustrated in Figure 3.54(b). It should be observed that the $\phi_{i-1}^{(e)}(x)$ function assumes 1 and 0 values when calculated at the nodal coordinates $x_{i-1}^{(e)}$ and $x_i^{(e)}$, respectively, analogously, for function $\phi_i^{(e)}(x)$.

We can imagine these functions printed on a stamp, so that when placed in the positions of each element of the bar mesh of Figure 3.54(a), they reconstitute the global functions. This procedure is called superposition or assembling of the local element functions to obtain the global interpolation functions.

The axial displacement and strain approximations of element e are given, in terms of the two local shape functions, as

$$
\begin{aligned}
u_{x_2}^{(e)}(x) &= a_{i-1}^{(e)}\phi_{i-1}^{(e)}(x) + a_i^{(e)}\phi_i^{(e)}(x) \\
&= \begin{bmatrix} \phi_{i-1}^{(e)} & \phi_i^{(e)} \end{bmatrix} \begin{Bmatrix} a_{i-1}^{(e)} \\ a_i^{(e)} \end{Bmatrix} = [\bar{N}^{(e)}]\{a^{(e)}\},
\end{aligned}
\tag{3.124}
$$

$$
\begin{aligned}
\varepsilon_{xx_2}^{(e)}(x) &= \frac{du_{x_2}^{(e)}(x)}{dx} = \begin{bmatrix} \dfrac{d\phi_{i-1}^{(e)}(x)}{dx} & \dfrac{d\phi_i^{(e)}(x)}{dx} \end{bmatrix} \begin{Bmatrix} a_{i-1}^{(e)} \\ a_i^{(e)} \end{Bmatrix} \\
&= \frac{1}{h^{(e)}}\begin{bmatrix} -1 & 1 \end{bmatrix}\begin{Bmatrix} a_{i-1}^{(e)} \\ a_i^{(e)} \end{Bmatrix} = [\bar{B}^{(e)}]\{a^{(e)}\},
\end{aligned}
\tag{3.125}
$$

with $[\bar{N}^{(e)}]$ and $[\bar{B}^{(e)}]$ the element matrices of shape functions and their first derivatives, respectively. The normal stress on each element is calculated using Hooke's law, that is,

$$
\sigma_{xx_2}^{(e)}(x) = E^{(e)}\varepsilon_{xx_2}^{(e)}(x).
\tag{3.126}
$$

Calculating the axial displacement given in (3.124) at the nodal coordinates $x = x_{i-1}^{(e)}$ and $x = x_i^{(e)}$, we have

$$
u_{x_2}^{(e)}(x_{i-1}^{(e)}) = a_{i-1}^{(e)}\phi_{i-1}^{(e)}(x_{i-1}^{(e)}) + a_i^{(e)}\phi_i^{(e)}(x_{i-1}^{(e)}) = a_{i-1}^{(e)},
\tag{3.127}
$$

$$
u_{x_2}^{(e)}(x_i^{(e)}) = a_{i-1}^{(e)}\phi_{i-1}^{(e)}(x_i^{(e)}) + a_i^{(e)}\phi_i^{(e)}(x_i^{(e)}) = a_i^{(e)}.
\tag{3.128}
$$

Thus, the approximation coefficients reduce to the nodal axial displacements, that is,

$$
a_{i-1}^{(e)} = u_{x_2}^{(e)}(x_{i-1}^{(e)}) = \bar{u}_{x_{i-1}}^{(e)},
$$

$$
a_i^{(e)} = u_{x_2}^{(e)}(x_i^{(e)}) = \bar{u}_{x_i}^{(e)}.
$$

This characteristic comes from the collocation property of the defined elementary interpolation functions, which assume values 0 and 1 when computed at the nodal coordinates of the element.

The stiffness matrix for the bar element is determined as

$$
\begin{aligned}
[\bar{K}^{(e)}] &= \int_{x_{i-1}^{(e)}}^{x_i^{(e)}} [\bar{B}^{(e)}]^T[D][\bar{B}^{(e)}]A\,dx = \frac{E^{(e)}A^{(e)}}{(h^{(e)})^2}\begin{bmatrix} -1 \\ -1 \end{bmatrix}\begin{bmatrix} -1 & -1 \end{bmatrix}\underbrace{\int_{x_{i-1}^{(e)}}^{x_i^{(e)}} dx}_{h^{(e)}} \\
&= \frac{E^{(e)}A^{(e)}}{h^{(e)}}\begin{bmatrix} 1 & -1 \\ -1 & 1 \end{bmatrix}.
\end{aligned}
\tag{3.129}
$$

This is the same expression previously obtained in (3.116). However, in this case, the stiffness matrix is for a single bar element, with the integration limits between the nodal coordinates $(x_{i-1}^{(e)}, x_i^{(e)})$ of each element.

The load vector of the element due to the distributed load $q_x(x)$ is calculated as

$$
\{\bar{f}^{(e)}\} = \int_{x_{i-1}^{(e)}}^{x_i^{(e)}} q_x(x)[\bar{N}^{(e)}]^T dx = \int_{x_{i-1}^{(e)}}^{x_i^{(e)}} q_x(x)\begin{Bmatrix} \phi_{i-1}^{(e)}(x) \\ \phi_i^{(e)}(x) \end{Bmatrix} dx.
\tag{3.130}
$$

The term $\{\bar{f}^{(e)}\}$ is denominated as the vector of equivalent nodal loads due to the distributed load of intensity $q_x(x)$, because the continuous load along the element is replaced by the nodal values. For $q_x(x) = p_0$ constant, the element nodal load vector is

$$\{\bar{f}^{(e)}\} = \frac{p_0 h^{(e)}}{2} \left\{ \begin{array}{c} 1 \\ 1 \end{array} \right\}. \tag{3.131}$$

The system of equations of the element for a constant distributed load p_0 is

$$\frac{E^{(e)} A^{(e)}}{h^{(e)}} \left[\begin{array}{cc} 1 & -1 \\ -1 & 1 \end{array} \right] \left\{ \begin{array}{c} \bar{u}_{x_{i-1}}^{(e)} \\ \bar{u}_{x_i}^{(e)} \end{array} \right\} = \frac{p_0 h^{(e)}}{2} \left\{ \begin{array}{c} 1 \\ 1 \end{array} \right\}. \tag{3.132}$$

In compact notation,

$$[\bar{K}^{(e)}]\{\bar{u}^{(e)}\} = \{\bar{f}^{(e)}\}. \tag{3.133}$$

The global stiffness matrix and global load vector for the bar are obtained by the superposition process illustrated in the next example.

The square of the L_2 error norm on each element is calculated as

$$||e^{(e)}||_{L_2} = \int_{x_{i-1}^{(e)}}^{x_i^{(e)}} \left(u_x(x) - u_{x_2}^{(e)}(x) \right)^2 dx. \tag{3.134}$$

The relative error of the approximated solution is obtained by adding the error norm on each element and dividing it by the L_2 norm of the exact solution, that is,

$$||e_r||_{L_2} = \frac{\sqrt{\sum_{e=1}^{N_{el}} ||e^{(e)}||_{L_2}}}{||u_x||_{L_2}}. \tag{3.135}$$

Example 3.26 *Apply the FEM to Example 3.21 using a uniform mesh with two linear elements with length $h^{(e)} = \dfrac{L}{2}$ and three nodes.*

The stiffness matrix for each element $e = 1, 2$ is given from (3.129) by

$$[\bar{K}^{(e)}] = 2\frac{EA}{L} \left[\begin{array}{cc} 1 & -1 \\ -1 & 1 \end{array} \right].$$

The global stiffness matrix for the bar has rank 3 and initially with all coefficients equal to zero, that is,

$$[\bar{K}] = \left[\begin{array}{ccc} 0 & 0 & 0 \\ 0 & 0 & 0 \\ 0 & 0 & 0 \end{array} \right].$$

The coefficients of the stiffness matrix of each element should be summed in the rows and columns of the global matrix, corresponding to the node numbers of each element. Thus, the coefficients of the stiffness matrix of element 1 are added to rows and columns 1 and 2 of the global stiffness matrix of the bar. Thus,

$$[\bar{K}] = 2\frac{EA}{L} \left[\begin{array}{ccc} 1 & -1 & 0 \\ -1 & 1 & 0 \\ 0 & 0 & 0 \end{array} \right].$$

Similarly, the coefficients of the stiffness matrix of element 2 are added to rows and column 2 and 3 of the global stiffness matrix, that is,

$$[\bar{K}] = 2\frac{EA}{L} \left[\begin{array}{ccc} 1 & -1 & 0 \\ -1 & 2 & -1 \\ 0 & -1 & 1 \end{array} \right].$$

Notice there is a sum of the coefficients of the elemental matrices in the position (2,2) of the global matrix, because both elements share the second node.

The load vectors for both elements are, respectively,

$$\{\bar{f}^{(1)}\} = \{ \ 0.363 + \bar{R}_{x_1} \quad 0.637 \ \}^T \ \text{ and } \ \{\bar{f}^{(2)}\} = \{ \ 0.637 \quad 0.363 + \bar{R}_{x_3} \ \}^T,$$

with $\bar{R}_{x_1}$ and $\bar{R}_{x_3}$ being the support reactions at $x = 0$ and $x = 1$ m, respectively. Using the same superposition procedure for the load vectors of the elements, we obtain the following global load vector:

$$\{\bar{f}\} = \left\{ \begin{array}{c} 0.363 + \bar{R}_{x_1} \\ 1.273 \\ 0.363 + \bar{R}_{x_3} \end{array} \right\}.$$

The system of equations to be solved is

$$2\frac{EA}{L} \begin{bmatrix} 1 & -1 & 0 \\ -1 & 2 & -1 \\ 0 & -1 & 1 \end{bmatrix} \left\{ \begin{array}{c} \bar{u}_{x_1} \\ \bar{u}_{x_2} \\ \bar{u}_{x_3} \end{array} \right\} = \left\{ \begin{array}{c} 0.363 + \bar{R}_{x_1} \\ 1.273 \\ 0.363 + \bar{R}_{x_3} \end{array} \right\}.$$

The coefficients of the first and third rows multiply the zero displacements $\bar{u}_{x_1} = u_x(x = 0) = 0$ and $\bar{u}_{x_3} = u_x(x = 1) = 0$, respectively. Thus, the resulting system of equations after applying the boundary conditions is

$$-2\frac{EA}{L}\bar{u}_{x_2} = 0.363 + \bar{R}_{x_1},$$

$$4\frac{EA}{L}\bar{u}_{x_2} = 1.273,$$

$$-2\frac{EA}{L}\bar{u}_{x_2} = 0.363 + \bar{R}_{x_3}.$$

The first and third equations contain two unknowns, $(\bar{u}_{x_2}, \bar{R}_{x_1})$ and $(\bar{u}_{x_2}, \bar{R}_{x_2})$, and cannot be solved. Using the second equation, we obtain $\bar{u}_{x_2} = 3.183 \times 10^{-5}$ m for $EA = 10^4$ N and $L = 1$ m. The support reactions are calculated from the first and third equations, resulting in $\bar{R}_{x_1} = \bar{R}_{x_3} = 0.6367$ N. It is observed that the application of zero displacement boundary conditions is equivalent to eliminating rows and columns 1 and 3 of the system of equations. After solving this system, the remaining reactions are determined.

The value of the axial nodal displacement is

$$\{\bar{u}\} = \{ \ 0 \quad 3.183 \times 10^{-5} \quad 0 \ \}^T \text{ m.}$$

From (3.122), (3.123), and (3.124), the local solutions for each finite element are, respectively,

$$\begin{aligned} u_{x_2}^{(1)}(x) &= \bar{u}_{x_1}\phi_1^{(1)}(x) + \bar{u}_{x_2}\phi_1^{(2)}(x) \\ &= (0)\left(1 - \frac{2}{L}x\right) + (3.183 \times 10^{-5})\left(\frac{2}{L}x\right) = 6.366 \times 10^{-5}x, \end{aligned}$$

$$\begin{aligned} u_{x_2}^{(2)}(x) &= \bar{u}_{x_2}\phi_1^{(2)}(x) + \bar{u}_{x_3}\phi_2^{(2)}(x) \\ &= (3.183 \times 10^{-5})\left(2 - \frac{2}{L}x\right) + (0)\left(\frac{2}{L}x - 1\right) = 6.366 \times 10^{-5}(1 - x). \end{aligned}$$

The normal strains in the elements are obtained from (3.125) or by the derivates of the previous expressions. Thus,

$$\varepsilon_{xx_2}^{(1)} = 6.366 \times 10^{-5},$$

$$\varepsilon_{xx_2}^{(2)} = -6.366 \times 10^{-5}.$$

The normal stress is given by Hooke's law in (3.126). For E = 10^8 Pa, we have

$$\sigma_{xx_2}^{(1)} = 6366.2 \text{ Pa},$$
$$\sigma_{xx_2}^{(2)} = -6366.2 \text{ Pa}.$$

The square of the L_2 error norm on each element is calculated with equation (3.134) and

$$||e_1||_{L_2} = ||e_2||_{L_2} = 1.153 \times 10^{-11}.$$

The relative percent error is calculated using (3.135) and

$$||e_r||_{L_2} = 21.34\%.$$

Figure 3.55(a) shows the exact and approximated solutions with five elements. Note that, due to the collocation property of the interpolation functions (3.111), the nodal displacements coincide with the considered exact solution calculated at the nodal coordinates, as indicated by circles.

Figure 3.55(b) illustrates the sparsity profile of the global stiffness matrix. Observe that, unlike dense matrices of the previous examples, the obtained global stiffness matrix with linear elements has nonzero coefficients in only three diagonals and is called three-diagonal.

File barsolapex6.m implements the solution of this example with the FEM for an arbitrary number of linear elements.

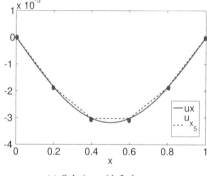

(a) Solution with 5 elements.

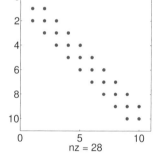

(b) Sparsity profile of the global stiffness matrix with 10 elements.

Figure 3.55 Example 3.26: FEM approximation with one-dimensional linear interpolation functions and sparsity profile of the global stiffness matrix.

□

Figure 3.51 illustrates the convergence of the approximated solution for meshes with 3 to 10 nodes. This is equivalent to using 3 to 10 global functions in the previous methods. Observe that the convergence rate obtained with the linear elements is lower than those obtained in other methods, which used interpolation functions with increasing orders. Thus, the FEM with linear elements has a linear convergence rate proportional to the number of nodes or global linear functions. The other methods achieved a spectral convergence rate, in which the error drops at least two orders of magnitude, when adding a new higher-order interpolation function. It will be shown later how to define, in a simpler way, high-order polynomial functions for the FEM, allowing us to achieve spectral convergence rates for problems with smooth solutions, as the examples considered here so far.

Increasing the number of nodes and mesh elements, for a constant interpolation order, defines the *h* version of the FEM. Keeping the mesh fixed and increasing the order of the interpolation functions defines the *p* version of the FEM. Refining the mesh elements, while simultaneously increasing the order of the interpolation functions, gives rise to the *hp*-FEM version [35].

3.16 ANALYSIS OF TRUSSES

One of the main applications of bar elements is in truss structures, as illustrated in Figure 1.6. It is assumed that the bars are connected by pins which transmit loads. Thus, the bar elements generally have only concentrated forces on the nodes and linear displacement fields. The use of linear interpolation functions is appropriate for the computational analysis of trusses.

The following example shows the design of bar trusses using the method of nodes to determine the axial forces in the bar elements.

Example 3.27 *The truss of Figure 3.56(a) consists of bars AB and BC, which are articulated at the ends. Design the bars for the admissible normal stresses for traction and compression equal to* 120 MPa *and* 80 MPa, *respectively.*

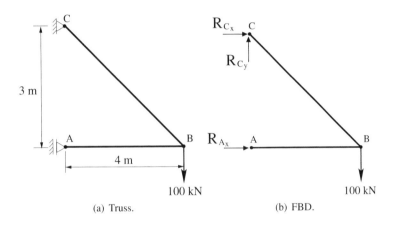

(a) Truss.   (b) FBD.

Figure 3.56 Example 3.27: design of truss bars.

Initially, we calculate the support reaction forces using the FBD of Figure 3.56(b). Thus,

- $\sum f_x = 0:\ R_{Ax} + R_{Cx} = 0;$
- $\sum F_y = 0:\ R_{Cy} - 100 = 0 \rightarrow R_{Cy} = 100.0$ kN;
- $\sum M_{z_A} = 0:\ -(100)(4) - 3R_{Cx} = 0 \rightarrow R_{Cx} = -1333.3$ kN.

From the first equilibrium condition, we have $R_{Ax} = 1333,3$ kN.

The forces in the AB and BC bars are $F_{AB} = -1333.3$ kN *and* $F_{BC} = 1666.7$ kN. *They are under compression and traction, respectively. The design of the bars is calculated as*

$$A_{AB} = \frac{F_{AB}}{\bar{\sigma}_c} = \frac{-1333.3 \times 10^3}{-80 \times 10^6} = 167.0 \text{ cm}^2,$$

$$A_{BC} = \frac{F_{BC}}{\bar{\sigma}_t} = \frac{1666.7 \times 10^3}{120 \times 10^6} = 139.0 \text{ cm}^2.$$

□

The considered linear bar finite element is in the horizontal position along the x axis of the adopted reference system. The BC bar in Figure 3.56 is inclined. Hence, a coordinate transformation must be defined to obtain the main variables of the element for an arbitrary orientation in the global xyz system. It is assumed here that the geometry of the bar is described by a straight line.

In the general case, the global reference system results in an inclined bar element, as illustrated in Figure 3.57. It is observed that $\bar{x}$ is the local reference system of the bar. Therefore, we need to perform a coordinate transformation between the local $\bar{x}$ and global xyz reference systems. In the global system, each node has three displacement degrees of freedom indicated by u_{x_i}, u_{y_i}, u_{z_i} ($i = 1, 2$), respectively, in the x, y, and z directions.

The length $h^{(e)}$ of the bar element can be obtained from the global coordinates of nodes 1 and 2, i.e., (x_1, y_1, z_1) and (x_2, y_2, z_2), respectively. Thus,

$$h^{(e)} = \sqrt{(x_2 - x_1)^2 + (y_2 - y_1)^2 + (z_2 - z_1)^2}. \tag{3.136}$$

The following trigonometric relations for the direction cosines l_x, l_y, and l_z, are obtained from Figure 3.57:

$$l_x = \frac{x_2 - x_1}{h^{(e)}}, \quad l_y = \frac{y_2 - y_1}{h^{(e)}}, \quad l_z = \frac{z_2 - z_1}{h^{(e)}}. \tag{3.137}$$

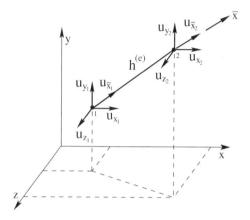

Figure 3.57 Bar element in the xyz coordinate system.

It is also verified that

$$
\begin{aligned}
\bar{u}_{x_1}^{(e)} &= l_x u_{x_1}^{(e)} + l_y u_{y_1}^{(e)} + l_z u_{z_1}^{(e)}, \\
\bar{u}_{x_2}^{(e)} &= l_x u_{x_2}^{(e)} + l_y u_{y_2^{(e)}} + l_z u_{z_2}^{(e)}.
\end{aligned}
$$

In matrix notation, we have

$$
\left\{ \begin{array}{c} \bar{u}_{x_1}^{(e)} \\ \bar{u}_{x_2}^{(e)} \end{array} \right\} =
\left[\begin{array}{cccccc} l_x & l_y & l_z & 0 & 0 & 0 \\ 0 & 0 & 0 & l_x & l_y & l_z \end{array} \right]
\left\{ \begin{array}{c} u_{x_1}^{(e)} \\ u_{y_1}^{(e)} \\ u_{z_1}^{(e)} \\ u_{x_2}^{(e)} \\ u_{y_2}^{(e)} \\ u_{z_2}^{(e)} \end{array} \right\}. \tag{3.138}
$$

In compact form,

$$\{\bar{u}^{(e)}\} = [T]\{u^{(e)}\}, \tag{3.139}$$

where $[T]$ is the transformation matrix between the global and local coordinate systems.

Substituting (3.139) in (3.133) and multiplying it by $[T]^T$ to keep the symmetry of the stiffness matrix, we obtain

$$[T]^T [\bar{K}^{(e)}][T]\{u^{(e)}\} = [T]^T \{\bar{f}^{(e)}\}.$$

or,

$$[K^{(e)}]\{u^{(e)}\} = \{f^{(e)}\}.$$

Hence, the stiffness matrix $[K^{(e)}]$ of the bar element written in the global reference system is

$$[K^{(e)}] = [T]^T[\bar{K}^{(e)}][T] = \begin{bmatrix} l_x & 0 \\ l_y & 0 \\ l_z & 0 \\ 0 & l_x \\ 0 & l_y \\ 0 & l_z \end{bmatrix} \frac{E^{(e)}A^{(e)}}{h^{(e)}} \begin{bmatrix} 1 & -1 \\ -1 & 1 \end{bmatrix} \begin{bmatrix} l_x & l_y & l_z & 0 & 0 & 0 \\ 0 & 0 & 0 & l_x & l_y & l_z \end{bmatrix}.$$

Developing the indicated multiplication, we have

$$[K^{(e)}] = \frac{E^{(e)}A^{(e)}}{h^{(e)}} \begin{bmatrix} [K_0] & -[K_0] \\ -[K_0] & [K_0] \end{bmatrix}, \tag{3.140}$$

with

$$[K_0] = \begin{bmatrix} l_x^2 & l_x l_y & l_x l_z \\ l_x l_y & l_y^2 & l_y l_z \\ l_x l_z & l_y l_z & l_z^2 \end{bmatrix}.$$

In an expanded form

$$[K^{(e)}] = \frac{E^{(e)}A^{(e)}}{h^{(e)}} \begin{bmatrix} l_x^2 & l_x l_y & l_x l_z & -l_x^2 & -l_x l_y & -l_x l_z \\ l_x l_y & l_y^2 & l_y l_z & -l_x l_y & -l_y^2 & -l_y l_z \\ l_x l_z & l_y l_z & l_z^2 & -l_x l_z & -l_y l_z & -l_z^2 \\ -l_x^2 & -l_x l_y & -l_x l_z & l_x^2 & l_x l_y & l_x l_z \\ -l_x l_y & -l_y^2 & -l_y l_z & l_x l_y & l_y^2 & l_y l_z \\ -l_x l_z & -l_y l_z & -l_z^2 & l_x l_z & l_y l_z & l_z^2 \end{bmatrix}.$$

Analogously, the element load vector $\{f^{(e)}\}$ for a distributed load of constant intensity p_0 is expressed in the global reference system as

$$\{f^{(e)}\} = [T]^T\{\bar{f}^{(e)}\} = \frac{p_0 h^{(e)}}{2} \begin{bmatrix} l_x & 0 \\ l_y & 0 \\ l_z & 0 \\ 0 & l_x \\ 0 & l_y \\ 0 & l_z \end{bmatrix} \begin{Bmatrix} 1 \\ 1 \end{Bmatrix} = \frac{p_0 h^{(e)}}{2} \begin{Bmatrix} l_x \\ l_y \\ l_z \\ l_x \\ l_y \\ l_z \end{Bmatrix}. \tag{3.141}$$

Substituting (3.139) in (3.125), we obtain the normal strain expressed in terms of the global displacements $(u_{x_1}, u_{y_1}, u_{z_1})$ and $(u_{x_2}, u_{y_2}, u_{z_2})$ as

$$
\varepsilon_{xx_2}^{(e)} = \frac{1}{h^{(e)}} \begin{bmatrix} -1 & 1 \end{bmatrix} \begin{bmatrix} l_x & l_y & l_z & 0 & 0 & 0 \\ 0 & 0 & 0 & l_x & l_y & l_z \end{bmatrix} \begin{Bmatrix} u_{x_1}^{(e)} \\ u_{y_1}^{(e)} \\ u_{z_1}^{(e)} \\ u_{x_2}^{(e)} \\ u_{y_2}^{(e)} \\ u_{z_2}^{(e)} \end{Bmatrix}
$$

$$
= \frac{1}{h^{(e)}} \begin{bmatrix} -l_x & -l_y & -l_z & l_x & l_y & l_z \end{bmatrix} \begin{Bmatrix} u_{x_1}^{(e)} \\ u_{y_1}^{(e)} \\ u_{z_1}^{(e)} \\ u_{x_2}^{(e)} \\ u_{y_2}^{(e)} \\ u_{z_2}^{(e)} \end{Bmatrix}
$$

$$
= \begin{bmatrix} B^{(e)} \end{bmatrix} \{u^{(e)}\}, \tag{3.142}
$$

with $\begin{bmatrix} B^{(e)} \end{bmatrix}$ the strain-displacement matrix of the element expressed in the global coordinate system. The element stress $\sigma_{xx_2}^{(e)}$ can be calculated using the constitutive equation (3.126).

In the case of a two-dimensional truss, we need simply to remove the rows and columns corresponding to the u_{z_1} and u_{z_2} displacements in the z direction. Therefore, the stiffness matrix, the equivalent nodal force vector, and the strain, given, respectively, in the expressions (3.140) to (3.142), reduce to

$$
[K^{(e)}] = \frac{E^{(e)} A^{(e)}}{h^{(e)}} \begin{bmatrix} l_x^2 & l_x l_y & -l_x^2 & -l_x l_y \\ l_x l_y & l_y^2 & -l_x l_y & -l_y^2 \\ -l_x^2 & -l_x l_y & l_x^2 & l_x l_y \\ -l_x l_y & -l_y^2 & l_x l_y & l_y^2 \end{bmatrix}, \tag{3.143}
$$

$$
\{f^{(e)}\} = \frac{p_0 h^{(e)}}{2} \begin{Bmatrix} l_x & l_y & l_x & l_y \end{Bmatrix}^T, \tag{3.144}
$$

$$
\varepsilon_{xx}^{(e)} = \frac{1}{h^{(e)}} \begin{bmatrix} -l_x & -l_y & l_x & l_y \end{bmatrix} \begin{Bmatrix} u_{x_1}^{(e)} \\ u_{y_1}^{(e)} \\ u_{x_2}^{(e)} \\ u_{y_2}^{(e)} \end{Bmatrix}. \tag{3.145}
$$

Example 3.28 *Consider the truss of Figure 3.58(a) consisting of three bars. Determine the normal stresses and strains for each bar element and Young's modulus equals to* 200 GPa.

The finite element model used is shown in Figure 3.58(b), where the node numbers, elements, and degrees of freedom are indicated. Table 3.1 summarizes the length, cross-section area, direction cosines, nodes, and degrees of freedom for each element.

The unrestricted degrees of freedom are numbered first and followed by those ones with boundary conditions.

From expression (3.143) and the parameters given in Table 3.1, the stiffness matrices for each element are, respectively,

$$
[K^{(1)}] = 1.0 \times 10^7 \begin{bmatrix} 0.0 & 0.0 & 0.0 & 0.0 \\ 0.0 & 5.0 & 0.0 & -5.0 \\ 0.0 & 0.0 & 0.0 & 0.0 \\ 0.0 & -5.0 & 0.0 & 5.0 \end{bmatrix},
$$

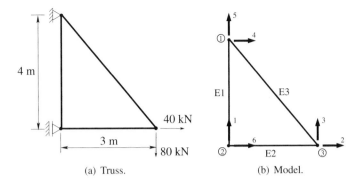

Figure 3.58 Example 3.28: truss with three bars.

e	$h^{(e)}\,[m]$	$A\,[cm^2]$	l_x	l_y	Nodes	DOFs
1	4.0	10.0	0.0	1.0	1.2	4.5,6,1
2	3.0	6.0	1.0	0.0	2.3	6,1,2,3
3	5.0	12.5	0.6	0.8	1.3	4.5,2,3

Table 3.1
Example 3.28: element parameters.

$$[K^{(2)}] = 1.0 \times 10^7 \begin{bmatrix} 4.0 & 0.0 & -4.0 & 0.0 \\ 0.0 & 0.0 & 0.0 & 0.0 \\ -4.0 & 0.0 & 4.0 & 0.0 \\ 0.0 & 0.0 & 0.0 & 0.0 \end{bmatrix},$$

$$[K^{(3)}] = 1.0 \times 10^7 \begin{bmatrix} 1.8 & 2.4 & -1.8 & -2.4 \\ 2.4 & 3.2 & -2.4 & -3.2 \\ -1.8 & -2.4 & 1.8 & 2.4 \\ -2.4 & 3.2 & 2.4 & 3.2 \end{bmatrix}.$$

The global stiffness matrix of the truss has rank 6. By making the sum of the element matrix coefficients in the rows and columns of the global matrix for the degrees of freedom indicated in Table 3.1, we obtain the following stiffness matrix for the truss:

$$[K] = 1.0 \times 10^7 \begin{bmatrix} 5.0 & 0.0 & 0.0 & 0.0 & -5.0 & 0.0 \\ 0.0 & 5.8 & 2.4 & -1.8 & -2.4 & -4.0 \\ 0.0 & 2.4 & 3.2 & -2.4 & -3.2 & 0.0 \\ 0.0 & -1.8 & -2.4 & 1.8 & 2.4 & 0.0 \\ -5.0 & -2.4 & -3.2 & 2.4 & 8.2 & 0.0 \\ 0.0 & -4.0 & 0.0 & 0.0 & 0.0 & 4.0 \end{bmatrix}.$$

The global load vector is analogously obtained as

$$\{f\} = \{\,0 \quad 40000 \quad -80000 \quad R_4 \quad R_5 \quad R_6\,\}^T,$$

with R_4, R_5, and R_6 the support reactions in the degrees of freedom 4, 5, and 6, respectively.

Due to the supports, the displacements of degrees of freedom 4, 5, and 6 are zero. Therefore, rows and columns 4, 5, and 6 of the global stiffness matrix and load vector are eliminated and the following system of equations is obtained:

$$[K] = 1.0 \times 10^7 \begin{bmatrix} 5.0 & 0.0 & 0.0 \\ 0.0 & 5.8 & 2.4 \\ 0.0 & 2.4 & 3.2 \end{bmatrix} \begin{Bmatrix} u_{y_1} \\ u_{x_2} \\ u_{y_3} \end{Bmatrix} = \begin{Bmatrix} 0 \\ 40000 \\ -80000 \end{Bmatrix}.$$

Nodal displacements u_{y_1}, u_{x_2}, and u_{y_3} are determined by solving the previous system of equations. The complete displacement vector, including the boundary conditions, is given by

$$\{u\} = \{ \ 0.0000 \quad 0.0025 \quad -0.0044 \quad 0.0000 \quad 0.0000 \quad 0.0000 \ \}^T.$$

The support reaction forces are obtained multiplying the matrix corresponding to rows 4, 5, and 6 and columns 1, 2, and 3 of $[K]$ by the displacements u_{y_1}, u_{x_2}, and u_{y_3}. Thus,

$$\begin{Bmatrix} R_4 \\ R_5 \\ R_6 \end{Bmatrix} = 1.0 \times 10^7 \begin{bmatrix} 0.0 & -1.8 & -2.4 \\ -5.0 & -2.4 & -3.2 \\ 0.0 & -4.0 & 0.0 \end{bmatrix} \begin{Bmatrix} 0.0000 \\ 0.0025 \\ -0.0044 \end{Bmatrix} = \begin{Bmatrix} 60000 \\ 80000 \\ -100000 \end{Bmatrix}.$$

The normal stresses and strains for each element are calculated using equations (3.145) and (3.126), respectively. The values are listed in Table 3.2.

e	$\varepsilon_{xx}^{(e)} \times 10^{-4}$	$\sigma_{xx}^{(e)} \ [MPa]$
1	0.00	0.00
2	8.33	166.67
3	-4.00	-80.00

Table 3.2
Example 3.28: element normal strains and stresses.

Suppose that the tension and compression admissible stresses are $\bar{\sigma}_t = 120$ MPa and $\bar{\sigma}_c = 70$ MPa, respectively. In this case, elements 2 and 3 must be redesigned and the new areas $A_i^{(e)}$ are determined from the expression of the axial force $F^{(e)} = \sigma_{xx}^{(e)} A^{(e)}$ as

$$A_i^{(e)} = \frac{F^{(e)}}{\bar{\sigma}} = \frac{\sigma_{xx}^{(e)}}{\bar{\sigma}} A^{(e)}.$$

The i subindex in $A_i^{(e)}$ indicates the design iteration. This occurs because generally, when calculating the new areas, the distribution of forces in the bars also changes, and the stresses may still be higher than the admissible values. Thus, an iterative design process is required. In this example, only one design iteration is necessary, and the final areas of bars 2 and 3 are $A^{(2)} = 8.33$ cm^2 and $A^{(3)} = 14.29$ cm^2, respectively.

The MATLAB programs truss2d.m and truss2ddesign.m perform the analysis and design of plane trusses.

□

Example 3.29 *Solve the truss of Example 2.8 using the linear bar element.*
    The stiffness matrices for each element are given from expression (3.143) by

$$[K^{(1)}] = 8000 \begin{bmatrix} 1 & 0 & -1 & 0 \\ 0 & 0 & 0 & 0 \\ -1 & 0 & 1 & 0 \\ 0 & 0 & 0 & 0 \end{bmatrix},$$

$$[K^{(2)}] = 9600 \begin{bmatrix} \frac{9}{25} & -\frac{12}{25} & -\frac{9}{25} & \frac{12}{25} \\ -\frac{12}{25} & \frac{16}{25} & \frac{12}{25} & -\frac{16}{25} \\ -\frac{9}{25} & \frac{12}{25} & \frac{9}{25} & -\frac{12}{25} \\ \frac{12}{25} & -\frac{16}{25} & -\frac{12}{25} & \frac{16}{25} \end{bmatrix},$$

$$[K^{(3)}] = 18000 \begin{bmatrix} 0 & 0 & 0 & 0 \\ 0 & 1 & 0 & -1 \\ 0 & 0 & 0 & 0 \\ 0 & -1 & 1 & 1 \end{bmatrix}.$$

Doing the assembly procedure of the element matrices and applying the boundary conditions, the system of equations relative to the free degrees of freedom is given by

$$\begin{bmatrix} 11456 & -4608 \\ -4608 & 24144 \end{bmatrix} \begin{Bmatrix} u_{x_1} \\ u_{y_2} \end{Bmatrix} = \begin{Bmatrix} 5 \\ -3 \end{Bmatrix},$$

with u_{x_1} and u_{y_2} the displacements of degrees of freedom of node B in the horizontal and vertical directions, respectively.

Note that the use of linear bar elements resulted in the same system of equations obtained with the application of the PVW in Example 2.8. This was expected to happen, because equation (2.26) is compatible with a linear axial displacement field.

□

3.17 FINAL COMMENTS

In this chapter, the variational formulation of the bar mechanical model was presented. This was the first deformable structural element considered in this book. It was also the first time that the variational formulation steps were presented. These same steps will be used in the formulation of other models which will be discussed in the following chapters.

An important aspect is that all features of the bar model were obtained from the kinematic definition and the concept of strain internal work. Thus, it is expected that the reader would be able to understand all the hypotheses of the mechanical model.

The use of singularity functions allows the solution of problems with concentrated loads applied along the bar length and the integration of discontinuous loading functions. In a more formal way, the Riemann integration is changed to Lebesgue, but these aspects will not be considered in this book. Interested readers can see the references [45, 40].

Starting from Section 3.15, the concept of bar approximation was introduced using weighted residue methods. Particularly, the FEM, based on the Galerkin method, will be adopted in the next chapters, and systematic construction procedures of interpolation functions will be presented. Furthermore, several programs in MATLAB will also be used to assist in the concepts assimilation.

3.18 PROBLEMS

1. A steel wire with length 7 m cannot elongate more than 80 mm when subjected to a tensile load of 8 kN. Using $E = 210$ GPa, calculate:
   a. The minimum diameter of the wire
   b. The respective normal force
2. A test specimen with 13 mm of diameter elongates 0.22 mm in 200 mm of length when subjected to a 30 kN tensile force. If the body is in the elastic range, what is the value of the elasticity modulus?
3. The rods BC and DE of Figure 3.59 are made of steel ($E = 200$ GPa) and have uniform retangular cross-sections of 6.2×25.5 mm. Calculate the maximum force P that can be applied at point A, in such way that the deflection at this point will not exceed 0.25 mm.

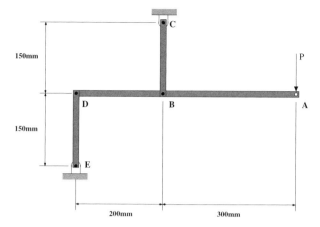

Figure 3.59 Problem 3.

4. A rigid bar AD is supported in A and C, as indicated in Figure 3.60, by two steel wires with 1.6 mm diameter ($E = 200$ GPa), and a support in D. Assume that the wires are initially straight. Calculate:

   a. The additional traction on each wire when a force P with intensity 530 N is applied at B

   b. The corresponding deflection at point B

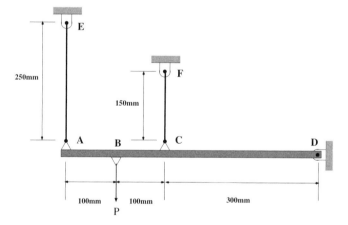

Figure 3.60 Problem 4.

5. In a tensile test, a steel bar with diameter 20 mm and length 200 mm is subjected to a force F with intensity 50 kN, as illustrated in Figure 3.61. Find:

   a. The elongation in the central region of the bar

   b. The elongation at $1/4$ from the superior end

   c. The diameter variation of the bar

6. Find the force required for the bar right end to touch the support at B of Figure 3.62.

7. A steel bar with diameter 60 mm and length 150 mm is axially compressed by a 200 kN force. Find the variation of the cross-section caused by this force with $E = 210$ GPa and $\nu = 0.3$.

8. An aparatus is necessary to maintain the cross-section of a circular steel bar with length 20 mm. If this bar is heated to a temperature of 200°C, what will be the necessary force to

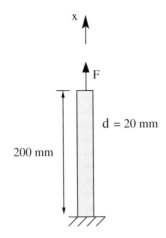

Figure 3.61 Problem 5.

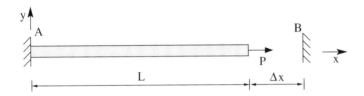

Figure 3.62 Problem 6.

maintain initial cross-section dimensions? Assume that the elasticity modulus of the steel is $E = 210$ GPa and $\alpha = 11.7 \times 10^{-6}/°C$.

9. Write the indicated boundary conditions for each of the bars in Figure 3.63.
10. For the bars illustrated in Figure 3.63, write the load equation and the necessary differential equation to solve the problem.
11. For the steel bars given in Figure 3.64, determine:
    a. The load equation
    b. The boundary conditions
    c. The normal force and the axial displacement diagrams
    d. The support reactions
    e. The maximum normal stress
12. Design
    a. The bar with circular cross-section and loading shown in Figure 3.63(c) for $E = 210$ GPa, and $\bar{\sigma} = 320$ MPa
    b. The bar with rectangular cross-section and the loading shown in Figure 3.63(f) for a base b, height $h = 2b$, $E = 210$ GPa, and $\bar{\sigma} = 240$ MPa
13. A bar composed of two cylindrical parts (see Figure 3.65) is fixed in both ends. The AB segment is made of steel ($E = 210$ GPa; $\alpha = 11.7 \times 10^{-6}/°C$) and the BC segment is made of aluminum ($E = 75$ GPa; $\alpha = 23.6 \times 10^{-6}/°C$). The system is unstressed in the initial configuration. Find:
    a. The normal stresses in segments AB and BC for a temperature variation of $100°C$
    b. The corresponding deflection at the interface point B
14. Figure 3.66 illustrates a brass tube ($E = 105$ GPa; $\alpha = 20.9 \times 10^{-6}/°C$ and $A = 315$ mm²), filled with a steel bar ($E = 210$ GPa; $\alpha = 11.7 \times 10^{-6}/°C$ and $A = 1650$ mm²). The initial

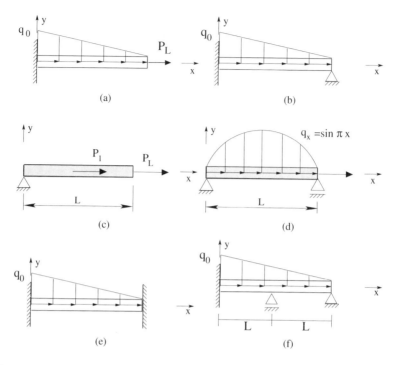

Figure 3.63 Problems 9 and 10.

temperature is 25°C. Considering only axial strains, determine the stress in the tube and in the steel core when the temperature is increased to 200°C.

15. Use the FEM to solve the bar of Figures 3.64(a) and 3.64(d) using four bar element meshes. Compare the analytical solutions with the numerical ones and calculate the approximation errors. Implement MATLAB programs to solve the indicated problems.

16. Design the trusses of Figure 3.67 using the truss2ddesign.m program. The bars are made of steel and adopt $A = 1$ cm^2 as the initial areas.

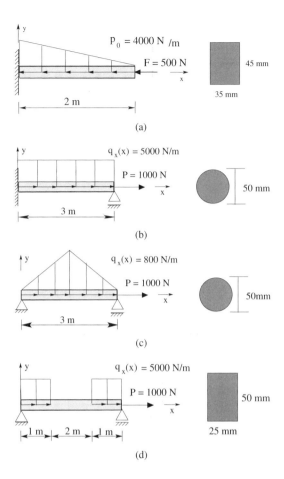

Figure 3.64 Problem 11.

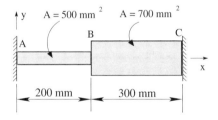

Figure 3.65 Problem 13.

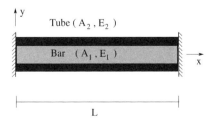

Figure 3.66 Problem 14.

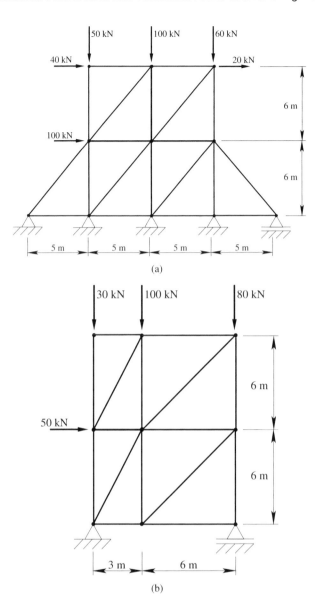

Figure 3.67   Problem 16.

4 FORMULATION AND APPROXIMATION OF SHAFTS

4.1 INTRODUCTION

This chapter presents the mechanical model called a shaft. It has a circular cross-section and is subjected to the action of torsion, characterized by rotations of the cross-sections about the longitudinal axis x.

As in the case of bars, the longitudinal dimension of the shaft is predominant. This allows the treatment of the shaft model by a one-dimensional mathematical model, analyzing its behavior along the longitudinal axis x, as shown in Figure 4.1. The adopted coordinate system is also illustrated and is the same used for the bar model, with the origin located at the geometric center (GC) of the cross-section.

The interest in the study of shafts is related only to kinematic actions causing the torsion of the cross-sections about the longitudinal dimension. In the variational approach, the formulation of the circular torsion model for shafts follows the same steps as the bar case. It is noted that shafts are presented practically in all mechanical systems, thus emphasizing the importance of this model.

In this chapter, we assume the case of small displacements and strains and the constitutive equation for isotropic linear elastic materials given by Hooke's law.

The FEM is applied to the approximation of shaft model. The construction of interpolation functions using Lagrange and Jacobi polynomials is then generalized, and the polynomials are defined in normalized local coordinate systems. The concepts of mapping between the local and global reference systems, numerical integration, and derivatives by collocation are also presented.

4.2 KINEMATICS

The following assumptions are adopted for the kinematic displacement actions for the torsion of circular cross-sections:

- The cross-sections remain plane and normal to the longitudinal axis, as in the bar model. It is further assumed that the parallel cross-sections remain at a constant distance to each other, that is, no longitudinal strains occur. This effect will appear in the torsion of noncircular sections, causing the warping of cross-sections (see Section 9.8).
- The kinematic actions are characterized by the rotation of the cross-sections about x. Therefore, each cross-section has a constant rigid rotation, as shown in Figure 4.1(a). This assumption means that the imaginary plane AO_1O_2B, illustrated in Figure 4.1(b) displaces to $A'O_1O_2B'$, after the rotation. As will be seen, the displacements of the points in the cross-section increase linearly from zero at the center of the section, and reach the maximum value at the boundary.

As each cross-section has a rigid rotation represented by θ_x about the longitudinal axis, it is constant for all points of any section x. Hence, θ_x is a scalar function of the x coordinate only, and we can write $\theta_x = \theta_x(x)$. It is called angle of twist or torsion angle.

The position of a point P in the cross-section x is given by the (y, z) coordinates, which can be written in terms of the angle β and the radial coordinate r, illustrated in Figure 4.2, as

$$\begin{aligned} y &= r\cos\beta, \\ z &= r\sin\beta, \end{aligned} \tag{4.1}$$

127

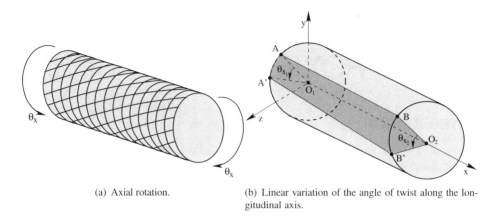

(a) Axial rotation.

(b) Linear variation of the angle of twist along the longitudinal axis.

Figure 4.1 Angle of twist.

where $r = \sqrt{y^2 + z^2}$ and $\tan\beta = \dfrac{z}{y}$.

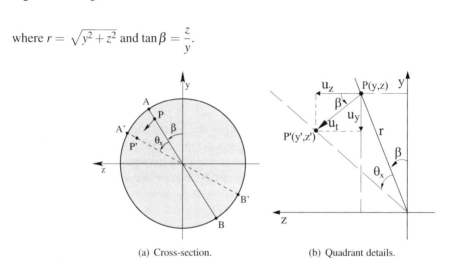

(a) Cross-section.

(b) Quadrant details.

Figure 4.2 Circular torsion kinematics.

Due to the rotation θ_x, point P displaces to its final position P' with coordinates (y', z') given by [see Figure 4.2(b)]

$$
\begin{aligned}
y' &= r\cos(\beta + \theta_x), \\
z' &= r\sin(\beta + \theta_x).
\end{aligned}
\tag{4.2}
$$

When point P displaces to the position P', there are the displacement components u_y and u_z, respectively, in the directions y and z of the reference system. These components are given by the difference between the final (y', z') and initial (y, z) coordinates [see Figure 4.2(b)], that is,

$$
\begin{aligned}
u_y &= y' - y, \\
u_z &= z' - z.
\end{aligned}
$$

Substituting equations (4.2) in the previous expression, we have

$$
\begin{aligned}
u_y &= r\cos(\beta + \theta_x) - y = r\cos\beta\cos\theta_x - r\sin\beta\sin\theta_x - y, \\
u_z &= r\sin(\beta + \theta_x) - z = r\sin\beta\cos\theta_x + r\cos\beta\sin\theta_x - z.
\end{aligned}
$$

From equation (4.1), we have $r\cos\beta = y$ and $r\sin\beta = z$. Thus,

$$
\begin{aligned}
u_y &= y\cos\theta_x - z\sin\theta_x - y, \\
u_z &= z\cos\theta_x + y\sin\theta_x - z.
\end{aligned}
$$

Assuming the case of small rotations, which means small θ_x, we can use the approximations $\cos\theta_x \approx 1$ and $\sin\theta_x \approx \theta_x$. Therefore, the above expressions reduce to

$$
\begin{aligned}
u_y &= y(1) - z(\theta_x) - y, \\
u_z &= z(1) + y(\theta_x) - z,
\end{aligned}
$$

that is,

$$
\begin{aligned}
u_y &= -z\theta_x, \\
u_z &= y\theta_x.
\end{aligned}
\tag{4.3}
$$

Hence, due to the rotation θ_x at the cross-section x, each point P with coordinates y and z has the displacement components u_y and u_z. Since there is no warping of the cross-section, the displacement component in direction x is zero, i.e., $u_x = 0$.

Note also that u_y and u_z vary linearly with the z and y coordinates, as illustrated in Figures 4.3(b) and 4.3(e) to a solid circular section with diameter d. In this case, the displacements u_y and u_z are zero in the center of the cross-section and reach the maximum value at the section boundary, where $\sqrt{y^2 + z^2} = \dfrac{d}{2}$. The negative sign in u_y is compatible with the direction of the rotation. When rotating the section about the positive direction of the x axis, point P moves downwards, opposite to the positive direction of y axis of the adopted reference system [see Figure 4.3(a)]. If the rotation θ_x is negative, or it is in the opposite to the positive direction of the x axis, point P moves upwards, causing a positive value for u_y and a negative one for u_z [see Figure 4.3(d)].

Figures 4.3(c) and 4.3(f) consider the case of a shaft with a hollow circular cross-section with internal d_i and external d_e diameters. In this case, the displacement components u_y and u_z have minimum values in the internal diameter, where $\sqrt{y^2 + z^2} = \dfrac{d_i}{2}$, and maximum at the section boundary, where $\sqrt{y^2 + z^2} = \dfrac{d_e}{2}$.

According to the adopted kinematic hypothesis, the angle of torsion θ_x is constant for a given cross-section, but it varies from one section to another, which can be denoted as $\theta_x = \theta_x(x)$. The displacement components given in (4.3) are valid for a point P with (y, z) coordinates in the x section. To indicate the kinematics for all points of the shaft, we can rewrite expression (4.3) by explicitly including the dependence of θ_x with the coordinate x of the section, that is, $\theta_x = \theta_x(x)$. Thus,

$$
\begin{aligned}
u_y(x, y, z) &= -z\theta_x(x), \\
u_z(x, y, z) &= y\theta_x(x).
\end{aligned}
\tag{4.4}
$$

The coordinate x allows us to find the section and the coordinates y and z indicate the point P in section x.

Recalling that the displacement in the longitudinal direction is zero, that is, $u_x(x, y, z) = 0$, it follows that the kinematics of a circular shaft is given by the vector field $\mathbf{u}(x, y, z)$ with the following components

$$
\mathbf{u}(x, y, z) = \left\{ \begin{array}{c} u_x(x, y, z) \\ u_y(x, y, z) \\ u_z(x, y, z) \end{array} \right\} = \left\{ \begin{array}{c} 0 \\ -z\theta_x(x) \\ y\theta_x(x) \end{array} \right\}.
\tag{4.5}
$$

The denomination vector field is employed to $\mathbf{u}(x, y, z)$, because the kinematics for each point of the shaft is described by a vector whose components depend on its (x, y, z) coordinates. As the shaft is continuous, there is an infinite number of points, that is, there is also an infinite number of vectors

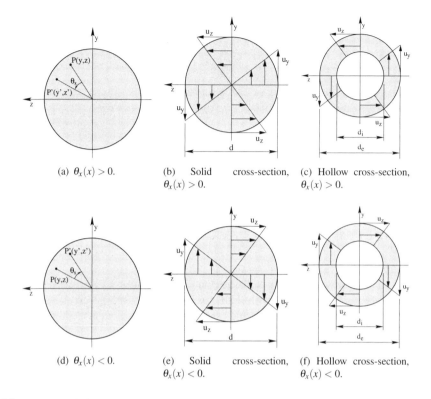

(a) $\theta_x(x) > 0.$ **(b)** Solid cross-section, $\theta_x(x) > 0.$ **(c)** Hollow cross-section, $\theta_x(x) > 0.$

(d) $\theta_x(x) < 0.$ **(e)** Solid cross-section, $\theta_x(x) < 0.$ **(f)** Hollow cross-section, $\theta_x(x) < 0.$

Figure 4.3 Transversal displacement components in the circular torsion.

describing the kinematics of the axis. For this reason, we use the concept of vector field to describe these infinite vectors.

From the kinematics given in (4.5), we can define the tangential displacement field $\mathbf{u}_t(x,y,z)$ for each point P of the shaft with (x,y,z) coordinates [see Figure 4.2(b)]. To do this, we only need to sum the vector components $u_y(x,y,z)$ and $u_z(x,y,z)$, that is,

$$\mathbf{u}_t(x,y,z) = u_y(x,y,z)\mathbf{e}_j + u_z(x,y,z)\mathbf{e}_k, \qquad (4.6)$$

with $\mathbf{e}_j$ and $\mathbf{e}_k$ the unit vectors in the y and z directions, respectively. The norm of $\mathbf{u}_t(x,y,z)$, indicated as $u_t(x,y,z)$, is obtained substituting $u_y(x,y,z)$ and $u_z(x,y,z)$ given in (4.5). Thus,

$$u_t(x,y,z) = \left(\sqrt{y^2+z^2} \right) \theta_x(x). \qquad (4.7)$$

Recalling that $r = \sqrt{y^2+z^2}$, we have

$$u_t(x,y,z) = r\theta_x(x). \qquad (4.8)$$

The tangential displacement u_t can be obtained directly from Figure 4.2(a) employing a polar coordinate system in the shaft cross-section. In this case, the position of P is expressed in terms of the radial r and circumferential β coordinates. Therefore, this position is given by the CP arc, which is equal to $r\beta$ for a small β. Similarly, the final position of P' is given by the CP' arc, that is, $r(\beta + \theta_x)$ for small θ_x. Therefore, the tangential displacement is simply the difference between the initial P and final P' positions

$$u_t = r(\beta + \theta_x) - r\beta,$$

that is,

$$u_t(x,r) = r\theta_x(x). \tag{4.9}$$

Again, $u_t(x)$ varies linearly in the cross-section and is zero at the center and maximum at the section boundary. Figure 4.4 illustrates the behavior of the tangential displacement for positive and negative angles of twist; analogously to the case of hollow cross-sections.

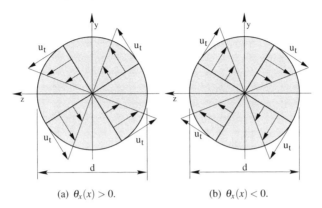

(a) $\theta_x(x) > 0$. (b) $\theta_x(x) < 0$.

Figure 4.4 Tangential displacement behavior in the shaft cross-section.

The kinematic constraints are represented by supports. In Figure 4.5, the shaft is fixed at $x = 0$ end. This implies that the angle of twist is zero in this section, namely $\theta_x(x = 0) = 0$. Despite the use of the clamp symbol to represent the support in Figure 4.5(a), the only valid restriction is $\theta_x(x = 0) = 0$, because this example considers only the shaft torsion.

Example 4.1 *Consider the shaft of length L fixed at the $x = 0$ end and subjected to an axial rotation θ at the free end ($x = L$), as shown in Figure 4.5(a).*
The function $\theta_x(x)$ that gives the angle of twist of each shaft section of this example is given by

$$\theta_x(x) = \frac{\theta}{L}x.$$

This function will be later determined by the solution of equations which will be derived at the end of the shaft formulation.

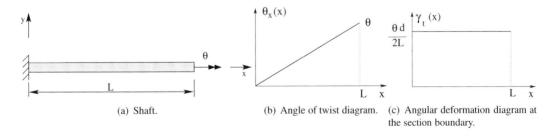

(a) Shaft. (b) Angle of twist diagram. (c) Angular deformation diagram at the section boundary.

Figure 4.5 Example 4.1: shaft with a rotation at the free end.

The diagram of the twist angle is a plot of the function $\theta_x(x)$ and provides the value of the angle for each x section. The diagram of this example is illustrated in Figure 4.5(b). It is observed that $\theta_x(x)$ satisfies the zero rotation kinematic constraint at $x = 0$, that is, $\theta_x(x = 0) = 0$. Moreover, $\theta_x(x)$ is a linear continuous function.

□

4.3 STRAIN MEASURE

In the case of the bar model, the longitudinal strain component $\varepsilon_{xx}(x)$ is related to stretching and shortening actions $u_x(x)$ of the bar sections. Thus, ε_{xx} is associated with a specific variation of the bar length.

In the case of circular torsion, the kinematics is described by the angle of twist $\theta_x(x)$, giving rise to a linear variation along the shaft length for small values of $\theta_x(x)$ [see Figure 4.2(a)]. Thus, for the circular torsion model, the strain measure will be angular, due to the variation of θ_x along the cross-sections, and not longitudinal as in the bar model. It is noted that if all sections have the same θ_x rotation, the shaft will have a rigid body rotation about the x axis, as will be seen in next section, so there is no strain on the shaft.

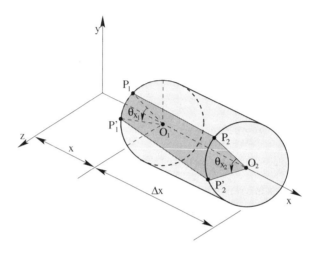

Figure 4.6 Strain measure analysis in circular torsion.

To characterize the shaft strain measure due to the angle of twist $\theta_x(x)$, the u_y and u_z displacements are compared at two points located in cross-sections with distinct angles of twists. For this purpose, consider Figure 4.6 which illustrates sections located at x and $x + \Delta x$ from the origin of the reference system. Let the points P_1 and P_2 with coordinates (x, y, z) and $(x + \Delta x, y, z)$, respectively. Thus, these points have the same cross-section y and z coordinates before the torsion action. Due to the torsion, sections x and $x + \Delta x$ present rigid body rotations, θ_{x_1} and θ_{x_2}, respectively, with $\theta_{x_2} > \theta_{x_1}$. In this case, points P_1 and P_2 assume the final positions P_1' and P_2', respectively. The transversal displacements u_y and u_z of these points are given from (4.5), respectively, by

$$u_y(x,y,z) = -z\theta_x(x) = -z\theta_{x_1} \qquad u_z(x,y,z) = y\theta_x(x) = y\theta_{x_1}$$
$$u_y(x+\Delta x,y,z) = -z\theta_x(x+\Delta x) = -z\theta_{x_2} \qquad u_z(x+\Delta x,y,z) = y\theta_x(x+\Delta x) = y\theta_{x_2} \, .$$

The relative transversal displacements between points P_1 and P_2 are given by

$$\Delta u_y = u_y(x+\Delta x,y,z) - u_y(x,y,z) = -z(\theta_{x_2} - \theta_{x_1}) = -z\Delta\theta_x,$$
$$\Delta u_z = u_z(x+\Delta x,y,z) - u_z(x,y,z) = y(\theta_{x_2} - \theta_{x_1}) = y\Delta\theta_x,$$

with $\Delta\theta_x = \theta_{x_2} - \theta_{x_1}$ the variation of the angle of twist at $x + \Delta x$ and x sections. These displacements are illustrated in Figure 4.7(a), taking the projections of the displacements of P_1 and P_2 in one section.

The absolute values of Δu_y and Δu_z do not allow us to identify if they are large or small. For example, if they are equal to 1 mm and the $\Delta x = 10$ mm, the ratio $1/10 = 0.1$ indicates that Δu_y and Δu_z represent 10% of the distance between the sections and hence the displacements can be

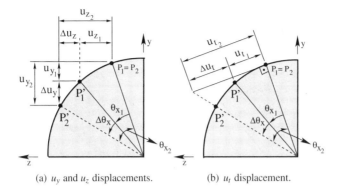

(a) u_y and u_z displacements.   (b) u_t displacement.

Figure 4.7 Relative transversal displacement components in the circular torsion.

considered large. For $\Delta x = 100$ mm, Δu_y and Δu_z represent 1% of the distance between the sections and are therefore small displacements. Thus, the specific relative transversal displacements are more representative of the displacement intensity. They are defined by

$$\frac{\Delta u_y}{\Delta x} = -z\frac{\Delta \theta_x}{\Delta x},$$

$$\frac{\Delta u_z}{\Delta x} = y\frac{\Delta \theta_x}{\Delta x}.$$

Due to the continuity hypothesis, the Δx distance can be as small as desired. Taking the limit $\Delta x \to 0$, we have the definition of the angular strain components at point P as

$$\gamma_{xy} = \lim_{\Delta x \to 0}\frac{u_y(x+\Delta x, y, z) - u_y(x, y, z)}{\Delta x} = \lim_{\Delta x \to 0}\frac{\Delta u_y}{\Delta x} = \frac{\partial u_y(x, y, z)}{\partial x}, \tag{4.10}$$

$$\gamma_{xz} = \lim_{\Delta x \to 0}\frac{u_z(x+\Delta x, y, z) - u_z(x, y, z)}{\Delta x} = \lim_{\Delta x \to 0}\frac{\Delta u_z}{\Delta x} = \frac{\partial u_z(x, y, z)}{\partial x}. \tag{4.11}$$

Therefore, associated to the transversal displacement components u_y and u_z, we have the angular or shear strain components γ_{xy} and γ_{xz}. The letter γ is used to indicate an angular strain, while ε represents a longitudinal strain. The x index represents the normal direction to the cross-section plane where the strain occurs. Indices y and z indicate the directions of the strain components. Thus, γ_{xy} is the angular strain at a point located on the x plane in the y axis direction. Analogously, γ_{xz} is an angular strain component in a point on the x plane in the z axis direction.

These strain components can be written in terms of the angle of twist, by substituting u_y and u_z given in (4.4). Thus,

$$\gamma_{xy}(x,z) = \frac{\partial u_y(x, y, z)}{\partial x} = \frac{\partial}{\partial x}(-z\theta_x(x)) = -z\frac{d\theta_x(x)}{dx}, \tag{4.12}$$

$$\gamma_{xz}(x,y) = \frac{\partial u_z(x, y, z)}{\partial x} = \frac{\partial}{\partial x}(y\theta_x(x)) = y\frac{d\theta_x(x)}{dx}. \tag{4.13}$$

Supposing the dimensions in meters and the angle of twist in radians, we observe that γ_{xy} has the following SI units

$$[\gamma_{xy}] = m\frac{rad}{m} = rad,$$

that is, the γ_{xy} component is also given in radians and represents an angular strain.

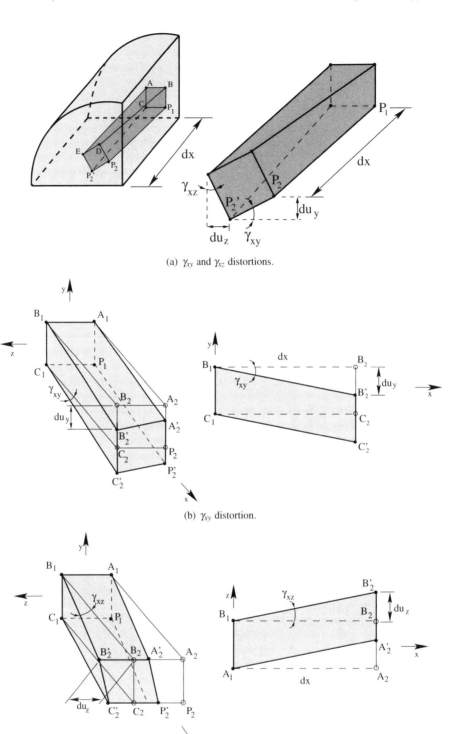

(a) γ_{xy} and γ_{xz} distortions.

(b) γ_{xy} distortion.

(c) γ_{xz} distortion.

Figure 4.8   Interpretation of the γ_{xy} and γ_{xz} strain components.

Figure 4.8 illustrates the distortion components. For this purpose, two area elements are considered around points P_1 and P_2 for the x and $x + dx$ sections. It is assumed that the angle of twist in section x is zero, and consequently the area element around P_1 does not have any distortion. Hence, the angle of twist in $x + dx$ corresponds to the $d\theta_x$ differential. The volume element obtained when considering the two area elements, with the distortion components, are illustrated in Figure 4.8(a). Figure 4.8(b) shows only the γ_{xy} distortion component. In this case, the area element $P_2A_2B_2C_2$ has only one displacement in the y direction, denoted by du_y. The $B_1B_2B_2'$ angle indicates the γ_{xy} distortion. From Figure 4.8(b) and equation (4.12), it is observed that

$$du_y = \gamma_{xy}dx = -zd\theta_x.$$

Similarly, Figure 4.8(c) shows only the γ_{xz} distortion component. In this case, the $P_2A_2B_2C_2$ area element has only a displacement in the z direction denoted by du_z. The $B_1B_2B_2'$ angle indicates the γ_{xz} distortion. From Figure 4.8(c) and equation (4.13), we observe that

$$du_z = \gamma_{xz}dx = yd\theta_x.$$

It is noted that before torsion, the angles between the edges of the area element were straight. After torsion, these angles are no longer straight, and the variation is given by the distortion components γ_{xy} and γ_{xz}.

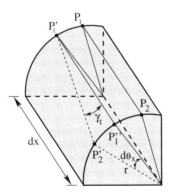

Figure 4.9 Tangential distortion component γ_t in a shaft.

Analogously to the tangential displacement field, the tangential angular or shear strain component is also defined and denoted by γ_t. In this case, the tangential displacements of points P_1 and P_2 are given by equation (4.9). Figure 4.7(b) shows the projection of these displacements in a single section. The specific relative tangential displacement is given by

$$\frac{\Delta u_t}{\Delta x} = r\frac{\Delta\theta_x}{\Delta x}.$$

Thus, the tangential strain of point P_1 is determined taking the limit for $\Delta x \to 0$ as

$$\gamma_t(x,r) = \lim_{\Delta x \to 0}\frac{u_t(x+\Delta x,r) - u_t(x,r)}{\Delta x} = \lim_{\Delta x \to 0}\frac{r\theta_{x_2} - r\theta_{x_1}}{\Delta x} = r\lim_{\Delta x \to 0}\frac{\Delta\theta_x}{\Delta x}. \tag{4.14}$$

Hence, employing the derivative definition, we have the final expression for the tangential strain γ_t, that is,

$$\gamma_t(x,r) = r\frac{d\theta_x(x)}{dx}. \tag{4.15}$$

The tangential strain varies linearly with the radial coordinate r in the cross-section of the shaft. This is expected, because the tangential displacement field (4.9) also varies linearly in the cross-section. The term shear comes from the physical effect of friction between two sections which are close to each other.

Figure 4.9 illustrates the tangential distortion component. After the θ_x and $\theta_x + d\theta_x$ rotations, points P_1 and P_2 assume the P_1' and P_2' positions, respectively. Point P_1'' indicates the projection of P_1' in the section $x + dx$. The arc length $P_1'P_1''P_2'$ indicates the tangential displacement du_t for small values of $d\theta_x$. From Figure 4.9, it is observed that

$$du_t = \gamma_t dx = r\theta_x.$$

Example 4.2 *Consider the shaft of Example 4.1. The angular tangential strain γ_t can be obtained by taking the derivative of the torsion angle expression $\theta_x(x) = \dfrac{\theta}{L}x$, that is,*

$$\gamma_t(x,r) = r\frac{\theta}{L}.$$

It is then observed that the strain is constant in all sections of the considered shaft for points with the same radial coordinate r, as illustrated in Figure 4.5(c) for points of the section boundaries with $r = \dfrac{d}{2}$.

☐

4.4   RIGID ACTIONS

The rigid actions are obtained by imposing that the shear strain components in the shaft are zero, that is,

$$
\begin{aligned}
\gamma_{xy}(x,z) &= -z\frac{d\theta_x(x)}{dx} = 0, \\
\gamma_{xz}(x,y) &= y\frac{d\theta_x(x)}{dx} = 0.
\end{aligned}
$$

The only condition that satisfies the previous relations is $\dfrac{d\theta_x(x)}{dx} = 0$ for any cross-section of the shaft. That implies that $\theta_x(x) = \theta_x$ must be a constant rotation for all sections. Thus, plane AO_1O_2B has only a rigid rotation with zero strain measure, as illustrated in Figure 4.10. Thus, the rigid actions are those where all cross-sections have the same rigid rotation around the x axis.

Example 4.3 *Consider the shaft of Example 4.1. It is fixed at the $x = 0$ end which gives $\theta_x(x = 0) = 0$. Thus, as the rigid rotation is constant for all sections of the shaft, and the kinematic constraint must be satisfied, we have $\theta_x = 0$.*

☐

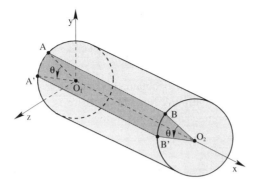

Figure 4.10 Rigid rotation in circular torsion.

4.5 DETERMINATION OF INTERNAL LOADS

Consider a shaft of length L and the differential volume $dV = dAdx$ illustrated in Figure 4.8(a). After torsion, the volume element has the transversal displacements du_y and du_z, as shown in Figure 4.8(a). Intuitively, it is known that the larger the transversal displacements, the larger will be the respective strain components and the work or energy needed to apply a torsion action in the shaft. Similarly to the bar model, the strain internal work W_i is a function of the γ_{xy} and γ_{xz} strain components, that is,

$$W_i = f(\gamma_{xy}, \gamma_{xz}). \tag{4.16}$$

In order to evaluate it, we need to define the internal loads presented in the deformed shaft.

For the differential volume dV, the internal work density dw_i is considered, i.e., the ratio of the strain internal work dW_i and the volume dV given by

$$dw_i = \frac{dW_i}{dV} = g(\gamma_{xy}, \gamma_{xz}). \tag{4.17}$$

Due to the proportionality between the internal work and strain measures, the relation between them is linear and the previous equation can be written as

$$dw_i = \frac{dW_i}{dV} = \tau_{xy}\gamma_{xy} + \tau_{xz}\gamma_{xz}, \tag{4.18}$$

with τ_{xy} and τ_{xz} the internal loads in the dV element. Substituting (4.12) and (4.13) in the previous expression, we obtain

$$dw_i = \tau_{xy}\left(-z\frac{d\theta_x(x)}{dx}\right) + \tau_{xz}\left(y\frac{d\theta_x(x)}{dx}\right). \tag{4.19}$$

Considering the dimensional analysis of the previous expression in the International System of Units (SI), we have

$$\frac{Nm}{m^3} = \frac{N}{m^2}m\frac{rad}{m} + \frac{N}{m^2}m\frac{rad}{m}.$$

Therefore, to represent a work density in the right-hand side of equation (4.19), the internal loads must necessarily be expressed as force per area, which are known as stress.

Thus, the stress components τ_{xy} and τ_{xz} represent the internal loads in the shaft. As the strain components γ_{xy} and γ_{xz} have an angular or shearing nature, τ_{xy} and τ_{xz} are called shearing stresses acting on plane x in the y and z directions, respectively. Each of the strain components is associated to its respective stress component. Expression $\tau_{xy}\gamma_{xy} + \tau_{xz}\gamma_{xz}$ represents the internal strain work density at any point P of the shaft.

From equations (4.18) and (4.19), the internal strain work in the volume element dV is given by

$$dW_i = \left[\tau_{xy} \left(-z \frac{d\theta_x(x)}{dx} \right) + \tau_{xz} \left(y \frac{d\theta_x(x)}{dx} \right) \right] dV. \tag{4.20}$$

The strain internal work for the shaft is obtained by the work summation of all differential elements of the shaft. Due to the continuity hypothesis, there is an infinite number of differential elements, and an infinity sum is represented by the following Riemann integral on the volume of the shaft:

$$
\begin{aligned}
W_i &= \int_V dW_i = \int_V \left[\tau_{xy}(x,y,z) \left(-z \frac{d\theta_x(x)}{dx} \right) + \tau_{xz}(x,y,z) \left(y \frac{d\theta_x(x)}{dx} \right) \right] dV \\
&= \int_V \left[-z\tau_{xy}(x,y,z) + y\tau_{xz}(x,y,z) \right] \frac{d\theta_x(x)}{dx} dV.
\end{aligned}
$$

Due to the assumed kinematics for the shaft, the angle of twist varies only from one section to another, that is, $\theta_x = \theta_x(x)$. Thus, we can express the previous volume integral as the product of an area integral in the cross-section and another along the shaft length. Therefore,

$$W_i = \int_0^L \left(\int_A \left[-z\tau_{xy}(x,y,z) + y\tau_{xz}(x,y,z) \right] dA \right) \frac{d\theta_x(x)}{dx} dx. \tag{4.21}$$

The area integral results in the moment along the x axis. Supposing that the length units are given in m and the stresses in N/m^2, the integrand is then given by N/m units, which results in Nm after the integration, that is, moment units. To interpret this integral, consider Figure 4.11(a) which shows an infinitesimal area element dA, around point P, with distances y and z, respectively, from the z and y axes of the adopted reference system. Note that τ_{xy} and τ_{xz} are the stress components on the x plane (the x axis is normal to the cross-sections) in the directions of y and z axes, respectively. On the other hand, the products $dF_y = -\tau_{xy}dA$ and $dF_z = \tau_{xz}dA$ indicate the resultant of internal forces in the area element in the $-y$ and z directions, respectively. Similarly, the products $-z\tau_{xy}dA$ and $y\tau_{xz}dA$ represent the moments in the positive direction of the x axis relative to the geometric center of the section. Thus, the moment in the x direction for the area element dA is given by

$$dM_x = \left[-z\tau_{xy}(x,y,z) + y\tau_{xz}(x,y,z) \right] dA. \tag{4.22}$$

Summing the moments for an infinite number of area elements, we have the moment in the x direction for the cross-section, that is,

$$M_x(x) = \int_A \left[-z\tau_{xy}(x,y,z) + y\tau_{xz}(x,y,z) \right] dA. \tag{4.23}$$

This is called longitudinal or twisting moment of the cross-section. The twisting moment may vary for each x section of the shaft, analogously to the angle of twist $\theta_x(x)$.

Substituting (4.23) in (4.21), we can write the strain internal work expression as

$$W_i = \int_0^L M_x(x) \frac{d\theta_x(x)}{dx} dx. \tag{4.24}$$

Analogously to the bar case, we can integrate the internal work expression by parts which gives

$$
\begin{aligned}
W_i &= -\int_0^L \frac{dM_x(x)}{dx} \theta_x(x) \, dx + M_x(x)\theta_x(x)\big|_0^L \\
&= -\int_0^L \frac{dM_x(x)}{dx} \theta_x(x) \, dx + \left[M_x(L)\theta_x(L) - M_x(0)\theta_x(0) \right]. \tag{4.25}
\end{aligned}
$$

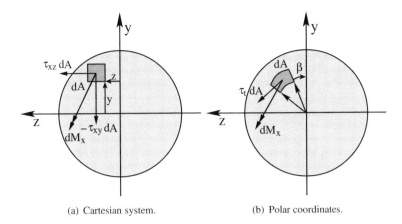

(a) Cartesian system.   (b) Polar coordinates.

Figure 4.11   Resultant in terms of the twisting moment in the cross-section of the shaft.

Thus, the internal loads which are compatible with the torsion kinematics are characterized by concentrated twisting moments $M_x(L)$ and $M_x(0)$ at the shaft ends, and a distributed twisting moment $\dfrac{dM_x(x)}{dx}$ along the shaft length. These loads are illustrated in the FBD of Figure 4.12(a), using the signs indicated in (4.25).

Figure 4.9 illustrates a volume element before and after torsion, considering the kinematics described by the tangential displacement u_t. The internal work may be written in terms of the angular strain component γ_t, that is,

$$W_i = \int_V \tau_t(x,r)\gamma_t(x,r)dV, \tag{4.26}$$

where $\tau_t(x,r)$ is the shear stress in the tangential direction of a point with polar coordinates (r,β) in the x cross-section. Substituting (4.15), the previous expression can be written analogously to (4.21) as

$$W_i = \int_0^L \left(\int_A r\tau_t(x,r)dA \right) \frac{d\theta_x(x)}{dx}dx. \tag{4.27}$$

In this case, the twisting moment is given by the area integral in terms of the tangential shear stress, that is,

$$M_x(x) = \int_A r\tau_t(x,r)dA, \tag{4.28}$$

as illustrated in Figure 4.11(b). It is observed that the term $dF_t = \tau_t(x,r)dA$ represents the tangential force in the dA element, while $dM_x = r\tau_t(x,r)dA$ is the twisting moment in the same element, relative to the geometrical center of the section. Substituting (4.28) in (4.27), we obtain the same expression (4.24) for the strain internal work. Again, integration by parts results in expression (4.25).

4.6   DETERMINATION OF EXTERNAL LOADS

The internal loads in a deformed shaft are given in equation (4.25). Thus, the external loads that may be applied to a shaft are those ones that can be balanced by the internal loads which are compatible with the shaft kinematics. Based on that, we have the following external loads in a shaft as shown in Figure 4.12(b):

-   $m_x(x)$: distributed torque along the shaft length, having units of Nm/m in the SI;
-   T_0 and T_L: concentrated torques at the $x = 0$ and $x = L$ shaft ends.

Hence, to characterize the external loads which are compatible with the internal ones, and consequently with the shaft kinematics free of any constraints, the terms of the internal work expression are used for an angle of twist θ_x. The work done by the internal twisting moments $M_x(L)\theta_x(L)$ and $M_x(0)\theta_x(0)$ at the shaft ends balances the respective work of the external loads $T_L\theta_x(L)$ and $T_0\theta_x(0)$. Besides that, the internal distributed twisting moment $\dfrac{dM_x(x)}{dx}$ balances the external distributed torque density $m_x(x)$. The work $\int_0^L m_x(x)\theta_x(x)\,dx$, associated to an arbitrary rotation $\theta_x(x)$, balances the internal work $\int_0^L \dfrac{dM_x(x)}{dx}\theta_x(x)\,dx$.

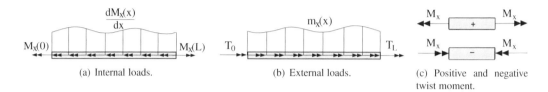

(a) Internal loads.   (b) External loads.   (c) Positive and negative twist moment.

Figure 4.12   Free body diagrams for the internal and external loads in a shaft and positive and negative twisting moments.

Finally, we obtain the following external work expression for an angle of twist $\theta_x(x)$:

$$W_e = T_0\theta_x(0) + T_L\theta_x(L) + \int_0^L m_x(x)\theta_x(x)\,dx. \tag{4.29}$$

4.7   EQUILIBRIUM

Consider a shaft in the equilibrium deformed configuration. To evaluate this equilibrium state, we introduce a virtual torsion action $\delta\theta_x(x)$ from the deformed position that satisfies the kinematic restrictions of the shaft. If the shaft is in equilibrium in this configuration, the internal and external works associated with any virtual action $\delta\theta_x(x)$ must be equal, that is,

$$\delta W_e = \delta W_i.$$

Substituting (4.25) and (4.29) in the previous PVW postulate, we obtain

$$[M_x(0) + T_0]\,\delta\theta_x(0) + [-M_x(L) + T_L]\,\delta\theta_x(L) - \int_0^L \left[\frac{dM_x(x)}{dx} + m_x(x)\right]\delta\theta_x(x)\,dx = 0. \tag{4.30}$$

This equation is valid for any arbitrary virtual action $\delta\theta_x(x)$, if the terms in the brackets are simultaneously zero, i.e.,

$$\begin{cases} \dfrac{dM_x(x)}{dx} + m_x(x) = 0 & \text{in } x \in (0,L) \\ M_x(0) = -T_0 & \text{in } x = 0 \\ M_x(L) = T_L & \text{in } x = L \end{cases} \tag{4.31}$$

The above expression defines the local equilibrium form of the circular torsion model, free of any kinematic constraints. There is the differential equation in terms of the twisting moment and two boundary conditions. This set (differential equation + boundary conditions) defines the boundary value problem (BVP) of equilibrium for the circular torsion. The boundary conditions indicate the positive and negative directions of the twist moment at the shaft ends as illustrated in Figure 4.12(c).

Solving the differential equation, the twist moment $M_x(x)$ along the shaft length is obtained. We can plot the diagram for the twisting moment $M_x(x)$, analogously to the diagram of the normal force $N_x(x)$ for the bar case.

If $\delta\theta_x(x)$ is a rigid virtual action, then the internal work is zero. In this case, the PVW states that for any rigid virtual action $\delta\theta_x(x)$, the external work given in (4.29) is zero for a shaft in equilibrium, that is,

$$T_0\delta\theta_x(0) + T_L\delta\theta_x(L) + \int_0^L m_x(x)\delta\theta_x(x)\,dx = 0. \tag{4.32}$$

The rigid actions for the shaft are constant rotations about the x axis. Thus, we have $\delta\theta_x(x) = \delta\theta_x = $ cte and substituting in the previous expression

$$\left(T_0 + T_L + \int_0^L m_x(x)\,dx\right)\delta\theta_x = 0.$$

This is the rigid equilibrium condition for the shaft and states that the resultant of external torques in the x direction must be zero, that is,

$$\sum m_x = T_0 + T_L + \int_0^L m_x(x)\,dx = 0. \tag{4.33}$$

The integral term is called the concentrated torque equivalent to the $m_x(x)$ distributed torque. Therefore, in terms of static equilibrium, it is possible to replace the distributed torque $m_x(x)$ by its equivalent concentrated torque.

It should be noted that the model of circular torsion is algebraically similar to the bar model. For its solution, the same integration procedure presented in Chapter 3 may be applied as illustrated in the following examples.

Example 4.4 *Plot the twisting moment diagram for the shaft illustrated in Figure 4.13(a), subjected to a concentrated torque at the free end. Use the integration of the equilibrium differential equation for the twisting moment.*

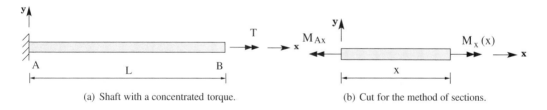

(a) Shaft with a concentrated torque.　　　　　(b) Cut for the method of sections.

Figure 4.13 Example 4.4: shaft subjected to a concentrated torque T at the free end.

- *Distributed load equation*
  As there is no distributed axial moment applied to the shaft, we have $m_x(x) = 0$.
- *Boundary conditions*
  The effect of the concentrated torque T acting at the free end $(x = L)$ is to rotate the axis in the positive direction of x, and therefore the torque is such that $M_x(x = L) = T$. The shaft is fixed at the $x = 0$ end and the angle of twist is zero, that is, $\theta_x(x = 0) = 0$. However, in this example, the differential equation of the twisting moment is integrated and the latter boundary condition is not useful for the solution of this problem. The use of boundary conditions in terms of the angle of twist is illustrated later.

- *Integration of the differential equation*
  The integration of the equilibrium differential equation in terms of the twisting moment results in

$$\frac{dM_x(x)}{dx} = -m_x(x) = 0 \rightarrow M_x(x) = C_1,$$

  where C_1 is the integration constant.
- *Determination of the integration constant*
  To calculate C_1, we apply the boundary condition in terms of the twisting moment. Thus,

$$M_x(x = L) = C_1 = T \rightarrow C_1 = T.$$

- *Final equation*
  The final equation for the twisting moment is obtained substituting C_1, that is,

$$M_x(x) = T.$$

- *Twisting moment diagram*
  The diagram for the twisting moment is illustrated in Figure 4.14 for $T = 100$ Nm and $L = 1$ m. It is noted that the twisting moment is constant for all sections of the shaft.

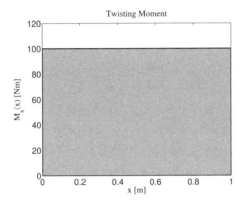

Figure 4.14   Example 4.4: twisting moment diagram.

- *Support reaction*
  It is initially assumed that the reaction support, expressed as the moment M_{Ax}, is in the positive direction of x. There are two ways of determining the reaction moment. The simplest one is to use the boundary condition at $x = 0$ of the equilibrium BVP given in (4.31). In this case, the reaction M_{Ax} is the T_0 torque and

$$M_{Ax} = -M_x(x = 0) = -100 \text{ Nm}.$$

  As the value of M_{Ax} is negative, the reaction points leftwards and therefore is contrary to the positive direction of the x axis.
  The second way to calculate the reaction moment is to employ the rigid body equilibrium equation given in (4.33). Hence,

$$\sum m_x = 0: \ M_{Ax} + T + \int_0^L (0) \, dx = 0 \rightarrow M_{Ax} = -T = -100 \text{ Nm}.$$

  The interpretation of the twisting moment diagram with the method of sections is illustrated in Figure 4.13(b). A cut is taken in any section x and the left portion of the shaft is isolated.

It is observed that this section should have the twisting moment $M_x(x)$ that balances the support reaction. Taking the equilibrium of moments in the x direction, we have

$$\sum m_x = -M_{Ax} + M_x(x) = 0 \rightarrow M_x(x) = 100 \text{ Nm}.$$

□

Example 4.5 *Plot the twisting moment diagram for the shaft illustrated in Figure 4.15(a), subjected to a distributed external torque of constant intensity t_0, by the integration of the differential equation.*

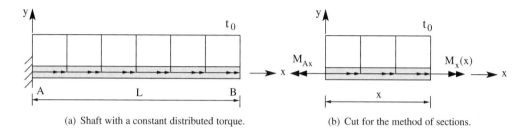

(a) Shaft with a constant distributed torque.　　(b) Cut for the method of sections.

Figure 4.15 Example 4.5: shaft subjected to a distributed torque of constant intensity.

- *Distributed load equation*
  As the shaft is subjected to a distributed torque of constant intensity t_0 in the positive direction of x, we have $m_x(x) = t_0$.
- *Boundary conditions*
  The shaft is fixed at $x = 0$ and the angle of twist is consequently zero. Thus, $\theta_x(x = 0) = 0$. As there is no concentrated torque applied at the free end, the twisting moment is zero, that is, $M_x(x = L) = 0$.
- *Integration of the differential equation*
  Integrating the equilibrium differential equation in terms of the twisting moment, we have

$$\frac{dM_x(x)}{dx} = -m_x(x) = -t_0 \rightarrow M_x(x) = -t_0 x + C_1,$$

with C_1 the integration constant.
- *Determination of the integration constant*
  The boundary condition in terms of the twisting moment is used to find C_1, that is,

$$M_x(x = L) = -t_0(L) + C_1 = 0 \rightarrow C_1 = t_0 L.$$

- *Final equation*
  The final expression for the twisting moment is obtained replacing the C_1 integration constant, that is,

$$M_x(x) = -t_0(x - L).$$

- *Twisting moment diagram*
  The twisting moment varies linearly. Its diagram is shown in Figure 4.16 for $L = 2$ m and $t_0 = 100 \text{ Nm/m}$.
- *Support reaction*
  The support reaction moment M_{Ax} at the $x = 0$ end of the shaft is given by the boundary condition, namely $M_{Ax} = -M_x(x = 0) = -t_0 L = -200$ Nm. Thus, the direction of the reaction is leftwards, and therefore contrary to the positive direction of the x axis. This value can also be obtained directly from the twisting moment diagram.

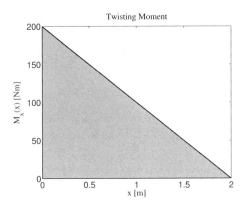

Figure 4.16   Example 4.5: twisting moment diagram.

Another way of calculating the reaction moment is by the equation of rigid body equilibrium given in (4.33), i.e.,

$$\sum m_x = 0: \ M_{Ax} + \int_0^L t_0 \, dx = 0 \rightarrow M_{Ax} = -t_0 L = -200 \text{ Nm}.$$

The interpretation of the twisting moment diagram using the method of sections is illustrated in Figure 4.15(b). A cut is made in a generic section x and the segment between 0 and x is isolated. It is observed that there must be a twisting moment $M_x(x)$ in this section which balances the support reaction and the distributed torque. Considering the balance of moments in the x direction, we have

$$\sum m_x = -M_{Ax} + t_0 x + M_x(x) = 0 \rightarrow M_x(x) = 200 - 100x = 100(2-x).$$

Thus, we recover the same expression as the one obtained by integrating the differential equation.
□

Example 4.6 *Plot the twisting moment diagram for the shaft illustrated in Figure 4.17, subjected to a linear distributed torque, by integrating the differential equation.*

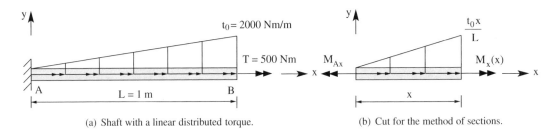

(a) Shaft with a linear distributed torque.   (b) Cut for the method of sections.

Figure 4.17   Example 4.6: shaft subjected to a distributed torque of linear intensity.

- *Distributed load equation*
  The expression for the linear distributed torque is $m_x(x) = \frac{t_0}{L}x$. Note that $m_x(x=0) = 0$ and $m_x(x=L) = t_0 = 2000 \text{ Nm/m}$.
- *Boundary conditions*
  As the shaft is fixed at $x = 0$, the angle of twist is zero and thus $\theta_x(x=0) = 0$. Due to the concentrated torque applied to the free end, the twisting moment is such that $M_x(x=L) = T = 500 \text{ Nm}$.

- *Integration of the differential equation*
  Integrating the equilibrium differential equation in terms of the twisting moment, we have

$$\frac{dM_x(x)}{dx} = -m_x(x) = -\frac{t_0}{L}x \rightarrow M_x(x) = -\frac{t_0}{2L}x^2 + C_1.$$

- *Determination of the integration constant*
  Applying the boundary condition in terms of the twisting moment, we have

$$M_x(x = L) = -\frac{t_0}{2L}(L)^2 + C_1 = T \rightarrow C_1 = T + \frac{t_0 L}{2}.$$

- *Final equation*
  The final equation for the twisting moment is obtained substituting the C_1 integration constant, that is,

$$M_x(x) = -\frac{t_0}{2L}x^2 + \left(T + \frac{t_0 L}{2}\right).$$

- *Twisting moment diagram*
  Considering $L = 1$ m, $t_0 = 2000$ Nm/m, and $T = 500$ Nm, we have

$$M_x(x) = -500(2x^2 - 3).$$

Note that due to the linear distributed torque, the twisting moment function is a parabola. The twisting moment diagram is illustrated in Figure 4.18.

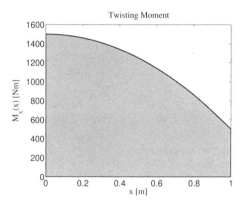

Figure 4.18 Example 4.6: twisting moment diagram.

- *Support reaction*
  It is initially assumed that the support reaction M_{Ax} is oriented in the positive direction of x. In this case, the support reaction M_{Ax} at the $x = 0$ shaft end is given by the boundary condition, that is, $M_{Ax} = -M_x(x = 0) = -1500$ Nm. Thus, the direction of the reaction is leftwards, and therefore contrary to the positive direction of x. This value can also be obtained directly from the diagram.
  Another way of calculating the reaction is to use the rigid body equilibrium equation given in (4.33), that is,

$$\sum m_x = 0 : M_{Ax} + T + \int_0^L \frac{t_0}{L}x \, dx = 0 \rightarrow M_{Ax} = -T - \frac{t_0 L}{2} = -1500 \text{ Nm.}$$

The interpretation of the twisting moment diagram using the method of sections is illustrated in Figure 4.17(b). A cut is made in a generic section x of the shaft and the left *segment is isolated. It is observed that a twisting moment $M_x(x)$ must exist in this section, which balances the support reaction and the distributed torque. Taking the balance of moments in the x direction, we have*

$$\sum m_x = -M_{Ax} + \int_0^x \frac{t_0}{L}x\,dx + M_x(x) = 0 \rightarrow M_x(x) = -500(2x^2 - 3),$$

and thus recovering the same expression as the one obtained by the integration of the differential equation.

□

4.8 MATERIAL BEHAVIOR

Similarly to the tensile test, the torsion test can be performed to characterize the behavior of a material when subjected to torsion. For this purpose, a specimen of length L with circular cross-section of constant radius R is used. It is fixed at one end and successively increasing values of external torque T are applied at the other end, measuring the angle of twist θ_x, as illustrated in Figure 4.19.

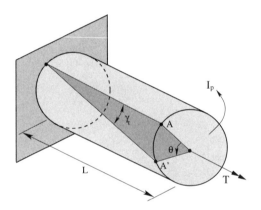

Figure 4.19   Specimen subjected to a torsion test.

The tangential angular strain at the boundary of the cross-section is determined by integrating equation (4.15). For an angle of twist θ_x, we have

$$\int_0^L \gamma_t\,dx = \int_0^{\theta_x} R\,d\theta_x.$$

As the γ_t distortion is constant along the shaft length, we have

$$\gamma_t = \frac{R\theta_x}{L}. \tag{4.34}$$

Another way to obtain this expression is considering the arc length AA' illustrated in Figure 4.19. For small strains, the following geometrical relation is valid:

$$\gamma_t L = R\theta_x \rightarrow \gamma_t = \frac{R\theta_x}{L}. \tag{4.35}$$

A material property called transversal elastic modulus, denoted by G, is defined as the angular coefficient of the linear regime of the torsion test diagram. The equation of the line for the elastic regime is given by

$$\tau_t = G\gamma_t. \tag{4.36}$$

This equation is known as Hooke's law for the case of torsion. It is noted that the same relationship is valid when taking the Cartesian shear components of stress (τ_{xy}, τ_{xz}) and strain (γ_{xy}, γ_{xz}), that is,

$$\tau_{xy} = G\gamma_{xy} \quad \text{and} \quad \tau_{xz} = G\gamma_{xz}. \tag{4.37}$$

As only the concentrated torque T is applied, the twisting moment for each cross-section x is constant with intensity of $M_x(x) = T$, as shown in Example 4.4. Replacing (4.36) in the twisting moment expression (4.28), we have

$$T = M_x(x) = \int_A G\gamma_t r \, dA = \frac{G\theta_x}{L} \int_A r^2 \, dA = \frac{GI_p}{L} \theta_x, \tag{4.38}$$

with I_p the polar moment of area of the circular cross-section.

The term

$$k_t = \frac{GI_p}{L}$$

is called torsional stiffness of the shaft. Analogous to the bar, a shaft of linear elastic material, subjected to a torque T at the end, behaves as a torsional spring with an elastic constant k_t.

From equation (4.38), we have

$$\frac{G\theta_x}{L} = \frac{T}{I_p}.$$

Substituting the previous equation in Hooke's law gives

$$\tau_t = \frac{T}{I_p} r, \tag{4.39}$$

which indicates that the shear stress has a linear variation in the cross-section for the elastic regime of the material.

For each value of the applied torque T, the angle of twist θ_x is measured. Then, the distortion γ_t and shear stress τ_t are calculated employing equations (4.35) and (4.39), respectively. A $\tau_t \times \gamma_t$ plot is called torsion test diagram. The behavior of this diagram for a ductile material is illustrated in Figure 4.20.

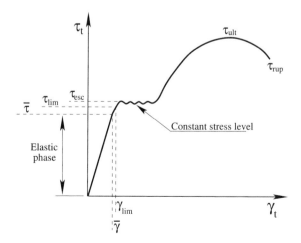

Figure 4.20 Torsion test diagram.

Stress values in the $0 \leq \tau_t \leq \tau_{\text{lim}}$ range characterize the elastic phase of the material, with τ_{lim} the proportionality limit shear stress. In general, shafts should be designed in such way they remain in the elastic regime; that is, the maximum shear stress τ_t^{max} must be in the $0 \leq \tau_t^{max} \leq \tau_{lim}$ range. It becomes difficult to accurately locate the value of the limit stress τ_{lim} from the torsion test. To overcome this limitation, the admissible shear stress $\bar{\tau}$ is employed to define the elastic phase. The $\bar{\tau}$ stress is given as the ratio of the yield shear stress τ_y and a safety coefficient k, that is,

$$\bar{\tau} = \frac{\tau_y}{k}. \tag{4.40}$$

Thus, the elastic range is then characterized by values of stress in the $0 \leq \tau \leq \bar{\tau}$ interval. The maximum shear stress in the torsion test is called ultimate shear stress τ_{ult}, while the stress at which rupture occurs is called rupture shear stress and is indicated by τ_{rup}.

It is verified experimentally that the yielding normal and shear stresses, σ_y and τ_y, are related to each other as

$$\tau_y \approx [0.55 \text{ to } 0.60] \, \sigma_y. \tag{4.41}$$

By dividing both sides of the previous equation by the safety coefficient k, we have

$$\frac{\tau_y}{k} \approx [0.55 \text{ to } 0.60] \, \frac{\sigma_y}{k},$$

that is,

$$\bar{\tau} \approx [0.55 \text{ to } 0.60] \, \bar{\sigma}. \tag{4.42}$$

As will be seen in Chapter 8, the following relation is valid between the normal and shear admissible stresses:

$$\bar{\tau} = 0.57\bar{\sigma}. \tag{4.43}$$

From the Poisson ratio v, we have the following relation between the longitudinal E and shear G elastic modulus:

$$G = \frac{E}{2(1+v)}. \tag{4.44}$$

For instance, taking Young's modulus and the Poisson ratio for the steel (210×10^9 N/m^2 and $v = 0.3$), the shear modulus for the steel has the following value:

$$G = \frac{210 \times 10^9}{2(1+0.3)} = 80.8 \times 10^9 \text{ N/m}^2 = 80.8 \text{ GPa}.$$

4.9   APPLICATION OF THE CONSTITUTIVE EQUATION

After defining Hooke's law (4.36) for a homogeneous isotropic linear elastic material subjected to torsion, the last step of the variational formulation can be taken, that is, the application of the constitutive relation in the equilibrium equation (4.31).

For this purpose, equation (4.15) is substituted in (4.36) resulting in

$$\tau_t(x, r) = G(x) \frac{d\theta_x(x)}{dx} r. \tag{4.45}$$

Substituting now the previous expression in (4.28), we have

$$M_x(x) = G(x) \frac{d\theta_x(x)}{dx} \int_A r^2 dA. \tag{4.46}$$

The previous integral represents the polar moment of area $I_p(x)$ for the cross-section x of the shaft. The SI unit for I_p is m^4. For a circular cross-section of diameter d, we have

$$I_p = \frac{\pi d^4}{32}. \tag{4.47}$$

Remembering that the radial r and Cartesian (y, z) coordinates are related by $r^2 = y^2 + z^2$, we have

$$I_p = \int_A r^2 \, dA = \int_A y^2 \, dA + \int_A z^2 \, dA = I_z + I_y, \tag{4.48}$$

with I_y and I_z the moments of area of the circular cross-section relative to the y and z axes of the Cartesian reference system, that is,

$$I_y = I_z = \frac{\pi d^4}{64}. \tag{4.49}$$

For a hollow circular cross-section with internal d_i and external d_e diameters, the polar moment of area is given by

$$I_p = \frac{\pi(d_e^4 - d_i^4)}{32}. \tag{4.50}$$

The final expression for the twisting moment of a shaft of elastic material according to Hooke's law is given by

$$M_x(x) = G(x)I_p(x)\frac{d\theta_x(x)}{dx}. \tag{4.51}$$

Replacing this equation in the differential equation (4.31) of the twisting moment, we obtain the differential equation for the shaft in terms of the angle of twist $\theta_x(x)$

$$\frac{d}{dx}\left(G(x)I_p(x)\frac{d\theta_x(x)}{dx}\right) + m_x(x) = 0. \tag{4.52}$$

For a shaft with constant cross-section and one material, we have $I_p(x) = I_p$ and $G(x) = G$. Thus, the previous differential equation can be simplified as

$$GI_p\frac{d^2\theta_x(x)}{dx^2} + m_x(x) = 0. \tag{4.53}$$

Therefore, for the case of linear elastic isotropic materials, we obtain a second-order differential equation, which must be integrated twice to obtain the function describing the angle of twist $\theta_x(x)$ of the shaft. The first integration provides the twisting moment $M_x(x) = G(x)I_p(x)\dfrac{d\theta_x(x)}{dx}$. The boundary conditions can now be given in terms of concentrated torques and the supports at the shaft ends, as shown in Figure 4.21.

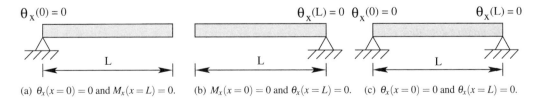

(a) $\theta_x(x = 0) = 0$ and $M_x(x = L) = 0$.   (b) $M_x(x = 0) = 0$ and $\theta_x(x = L) = 0$.   (c) $\theta_x(x = 0) = 0$ and $\theta_x(x = L) = 0$.

Figure 4.21   Boundary conditions in terms of the angle of twist and the twisting moment.

From (4.51), we have

$$\frac{d\theta_x(x)}{dx} = \frac{M_x(x)}{G(x)I_p(x)}, \tag{4.54}$$

which, when substituted in (4.45), obtains the tangential shear stress expression in terms of the twisting moment

$$\tau_t(x,r) = \frac{M_x(x)}{I_p(x)}r. \tag{4.55}$$

Thus, there is a linear variation of the shear stress in the shaft cross-sections in terms of the radial coordinate r of the points. This characteristic shows that the shear stress distribution is kinematically compatible with the tangential displacement field, which also varies linearly with the r radial coordinate of the points of the cross-section. The maximum shear stress occurs at the section boundary where $r = \frac{d}{2}$. Figure 4.22 illustrates the linear behavior of the tangential shear strain.

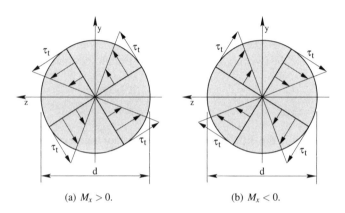

(a) $M_x > 0$.   (b) $M_x < 0$.

Figure 4.22   Distribution of tangential shear stresses in the shaft cross-sections.

Similar equations are obtained taking the Cartesian components of stress (τ_{xy}, τ_{xz}) and strain (γ_{xy}, γ_{xz}). We employ simply the expressions in (4.37) along with (4.12), (4.13), and (4.54) to obtain

$$\tau_{xy}(x,y,z) = -\frac{M_x(x)}{I_p(x)}z, \tag{4.56}$$

$$\tau_{xz}(x,y,z) = \frac{M_x(x)}{I_p(x)}y. \tag{4.57}$$

Example 4.7 *Derive the expression of the angle of twist for the shaft of Example 4.1, subjected to the angle of twist θ at the free end.*

To solve this example, the previous integration procedure is applied to equation (4.53).

- *Distributed load equation*
  As there is no distributed torque applied to the shaft, we have $m_x(x) = 0$.
- *Boundary conditions*
  Due to the rotation θ at the free shaft end $x = L$, we have the boundary condition $\theta_x(x = L) = \theta$. As the shaft is fixed at the $x = 0$ end, the angle of twist is zero and $\theta_x(x = 0) = 0$.
- *Integration of the differential equation*
  We consider the differential equation in terms of the angle of twist. The first indefinite integration results in the twisting moment expression

$$M_x(x) = GI_p\frac{d\theta_x(x)}{dx} = -m_x(x) = 0 \rightarrow M_x(x) = C_1.$$

The second integration gives the expression for the angle of twist, that is,

$$\theta_x(x) = \frac{1}{GI_p}\left(C_1 x + C_2\right),$$

with C_1 and C_2 the integration constants.

- *Determination of the integration constants*
  To calculate C_1 and C_2, the boundary conditions in terms of the angle of twist are employed. Thus,

$$\theta_x(x=0) = \frac{1}{GI_p}\left(C_1(0) + C_2\right) = 0 \rightarrow C_2 = 0,$$

$$\theta_x(x=L) = \frac{C_1(L)}{GI_p} = \theta \rightarrow C_1 = \frac{GI_p\theta}{L}.$$

- *Final equations*
  Replacing C_1 and C_2, we obtain the final equations for the twisting moment and the angle of twist, that is,

$$M_x(x) = \frac{GI_p\theta}{L},$$

$$\theta_x(x) = \frac{\theta}{L}x.$$

Hence, the expression of the angle of twist given in Example 4.1 is recovered.

□

Example 4.8 *Plot the diagrams of the twisting moment and angle of twist for the simply supported shaft of Figure 4.15, subjected to the constant distributed axial torque t_0, integrating the equilibrium differential equation.*

This is a hyperstatic shaft and the differential equation in terms of the kinematics must be considered.

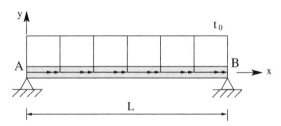

Figure 4.23 Example 4.8: simply supported shaft subjected to a constant distributed torque.

- *Distributed load equation*
  As the shaft is subjected to the constant distributed torque t_0 in the positive direction of x, we have $m_x(x) = t_0$.
- *Boundary conditions*
  As the shaft is supported at $x = 0$ and $x = L$, the angle of twist is zero at these ends. Thus, $\theta_x(x=0) = 0$ and $\theta_x(x=L) = 0$.
- *Integration of the differential equation*
  Integrating the differential equation in terms of the angle of twist, we obtain, respectively, the expressions for the twisting moment and the angle of twist.

$$GI_p\frac{d\theta_x(x)}{dx} = -m_x(x) = -t_0 \rightarrow M_x(x) = -t_0x + C_1,$$

$$\theta_x(x) = \frac{1}{GI_p}\left(-\frac{t_0}{2}x^2 + C_1x + C_2\right),$$

with C_1 and C_2 the integration constants.

- *Determination of the integration constants*
  In order to calculate C_1 and C_2, the boundary conditions in terms of the angle of twist are used, that is,

$$\theta_x(x=0) = \frac{1}{GI_p}\left(-\frac{t_0}{2}(0)^2 + C_1(0) + C_2\right) = 0 \rightarrow C_2 = 0,$$

$$\theta_x(x=L) = \frac{1}{GI_p}\left(-\frac{t_0}{2}(L)^2 + C_1(L)\right) = 0 \rightarrow C_1 = \frac{t_0}{2}L.$$

- *Final equations*
  The final expressions for the twisting moment and angle of twist are obtained substituting the C_1 and C_2 integration constants, that is,

$$M_x(x) = -\frac{t_0}{2}(2x - L),$$

$$\theta_x(x) = -\frac{t_0}{2GI_p}\left(x^2 - Lx\right).$$

- *Twisting moment and angle of twist diagrams*
  The twisting moment varies linearly and the angle of twist varies as a parabola along the shaft length. As the sign of the term x^2 is negative, the concavity of the parabola points downwards. The twisting moment and angle of twist diagrams are shown in Figure 4.24 for $L = 2$ m, $t_0 = 10$ Nm/m and $GI_p = 10^6$ Nm^2.

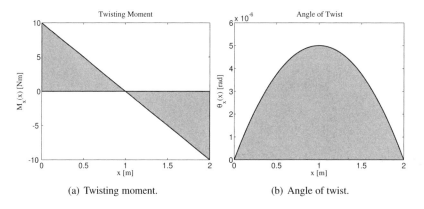

(a) Twisting moment.   (b) Angle of twist.

Figure 4.24 Example 4.8: diagrams for the twisting moment and angle of twist.

- *Support reactions*
  In this case, the support reactions M_{Ax} and M_{Bx} at the $x = 0$ and $x = 2$ m shaft ends are given by the boundary conditions of the BVP, that is,

$$M_{Ax} = -M_x(x=0) = -10 \, \text{Nm}$$

and

$$M_{Bx} = M_x(x=2) = -10 \, \text{Nm}.$$

Thus, the directions of the two reactions are leftwards and therefore contrary to the positive direction of the x axis. These values can also be obtained directly from the twisting moment diagram.

In this case, the rigid body equilibrium equation given in (4.33) cannot be applied, since the problem is statically indeterminate. Using the equilibrium equation, we have

$$M_{Ax} + M_{Bx} + \int_0^L t_0 \, dx = 0.$$

Since there are two unknowns and one rigid body equilibrium equation, the problem is indeterminate.

It should be noted that the same solution procedure was also applied to a statically indeterminate problem. This comes from the fact that the balance between internal and external loads is already represented in the differential equation (4.53).

□

Example 4.9 *Consider a steel tube with internal and external diameters equal to 5 cm and 8 cm, respectively, enclosed by an aluminum tube with 1 cm of thickness. Both metals are rigidly connected. Calculate the maximum shear stresses in the steel and aluminum tubes, due to an applied external torque of $T = 7500$ Nm, and considering $G_1 = 120$ GPa for the steel, and $G_2 = 70$ GPa to the aluminum.*

From the equilibrium, the summation of twisting moments on the steel tube (T_1) and in the aluminum tube (T_2) are equal to T, that is,

$$T = T_1 + T_2 = 7500.$$

The angles of twist for the steel and aluminum tubes are θ_1 and θ_2, respectively. Due to the kinematic compatibility, these angles of twist must be equal. Thus,

$$\theta_1 = \theta_2,$$

with

$$\theta_1 = \frac{T_1 L}{G_1 I_{p_1}} \quad and \quad \theta_2 = \frac{T_2 L}{G_2 I_{p_2}}.$$

Hence, for $\theta_1 = \theta_2$, we have

$$T_1 = T_2 \frac{G_1 I_{p_1}}{G_2 I_{p_2}}.$$

The polar moments of area are given by

$$I_{p_1} = \frac{\pi}{32}(8^4 - 5^4) = 3.41 \times 10^{-6} \, \text{m}^4 \quad and \quad I_{p_2} = \frac{\pi}{32}(9^4 - 8^4) = 24.65 \times 10^{-6} \, \text{m}^4.$$

Thus,

$$T_1 = T_2 \frac{(120 \times 10^9)(3.41 \times 10^{-6})}{(70 \times 10^9)(24.65 \times 10^{-6})} = 0.237 T_2.$$

Then, we have a system consisting of two equations and two unknowns, given by

$$\begin{cases} T_1 + T_2 & = \quad 7500 \\ T_1 - 0.237 T_2 & = \quad 0 \end{cases}.$$

Solving the previous system of equation, we have $T_1 = 1397.2$ Nm and $T_2 = 5891.6$ Nm.
The maximum shear stresses are now calculated for the maximum radius of each one of the tubes. Thus,

$$\tau_1^{max} = \frac{T_1 d_1}{2 I_{p_1}} = \frac{(1397.2)(8 \times 10^{-2})}{(2)(3.41 \times 10^{-6})} = 16.40 \, \text{MPa},$$

$$\tau_2^{max} = \frac{T_2 d_2}{2 I_{p_2}} = \frac{(5891.6)(9 \times 10^{-2})}{(2)(24.65 \times 10^{-6})} = 10.75 \, \text{MPa}.$$

□

4.10 DESIGN AND VERIFICATION

Designing a shaft means calculating the minimum diameter of the cross-section such that it remains in the elastic regime. The design considered here employs the maximum shear stress of the shaft. Similarly to the bar model, we consider the following steps in the design of shafts based on the maximum shear stress:

1. Calculate the function and plot the diagram of the twisting moment $M_x(x)$ by integrating the differential equation (4.31) for statically determinate problems, and equation (4.52) for statically indeterminate problems.
2. Based on this diagram, find the critical section, i.e., the section subjected to the largest absolute twisting moment, denoted by $M_x^{\max}$.
3. From expression (4.55), the maximum shear stress $\tau_t^{\max}$ is on the boundary of the critical cross-section. For a circular section, $r = \dfrac{d}{2}$ and therefore

$$\tau_t^{\max} = \frac{M_x^{\max}}{I_p}\frac{d}{2}. \tag{4.58}$$

As the cross-section dimensions are still unknown, the terms involving the diameter are grouped in the torsional strength modulus W_x, given by

$$W_x = I_p\frac{2}{d} = \frac{\pi d^3}{16}. \tag{4.59}$$

Note that the SI unit for W_x is m^3.
Based on that definition, we can rewrite expression (4.58) as

$$\tau_t^{\max} = \frac{M_x^{\max}}{W_x}. \tag{4.60}$$

4. The shaft remains in the elastic regime if the maximum shear stress is below the admissible shear stress $\bar{\tau}$ of the material, i.e.,
$$\tau_t^{\max} \leq \bar{\tau}. \tag{4.61}$$

The minimum torsional strength modulus is obtained by taking the equality of the previous expression. Therefore,

$$W_x = \frac{M_x^{\max}}{\bar{\tau}}. \tag{4.62}$$

Once W_x is known, the diameter d can be calculated as

$$d = \left(\frac{16W_x}{\pi}\right)^{1/3}. \tag{4.63}$$

5. Using standardized tables, we select the cross-section whose diameter is larger than or equal to the minimum calculated diameter.

In the case of verification, the dimensions of the cross-section are known and it is necessary to check if the shaft remains in the elastic range when subjected to a certain loading. For this purpose, the maximum shear stress $\tau_t^{\max}$ is calculated using (4.60). With this maximum stress, we check if it is less than the admissible shear stress as indicated in equation (4.61). In this case, the shaft remains in the elastic regime. If condition (4.61) is not satisfied, the shaft must be redesigned applying the above procedure.

Example 4.10 *Calculate the diameter of the steel shaft of Example 4.6 for $\bar{\tau} = 120$ MPa.*

The critical section is the one with the largest absolute value of the twisting moment. From Figure 4.18, $x = 0$ is the critical session and $M_x^{max} = 1500$ Nm.

The next step is to calculate the maximum shear stress corresponding to M_x^{max}. For this purpose, the following expression is used:

$$\tau_t^{max} = \frac{M_x^{max}}{W_x} = \frac{1500}{W_x}.$$

To keep the shaft in the elastic regime, this stress must not exceed the admissible shear stress $\bar{\tau}_t$ of the material, that is,

$$\bar{\tau}_t = \tau_t^{max} = \frac{M_x^{max}}{W_x} = \frac{1500}{W_x}.$$

Thus, we have the following value for the torsional strength modulus:

$$W_x = \frac{M_x^{max}}{\bar{\tau}_t} = \frac{1500}{120 \times 10^6} = 1.25 \times 10^{-5} \text{ m}^3.$$

For a circular section, $W_x = \pi d^3/16$ and thus the diameter of the shaft is

$$d = \left(\frac{16 W_x}{\pi}\right)^{1/3} = 4.0 \text{ cm}.$$

This is the minimum diameter to maintain the shaft in the elastic regime. Using a table of standard profiles, we select a shaft with diameter greater than or equal to 4.0 cm.
□

Example 4.11 *A circular shaft of a rotor must transmit 10 hp of power, operating at 3600 rpm. For the admissible shear stress of 70 MPa, calculate the minimum diameter of the shaft.*

Initially, the power P of the rotor is expressed in Watts and its angular velocity f in Hz, that is,

$$P = 10 \text{ hp}\frac{746 \text{ Nm/s}}{1 \text{ hp}} = 7460 \text{ N}\frac{\text{m}}{\text{s}} = 7460 \text{ W},$$

$$f = 3600\frac{1 \text{ Hz}}{60 \text{ rpm}} = 60 \text{ Hz} = 60 \text{ s}^{-1}.$$

The torque can be calculated using the following relation between the power and the angular velocity of the shaft:

$$T = \frac{P}{2\pi f} = \frac{7460}{(2\pi)(60)} = 19.80 \text{ Nm}.$$

From the torque and the maximum shear stress of the shaft, the minimum diameter d is calculated as

$$\tau_t^{max} = 70 = \frac{Td}{2I_p} = 19.80\frac{16}{\pi d^3} \rightarrow d = 11.30 \text{ mm}.$$

□

4.11 SINGULARITY FUNCTIONS FOR EXTERNAL LOADING REPRESENTATION

Similarly to the bar case, the external loads compatible with the circular torsion kinematics are given by a continuous function $m_x(x)$, representing the distributed torque applied along the shaft length, and the concentrated torques T_0 and T_L, acting at the shaft ends. Thus, concentrated torques in the internal sections of the shaft are not compatible with the presented model.

The use of concentrated torques along the shaft length requires the use of singularity functions for $m_x(x)$. For example, the concentrated torques shown in Figure 4.25(a) are denoted in the distributed loading function as

$$m_x(x) = T_1 < x - \frac{L}{3} >^{-1} + T_2 < x - \frac{2L}{3} >^{-1}.$$

The exponent -1 is used again for concentrated torques, analogously to the case of axial concentrated forces for bars. The above expression gives the distributed torque to each of the three segments of interest in the study of the shaft as

$$m_x(x) = \begin{cases} 0 & 0 \leq x < L/3 \\ T_1 \left(x - \frac{L}{3}\right)^{-1} & L/3 \leq x < 2L/3 \\ T_1 \left(x - \frac{L}{3}\right)^{-1} + T_2 \left(x - \frac{2L}{3}\right)^{-1} & 2L/3 \leq x < L \end{cases}.$$

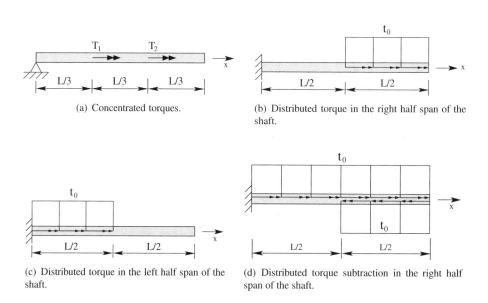

(a) Concentrated torques.

(b) Distributed torque in the right half span of the shaft.

(c) Distributed torque in the left half span of the shaft.

(d) Distributed torque subtraction in the right half span of the shaft.

Figure 4.25   Singularity functions for the circular torsion.

The singularity function notation is also useful when representing distributed loadings applied to parts of the shaft. Consider the distributed torque with constant intensity t_0 applied to the second half of Figure 4.25(b). In this case, the function for the distributed torque is written as

$$m_x(x) = t_0 < x - \frac{L}{2} >^0.$$

As $\left\langle x - \frac{L}{2} \right\rangle^0 = 0$ for $x < \frac{L}{2}$ and $\left\langle x - \frac{L}{2} \right\rangle^0 = 1$ for $x \geq \frac{L}{2}$, we have

$$m_x(x) = \begin{cases} 0 & 0 \leq x < \frac{L}{2} \\ t_0 & \frac{L}{2} \leq x < L \end{cases}.$$

Consquently, we have the correct representation of the distributed torque on the shaft.

For a constant distributed torque t_0 acting in the first half of the shaft, as illustrated in Figure 4.25(c), the $m_x(x)$ function is given by

$$m_x(x) = t_0 < x - 0 >^0 - t_0 < x - \frac{L}{2} >^0.$$

The first term indicates that t_0 acts along the entire shaft, because $t_0 < x - 0 >^0 = t_0$ for $x \geq 0$. As there is no distributed torque in the second half of the shaft, we should subtract $t_0 < x - \frac{L}{2} >^0 = t_0$ for $x \geq \frac{L}{2}$, observing that $t_0 < x - \frac{L}{2} >^0 = 0$ for $x < \frac{L}{2}$. Thus, expanding the $m_x(x)$ function, we have

$$m_x(x) = \begin{cases} t_0(x-0)^0 = t_0 & 0 \leq x < \frac{L}{2} \\ t_0(x-0)^0 - t_0 \left(x - \frac{L}{2}\right)^0 = 0 & \frac{L}{2} \leq x < L \end{cases}.$$

Hence, the distributed torque can be denoted in this case as

$$m_x(x) = t_0 - t_0 < x - \frac{L}{2} >^0.$$

Example 4.12 *The simply supported shaft shown in Figure 4.26 is loaded by two twisting moments M_1 and M_2. Calculate the support reactions.*

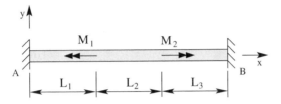

Figure 4.26   Example 4.12: shaft with concentrated torques.

The shaft is statically indeterminate, because there are two unknown support reactions M_{A_x} and M_{B_x}, at the A and B ends, and only one equilibrium equation, $\sum m_x = 0$. Thus, the second-order differential equation in terms of the angle of twist must be integrated.

- *Distributed load equation*
  The expression for the distributed torque is given by

$$m_x(x) = -M_1 < x - L_1 >^{-1} + M_2 < x - (L_1 + L_2) >^{-1}.$$

- *Boundary conditions*
  As the shaft is supported in both ends, the respective angles of twist are zero and hence $\theta_x(x=0) = 0$ and $\theta_x(x=L) = 0$.
- *Integration of the differential equation*
  The differential equation in terms of the angle of twist is considered

$$GI_p \frac{d^2 \theta_x(x)}{dx^2} = -m_x(x) = M_1 < x - L_1 >^{-1} - M_2 < x - (L_1 + L_2) >^{-1}.$$

The first integration results in the twisting moment expression

$$M_x(x) = GI_p \frac{d\theta_x(x)}{dx} = M_1 < x - L_1 >^0 - M_2 < x - (L_1 + L_2) >^0 + C_1$$

The second integration results in the expression of the angle of twist

$$\theta_x(x) = \frac{1}{GI_p} \left(M_1 < x - L_1 >^1 - M_2 < x - (L_1 + L_2) >^1 + C_1 x + C_2 \right).$$

- *Determination of the integration constants*
  Integration constants C_1 and C_2 are determined using the boundary conditions in terms of the angles of twist at the shaft ends. Thus,

$$\theta_x(x=0) = \frac{1}{GI_p}\left((0)+(0)+C_1(0)+C_2\right) = 0 \rightarrow C_2 = 0,$$

$$\theta_x(x=L) = \frac{1}{GI_p}\left(M_1(L-L_1)^1 - M_2(L-L_2)^1 + C_1(L)\right) = 0.$$

For $L = L_1 + L_2 + L_3$, we have

$$C_1 = \frac{M_2 L_3 - M_1(L_2 + L_3)}{(L_1 + L_2 + L_3)}.$$

- *Final equations*
  Substituting the C_1 and C_2 integration constants, the final equations for the twisting moment and angle of twist are given, respectively, by

$$M_x(x) = M_1 <x - L_1>^0 - M_2 <x - (L_1 + L_2)>^0 + \frac{M_2 L_3 - M_1(L_2 + L_3)}{(L_1 + L_2 + L_3)},$$

$$\theta_x(x) = \frac{1}{GI_p}\left(M_1 <x - L_1>^1 - M_2 <x - (L_1 + L_2)>^1 + \frac{M_2 L_3 - M_1(L_2 + L_3)}{(L_1 + L_2 + L_3)}x\right).$$

- The M_{A_x} and M_{B_x} support reactions are determined from the boundary conditions of the BVP in terms of the twisting moment as

$$M_{A_x} = -M_x(x=0) = (0) - (0) + \frac{M_2 L_3 - M_1(L_2 + L_3)}{(L_1 + L_2 + L_3)},$$

$$M_{B_x} = M_x(x=L) = M_1 - M_2 + \frac{M_2 L_3 - M_1(L_2 + L_3)}{(L_1 + L_2 + L_3)}.$$

Hence,

$$M_{A_x} = \frac{M_1(L_2 + L_3)}{L} - \frac{M_2 L_3}{L} \quad and \quad M_{B_x} = \frac{M_1 L_1}{L} - \frac{M_2(L_1 + L_2)}{L}.$$

□

Example 4.13 *Consider the shaft shown in Figure 4.27 subjected to the indicated loading. Design the shaft assuming a circular section with a diameter d and a hollow circular section with internal and external diameters d_i and d_e, with $d_i/d_e = 0.7$. Compare the final mass and angles of twist for both designed shafts. Use $L = 2$ m, $t_0 = 2000$ Nm/m, $T = 500$ Nm, and $\bar{\tau} = 50$ MPa.*

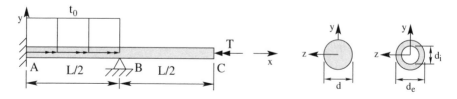

Figure 4.27   Example 4.13: shaft with solid and hollow cross-sections.

- *Distributed load equation*

  This is a statically indeterminate problem, and there is a support in the middle span of the shaft. To solve this problem, we include the M_{Bx} reaction at the B support in the distributed load equation. We employ also the additional constraint of zero angle of twist at the $x = \frac{L}{2}$ section.

  The external distributed torque in the shaft is indicated using the singular function notation as

  $$m_x(x) = t_0 - t_0 < x - \frac{L}{2} >^0 + M_{Bx} < x - \frac{L}{2} >^{-1}.$$

  The second term is subtracted to indicate that the distributed torque is zero at the second half of the shaft, analogously to Figure 4.25(c).

- *Boundary conditions and additional constraint*

  As the shaft is supported at $x = 0$ section, the angle of twist is zero, i.e., $\theta_x(x = 0) = 0$. The twisting moment at the right end must be equal to the negative external torque, that is, $M_x(x = L) = -T$. Furthermore, due to the support at the middle of the shaft, we have the additional constraint of zero angle of twist at this section, namely $\theta_x\left(x = \frac{L}{2}\right) = 0$. This constraint provides an additional equation, allowing the calculation of the two integration constants and the support reaction in B.

- *Integration of the differential equation*

  As the problem is statically indeterminate, the integration procedure is applied to the differential equation in terms of the angle of twist. Thus,

  $$GI_p \frac{d^2\theta_x(x)}{dx^2} = -m_x(x) = -t_0 + t_0 < x - \frac{L}{2} >^0 - M_{Bx} < x - \frac{L}{2} >^{-1}.$$

  The first integration gives the twisting moment expression

  $$M_x(x) = GI_p \frac{d\theta_x(x)}{dx} = -t_0 x + t_0 < x - \frac{L}{2} >^1 - M_{Bx} < x - \frac{L}{2} >^0 + C_1.$$

  The second integration results in the angle of twist expression

  $$\theta_x(x) = \frac{1}{GI_p}\left(-\frac{t_0}{2}x^2 + \frac{t_0}{2} < x - \frac{L}{2} >^2 - M_{Bx} < x - \frac{L}{2} >^1 + C_1 x + C_2\right).$$

- *Determination of the integration constants and the M_{Bx} reaction*

  Applying the boundary conditions and the additional constraint, we obtain the following expressions relating the integration constants C_1 and C_2 and the support reaction M_{Bx}:

  $$\theta_x(x = 0) = \frac{1}{GI_p}\left[-\frac{t_0}{2}(0) + \frac{t_0}{2}(0) - M_{Bx}(0) + C_1(0) + C_2\right] = 0$$

  $$\rightarrow \quad C_2 = 0,$$

  $$M_x(x = L) = -t_0 L + t_0\left(L - \frac{L}{2}\right)^1 - M_{Bx}\left(L - \frac{L}{2}\right)^0 + C_1 = -T$$

  $$\rightarrow \quad -M_{Bx} + C_1 = -T + t_0\frac{L}{2},$$

  $$\theta_x\left(x = \frac{L}{2}\right) = \frac{1}{GI_p}\left[-\frac{t_0}{2}\frac{L}{2} + \frac{t_0}{2}\left(\frac{L}{2} - \frac{L}{2}\right)^1 - M_{Bx}\left(\frac{L}{2} - \frac{L}{2}\right)^1 + C_1\frac{L}{2}\right] = 0$$

  $$\rightarrow \quad C_1 = \frac{t_0 L}{4}.$$

  Substituting C_1 in the second expression above, we obtain

  $$M_{Bx} = T - \frac{t_0 L}{4}.$$

- *Final equations*
  Substituting the integration constants and the support reaction, we obtain the final equations for the twisting moment and angle of twist, that is,

$$M_x(x) = -t_0 x + t_0 < x - \frac{L}{2} >^1 - \left(T - \frac{t_0 L}{4} \right) < x - \frac{L}{2} >^0 + \frac{t_0 L}{4},$$

$$\theta_x(x) = \frac{1}{GI_p} \left[-\frac{t_0}{2} x + \frac{t_0}{2} < x - \frac{L}{2} >^1 - \left(T - \frac{t_0 L}{4} \right) < x - \frac{L}{2} >^1 + \frac{t_0 L}{4} x \right].$$

- *Twisting moment diagram*
  Substituting the given values, the twisting moment expression is

$$M_x(x) = -2000x + 2000 < x - 1 >^1 + 500 < x - 1 >^0 + 1000.$$

To plot the diagram, the previous expression for the twisting moment is written for both shaft segments. Thus,

$$M_x(x) = \begin{cases} -2000x + 1000 & 0 \le x < 1 \\ -2000x + 2000(x - 1) + 500 + 1000 = -500 & 1 \le x \le 2 \end{cases}.$$

Calculating the previous expression for each of the limit sections of both segments, we obtain $M_x(x \to 0^+) = 1000$ Nm, $M_x(x \to 1^-) = -1000$ Nm, $M_x(x \to 1^+) = -500$ Nm, and $M_x(x \to 2^-) = -500$ Nm. The twisting moment diagram is illustrated in Figure 4.28.

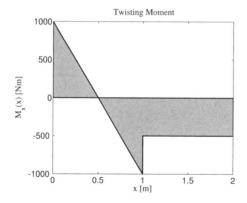

Figure 4.28 Example 4.13: twisting moment diagram.

- *Design*
  In the design procedure, the critical section is considered, that is, the section with the largest absolute value for the twisting moment. In this case, the critical sections are $x = 0$ and $x = 1$ m, with $M_x^{max} = 1000$ Nm.
  In the design, the maximum shear stress must be equal to the admissible shear stress of the material. Thus,

$$\tau^{max} = \frac{M_x^{max}}{W_x} = \bar{\tau}.$$

Hence, the minimum torsional strength modulus to maintain the shaft in the elastic range is

$$W_x = \frac{1000}{100 \times 10^6} = 1.0 \times 10^{-5} \text{ m}^3.$$

For a circular section with diameter d, $W_x = \dfrac{\pi d^3}{16}$. Thus,

$$d = \left(\frac{16W_x}{\pi}\right)^{\frac{1}{3}} = 4.67 \text{ cm.}$$

For a hollow circular section with internal and external diameters d_i and d_e, the torsional strength modulus is given by

$$W_x = \frac{I_p}{\frac{d_e}{2}} = \frac{\frac{\pi}{32}(d_e^4 - d_i^4)}{\frac{d_e}{2}} = \frac{\pi}{16}\frac{d_e^4 - d_i^4}{d_e}.$$

Substituting $\dfrac{d_i}{d_e} = 0.7$, we determine the external diameter d_e as

$$d_e = \left[\frac{16W_x}{\pi(1 - 0.7^4)}\right]^{\frac{1}{3}} = 5.12 \text{ cm.}$$

Hence, $d_i = 3.58$ cm.

The masses of the shafts with solid and hollow cross-sections, denoted as m_s and m_h, are given, respectively, by $m_s = \rho V_s$ and $m_h = \rho V_h$, where ρ is the material density; V_s and V_h are the shaft volumes. Thus, the relation between the masses is the following:

$$\frac{m_s}{m_h} = \frac{V_s}{V_h} = \frac{L(\frac{\pi}{4})d^2}{L(\frac{\pi}{4})(d_e^2 - d_i^2)} = \frac{d^2}{(d_e^2 - d_i^2)} = 1.63.$$

Thus, the mass of the shaft with the solid section is about 63% larger than the mass of the hollow section.

The shafts with solid and hollow sections have the same expressions of the angle of twist. However, the polar moments of area are different for them. The relation between the angles of twist is given by

$$\frac{\theta_{x_s}}{\theta_{x_h}} = \frac{I_{p_h}}{I_{p_s}} = \frac{d_e^4 - d_i^4}{d^4} = 1.10.$$

Thus, not only the mass of the solid shaft is larger than the hollow shaft, its angle of twist is also 10% larger.

□

Example 4.14 *Consider the shaft of Figure 4.29 fixed at end E and subjected to torques at sections B and D. Calculate the twist angle at point A. The cross-section of the AC segment is solid and has a diameter of 25 mm. Segment CE has a hollow section with external and internal diameters equal to 50 mm and 25 mm, respectively. Use $G = 80$ GPa, $T_1 = 150$ Nm, and $T_2 = 1000$ Nm.*

The cross-section of the shaft is not uniform in this example. There are two polar moments of area I_{p_1} and I_{p_2}, corresponding to the AC and CE segments. To solve the problem, the shaft is split in two segments, AC and CE, as illustrated in Figure 4.30. The differential equation in terms of the angle of twist must be integrated for each of the segments. The kinematic compatibility at the interface of segments must also be considered. Therefore, at the interface of segments, we have the same angle of twist θ_1 and twisting moment M_{x_1}.

The AC segment of the shaft is considered first ($0 < x \leq 450$ mm). In this case, the unknowns are the integration constants C_1 and C_2 and the angle of twist θ_1 at the interface of both segments. The previous integration procedure is applied to each segment of the shaft.

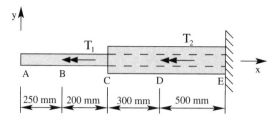

Figure 4.29   Example 4.14: shaft with nonuniform cross-sections.

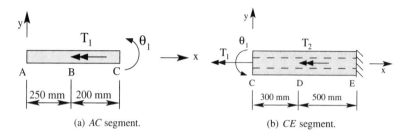

(a) *AC* segment.   (b) *CE* segment.

Figure 4.30   Example 4.12: shaft segments.

- *Distributed load equation*

$$m_{x_1}(x) = -T_1 < x - 250 >^{-1}.$$

- *Boundary conditions*

$$M_{x_1}(x = 0) = 0 \quad and \quad \theta_{x_1}(x = 450) = \theta_1.$$

- *Integration of the differential equation*

$$GI_{p_1} \frac{d^2\theta_{x_1}(x)}{dx^2} = -m_{x_1}(x) = T_1 < x - 250 >^{-1}.$$

The first integration results in the twisting moment in the AC segment, that is,

$$M_{x_1}(x) = T_1 < x - 250 >^0 + C_1.$$

The second integration gives the angle of twist

$$\theta_{x_1}(x) = \frac{1}{GI_{p_1}}\left(T_1 < x - 250 >^1 + C_1 x + C_2\right).$$

- *Determination of C_1 and C_2*
  Applying the boundary conditions, we determine the integration constants

$$M_{x_1}(x = 0) = (0) + C_1 = 0 \rightarrow C_1 = 0,$$

$$\theta_{x_1}(x = 450) = \theta_1 = \frac{1}{GI_{p_1}}\left[T_1(450 - 250)^1 + C_2\right] \rightarrow C_2 = GI_{p_1}\theta_1 - 200T_1.$$

- *Final equations*
  Substituting C_1 and C_2, the equations for the twisting moment and angle of twist are, respectively,

$$M_{x_1}(x) = T_1 < x - 250 >^0,$$

$$\theta_{x_1}(x) = \frac{1}{GI_{p_1}}\left(T_1 < x - 250 >^1 + GI_{p_1}\theta_1 - 200T_1\right).$$

The segment CE (450 mm $< x \leq$ 1250 mm) of the shaft is considered now. In this case, the unknowns are the integration constants C_3 and C_4, as well as the angle of twist θ_1 at the interface.

- Distributed load equation

$$m_{x_2}(x) = -T_2 < x - 750 >^{-1}.$$

- Boundary conditions

$$M_{x_2}(x = 450) = T_1 \quad and \quad \theta_{x_2}(x = 1250) = 0.$$

- Integration of the differential equation

$$GI_{p2} \frac{d^2 \theta_{x_2}(x)}{dx^2} = -m_{x_2}(x) = T_2 < x - 750 >^{-1}.$$

The first integration results in the twisting moment in the CE segment

$$M_{x_2}(x) = T_2 < x - 750 >^0 + C_3.$$

The second integration results in the angle of twist

$$\theta_{x_2}(x) = \frac{1}{GI_{p2}} \left(T_2 < x - 750 >^1 + C_3 x + C_4 \right).$$

- Determination of C_3 and C_4
  The integration constants are calculated applying the boundary conditions, that is,

$$
\begin{aligned}
M_{x_2}(x = 450) &= (0) + C_3 = T_1 \rightarrow C_3 = T_1, \\
\theta_{x_2}(x = 1250) &= \frac{1}{GI_{p2}} [T_2(1250 - 750) + T_1(1250) + C_4] = 0 \\
&\rightarrow C_4 = -1250T_1 - 500T_2.
\end{aligned}
$$

- Final equations
  Substituting C_3 and C_4, the equations for the twisting moment and angle of twist are determined, respectively, by

$$M_{x_2}(x) = T_2 < x - 750 >^0 + T_1,$$

$$\theta_{x_2}(x) = \frac{1}{GI_{p2}} \left(T_2 < x - 750 >^1 + T_1 x - 1250T_1 - 500T_2 \right).$$

The angle of twist θ_1 can now be calculated taking the previous equation for $x = 450$ mm. Hence,

$$\theta_1 = \theta_{x_2}(x = 450) = \frac{1}{GI_{p2}} [(0) + T_1(450) - 1250T_1 - 500T_2] = \frac{1}{GI_{p2}} (-800T_1 - 500T_2).$$

Substituting θ_1 in the equation of the angle of twist for the AC segment, we obtain the following expression for $\theta_{x_1}(x)$:

$$\theta_{x_1}(x) = \frac{T_1}{GI_{p1}} \left(< x - 250 >^1 - 200 \right) + \frac{1}{GI_{p2}} (-800T_1 - 500T_2).$$

The polar moments of area are given by

$$I_{p1} = \frac{\pi d^4}{32} = \frac{\pi (25)^4}{32} = 38.3 \times 10^3 \text{ mm}^4,$$

$$I_{p2} = \frac{\pi}{32}(d_e^4 - d_i^4) = \frac{\pi}{32}(50^4 - 25^4) = 575.0 \times 10^3 \text{ mm}^4.$$

Substituting the numerical values, the angle of twist at A is determined as

$$
\begin{aligned}
\theta_{x_1}(x=0) &= -\frac{(200)(150 \times 10^3)}{(80 \times 10^3)(38.3 \times 10^3)} - \frac{\left[(800)(150 \times 10^3) + (500)(1000 \times 10^3)\right]}{(80 \times 10^3)(575.0 \times 10^3)} \\
&= -0.0233 \text{ rad.}
\end{aligned}
$$

Another way of determining the angle of twist at A is summing the angles of twist for each of the shaft segments using equation (4.38). From the twisting moment diagram illustrated in Figure 4.31, we have

$$
\begin{aligned}
\theta_{x_1}(x=0) &= \theta_{x_{AB}} + \theta_{x_{BC}} + \theta_{x_{CD}} + \theta_{x_{DE}} \\
&= \frac{M_{x_{AB}}L_{AB}}{I_{P_{AB}}G} + \frac{M_{x_{BC}}L_{BC}}{I_{P_{BC}}G} + \frac{M_{x_{CD}}L_{CD}}{I_{P_{CD}}G} + \frac{M_{x_{DE}}L_{DE}}{I_{P_{DE}}G} \\
&= -\frac{(0)(250)}{(38.3 \times 10^3)(80000)} - \frac{(150000)(200)}{(38.3 \times 10^3)(80000)} \\
&\quad - \frac{(150000)(300)}{(575.0 \times 10^3)(80000)} - \frac{(1150000)(500)}{(575.0 \times 10^3)(80000)} = -0.0233 \text{ rad.}
\end{aligned}
$$

The sign is negative because the angle of twist at $x = 0$ is in the negative direction of the x axis, according to the applied torques.

As the shaft is isostatic, the support is at the right end and no torque is applied at the left end, it is known that the twisting moment at the interface of the two segments of the shaft is equal to T_1. Thus, we may use T_1 directly as a boundary condition to the second shaft segment.

In Example 3.18, the boundary condition was applied in terms of the normal force $N_{x_1}(x = L_1) = N_1$ at the interface of the segments for the AB bar. The kinematic boundary condition $u_{x_1}(x = L_1) = u_1$ was applied in the second bar segment. In this problem, a kinematic boundary condition was applied to the AC segment, and a twisting moment boundary condition was applied to the CE segment. The general conclusion is that we always need a kinematic boundary condition and another in terms of the loading at the interface of the elements to achieve a correct kinematic and loading couplings at the interface.

$\square$

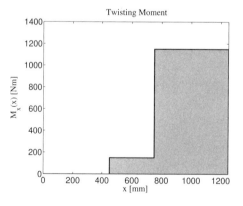

Figure 4.31 Example 4.14: twisting moment diagram.

4.12 SUMMARY OF THE VARIATIONAL FORMULATION OF SHAFTS

A summary of the variational formulation of circular torsion is presented in this section.

The set of possible kinematic actions $\mathscr{V}$ consists of the displacement field $\mathbf{u}(x,y,z)$ given in (4.5), where $\theta_x(x)$ is a smooth function of x. Thus,

$$\mathscr{V} = \{\mathbf{u}, u_x = 0,\ u_y(x,z) = -z\theta_x(x),\ u_z(x,y) = y\theta_x(x) \text{ and } \theta_x(x) \text{ is a smooth function}\}. \quad (4.64)$$

For a shaft free of kinematic restrictions, all elements $\mathbf{u} \in \mathscr{V}$ are also admissible actions because there are no supports at the ends. When a constraint is present, only the subset Kin_v of $\mathscr{V}$, formed by the functions that respect the kinematic constraints, constitutes the admissible displacement actions.

The set of the strain actions $\mathscr{W}$, which are compatible with the circular torsion kinematics, consists of the continuous functions $\gamma_{xy}(x,z)$ and $\gamma_{xz}(x,y)$. The strain operator $\mathscr{D} = \dfrac{\partial}{\partial x}$ relates the sets of possible displacement and strain actions $\mathscr{V}$ and $\mathscr{W}$. Therefore, applying $\mathscr{D}$ to a displacement action $\mathbf{u}(x,y,z) = \begin{bmatrix} u_y(x,z) \\ u_z(x,y) \end{bmatrix}$, we have the respective strain components, that is,

$$\mathscr{D}\mathbf{u} = \frac{\partial}{\partial x}\begin{bmatrix} u_y(x,z) \\ u_z(x,y) \end{bmatrix} = \begin{bmatrix} \gamma_{xy}(x,z) \\ \gamma_{xz}(x,y) \end{bmatrix}.$$

The operator $\mathscr{D}$ can also be indicated as

$$\mathscr{D}:\quad \mathscr{V} \to \mathscr{W}$$
$$\begin{bmatrix} 0 \\ u_y(x,z) \\ u_z(x,y) \end{bmatrix} \to \frac{\partial}{\partial x}\begin{bmatrix} 0 \\ u_y(x,z) \\ u_z(x,y) \end{bmatrix} = \begin{bmatrix} 0 \\ \gamma_{xy}(x,z) \\ \gamma_{xz}(x,y) \end{bmatrix}. \quad (4.65)$$

Considering the tangential displacement component $u_t(x,r)$, the set of compatible strain actions $\mathscr{W}$ is constituted of continuous functions $\gamma_t(x,r)$ which represents the tangential strain. In this case, the strain operator $\mathscr{D}$ is also given by $\mathscr{D} = \dfrac{d}{dx}$ such that

$$\mathscr{D}:\quad \mathscr{V} \to \mathscr{W}$$
$$u_t(x,r) = r\theta_x(x) \to \gamma_t(x,r) = r\frac{d\theta_x(x)}{dx}. \quad (4.66)$$

The subset $\mathscr{N}(\mathscr{D})$ of rigid actions is composed of the displacement actions given in (4.5) with θ_x constant. Hence, the set $\mathscr{N}(\mathscr{D})$ is defined as

$$\mathscr{N}(\mathscr{D}) = \{\mathbf{u};\mathbf{u} \in \mathscr{V} \mid \theta_x(x) = \theta_x \text{ constant}\}. \quad (4.67)$$

The internal work allows the association of the $\gamma_{xy}(x,z)$ and $\gamma_{xz}(x,y)$ strain components with the respective $\tau_{xy}(x,z)$ and $\tau_{xz}(x,y)$ stress components, representing the internal loads in the shaft. As the strains are angular, $\tau_{xy}(x,z)$ and $\tau_{xz}(x,y)$ are called shear stresses. They act at each point of the x plane, in the y and z directions, respectively. As there are two strain components, each one is associated to its respective stress component. Relation $\tau_{xy}(x,z)\gamma_{xy}(x,z) + \tau_{xz}(x,y)\gamma_{xz}(x,y)$ represents the internal work density for each point of the shaft. These contributions must be added for each point. As the shaft is continuous, that is, it has an infinite number of points, the summation is written as a volume integral, that is,

$$W_i = \int_V [\tau_{xy}(x,z)\gamma_{xy}(x,z) + \tau_{xz}(x,y)\gamma_{xz}(x,y)]\,dV. \quad (4.68)$$

Substituting the expressions for the strain components given in (4.13), we obtain

$$
\begin{aligned}
W_i &= \int_V \left[\tau_{xy}(x,z) \left(-z\frac{d\theta_x(x)}{dx} \right) + \tau_{xz}(x,y) \left(y\frac{d\theta_x(x)}{dx} \right) \right] dV \\
&= \int_V \left[-z\tau_{xy}(x,z) + y\tau_{xz}(x,z) \right] \frac{d\theta_x(x)}{dx} dV.
\end{aligned}
$$

The previous volume integral can be rewritten as the product of the integrals along the length and cross-section area as follows

$$
W_i = \int_0^L \left(\int_A \left[-z\tau_{xy}(x,z) + y\tau_{xz}(x,y) \right] dA \right) \frac{d\theta_x(x)}{dx} dx. \tag{4.69}
$$

The area integral represents the twisting moment

$$
M_x(x) = \int_A \left[-z\tau_{xy}(x,z) + y\tau_{xz}(x,y) \right] dA. \tag{4.70}
$$

Therefore, the set $\mathscr{W}'$ of internal loads is constituted of continuous scalar functions $M_x(x)$, characterizing the twisting moment in each cross-section of the x axis.

From equation (4.31), the equilibrium operator $\mathscr{D}^*$ between the external and internal loadings is defined. This operator can be written as

$$
\mathscr{D}^* M_x(x) = \begin{cases} -\dfrac{d}{dx} M_x(x) & \text{at } x \in (0,L) \\ - M_x(x)|_{x=0} & \text{at } x = L \\ M_x(x)|_{x=L} & \text{at } x = 0 \end{cases} . \tag{4.71}
$$

The operator $\mathscr{D}^*$ maps the sets of internal and external loadings $\mathscr{W}'$ and $\mathscr{V}'$. The set of external loads $\mathscr{V}'$ is characterized by the continuous scalar function $m_x(x)$, indicating the distributed torque along the length, and the concentrated torques T_0 and T_L at the shaft ends and treated as boundary conditions of the model. Therefore, we denote the equilibrium operator $\mathscr{D}^*$ as

$$
\mathscr{D}^* : \ \mathscr{W} \to \mathscr{V}'
$$

$$
M_x(x) \to \mathscr{D}^* M_x(x) = \begin{cases} -\dfrac{d}{dx} M_x(x) = m_x(x) & \text{at } x \in (0,L) \\ - M_x(x)|_{x=0} = T_0 & \text{at } x = L \\ M_x(x)|_{x=L} = T_L & \text{at } x = 0 \end{cases} . \tag{4.72}
$$

The PVW represents the balance of internal and external works for any virtual action $\delta\theta_x(x)$ from the deformed position of the shaft. Thus, the external loads in the shaft must be such that the work done by them for a virtual rotation $\delta\theta_x(x)$ balances the work of the internal loads to the same virtual action.

Figure 4.32 illustrates the variational formulation of the circular torsion model.

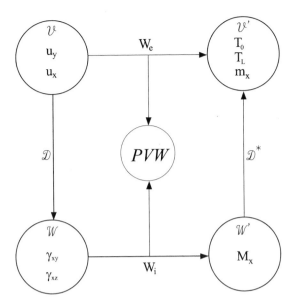

Figure 4.32 Variational formulation for the circular torsion model.

4.13 APPROXIMATED SOLUTION

Analogously to the bar model, the weak form of the circular torsion model is obtained by multiplying the differential equation given in (4.52) by the test function $v(x)$ and integrating in the shaft length L, that is,

$$\int_0^L \frac{d}{dx}\left(G(x)I_p(x)\frac{d\theta_x(x)}{dx}\right)v(x)dx = -\int_0^L m_x(x)v(x)dx.$$

Integrating the above expression by parts and using the boundary conditions given in the BVP (4.31), we obtain the following equilibrium integral equation or weak form for the circular torsion:

$$\int_0^L G(x)I_p(x)\frac{d\theta_x(x)}{dx}\frac{dv(x)}{dx}dx = \int_0^L m_x(x)v(x)dx + T_L v(L) + T_0 v(0). \tag{4.73}$$

The angle of twist $\theta_x(x)$ and the test function $v(x)$ are approximated with the Galerkin method by the following linear combinations of the global interpolation functions $\{\Phi_i(x)\}_{i=1}^N$:

$$\theta_{x_N}(x) = \sum_{i=1}^N a_i\Phi_i(x) \quad \text{and} \quad v_N(x) = \sum_{j=1}^N b_j\Phi_j(x).$$

Substituting the previous expressions in (4.73), we find, analogous to (3.108), the following system of equations

$$\sum_{i,j=1}^n \left(\int_0^L G(x)I_p(x)\Phi_i'(x)\Phi_j'(x)dx\right)a_j = \int_0^L m_x(x)\Phi_i(x)dx + T_L\Phi_i(L) + T_0\Phi_i(0). \tag{4.74}$$

Assuming $G(x) = G$ and $I_p(x) = I_p$ constants, we have the expanded form

$$
GI_p \int_0^L
\begin{bmatrix}
\Phi_1'(x)\Phi_1'(x) & \cdots & \Phi_1'(x)\Phi_j'(x) & \cdots & \Phi_1'(x)\Phi_N'(x) \\
\vdots & \cdots & \vdots & \cdots & \vdots \\
\Phi_i'(x)\Phi_1'(x) & \cdots & \Phi_i'(x)\Phi_j'(x) & \cdots & \Phi_i'(x)\Phi_N'(x) \\
\vdots & \vdots & \vdots & \vdots & \vdots \\
\Phi_N'(x)\Phi_1'(x) & \cdots & \Phi_N'(x)\Phi_j'(x) & \cdots & \Phi_N'(x)\Phi_N'(x)
\end{bmatrix}
dx
\begin{Bmatrix}
a_1 \\ \vdots \\ a_i \\ \vdots \\ a_N
\end{Bmatrix}
=
$$

$$
\int_0^L
\begin{Bmatrix}
m_x(x)\Phi_1(x) \\ \vdots \\ m_x(x)\Phi_i(x) \\ \vdots \\ m_x(x)\Phi_N(x)
\end{Bmatrix}
dx +
\begin{Bmatrix}
T_L\Phi_1(L) + T_0\Phi_1(0) \\ \vdots \\ T_L\Phi_i(L) + T_0\Phi_i(0) \\ \vdots \\ T_L\Phi_N(L) + T_0\Phi_N(0)
\end{Bmatrix},
\tag{4.75}
$$

or

$$
[K]\{a\} = \{f\},
$$

where $[K]$ is the stiffness matrix, $\{a\}$ the vector of approximation coefficients, and $\{f\}$ the load vector.

We can partition the shaft of length L in several finite elements of length $h^{(e)}$. Considering the two-node finite element, the approximations of the angle of twist and the tangential distortion in element e are given in terms of the local shape functions given in (3.122) and (3.123) as

$$
\begin{aligned}
\theta_{x_2}^{(e)}(x) &= \theta_{x_{i-1}}^{(e)}\phi_{i-1}^{(e)}(x) + \theta_{x_i}^{(e)}\phi_i^{(e)}(x) \\
&= \begin{bmatrix} \phi_{i-1}^{(e)}(x) & \phi_i^{(e)}(x) \end{bmatrix}
\begin{Bmatrix} \theta_{x_{i-1}}^{(e)} \\ \theta_{x_i}^{(e)} \end{Bmatrix} = [N^{(e)}]\{\theta_x^{(e)}\},
\end{aligned}
\tag{4.76}
$$

$$
\begin{aligned}
\gamma_{t_2}^{(e)}(x,r) &= r\frac{d\theta_{x_2}^{(e)}(x)}{dx} = r\begin{bmatrix} \dfrac{d\phi_{i-1}^{(e)}(x)}{dx} & \dfrac{d\phi_i^{(e)}(x)}{dx} \end{bmatrix}
\begin{Bmatrix} \theta_{x_{i-1}}^{(e)} \\ \theta_{x_i}^{(e)} \end{Bmatrix} \\
&= \frac{r}{h^{(e)}}\begin{bmatrix} -1 & 1 \end{bmatrix}
\begin{Bmatrix} \theta_{x_{i-1}}^{(e)} \\ \theta_{x_i}^{(e)} \end{Bmatrix} = [B^{(e)}]\{\theta_x^{(e)}\},
\end{aligned}
\tag{4.77}
$$

with $\theta_{x_{i-1}}^{(e)}$ and $\theta_{x_i}^{(e)}$ the angles of twist for each node; $[N^{(e)}]$ and $[B^{(e)}]$ are the element matrices whose coefficients are the shape functions and their first derivatives, respectively. The tangential shear stress in the element is calculated using Hooke's law given in (4.36), i.e.,

$$
\tau_{t_2}^{(e)}(x,r) = G^{(e)}\gamma_{t_2}^{(e)}(x,r).
\tag{4.78}
$$

The stiffness matrix of the shaft element is obtained as

$$
\begin{aligned}
[K^{(e)}] &= \int_{x_{i-1}^{(e)}}^{x_i^{(e)}} \int_A [B^{(e)}]^T[D][B^{(e)}]\,dA\,dx \\
&= \frac{G^{(e)}}{(h^{(e)})^2}\begin{bmatrix} -1 \\ 1 \end{bmatrix}\begin{bmatrix} -1 & 1 \end{bmatrix}\underbrace{\int_{x_{i-1}^{(e)}}^{x_i^{(e)}} dx}_{h^{(e)}}\underbrace{\int_A r^2\,dA}_{I_p^{(e)}} \\
&= \frac{G^{(e)}I_p^{(e)}}{h^{(e)}}\begin{bmatrix} 1 & -1 \\ -1 & 1 \end{bmatrix}.
\end{aligned}
\tag{4.79}
$$

The equivalent nodal load vector to the distributed torque $m_x(x)$ of the element is

$$\{f^{(e)}\} = \int_{x_{i-1}^{(e)}}^{x_i^{(e)}} m_x(x)\{N^{(e)}\}^T dx = \int_{x_{i-1}^{(e)}}^{x_i^{(e)}} m_x(x) \left\{ \begin{array}{c} \phi_{i-1}^{(e)}(x) \\ \phi_i^{(e)}(x) \end{array} \right\} dx. \tag{4.80}$$

For a constant intensity $m_x(x) = t_0$, the equivalent nodal load vector in element e reduces to

$$\{f^{(e)}\} = \frac{t_0 h^{(e)}}{2} \left\{ \begin{array}{c} 1 \\ 1 \end{array} \right\}. \tag{4.81}$$

Therefore, the distributed moment transforms into two concentrated torques with the same intensity $\dfrac{t_0 h^{(e)}}{2}$, applied to the nodes.

The final system of equations in element e for $m_x(x) = t_0$ is given by

$$\frac{G^{(e)} I_p^{(e)}}{h^{(e)}} \begin{bmatrix} 1 & -1 \\ -1 & 1 \end{bmatrix} \left\{ \begin{array}{c} \theta_{x_{i-1}}^{(e)} \\ \theta_{x_i}^{(e)} \end{array} \right\} = \frac{t_0 h^{(e)}}{2} \left\{ \begin{array}{c} 1 \\ 1 \end{array} \right\}. \tag{4.82}$$

In a compact notation,

$$[K^{(e)}]\{\theta_x^{(e)}\} = \{f^{(e)}\}. \tag{4.83}$$

The stiffness matrix and global load vector for the shaft are obtained by the assembly procedure illustrated in the next example. As the angle of twist is along the x axis, its representation is independent of the element orientation. Hence, there is no need to apply the coordinate transformation matrix as in the bar case.

In the previous chapter, we defined the L_2 norm of the approximation error in equation (3.134). We can also calculate the error by the energy norm, denoted as $||e||_E$. This error is given for a shaft finite element by

$$||e^{(e)}||_E = \sqrt{\int_{x_{i-1}^{(e)}}^{x_i^{(e)}} G^{(e)} I_p^{(e)} \left[\left(\frac{d\theta_x(x)}{dx} \right)^2 - \left(\frac{d\theta_{x_N}(x)}{dx} \right)^2 \right] dx}. \tag{4.84}$$

The relative energy error of the approximated solution for a mesh with N_{el} elements is obtained, analogously to equation (3.135), as

$$||e_r||_E = \frac{\sqrt{\sum_{e=1}^{N_{el}} ||e^{(e)}||_E}}{||\theta_x||_E}. \tag{4.85}$$

The energy norm of the analytical solution is calculated as

$$||\theta_x||_E = \sqrt{\int_0^L G I_p \left(\frac{d\theta_x(x)}{dx} \right)^2 dx}. \tag{4.86}$$

Example 4.15 *Apply the FEM to Example 4.8 for an uniform mesh with two linear elements of length $h^{(e)} = \frac{L}{2}$ and three nodes.*

The stiffness matrix for each element $e = 1, 2$ is given from (4.79) by

$$[K^{(e)}] = 2\frac{GI_p}{L} \begin{bmatrix} 1 & -1 \\ -1 & 1 \end{bmatrix}.$$

The global stiffness matrix of the shaft has a rank 3 and initially all coefficients are zero, that is,

$$[K] = \begin{bmatrix} 0 & 0 & 0 \\ 0 & 0 & 0 \\ 0 & 0 & 0 \end{bmatrix}.$$

The assembly procedure is performed for the coefficients of the elemental stiffness matrices into the rows and columns of the global stiffness matrix corresponding to the node numbers of each element. Thus, we add the coefficients of the stiffness matrix of element 1 into rows and columns 1 and 2 of the global stiffness. Therefore,

$$[K] = 2\frac{GI_p}{L} \begin{bmatrix} 1 & -1 & 0 \\ -1 & 1 & 0 \\ 0 & 0 & 0 \end{bmatrix}.$$

Analogously, the coefficients of the stiffness matrix of element 2 are added into rows and columns 2 and 3 of the global stiffness matrix, that is,

$$[K] = 2\frac{GI_p}{L} \begin{bmatrix} 1 & -1 & 0 \\ -1 & 2 & -1 \\ 0 & -1 & 1 \end{bmatrix}.$$

The load vectors of the elements are, respectively,

$$\{f^{(1)}\} = \{\ 5 + M_{x_1} \quad 5\ \}^T \quad and \quad \{f^{(2)}\} = \{\ 5 \quad 5 + M_{x_3}\ \}^T,$$

with M_{x_1} and M_{x_3} the reaction torques at $x = 0$ and $x = 2$ m, respectively. Doing the same assembly procedure for the element load vectors, we have the following global load vector:

$$\{f\} = \{\ 5 + M_{x_1} \quad 10 \quad 5 + M_{x_3}\ \}^T.$$

The system of equations to be solved is

$$10^6 \begin{bmatrix} 1 & -1 & 0 \\ -1 & 2 & -1 \\ 0 & -1 & 1 \end{bmatrix} \begin{Bmatrix} \theta_{x_1} \\ \theta_{x_2} \\ \theta_{x_3} \end{Bmatrix} = \begin{Bmatrix} 5 + M_{x_1} \\ 10 \\ 5 + M_{x_3} \end{Bmatrix}.$$

Rows and columns 1 and 3 can be eliminated due to the supports at the shaft ends. Thus, the following equation is evaluated to θ_{x_2}:

$$2 \times 10^6 \theta_{x_2} = 10 \to \theta_{x_2} = 5 \times 10^{-6} \text{ rad.}$$

The nodal vector for the angle of twist is

$$\{\theta_x\} = \{\ 0 \quad 5 \times 10^{-6} \quad 0\ \}^T \text{ rad.}$$

The reaction torques are obtained from rows 1 and 3 of the previous system of equations as

$$\begin{Bmatrix} M_{x_1} \\ M_{x_3} \end{Bmatrix} = 10^6 \begin{bmatrix} 1 & -1 & 0 \\ 0 & -1 & 1 \end{bmatrix} \begin{Bmatrix} 0 \\ 5 \times 10^{-6} \\ 0 \end{Bmatrix} - \begin{Bmatrix} 5 \\ 5 \end{Bmatrix} = \begin{Bmatrix} -10 \\ -10 \end{Bmatrix} \text{ Nm.}$$

From (3.122), (3.123) and (4.76), the local solutions of the twist angle for each finite element are, respectively,

$$\theta_{x_2}^{(1)}(x) = \theta_{x_1}\phi_1^{(1)}(x) + \theta_{x_2}\phi_2^{(1)}(x)$$

$$= (0)\left(1 - \frac{2}{L}x\right) + (5 \times 10^{-6})\left(\frac{2}{L}x\right) = 5 \times 10^{-6}x,$$

$$\theta_{x_2}^{(2)}(x) = \theta_{x_2}\phi_1^{(2)}(x) + \theta_{x_3}\phi_2^{(2)}(x)$$

$$= (5 \times 10^{-6})\left(2 - \frac{2}{L}x\right) + (0)\left(\frac{2}{L}x - 1\right) = 5 \times 10^{-6}(2 - x).$$

The tangential distortion in the elements is obtained from (4.77) or the derivatives of the previous expressions. The maximum distortion in the elements occurs to $r = \frac{d}{2} = 5$ cm. Hence,

$$\gamma_{t_2}^{(1)} = 2.5 \times 10^{-7} \text{ rad,}$$

$$\gamma_{t_2}^{(2)} = -2.5 \times 10^{-7} \text{ rad.}$$

The shear stresses are given by Hooke's law (4.78). For $G = 10^9$ Pa, we have

$$\tau_{t_2}^{(1)} = 250 \, Pa,$$

$$\tau_{t_2}^{(2)} = -250 \, Pa.$$

The square of the L_2 and energy error norms for each element are calculated by equations (3.134) and (4.84), respectively. Therefore,

$$||e^{(1)}||_{L_2} = ||e^{(2)}||_{L_2} = 8.333 \times 10^{-13}$$

$$||e^{(1)}||_E = ||e^{(2)}||_E = 8.333 \times 10^{-6}.$$

The respective relative percentual errors are calculated using (3.135) and (4.85). Thus,

$$||e_r||_{L_2} = 25.0\% \quad and \quad ||e_r||_E = 50.0\%.$$

The energy norm error is much larger than the L_2 norm error. This is justified because the energy norm is based on the derivatives of the approximated solution, which has a larger approximation error, as illustrated in Figure 4.33.

The exact solution and the approximated one with four linear elements are illustrated in Figure 4.33(a). Again, the nodal values coincide with the exact solution calculated at the nodal coordinates, as indicated by the circles. On the other hand, Figure 4.33(b) shows the derivatives of the analytical and approximated solutions. While the analytical derivative is a continuous function, the approximated solution is discontinuous on the nodal coordinates. This characteristic is related to the construction procedure for the interpolation functions which uses polynomials by parts. Hence, the approximation derivatives obtained with the FEM, as for strains and stresses, will generally be discontinuous at the elemental interfaces. Smoothing procedures can be used to obtain a continuous representation of the strain components [53].

File shaftsolapex1.m implements the solution of this example with the FEM for an arbitrary number of linear interpolation functions.

□

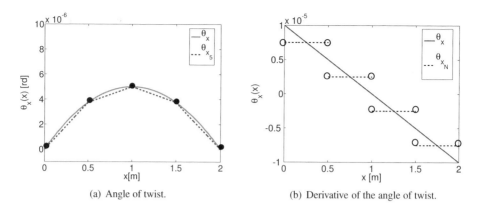

(a) Angle of twist.   (b) Derivative of the angle of twist.

Figure 4.33 Example 4.15: FEM approximations with four linear elements.

4.14 MATHEMATICAL ASPECTS OF THE FEM

An important aspect of the FEM, and indeed for any approximation numerical method of BVPs, is to determine the error and convergence rate of the approximated solution. The error norms used in the previous example are called *a posteriori*, because we can calculate the approximation error using a known analytical solution. Another type of error, called *a priori*, is determined from the characteristics of the mesh, approximation space, and smoothness of the load term. In the case of linear finite elements and one-dimensional problems, the *a priori* estimate of the approximation error in the energy norm is given by [4, 35, 48]

$$||e||_E \leq Ch, \tag{4.87}$$

where C is a constant that depends on the right-hand side of the BVP and h is the characteristic mesh element size.

It is observed in the previous expression that the approximation error in the h version of the FEM is proportional to the mesh size, i.e., if the element size is halved, the error should decrease twice. Hence, the convergence rate is algebraic. As the previous example has a smooth analytical solution, it is possible to check if such convergence rate is achieved. For this purpose, uniform meshes with 2, 4, 8, and 16 elements are considered, with characteristic element sizes $h^{(e)} = 1$, $h^{(e)} = \frac{1}{2}$, $h^{(e)} = \frac{1}{4}$, and $hV = \frac{1}{8}$, respectively. Using expression (4.87), the decreasing rate of the approximation error between two consecutive meshes $i-1$ and i is given by

$$\eta = \frac{||e||_E^{h_{i-1}}}{||e||_E^{h_i}} = \frac{Ch_{i-1}}{Ch_i} = 2.$$

Thus, the approximation error is expected to reduce by a factor of 2 when the element sizes are halved. Using the MATLAB script shaftsolapex1.m with the four considered meshes, this convergence rate is verified. Figure 4.34 illustrates, in a logarithmic scale, the behavior of the relative error in the L_2 and energy norms in terms of the number of elements. The error convergence rate in the energy norm is 2. On the other hand, the error drops four times for the L_2 norm. This fact is related to the use of derivatives of the approximation when calculating the error in the energy norm. The *a priori* and *a posteriori* convergence rates in the energy norm coincide in this example.

Although the bar and shaft models are very distinct physically, their strong and weak forms have the same mathematical structure and share several common properties. It is interesting to use a more abstract notation based on operators to denote their BVPs and respective weak forms.

The strong forms of the considered models in this book can be denoted as

$$\mathscr{A}u = f, \tag{4.88}$$

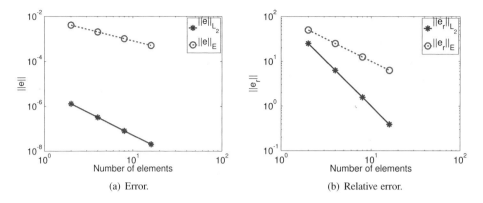

Figure 4.34 Example 4.15: approximation errors in the energy and L_2 norms.

with $\mathscr{A}$ a differential operator, u the solution field of the considered problem, and f the independent term field. For bars and shafts, we have, respectively,

$$\mathscr{A} = EA\frac{d^2}{dx^2}, \ u = u_x(x), \ f = -q_x(x), \tag{4.89}$$

$$\mathscr{A} = GI_p\frac{d^2}{dx^2}, \ u = \theta_x(x), \ f = -m_x(x). \tag{4.90}$$

It is possible to define the domain $D_{\mathscr{A}}$ of the solution u of the strong form. This is the set of functions that has enough derivatives, such that the differential operator can be applied, and satisfy the boundary conditions of the problem. As the bar and shaft models have operators with second-order derivatives, we define

$$D_{\mathscr{A}} = \{u \mid u \in C^2(0,L), u \text{ satisfies the boundary conditions}\}, \tag{4.91}$$

where $C^2(0,L)$ is the set of functions with continuous second-order derivatives in the open interval $(0,L)$, with L the length of the structural element. This definition of $D_{\mathscr{A}}$ is valid for all the examples presented here with continuous loading functions.

Example 4.16 *For the shaft of Example 4.8, the solution domain of the strong form is given by*

$$D_{\mathscr{A}} = \{\theta_x(x) \mid \theta_x(x) \in C^2(0,L), \theta_x(0) = \theta_x(L) = 0\}.$$

$\square$

Analogously, the weak forms of the problems treated here can be written as

$$a(u,v) = f(v), \tag{4.92}$$

with $a(u,v)$ and $f(v)$ the bilinear and linear forms, respectively. For the bar and shaft models, we have

$$a(u_x,v) = \int_0^L EA\frac{du_x(x)}{dx}\frac{dv(x)}{dx}dx \ \text{ and } \ f(v) = \int_0^L q_x(x)v(x)dx + P_L v(L) + P_0 v(0), \tag{4.93}$$

$$a(\theta_x,v) = \int_0^L EA\frac{d\theta_x(x)}{dx}\frac{dv(x)}{dx}dx \ \text{ and } \ f(v) = \int_0^L m_x(x)v(x)dx + T_L v(L) + T_0 v(0). \tag{4.94}$$

As the previous weak forms have only first derivatives, their solution belong to the set $C^1(0,L)$ of functions with continuous first-order derivatives. Actually, we can be more restrictive and use the set $C^1_{cp}(0,L)$ of continuous functions with continuous piecewise first-order derivatives, which is a broader set than $C^1(0,L)$. The shape functions illustrated in Figure 3.54 belong to the $C^1_{cp}(0,L)$ set. Precisely for this reason, the derivative of the approximated solution shown in Figure 4.33(b) is discontinuous at the nodal coordinates.

For $u = v$, the bilinear form $a(u,u)$ represents the strain energy in structural mechanics. For the circular torsion model, the strain energy and the energy norm are related based on (4.86) as

$$a(u,u) = ||\theta_x||^2_E = \int_0^L GI_p \left(\frac{d\theta_x(x)}{dx} \right)^2 dx. \tag{4.95}$$

Generally, the energy norm of the function u can be defined in terms of the bilinear form as

$$||u||_E = \sqrt{a(u,u)}. \tag{4.96}$$

Until now, only elemental linear interpolation functions defined in the Cartesian coordinate system were considered. To make it simpler to increase the order of the interpolation functions and their extension for two- and three-dimensional domains, it is interesting to use polynomial bases defined in the local normalized interval $[-1,1]$. This requires the definition of global and local coordinate systems for the elements and mapping procedures between these systems. Furthermore, the coefficients of the element matrices and vectors are more conveniently calculated by numerical integration and derivative by collocation. These aspects will be discussed in the following sections.

4.15 LOCAL COORDINATE SYSTEMS

When using the FEM for structural analysis, a global Cartesian reference system is employed to input the nodal coordinates of the mesh. For one-dimensional elements of bars, shafts, and beams, which may be arbitrarily oriented in the global system, it is convenient to have a local Cartesian coordinate system attached to the element. A transformation matrix to mapping the local and global systems is defined to orientate the element in the global system, as given in equation (3.138) for bars.

Another local coordinate system is defined for the element in the $[-1,1]$ or $[0,1]$ intervals. It makes easier the construction of shape functions and procedures of numerical integration and derivative by collocation. This local system is called normalized. In this section, the normalized local systems in the $[-1,1]$ and $[0,1]$ intervals are discussed.

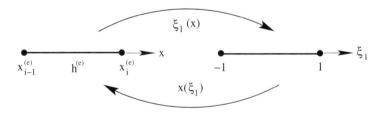

Figure 4.35 Elemental local Cartesian and normalized coordinate systems.

Consider the local Cartesian system x for the two-node element with nodal coordinates $x^{(e)}_{i-1}$ and $x^{(e)}_i$. Consider also the normalized local system ξ_1 in the closed interval $[-1,1]$. Both systems are illustrated in Figure 4.35. It is interesting to determine a transformation between these two coordinate systems. As the element geometry is a straight line, a linear transformation $x(\xi_1)$ is assumed, i.e.,

$$x(\xi_1) = a\xi_1 + b.$$

The coefficients a and b are determined using the fact that the previous transformation should be one-to-one. Therefore, each coordinate ξ_1 must be mapped onto only one coordinate x and vice versa. Thus, the local coordinates -1 and 1 should be transformed to the Cartesian coordinates $x_{i-1}^{(e)}$ and $x_i^{(e)}$, respectively, i.e.,

$$\begin{aligned} x(-1) &= a(-1) + b = x_{i-1}^{(e)}, \\ x(1) &= a(1) + b = x_i^{(e)}. \end{aligned}$$

Solving the previous system of equations, the coefficients a and b are determined as

$$a = \frac{1}{2}(x_i^{(e)} - x_{i-1}^{(e)}) = \frac{h^{(e)}}{2} \quad \text{and} \quad b = \frac{1}{2}(x_{i-1}^{(e)} + x_i^{(e)}).$$

Hence, the final transformation is given by

$$x(\xi_1) = \frac{1}{2}(x_i^{(e)} - x_{i-1}^{(e)})\xi_1 + \frac{1}{2}(x_{i-1}^{(e)} + x_i^{(e)}) = \frac{1}{2}(1 - \xi_1)x_{i-1}^{(e)} + \frac{1}{2}(1 + \xi_1)x_i^{(e)}. \tag{4.97}$$

As the previous equation is linear in ξ_1, the inverse transformation can be explicitly determined and

$$\xi_1(x) = \frac{2x - (x_{i-1}^{(e)} + x_i^{(e)})}{x_i^{(e)} - x_{i-1}^{(e)}} = \frac{2x - (x_{i-1}^{(e)} + x_i^{(e)})}{h^{(e)}}. \tag{4.98}$$

Example 4.17 *For $x_{i-1}^{(e)} = -10$ and $x_i^{(e)} = 20$, the previous transformations reduces to*

$$\xi_1(x) = \frac{2}{30}x - \frac{1}{3} \quad \text{and} \quad x(\xi_1) = 15\xi_1 + 5.$$

For $x = 5$, we calculate $\xi_1 = 0$ from the first transformation; for $\xi_1 = 0$, we recover $x = 5$ from the second transformation.
□

Normalized natural coordinates L_1 and L_2 are also employed in the interval $[0, 1]$, as shown in Figure 4.15. Consider a given point P located in the two-node element with $x_{i-1}^{(e)}$ and $x_i^{(e)}$ coordinates. The distances from P to the element ends are denoted by l_1 and l_2. The following relation is valid for the length l of the element:

$$l_1 + l_2 = l.$$

By dividing both sides by l, we have

$$\frac{l_1}{l} + \frac{l_2}{l} = 1.$$

Defining

$$L_1 = \frac{l_1}{l} \quad \text{and} \quad L_2 = \frac{l_2}{l}, \tag{4.99}$$

the previous relation can be rewritten as

$$L_1 + L_2 = 1. \tag{4.100}$$

Therefore, L_1 and L_2 define the natural coordinate system, which allows us to locate any point P. If P is at the left end of the element, we have $L_1 = 0$ and $L_2 = 1$; if P is on the right end, we have $L_1 = 1$ and $L_2 = 0$. As the element is one-dimensional, one of the coordinates is dependent. For example, from (4.100), we have

$$L_2 = 1 - L_1. \tag{4.101}$$

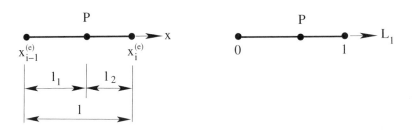

Figure 4.36 Local natural coordinate system.

The linear transformations between the normalized systems are illustrated in Figure 4.37 and given by

$$\xi_1(L_1) = 2L_1 - 1, \tag{4.102}$$

$$L_1(\xi_1) = \frac{1}{2}(1 + \xi_1). \tag{4.103}$$

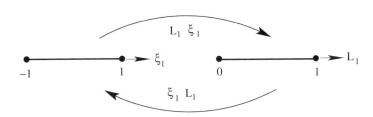

Figure 4.37 Transformation between the normalized local systems $[-1, 1]$ and $[0, 1]$.

The advantages of using normalized local systems are:

- Make simpler the construction of polynomial shape functions for the finite elements
- Use fixed integration limits to calculate the element matrix and vector coefficients
- Allow the use of distorted elements in two- and three-dimensional finite element meshes

4.16 ONE-DIMENSIONAL SHAPE FUNCTIONS

In this section, the construction of nodal and modal one-dimensional shape functions for any order is presented, using, respectively, Lagrange and Jacobi polynomials defined in the local normalized systems discussed in the previous section.

4.16.1 NODAL BASIS

Consider a set of $P_1 + 1$ points with coordinates ξ_{1_p} ($p = 0, \ldots, P_1$), illustrated in Figure 4.38(a) for the local coordinate system ξ_1. The Lagrange polynomial $l_p^{(P_1)}(\xi_1)$ of order P_1 associated to the coordinate ξ_{1_p} is given by [54, 18]

$$l_p^{(P_1)}(\xi_1) = \prod_{q=0(p \neq q)}^{P_1} \frac{(\xi_1 - \xi_{1_q})}{(\xi_{1_p} - \xi_{1_q})}$$

$$= \frac{(\xi_1 - \xi_{1_0}) \cdots (\xi_1 - \xi_{1_{p-1}})(\xi_1 - \xi_{1_{p+1}}) \cdots (\xi_1 - \xi_{1_{P_1}})}{(\xi_{1_p} - \xi_{1_0}) \cdots (\xi_{1_p} - \xi_{1_{p-1}})(\xi_{1_p} - \xi_{1_{p+1}}) \cdots (\xi_{1_p} - \xi_{1_{P_1}})}.$$

(4.104)

The zero-order polynomial is such that $l_p^{(0)}(\xi_1) = 1$ for any p. The term ξ_{1_p} is defined as roots, nodal coordinates, or collocation points of the Lagrange polynomials.

(a) ξ_1.   (b) L_1.

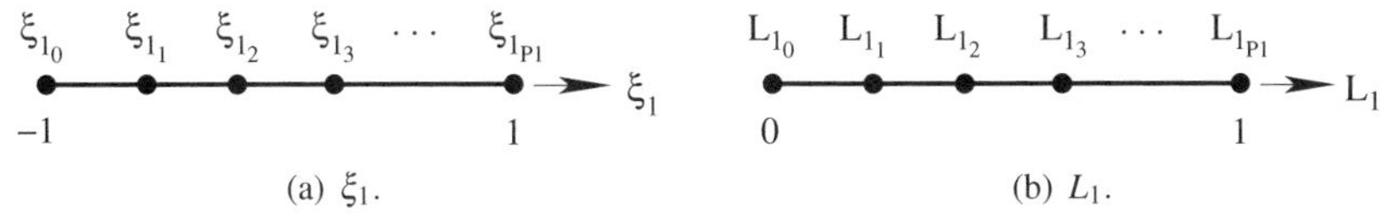

Figure 4.38 Nodes in the local coordinate systems ξ_1 and L_1.

Using the previous definition, the Lagrange polynomials have the following collocation property:

$$l_p^{(P_1)}(\xi_{1_q}) = \delta_{pq},$$

(4.105)

where the Kronecker's delta is defined by

$$\delta_{pq} = \left\{ \begin{array}{ll} 1 & p = q \\ 0 & p \neq q \end{array} \right. .$$

(4.106)

Therefore, the Lagrange polynomial $l_p^{(P_1)}(\xi_1)$ assumes the unit value when calculated on the coordinate ξ_{1_p} of the collocation point which is associated. When $l_p^{(P_1)}(\xi_1)$ is calculated on the coordinates of any other collocation point, its value is zero. This property can be easily observed substituting $\xi_1 = \xi_{1_p}$ and, for instance, $\xi_1 = \xi_{1_0}$ ($p \neq 0$) in equation (4.104), i.e.,

$$l_p^{(P_1)}(\xi_{1_p}) = \frac{(\xi_{1_p} - \xi_{1_0}) \cdots (\xi_{1_p} - \xi_{1_{p-1}})(\xi_{1_p} - \xi_{1_{p+1}}) \cdots (\xi_{1_p} - \xi_{1_{P_1}})}{(\xi_{1_p} - \xi_{1_0}) \cdots (\xi_{1_p} - \xi_{1_{p-1}})(\xi_{1_p} - \xi_{1_{p+1}}) \cdots (\xi_{1_p} - \xi_{1_{P_1}})} = 1,$$

$$l_p^{(P_1)}(\xi_{1_0}) = \frac{(\xi_{1_0} - \xi_{1_0}) \cdots (\xi_{1_0} - \xi_{1_{p-1}})(\xi_{1_0} - \xi_{1_{p+1}}) \cdots (\xi_{1_0} - \xi_{1_{P_1}})}{(\xi_{1_p} - \xi_{1_0}) \cdots (\xi_{1_p} - \xi_{1_{p-1}})(\xi_{1_p} - \xi_{1_{p+1}}) \cdots (\xi_{1_p} - \xi_{1_{P_1}})} = 0.$$

The local shape functions $\phi_p(\xi_1)$ associated to the p ($p = 0, \ldots, P_1$) nodes of the one-dimensional finite elements are indeed the Lagrange polynomials, that is,

$$\phi_p^{(e)}(\xi_1) = l_p^{(P_1)}(\xi_1).$$

(4.107)

Example 4.18 *Figures 4.39 to 4.41 illustrate the one-dimensional Lagrangian elements with two, three, and four nodes with equally spaced collocation coordinates.*

The local shape functions for the two-node linear element are determined from (4.104) as

$$\phi_0^{(e)}(\xi_1) = \frac{\xi_1 - 1}{(-1) - 1} = \frac{1}{2}(1 - \xi_1),$$

$$\phi_1^{(e)}(\xi_1) = \frac{\xi_1 - (-1)}{1 - (-1)} = \frac{1}{2}(1 + \xi_1).$$

For the three-node second-order element, the local shape functions are given by

$$\phi_0^{(e)}(\xi_1) = \frac{(\xi_1 - 0)(\xi_1 - 1)}{[(-1) - 0][(-1) - 1]} = -\frac{1}{2}\xi_1(1 - \xi_1),$$

$$\phi_1^{(e)}(\xi_1) = \frac{[\xi_1 - (-1)](\xi_1 - 1)}{[0 - (-1)](0 - 1)} = 1 - \xi_1^2,$$ (4.108)

$$\phi_2^{(e)}(\xi_1) = \frac{[\xi_1 - (-1)](\xi_1 - 0)}{[1 - (-1)](1 - 0)} = \frac{1}{2}\xi_1(1 + \xi_1).$$

The local shape functions for the third-order element are obtained from (4.104) as

$$\phi_0^{(e)}(\xi_1) = \frac{[\xi_1 - (-\frac{1}{3})](\xi_1 - \frac{1}{3})(\xi_1 - 1)}{[-1 - (-\frac{1}{3})](-1 - \frac{1}{3})(-1 - 1)} = \frac{1}{16}(1 - \xi_1)(9\xi_1^2 - 1),$$

$$\phi_1^{(e)}(\xi_1) = \frac{[\xi_1 - (-1)](\xi_1 - \frac{1}{3})(\xi_1 - 1)}{[-\frac{1}{3} - (-1)](-\frac{1}{3} - \frac{1}{3})(-\frac{1}{3} - 1)} = \frac{9}{16}(1 - \xi_1^2)(1 - 3\xi_1),$$

$$\phi_2^{(e)}(\xi_1) = \frac{[\xi_1 - (-1)][\xi_1 - (-\frac{1}{3})](\xi_1 - 1)}{[\frac{1}{3} - (-1)][\frac{1}{3} - (-\frac{1}{3})](-\frac{1}{3} - 1)} = \frac{9}{16}(1 - \xi_1^2)(1 + 3\xi_1),$$ (4.109)

$$\phi_3^{(e)}(\xi_1) = \frac{[\xi_1 - (-1)][\xi_1 - (-\frac{1}{3})](\xi_1 - \frac{1}{3})}{[1 - (-1)][1 - (-\frac{1}{3})](1 - \frac{1}{3})} = \frac{1}{16}(1 + \xi_1)(9\xi_1^2 - 1).$$

Note that the number of shape functions is always $P_1 + 1$ and all functions have the same order P_1.

□

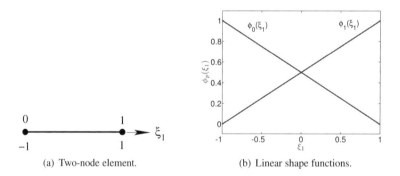

(a) Two-node element.   (b) Linear shape functions.

Figure 4.39   Example 4.18: one-dimensional Lagrange shape functions for a two-node element.

Considering the $P_1 + 1$ collocation points in the natural coordinate system L_1, illustrated in Figure 4.38(b), expression (4.104) for the Lagrange polynomials is written as

$$l_p^{(P_1)}(L_1) = \prod_{q=0, p \neq q}^{P_1} \frac{(L_1 - L_{1_q})}{(L_{1_p} - L_{1_q})}$$

$$= \frac{(L_1 - L_{1_0}) \ldots (L_1 - L_{1_{p-1}})(L_1 - L_{1_{p+1}}) \ldots (L_1 - L_{1_{P_1}})}{(L_{1_p} - L_{1_0}) \ldots (L_{1_p} - L_{1_{p-1}})(L_{1_p} - L_{1_{p+1}}) \ldots (L_{1_p} - L_{1_{P_1}})}.$$

(4.110)

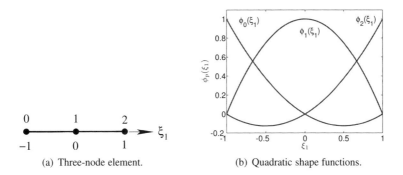

Figure 4.40 Example 4.18: one-dimensional Lagrange shape functions for the three-node element.

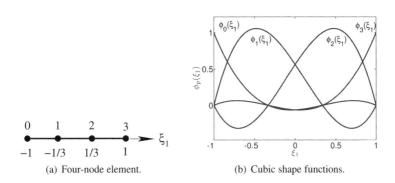

Figure 4.41 Example 4.18: one-dimensional Lagrange shape functions for the four-node element.

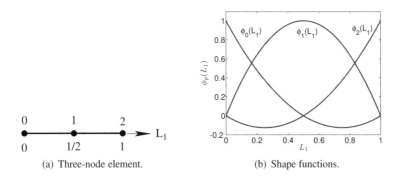

Figure 4.42 Example 4.19: one-dimensional Lagrange shape functions for a three-node element in the L_1 natural system.

Example 4.19 *For the second-order element with equally spaced nodes in the natural coordinate system L_1 shown in Figure 4.42, the local shape functions are determined from (4.110) as*

$$\phi_0^{(e)}(L_1) = \frac{(L_1 - \frac{1}{2})(L_1 - 1)}{(0 - \frac{1}{2})(0 - 1)} \quad = \quad -(2L_1 - 1)(1 - L_1),$$

$$\phi_1^{(e)}(L_1) = \frac{(L_1 - 0)(L_1 - 1)}{(\frac{1}{2} - 0)(\frac{1}{2} - 1)} \quad = \quad 4L_1(1 - L_1),$$

$$\phi_2^{(e)}(L_1) = \frac{(L_1 - 0)(L_1 - \frac{1}{2})}{(1 - 0)(1 - \frac{1}{2})} \quad = \quad L_1(2L_1 - 1).$$

We can also obtain the shape functions in the L_1 coordinate system substituting (4.102) in (4.104). For example, taking the shape function ϕ_1 for the quadratic element given in (4.109), we have

$$\phi_1^{(e)}(L_1) = \phi_1^{(e)}[\xi_1(L_1)] = 1 - (2L_1 - 1)^2 = 4L_1(1 - L_1).$$

□

The following expression can be employed to denote the vertex ($p = 0$ and $p = P_1 + 1$) and internal ($0 < p < P_1$) shape functions:

$$\phi_p^{(e)}(\xi_1) = \begin{cases} \frac{1}{2}(1 - \xi_1)L_{p,P_1}^{(P_1-1)}(\xi_1) & p = 0 \\ \frac{1}{2}(1 + \xi_1)L_{p,0}^{(P_1-1)}(\xi_1) & p = P_1 \\ \frac{1}{4}(1 - \xi_1)(1 + \xi_1)L_p^{(P_1-2)}(\xi_1) & 0 < p < P_1 \end{cases}, \qquad (4.111)$$

with

$$L_{p,l}^{(P_1-1)}(\xi_1) = -\frac{\prod_{q=0,q\neq\{p,l\}}^{P_1}(\xi_1 - \xi_{1q})}{\prod_{q=0,q\neq\{p,l\}}^{P_1}(\xi_{1p} - \xi_{1q})}, \qquad (4.112)$$

$$L_p^{(P_1-2)}(\xi_1) = -4\frac{\prod_{q=1,q\neq p}^{P_1-1}(\xi_1 - \xi_{1q})}{\prod_{q=1,q\neq p}^{P_1-1}(\xi_{1p} - \xi_{1q})}. \qquad (4.113)$$

Example 4.20 *The idea of the previous equations is to factorize the $(1 \pm \xi_1)$ monomials in the Lagrange polynomials expressions. This allows us to obtain similar expressions for the nodal and modal bases and reduce the number of integration points, as will be seen later.*

Consider the third-order shape functions given in (4.110). In this case, we have

$$L_{0,3}^{(2)}(\xi_1) = L_{3,0}^{(2)}(\xi_1) = \frac{1}{8}(9\xi_1^2 - 1), \quad L_1^{(1)}(\xi_1) = \frac{9}{4}(1 - 3\xi_1), \quad \text{and} \quad L_2^{(1)}(\xi_1) = \frac{9}{4}(1 + 3\xi_1).$$

□

In natural coordinates, equation (4.111) is given by

$$\phi_p^{(e)}(L_1) = \begin{cases} (1 - L_1)L_{p,P_1}^{(P_1-1)}(2L_1 - 1) & p = 0 \\ L_1 L_{p,0}^{(P_1-1)}(2L_1 - 1) & p = P_1 \\ (1 - L_1)L_1 L_p^{(P_1-2)}(2L_1 - 1) & 0 < p < P_1 \end{cases}, \qquad (4.114)$$

with

$$L_{p,l}^{(P_1-1)}(L_1) = -\frac{\prod_{q=0,q\neq\{p,l\}}^{P_1}(L_1 - L_{1q})}{\prod_{q=0,q\neq\{p,l\}}^{P_1}(L_{1p} - L_{1q})}, \qquad (4.115)$$

$$L_p^{(P_1-2)}(L_1) = -4\frac{\prod_{q=1,q\neq p}^{P_1-1}(L_1 - L_{1q})}{\prod_{q=1,q\neq p}^{P_1-1}(L_{1p} - L_{1q})}. \qquad (4.116)$$

4.16.2 MODAL BASIS

Due to the Lagrange polynomial definition (4.104), the shape functions given in (4.107) are directly related to the element nodes and thus constitute a nodal basis.

In the case of a modal basis, there are only the vertex nodes for the element. The shape functions starting from the second-order are not associated to nodal variables. For this purpose, we should employ a polynomial basis whose definition is not dependent on nodal coordinates of the element. Orthogonal Jacobi polynomials are used here.

The one-dimensional Jacobi polynomials of order n, denoted by $P_n^{\alpha_1,\beta_1}(\xi_1)$, are a family of polynomial solutions to the Sturm-Liouville singular problem [35]. These polynomials are orthogonal in the $[-1,1]$ interval, with respect to the weight function $(1-\xi_1)^{\alpha_1}(1+\xi_1)^{\beta_1}$ $(\alpha_1,\beta_1 > -1; \alpha_1,\beta_1 \in \Re)$, such that [17]

$$\int_{-1}^{1}(1-\xi_1)^{\alpha_1}(1+\xi_1)^{\beta_1}P_n^{\alpha_1,\beta_1}(\xi_1)P_m^{\alpha_1,\beta_1}(\xi_1)d\xi_1 = C\delta_{nm}, \quad \xi_1 \in [-1,1], \tag{4.117}$$

with the constant C given by

$$C = \frac{2^{\alpha_1+\beta_1+1}}{2n+\alpha_1+\beta_1+1}\frac{\Gamma(n+\alpha_1+1)\Gamma(n+\beta_1+1)}{n!\Gamma(n+\alpha_1+\beta_1+1)}.$$

The Γ function for integer numbers n reduces to the factorial, that is, $\Gamma(n)=n!$.

The previous orthogonality propery can be mapped to the interval $[0,1]$ of the natural coordinates L_1 substituting (4.102) in (4.117). Thus,

$$\int_{0}^{1}(1-L_1)^{\alpha_1}L_1^{\beta_1}P_n^{\alpha_1,\beta_1}(2L_1-1)P_m^{\alpha_1,\beta_1}(2L_1-1)dL_1 = D\delta_{nm}, \quad L_1 \in [0,1], \tag{4.118}$$

and

$$D = \frac{1}{2^{\alpha_1+\beta_1+1}}C.$$

The Legendre and Chebyshev polynomials are particular cases of Jacobi polynomials and obtained for the weights $(\alpha_1 = \beta_1 = 0)$ and $\left(\alpha_1 = \beta_1 = -\frac{1}{2}\right)$, respectively.

Example 4.21 *Consider the following integral with the multiplication of a third-order Jacobi polynomial $P_3^{\alpha_1,\beta_1}(\xi_1)$ and a second-order polynomial $f(\xi_1)$, that is,*

$$I = \int_{-1}^{1}(1-\xi_1)^{\alpha_1}(1+\xi_1)^{\beta_1}P_3^{\alpha_1,\beta_1}(\xi_1)f(\xi_1)d\xi_1.$$

Because $f(\xi_1)$ is a second-order polynomial function, we can represent it exactly by the following linear combination of Jacobi polynomials:

$$f(\xi_1) = a_0P_0^{\alpha_1,\beta_1}(\xi_1)+a_1P_1^{\alpha_1,\beta_1}(\xi_1)+a_2P_2^{\alpha_1,\beta_1}(\xi_1).$$

Substituting $f(\xi_1)$ in the expression for I and using the orthogonality equation (4.117), we have

$$\begin{aligned}I &= a_0\int_{-1}^{1}(1-\xi_1)^{\alpha_1}(1+\xi_1)^{\beta_1}P_3^{\alpha_1,\beta_1}(\xi_1)P_0^{\alpha_1,\beta_1}(\xi_1)d\xi_1\\
&+ a_1\int_{-1}^{1}(1-\xi_1)^{\alpha_1}(1+\xi_1)^{\beta_1}P_3^{\alpha_1,\beta_1}(\xi_1)P_1^{\alpha_1,\beta_1}(\xi_1)d\xi_1\\
&+ a_2\int_{-1}^{1}(1-\xi_1)^{\alpha_1}(1+\xi_1)^{\beta_1}P_3^{\alpha_1,\beta_1}(\xi_1)P_2^{\alpha_1,\beta_1}(\xi_1)d\xi_1\\
&= 0.\end{aligned}$$

□

The following recurrence relations can be used to calculate the Jacobi polynomials [17, 35]:

$$
\begin{aligned}
P_0^{\alpha_1,\beta_1}(\xi_1) &= 1, \\
P_1^{\alpha_1,\beta_1}(\xi_1) &= \frac{1}{2}[\alpha_1 - \beta_1 + (\alpha_1 + \beta_1 + 2)\xi_1], \\
a_n^1 P_{n+1}^{\alpha_1,\beta_1}(\xi_1) &= (a_n^2 + a_n^3 \xi_1)P_n^{\alpha_1,\beta_1}(\xi_1) - a_n^4 P_n^{\alpha_1,\beta_1}(\xi_1),
\end{aligned}
\tag{4.119}
$$

where

$$
\begin{aligned}
a_n^1 &= 2(n+1)(n+\alpha_1+\beta_1+1)(2n+\alpha_1+\beta_1), \\
a_n^2 &= (2n+\alpha_1+\beta_1+1)(\alpha_1^2-\beta_1^2), \\
a_n^3 &= (2n+\alpha_1+\beta_1)(2n+\alpha_1+\beta_1+1)(2n+\alpha_1+\beta_1+2), \\
a_n^4 &= 2(n+\alpha_1)(n+\beta_1)(2n+\alpha_1+\beta_1+2).
\end{aligned}
\tag{4.120}
$$

Analogously, the derivatives of the Jacobi polynomials have the following recurrence relation:

$$
b_n^1(\xi_1)\frac{d}{d\xi_1}P_n^{\alpha_1,\beta_1}(\xi_1) = b_n^2(\xi_1)P_n^{\alpha_1,\beta_1}(\xi_1) + b_n^3(\xi_1)P_{n-1}^{\alpha_1,\beta_1}(\xi_1),
\tag{4.121}
$$

with

$$
\begin{aligned}
b_n^1(\xi_1) &= (2n+\alpha_1+\beta_1)(1-\xi_1^2), \\
b_n^2(\xi_1) &= n(\alpha_1-\beta_1-(2n+\alpha_1-\beta_1)\xi_1), \\
b_n^3(\xi_1) &= 2(n+\alpha_1)(n+\beta_1).
\end{aligned}
\tag{4.122}
$$

From the previous definitions, the modal shape functions can be defined in the ξ_1 and L_1 coordinate systems, analogously to equations (4.111) and (4.114), respectively, as

$$
\phi_p^{(e)}(\xi_1) =
\begin{cases}
\frac{1}{2}(1-\xi_1) & p = 0 \\
\frac{1}{2}(1+\xi_1) & p = P_1 \\
\frac{1}{4}(1-\xi_1)(1+\xi_1)P_{p-2}^{\alpha_1,\beta_1}(\xi_1) & 0 < p < P_1
\end{cases},
\tag{4.123}
$$

$$
\phi_p^{(e)}(L_1) =
\begin{cases}
1-L_1 & p = 0 \\
L_1 & p = P_1 \\
(1-L_1)L_1 P_{p-2}^{\alpha_1,\beta_1}(2L_1 - 1) & 0 < p < P_1
\end{cases}.
\tag{4.124}
$$

Note that the vertex shape functions are the same functions of the linear Lagrangian element. The terms $(1-\xi_1)(1+\xi_1)$ and $(1-L_1)L_1$ make the internal functions zero at the element ends, as will be seen in the next example.

Example 4.22 *The local shape functions for the second-, third- and fourth-order elements and* $\alpha_1 = \beta_1 = 1$ *are illustrated in Figure 4.43.*

The internal mode for the second-order element is given from (4.123) by

$$
\phi_1^{(e)}(\xi_1) = \frac{1}{4}(1-\xi_1)(1+\xi_1).
$$

Analogously, the two internal modes for the third-order element are

$$
\begin{aligned}
\phi_1^{(e)}(\xi_1) &= \frac{1}{4}(1-\xi_1)(1+\xi_1), \\
\phi_2^{(e)}(\xi_1) &= \frac{1}{2}(1-\xi_1)(1+\xi_1)\xi_1.
\end{aligned}
$$

The three internal modes for the fourth-order element are obtained from (4.123) and given by

$$\phi_1^{(e)}(\xi_1) = \frac{1}{4}(1-\xi_1)(1+\xi_1),$$

$$\phi_2^{(e)}(\xi_1) = \frac{1}{2}(1-\xi_1)(1+\xi_1)\xi_1,$$

$$\phi_3^{(e)}(\xi_1) = -\frac{1}{16}(1-\xi_1)^2(1+\xi_1)(23+5\xi_1).$$

A hierarchy of shape functions is observed, because the set of functions with order P_1 is included in the set of functions with order $P_1 + 1$.

The weights α_1 and β_1 can be selected to zero the coefficients of the elemental matrices, as will be discussed in the next section.

□

The collocation property of the vertex linear functions and the fact that the internal functions have zero values at the element ends make the application of boundary conditions easier.

4.16.3 SCHUR'S COMPLEMENT

We can partition the element system of equations

$$[K^{(e)}]\{a^{(e)}\} = \{f^{(e)}\}$$

in terms of the blocks relative to the vertex (v) and internal (i) functions as

$$\begin{bmatrix} \left[K_{vv}^{(e)}\right] & \left[K_{vi}^{(e)}\right] \\ \left[K_{vi}^{(e)}\right]^T & \left[K_{ii}^{(e)}\right] \end{bmatrix} \left\{ \begin{array}{c} \{a_v^{(e)}\} \\ \{a_i^{(e)}\} \end{array} \right\} = \left\{ \begin{array}{c} \{f_v^{(e)}\} \\ \{f_i^{(e)}\} \end{array} \right\}. \tag{4.125}$$

In expanded form, the previous expression results in

$$\begin{array}{c} \left[K_{vv}^{(e)}\right]\{a_v^{(e)}\} + \left[K_{vi}^{(e)}\right]\{a_i^{(e)}\} = \{f_v^{(e)}\} \\ \left[K_{vi}^{(e)}\right]^T\{a_v^{(e)}\} + \left[K_{ii}^{(e)}\right]\{a_i^{(e)}\} = \{f_i^{(e)}\} \end{array}. \tag{4.126}$$

The following expression for $\{a_i^{(e)}\}$ is obtained from the latter equation:

$$\{a_i^{(e)}\} = \left[K_{ii}^{(e)}\right]^{-1}\left(\{f_i^{(e)}\} - \left[K_{vi}^{(e)}\right]^T\{a_v^{(e)}\}\right). \tag{4.127}$$

Substituting this equation into the first equation in (4.126), we have an expression in terms of the vertex coefficients only, that is,

$$\left(\left[K_{vv}^{(e)}\right] - \left[K_{vi}^{(e)}\right]\left[K_{ii}^{(e)}\right]^{-1}\left[K_{vi}^{(e)}\right]^T\right)\{a_v^{(e)}\} = \{f_v^{(e)}\} - \left[K_{vi}^{(e)}\right]\left[K_{ii}^{(e)}\right]^{-1}\{f_i^{(e)}\}. \tag{4.128}$$

In a compact form, we have

$$\left[K_S^{(e)}\right]\{a_v^{(e)}\} = \{f_S^{(e)}\}, \tag{4.129}$$

with

$$\left[K_S^{(e)}\right] = \left[K_{vv}^{(e)}\right] - \left[K_{vi}^{(e)}\right]\left[K_{ii}^{(e)}\right]^{-1}\left[K_{vi}^{(e)}\right]^T, \tag{4.130}$$

$$\{f_S^{(e)}\} = \{f_v^{(e)}\} - \left[K_{vi}^{(e)}\right]\left[K_{ii}^{(e)}\right]^{-1}\{f_i^{(e)}\}. \tag{4.131}$$

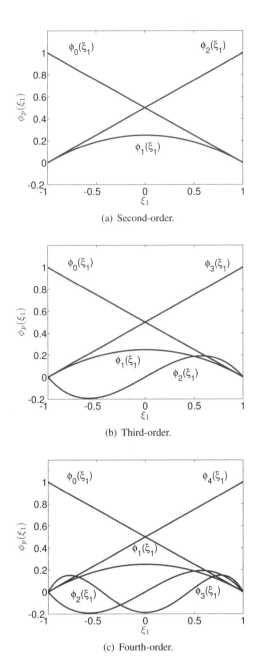

(a) Second-order.

(b) Third-order.

(c) Fourth-order.

Figure 4.43 Example 4.22: modal shape functions for one-dimensional elements of second-, third-, and fourth-orders with $\alpha_1 = \beta_1 = 1$.

The advantage of using the above system of equations is that it is written only in terms of the vertex coefficients, which decreases the number of final equations to be solved. In addition, there is a decrease of the condition number of matrices, as will be seen in the next section. The previous procedure is called Schur's complement or substructuring. Once the vertex coefficients are obtained, the internal coefficients are calculated using (4.127).

4.16.4 SPARSITY AND NUMERICAL CONDITIONING

A fundamental aspect of the FEM, especially in the case of high-order approximations, is the choice of the basis of shape functions. This choice affects the conditioning of matrices and therefore the efficiency of the numerical analysis.

The numerical efficiency is related not only to the computational cost to calculate the elemental matrices but also to the solution of the global system of equations. In addition, the basis affects the number of nonzero coefficients of the matrices, which in turn may have implications in computational cost.

The mass matrix coefficients for a bar element in the local system ξ_1 are given by [18, 35]

$$m_{pq} = \rho A \int_{-1}^{1} \phi_p(\xi_1)\phi_q(\xi_1)d\xi_1, \ 0 \le p,q \le P_1, \tag{4.132}$$

with ρ the material density and A the cross-section area.

The coefficients of the mass matrix corresponding to the internal modes can be rewritten employing Jacobi polynomials with $\alpha_1 = \beta_1 = 1$ as

$$m_{pq} = \rho A \int_{-1}^{1} \phi_p(\xi_1)\phi_q(\xi_1)d\xi_1 = \frac{\rho A}{4} \int_{-1}^{1} (1-\xi_1)(1+\xi_1)P_{p-2}^{1,1}(\xi_1)\phi_q(\xi_1)d\xi_1. \tag{4.133}$$

The term $(1-\xi_1)(1+\xi_1)$ corresponds to the weight function of the orthogonality relation (4.117) of the Jacobi polynomials for $\alpha_1 = \beta_1 = 1$. Because $\phi_q(\xi_1)$ $(0 < q < P_1)$ denotes the internal functions, its order is $q+1$. Therefore, the above integral is zero when $p-2 > q-2$, which implies that $p > q+4$. Due to the symmetry of the matrix, there are two upper and two lower diagonals, and thus making the mass matrix pentadiagonal, as shown in Figure 4.44(a) for $P_1 = 10$. In particular, the internal coefficients of two of these diagonals are zero.

The blocks relative to the vertex and internal functions are denoted as $[M_{vv}]$ and $[M_{ii}]$, respectively, while the coupled blocks are denoted by $[M_{vi}]$ and $[M_{iv}]$. The mass matrix after the Schur complement is shown in Figure 4.44(b), with the coefficients of the coupling blocks $[M_{vi}]$ and $[M_{iv}]$ zero.

The coefficients of the bar stiffness matrix in the local system ξ_1 are given by

$$k_{pq} = EA \int_{-1}^{1} \phi_{p,\xi_1}(\xi_1)\phi_{q,\xi_1}(\xi_1)d\xi_1,$$

with $0 \le p,q \le P_1$, and ϕ_{p,ξ_1} the derivative of ϕ_p with respect to ξ_1.

The coefficients corresponding to internal modes can be written after integration by parts as [35]

$$\begin{aligned}
k_{pq} &= EA \int_{-1}^{1} \phi_{p,\xi_1}(\xi_1)\phi_{q,\xi_1}(\xi_1)d\xi_1 \\
&= -\frac{EA}{4} \int_{-1}^{1} (1-\xi_1)(1+\xi_1)P_{p-2}^{1,1}(\xi_1)\frac{d^2\phi_q(\xi_1)}{d^2\xi_1}d\xi_1 \\
&\quad + \frac{EA}{4}(1-\xi_1)(1+\xi_1)P_{p-2}^{1,1}(\xi_1)\phi_{q,\xi_1}\Big|_{-1}^{1} \\
&= -\frac{EA}{4} \int_{-1}^{1} (1-\xi_1)(1+\xi_1)P_{p-2}^{1,1}(\xi_1)\frac{d^2\phi_q(\xi_1)}{d^2\xi_1}d\xi_1. \tag{4.134}
\end{aligned}$$

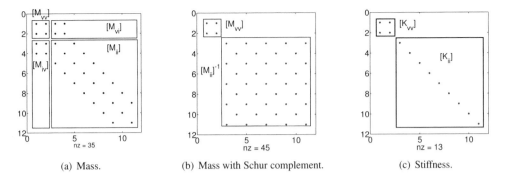

(a) Mass.   (b) Mass with Schur complement.   (c) Stiffness.

Figure 4.44   Sparsity profiles of the mass and stiffness matrices for the Jacobi basis with $\alpha_1 = \beta_1 = 1$.

The boundary term is zero, because the internal shape functions are zero at the element boundaries. Using the orthogonality relation (4.117) of the Jacobi polynomials with $\alpha_1 = \beta_1 = 1$, and noting that $\dfrac{d^2}{d^2 \xi_1} \phi_q(\xi_1)$ is a polynomial of $q - 1$ order, the internal coefficients of the stiffness matrix are zero for $p > q$. Therefore, this matrix is diagonal, as illustrated in Figure 4.44(c) for $P_1 = 10$. The coupling blocks $[K_{vi}]$ and $[K_{iv}]$ are zero, and the same sparsity profile is obtained after the Schur complement. The mass and stiffness matrices are always dense for a Lagrangian basis.

The numerical conditioning provides information about the linear independence of the basis and the number of iterations required for the solution of the systems of equations when using iterative methods. The spectral condition number of the matrices is calculated as [2]

$$\kappa = \frac{\lambda_{\max}}{\lambda_{\min}}, \tag{4.135}$$

with $\lambda_{\max}$ the largest eigenvalue and $\lambda_{\min}$ the lowest nonzero eigenvalue of the matrix.

Figures 4.45(a) and 4.45(b) show the behavior of the condition numbers of the mass and stiffness matrices for the bar with $P_1 = 1, \ldots, 10$ calculated using Lagrange and Jacobi with $\alpha_1 = \beta_1 = 1$ bases. A better conditioning of the Lagrange polynomials is observed for the mass matrix. The Jacobi polynomials have better condition numbers for the stiffness matrix. The condition numbers of the same matrices after applying the Schur complement are also illustrated. It is observed that the mass and stiffness matrices with Lagrange and Jacobi bases have the same condition numbers and they are much smaller when compared to the values calculated for the standard matrices. The results in this section were obtained with $\rho = A = E = 1$.

A fundamental aspect of any polynomial basis is to generate complete approximations. A polynomial $p_n(\xi_1)$ of order n, expressed in the local system ξ_1, is complete if it has all monomials ξ_1^i $(i = 0, \ldots, n)$, that is,

$$p_n(\xi_1) = a_0 + a_1 \xi_1 + a_2 \xi_1^2 + \ldots + + a_n \xi_1^n = \sum_{i=0}^{n} a_i \xi_1^i. \tag{4.136}$$

Thus, for instance, function $f(\xi_1) = 1 + 3\xi_1 - 5\xi_1^2$ is a complete second-order polynomial. On the other hand, $g(\xi_1) = 1 + 2\xi_1^2$ is not complete, because it misses the term relative to ξ_1. The derivative of the previous expression is

$$p_n'(\xi_1) = a_1 + 2a_2 \xi_1 + \ldots + + n a_n \xi_1^n = \sum_{i=1}^{n} i a_i \xi_1^{i-1}. \tag{4.137}$$

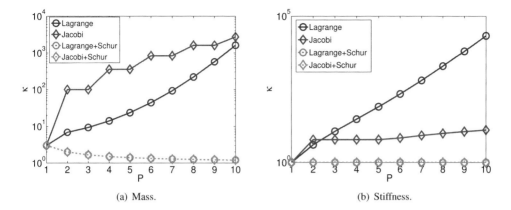

Figure 4.45 Numerical conditioning of the mass and stiffness matrices for one-dimensional elements with Lagrange and Jacobi bases, $P_1 = 1, \ldots, 10$.

Note that the previous polynomial is not complete, because it misses the term relative to the constant coefficient.

Since the mass matrix coefficients are given by the integration of the product of shape functions, and assuming that the basis is complete, all eigenvalues are positive and the matrix is called positive-definite. On the other hand, as the coefficients of the stiffness matrix are given from the integration of the product of shape function derivatives, it has one zero eigenvalue, because the constant term is not represented by the polynomial approximations of the derivatives. Thus, the obtained stiffness matrix is positive semi- definite. This fact justifies the use of λ_{min} as the first nonzero eigenvalue in (4.135).

4.17 MAPPING

As shown in the previous section, the use of normalized local coordinate systems allows a simpler definition of one-dimensional nodal and modal shape functions using Lagrange and Jacobi polynomial bases, respectively. However, the stiffness matrix coefficients must still be calculated using the derivatives of the interpolation functions with respect to the global coordinate x.

It may be observed in equation (4.97) that the terms in ξ_1 are the linear Lagrange shape functions. Therefore, we can employ the shape functions to interpolate not only the primary quantity of the considered problem but also the geometry.

As indicated in (4.97), the geometry of elements with straight sides can be interpolated using the linear shape functions only. Taking the derivative of (4.97), we have the following Jacobian matrix:

$$[J] = \frac{dx(\xi_1)}{d\xi_1} = \frac{1}{2}(x_i^{(e)} - x_{i-1}^{(e)}) = \frac{h^{(e)}}{2}. \tag{4.138}$$

From the previous expression, the following relation between the differentials of the global and local reference systems is obtained:

$$dx = \frac{h^{(e)}}{2}d\xi_1. \tag{4.139}$$

Substituting it in the equations of the shaft mass and stiffness matrices allows their representation in the local coordinate system. Thus,

$$[M^{(e)}] = \rho^{(e)}A^{(e)} \int_{x_{i-1}}^{x_i} [N^{(e)}]^T[N^{(e)}]dx = \rho^{(e)}A^{(e)} \int_{-1}^{1} [N^{(e)}]^T[N^{(e)}]|J|d\xi_1, \tag{4.140}$$

$$[K^{(e)}] = I_p^{(e)} \int_{x_{i-1}}^{x_i} [B^{(e)}]^T [D][B^{(e)}]dx = G^{(e)} I_p^{(e)} \int_{-1}^{1} [B^{(e)}]^T [B^{(e)}]|J|d\xi_1, \qquad (4.141)$$

with

$$|J| = |\det[J]| = \frac{h^{(e)}}{2}. \qquad (4.142)$$

Note that the integration limits are now fixed and equal to -1 and 1. However, the Jacobian of the local-global transformation must be also integrated.

The derivatives of the interpolation functions relative to the global variable x, used in the strain matrix $[B^{(e)}]$, are calculated in terms of the local shape function derivatives observing that

$$\phi_i(x) = \phi_i[x(\xi_1)].$$

Using the chain rule, we have

$$\frac{d\phi_i[x(\xi_1)]}{d\xi_1} = \frac{d\phi_i[x(\xi_1)]}{dx}\frac{dx(\xi_1)}{d\xi_1} = \frac{h^{(e)}}{2}\frac{d\phi_i[x(\xi_1)]}{dx}. \qquad (4.143)$$

Thus, the derivative of the shape function ϕ_i relative to x is given by

$$\frac{d\phi_i[x(\xi_1)]}{dx} = \frac{2}{h^{(e)}}\frac{d\phi_i(\xi_1)}{d\xi_1}. \qquad (4.144)$$

Substituting (4.142) and (4.144) in (4.141), the stiffness matrix for the shaft element is written as

$$[K^{(e)}] = \frac{2G^{(e)}I_p^{(e)}}{h^{(e)}} \int_{-1}^{1} [B_{\xi_1}^{(e)}]^T [B_{\xi_1}^{(e)}]d\xi_1. \qquad (4.145)$$

The matrix of the local shape function derivatives relative to ξ_1 is given by

$$\left[B_{\xi_1}^{(e)}\right] = \left[\begin{array}{cccc} \phi_{0,\xi_1} & \phi_{1,\xi_1} & \cdots & \phi_{P_1,\xi_1} \end{array}\right]. \qquad (4.146)$$

The load vector for the shaft element, expressed in the local coordinate system, is

$$\{f^{(e)}\} = \int_{-1}^{1} m_x(x)[N^{(e)}]^T dx = \frac{h^{(e)}}{2} \int_{-1}^{1} m_x(x(\xi_1))[N^{(e)}]^T d\xi_1. \qquad (4.147)$$

Example 4.23 *Consider the shaft of Example 4.15, but now subjected to a distributed moment given by* $m_x(x) = \sin\left(\frac{\pi}{2}x\right)$ *with the* $x = L$ *end free. The analytical solution for the angle of twist is*

$$\theta_x(x) = \left(\frac{2}{\pi}\right)^2 \frac{1}{GI_p} \sin\left(\frac{\pi}{2}x\right).$$

This problem is approximated with the h refinement using four linear meshes with 2, 4, 8, and 16 free degrees of freedom. For the p refinement, a mesh with two elements of orders $P = 1, 2, 4, 8$ was used, reaching the same number of unknowns used for the h refinement. Figure 4.46 illustrates the relative error in the energy norm in terms of the number of degrees of freedom.

The hp refinement is also employed with the mesh of four elements and orders $P = 1, 2, 4, 8$, reaching 4, 6, 10, and 18 degrees of freedom, respectively. The rate of decrease of the error is larger than the p version for the first three refinements. For $P = 8$, the error is zero.

An algebraic convergence rate is obtained with the h refinement. The error reduces twice when the size of the elements is halved, according to the a priori estimate given in (4.87). On the other hand, the error decreases 10, 300, and 2700 times for each of the p refinements. In the case of the

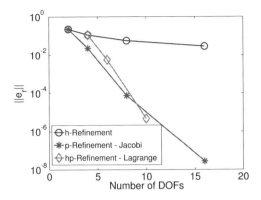

Figure 4.46 Example 4.23: relative errors in the energy norm for the *h*, *p*, and *hp* refinements.

first two hp refinements, the error drops about 200 and 1200 times. Thus, the convergence rates are spectral for case the p and hp refinements.

File shaftsolapex2.m implements the solution of this example and allows the variation of the number of elements and the order of the Lagrange and Jacobi shape functions. An important aspect to be observed is that the x coordinate in the distributed load intensity [see equation (4.147)] is interpolated in each element, using equation (4.97) and the vertex coordinates of the element.

□

We can employ the same interpolation order for the variable of interest and the element geometry, defining the isoparametric mapping. The *x* coordinate is interpolated isoparametrically using the global coordinates $x_i^{(e)}$ of element nodes, as

$$x(\xi_1) = \sum_{i=0}^{P_1} \phi_i^{(e)}(\xi_1)x_i^{(e)}. \tag{4.148}$$

The previous expression allows the use of curved elements, as illustrated in the following example.

Example 4.24 *Consider the quarter of circumference with unit radius illustrated in Figure 4.47, with nodes 0, 1, and 2 indicated. We want to determine the expressions that interpolate the (x,y) global coordinates of the points along the arc of circle.*

The general equation $(x-a)^2 + (y-b)^2 = r^2$ for a circle of radius r and center (a,b) is a second-order expression. It can be mapped to the second-order one-dimensional element, as shown in Figure 4.47. The global (x,y) coordinates are interpolated using the shape functions (4.109) and the mapping (4.148). Therefore,

$$
\begin{aligned}
x(\xi_1) &= \phi_0^{(e)}(\xi_1)x_0^{(e)} + \phi_1^{(e)}(\xi_1)x_1^{(e)} + \phi_2^{(e)}(\xi_1)x_2^{(e)} \\
&= -\frac{1}{2}\xi_1(1-\xi_1)x_0^{(e)} + \frac{1}{2}\xi_1(1+\xi_1)x_1^{(e)} + (1-\xi_1^2)x_2^{(e)}, \\
y(\xi_1) &= \phi_0^{(e)}(\xi_1)y_0^{(e)} + \phi_1^{(e)}(\xi_1)y_1^{(e)} + \phi_2^{(e)}(\xi_1)y_2^{(e)} \\
&= -\frac{1}{2}\xi_1(1-\xi_1)y_0^{(e)} + \frac{1}{2}\xi_1(1+\xi_1)y_1^{(e)} + (1-\xi_1^2)y_2^{(e)}.
\end{aligned}
$$

Substituting the global coordinates of nodes 0 to 2, given in Figure 4.47, and doing some simplifications, we obtain the expressions that interpolate the global coordinates along the arc, that is,

$$x(\xi_1) = -0.2071\xi_1^2 + 0.5000\xi_1 + 0.7071,$$
$$y(\xi_1) = -0.2071\xi_1^2 - 0.5000\xi_1 + 0.7071.$$

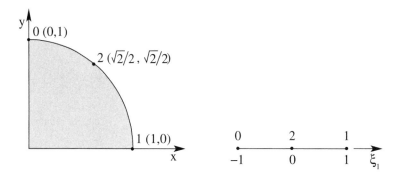

Figure 4.47   Example 4.24: local to global transformation of a circle arc.

Substituting $\xi_1 = -1$, $\xi_1 = 0$, and $\xi_1 = 1$ in the previous expressions, we obtain, respectively, the global coordinates (x,y) of nodes 0 to 2, that is, $(1,0)$, $(\frac{\sqrt{2}}{2}, \frac{\sqrt{2}}{2})$, and $(0,1)$.

□

The Jacobian of the local to global transformation for a curved element is given, in a general form, by

$$[J] = \frac{dx(\xi_1)}{d\xi_1} = \sum_{i=0}^{P_1} \frac{d\phi_i^{(e)}(\xi_1)}{d\xi_1} x_i^{(e)} = \underbrace{\left[\begin{array}{cccc} \phi_{0,\xi_1}^{(e)} & \phi_{1,\xi_1}^{(e)} & \cdots & \phi_{P_1,\xi_1}^{(e)} \end{array} \right]}_{[B_{\xi_1}^{(e)}]} \begin{bmatrix} x_0^{(e)} \\ x_1^{(e)} \\ \vdots \\ x_{P_1}^{(e)} \end{bmatrix}. \tag{4.149}$$

The derivatives of the shape functions relative to x are obtained by the chain rule, i.e.,

$$\frac{d\phi_i^{(e)}[x(\xi_1)]}{d\xi_1} = \frac{d\phi_i^{(e)}[x(\xi_1)]}{dx} \frac{dx[x(\xi_1)]}{d\xi_1} = [J] \frac{d\phi_i^{(e)}[x(\xi_1)]}{dx}. \tag{4.150}$$

The global derivative of $\phi_i^{(e)}$ is obtained by the inverse of the Jacobian matrix

$$\frac{d\phi_i^{(e)}[x(\xi_1)]}{dx} = [J]^{-1} \frac{d\phi_i^{(e)}(\xi_1)}{d\xi_1}. \tag{4.151}$$

As mentioned before, the internal modes of a modal basis are not associated to element nodes. Hence, it is not possible to apply expression (4.148) directly. It is necessary first to determine a set of coordinates for the internal modes, such that the modal isoparametric mapping $x_m(\xi_1)$, obtained with (4.148), provides the same results when using the nodal basis mapping $x_n(\xi_1)$. Hence,

$$x_m(\xi_1) = x_n(\xi_1), \tag{4.152}$$

where

$$x_n(\xi_1) = \sum_{i=0}^{P_1} \phi_{n_i}^{(e)}(\xi_1) x_{n_i}^{(e)}, \tag{4.153}$$

$$x_m(\xi_1) = \sum_{i=0}^{P_1} \phi_{m_i}^{(e)}(\xi_1) x_{m_i}^{(e)}. \tag{4.154}$$

In the previous expression, $\{\phi_{n_i}^{(e)}(\xi_1)\}_{i=0}^{P_1}$ and $\{\phi_{m_i}^{(e)}(\xi_1)\}_{i=0}^{P_1}$ are the nodal and modal bases, respectively; $x_{n_i}^{(e)}$ and $x_{m_i}^{(e)}$ are the respective nodal and modal coordinates, associated to the vertex and internal modes of element e. The local vertex coordinates are the same for both bases, that is, $\xi_0 = -1$ and $\xi_{P_1} = 1$.

Substituting (4.154) in (4.152), we have

$$\sum_{i=0}^{P_1} \phi_{m_i}^{(e)}(\xi_1) x_{m_i}^{(e)} = x_n(\xi_1),$$

The Galerkin method is employed to determine the coordinates x_{m_i} for the element internal modes. Thus, the previous expression is multiplied by the test function $v(\xi_1)$, and integrated in the local interval $[-1, 1]$. Therefore,

$$\left(\sum_{i=0}^{P_1} \int_{-1}^{1} \phi_{m_i}^{(e)}(\xi_1) v(\xi_1) |J| d\xi_1 \right) x_{m_i}^{(e)} = \int_{-1}^{1} x_n(\xi_1) v(\xi_1) |J| d\xi_1. \tag{4.155}$$

The test function is written as the linear combination of the modal basis as

$$v(\xi_1) = \sum_{j=0}^{P_1} b_{m_j} \phi_{m_j}^{(e)}(\xi_1).$$

Substituting this equation in (4.155) and after some simplifications, we obtain the following system of equations for element e:

$$\left(\sum_{i=0}^{P_1} \int_{-1}^{1} \phi_{m_i}^{(e)}(\xi_1) \phi_{m_j}^{(e)}(\xi_1) |J| d\xi_1 \right) x_{m_i}^{(e)} = \int_{-1}^{1} x_n(\xi_1) \phi_{m_j}^{(e)}(\xi_1) |J| d\xi_1, \quad j = 1, \dots, m, \tag{4.156}$$

where the unknowns are the vertex and internal modal coordinates $x_{m_i}^{(e)}$ of the element.

The left side of the above equation is a modal mass matrix $[M_m^{(e)}]$ for element e. The right-hand side is the independent term, denoted by $\{f_m^{(e)}\}$. The previous system of equations is rewritten, in a compact form, as

$$[M_m^{(e)}]\{x_m^{(e)}\} = \{f_m^{(e)}\}. \tag{4.157}$$

The above expression defines a projection problem. Generally, given any function, in this case $x_n(\xi_1)$, we want to determine the coefficients of the linear combination, in this case $x_{m_i}^{(e)}$, which represent a given function in the approximation space used to construct the mass matrix, that is, the modal basis $\{\phi_{m_i}^{(e)}(\xi_1)\}_{i=1}^{P_1}$.

To ensure the continuity of the nodal coordinates between elements, we can employ the Gauss-Lobatto-Legendre collocation points (see Section 4.18) to construct the Lagrange polynomials used in the mapping $x_n(\xi_1)$.

Example 4.25 *Consider the arc of circumference of Example 4.24. Calculate the coordinates associated to the modal basis given in (4.123) with $\alpha = \beta = 1$.*

The expanded form of the system of equations (4.157) is given by

$$\left(\int_{-1}^{1} \begin{bmatrix} \frac{1}{4}(1-\xi_1)^2 & \frac{1}{4}(1-\xi_1)(1+\xi_1) & \frac{1}{8}(1-\xi_1)^2(1+\xi_1) \\ \frac{1}{4}(1-\xi_1)(1+\xi_1) & \frac{1}{4}(1+\xi_1)^2 & \frac{1}{8}(1-\xi_1)(1+\xi_1)^2 \\ \frac{1}{8}(1-\xi_1)^2(1+\xi_1) & \frac{1}{8}(1-\xi_1)(1+\xi_1)^2 & \frac{1}{16}(1-\xi_1)^2(1+\xi_1)^2 \end{bmatrix} |J|d\xi_1 \right)$$

$$\left\{ \begin{array}{c} x_{0_m} \\ x_{1_m} \\ x_{2_m} \end{array} \right\} = \int_{-1}^{1} x_n(\xi_1) \left\{ \begin{array}{c} \frac{1}{2}(1-\xi_1) \\ \frac{1}{2}(1+\xi_1) \\ \frac{1}{4}(1-\xi_1^2) \end{array} \right\} |J|d\xi_1.$$

The nodal mapping and the respective Jacobian are obtained from Example 4.24. Thus,

$$x_n(\xi_1) = -0.2071\xi_1^2 + 0.5000\xi_1 + 0.7071,$$

$$|J| = \frac{dx_n(\xi_1)}{d\xi_1} = -0.4142\xi_1 + 0.5000.$$

Substituting these relations in the previous system of equations and doing the indicated operations, we have

$$\begin{bmatrix} 0.9428 & 0.1667 & 0.1443 \\ 0.1667 & -0.2761 & 0.0224 \\ 0.1443 & 0.0224 & 0.0333 \end{bmatrix} \left\{ \begin{array}{c} x_{0_m} \\ x_{1_m} \\ x_{2_m} \end{array} \right\} = \left\{ \begin{array}{c} 0.0529 \\ 0.0586 \\ 0.3886 \end{array} \right\}.$$

The solution is $\{x_m^{(e)}\} = \{0.0000 \ 1.0000 \ 0.8284\}^T$. Hence, the two coordinates at the ends of the circle arc are recovered and the internal mode coordinate is determined.

Substituting these coordinates and the modal functions in (4.148), the same expression for the nodal mapping of Example 4.24 is recovered, that is,

$$x_m(\xi_1) = x_n(\xi_1) = -0.2071\xi_1^2 + 0.5000\xi_1 + 0.7071.$$

File shaftsolapex3.m implements the solution of this example using nodal and modal bases.
□

Note that the obtained value to the coordinate associated to the internal mode in the previous example does not belong to the quarter of circle of unit radius. The coordinates are denominated physical and transformed when they are calculated using nodal and modal mappings, respectively.

Generally, given a function $u[x(\xi_1)]$ and its interpolation in the element e with a modal basis

$$u[x(\xi_1)] = \sum_{i=0}^{P_1} a_{m_i} \phi_{m_i}^{(e)}(\xi_1), \tag{4.158}$$

the term $u[x(\xi_1)]$ is denominated physical space representation. On the other hand, a_{m_i} are the coefficients of the representation of u in the transformed space.

To obtain the coefficients of the nodal approximation from the preceding equation, we need to calculate it on the coordinates of the collocation points. This comes from the collocation property of the Lagrange polynomials. This is called backward operation.

On the other hand, the forward transformation to obtain the modal coefficients from a physical representation $u[x(\xi_1)]$, is given by a projection problem, analogous to (4.157), i.e.,

$$[M_m^{(e)}]\{a_m^{(e)}\} = \{f_m^{(e)}\}, \tag{4.159}$$

with $\{a_m^{(e)}\}$ the vector of coefficient of the modal approximation and the right-hand side is

$$f_{m_i}^{(e)} = \int_{-1}^{1} u[x(\xi_1)] \phi_{m_i}^{(e)}(\xi_1)|J|d\xi_1, \quad i = 0, \dots, P_1.$$

4.18 NUMERICAL INTEGRATION

The coefficients of the finite element matrices and vectors are calculated more conveniently by numerical integration. Generally, the integral of a polynomial function $u(\xi_1)$ is calculated numerically as

$$I = \int_{-1}^{1} u(\xi_1)d\xi_1 = \sum_{i=0}^{Q_1-1} w_{1_i}u(\xi_{1_i}) + R(u), \tag{4.160}$$

with ξ_{1_i} and w_{1_i} the coordinates and weights of the Q_1 integration points and $R(u)$ the approximation error term of the quadrature procedure.

The number of points is selected such that the cost of the numerical integration procedure is minimized and the error is zero. In this sense the Gauss-Legendre quadrature is the most common procedure for the FEM. In fact, it is a particular case of the Gauss-Jacobi quadrature, which will be presented here according to [35].

Let $u(\xi_1)$ be a polynomial function of order $2Q_1 - k$, with k an integer related to the distribution of the quadrature points, as will be seen later. We want to integrate $u(\xi_1)$ with respect to the weighting function $(1 - \xi_1)^{\alpha}(1 + \xi_1)^{\beta}$, that is,

$$I = \int_{-1}^{1} (1 - \xi_1)^{\alpha}(1 + \xi_1)^{\beta}u(\xi_1)d\xi_1. \tag{4.161}$$

As $u(\xi_1)$ is a polynomial function, we can write it as the following linear combination of Lagrange polynomials $l_i^{(Q_1-1)}(\xi_1)$ through the Q_1 integration points $\xi_{1_i}^{\alpha,\beta}$, i.e.,

$$u(\xi_1) = \sum_{i=0}^{Q_1-1} u(\xi_{1_i}^{\alpha,\beta})l_i^{(Q_1-1)}(\xi_1) + s(\xi_1)r(\xi_1). \tag{4.162}$$

The coefficients $u(\xi_{1_i}^{\alpha,\beta})$ are used due to the collocation property of the Lagrange polynomials; $s(\xi_1)$ is a polynomial of order Q_1 with roots on the integration points $\xi_{1_i}^{\alpha,\beta}$, that is, $s(\xi_{1_i}^{\alpha,\beta}) = 0$; $r(\xi_1)$ is the remainder polynomial with order $Q_1 - k$. Hence, the previous expression represents a polynomial of order $2Q_1 - k$.

Substituting (4.162) in (4.161), we have

$$
\begin{aligned}
I &= \int_{-1}^{1} (1 - \xi_1)^{\alpha}(1 + \xi_1)^{\beta} \left(\sum_{i=0}^{Q_1-1} u(\xi_{1_i}^{\alpha,\beta})l_i^{(Q_1-1)}(\xi_1) + s(\xi_1)r(\xi_1) \right) d\xi_1 \\
&= \sum_{i=0}^{Q_1-1} u(\xi_{1_i}^{\alpha,\beta}) \underbrace{\left(\int_{-1}^{1} (1 - \xi_1)^{\alpha}(1 + \xi_1)^{\beta}l_i^{(Q_1-1)}(\xi_1)d\xi_1 \right)}_{w_{1_i}^{\alpha,\beta}} \\
&\quad + \underbrace{\left(\int_{-1}^{1} (1 - \xi_1)^{\alpha}(1 + \xi_1)^{\beta}s(\xi_1)r(\xi_1)d\xi_1 \right)}_{R(u)} \\
&= \sum_{i=0}^{Q_1-1} w_{1_i}^{\alpha,\beta}u(\xi_{1_i}^{\alpha,\beta}) + R(u),
\end{aligned} \tag{4.163}
$$

with $w_{1_i}^{\alpha,\beta}$ and $\xi_{1_i}^{\alpha,\beta}$ the weights and roots of the integration points, respectively; $R(u)$ is the approximation error term of the Gauss-Jacobi quadrature rule.

The weights $w_{1_i}^{\alpha,\beta}$ are defined by the integration of the Lagrange polynomials, multiplied by the weight function, that is,

$$w_{1_i}^{\alpha,\beta} = \int_{-1}^{1} (1 - \xi_1)^{\alpha}(1 + \xi_1)^{\beta} l_i^{(Q_1 - 1)}(\xi_1) d\xi_1. \tag{4.164}$$

We want to determine the $s(\xi_1)$ function, such that the approximation error term $R(u)$ is zero, which results in an exact integration. For $s(\xi_1) = P_{Q_1}^{\alpha,\beta}(\xi_1)$ and $k = 1$, we have

$$R(u) = \int_{-1}^{1} (1 - \xi_1)^{\alpha}(1 + \xi_1)^{\beta} P_{Q_1}^{\alpha,\beta}(\xi_1) r(\xi_1) d\xi_1 = 0,$$

because $r(\xi_1)$ is a polynomial of order $Q_1 - 1$ and $s(\xi_1)$ of order Q_1 (see Example 4.21).

This choice defines the Gauss-Jacobi quadrature and the coordinates $\xi_{1_i}^{\alpha,\beta}$ are the roots of $P_{Q_1}^{\alpha,\beta}(\xi_1)$, which are obtained with the Newton-Raphson method, as indicated in[35] and used in Example 4.26.

For some cases, the coordinates $\xi_1 = \pm 1$ are employed as quadrature points. For instance, their use is convenient in the application of boundary conditions. This choice defines the Gauss-Radau and Gauss-Lobatto quadratures. The Gauss ($k = 1$), Gauss-Radau ($k = 2$), and Gauss-Lobatto ($k = 3$) quadratures integrate exactly the polynomials of orders $2Q_1 - 1$, $2Q_1 - 2$, and $2Q_1 - 3$, respectively. Figure 4.48 illustrates the three considered distributions of points. For $\alpha = \beta = 0$, we have the Gauss-Legendre (GL), Gauss-Radau-Legendre (GRL), and Gauss-Lobatto-Legendre (GLL) quadratures, which are commonly used in the FEM.

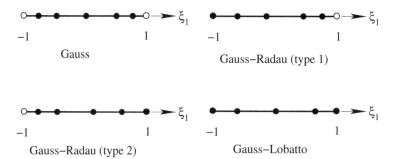

Figure 4.48   Gauss ($k = 1$), Gauss-Radau ($k = 2$), and Gauss-Lobatto ($k = 3$) quadratures.

The required number of points to integrate exactly a polynomial function of order P_1 using the Gauss, Gauss-Radau, and Gauss-Lobatto quadratures are, respectively,

$$2Q_1 - 1 \;=\; P_1 \rightarrow Q_1 = \frac{1}{2}(P_1 + 1), \tag{4.165}$$

$$2Q_1 - 2 \;=\; P_1 \rightarrow Q_1 = \frac{1}{2}(P_1 + 2), \tag{4.166}$$

$$2Q_1 - 3 \;=\; P_1 \rightarrow Q_1 = \frac{1}{2}(P_1 + 3). \tag{4.167}$$

Note that because the Radau and Lobatto quadratures use the vertex coordinates, they require a larger number of points to integrate the same polynomial function of order P_1.

Example 4.26 *Consider the following function:*

$$u(\xi_1) = 1 - \xi_1^2 = (1 - \xi_1)(1 + \xi_1), \quad \xi_1 \in [-1, 1].$$

The exact integration is

$$I = \int_{-1}^{1} u(\xi_1) d\xi_1 = \int_{-1}^{1} (1 - \xi_1^2) d\xi_1 = \xi_1 - \frac{1}{3}\xi_1^3 \Big|_{-1}^{1} = \frac{4}{3}.$$

Table 4.1 lists the coordinates and weights of the Gauss-Legendre (GL), Gauss-Radau-Legendre (GRL), and Gauss-Lobatto-Legendre (GLL) quadrature rules to integrate exactly the given function. The required number of points is 2, 2, and 3 points, respectively.

Since $u(\xi_1)$ is the weight function for $\alpha_1 = \beta_1 = 1$, the value of the integral is equal to the weight for the Gauss-Jacobi (GJ) quadrature (see last row of Table 4.1). There is no need to do the sum indicated in (4.163), as presented in [11].

The Newton-Raphson procedure for the coordinate calculation is implemented in the jacobi-root.m file. This procedure uses the jacobi-polynomials.m and D-jacobi-polynomials.m files, which implement the recurrence relations (4.120) and (4.121), respectively. The calculation of the weights, according to equation (4.163), is implemented in jacobi-weights.m.

Quadrature	ξ_1	w_1
GL ($\alpha = \beta = 0$)	-0.5774	1
	0.5774	1
GRL ($\alpha = \beta = 0$)	-1.0000	1/2
	0.3333	3/2
GLL ($\alpha = \beta = 0$)	-1.0000	1/3
	0.0000	4/3
	1.0000	1/3
GJ ($\alpha = 1, \beta = 1$)	0.0000	4/3

Table 4.1
Example 4.26: coordinates and weights of the integration points for different quadratures.

☐

Example 4.27 *All MATLAB programs presented so far have used symbolic manipulation. That means that symbolic variables are employed and differentiation and integration procedures are performed analytically. Although quite interesting for didactic purposes, symbolic manipulation is slow in terms of computational time to run the scripts.*

File shaftsolapex4.m implements the solution of Example 4.23 using numerical integration and Jacobi polynomials. The coordinates and weights of the integration points are calculated by the GetPointsWeight1D.m function. It has as input arguments α_1, β_1, the integrand order to determine the number of points, and the quadrature type (Gauss-Jacobi (GJ), Gauss-Radau-Jacobi (GRJ), or Gauss-Lobatto-Jacobi (GLJ)).

Figure 4.49 illustrates the same refinements considered in Figure 4.46 and introduces the hp refinement, similar to the previous one, but with an eight element mesh. The number of degrees of freedom is 4, 16, 32, and 64, while the rates of decrease of errors are of 40, 490, and 10^8 times.

☐

The coordinates of the Gauss-Jacobi quadrature can be used as the nodes or collocation points of the Lagrange polynomials. In particular, the Gauss-Lobatto-Legendre quadrature has been employed with Lagrange polynomials, allowing the generation of a polynomial basis with better condition numbers for the element mass matrices. Furthermore, the nodal interpolation functions become less oscillatory when compared with the functions generated with equally spaced collocation points.

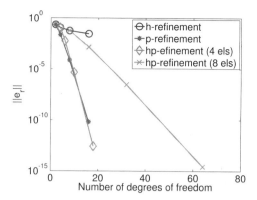

Figure 4.49 Example 4.27: relative errors in the energy norm for the h, p, and hp refinements.

The number of points for the exact integration of the mass matrix coefficients given in (4.140), with integrand order $2P_1$ and using the Gauss-Lobatto-Legendre quadrature, is given by (4.167). Therefore,

$$Q_1 = \frac{1}{2}(2P_1 + 3) = P_1 + \frac{3}{2} \rightarrow Q_1 = P_1 + 2.$$

When using $P_1 + 2$ integration points, the element mass matrix is dense.

However, when using $P_1 + 1$ Gauss-Lobatto-Legendre quadrature points simultaneously as the nodes of the Lagrange polynomials and the integration coordinates, we obtain a diagonal mass matrix. This fact is due to the collocation property of the Lagrange polynomials. Therefore, for the coefficients of the local mass matrix, we have

$$M_{pq}^{(e)} = \int_{-1}^{1} \phi_p^{(e)}(\xi_1)\phi_q^{(e)}(\xi_1)d\xi_1 = \sum_{i=0}^{P_1} w_{1_i}^{0,0}\phi_p^{(e)}(\xi_{1_i})\phi_q^{(e)}(\xi_{1_i}) = \sum_{i=0}^{P_1} w_{1_i}^{0,0}\delta_{pi}\delta_{qi} = w_{1_p}^{0,0}\delta_{pq}. \quad (4.168)$$

The missed integration point makes inexact the coefficients of the diagonal of the mass matrix.

The coefficients of the main diagonal of the mass matrix, using reduced integration, are equal to the sum of the row coefficients of the consistently integrated mass matrix, as can be seen for row p,

$$\sum_{q=0}^{P_1} M_{pq}^{(e)} = \sum_{q=0}^{P_1} \int_{-1}^{1} \phi_p^{(e)}(\xi_1)\phi_q^{(e)}(\xi_1)d\xi_1 = \int_{-1}^{1} \phi_p^{(e)}(\xi_1)\left(\sum_{q=0}^{P_1}\phi_q^{(e)}(\xi_1)\right)d\xi_1.$$

The sum of all Lagrange polynomials is equal to 1. Hence,

$$\sum_{q=0}^{P_1} M_{pq}^{(e)} = \int_{-1}^{1} \phi_p^{(e)}(\xi_1)d\xi_1 = w_{1_p}^{0,0}.$$

Equation (4.164) was used for the weights w_{1_p} with $\alpha = \beta = 0$. Hence, the main diagonal coefficients of the local mass matrix are equal to the weights of the collocation points.

4.19 COLLOCATION DERIVATIVE

Consider the polynomial function $u(\xi_1)$ with order P_1 and $\xi_1 \in [-1, 1]$. We can write it exactly as the linear combination of $Q_1 \geq P_1 + 1$ Lagrange polynomials $l_p^{(Q_1)}(\xi_1)$ as

$$u(\xi_1) = \sum_{i=0}^{Q_1-1} u(\xi_{1_i})l_i^{(Q_1-1)}(\xi_1), \quad (4.169)$$

with ξ_{1_i} the collocation points.

The derivative of $u(\xi_1)$ is given by

$$\frac{du(\xi_1)}{d\xi_1} = \sum_{i=0}^{Q_1-1} u(\xi_{1_i}) \frac{dl_i^{(Q_1-1)}(\xi_1)}{d\xi_1}. \tag{4.170}$$

Generally, it is required to calculate the previous equation at the ξ_{1_j} coordinates of the integration or collocation points, that is,

$$\frac{du(\xi_{1_j})}{d\xi_1} = \sum_{i=0}^{Q_1-1} u(\xi_{1_i}) \frac{dl_i^{(Q_1-1)}(\xi_{1_j})}{d\xi_1} = \sum_{i=0}^{Q_1-1} d_{ij} u(\xi_{1_i}), \tag{4.171}$$

with

$$d_{ij} = \frac{dl_i^{(Q_1-1)}(\xi_{1_j})}{d\xi_1}. \tag{4.172}$$

We can denote the Lagrange polynomials in the following alternative form [24]:

$$l_i^{(Q_1-1)}(\xi_1) = \frac{h_{Q_1}(\xi_1)}{h'_{Q_1}(\xi_{1_i})(\xi_1 - \xi_{1_i})}, \tag{4.173}$$

with the nodal polynomial $h_{Q_1}(\xi_1)$ given by

$$h_{Q_1}(\xi_1) = \prod_{i=0}^{Q_1}(\xi_1 - \xi_{1_i}). \tag{4.174}$$

The derivative of (4.173) is

$$\frac{dl_i^{(Q_1-1)}(\xi_1)}{d\xi_1} = \frac{h'_{Q_1}(\xi_1)(\xi_1 - \xi_{1_i}) - h_{Q_1}(\xi_1)}{h'_{Q_1}(\xi_{1_i})(\xi_1 - \xi_{1_i})^2}. \tag{4.175}$$

As both numerator and denominator of the previous expression tend to zero when $\xi_1 \to \xi_{1_i}$, L'Hôpital's rule is employed to solve this indetermination, that is,

$$\lim_{\xi_1 \to \xi_{1_i}} \frac{dl_i^{(Q_1-1)}(\xi_1)}{d\xi_1} = \lim_{\xi_1 \to \xi_{1_i}} \frac{h''_{Q_1}(\xi_1)}{2h'_{Q_1}(\xi_1)} = \frac{h''_{Q_1}(\xi_{1_i})}{2h'_{Q_1}(\xi_{1_i})}.$$

The term d_{ij} has the following final form:

$$d_{ij} = \begin{cases} \dfrac{h'_{Q_1}(\xi_{1_j})}{h'_{Q_1}(\xi_{1_i})} \dfrac{1}{(\xi_1 - \xi_{1_i})} & i \neq j \\[4mm] \dfrac{h''_{Q_1}(\xi_{1_i})}{2h'_{Q_1}(\xi_{1_i})} & i = j \end{cases}. \tag{4.176}$$

This is the general expression for the derivatives of the Lagrange polynomials calculated at the integration points. Specific expressions can be obtained for h'_{Q_1} and h''_{Q_1}, depending on the considered quadrature.

The most employed expressions use the Gauss-Legendre quadrature points. The roots of the Jacobi polynomials are denoted by $\xi_{1_i}^{\alpha,\beta}$. The derivative matrices d_{ij} for three quadrature rules are presented below according to [35].

1. Gauss-Legendre: in this case, the collocation points are the roots of the Legendre polynomial $P_{Q_1}^{0,0}(\xi_1)$, that is, $\xi_{1_i} = \xi_{1_i}^{0,0}$ for $0 \leq i \leq Q_1 - 1$. The d_{ij} coefficients are

$$
d_{ij} = \begin{cases} \dfrac{\left[P_{Q_1}^{0,0}(\xi_{1_i})\right]'}{\left[P_{Q_1}^{0,0}(\xi_{1_j})\right]'(\xi_{1_i} - \xi_{1_j})} & i \neq j, 0 \leq i,j \leq Q_1 - 1 \\[4mm] \dfrac{\xi_{1_i}}{1 - \xi_{1_i}^2} & i = j \end{cases} \tag{4.177}
$$

2. Gauss-Radau-Legendre: the collocation points are $\xi_{1_0} = -1$ and the roots of the Jacobi polynomial $P_{Q_1-1}^{0,1}(\xi_1)$, that is, $\xi_{1_i} = \xi_{1_i}^{0,1}$ with $1 \leq i \leq Q_1 - 1$. The d_{ij} coefficients are

$$
d_{ij} = \begin{cases} -\dfrac{1}{4}(Q_1 - 1)(Q_1 + 1) & i = j = 0 \\[4mm] \dfrac{P_{Q_1-1}^{0,0}(\xi_{1_i})}{P_{Q_1-1}^{0,0}(\xi_{1_j})} \dfrac{1 - \xi_{1_j}}{1 - \xi_{1_i}} \dfrac{1}{\xi_{1_i} - \xi_{1_j}} & i \neq j, 0 \leq i,j \leq Q_1 - 1 \\[4mm] \dfrac{1}{2(1 - \xi_{1_i})} & i = j, 1 \leq i \leq Q_1 - 1 \end{cases} \tag{4.178}
$$

3. Gauss-Lobatto-Legendre: the collocation points are $\xi_{1_0} = -1$, $\xi_{1_{Q_1-1}} = 1$ and the roots of the Jacobi polynomial $P_{Q_1-2}^{1,1}(\xi_1)$, that is, $\xi_{1_i} = \xi_{1_i}^{1,1}$ with $1 \leq i,j \leq Q_1 - 2$. The d_{ij} coefficients are:

$$
d_{ij} = \begin{cases} -\dfrac{1}{4}Q_1(Q_1 - 1) & i = j = 0 \\[4mm] \dfrac{P_{Q_1-1}^{0,0}(\xi_{1_i})}{P_{Q_1-1}^{0,0}(\xi_{1_j})} \dfrac{1}{\xi_{1_i} - \xi_{1_j}} & i \neq j, 0 \leq i,j \leq Q_1 - 1 \\[4mm] 0 & i = j, 1 \leq i \leq Q_1 - 2 \\[4mm] \dfrac{1}{4}Q_1(Q_1 - 1) & i = j = Q_1 - 1 \end{cases} \tag{4.179}
$$

Example 4.28 *Consider the function $f(\xi_1) = 5\xi_1^8$ with $\xi_1 \in [-1,1]$. The analytical derivative is $f'(\xi_1) = 40\xi_1^7$. We want to evaluate the collocation derivative $f_c'(\xi_1)$ using $Q_1 = 8,9,10$ and the Gauss-Legendre quadrature. The average error will be calculated as*

$$
e = \frac{1}{Q_1} \sum_{i=0}^{Q_1-1} |f'(\xi_{1_i}) - f_c'(\xi_{1_i})|.
$$

The obtained values are 0.6527, $5.4054398 \times 10^{-15}$, and 1.50029×10^{-15}, respectively, for 8, 9, and 10 collocation points.

□

Example 4.29 *Consider now the function $g(x) = 5x^8$ with $x \in [2,8]$. Its analytical derivative is $g'(x) = 40x^7$. We want to evaluate the collocation derivative $g_c'(\xi_1)$ using $Q_1 = 8,9,10$ and the Gauss-Legendre quadrature.*

Considering the mapping given in (4.97), the local coordinates of the collocation points are transformed to the $[2,8]$ interval. By the chain rule, we have

$$\frac{dg\left[x(\xi_1)\right]}{d\xi_1} = \frac{dg\left[x(\xi_1)\right]}{dx}\frac{dx(\xi_1)}{d\xi_1} = [J]\frac{dg\left[x(\xi_1)\right]}{dx}.$$

Taking the inverse of the Jacobian, we obtain the derivative of $g(x)$ relative to x, that is,

$$\frac{dg\left[x(\xi_1)\right]}{dx} = [J]^{-1}\frac{dg\left[x(\xi_1)\right]}{d\xi_1}.$$

The calculated derivative at the $j = 0,\ldots,Q_1 - 1$ collocation points is

$$\frac{dg\left[x(\xi_{1_j})\right]}{dx} = [J]^{-1}\sum_{i=0}^{Q_1-1} d_{ij}g\left[x(\xi_{1_i})\right].$$

The obtained errors are 1.42740×10^3, 2.72669×10^{-8}, and 1.02085×10^{-9} for 8, 9, and 10 collocation points, respectively.

MATLAB file shaftsolapex5.m implements the collocation derivative procedure to functions f and g of the two last examples.

□

The modal bases can be expressed in terms of the Lagrange polynomials. In this way, the collocation derivative procedure can be also used for modal expansions.

4.20 FINAL COMMENTS

In this chapter, the variational formulation of circular torsion of shafts was presented. Again, the same sequence of steps was applied to obtain the shaft model, based on the kinematics definition and the concept of strain internal work. The use of singularity functions allowed the solution of problems with concentrated loads applied along the shaft length. After this chapter, it is expected that the reader has become familiar with the steps of the variational formulation procedure of mechanical models.

Regarding approximation, several fundamental concepts of the FEM were introduced, such as local coordinate systems, construction of shape functions, mapping, numerical integration, and collocation derivative. More formal aspects, such as the representation of the BVP and weak forms using operators, *a priori* error estimates, and refinement strategies for low and high orders were also considered. Furthermore, several programs in MATLAB illustrated the presented concepts. Other aspects of the FEM will be considered in the following chapters.

4.21 PROBLEMS

1. A circular shaft with diameter of 50 mm is subjected to the constant twisting moment $T = 20000$ Nm applied at the free end. Considering a length of 5 m and an angle of twist $\theta_x = 6°$, what is the shear modulus G?

2. Consider the hollow cross-section shaft with an internal diameter of 5 cm and external diameter of 8 cm. The maximum shear stress is $\tau_t = 80$ MPa. What is the stress in the internal fibers of the shaft?

3. A shaft must transmit 7500 hp of power at 3600 rpm. The admissible shear stress is 80 MPa and the external/internal diameter ratio is equal to 2. What is the external diameter for $G = 80$ GPa?

4. What are the advantages of using a shaft with hollow section compared to a solid section for a given load?
5. Suppose that the shear modulus for a given material is unknown, but the Young's modulus was obtained by the tensile test. Furthermore, the Poisson ratio is known and equal to 0.30. How can we use these values and solve the circular torsion problem?
6. Given the shafts shown in Figure 4.50, indicate the boundary conditions and load equations for the circular torsion model.

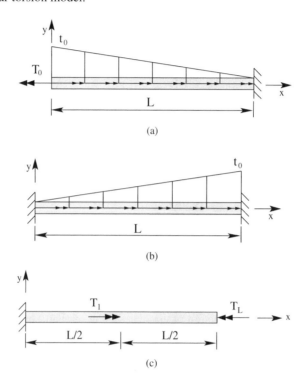

(a)

(b)

(c)

Figure 4.50   Problem 6.

7. Design the shafts of Figure 4.51 given
   a. $G = 70$ GPa and $\tau_t = 60$ MPa for 4.51(a);
   b. $G = 27.6$ GPa and $\tau_t = 100$ MPa for 4.51(b).
8. Consider a solid steel shaft with diameter $d_1 = 5$ cm, enclosed by an aluminum tube which is 3 cm thick [see Figure 4.52]. Both metals are rigidly connected to each other. Calculate the maximum shear stresses in the steel and aluminum for the torque $T = 2000$ Nm applied to the free end. Plot the stresses in the cross-section. Consider $G_1 = 120$ MPa for steel and $G_2 = 70$ MPa for aluminum.
9. The shaft of Figure 4.53 is statically indeterminate. Find the angle of twist at the interface of the segments with smaller and larger cross-sections (point C). Plot also the twisting moment diagram for the shaft. Assume that the diameter of the smaller section is 25 mm. The larger cross-section has an internal diameter of 25 mm and outer diameter 50 mm. The torques are $T_1 = 300$ Nm and $T_2 = 2000$ Nm and $G = 75$ GPa.
10. An aluminum tube ($E_1 = 70$ GPa and $v = 0.25$) has a steel core ($E_2 = 210$ GPa and $v = 0.3$). The cross-section is shown in Figure 4.54. The shaft has 2 m of length. The left end is fixed and the right end is free. A concentrated torque is applied at the free end. What is the maximum shear stress at this end? Plot the shear stress distribution.

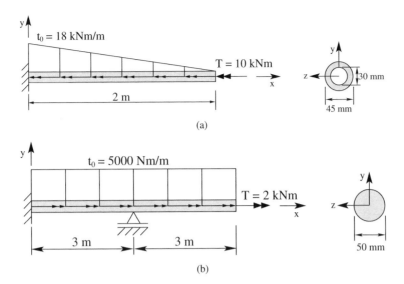

(a)

(b)

Figure 4.51 Problem 7.

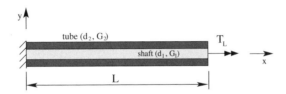

Figure 4.52 Problem 8.

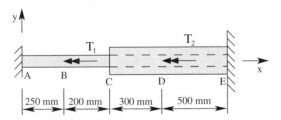

Figure 4.53 Problem 9.

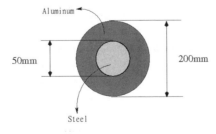

Figure 4.54 Problem 10.

11. A circular shaft with 2 m of length is fixed in one of the ends and free at the other, and is subjected to a distributed load as illustrated in Figure 4.55. The torsional stiffness $\dfrac{GI_p}{L}$ of the shaft is contant. Calculate the angle of twist at the free end due to the distributed torque.

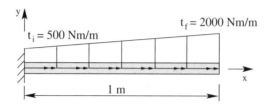

Figure 4.55 Problem 11.

12. Consider the function $f(x) = \sin(\pi x)$ for $x \in [0, \pi]$. Calculate the error for the numerical integration of f for $P = 1, \ldots, 10$, using the Gauss-Legendre, Gauss-Legendre-Radau, and Gauss-Legendre-Lobatto quadratures.

13. Implement MATLAB programs to solve the shaft of Figure 4.51. Calculate the approximation errors in the L_2 and energy norms.

14. Calculate the collocation derivative for function $f(x) = 3\sin(\pi x)$, $x \in [0, 4]$ using 4 to 12 collocation points. Plot the average error in terms of the number of collocation points.

5 FORMULATION AND APPROXIMATION OF BEAMS

5.1 INTRODUCTION

Similarly to the cases of bars and shafts, a beam is a one-dimensional mechanical element for which the length is much larger than the cross-section dimensions. The Cartesian reference system is located at the geometric center (GC) of the left end cross- section, as illustrated in Figure 5.1.

The main interest is related to displacement actions that cause bending of beams, as shown in Figure 5.2 for a positive bending action about the z axis. In this case, there is a length increase of the longitudinal fibers located below the x axis of the reference system, while the lengths of fibers above the x axis decrease.

The analysis of beams is quite common in real engineering problems, and the study of the beam models is of fundamental importance. For this purpose, the Euler-Bernoulli and Timoshenko beam models are considered. The basic difference between these models is related to the fact that the Euler-Bernoulli formulation does not consider the shear strains presented in the cross-sections and longitudinal fibers of the beam. To include this effect, we consider the Timoshenko model discussed in the next chapter. The following sections present the Euler-Bernoulli beam model.

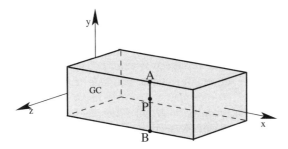

Figure 5.1 Coordinate system located at the cross-section geometrical center (GC) at the left end of the beam.

The classical Euler-Bernoulli model considers prismatic uniform (constant cross-section) beams, with the longitudinal length as the predominant dimension. Only cross-sections which are symmetric about the y axis are considered, with bending along the plane defined by the z axis. The bending actions are characterized by transversal displacements in the y direction, associated to rotations of the cross-sections about the z axis, according to the coordinate system shown in Figure 5.1.

Beams with bending in y and z planes simultaneously are called composite bending beams and will be studied in Chapter 7. In this case, the bending actions about the y axis are characterized by transversal displacements in z and rotations in y.

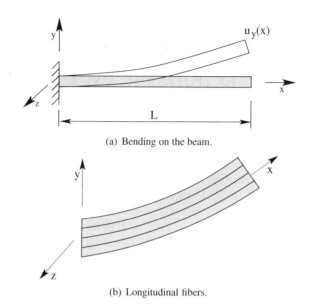

(a) Bending on the beam.

(b) Longitudinal fibers.

Figure 5.2   Positive bending action of a beam about the z axis of the reference system.

5.2   KINEMATICS

The kinematic hypothesis of the Euler-Bernoulli model assumes that the possible displacement actions are such that the cross-sections remain planar, nondeformed and orthogonal to the longitudinal axis x of the beam. This assumption is illustrated in Figure 5.3 for the section AB, which is at a distance of x units from the origin of the reference system. After the bending action, section AB is at the position indicated by $A''B''$ but remains planar, nondistorted, and orthogonal to the tangent of beam deformed axis x.

Therefore, the possible displacement actions make each cross-section x displace rigidly by $u_y(x)$ in y direction (constant for all points of the section). There is also a rigid rotation about the z axis, denoted by $\theta_z(x)$, as illustrated in Figure 5.3 for the section AB.

Initially, section AB takes the position $A'B'$, due to the rigid transversal displacement $u_y(x)$ in the y direction. From this position, there is a rigid rotation of angle $\theta_z(x)$ about the z axis, and the section rotates until it reaches the final position $A''B''$. Due to this rotation, section $A'B'$ has the displacement u_x in the longitudinal direction and the Δu_y displacement in y, as shown in Figure 5.4.

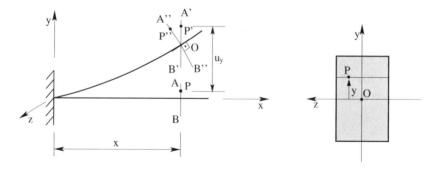

Figure 5.3   Kinematics of the Euler-Bernoulli beam.

From Figure 5.4, the axial displacement u_x of any point P, which is far y units from the cross-

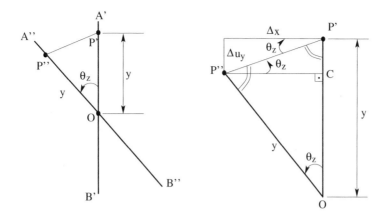

Figure 5.4 Detail of the cross-section after bending of the Euler-Bernoulli beam.

section geometrical center O, is given by the difference between the final and initial positions of this point, that is,

$$u_x = (x - \Delta x) - x = -\Delta x.$$

The following trigonometric relations are valid for the triangles $P''CP'$ and OCP'', respectively,

$$\sin \theta_z = \frac{\Delta x}{y} = -\frac{u_x}{y}, \tag{5.1}$$

$$\tan \theta_z = \frac{\Delta u_y}{\Delta x}. \tag{5.2}$$

The distances $\overline{OP}$, $\overline{OP'}$, and $\overline{OP''}$ are all equal to y, because there is ny deformation of the cross-section.

In the case of small displacements, we have θ_z and Δx also small. Based on this hypothesis, the following simplifications are valid to the previous equations:

$$\sin \theta_z \approx \theta_z \quad \text{and} \quad \tan \theta_z \approx \theta_z.$$

Using these simplifications in (5.1) and (5.2), and taking the limit to $\Delta x \to 0$ in (5.2), we have

$$u_x(x,y) = -y\theta_z(x), \tag{5.3}$$

$$\theta_z(x) = \lim_{\Delta x \to 0} \frac{\Delta u_y}{\Delta x} = \frac{du_y(x)}{dx}. \tag{5.4}$$

Combining these expressions, the final equation for the axial displacement of the points of the cross-sections for the Euler-Bernoulli beam is

$$u_x(x,y) = -y\frac{du_y(x)}{dx}. \tag{5.5}$$

Therefore, due to the bending action, the axial displacement $u_x(x,y)$ varies linearly in the cross-section. Note that $\theta_z(x) = \dfrac{du_y(x)}{dx}$ represents the rotation or bending angle of the cross-section x about the z axis of the reference system. The negative sign in $u_x(x,y)$ comes from the fact that when the rotation is positive, that is, $\theta_z(x) > 0$, the longitudinal displacement is in the negative direction of axis x of the reference system for the points situated above the GC of the section, as can be seen in Figure 5.5(b). Figure 5.5 illustrates the behavior of the axial displacement $u_x(x,y)$ for positive $[\theta_z(x) > 0]$ and negative $[\theta_z(x) < 0]$ rotations.

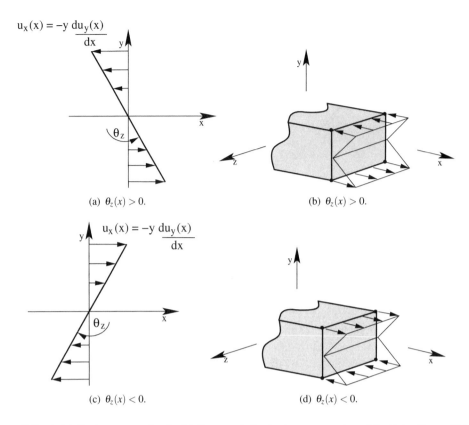

Figure 5.5 Axial displacement $u_x(x,y)$ with linear variation in the cross-section of the Euler-Bernoulli beam.

Example 5.1 *Consider the cantilever beam of length L, clamped at the left end ($x = 0$) and sub-jected to a δ transversal displacement at the free end ($x = L$), as illustrated in Figure 5.6.*

The transversal displacement $u_y(x)$ of the beam sections of this example is

$$u_y(x) = \frac{\delta}{2L^3}(3L - x)x^2.$$

This function will be determined later with the solution of the equations which will be derived in the final part of the beam formulation (see Example 5.7).

Analogously, the function describing the rotation of the beam cross-sections is obtained by the derivative of the previous function for the transversal displacement. Hence,

$$\theta_z(x) = \frac{du_y(x)}{dx} = \frac{\delta}{2L^3}(6L - 3x)x.$$

The transversal displacement diagram is a plot of the function $u_y(x)$ and gives the value of the transversal displacement for the points of a given section x; analogously for the $\theta_z(x)$ rotation. The displacement and rotation diagrams of this example are illustrated in Figures 5.23(a) and 5.23(b). It is observed that $u_y(x)$ and $\theta_z(x)$ satisfy, respectively, the kinematic constraints of zero displacement and rotation at $x = 0$, that is, $u_y(x = 0) = 0$ and $\theta_z(x = 0) = 0$. Furthermore, $u_y(x)$ and $\theta_z(x)$ are continuous functions.

□

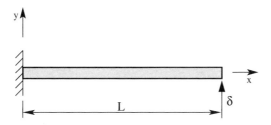

Figure 5.6 Example 5.1: cantilever beam with a deflection at the free end.

5.3 STRAIN MEASURE

The kinematics of the Euler-Bernoulli beam is such that the cross-sections have the axial displacement $u_x(x,y)$ which varies linearly with the y coordinate of each point. Similarly to the bar case, the strain measure associated with $u_x(x,y)$ is the specific longitudinal strain $\varepsilon_{xx}(x)$.

The procedure for determining the expression of $\varepsilon_{xx}(x)$ is analogous to the bar and shaft models. Therefore, the variation of the axial displacements of points located in two arbitrary sections of the beam is considered.

Figure 5.7 illustrates the sections AB and CD, respectively, far x and $x+\Delta x$ units from the origin of the beam reference system. Due to the bending action, points P and Q of these sections, which are y units far from the GC, have $u_x(x,y)$ and $u_x(x+\Delta x)$ axial displacements. These displacements are given in terms of the $u_y(x)$ and $u_y(x+\Delta x, y)$ transversal displacements, respectively, by

$$u_x(x,y) = -y\frac{du_y(x)}{dx} = -y\theta_z(x),$$

$$u_x(x+\Delta x, y) = -y\frac{du_y(x+\Delta x)}{dx} = -y\theta_z(x+\Delta x).$$

The relative axial displacement between points Q and P is determined by the difference of the previous expressions

$$\Delta u_x = u_x(x+\Delta x, y) - u_x(x,y) = -y\frac{d}{dx}[u_y(x+\Delta x, y) - u_y(x,y)]$$

$$= -y\frac{d}{dx}(\Delta u_y) = -y\Delta\left(\frac{du_y}{dx}\right) = -y\Delta\theta_z.$$

The specific relative axial displacement is obtained by the ratio of the above equation and Δx, that is,

$$\frac{\Delta u_x}{\Delta x} = -y\frac{\Delta\theta_z}{\Delta x}.$$

The specific strain $\varepsilon_{xx}(x,y)$ is defined by taking the limit of the previous equation for Δx going to zero, that is,

$$\varepsilon_{xx}(x,y) = \lim_{\Delta x\to 0}\frac{\Delta u_x}{\Delta x} = \frac{\partial u_x(x,y)}{\partial x} = -y\frac{d\theta_z(x)}{dx}.$$

Hence, the strain measure $\varepsilon_{xx}(x,y)$ for the beam is given by

$$\varepsilon_{xx}(x,y) = -y\frac{d\theta_z(x)}{dx} = -y\frac{d}{dx}\left(\frac{du_y(x)}{dx}\right) = -y\frac{d^2u_y(x)}{dx^2}. \tag{5.6}$$

The bar strain measure $\varepsilon_{xx}(x)$ is constant for all points of the cross-section. On the other hand, the longitudinal strain measure for the beam varies linearly according to the coordinate y of the points. This behavior is analogous to the axial displacement $u_x(x,y)$, as illustrated in Figure 5.5.

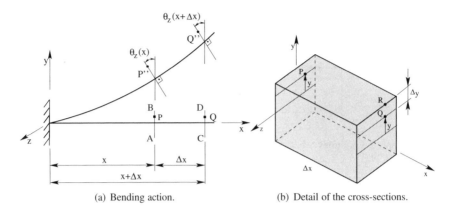

(a) Bending action.   (b) Detail of the cross-sections.

Figure 5.7   Strain measure in the Euler-Bernoulli beam.

Due to the transversal displacement $u_y(x)$ in the y direction, there is the transversal strain component $\varepsilon_{yy}(x)$ associated to $u_y(x)$. To characterize it, we take the difference between the transversal displacement of two points Q and R, which are on the same section x and far Δy from each other, as shown in Figure 5.7(b). Therefore, the strain component $\varepsilon_{yy}(x)$ is given by the limit of Δy going to zero, of the following specific relative transversal displacement:

$$\varepsilon_{yy}(x) = \lim_{\Delta y \to 0} \frac{\Delta u_y}{\Delta y} = \frac{du_y(x)}{dy}. \tag{5.7}$$

However, the transversal displacement $u_y(x)$ is constant for all points of cross-section x. Consequently, the relative displacement Δu_y is zero and $\dfrac{du_y(x)}{dy} = 0$. Hence, the specific transversal strain $\varepsilon_{yy}(x)$, associated to $u_y(x)$ is zero. This result is consistent with the hypothesis that there is no cross-section deformation of the Euler-Bernoulli model.

Similarly, the shear strain component $\gamma_{xy}(x)$ is zero, because the cross-sections remain planar and orthogonal to the longitudinal axis of the beam. Remember that the x index stands for the plane in which the strain measure is present (in this case, the x axis is normal to the cross-sections of the beam), while the y index indicates the strain measure direction. Thus,

$$\gamma_{xy}(x) = 0. \tag{5.8}$$

Analogous to the bar, we conclude that the only nonzero strain measure component for the beam is ε_{xx}.

Example 5.2 *Consider the beam of Example 5.1. The specific normal strain ε_{xx} can be obtained using equation (5.6), that is,*

$$\varepsilon_{xx}(x,y) = -y\frac{d\theta_z(x)}{dx} = \frac{3\delta}{L^3}(L-x)\,y.$$

For a rectangular cross-section of height h, the strain measure ε_{xx} is zero for all points located at the GC of the cross-section, because, for this case, $y = 0$. Similarly, the maximum ε_{xx} is in the upper edge with $y = \frac{h}{2}$. The minimum value is in the lower edge with $y = -\frac{h}{2}$. Figure 5.23(e) shows the maximum strain measure of a beam with rectangular cross-section.

□

5.4 RIGID ACTIONS

As mentioned before, the strain component $\varepsilon_{xx}(x,y)$ is zero for a rigid action. Therefore,

$$\varepsilon_{xx}(x,y) = -y\frac{d^2 u_y(x)}{dx^2} = 0, \quad x \in (0,L). \tag{5.9}$$

The previous relation can be rewritten as

$$\varepsilon_{xx}(x,y) = -y\frac{d}{dx}\left(\frac{du_y(x)}{dx}\right) = -y\frac{d\theta_z(x)}{dx} = 0.$$

In order to satisfy the previous conditions, the displacement u_y or rotation $\theta_z(x) = \dfrac{du_y(x)}{dx}$ must be constant for all cross-sections of the beam. This implies that the rigid actions for the beam are a translation in the y direction and a rotation about the z axis, as illustrated in Figure 5.8.

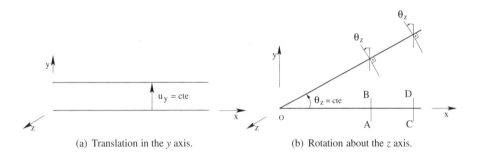

(a) Translation in the y axis.　　　　　(b) Rotation about the z axis.

Figure 5.8　Rigid actions for the Euler-Bernoulli beam.

Example 5.3 *The beam of Example 5.1 is clamped at the $x = 0$ end. Consequently, $u_y(x = 0) = 0$ and $\theta_z(x = 0) = 0$. Therefore, since the displacement and rigid rotation are constant for all cross-sections of the beam and the kinematic constraints must be satisfied, the rigid transversal displacement and rigid rotation are respectively $u_y = 0$ and $\theta_z = 0$.*

□

5.5 DETERMINATION OF INTERNAL LOADS

Analogous to the bar model, consider the beam of length L and the differential volume element $dV = dAdx$ about an arbitrary point, in the nondeformed configuration, illustrated in Figure 5.9(a). The beam has a final length L' after the bending action and the volume element elongated by the axial displacement du_x, as shown in Figure 5.9(b).

Intuitively, it is known that the larger the magnitude of the bending action on the beam, the larger the work or energy required to do it. Thus, the strain internal work for the the beam is a function of the strain measure, that is,

$$W_i = f(\varepsilon_{xx}). \tag{5.10}$$

To quantify W_i, it is necessary to obtain the internal loads in the deformed beam.

In the case of the volume element dV, we consider the internal work density dw_i, that is, the strain internal work dW_i by units of volume dV, given by

$$dw_i = \frac{dW_i}{dV} = g(\varepsilon_{xx}). \tag{5.11}$$

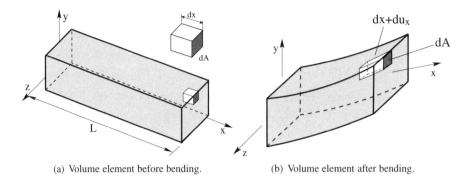

(a) Volume element before bending.   (b) Volume element after bending.

Figure 5.9 Differential volume elements before and after bending of the Euler-Bernoulli beam.

Due to the proportionality between the internal work and the strain measure, the relation between both quantities is linear. Thus, the previous equation can be written as

$$dw_i = \frac{dW_i}{dV} = \sigma_{xx}\varepsilon_{xx}, \tag{5.12}$$

with σ_{xx} the internal loads in the volume element dV.

Considering a dimensional analysis of the previous expression in the International System of Units (SI), we have

$$\frac{\mathrm{Nm}}{\mathrm{m}^3} = \frac{\mathrm{N}}{\mathrm{m}^2}\frac{\mathrm{m}}{\mathrm{m}}.$$

Therefore, for the left side to represent work density, the internal loads must be expressed as a load density by units of area, called stress. The term σ_{xx} is called normal stress, because it is present in the points of the cross-sections in the normal or perpendicular direction, i.e., the x plane in the x direction, as shown in Figure 5.10. Positive values indicate tractive normal stresses, while negative values indicate compressive normal stresses.

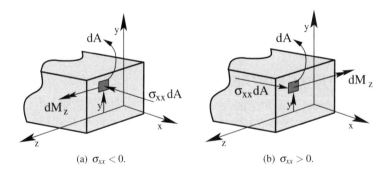

(a) $\sigma_{xx} < 0$.   (b) $\sigma_{xx} > 0$.

Figure 5.10 Bending moment in the cross-section of the Euler-Bernoulli beam.

Substituting (5.6) in (5.12), we obtain

$$dW_i = \sigma_{xx}\left(-y\frac{d^2u_y(x)}{dx^2}\right)dV. \tag{5.13}$$

The strain internal work of the beam is determined by the sum of the contributions of all volume differential elements of the beam. Due to the continuity hypothesis, there is an infinite number of differential elements, and an infinite summation is represented by the following Riemann integral:

$$W_i = \int_V dW_i = \int_V \sigma_{xx}(x,y)\varepsilon_{xx}(x,y)dV = \int_V \sigma_{xx}(x,y)\left(-y\frac{d^2u_y(x)}{dx^2}\right)dV. \tag{5.14}$$

Due to the geometrical hypothesis of the beam, the above volume integral can be decomposed in the product of two integrals: one along the length L and the other in the cross-section area A of the beam. Thus,

$$W_i = \int_0^L \left(-\int_A y\sigma_{xx}(x,y)dA\right)\frac{d^2u_y(x)}{dx^2}dx. \tag{5.15}$$

The area integral results in a moment in the z direction of the adopted reference system, as illustrated in Figure 5.10 for an infinitesimal area element dA. The term $dN_x = \sigma_{xx}(x,y)dA$ represents the normal force in the x direction acting on the area element dA. Multiplying this force by the distance y, we have a moment in the negative direction of z. Hence,

$$dM_z = -y\sigma_{xx}(x,y)\,dA.$$

The moment for the cross-section x is given by the summation of the dM_z contributions for infinite area elements of the section. This summation is given by the following area integral:

$$M_z(x) = \int_A dM_z = -\int_A y\sigma_{xx}(x,y)\,dA. \tag{5.16}$$

As the moment is associated to the bending on the beam, it is called bending moment relative to the z axis of the reference system and denoted by $M_z(x)$.

Substituting (5.16) in (5.15), the internal strain work expression becomes

$$W_i = \int_0^L M_z(x)\frac{d^2u_y(x)}{dx^2}dx. \tag{5.17}$$

The transversal displacement $u_y(x)$ is derived twice in the internal work expression. To obtain a relation containing $u_y(x)$ only, we should integrate (5.17) by parts twice, that is,

$$
\begin{aligned}
W_i &= \int_0^L M_z(x)\frac{d^2u_y(x)}{dx^2}dx \\
&= -\int_0^L \frac{dM_z(x)}{dx}\frac{du_y(x)}{dx}\,dx + M_z(x)\frac{du_y(x)}{dx}\bigg|_0^L \\
&= \int_0^L \frac{d^2M_z(x)}{dx^2}u_y(x)\,dx + M_z(x)\frac{du_y(x)}{dx}\bigg|_0^L - \frac{dM_z(x)}{dx}u_y(x)\bigg|_0^L.
\end{aligned}
\tag{5.18}
$$

There is an integral expression along the beam length and two boundary terms. The derivative $\frac{dM_z(x)}{dx}$ represents an internal force, because if the bending moment has Nm units and the length m units, the previous derivative results in N, that is,

$$\left[\frac{dM_z(x)}{dx}\right] = \frac{\text{Nm}}{\text{m}} = \text{N}.$$

As the moment is in the z direction and the length of the beam in x, we have a transversal force in the y direction, that is, parallel to the vertical direction in each cross-section of the beam. This force is denoted by $V_y(x)$ and called shear force in y, because it has a shear effect on each cross-section. Therefore, the following relation is valid for the bending moment and shear force in the beam cross-sections

$$V_y(x) = \frac{dM_z(x)}{dx}. \tag{5.19}$$

The term $\dfrac{d^2 M_z(x)}{dx^2}$ represents a force by units of length, that is, a distributed internal load acting along the beam length. To verify this fact, we should check the units of $\dfrac{d^2 M_z(x)}{dx^2}$, that is,

$$\left[\frac{d^2 M_z(x)}{dx^2} \right] = \frac{\text{Nm}}{\text{m}^2} = \frac{\text{N}}{\text{m}}.$$

Also notice that analogously to the shear force, this distributed load acts in the vertical direction, that is, in the y axis of the beam reference system.

Recalling that the rotation is given as $\theta_z(x) = \dfrac{d u_y(x)}{dx}$ and substituting expression (5.19) in (5.18), we obtain the final equation for the strain internal work of the beam

$$W_i = \int_0^L \frac{d^2 M_z(x)}{dx^2} u_y(x)\, dx + M_z(x)\theta_z(x)\big|_0^L - V_y(x) u_y(x)\big|_0^L. \tag{5.20}$$

This expression can be expanded as

$$W_i = \int_0^L \frac{d^2 M_z(x)}{dx^2} u_y(x)\, dx + M_z(L)\theta_z(L) - M_z(0)\theta_z(0) - V_y(L) u_y(L) + V_y(0) u_y(0). \tag{5.21}$$

Therefore, the internal loads which are compatible with the beam kinematics are concentrated shear forces and bending moments at the ends, and a transversal distributed force along the length. These loads are illustrated in the FBD of Figure 5.9(a). The directions of the loads are in accordance to the signs of equation (5.21).

5.6   DETERMINATION OF EXTERNAL LOADS

The internal loads for a deformed beam are given in equation (5.21). The respective external loads are those ones that can be balanced by the internal loads compatible with the beam kinematics. Hence, we have the following external loads for beams as illustrated in Figure 5.9(b):

- $q_y(x)$: transversal distributed load function in the y direction applied along the beam length with typical units N/m;
- V_{y0} and V_{yL}: concentrated transversal forces in the y direction applied to the $x = 0$ and $x = L$ ends;
- M_{z0} and M_{zL}: concentrated moments in the z direction applied to the $x = 0$ and $x = L$ ends.

Analogous to (5.21), the external work W_e for any bending action is

$$W_e = \int_0^L q_y(x) u_y(x)\, dx + M_{zL}\theta_z(L) + M_{z0}\theta_z(0) + V_{yL} u_y(L) + V_{y0} u_y(0) u_y(0). \tag{5.22}$$

Notice that the external loads in the FBD of Figure 5.9(b) are indicated in the positive direction of the axes of the reference system.

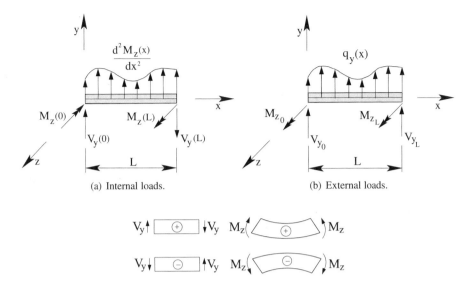

(a) Internal loads.   (b) External loads.

(c) Positive and negative shear forces and bending
moments.

Figure 5.11   Free body diagrams of the internal and external loads in the Euler-Bernoulli beam, positive and
negative shear forces, and bending moments.

5.7   EQUILIBRIUM

The PVW establishes the balance of internal and external works for any virtual action $\delta\mathbf{u}(x,y) = \{\delta u_x(x,y)\ \delta\theta_z(x)\}^T$, from the deformed equilibrium position of the beam. Based on that, it was possible to characterize the external loads which are compatible with the internal loads given in (5.21).

The external loads are given by the transversal forces V_{y_0} and V_{y_L} and the moments M_{z_0} and M_{z_L}, applied to the $x = 0$ and $x = L$ beam ends, along with the distributed transversal load $q_y(x)$. It is observed that, for a virtual action $\delta\mathbf{u}(x,y) = \{\delta u_x(x,y)\ \delta\theta_z(x)\}^T$, the terms $V_y(L)\,\delta u_y(L)$ and $V_y(0)\,\delta u_y(0)$ of the internal work balance the external work terms $V_{y_L}\delta u_y(L)$ and $V_{y_0}\delta u_y(0)$, respectively. Analogously, the internal work terms $M_z(L)\delta\theta_z(L)$ and $M_z(0)\delta\theta_z(0)$ balance the external work terms $M_{z_L}\delta\theta_z(L)$ and $M_{z_0}\delta\theta_z(0)$, respectively. Finally, the work done by the internal transversal load $\int_0^L \frac{d^2 M_z(x)}{dx^2}\delta u_y(x)dx$ must be equal to the work $\int_0^L q_y(x)\delta u_y(x)dx$ done by the external distributed load.

Substituting (5.21) and (5.22) in the PVW statement (3.18) and rearranging the terms, we have

$$\int_0^L \left[\frac{d^2 M_z(x)}{dx^2} - q_y(x)\right]\delta u_y\, dx$$
$$+\ \left[-M_z(L) + M_{z_L}\right]\delta\theta_z(L) + \left[M_z(0) + M_{z_0}\right]\delta\theta_z(0)$$
$$+\ \left[V_y(L) + V_{y_L}\right]\delta u_y(L) + \left[-V_y(0) + V_{y_0}\right]\delta u_y(0) = 0.$$

In order to satisfying the above equation for any arbitrary virtual action, all terms inside the brackets must be simultaneously zero, that is,

$$
\begin{cases}
\dfrac{d^2 M_z(x)}{dx^2} = q_y(x) & \text{in } x \in (0, L) \\
V_y(0) = V_{y_0} & \text{in } x = 0 \\
V_y(L) = -V_{y_L} & \text{in } x = L \\
M_z(0) = -M_{z_0} & \text{in } x = 0 \\
M_z(L) = M_{z_L} & \text{in } x = L
\end{cases}
\tag{5.23}
$$

The previous expression defines the local form or the boundary value problem (BVP) of equilibrium of the unconstrained beam. This BVP consists of the second-order differential equation in terms of the bending moment and four boundary conditions in terms of shear forces and bending moments at the beam ends. The solution of the BVP results in the functions for the shear force $V_y(x)$ and bending moment $M_z(x)$.

A positive value of $M_z(x)$ indicates that the fibers of the beam below the x axis are under traction, while the fibers above the x axis are under compression, as shown in Figure 5.2. A negative $M_z(x)$ indicates that the fibers below the x axis are under compression, and the fibers above are under traction. The signs of the shear force and bending moment are illustrated in Figure 5.11(c). When the beam "smiles", bending will be positive. The positive shear force between two sections of a beam makes it rotate clockwise (negative z), while negative shear forces indicate a counterclockwise rotation (positive z).

The plots of the functions $V_y(x)$ and $M_z(x)$ are known as shear force and bending moment diagrams for the beam model.

The differential equation in terms of the bending moment given in (5.23) can be rewritten as

$$
\frac{d}{dx}\left(\frac{dM_z(x)}{dx}\right) = q_y(x).
\tag{5.24}
$$

Substituting expression (5.19) in the previous equation, we obtain the differential equation of equilibrium in terms of the shear force, that is,

$$
\frac{dV_y(x)}{dx} = q_y(x).
\tag{5.25}
$$

Hence, we can solve the second-order differential equation (5.24) in terms of the bending moment or the system of two first-order differential equations (5.19) and (5.25).

If $(\delta u_y, \delta \theta_z)$ is a virtual rigid action, the strain measure and consequently the internal work are both zero. In this case, the PVW establishes that for any rigid virtual action of a beam in equilibrium, the external work (5.22) is zero, that is,

$$
\delta W_e = \int_0^L q_y(x)\,\delta u_y(x)\,dx + M_{z_L}\delta\theta_z(L) + M_{z_0}\delta\theta_z(0) + V_{y_L}\delta u_y(L) + V_{y_0}\delta u_y(0) = 0.
\tag{5.26}
$$

The rigid actions for the beam are the constant transversal displacements and the constant rotations in the y and z directions, respectively. Thus, we have $\delta u_y(x) = \delta u_y = \text{cte}$ and $\delta\theta_z(x) = \delta\theta_z = \text{cte}$. Substituting them in the previous expression, we have

$$
\delta W_e = \left(\int_0^L q_y(x)\,dx + V_{y_L} + V_{y_0}\right)\delta u_y + \left(M_{z_L} + M_{z_0}\right)\delta\theta_z = 0.
$$

Based on this equation, the rigid equilibrium conditions of the beam are recovered, which state that the resultants of external forces in y direction and moments in z direction must be zero, that is,

$$
\int_0^L q_y(x)\,dx + V_{y_L} + V_{y_0} = 0,
\tag{5.27}
$$

$$
M_{z_L} + M_{z_0} = 0.
\tag{5.28}
$$

The term resulting of the integral of $q_y(x)$ along the beam length is the transversal concentrated force which is equivalent to the transversal distributed load. Hence, in a strict equilibrium point of view, $q_y(x)$ is transformed into a concentrated force, acting at the geometric center of the beam segment where the distributed force is applied.

The two previous equations are equivalent, respectively, to the following Newton equilibrium conditions:

$$\sum f_y = 0,$$

$$\sum m_z = 0.$$

In the case of the second condition, we should take a convenient point to calculate the resultant of moments of all external loads on the beam.

Before including the material behavior, some examples are presented to determine the functions for the shear force and bending moment using the integration of the BVP given in (5.23).

Example 5.4 *Plot the shear force and bending moment diagrams for the simply supported beam illustrated in Figure 5.12, by integrating the differential equation given in (5.23).*

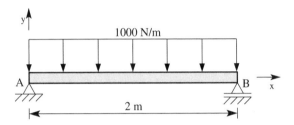

Figure 5.12   Example 5.4: beam subjected to a constant distributed load.

- *Distributed load equation*
  Due to the transversal distributed load applied to the beam in the negative y direction, we have $q_y(x) = -1000$ N/m.
- *Boundary conditions*
  As the beam is supported at both ends, the transversal displacements are zero, that is, $u_y(x = 0) = u_y(x = 2) = 0$. However, as the considered beam is isostatic, the differential equation in terms of the bending moment is integrated and the kinematic boundary conditions are not used for the solution of this problem. Subsequently, the use of kinematic boundary conditions in terms of transversal displacements and rotations will be presented. There is no moment applied to beam ends. Consenquently, the bending moments are zero, i.e, $M_z(x = 0) = 0$ and $M_z(x = 2) = 0$.
  The boundary conditions in terms of the shear force, that is, $V_y(x = 0) = R_{Ay}$ and $V_y(x = 0) = R_{By}$, cannot be used, because the support reactions R_{Ay} and R_{By} are still unknown. At the end of the solution procedure, the reactions are obtained naturally, as the balance is already represented in the BVP (5.23).
- *Integration of the differential equation*
  The differential equation of equilibrium in terms of the bending moment is considered, that is,

$$\frac{d^2 M_z(x)}{dx^2} = q_y(x) = -1000.$$

The first integration results in the shear force expression

$$V_y(x) = \frac{dM_z(x)}{dx} = -1000x + C_1.$$

The second integration results in the bending moment expression

$$M_z(x) = -500x^2 + C_1 x + C_2,$$

with C_1 and C_2 the integration constants.

• *Determination of the integration constants*
To calculate C_1 and C_2, the boundary conditions in terms of the bending moment are used. Thus,

$$M_z(x = 0) = -500(0) + C_1(0) + C_2 = 0 \rightarrow C_2 = 0,$$
$$M_z(x = 2) = -500(2)^2 + C_1(2) + 0 = 0 \rightarrow C_1 = 1000.$$

• *Final equations*
Substituting C_1 and C_2, we obtain the final equations for the shear force and bending moment, respectively,

$$V_y(x) = -1000x + 1000 \quad and \quad M_z(x) = -500x^2 + 1000x.$$

• *Diagrams*
The diagrams of shear force and bending moments are illustrated in Figure 5.13. Since the shear force is zero at $x = 1.5$ m, the bending moment function have an inflection point and may assume a maximum or minimum value. In general, if the shear force is zero in any section, the bending moment must be calculated there. In this case, $M_z(x = 1.5) = 500$ Nm is the maximum bending moment for the beam.

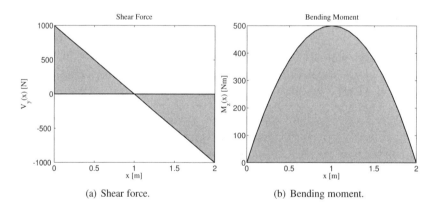

(a) Shear force.   (b) Bending moment.

Figure 5.13   Example 5.4: shear force and bending moment diagrams.

• *Support reactions*
It is initially assumed that the support reactions R_{Ay} and R_{By} are in the y positive direction. There are two ways to calculate the reactions. The simplest one is to use the boundary conditions of the equilibrium BVP (5.23) in terms of the shear force at $x = 0$ and $x = 3$ m. In this case, R_{Ay} and R_{By} are V_{y0} and V_{yL}, respectively. Thus,

$$R_{Ay} = V_y(x = 0) = 1000 \text{ N} \quad and \quad R_{By} = -V_y(x = L) = 1000 \text{ N}.$$

Hence, as the values for R_{Ay} and R_{By} are positive, the initially assumed directions for the support reactions are correct.

The second option of calculating the reactions is employing the rigid body equilibrium equations given in (5.27) and (5.28). Thus, considering the resultant of forces in the y direction and moments in the z direction for the reference point A, we have, respectively,

$$\sum f_y = 0 : \ R_{Ay} + R_{By} + \int_0^2 1000 \, dx = 0 \rightarrow R_{Ay} + R_{By} = 2000,$$

$$\sum m_{z_A} = 0 : \ (2)(R_{By}) - \left(\int_0^2 1000 \, dx \right)(1) = 0.$$

Thus, $R_{By} = 1000$ N and $R_{Ay} = 1000$ N. These values can also be obtained directly from the shear force diagram.

The interpretation of the shear force and bending moment diagrams, based on the method of sections, is illustrated in Figure 5.14. A cut is made in a generic section x and the left segment of the beam is analyzed. The shear force $V_y(x)$ and the bending moment $M_z(x)$ must balance the applied external forces and moments in the considered beam segment. Doing the balance of forces in y direction and moments in z, taking point A as reference, we have

$$\sum f_y = 0 : \ R_{Ay} - 1000x - V_y(x) = 0,$$

$$\sum m_{z_A} = 0 : \ -(R_{Ay})(x) + (1000x) \left(\frac{x}{2} \right) + M_z(x) = 0.$$

Hence,

$$V_y(x) = -1000x + 1000 \quad and \quad M_z(x) = -500x^2 + 1000x.$$

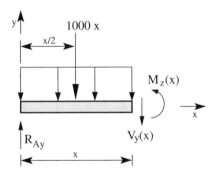

Figure 5.14   Example 5.4: FBD used in the method of sections.

Note that the previous expressions are the same ones obtained by the integration of the differential equation of equilibrium. The method of sections can be applied only to statically determinate problems.

☐

Example 5.5 *Plot the shear force and bending moment diagrams for the cantilever beam illustrated in Figure 5.15, by integrating the differential equation given in (5.23).*

- *Distributed load equation*
  Due to the linear distributed transversal load applied on the beam in the y negative direction, we have $q_y(x) = -q_0(1 - \frac{x}{L}) = -500(2 - x)$. Observe that $q_y(x = 0) = -1000$ N/m and $q_y(x = 2) = 0$.

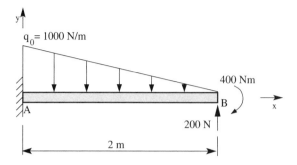

Figure 5.15 Example 5.5: beam subjected to a distributed linear load.

- *Boundary conditions*
  This is a cantilever beam, because it is clamped at $x = 0$ and free at the $x = 2$ m. Due to the application of the transversal force of 200 N and the moment of 400 Nm at the free end, the boundary conditions are expressed in terms of the shear force and bending moment. Thus, $V_y(x = 2) = -200$ N and $M_z(x = 2) = -400$ Nm. The transversal displacement and rotation are zero at the clamped end, that is, $u_y(x = 0) = 0$ and $\theta_z(x = 0) = 0$. However, as the beam is isostatic, the differential equation in terms of the bending moment is integrated and the kinematic boundary conditions are not used to solve this problem. Posteriorly, the use of kinematic boundary conditions in terms of the transversal displacements and rotations is considered.
- *Integration of the differential equation*
  The differential equation of equilibrium in terms of the bending moment is used, that is,

$$\frac{d^2 M_z(x)}{dx^2} = q_y(x) = -500(2 - x).$$

The first integration gives the shear force expression

$$V_y(x) = \frac{dM_z(x)}{dx} = 250x^2 - 1000x + C_1.$$

The second integration results in the bending moment expression

$$M_z(x) = \frac{250}{3}x^3 - 500x^2 + C_1 x + C_2,$$

with C_1 and C_2 the integration constants.
- *Determination of the integration constants*
  In order to calculate C_1 and C_2, the boundary conditions in terms of the shear force and bending moment are used. Thus,

$$V_y(x = 2) = 250(4) - 1000(2) + C_1 = -200 \rightarrow C_1 = 800,$$

$$M_z(x = 2) = \frac{250}{3}(8) - 500(4) + 900(2) + C_2 = -400 \rightarrow C_2 = -\frac{2000}{3}.$$

- *Final equations*
  Substituting C_1 and C_2, we obtain the final equations for the shear force and bending moment, respectively,

$$V_y(x) = 250x^2 - 1000x + 800 \quad and \quad M_z(x) = \frac{250}{3}x^3 - 500x^2 + 800x - \frac{2000}{3}.$$

- *Diagrams*
  The shear force and bending moment diagrams are illustrated in Figure 5.16. Due to the linear distributed load, the diagrams for the shear force and bending moment vary as parabolic and cubic functions, respectively.
  To determine the section with zero shear force, we have

  $$V_y(x) = 250x^2 - 1000x + 800 = 0,$$

  which results in $x = 1.10$ m and $x = 2.89$ m. As the length is $L = 2$ m, we adopt $x = 1.10$ m. The bending moment in this section is $M_z(x = 1.10) = -280.74$ Nm.

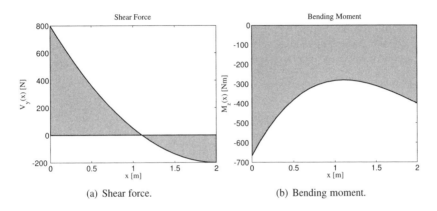

(a) Shear force. (b) Bending moment.

Figure 5.16 Example 5.5: shear force and bending moment diagrams.

- *Support reactions*
  It is initially assumed that the support reactions R_{Ay} and M_{Az} are in the positive directions of y and z, respectively. There are two ways of calculating these reactions. The simplest one is to use the boundary conditions in terms of the shear force and bending moment at $x = 0$ for the equilibrium BVP given in (5.23). In this case, R_{Ay} and M_{Az} are V_{y_0} and M_{z_0}, respectively. Thus,

  $$R_{Ay} = V_y(x = 0) = 800 \text{ N} \quad and \quad M_{Az} = -M_z(x = 0) = 666.7 \text{ Nm.}$$

  Hence, as the values of R_{Ay} and M_{Az} are positive, the initially assumed directions of the support reactions are correct.
  The second way of calculating the reactions is employing the rigid body equilibrium equations given in (5.27) and (5.28). Considering the resultant of forces in the y direction and moments in z, and taking point A as reference, we have, respectively,

  $$\sum f_y = 0 : \ R_{Ay} + \int_0^2 [-500(2-x)]dx + 200 = 0,$$

  $$\sum m_{z_A} = 0 : \ M_{Az} - \left(\int_0^2 [-500(2-x)]dx \right)(0.6667) + (2)(200) + 400 = 0.$$

  Thus, $R_{Ay} = 800$ N and $M_{Az} = 666.7$ Nm. These values can also be directly obtained from the shear force and bending moment diagrams.
  The interpretation of the shear force and bending moment diagrams based on the method of sections is illustrated in Figure 5.17. A cut is made in the generic section x and the left segment of the beam is analyzed. The shear force $V_y(x)$ and bending moment $M_z(x)$

must balance the external forces and moments in the considered beam segment. Taking the balance of forces in y direction and moments in z, for the reference point A, we have, respectively,

$$\sum f_y = 0 : R_{Ay} - \int_0^x 500(2-x)\,dx - V_y(x) = 0,$$

$$\sum m_{z_A} = 0 : M_{Az} - \left(\int_0^x 500(2-x)\,dx\right)x + M_z(x) = 0.$$

Thus,

$$V_y(x) = 250x^2 - 1000x + 800 \quad and \quad M_z(x) = -\frac{250}{3}x^3 + 500x^2 + 800x - \frac{2000}{3}.$$

The previous expressions are the same ones obtained from the integration of the equilibrium differential equation. The method of sections can be applied only to statically determinate problems.

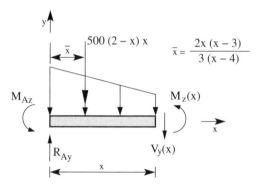

Figure 5.17 Example 5.5: FBD used in the method of sections.

□

Example 5.6 *Plot the shear force and bending moment diagrams for the beam illustrated in Figure 5.18, integrating the differential equation given in (5.23).*

There is a hinge at the central section of the beam. This type of restriction avoids the transmission of bending moments from one beam part to other one. The hinge will give rise to an additional restriction of zero bending moment in the beam center. This will allow the solution of this hyperstatic problem, because there are three unknowns (R_{Ay}, M_{Az}, and R_{Cy}) and only two equilibrium conditions ($\sum f_y = 0$ and $\sum m_z = 0$).

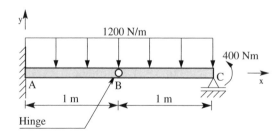

Figure 5.18 Example 5.6: beam with a hinge subjected to a constant distributed load.

- *Distributed load equation*
  Due to the constant distributed transversal load applied to the beam in the negative direction of y, we have $q_y(x) = -1200 \, \text{N/m}$.
- *Boundary conditions*
  The beam is clamped in the $x = 0$ *section. Consequently, the transversal displacement and the rotation are zero, that is,* $u_y(x = 0) = 0$ *and* $\theta_z(x = 0) = 0$. *Due to the simple support at* $x = 2$ *m, the transversal displacement in this section is also zero, that is,* $u_y(x = 2) = 0$. *However, the bending moment differential equation is still considered, and these boundary conditions are not used in this problem. At the* $x = 2$ *m end, due to the moment, we have* $M_z(x = 2) = 400 \, \text{Nm}$.
- *Additional constraint*
  The hinge gives an additional constraint of zero bending moment at $x = 1$ *m, i.e.,* $M_z(x = 1) = 0$. *This condition is called additional constraint or overhang, because the hinge is located at the middle section, and not at the beam ends.*
- *Integration of the differential equation*
  The differential equation of equilibrium in terms of the bending moment is considered, that is,

$$\frac{d^2 M_z(x)}{dx^2} = q_y(x) = -1200.$$

The first integration gives the shear force expression

$$V_y(x) = \frac{dM_z(x)}{dx} = -1200x + C_1.$$

The second integration results in the bending moment expression

$$M_z(x) = -600x^2 + C_1 x + C_2,$$

with C_1 *and* C_2 *the integration constants.*
- *Determination of the integration constants*
  The boundary conditions in terms of the bending moment and the hinge additional constraint are used to calculate C_1 *and* C_2. *Therefore,*

$$M_z(x = 2) = -(600)(4) + 2C_1 + C_2 = 400 \rightarrow 2C_1 + C_2 = 2800,$$

$$M_z(x = 1) = -600 + C_1 + C_2 = 0 \rightarrow C_1 + C_2 = 600.$$

Solving the system defined by the two previous equations, we have $C_1 = 2200$ *and* $C_2 = -1600$.
- *Final equations*
  Replacing C_1 *and* C_2, *we obtain the final equations for the shear force and bending moment, respectively, given by*

$$V_y(x) = -1200x + 2200 \quad and \quad M_z(x) = -600x^2 + 2200x - 1600.$$

- *Diagrams*
  The shear force and bending moment diagrams are illustrated in Figure 5.19. It is observed that the moment is zero at the hinge. The section where the shear force is zero is obtained as

$$V_y(x) = -1200x + 2200 = 0 \rightarrow x = 1.83 \, \text{m}.$$

The bending moment in this section is $M_z(x) = 416.67 \, \text{Nm}$.

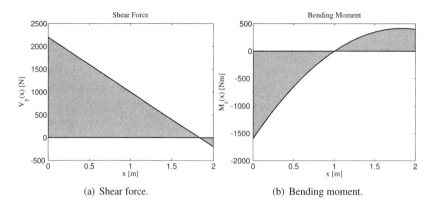

(a) Shear force.   (b) Bending moment.

Figure 5.19   Example 5.6: shear force and bending moment diagrams.

- *Support reactions*
  It is initially assumed that the support reactions R_{Ay} and R_{Cy} are in the y positive direction and the moment M_{Az} is in the positive direction of z. Because the problem is statically indeterminate, the boundary conditions of the equilibrium BVP (5.23) must be considered. The R_{Ay} and R_{Cy} reactions are, respectively, V_{y0} and V_{yL}, while M_{Az} represents M_{z0}. Thus,

$$
\begin{aligned}
R_{Ay} &= V_y(x=0) = 2200 \text{ N}, \\
R_{Cy} &= -V_y(x=2) = 200 \text{ N}, \\
M_{Az} &= -M_z(x=0) = 1600 \text{ Nm}.
\end{aligned}
$$

  As the calculated reactions are positive, the initially assumed directions are correct. It is observed that the same procedure to determine the support reactions using the BVP boundary conditions can also be used to hyperstatic problems.

□

5.8   APPLICATION OF THE CONSTITUTIVE EQUATION

For an isotropic homogeneous linear elastic material, the normal stress component $\sigma_{xx}(x,y)$ is related to the longitudinal strain component $\varepsilon_{xx}(x,y)$ by the longitudinal elastic modulus $E(x)$, that is,

$$
\sigma_{xx}(x,y) = E(x)\varepsilon_{xx}(x,y). \tag{5.29}
$$

Substituting (5.6) in the previous expression, we have

$$
\sigma_{xx}(x,y) = -E(x)\frac{d^2u_y(x)}{dx^2}y. \tag{5.30}
$$

Replacing the above expression in (5.16), the bending moment $M_z(x)$ is rewritten as

$$
M_z(x) = -\int_A \left(-E(x)\frac{d^2u_y(x)}{dx^2}y^2 \right) dA = E(x)\frac{d^2u_y(x)}{dx^2} \int_A y^2 \, dA.
$$

Recalling that

$$
I_z(x) = \int_A y^2 \, dA \tag{5.31}
$$

is the moment of area of the cross-section x relative to the z axis, the final expression for the bending moment for a beam with a Hookean material is

$$M_z(x) = E(x)I_z(x)\frac{d^2u_y(x)}{dx^2}. \tag{5.32}$$

The moments of area for circular cross-section with diameter d and for rectangular section with base b and height h are given, respectively, by

$$I_z = \frac{\pi d^4}{64}, \tag{5.33}$$

$$I_z = \frac{bh^3}{12}. \tag{5.34}$$

As the shear force is the derivative of the bending moment, the following expression is valid:

$$V_y(x) = \frac{d}{dx}\left(E(x)I_z(x)\frac{d^2u_y(x)}{dx^2}\right).$$

For a beam with one material and constant cross-section, we have $E(x) = E$ and $I_z(x) = I_z$. Hence, the previous equation simplifies to

$$V_y(x) = EI_z\frac{d^3u_y(x)}{dx^3}. \tag{5.35}$$

Substituting $M_z(x)$ given in (5.32) in the differential equation in terms of the bending moment indicated in (5.23), we have

$$\frac{d^2}{dx^2}\left(E(x)I_z(x)\frac{d^2u_y(x)}{dx^2}\right) = q_y(x). \tag{5.36}$$

For $E(x) = E$ and $I_z(x) = I_z$, the previous expression reduces to

$$EI_z\frac{d^4u_y(x)}{dx^4} = q_y(x). \tag{5.37}$$

These last two expressions represent the fourth-order differential equation in terms of the beam kinematics. The solution gives the function for the transversal displacement or deflection $u_y(x)$ along the beam. The differential equation should be integrated four times, resulting, respectively, in expressions for the shear force $V_y(x) = EI_z\frac{d^3u_y(x)}{dx^3}$, bending moment $M_z(x) = EI_z\frac{d^2u_y(x)}{dx^2}$, rotation $\theta_z(x) = \frac{du_y(x)}{dx}$, and transversal displacement $u_y(x)$. The boundary conditions can now be given in terms of concentrated forces and moments and transversal displacements and rotations, as shown in Figure 5.20.

From (5.30), we obtain

$$\frac{d^2u_y(x)}{dx^2} = -\frac{\sigma_{xx}(x,y)}{E(x)y}, \tag{5.38}$$

which, when substituted in (5.32), results in the following expression for the normal stress at beam section x:

$$\sigma_{xx}(x,y) = -\frac{M_z(x)}{I_z(x)}y. \tag{5.39}$$

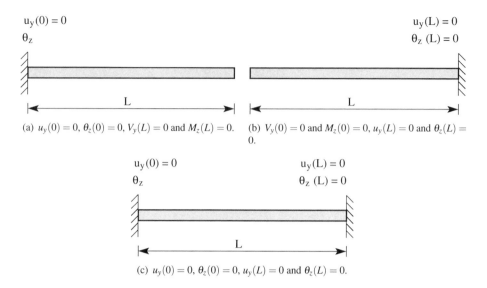

Figure 5.20   Examples of kinematic and force boundary conditions of the beam.

Analogously, from (5.29) and (5.39), we have that the longitudinal strain component ε_{xx} associated to σ_{xx} is given by

$$\varepsilon_{xx}(x,y) = \frac{\sigma_{xx}(x,y)}{E(x)} = -\frac{M_z(x)}{E(x)I_z(x)}y. \tag{5.40}$$

The normal stress and strain vary linearly with the y coordinate of the beam cross-section, reaching the maximum value on the section boundary. Depending on the sign of the bending moment, the fibers in the upper part of the beam will be under traction or compression, as shown in Figure 5.21.

As mentioned previously, the beam reference system is located at the geometric center of the cross-section. To confirm this fact, we need to determine the origin of the y and z axes in the cross-section. Because we consider pure bending only, the resultant of forces in the x direction of any section is zero, that is,

$$\sum f_x = \int_A \sigma_{xx}\, dA = 0. \tag{5.41}$$

Substituting (5.39) in the previous expression and observing that M_z and I_z vary only with coordinate x, we have

$$-\frac{M_z(x)}{I_z(x)}\int_A y\, dA = 0. \tag{5.42}$$

For the previous equation to be satisfied, the first moment of area for the z axis

$$M_{s_z}(x) = \int_A y\, dA \tag{5.43}$$

must be zero. Consequently, the reference system indicated in Figure 5.1 is located at the geometrical center of the beam cross-section.

The z axis is called neutral line of the cross-section and the union of these lines for each section defines the neutral surface, as indicated in Figure 5.22 for a rectangular section.

Example 5.7 *Consider the beam of Example 5.1 clamped at the left end and subjected to the displacement δ at the right end. Determine the expressions for the transversal displacement and rotation. For this purpose, the integration procedure is applied to the differential equation (5.37).*

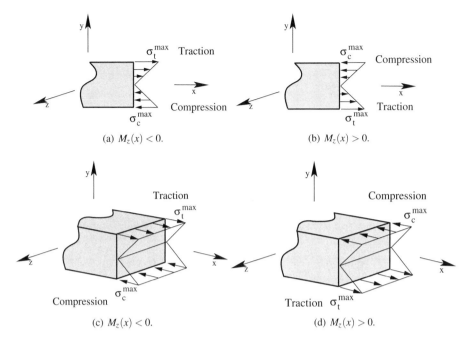

(a) $M_z(x) < 0.$ (b) $M_z(x) > 0.$

(c) $M_z(x) < 0.$ (d) $M_z(x) > 0.$

Figure 5.21   Normal stresses of traction and compression in the cross-section of the Euler-Bernoulli beam.

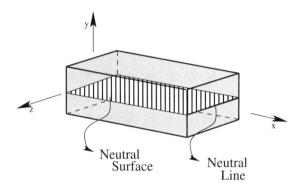

Figure 5.22   Neutral line and surface of the Euler-Bernoulli beam.

- *Distributed load equation*
  As there is no transversal distributed load applied to the beam, we have $q_y(x) = 0$.
- *Boundary conditions*
  Because the beam is clamped at $x = 0$, the transversal displacement and rotation are zero, that is, $u_y(x = 0) = 0$ and $\theta_z(x = 0) = 0$. The displacement δ at $x = L$ is such that $u_y(x = L) = \delta$. The bending moment at $x = L$ is zero, $M_z(x = L) = 0$, since there is no moment applied. As this example considers the differential equation in terms of the transversal displacement, all boundary conditions are used to obtain the solution.
- *Integration of the differential equation*
  Because we want to determine the expressions for the rotation and transversal displacement, the differential equation of equilibrium in terms of the transversal displacement is

considered, even though the problem is statically determinate. Thus,

$$EI_z \frac{d^4 u_y(x)}{dx^4} = q_y(x) = 0.$$

The first integration gives the shear force expression

$$V_y(x) = EI_z \frac{d^3 u_y(x)}{dx^3} = x + C_1.$$

The second integration results in the bending moment expression

$$M_z(x) = EI_z \frac{d^3 u_y(x)}{dx^3} = \frac{1}{2} x^2 + C_1 x + C_2.$$

The third integration results in the expression for the cross-section rotation

$$\theta_z(x) = \frac{d u_y(x)}{dx} = \frac{1}{EI_z} \left(\frac{1}{6} x^3 + \frac{1}{2} C_1 x^2 + C_2 x + C_3 \right).$$

And finally, the fourth integration gives the transversal displacement equation

$$u_y(x) = \frac{1}{EI_z} \left(\frac{1}{24} x^4 + \frac{1}{6} C_1 x^3 + \frac{1}{2} C_2 x^2 + C_3 x + C_4 \right),$$

with C_1 to C_4 the integration constants.
- *Determination of the integration constants*
The previous boundary conditions are used to calculate C_1 to C_4. Hence,

$$
\begin{aligned}
\theta_z(x=0) &= \frac{1}{EI_z} \left[\frac{1}{6}(0)^3 + \frac{1}{2} C_1 (0)^2 + C_2(0) + C_3 \right] \\
&\rightarrow C_3 = 0, \\
u_y(x=0) &= \frac{1}{EI_z} \left[\frac{1}{24}(0)^4 + \frac{1}{6} C_1 (0)^3 + \frac{1}{2} C_2 (0)^2 + C_3(0) + C_4 \right] = 0 \\
&\rightarrow C_4 = 0, \\
M_z(x=L) &= \frac{1}{2}(L)^2 + C_1(L) + C_2 = 0 \\
&\rightarrow LC_1 + C_2 = -\frac{L^2}{2}, \\
u_y(x=L) &= \frac{1}{EI_z} \left[\frac{1}{24}(L)^4 + \frac{1}{6} C_1(L)^3 + \frac{1}{2} C_2(L)^2 \right] = \delta \\
&\rightarrow \frac{L^3}{6} C_1 + \frac{L^2}{2} C_2 = EI_z \delta - \frac{L^4}{24}.
\end{aligned}
$$

Solving the system defined by the two previous equations, we determine

$$C_1 = -\frac{3EI_z}{L^3} \delta \quad and \quad C_2 = \frac{3EI_z}{L^2} \delta.$$

- *Final equations*
Substituting C_1 to C_4, we obtain the final equations for the shear force, bending moment,

rotation, and transversal displacement, which are given, respectively, by

$$V_y(x) = -\frac{3EI_z\delta}{L^3},$$

$$M_z(x) = -\frac{3EI_z\delta}{L^2}\left(\frac{x}{L}-1\right),$$

$$\theta_z(x) = -\frac{3\delta}{L^2}\left(\frac{x^2}{2L}-x\right),$$

$$u_y(x) = -\frac{3\delta}{L^2}\left(\frac{x^3}{6L}-\frac{x^2}{2}\right) = \frac{\delta}{2L^3}(3L-x)x^2.$$

- *Diagrams*
  Figure 5.23 illustrates the diagrams for the shear force, bending moment, rotation, transversal displacement, maximum normal strain and stress for $\delta = 10^{-4}$ m, $E = 210$ GPa, $L = 1$ m, with a rectangular cross-section of dimensions $b = 0.1$ m and $h = 0.2$ m. As the beam is under a positive bending along its length, the maximum strain and stress occur in the inferior edge of each cross-sections where $y = -0.1$ m.
- *Support reactions*
  It is assumed that the support reactions R_{Ay} and M_{Az} are in the positive direction of the y and z axes, respectively. Employing the boundary conditions of the equilibrium BVP (5.23), we have that R_{Ay} is V_{y0}, while M_{Az} represents M_{z0}. Thus,

$$R_{Ay} = V_y(x=0) = -\frac{3EI_z}{L^3}\delta \quad and \quad M_{Az} = -M_z(x=0) = \frac{3EI_z}{L^2}\delta.$$

☐

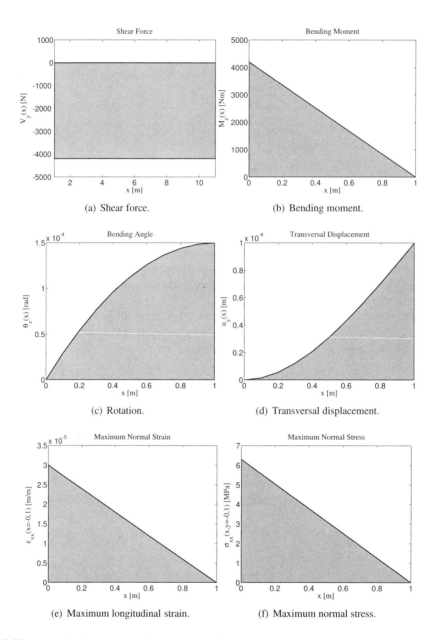

(a) Shear force.

(b) Bending moment.

(c) Rotation.

(d) Transversal displacement.

(e) Maximum longitudinal strain.

(f) Maximum normal stress.

Figure 5.23 Example 5.7: diagrams for the beam with a transversal displacement at the free end.

5.9 DESIGN AND VERIFICATION

Designing a beam means to determine the minimum cross-section dimensions, such that it remains in the elastic regime. The design considered here is based on the maximum normal stress in the beam. Analogously to the case of bars and shafts, we consider the following steps in the design for the maximum stress:

1. Obtain the functions and diagrams for the shear force $V_y(x)$ and the bending moment $M_z(x)$, integrating the differential equations (5.23) or (5.37) for statically determinate or indeterminate problems, respectively.
2. Based on these diagrams, the critical section is determined. This is the section with the largest absolute value for the bending moment, denoted by M_z^{max}.
3. Applying expression (5.39), the maximum normal stress σ_{xx}^{max} occurs on the boundary of critical section denoted by the coordinate $y = y^{max}$. Thus,

$$\sigma_{xx}^{max} = \frac{M_z^{max}}{I_z} y^{max}. \tag{5.44}$$

As the cross-section dimensions are still unknown, the terms of the previous expression related to these dimensions are grouped in the bending strength modulus W_z, given by

$$W_z = \frac{I_z}{y^{max}}. \tag{5.45}$$

By this manner, we can rewrite expression (5.44) as

$$\sigma_{xx}^{max} = \frac{M_z^{max}}{W_z}. \tag{5.46}$$

4. The beam remains in the elastic range if the maximum normal stress is less than or equal than the admissible normal stress $\bar{\sigma}$ of the material, that is,

$$\sigma_{xx}^{max} \leq \bar{\sigma}. \tag{5.47}$$

For materials with different values of admissible traction ($\bar{\sigma}_t$) and compression ($\bar{\sigma}_c$) stresses, $\bar{\sigma}$ is taken as the minimum value between them, that is, $\bar{\sigma} = \min(\bar{\sigma}_c, \bar{\sigma}_t)$.
The bending strength modulus is obtained taking the equality of the previous expression, resulting in

$$W_z = \frac{M_z^{max}}{\bar{\sigma}}. \tag{5.48}$$

Knowing W_z, the cross-section dimensions are determined. For instance, for the circular cross-section of diameter d, we have

$$W_z = \frac{\pi d^4/64}{d/2} = \frac{\pi d^3}{32} \rightarrow d = \left(\frac{32W_z}{\pi}\right)^{1/3}. \tag{5.49}$$

In the case of the rectangular section of base b and height h, we have

$$W_z = \frac{bh^3/12}{h/2} = \frac{bh^2}{6} \rightarrow bh^2 = 6W_z. \tag{5.50}$$

Knowing the relation between b and h, their values can be determined.

In the case of beam verification, the cross-section dimensions are known and we want to check if it remains in the elastic regime when subjected to a certain load. For this purpose, the maximum normal stress $\sigma_{xx}^{\max}$ is calculated using (5.44). With this maximum stress, we just verify if it is less than or equal to the material admissible stress, i.e.,

$$\sigma_{xx}^{\max} \leq \bar{\sigma}. \tag{5.51}$$

In this case, the beam remains in the elastic range. If this condition is not satisfied, the beam should be resized with the application of the previous procedure.

Example 5.8 *Design the beam of Example 5.4 for a circular cross-sections of diameter d, hollow circular cross-section with internal d_i and external d_e ($d_i/d_e = 0.8$) diameters and a rectangular cross-section of base b and height h ($h = 4b$). Adopt $\bar{\sigma} = 150$ MPa.*

The beam critical cross-section is $x = 1$ m with the bending moment $M_z^{\max} = 500$ Nm. Note that the shear force in this section is zero and consequently the bending moment has a maximum value.

The bending strength modulus is obtained from (5.48), that is,

$$W_z = \frac{500}{150 \times 10^6} = 3.33 \times 10^{-6}\ \mathrm{m}^3.$$

The diameter of the circular section is calculated using expression (5.49). Thus,

$$d = \left(\frac{32}{\pi} 3.33 \times 10^{-6}\right)^{1/3} = 3.24\ \mathrm{cm}.$$

For the hollow cross-section, we have $d_i/d_e = 0.8$. Hence,

$$W_z = \frac{\pi}{32} \frac{d_e^4 - d_i^4}{d_e} = \frac{\pi}{32} \frac{d_e^4 - (0.8 d_e)^4}{d_e}$$

and

$$d_e = \left[\frac{32}{\pi(1 - 0.8^4)} 3.75 \times 10^{-6}\right]^{1/3} = 3.86\ \mathrm{cm}.$$

Once d_e is determined, the internal diameter is calculated as

$$d_i = 0.8 d_e = 3.09\ \mathrm{cm}.$$

For the rectangular section, equation (5.50) gives

$$bh^2 = 6W_z = 2.00 \times 10^{-5}.$$

For $h = 4b$, we have $b = 1.08$ cm and $h = 4.32$ cm.

□

5.10 SINGULARITY FUNCTIONS FOR EXTERNAL LOADING REPRESENTATION

As stated previously, the compatible concentrated external loads with the bending kinematics are at the beam ends. However, when modeling physical systems, the use of concentrated loads along the beam length is very common. One way to include these loads is to use the singularity function notation.

Consider initially the following example of an isostatic beam where the shear force and bending moment expressions are obtained by the method of sections.

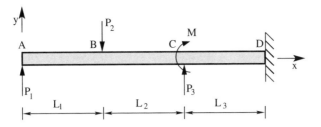

Figure 5.24 Example 5.9: cantilever beam with concentrated loads.

Example 5.9 *Obtain the equations for the shear force and bending moment for the cantilever beam illustrated in Figure 5.24.*

The three beam segments illustrated in Figure 5.25 are considered to determine the shear force and bending moment equations for the given beam.

Applying the equilibrium conditions in terms of the resultant of forces in y ($\sum f_y = 0$) and moments in z ($\sum m_{z_A} = 0$) for each beam segment, we obtain the following expressions for the shear force and bending moments for the beam:

$$V_y(x) = \begin{cases} P_1 & 0 \le x < L_1 \\ P_1 - P_2 & L_1 \le x < L_1 + L_2 \\ P_1 - P_2 + P_3 & L_1 + L_2 \le x < L_1 + L_2 + L_3 \end{cases}, \tag{5.52}$$

$$M_z(x) = \begin{cases} P_1 x & 0 \le x < L_1 \\ P_1 x - P_2(x - L_1) & L_1 \le x < L_1 + L_2 \\ P_1 x - P_2(x - L_1) + P_3(x - L_1 - L_2) + M & L_1 + L_2 \le x < L_1 + L_2 + L_3 \end{cases}. \tag{5.53}$$

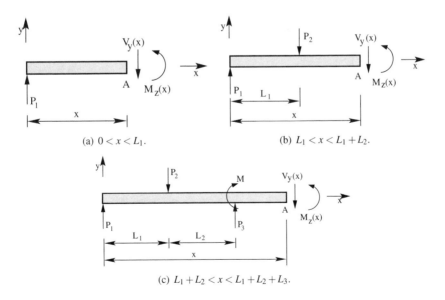

Figure 5.25 Example 5.9: beam segments.

The shear force and bending moment diagrams for $L_1 = L_2 = L_3 = 1$ m, $P_1 = 1$ kN, $P_2 = 2$ kN, $P_3 = 3$ kN, and $M = 1$ kNm are illustrated in Figure 5.26.

Some observations can be made from the obtained diagrams:

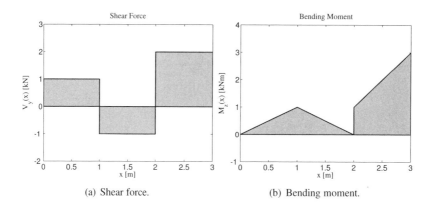

(a) Shear force.   (b) Bending moment.

Figure 5.26 Example 5.9: diagrams of shear force and bending moment.

- *The concentrated transversal forces P_1, P_2, and P_3 cause discontinuities in the shear force diagram, in the respective coordinates of the application points, and with values equal to the intensity of forces.*
- *The moment M also causes a discontinuity in the bending moment diagram in its point of application, with a value equal to its intensity. However, the moment does not affect the shear force diagram.*
- *To plot the shear force of this example, we may just follow the direction and intensity of the applied forces. Thus, at $x = 0$, as the force $P_1 = 1$ kN is applied upwards, there is a positive jump or discontinuity. Constant intensity remains until section $x = 1$ m, in which force $P_2 = 2$ kN causes a jump of two units. The diagram remains constant until section $x = 2$ m, where force $P_3 = 3$ kN gives a positive jump. There is another jump at $x = 3$ m, whichcorresponds to the support reaction R_{Dy}.*
- *Using relation (5.19), we can write the following expression:*

$$dM_z(x) = V_y(x)dx. \tag{5.54}$$

Integrating this equation between the arbitrary sections x_1 and x_2 results in

$$\int_{x_1}^{x_2} dM_z(x) = \int_{x_1}^{x_2} V_y(x)dx, \tag{5.55}$$

that is,

$$M_z(x_2) = \int_{x_1}^{x_2} V_y(x)dx + M_z(x_1). \tag{5.56}$$

Thus, the bending moment at x_2 is equal to the bending moment at x_1 plus the area of the shear force diagram between x_1 and x_2.

Using the previous expression between the sections where loads are applied, we have

$$\begin{aligned}
M_z(x=0) &= 0, \\
M_z(x=1) &= \int_0^1 (1)dx + 0 = 1 \text{ kNm}, \\
M_z(x=2) &= \int_1^2 (-1)dx + 1 = 0 \text{ kNm}, \\
M_z(x=3) &= \int_2^3 (2)dx + 1 = 3 \text{ kNm}.
\end{aligned}$$

Therefore, we can plot the bending moment diagram using the area of the shear force diagram between two sections of interest and the bending moment in the left section of the considered interval.

- *The support reactions in D can be obtained by equilibrium, resulting in $R_{Dy} = -(P_1 - P_2 + P_3)$ and $M_{Dz} = P_1(L_1 + L_2 + L_3) + P_2(L_2 + L_3) + P_3 L_3 + M$. Substituting the given values, we have $R_{Dy} = -2$ kN and $M_{Dz} = 3$ kNm.*
Using the boundary conditions of the equilibrium BVP (5.23), we have

$$R_{Dy} = -V_y(x = 3) = -2 \text{ kN} \quad and \quad M_{Dz} = M_z(x = 3) = 3 \text{ kNm}.$$

As expected, both procedures give the same values for the support reactions.

□

Expressions for the shear force and bending moment for each beam segment can be written in a unique form, using the Macaulay notation, as

$$\begin{aligned}
V_y(x) &= P_1 <x-0>^0 - P_2 <x-L_1>^0 + P_3 <x-L_2>^0, & (5.57) \\
M_z(x) &= P_1 <x-0>^1 - P_2 <x-L_1>^1 + P_3 <x-L_2>^1 \\
&+ M <x-(L_1+L_2)>^0. & (5.58)
\end{aligned}$$

The $<x-a>^n$ term has the following meaning:

$$<x-a>^n = \begin{cases} 0 & x < a \\ (x-a)^n & x \geq a \end{cases}. \qquad (5.59)$$

Using the previous definition in expressions (5.57) and (5.58), we recover the expressions for the shear force and bending moment given in (5.52) and (5.53), respectively.

From the operational calculus, we can identify $<x-a>^0$ as the generalized Heaviside function, illustrated in Figure 5.27(a), and given by

$$\theta(x-a) = \begin{cases} 0 & x < a \\ 1 & x \geq a \end{cases}. \qquad (5.60)$$

Thus, the $<x-a>^0$ term is equivalent to the Heaviside function, that is,

$$<x-a>^0 = \theta(x-a). \qquad (5.61)$$

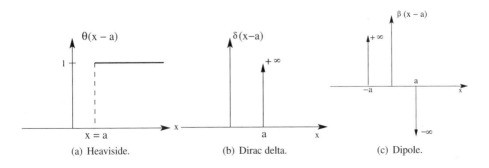

(a) Heaviside.   (b) Dirac delta.   (c) Dipole.

Figure 5.27 Generalized functions.

The Dirac's delta, illustrated in Figure 5.27(b), is another generalized function defined by

$$\delta(x-a) = \begin{cases} 0 & x \neq a \\ \infty & x = a \end{cases}.$$
(5.62)

The following property is valid for a function $f(x)$ using the Dirac delta:

$$\int_{-\infty}^{+\infty} f(x)\delta(x-a)dx = f(a).$$
(5.63)

The derivative of the Heaviside function results in the Dirac delta. Thus,

$$\delta(x-a) = \frac{d}{dx}\theta(x-a).$$
(5.64)

From (5.61), we have

$$\delta(x-a) = \frac{d}{dx}\theta(x-a) = <x-a>^{-1}.$$
(5.65)

The Dirac delta derivative results in the generalized dipole function illustrated in Figure 5.27(c)

$$\beta(x-a) = \frac{d}{dx}\delta(x-a) = <x-a>^{-2}.$$
(5.66)

We can use the previous functions to express the distributed load equation for the beam of Figure 5.24 as

$$q_y(x) = P_1 <x-0>^{-1} -P_2 <x-L_1>^{-1} +P_3 <x-(L_1+L_2)>^{-1} +M <x-(L_1+L_2)>^{-2}.$$
(5.67)

Observe that the moment M is indicated as positive even though it is in the negative direction of the z axis. The reason for this will be explained later.

Integrating the previous expression, we obtain the shear force expression

$$V_y(x) = P_1 <x-0>^{0} -P_2 <x-L_1>^{0} +P_3 <x-(L_1+L_2)>^{0} M <x-(L_1+L_2)>^{-1} +C_1.$$

The last term relative to the moment can be neglected, because, as observed above, it will not influence the behavior of the shear force.

Integrating the shear force expression, we obtain the bending moment equation

$$M_z(x) = P_1 <x-0>^{1} -P_2 <x-L_1>^{1} +P_3 <x-(L_1+L_2)>^{1} M <x-(L_1+L_2)>^{0} +C_1x+C_2.$$

From this point, the following definition can be used to integrate the $<x-a>^n$ term:

$$\int <x-a>^n dx = \begin{cases} <x-a>^{n+1} & n < 0 \\ \dfrac{<x-a>^{n+1}}{n+1} & n \geq 0 \end{cases}.$$
(5.68)

The three functions illustrated in Figure 5.27 do not satisfy the standard definition of functions and thus are called generalized. One way to obtain equations (5.61), (5.65), and (5.66) is taking the limit of series of continuous function. There are several series that may be used for this purpose. For instance [1],

$$f_n(x) = \sqrt{\frac{n}{\pi}}\exp(-nx^2).$$
(5.69)

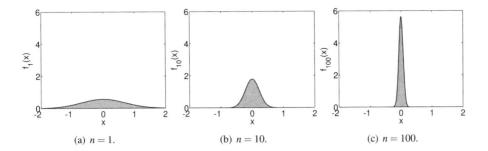

(a) $n = 1$.   (b) $n = 10$.   (c) $n = 100$.

Figure 5.28 Plots of functions of series $f_n(x)$.

As n increases, functions of the previous series become more concentrated about $x = 0$, as illustrated in Figure 5.28 for $n = 1, 10, 100$. Besides that, integration of $f_n(x)$, for any value of n, results in 1, that is,

$$\int_{-\infty}^{\infty} f_n(x)dx = 1. \tag{5.70}$$

The sequence of functions in (5.69) results in the Dirac delta taking the limit to $n \to \infty$. Thus,

$$\delta(x) = \lim_{n \to \infty} f_n(x). \tag{5.71}$$

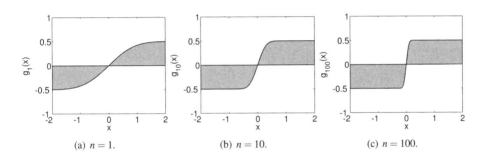

(a) $n = 1$.   (b) $n = 10$.   (c) $n = 100$.

Figure 5.29 Plots of the $g_n(x)$ functions.

The indefinite integration of (5.69) gives

$$g_n(x) = \int f_n(x)dx = -2n\sqrt{\frac{n}{\pi}}\exp(-nx^2)x.$$

Figure 5.29 illustrates $g_n(x)$ for $n = 1, 10, 100$. It is observed that when n increases, the sequence of functions $g_n(x)$ approximates the Heaviside function, that is,

$$\theta(x) = \lim_{n \to \infty} g_n(x). \tag{5.72}$$

Taking the derivative of $f_n(x)$, we have

$$h_n(x) = \frac{d}{dx}f_n(x) = -\sqrt{\frac{4n^3}{\pi}}x\exp(-nx^2). \tag{5.73}$$

Figure 5.30 illustrates the behavior of $h_n(x)$ for $n = 1, 10, 100$. Note that as n increases, the sequence of functions $h_n(x)$ approximates the dipole function, that is,

$$\beta(x) = \lim_{n \to \infty} h_n(x). \tag{5.74}$$

This function represents a binary of forces about $x = 0$, which is equivalent to a positive pure bending moment or couple in the clockwise direction. This is the reason why the moment was considered as positive in the loading equation (5.67).

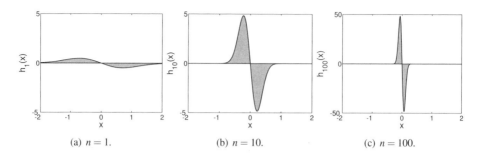

(a) $n = 1$.   (b) $n = 10$.   (c) $n = 100$.

Figure 5.30   Plots of the $h_n(x)$ function.

The notation of singularity functions will be used to represent the distributed load equation for beams, as illustrated below.

Example 5.10 *Indicate the expressions for the distributed load function $q_y(x)$ for the beams illustrated in Figures 5.31.*

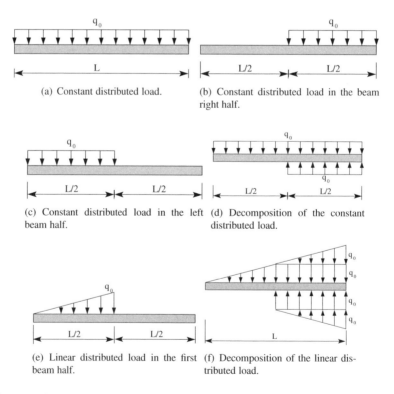

(a) Constant distributed load.

(b) Constant distributed load in the beam right half.

(c) Constant distributed load in the left beam half.

(d) Decomposition of the constant distributed load.

(e) Linear distributed load in the first beam half.

(f) Decomposition of the linear distributed load.

Figure 5.31   Example 5.10: beam loads indicated with singularity functions.

The load function expression for the beam of Figure 5.31(a) is

$$q_y(x) = -q_0 <x - 0>^0 = -q_0.$$

For the loading given in Figure 5.31(b), we have

$$q_y(x) = -q_0 <x-\frac{L}{2}>^0 = \begin{cases} 0 & 0<x<\frac{L}{2} \\ -q_0 & \frac{L}{2}\leq x<L \end{cases}.$$

For the beam of Figure 5.31(c), the load function is

$$q_y(x) = -q_0 <x-0>^0 +q_0 <x-\frac{L}{2}>^0.$$

As $<x-0>^0=1$ for $x \geq 0$, the first term applies the load along the entire beam length. Thus, we should add the second term from $x = \frac{L}{2}$ to zero the distributed load in the interval $\frac{L}{2} \leq x < L$, as shown in Figure 5.31(d). Thus,

$$q_y(x) = q_0 <x-0>^0 -q_0 <x-\frac{L}{2}>^0 = \begin{cases} -q_0 & 0<x<\frac{L}{2} \\ 0 & \frac{L}{2}\leq x<L \end{cases}.$$

For the beam of Figure 5.31(e), we have a similar behavior, because the first term in

$$q_y(x) = -\frac{q_0}{\frac{L}{2}} <x-0>^1 +q_0 <x-\frac{L}{2}>^0 +\frac{q_0}{\frac{L}{2}} <x-\frac{L}{2}>^1$$

makes the distributed load to be applied along the entire beam length. The second and third terms add, respectively, the constant and linear load components q_0, both starting at $x = \frac{L}{2}$, as illustrated in Figure 5.31(f). Thus,

$$q_y(x) = \begin{cases} -\dfrac{2q_0}{L}x & 0<x<\frac{L}{2} \\ -\dfrac{2q_0}{L}x+q_0+\dfrac{2q_0}{L}(x-\frac{L}{2})=0 & \frac{L}{2}\leq x<L \end{cases}.$$

□

The following examples use the notation of singularity functions to express the distributed load, kinematics, and internal loads for beams.

Example 5.11 *Plot the diagrams for the shear force and bending moment for the beam illustrated in Figure 5.32.*

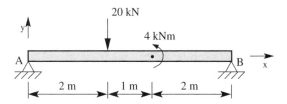

Figure 5.32 Example 5.11: simply supported beam with concentrated loads.

As the beam is statically determined, we use the second-order differential equation in terms of the bending moment. The boundary conditions for the shear force and bending moment are used. The load equation is denoted by the singularity function notation, since we have concentrated forces and moments along the beam length. The same previous integration procedure is applied in this example to solve the differential equation.

- *Distributed load equation*
  Using the singularity function notation, the distributed load equation on the beam is given by
  $$q_y(x) = -20000 < x - 2 >^{-1} -4000 < x - 3 >^{-2}.$$

- *Boundary conditions*
  As the beam is simply supported, the bending moments at the ends are zero. Thus, $M_z(x = 0) = 0$ and $M_z(x = 5) = 0$.

- *Integration of the differential equation*
  The differential equation of equilibrium in terms of the bending moment is considered
  $$\frac{d^2 M_z(x)}{dx^2} = q_y(x) = -20000 < x - 2 >^{-1} -4000 < x - 3 >^{-2}.$$

  From the integration of the above equation, the expressions for the shear force and bending moment are given, respectively, by
  $$V_y(x) = \frac{dM_z(x)}{dx^2} = -20000 < x - 2 >^0 -4000 < x - 3 >^{-1} +C_1,$$

  $$M_z(x) = -20000 < x - 2 >^1 -4000 < x - 3 >^0 +C_1 x + C_2.$$

- *Determination of the integration constants C_1 and C_2*
  Applying the boundary conditions, the integration constants C_1 and C_2 are determined as
  $$M_z(x = 0) = -20000(0) - 4000(0) + C_1(0) + C_2 = 0 \rightarrow C_2 = 0,$$

  $$M_z(x = 5) = -20000(5 - 2) - 4000 + C_1(5) = 0 \rightarrow C_1 = 12800.$$

- *Final equations*
  Substituting the integration constants C_1 and C_2, we have the final expressions for the shear force and bending moment given, respectively, by
  $$V_y(x) = -20000 < x - 2 >^0 +12800,$$

  $$M_z(x) = -20000 < x - 2 >^1 -4000 < x - 3 >^0 +12800x.$$

  The expressions of the shear force and bending moment for each of the three beam segments between the applied loads are
  $$V_y(x) = \begin{cases} 12800 & 0 \leq x < 2 \\ -7200 & 2 \leq x < 3 \\ -7200 & 3 \leq x \leq 5 \end{cases},$$

  $$M_z(x) = \begin{cases} 12800x & 0 \leq x < 2 \\ -7200x + 40000 & 2 \leq x < 3 \\ -7200x + 36000 & 3 \leq x \leq 5 \end{cases}.$$

- *Diagrams*
  Based on the final equations, the shear force and bending moment diagrams are illustrated in Figure 5.33.

- *Support reactions*
  The support reactions are determined from the boundary conditions in terms of the shear force as
  $$R_{Ay} = V_y(x = 0) = 12800 \text{ N} \quad and \quad R_{Dy} = -V_y(x = 5) = 7200 \text{ N}.$$

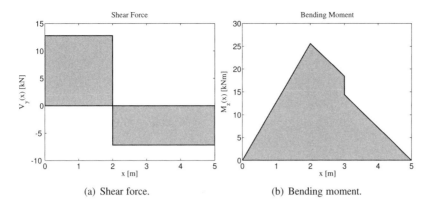

(a) Shear force.   (b) Bending moment.

Figure 5.33 Example 5.11: shear force and bending moment diagrams.

□

Example 5.12 *Consider the same beam of Example 5.6 but without the hinge, as shown in Figure 5.34. Plot the diagrams of shear force, bending moment, rotation, and transversal displacement.*

The beam is hyperstatic with three unknowns (R_{Ay}, M_{Az}, and R_{By}) and only two equilibrium equations ($\sum f_y = 0$ and $\sum m_z = 0$). As there is no hinge, we must necessarily integrate the differential equation (5.37) for the transversal displacement.

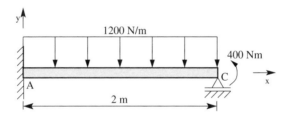

Figure 5.34 Example 5.12: beam subjected to a constant distributed load.

- *Distributed load equation*
  Due to the constant distributed transversal load applied in the negative y direction, we have $q_y(x) = -1200 \, \text{N/m}$.
- *Boundary conditions*
  The clamp at $x = 0$ makes the transversal displacement and rotation zero, that is, $u_y(x = 0) = 0$ and $\theta_z(x = 0) = 0$. The transversal displacement at the simple support is also zero and consequently $u_y(x = 2) = 0$. The bending moment is such that $M_z(x = 2) = 400 \, \text{Nm}$ due to the applied moment.
- *Integration of the differential equation*
  The equilibrium differential equation in terms of the transversal displacement is considered because the beam is hyperstatic. Thus,

$$EI_z \frac{d^4 u_y(x)}{dx^4} = q_y(x) = -1200.$$

The first integration results in the expression for the shear force

$$V_y(x) = EI_z \frac{d^3 u_y(x)}{dx^3} = -1200x + C_1.$$

The second integration gives the bending moment expression

$$M_z(x) = EI_z \frac{d^2 u_y(x)}{dx^2} = -600x^2 + C_1 x + C_2.$$

The third and fourth integrations give, respectively, the expressions for the rotation and transversal displacements

$$\theta_z(x) = \frac{1}{EI_z} \left(-200x^3 + \frac{C_1}{2} x^2 + C_2 x + C_3 \right),$$

$$u_y(x) = \frac{1}{EI_z} \left(-50x^4 + \frac{C_1}{6} x^3 + \frac{C_2}{2} x^2 + C_3 x + C_4 \right),$$

with C_1 to C_4 the integration constants.

- *Determination of the integration constants*
  In order to determine the constants C_1 to C_4, the given boundary conditions are used. Thus,

$$
\begin{aligned}
u_y(x=0) &= \frac{1}{EI_z}\left[-(50)(0)^4 + \frac{C_1}{6}(0)^3 + \frac{C_2}{2}(0)^2 + C_3(0) + C_4 \right] = 0 \\
&\rightarrow\quad C_4 = 0, \\
\theta_z(x=0) &= \frac{1}{EI_z}\left[-200(0)^3 + \frac{C_1}{2}(0)^2 + C_2(0) + C_3 \right] = 0 \\
&\rightarrow\quad C_3 = 0, \\
u_y(x=2) &= \frac{1}{EI_z}\left[-(50)(2)^4 + \frac{C_1}{6}(2)^3 + \frac{C_2}{2}(2)^2 \right] = 0 \\
&\rightarrow\quad 2C_1 + 3C_2 = 1200, \\
M_z(x=2) &= -(600)(2)^4 + 2C_1 + C_2 = 400 \\
&\rightarrow\quad 2C_1 + C_2 = 2800.
\end{aligned}
$$

Solving the system defined by the two previous equations, we determine $C_1 = 1800$ and $C_2 = -800$.

- *Final equations*
  Substituting C_1 to C_4, we obtain, respectively, the final equations for the shear force, bending moment, rotation, and transversal displacement

$$V_y(x) = -1200x + 1800,$$

$$M_z(x) = -600x^2 + 1800x - 800,$$

$$\theta_z(x) = \frac{1}{EI_z}\left(-200x^3 + 900x^2 - 800x \right),$$

$$u_y(x) = \frac{1}{EI_z}\left(-50x^4 + 300x^3 - 400x^2 \right).$$

- *Diagrams*
  The diagrams for the shear force, bending moment, rotation, and transversal displacement are illustrated in Figure 5.35. The shear force is zero for section $x = 1.5$ m. The respective bending moment is $M_z(x = 1.5) = 550$ Nm.
- *Support reaction*
  It is initially assumed that the support reactions R_{Ay} and R_{By} are in the y positive direction and the moment M_{Az} is in the z positive direction. Since the problem is statically indeterminate, we must employ the boundary conditions given in the equilibrium BVP (5.23).

In this case, R_{Ay} and R_{By} are V_{y0} and V_{yL}, respectively, while M_{Az} represents M_{z0}. Thus,

$$
\begin{aligned}
R_{Ay} &= V_y(x=0) = 1600 \text{ N}, \\
R_{By} &= -V_y(x=2) = 600 \text{ N}, \\
M_{Az} &= -M_z(x=0) = 800 \text{ Nm}.
\end{aligned}
$$

As the reactions were positive, the initially defined directions are correct.

The support reactions in A of Example 5.6 have larger values than those obtained in this example. As the bending moment is zero in the middle of the beam where the hinge is located, larger forces are supported at the left end clamp. In this example, the supports at both ends work more uniformly and contribute to the balance of external loads applied to the beam.

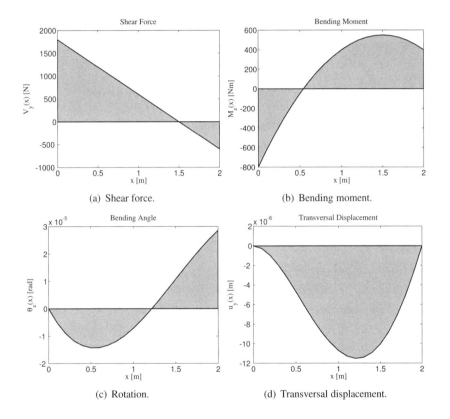

(a) Shear force.

(b) Bending moment.

(c) Rotation.

(d) Transversal displacement.

Figure 5.35 Example 5.12: diagrams of the shear force, bending moment, rotation, and transversal displacement.

□

Example 5.13 *Plot the diagrams of bending angle and transversal displacement for the beam with the hinge of Example 5.6.*

As the bending moment is zero at the hinge section, we have from equation (5.32)

$$
M_z(x) = EI_z \frac{d\theta_z(x)}{dx} = 0.
$$

Therefore, the rotation differential $d\theta_z(x)$ is also zero at the same section. Integrating this differential between the sections immediately before and after the hinge section $x = a$, we have

$$\Delta\theta_z = \int_{a^-}^{a^+} d\theta_z(x) = C, \tag{5.75}$$

with C a constant. Hence, there is a rotation discontinuity at the hinge, denoted by $\Delta\theta_z$, which must be determined by the solution procedure. For this purpose, the discontinuity is included as $EI_z\Delta\theta_z < x - a >^{-3}$ in the distributed load equation, as illustrated below.

- *Distributed load equation*
  Due to the constant distributed load applied to the beam in the negative y direction and the hinge discontinuity, we have

  $$q_y(x) = -1200 + EI_z\Delta\theta_z < x - 1 >^{-3}.$$

- *Boundary conditions*
  The transversal displacement and rotation are zero at the $x = 0$ end, that is, $u_y(x = 0) = 0$ and $\theta_z(x = 0) = 0$. At the simple support, the transversal displacement is also zero, that is, $u_y(x = 2) = 0$. Due to the moment applied at the $x = 2$ m end, we have $M_z(x = 2) = 400$ Nm.
- *Additional constraint*
  We have the additional restriction of zero moment at the hinge section, $x = 1$ m, that is, $M_z(x = 1) = 0$.
- *Integration of the differential equation*
  The differential equation of equilibrium in terms of the transversal displacement is considered, that is,

  $$EI_z\frac{d^4u_y(x)}{dx^4} = q_y(x) = -1200 + EI_z\Delta\theta_z < x - 1 >^{-3}.$$

  The first integration results in the shear force expression

  $$V_y(x) = EI_z\frac{d^3u_y(x)}{dx^3} = -1200x + EI_z\Delta\theta_z < x - 1 >^{-2} +C_1.$$

  The second integration gives the bending moment expression

  $$M_z(x) = EI_z\frac{d^2u_y(x)}{dx^2} = -600x^2 + EI_z\Delta\theta_z < x - 1 >^{-1} +C_1x+C_2,$$

  The third and fourth integrations result in, respectively, the rotation and transversal displacement equations

  $$\theta_z(x) = \frac{1}{EI_z}\left(-200x^3 + EI_z\Delta\theta_z < x - 1 >^0 +\frac{C_1}{2}x^2 + C_2x + C_3\right),$$

  $$u_y(x) = \frac{1}{EI_z}\left(-50x^4 + EI_z\Delta\theta_z < x - 1 >^1 +\frac{C_1}{6}x^3 + \frac{C_2}{2}x^2 + C_3x + C_4\right),$$

  with C_1 to C_4 the integration constants.
- *Determination of the integration constants*

To determine the C_1 to C_4 integration constants, the boundary conditions and the additional constraint are used. Thus,

$$u_y(x=0) = \frac{1}{EI_z}\left[-(50)(0)^4 + EI_z\Delta\theta_z(0) + \frac{C_1}{6}(0)^3 + \frac{C_2}{2}(0)^2 + C_3(0) + C_4\right] = 0$$

$$\rightarrow \quad C_4 = 0,$$

$$\theta_z(x=0) = \frac{1}{EI_z}\left[-200(0)^3 + EI_z\Delta\theta_z(0) + \frac{C_1}{2}(0)^2 + C_2(0) + C_3\right] = 0$$

$$\rightarrow \quad C_3 = 0,$$

$$u_y(x=2) = \frac{1}{EI_z}\left[-(50)(2)^4 + EI_z\Delta\theta_z(2-1)^0 + \frac{C_1}{6}(2)^3 + \frac{C_2}{2}(2)^2\right] = 0$$

$$\rightarrow \quad EI_z\Delta\theta_z + \frac{4}{3}C_1 + 2C_2 = 800,$$

$$M_z(x=2) = -(600)(2)^2 + 2C_1 + C_2 = 400 \rightarrow 2C_1 + C_2 = 2800,$$

$$M_z(x=1) = -(600)(1)^2 + (1)C_1 + C_2 = 0 \rightarrow C_1 + C_2 = 600.$$

Solving the system defined by the three previous equations, for $E = 210$ GPa and a rectangular section with $b = 0.1$ m and $h = 0.2$ m ($I_z = 6.67 \times 10^{-5}$ m^4), that is,

$$\begin{bmatrix} EI_z & \frac{4}{3} & 2 \\ 0 & 2 & 1 \\ 0 & 1 & 1 \end{bmatrix} \begin{Bmatrix} \Delta\theta_z \\ C_1 \\ C_2 \end{Bmatrix} = \begin{Bmatrix} 800 \\ 2800 \\ 600 \end{Bmatrix},$$

we determine $C_1 = 2200$ and $C_2 = -1600$ with $EI_z\Delta\theta_z = 1066.67$.

- *Final equations*
  Replacing C_1 to C_4 and $EI_z\Delta\theta_z$, we obtain the final equations, respectively, for the shear force, bending moment, rotation, and transversal displacement

$$V_y(x) = -1200x + 2200,$$

$$M_z(x) = -600x^2 + 2200x - 1600,$$

$$\theta_z(x) = \frac{1}{EI_z}\left(-200x^3 + 1066.67 <x-1>^0 + 1100x^2 - 1600x\right),$$

$$u_y(x) = \frac{1}{EI_z}\left(-50x^4 + 1066.67 <x-1>^1 + 366.67x^3 - 800x^2\right).$$

- *Diagrams*
  The diagrams of shear force and bending moment are the same ones obtained in Example 5.6. The diagrams of the rotation and transversal displacement are illustrated in Figure 5.36. There is discontinuity at the $x = 1$ m hinge section. In general, the displacements and rotations of the beam with the hinge are larger than those ones obtained in the previous example without the hinge.
- *Support reactions*
  The support reactions are the same of Example 5.6, that is,

$$\begin{aligned} R_{Ay} &= V_y(x=0) = 2200 \text{ N}, \\ R_{Cy} &= -V_y(x=2) = 200 \text{ N}, \\ M_{Az} &= -M_z(x=0) = 1600 \text{ Nm}. \end{aligned}$$

□

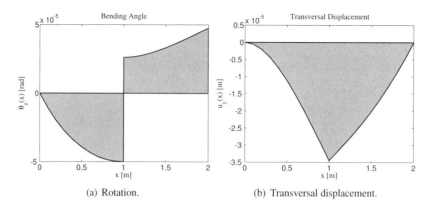

(a) Rotation.

(b) Transversal displacement.

Figure 5.36   Example 5.13: diagrams of rotation and transversal displacement.

Example 5.14 *Design the beams of Examples 5.6 and 5.12 to the maximum stress. Assume a circular cross-section and an admissible normal stress* $\bar{\sigma} = 100$ MPa.

For the beam with hinge, the critical section is $x = 0$ *and the maximum bending moment is* $M_z^{\max} = 1600$ Nm. *The bending strength modulus is calculated as*

$$W_z = \frac{M_z^{\max}}{\bar{\sigma}} = 1.6 \times 10^{-5} \text{ m}^3.$$

The diameter of the circular section is

$$d = \left(\frac{32}{\pi} W_z\right)^{1/3} = 2.52 \text{ cm}.$$

Analogous to the beam without hinge, the maximum bending moment is $M_z^{\max} = 800$ Nm *and the diameter of the circular section is* $d = 1.26$ cm.

Thus, the diameter of the beam with hinge is twice the one for the beam without hinge.
□

Example 5.15 *Consider the beam shown in Figure 5.37 which consists of two different cross-sections with material and geometric properties indicated by* (E_1, I_{z_1}) *and* (E_2, I_{z_2}), *respectively. Plot the diagrams of shear force, bending moment, rotation, and transversal displacement to* $L_1 = L_2 = 1$ m, $E_1 = 70$ GPa, $E_2 = 210$ GPa *and circular cross-sections of diameters* $d_1 = 15$ cm *and* $d_2 = 10$ cm. *Adopt* $q_0 = 1000$ N/m.

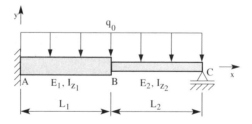

Figure 5.37   Example 5.15: beam with distinct cross-sections and materials.

Initially, the beam segment AB with $0 < x < L_1$ *is considered. The unknowns are the* C_1, C_2, C_3, *and* C_4 *integration constants, as well as the shear force* V_1 *and the bending moment* M_1 *at the interface of both segments, as illustrated in Figure 5.38(a).*

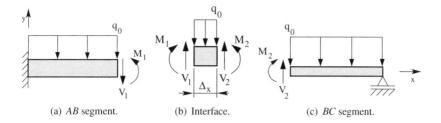

(a) *AB* segment.   (b) Interface.   (c) *BC* segment.

Figure 5.38 Example 5.12: FBDs for the *AB* and *BC* beam segments and the interface between them.

- *Distributed load equation*
  The expression for the distributed load is $q_y(x) = q_0$.
- *Boundary conditions*
  The kinematic boundary conditions are $u_{y_1}(x = 0) = 0$ and $\theta_{z_1}(x = 0) = 0$ due to the clamp at $x = 0$. The natural boundary conditions at the right end are $V_{y_1}(x = L_1) = V_1$ and $M_{z_1}(x = L_1) = M_1$.
- *Differential equation integration*
  As the problem is hyperstatic, the differential equation in terms of the kinematics is considered, that is,

$$E_1 I_{z_1} \frac{d^4 u_y(x)}{dx^4} = q_y(x) = -q_0.$$

The four successive integrations give, respectively, the expressions for the shear force, bending moment, rotation and transversal displacement, that is,

$$
\begin{aligned}
V_{y_1}(x) &= -q_0 x + C_1, \\
M_{z_1}(x) &= -\frac{q_0}{2} x^2 + C_1 x + C_2, \\
\theta_{z_1}(x) &= \frac{du_{y_1}(x)}{dx} = \frac{1}{E_1 I_{z_1}} \left(-\frac{q_0}{6} x^3 + C_1 \frac{x^2}{2} + C_2 x + C_3 \right), \\
u_{y_1}(x) &= \frac{1}{E_1 I_{z_1}} \left(-\frac{q_0}{24} x^4 + C_1 \frac{x^3}{6} + C_2 \frac{x^2}{2} + C_3 x + C_4 \right).
\end{aligned}
$$

Substituting the kinematic boundary conditions in the expressions for the rotation and transversal displacement, we obtain $C_3 = C_4 = 0$. Using the natural boundary conditions, we obtain the following two equations:

$$
\begin{aligned}
C_1 - q_0 L_1 &= V_1, & (5.76) \\
C_1 L_1 + C_2 - \frac{q_0}{2} L_1^2 &= M_1. & (5.77)
\end{aligned}
$$

Now consider the beam segment BC with $L_1 < x < L_1 + L_2$. In this case, the unknowns are the integration constants C_5, C_6, C_7, and C_8, as well as the shear force V_2 and the bending moment M_2 at the interface of the beam segments AB and BC, as illustrated in Figure 5.38(c).

- *Distributed load equation*
  The expression for the distributed load in this segment is $q_y(x) = -q_0$.
- *Boundary conditions*
  The boundary conditions in $x = L_1$ are $V_{y_2}(x = L_1) = V_2$ and $M_{z_2}(x = L_1) = M_2$. For the right end, we have the boundary conditions $u_{y_2}(x = L_1 + L_2) = 0$ and $M_{z_2}(x = L_1 + L_2) = 0$.

- *Differential equation integration*
  The differential equation in terms of the kinematics is considered

$$E_2 I_{z2} \frac{d^4 u_{y2}(x)}{dx^4} = -q_0.$$

The four successive integrations give, respectively, the expressions of the shear force, bending moment, rotation, and transversal displacement, that is,

$$
\begin{aligned}
V_{y2}(x) &= -q_0 x + C_5, \\
M_{z2}(x) &= -\frac{q_0}{2} x^2 + C_5 x + C_6, \\
\theta_{z2}(x) &= \frac{d u_{y2}(x)}{dx} = \frac{1}{E_2 I_{z2}} \left(-\frac{q_0}{6} x^3 + C_5 \frac{x^2}{2} + C_6 x + C_7 \right), \\
u_{y2}(x) &= \frac{1}{E_2 I_{z2}} \left(-\frac{q_0}{24} x^4 + C_5 \frac{x^3}{6} + C_6 \frac{x^2}{2} + C_7 x + C_8 \right).
\end{aligned}
$$

Substituting the boundary conditions in the previous expressions, the four following equations are determined:

$$-q_0 L_1 + C_5 = V_2, \tag{5.78}$$

$$-\frac{q_0}{2} L_1^2 + C_5 L_1 + C_6 = M_2, \tag{5.79}$$

$$(L_1 + L_2)C_5 + C_6 = \frac{q_0}{2}(L_1 + L_2)^2, \tag{5.80}$$

$$\frac{(L_1 + L_2)^3}{6} C_5 + \frac{(L_1 + L_2)^2}{2} C_6 + (L_1 + L_2)C_7 + C_8 = \frac{q_0}{24}(L_1 + L_2)^4. \tag{5.81}$$

The balance of loads at the interface of the two beam segments is now considered. For this purpose, a beam element of length Δx at the interface of two segments is considered, as shown in Figure 5.38(b). The equilibrium conditions are

$$\sum f_y = 0 : V_1 - V_2 - q_0 \Delta x = 0,$$

$$\sum m_z = 0 : -M_1 - V_1 \Delta x + q_0 \frac{\Delta x^2}{2} + M_2 = 0.$$

Taking the limit to $\Delta x \to 0$, we obtain the following expressions relating the shear force and the bending moment at the interface:

$$V_1 = V_2,$$

$$M_1 = M_2.$$

Substituting these expressions in (5.78) and (5.79) and using (5.76) and (5.77) we have

$$
\begin{aligned}
C_1 - q_0 L_1 &= C_5 - q_0 L_1, \\
C_1 L_1 + C_2 - \frac{q_0}{2} L_1^2 &= C_5 L_1 + C_6 - \frac{q_0}{2} L_1^2,
\end{aligned}
$$

or

$$
\begin{aligned}
C_1 - C_5 &= 0, \tag{5.82} \\
C_2 - C_6 &= 0. \tag{5.83}
\end{aligned}
$$

The kinematic compatibility at the interface between the two beam segments means that the transversal displacements u_{y_1} and u_{y_2} and the rotations θ_{z_1} and θ_{z_2} must be equal. Thus,

$$\theta_{z_1}(L_1) = \theta_{z_2}(L_1),$$

$$u_{y_1}(L_1) = u_{y_2}(L_1).$$

Substituting the previously obtained expressions for the transversal displacements and rotations, we have

$$\frac{1}{E_1 I_{z_1}}\left(-\frac{q_0}{6}L_1^3 + C_1\frac{L_1^2}{2} + C_2 L_1\right) = \frac{1}{E_2 I_{z_2}}\left(-\frac{q_0}{6}L_1^3 + C_5\frac{L_1^2}{2} + C_6 L_1 + C_7\right),$$

$$\frac{1}{E_1 I_{z_1}}\left(-\frac{q_0}{24}L_1^4 + C_1\frac{L_1^3}{6} + C_2\frac{L_1^2}{2}\right) = \frac{1}{E_2 I_{z_2}}\left(-\frac{q_0}{24}L_1^4 + C_5\frac{L_1^3}{6} + C_6\frac{L_1^2}{2} + C_7 L_1 + C_8\right),$$

or

$$\frac{L_1^2}{2}C_1 + L_1 C_2 - k\frac{L_1^2}{2}C_5 - kL_1 C_6 - kC_7 = (1-k)\frac{q_0}{6}L_1^3, \tag{5.84}$$

$$\frac{L_1^3}{6}C_1 + \frac{L_1^2}{2}C_2 - k\frac{L_1^3}{6}C_5 - k\frac{L_1^2}{2}C_6 - kL_1 C_7 - kC_8 = (1-k)\frac{q_0}{24}L_1^4, \tag{5.85}$$

with $k = \frac{E_1 I_{z_1}}{E_2 I_{z_2}}$.

Equations (5.80) to (5.85) result in the following system of equations:

$$\begin{bmatrix} 1 & 0 & -1 & 0 & 0 & 0 \\ 0 & 1 & 0 & -1 & 0 & 0 \\ \frac{L_1^2}{2} & L_1 & -k\frac{L_1^2}{2} & -kL_1 & -k & 0 \\ \frac{L_1^3}{6} & \frac{L_1^2}{2} & -k\frac{L_1^3}{6} & -k\frac{L_1^2}{2} & -kL_1 & -k \\ 0 & 0 & (L_1+L_2) & 1 & 0 & 0 \\ 0 & 0 & \frac{(L_1+L_2)^3}{6} & \frac{(L_1+L_2)^2}{2} & (L_1+L_2) & 1 \end{bmatrix} \begin{Bmatrix} C_1 \\ C_2 \\ C_5 \\ C_6 \\ C_7 \\ C_8 \end{Bmatrix} = \begin{Bmatrix} 0 \\ 0 \\ (1-k)\frac{q_0}{6}L_1^3 \\ (1-k)\frac{q_0}{24}L_1^4 \\ \frac{q_0}{2}(L_1+L_2)^2 \\ \frac{q_0}{24}(L_1+L_2)^4 \end{Bmatrix}.$$

Substituting the given values, we obtain $C_1 = 1279.67$, $C_2 = -559.35$, $C_5 = 1279.68$, $C_6 = -559.35$, $C_7 = 35.11$, and $C_8 = 8.92$.

The normal strains and stresses in each segment are calculated, respectively, as

$$\varepsilon_{xx_1}(x,y) = -y\frac{d^2 u_{y_1}(x)}{dx^2}, \quad \sigma_{xx_1}(x,y) = E_1 \varepsilon_{xx_1}(x,y),$$

$$\varepsilon_{xx_2}(x,y) = -y\frac{d^2 u_{y_2}(x)}{dx^2}, \quad \sigma_{xx_2}(x,y) = E_2 \varepsilon_{xx_2}(x,y).$$

The diagrams of the shear force, bending moment, rotation, transversal displacement, and longitudinal strains and normal stresses in the upper edge of the beam cross-sections are illustrated in Figure 5.39. Note that the internal loads and kinematics are continuous at the interface of segments. On the other hand, the normal strains and stresses are discontinuous at the interface, due to the change in the cross-section dimensions and material properties.

File beamexemp15.m implements the solution of this example using symbolic manipulation toolkit available in MATLAB.

☐

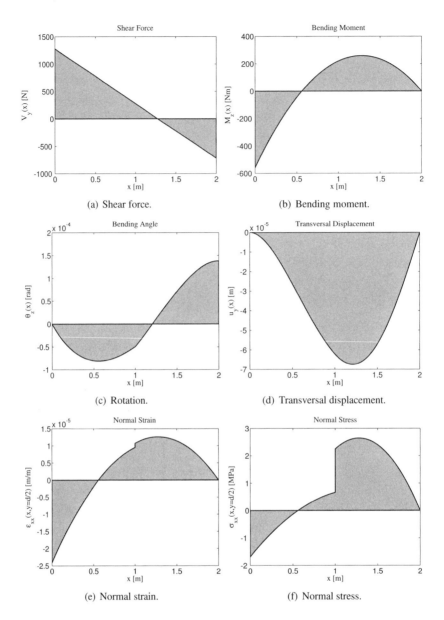

(a) Shear force.

(b) Bending moment.

(c) Rotation.

(d) Transversal displacement.

(e) Normal strain.

(f) Normal stress.

Figure 5.39 Example 5.15: diagrams of the shear force, bending moment, rotation, transversal displacement, and normal strains and stresses and at the upper edge of the cross-sections.

Example 5.16 *Consider the beam of Example 5.11, with the support at $x = 3$ m, as illustrated in Figure 5.40. Plot the shear force and bending moment diagrams.*

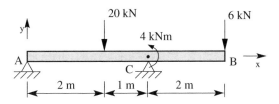

Figure 5.40 Example 5.16: beam supported in an internal section.

- *Distributed load equation*
  In this case, the support reaction R_{Cy} at C is included in the beam distributed load equation. Thus,

  $$q_y(x) = -20000 < x - 2 >^{-1} + R_{Cy} < x - 3 >^{-1} - 4000 < x - 3 >^{-2}.$$

- *Boundary conditions*
  As the beam is supported at $x = 0$, the bending moment and the transversal displacement are zero, that is, $M_z(x = 0) = 0$ and $u_y(x = 0) = 0$. The boundary conditions at the right end are in terms of the shear force and bending moment, that is, $V_y(x = 5) = 6000$ and $M_z(x = 5) = 0$.
- *Additional constraint*
  The transversal displacement is zero at C and $u_y(x = 3) = 0$. As the beam is isostatic, this condition will not be used for the solution of this example.
- *Integration of the differential equation*
  For the given isostatic beam, it is enough to consider the differential equation in terms of the bending moment. i.e.,

  $$\frac{d^2 M_z(x)}{dx^2} = q_y(x) = -20000 < x - 2 >^{-1} + R_{Cy} < x - 3 >^{-1} - 4000 < x - 3 >^{-2}.$$

  Integrating this differential equation, the expressions for the shear force and bending moment are given, respectively, by

  $$V_y(x) = -20000 < x - 2 >^0 + R_{Cy} < x - 3 >^0 - 4000 < x - 3 >^{-1} + C_1,$$

  $$M_z(x) = -20000 < x - 2 >^1 + R_{Cy} < x - 3 >^1 - 4000 < x - 3 >^0 + C_1 x + C_2.$$

- *Determination of the integration constants C_1 and C_2 and support reaction R_{Cy}*
  Applying the boundary conditions, the following expressions are determined:

  $$
  \begin{aligned}
  M_z(x = 0) &= -(2000)(0) + R_{Cy}(0) - (4000)(0) + C_1(0) + C_2 = 0 \\
  &\rightarrow C_2 = 0, \\
  V_y(x = 5) &= -20000 + R_{Cy} + C_1 = 6000 \\
  &\rightarrow C_1 + R_{Cy} = 26000, \\
  M_z(x = 5) &= -20000(5 - 2) + R_{Cy}(5 - 3) - 4000 + C_1(5) = 0 \\
  &\rightarrow 5C_1 + 2R_{Cy} = 64000.
  \end{aligned}
  $$

  Solving the system of equations with the two previous expressions, we obtain $C_1 = 4000$ and $R_{Cy} = 22000$ N.

- *Final equations*
  Substituting the integration constants C_1 and C_2 and the reaction force R_{Cy}, we have the final expressions for the shear force and bending moment given by

$$V_y(x) = -20000 < x - 2 >^0 + 26000 < x - 3 >^0 + 4000,$$

$$M_z(x) = -20000 < x - 2 >^1 + 26000 < x - 3 >^1 - 4000 < x - 3 >^0 + 4000x.$$

The expressions for the shear force and bending moment for each of the three beam segments between the applied loads are:

$$V_y(x) = \begin{cases} 4000 & 0 \leq x < 2 \\ -16000 & 2 \leq x < 3 \\ 6000 & 3 \leq x \leq 5 \end{cases},$$

$$M_z(x) = \begin{cases} 4000x & 0 \leq x < 2 \\ -16000x + 40000 & 2 \leq x < 3 \\ 6000x - 30000 & 3 \leq x \leq 5 \end{cases}.$$

- *Diagrams*
  Based on the final equations, the shear force and bending moment diagrams are illustrated in Figure 5.41.

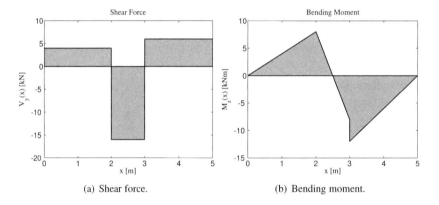

| (a) Shear force. | (b) Bending moment. |

Figure 5.41   Example 5.16: diagrams of shear force and bending moment.

- *Support reaction*
  The support reaction at $x = 0$ is determined from the shear force boundary condition as

$$R_{Ay} = V_y(x = 0) = 4000 \text{ N}.$$

□

Example 5.17 *The beam ABCD of Figure 5.42(a) holds an object with weight $P = 30$ kN using the rigid cable that passes through the pulley at B with radius $r = 30$ mm. Find the expressions of the shear force and bending moment and calculate the support reactions at A and D. Neglect the friction betweeen the cable and pulley, beam thickness, and vertical arm.*

One important point of this example relies on how to transfer the forces at the cable-pulley set to the beam. Special attention should be given to the forces that arise at points B and E. It is important to correctly define the free body diagram for the model.

Figure 5.42(b) illustrates the free body diagram used for the solution. Forces E_x and E_y arise from the decomposition of the cable traction force T. The moment M_B comes from the transference

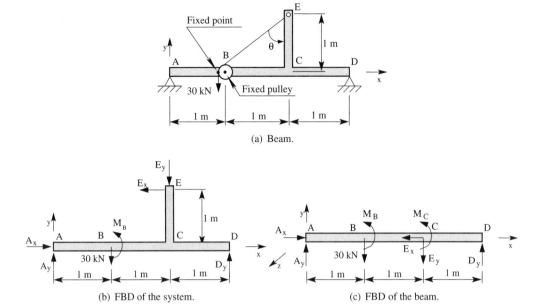

(a) Beam.

(b) FBD of the system.   (c) FBD of the beam.

Figure 5.42   Example 5.17: transmission of loads to the beam *ABCD*.

*of force P to the pulley center. Because the friction on the pulley is neglected, we have T = P.
Therefore,*

$$E_x = T\sin\theta = (30000)\sin(45^o) = 21213.2 \text{ N},$$

$$E_y = T\cos\theta = (30000)\cos(45^o) = 21213.2 \text{ N}.$$

*The transference of P to the pulley center gives rise to the moment M_B with counter-clockwise
direction and magnitude calculated as*

$$M_B = Pr = (30000)(30 \times 10^{-3}) = 900 \text{ Nm}.$$

*We need now to transfer forces on point E to point C as illustrated in Figure 5.42(c). This will
lead to the counter-clockwise moment M_C at point C with magnitude*

$$M_C = E_x(1) = (21213.2)(1) = 21213.2 \text{ Nm}.$$

*The axial forces are neglected. The shear force and bending moment diagrams will be obtained
using the previous integration procedure.*

· *Distributed load equation*

$$q_y(x) = -30000 < x - 1 >^{-1} -900 < x - 1 >^{-2} -21213.2 < x - 2 >^{-1}$$
$$-21213.2 < x - 2 >^{-2}.$$

· *Boundary conditions*
  *As the beam is simply supported, the bending moments at the ends are zero. Thus, $M_z(x = 0) = 0$ and $M_z(x = 3) = 0$. The kinematic boundary conditions $u_y(x = 0) = u_y(x = 3) = 0$
  will not be employed for the solution of this isostatic problem.*

- *Integration of the differential equation*
  The differential equation in terms of the bending moment is considered and

$$\frac{dM_z^2(x)}{dx^2} = q_y(x) = -30000 < x - 1 >^{-1} - 900 < x - 1 >^{-2}$$
$$-21213.2 < x - 2 >^{-1} - 21213.2 < x - 2 >^{-2}.$$

  Two indefinite integrals of this equation give the expressions for the shear force and bending moment, that is,

$$V_y(x) = -30000 < x - 1 >^0 - 900 < x - 1 >^{-1} - 21213.2 < x - 2 >^0$$
$$-21213.2 < x - 2 >^{-1} + C_1,$$
$$M_z(x) = -30000 < x - 1 >^1 - 900 < x - 1 >^0 - 21213.2 < x - 2 >^1$$
$$-21213.2 < x - 2 >^0 + C_1 x + C_2.$$

- *Determination of the integration constants*
  Applying the boundary conditions in terms of the bending moment, the integration constants C_1 and C_2 are determined as

$$M_z(x = 0) = -(30000)(0) - (900)(0) - (21213.2)(0) - (21213.2)(0)$$
$$+C_1(0) + C_2 = 0 \rightarrow C_2 = 0,$$
$$M_z(x = 3) = -(30000)(3 - 1)^1 - (900)(3 - 1)^0 - (21213.2)(3 - 2)^1$$
$$-21213.2 + C_1(3) = 0 \rightarrow C_1 = 34442.1.$$

- *Final equations*
  Substituting the integration constants C_1 and C_2, we have the following expressions for the shear force and bending moment, respectively,

$$V_y(x) = -30000 < x - 1 >^0 - 21213.2 < x - 2 >^0 + 34442.1,$$
$$M_z(x) = -30000 < x - 1 >^1 - 900 < x - 1 >^0 - 21213.2 < x - 2 >^1$$
$$-21213.2 < x - 2 >^0 + 34442.1x.$$

  The expressions for the shear force and bending moment for each of the three beam segments between the applied forces and momets are

$$V_y(x) = \begin{cases} 34442.1 & 0 \le x < 1 \\ 4442.1 & 1 \le x < 2 \\ -16771.1 & 2 \le x \le 3 \end{cases},$$

$$M_z(x) = \begin{cases} 34442.1x & 0 \le x < 1 \\ 4442.1x + 29100 & 1 < x \le 2 \\ -16771.1x + 50313.2 & 2 \le x \le 3 \end{cases}.$$

- *Diagrams*
  Using the final equations, the diagrams for the shear force and bending moment are illustrated in Figure 5.43.
- *Support reactions*
  The support reactions can be directly obtained from the shear force diagram and using the boundary conditions indicated in (5.23). Thus,

$$A_y = V_y(x = 0) = 34442.1 \text{ N and } D_y = -V_y(x = 3) = 16771.1 \text{ N}.$$

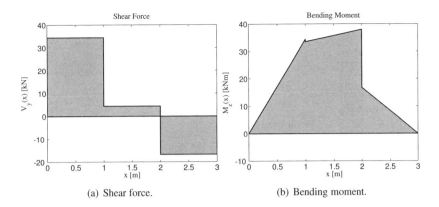

(a) Shear force. (b) Bending moment.

Figure 5.43 Example 5.17: diagrams for the shear force and bending moment.

Using the Newtonian equilibrium conditions, we have

$$\sum f_y = 0: \ A_y - 30000 + E_y + D_y = 0;$$

$$\sum m_{z_A} = 0: \ M_B - (30000)(1) - M_C + E_y(2) + D_y(3) = 0.$$

Substituting M_B, M_C, and E_y, we determine $D_y = 16771.1$ N and consequently $A_y = 34442.1$ N.

□

Example 5.18 *A steel wall with rectangular cross-section, illustrated in Figure 5.44(a), is supported on a rigid foundation and behaves as a dam. The wall has height $h = 2$ m and thickness $t = 0.3$ m. Calculate the maximum traction (σ_t) and compression (σ_c) stresses at the wall base for a water column height d. Consider the gravity acceleration $g = 9.81$ m/s^2, density $\rho = 1000$ kg/m^3 and steel Young's modulus $E = 210$ GPa.*

The first step is to construct the free body diagram as illustrated in Figure 5.44(b). From fluid mechanics, it is known that for a column height d, the pressure at the wall base is $\rho g d$, which represents a force density per units of area. It is then necessary to transform this area density to a force density per units of length (p_{unit}). Considering a wall with width $b = 1$ m, we have

$$p_{unit} = (\rho g d)b = 9.81d \times 10^3 \ \text{N/m}.$$

We can now apply the same previous integration procedure for the cantilever beam with a linear distributed load illustrated in Figure 5.44(b).

- *Distributed load equation*
  For the linear distributed load illustrated in Figure 5.44(b), we have

$$q_y(x) = \frac{9.81d \times 10^3}{d} <x-a>^1 = 9.81 \times 10^3 <x-a>^1.$$

- *Boundary conditions*
  As the problem is statically determinate, the boundary conditions are in terms of the shear force and bending moment. At the free end, we have $M_z(x = 0) = 0$ and $V_y(x = 0) = 0$.

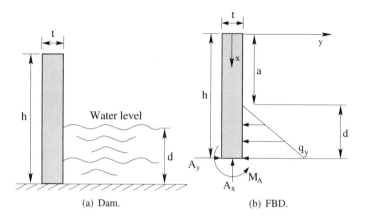

Figure 5.44   Example 5.18: beam model for a dam.

- *Integration of the differential equation*
  As the problem is isostatic, the differential equation of equilibrium in terms of the bending moment is considered

$$\frac{d^2 M_z(x)}{dx^2} = q_y(x) = 9.81 \times 10^3 <x-a>^1 .$$

The integration of the above differential equation results in the expressions for the shear force and bending moment given, respectively, by

$$V_y(x) = \frac{9.81 \times 10^3}{2} <x-a>^2 +C_1,$$

$$M_z(x) = \frac{9.81 \times 10^3}{6} <x-a>^3 +C_1 x + C_2.$$

- *Determination of the integration constants*
  Applying the boundary conditions at the free end of the beam, the integration constants C_1 and C_2 are calculated as

$$V_y(x=0) = 0 + C_1 = 0 \rightarrow C_1 = 0,$$

$$M_z(x=0) = 0 + 0 + C_2 = 0 \rightarrow C_2 = 0.$$

- *Final equations*
  Replacing C_1 and C_2, we have the final expressions for the shear force and bending moment, that is,

$$V_y(x) = \frac{9.81 \times 10^3}{2} <x-a>^2,$$

$$M_z(x) = \frac{9.81 \times 10^3}{6} <x-a>^3 .$$

- *Normal stress calculation for a column of height d*
  As a unit thickness was considered for the wall, the moment of area relative to the z axis is calculated as

$$I_z = \frac{bt^3}{12} = \frac{(1)(0.3)^3}{12} = 22.5 \times 10^{-4} \ \text{m}^4.$$

The traction and compression normal stresses at the base are located at $y = \pm 0.15$ m from the neutral line and are determined as

$$\sigma_t = \frac{M_z}{I_z}y = \frac{(9.81 \times 10^3)(0.15)}{(22.5 \times 10^{-4})(6)}(2-a)^3 = 872000 - 109000a^3 \text{ Pa},$$

$$\sigma_c = -\frac{M_z}{I_z}y = -\frac{(9.81 \times 10^3)(0.15)}{(22.5 \times 10^{-4})(6)}(2-a)^3 = -(872000 - 109000a^3) \text{ Pa}.$$

For $a = 1$ m, we have $\sigma_t = 761$ kPa and $\sigma_c = -761$ kPa.

☐

5.11 SUMMARY OF THE VARIATIONAL FORMULATION FOR THE EULER-BERNOULLI BEAM

The displacement bending actions are given by the vector field **u** with the following components:

$$\mathbf{u} = \left\{ \begin{array}{c} u_x(x,y) \\ u_y(x) \\ u_z(x) \end{array} \right\} = \left\{ \begin{array}{c} -y\dfrac{du_y(x)}{dx} \\ u_y(x) \\ 0 \end{array} \right\}. \tag{5.86}$$

The set of possible kinematic actions $\mathcal{V}$ is defined by

$$\mathcal{V} = \{\mathbf{u}|\, u_x(x,y) = -y\frac{du_y(x)}{dx},\, u_y(x),\, u_z(x) = 0\}. \tag{5.87}$$

For an unconstrained beam, the set of admissible actions coincides with $\mathcal{V}$.

The strain operator $\mathcal{D}$ is given by $\mathcal{D} = \dfrac{\partial}{\partial x}$. In the same way, the space $\mathcal{W}$ is the set of all scalar functions $\varepsilon_{xx}(x,y)$, called normal strains, and obtained by the derivative of the displacement actions $u_x(x,y) \in \mathcal{V}$. It is observed that the strain operator $\mathcal{D}: \mathcal{V} \longrightarrow \mathcal{W}$ relates the kinematics to the strain, that is,

$$\begin{array}{c} \mathcal{D}: \quad \mathcal{V} \longrightarrow \mathcal{W} \\ u_x(x,y) \longrightarrow \varepsilon_{xx}(x) = \mathcal{D}u_x(x,y) = -y\dfrac{d^2u_y(x)}{dx^2} \end{array}. \tag{5.88}$$

The set of all rigid actions in $\mathcal{V}$, that is, the actions $\mathbf{u} \in \mathcal{V}$ such that $\varepsilon_{xx} = \mathcal{D}u_x(x,y) = -y\dfrac{d^2u_y(x)}{dx^2} = 0$, defines the subset $\mathcal{N}(\mathcal{D})$ of rigid actions of the beam. This subset is formally defined as

$$\mathcal{N}(\mathcal{D}) = \left\{ \mathbf{u}(x) \in \mathcal{V} \mid u_y(x) = \text{cte or } \frac{du_y(x)}{dx} = \text{cte}, \mathcal{D}u_x(x,y) = -y\frac{d^2u_y(x)}{dx^2} = 0 \right\}, \tag{5.89}$$

that is, $\mathcal{N}(\mathcal{D})$ is the subset of all kinematic actions $\mathbf{u}$ of $\mathcal{V}$, such that $u_y(x)$ or $\dfrac{du_y(x)}{dx}$ are constants, implying that the normal strain $\varepsilon_{xx}(x,y)$ is zero.

In the case of the Euler-Bernoulli beam, the set $\mathcal{W}$ of internal loads is defined by continuous functions $M_z(x)$ representing the resultant in terms of the bending moment at the beam sections.

Operator $\mathcal{D}^*$ maps the vector spaces $\mathcal{W}$ of the internal loads and $\mathcal{V}'$ of the external loads. In this case, the vector space of external loads $\mathcal{V}'$ is characterized by a continuous scalar function $q_y(x)$, giving the distributed transversal load on the beam; transversal concentrated forces V_{y_0} and V_{y_L}; and moments M_{z_0} and M_{z_L} at the beam ends. These concentrated loads are treated as natural or force boundary conditions of the problem. Thus, $\mathscr{D}^*$ is denoted as

$$\mathscr{D}^* : \quad \mathscr{W} \to \mathscr{V}'$$

$$M_z(x) \to \mathscr{D}^* M_z(x) = \begin{cases} \dfrac{d^2 M_z(x)}{dx^2} & \text{in } x \in (0,L) \\[2mm] \dfrac{dM_z(x)}{dx} & \text{in } x = 0 \\[2mm] -\dfrac{dM_z(x)}{dx} & \text{in } x = L \\[2mm] -\left. M_z(x) \right|_{x=0} & \text{in } x = 0 \\[2mm] \left. M_z(x) \right|_{x=L} & \text{in } x = L \end{cases} \qquad (5.90)$$

From equation (5.23), the differential equilibrium operator $\mathscr{D}^*$ between the external and internal loads is defined by

$$\mathscr{D}^*(\cdot) = \begin{cases} \dfrac{d^2}{dx^2}(\cdot) & \text{in } x \in (0,L) \\[2mm] \dfrac{d}{dx}(\cdot) & \text{in } x = 0 \\[2mm] -\dfrac{d}{dx}(\cdot) & \text{in } x = L \\[2mm] -(\cdot)|_{x=0} & \text{in } x = 0 \\[2mm] (\cdot)|_{x=L} & \text{in } x = L \end{cases} \qquad (5.91)$$

The schematic form of the beam problem formulation is shown in Figure 5.45.

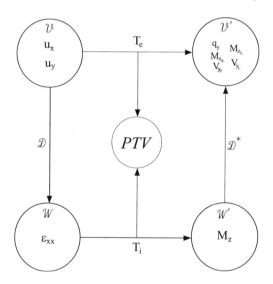

Figure 5.45   Variational formulation of the Euler-Bernoulli beam model.

5.12 BUCKLING OF COLUMNS

The above-mentioned design procedures consider only the stiffness and strength aspects of mechanical components. Another characteristic that should be considered in the design procedure of mechanical components is stability.

To illustrate this aspect, consider the cylindrical bars illustrated in Figure 5.46, of lengths L_1 and L_2 ($L_2 > L_1$) and same cross-section areas A, subjected to axial compressive forces P_1 and P_2, respectively.

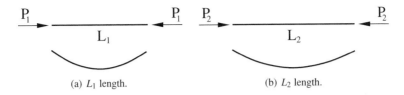

(a) L_1 length.　　　　　(b) L_2 length.

Figure 5.46　Bars with different lengths subjected to compressive axial forces.

For certain values of compressive forces, the bars instantly change the equilibrium configuration. This phenomenon is an example of stability. It is observed that the critical force $P_{2_{cr}}$ for buckling the longer bar, which consequently is more flexible, is smaller than $P_{1_{cr}}$ for the shorter bar.

Consider now the column of negligible weight shown in Figure 5.47(a) and subjected to an axial compressive force P. It is supported at the lower end in a torsional spring of stiffness coefficient k_t. Following the same procedure of Example 2.9 and assuming that the column is initially in the vertical position, the external work function is given in this case by

$$W_e(\theta) = P(L - L\cos\theta) - M_t\theta, \tag{5.92}$$

with M_t the reaction moment in the spring.

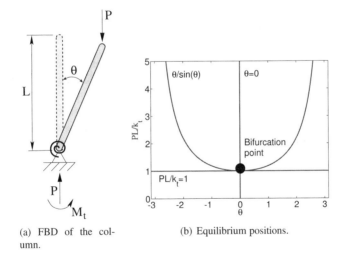

(a) FBD of the column.　　　(b) Equilibrium positions.

Figure 5.47　Vertical column for stability analysis.

The equilibrium condition of the lever is determined by imposing that the differential of the previous expression is zero, that is,

$$\delta W_e(\theta) = (PL\sin\theta - M_t)\delta\theta = 0. \tag{5.93}$$

As the virtual rotation $\delta\theta$ is arbitrary, the previous expression is satisfied if

$$PL\sin\theta = M_t. \tag{5.94}$$

One of the possible solutions for this equation is $\theta = 0$, which results in $\sin\theta = 0$ and $M_t = 0$. Assuming that the torsional spring is proportional, the relation between M_t and θ is $M_t = k_t\theta$. Thus, the above expression is rewritten as

$$\frac{PL}{k_t} = \frac{\theta}{\sin\theta}. \tag{5.95}$$

For $\frac{PL}{k_t} < 1$, there is only one solution with $\theta = 0$; for $\frac{PL}{k_t} > 1$, there are three possible solutions corresponding to the vertical position $\theta = 0$ of the column and rotations of an angle θ to the left and right of the vertical position. These solutions are shown in Figure 5.47(b). The intersection point of all solutions is called the bifurcation point.

For a small θ, $\sin\theta \approx \theta$, and equation (5.94) reduces to the following linear stability problem:

$$(k_t - PL)\theta = 0. \tag{5.96}$$

The solutions of the previous equation are $\theta = 0$ and $\frac{PL}{k_t} = 1$ for any value of θ, both shown in Figure 5.47(b). The $P_{cr} = P = \frac{k_t}{L}$ relation is called the critical buckling load and represents the lowest compressive force that causes instability of the column. These concepts will be considered in the next section for the Euler column. For more general structures, the above equation represents an eigenvalue problem, with (P, θ) an eigenpair comprising the value of critical force and the respective buckling mode.

5.12.1   EULER COLUMN

Consider a straight Euler column illustrated in Figure 5.48(a) with pinned supports at both ends. We wish to determine the lower compression force P, called Euler critical load, such that buckling occurs in the column.

The buckling modes depend not only on the geometric properties of the column but also on the intensity of the compressive force. It is assumed that the considered columns have only global buckling modes due to bending.

For cross-sections with two axes of symmetry (for instance, circular, rectangular, and an I-shaped), buckling can occur laterally in any direction. But for cross-sections with only one axis of symmetry, buckling occurs about the axis with the smallest moment of area. Thus, I_{min} is denoted as the smallest moment of area of the cross-section in the formulation presented below.

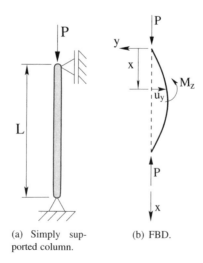

(a) Simply supported column.   (b) FBD.

Figure 5.48   Euler column.

The deformed geometry of the Euler column is illustrated in Figure 5.48(b). The bending moment in a generic section x of the beam is given by $M_z(x) = -Pu_y(x)$. Assuming that the column has the same material and constant cross-sections, we obtain the following differential equation substituting $M_z(x)$ into expression (5.32):

$$\frac{d^2u_y(x)}{dx^2} = -\frac{P}{EI_{min}}u_y(x). \tag{5.97}$$

Defining

$$\lambda^2 = \frac{P}{EI_{min}}, \tag{5.98}$$

we rewrite equation (5.97) as

$$\frac{d^2u_y(x)}{dx^2} + \lambda^2 u_y(x) = 0. \tag{5.99}$$

The above homogeneous differential equation defines a continuum eigenvalue problem for the transversal displacement function $u_y(x)$. The general solution of this equation is given by

$$u_y(x) = C_1\sin(\lambda x) + C_2\cos(\lambda x), \tag{5.100}$$

with C_1 and C_2 arbitrary constants to be determined using the boundary conditions $u_y(0) = u_y(L) = 0$. Hence,

$$
\begin{aligned}
u_y(x=0) &= C_1\sin(\lambda 0) + C_2\cos(\lambda 0) = 0 \rightarrow C_2 = 0,\\
u_y(x=L) &= C_1\sin(\lambda L) = 0.
\end{aligned}
$$

The last equation is satisfied with $C_1 = 0$. However, as $C_2 = 0$, such combination results in the trivial solution corresponding to the straight column. Another solution is obtained imposing $\sin(\lambda L) = 0$, which is achieved if

$$\lambda L = n\pi, \tag{5.101}$$

for $n \geq 1$ an integer number.

Substituting (5.98) in (5.101), we obtain the compressive forces that cause the column buckling as

$$P_n = \left(\frac{n\pi}{L}\right)^2 EI_{min}. \tag{5.102}$$

The respective buckling modes are given by

$$u_y(x) = C_1\sin\left(\frac{n\pi}{L}x\right), \tag{5.103}$$

where the C_1 is the indeterminate amplitude.

The smallest compression force is the most important in stability analysis. It is obtained for $n = 1$, and is called Euler critical buckling load and given by

$$P_{cr} = \frac{\pi^2 EI_{min}}{L^2}. \tag{5.104}$$

The critical buckling mode is obtained from (5.104) as

$$u_y(x) = C_1\sin\left(\frac{\pi}{L}x\right) \tag{5.105}$$

and corresponds to a half sinusoid.

Consider now the column illustrated in Figure 5.49(a) with one clamped end and the other supported. The bending moment in the beam generic section x is given by

$$M_z(x) = -Pu_y(x) + M_{Az}\left(1 - \frac{x}{L}\right),$$

where M_{Az} is the clamp reaction moment. The differential equation in terms of the bending moment is

$$\frac{d^2 u_y(x)}{dx^2} = -\frac{1}{EI_{min}}\left[-Pu_y(x) + M_{Az}\left(1 - \frac{x}{L}\right)\right].$$

(5.106)

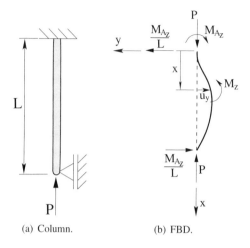

(a) Column.   (b) FBD.

Figure 5.49 Column with supported and clamped ends.

We can rewrite the previous equation in terms of λ as

$$\frac{d^2 u_y(x)}{dx^2} + \lambda^2 u_y(x) = \frac{\lambda^2 M_{Az}}{P}\left(1 - \frac{x}{L}\right).$$

(5.107)

This is a nonhomogeneous differential equation, whose solution is the sum of a homogeneous solution $u_y^h(x)$ and a particular solution $u_y^p(x)$, that is,

$$u_y(x) = u_y^h(x) + u_y^p(x).$$

The homogeneous solution is the same used in (5.100) and the particular solution is $u_y^p(x) = \frac{M_{Az}}{P}\left(1 - \frac{x}{L}\right)$. Thus,

$$u_y(x) = C_1 \sin(\lambda x) + C_2 \cos(\lambda x) + \frac{M_{Az}}{P}\left(1 - \frac{x}{L}\right).$$

(5.108)

Applying the boundary conditions for the considered column, we have

$$
\begin{aligned}
u_y(x=0) &= C_1 \sin(\lambda 0) + C_2 \cos(\lambda 0) + \frac{M_{Az}}{P} = 0 \rightarrow C_2 = -\frac{M_{Az}}{P}, \\
u_y(x=L) &= C_1 \sin(\lambda L) + C_2 \cos(\lambda L) = 0, \\
\frac{du_y(x=0)}{dx} &= C_1 \lambda \cos(\lambda 0) - C_2 \lambda \sin(\lambda 0) - \frac{M_{Az}}{PL} = 0 \rightarrow C_1 \lambda - \frac{M_{Az}}{PL} = 0.
\end{aligned}
$$

From the first and second equations, we obtain

$$C_1 = \frac{M_{Az}}{P \tan(\lambda L)},$$

which substituted in the third previous equation results in

$$\frac{M_{Az}}{P}\left(\frac{\lambda}{\tan(\lambda L)} - \frac{1}{L}\right) = 0.$$

As M_{Az} and P are arbitrary, the necessary condition to satisfy the previous equation results in the following transcedental equation:

$$\lambda L = \tan(\lambda L). \tag{5.109}$$

The smallest nonzero root of the above equation is

$$\lambda L = 4.493.$$

The critical buckling load for the considered column can be expressed analogously to (5.104) as

$$P_{cr} = \frac{2.05\pi^2 EI_{min}}{L^2}. \tag{5.110}$$

Adopting the same previous procedure, the critical load for the double-clamped and free end beams illustrated, respectively, in Figures 5.50(c) and 5.50(d), are given by

$$P_{cr} = \frac{4\pi^2 EI_{min}}{L^2} \quad \text{and} \quad P_{cr} = \frac{\pi^2 EI_{min}}{4L^2}. \tag{5.111}$$

Thus, the critical buckling load will be larger for the stiffer beams.

We can express the critical buckling load for columns with different boundary conditions using equation (5.104). For this purpose, the beam length L is replaced by the effective length L_e

$$L_e = k_e L, \tag{5.112}$$

with k_e the effective length factor. As illustrated in Figure 5.50, L_e is the distance between the inflections of the elastic curves of the columns. Hence, $k_e = 1$ for the standard simply supported beam; $k_e = 0.7$ for the column with clamped and supported ends; $k_e = 0.5$ for the double-clamped column, and $k_e = 2$ for the column with free ends. Based on that, (5.104) is rewritten as

$$P_{cr} = \frac{\pi^2 EI_{min}}{(k_e L)^2} = \frac{\pi^2 EI_{min}}{L_e^2}. \tag{5.113}$$

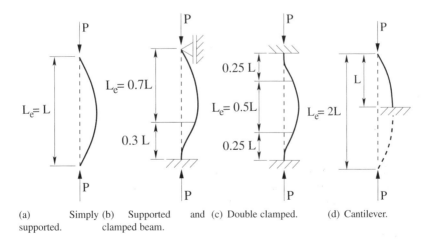

(a) Simply supported. (b) Supported and clamped beam. (c) Double clamped. (d) Cantilever.

Figure 5.50 Effective lengths of the column.

The obtained expressions for the compressive buckling force of the considered columns depend only on Young's modulus, which represents the stiffness of the material. There is no explicit dependence of the material strength in terms of the admissible stresses. However, obtained equations are valid only for columns in the elastic regime.

We can express the minimum moment of area of the section using the radius of gyration r_g as

$$I_{min} = A r_g^2.$$

Equation (5.113) is now rewritten as

$$P_{cr} = \frac{\pi^2 E I_{min}}{L_e^2} = \frac{\pi^2 E A}{(L_e/r_g)^2}. \tag{5.114}$$

Using this equation, the critical buckling normal stress is given by

$$\sigma_{cr} = \frac{P_{cr}}{A} = \frac{\pi^2 E}{(L_e/r_g)^2}. \tag{5.115}$$

The ratio L_e/r_g is called slenderness ratio of the column. We can plot a curve of the critical normal stress by the slenderness ratio of the column for different materials using the previous equation. Figure 5.51 illustrates the behavior of the critical normal stress for steel and aluminum columns, considering the longitudinal elasticity modulus equal to 210 GPa and 70 GPa, respectively.

It is also observed from the above expression that buckling always occurs about the axis with the smallest slenderness ratio, because this gives the smallest critical buckling load.

As the critical load expressions obtained above are valid for the elastic regime, we must determine the minimum slenderness ratio so that the column does not yield. Replacing the yield normal stresses of steel ($\sigma_{cr} = \sigma_y = 250$ MPa) and aluminum ($\sigma_{cr} = \sigma_y = 190$ MPa) in equation (5.115), we determine, respectively, the minimum slenderness ratios of 91 and 60.

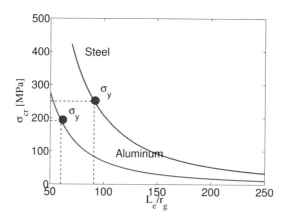

Figure 5.51   Critical normal stress in terms of the slenderness ratio for aluminum and steel columns.

Example 5.19 *Consider a simply supported steel column with length $L = 2$ m and cross-section with $A = 500$ cm^2, $I_y = 2200$ cm^4, and $I_z = 3000$ cm^4. Calculate the largest axial compressive force that can be applied to the column before the buckling occurs without yielding.*

As $I_{min} = I_y$, buckling will occur about the y axis of the column cross-section. The critical buckling load is calculated as

$$P_{cr} = \frac{\pi^2 E I_{min}}{L^2} = \frac{\pi^2 (210 \times 10^9)(2200 \times 10^{-4})}{2^2} = 11.4 \text{ MN}.$$

The normal critical buckling stress is determined as

$$\sigma_{cr} = \frac{P_{cr}}{A} = \frac{11.40 \times 10^6}{500 \times 10^{-4}} = 228.0 \text{ MPa}.$$

As this stress exceeds the steel yield stress $\sigma_y = 200$ MPa, the maximum compressive force must be calculated as

$$P = \sigma_y A = (200 \times 10^6)(500 \times 10^{-4}) = 10.0 \text{ MN}.$$

☐

Example 5.20 *Consider the same column of the previous example, but clamped at the ground and fixed in the upper end by a rope in z direction, as illustrated in Figure 5.52. Calculate the largest compressive force P which can be applied to the column to avoid buckling or yielding.*

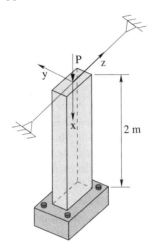

Figure 5.52   Example 5.20: clamped column at one of the ends.

It is a cantilever beam at the xy plane, and the buckling will occur about the z axis. The critical load is given from equation (5.113) by

$$P_{cr_z} = \frac{\pi^2 E I_z}{(k_e L)_z^2} = \frac{\pi^2 (210 \times 10^9)(3000 \times 10^{-4})}{4^2} = 3.9 \text{ MN}.$$

On the other hand, the column is supported in the z direction at plane xz, and the buckling will occur about the y axis. The critical load is given from equation (5.113) by

$$P_{cr_y} = \frac{\pi^2 E I_y}{(k_e L)_y^2} = \frac{\pi^2 (210 \times 10^9)(2200 \times 10^{-4})}{(2(0.7))^2} = 23.3 \text{ MN}.$$

Thus, when the intensity of P is increased, buckling will necessarily occur about the z axis. The critical normal stress is calculated as

$$\sigma_{cr} = \frac{P_{cr}}{A} = \frac{3.9 \times 10^6}{500 \times 10^{-4}} = 77.7 \text{ MPa}.$$

As the obtained value is below the steel yield stress, the column will buckle without yielding.

☐

5.13   APPROXIMATED SOLUTION

To obtain the weak form of the Euler-Bernoulli beam, we multiply the strong form (5.37) by the test function $v(x)$ and integrate between 0 and L, that is,

$$\int_0^L E I_z \frac{d^4 u_y(x)}{dx^4} v(x) dx = \int_0^L q_y(x) v(x) dx. \tag{5.116}$$

Integrating the left-hand side of the previous expressions twice by parts, and considering the definitions of bending moment (5.32) and shear force (5.35), we have

$$
\int_0^L EI_z \frac{d^4 u_y(x)}{dx^4} v(x) dx = - \int_0^L EI_z \frac{d^3 u_y(x)}{dx^3} \frac{dv(x)}{dx} dx + EI_z \frac{d^3 u_y(x)}{dx^3} v(x) \Big|_0^L
$$

$$
= \int_0^L EI_z \frac{d^2 u_y(x)}{dx^2} \frac{d^2 v(x)}{dx^2} dx + V_y(x) v(x) \Big|_0^L - EI_z \frac{d^2 u_y(x)}{dx^2} \frac{dv(x)}{dx} \Big|_0^L
$$

$$
= \int_0^L EI_z \frac{d^2 u_y(x)}{dx^2} \frac{d^2 v(x)}{dx^2} dx + V_y(x) v(x) \Big|_0^L - M_z(x) \frac{dv(x)}{dx} \Big|_0^L .
$$

Substituting this expression in (5.116) and using the boundary conditions given in the BVP (5.23), we have the beam weak form

$$
\int_0^L EI_z \frac{d^2 u_y(x)}{dx^2} \frac{d^2 v(x)}{dx^2} dx = \int_0^L q_y(x) v(x) dx + V_{y_L} v(L) + V_{y_0} v(0) + M_{z_L} \frac{dv(L)}{dx} + M_{z_0} \frac{dv(0)}{dx}, \quad (5.117)
$$

The first term of this expression involves second-order derivatives. Thus, when approximating the weak form, we should employ shape functions which have at least second-order piecewise continuous derivatives. This means that the functions and their first derivatives must be continuous. Hence, the transversal displacement and rotation of the Euler-Bernoulli beam are approximated. The cubic Hermite polynomials satisfy the continuity requirements for the beam approximation.

The simplest beam element has two nodes with two degrees of freedom per node, corresponding to the transversal displacement and rotation as illustrated in Figure 5.53. The nodal coordinates are indicated by x_1 and x_2. Transverse forces in $\bar{y}$ and moments in $\bar{z}$ can be applied at the element ends. A bar is used above the names of the degrees of freedom and loads to indicate that they are represented in the non-normalized local reference system $\bar{x}\bar{y}$.

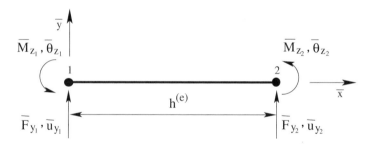

Figure 5.53   Beam element in the the local reference system $\bar{x}\bar{y}$.

The transversal displacement $u_y^{(e)}(x)$ and rotation $\theta_z^{(e)}(x)$ in element e can be approximated by the linear combinations of the four cubic Hermite polynomials and their first derivatives as

$$
u_{y_4}^{(e)}(x) = \sum_{i=1}^4 a_i \phi_i^{(e)}(x), \quad (5.118)
$$

$$
\theta_{z_4}^{(e)}(x) = \sum_{i=1}^4 a_i \phi_i^{(e)'}(x). \quad (5.119)
$$

The general expression for these polynomials and their first derivatives, for $i = 1,2,3,4$, are given, respectively, by

$$
\phi_i^{(e)}(x) = a_{0_i} + a_{1_i} x + a_{2_i} x^2 + a_{3_i} x^3, \quad (5.120)
$$

$$
\phi_i^{(e)'}(x) = a_{1_i} + 2a_{2_i} x + 3a_{3_i} x^2. \quad (5.121)
$$

The coefficients a_{i_j} $(i, j = 1, \ldots, 4)$ are determined using the following collocation properties:

$$
\begin{aligned}
\phi_1^{(e)}(x_1) = 1 \quad &\phi_1^{(e)}(x_2) = 0 \quad \phi_1^{(e)'}(x_1) = 0 \quad \phi_1^{(e)'}(x_2) = 0, \\
\phi_2^{(e)}(x_1) = 0 \quad &\phi_2^{(e)}(x_2) = 1 \quad \phi_2^{(e)'}(x_1) = 0 \quad \phi_2^{(e)'}(x_2) = 0, \\
\phi_3^{(e)}(x_1) = 0 \quad &\phi_3^{(e)}(x_2) = 0 \quad \phi_3^{(e)'}(x_1) = 1 \quad \phi_3^{(e)'}(x_2) = 0, \\
\phi_4^{(e)}(x_1) = 0 \quad &\phi_4^{(e)}(x_2) = 0 \quad \phi_4^{(e)'}(x_1) = 0 \quad \phi_4^{(e)'}(x_2) = 1.
\end{aligned}
$$

For instance, using the previous conditions for $\phi_1^{(e)}$ and $\phi_1^{(e)'}$ in (5.120) and (5.121), we have

$$
\begin{aligned}
a_{0_1}(1) + a_{1_1}(x_1) + a_{2_1}(x_1^2) + a_{3_1}(x_1^3) &= 1, \\
a_{0_1}(1) + a_{1_1}(x_2) + a_{2_1}(x_2^2) + a_{3_1}(x_2^3) &= 0, \\
a_{0_1}(0) + a_{1_1}(1) + a_{2_1}(2x_1) + a_{3_1}(3x_1^2) &= 0, \\
a_{0_1}(0) + a_{1_1}(1) + a_{2_1}(2x_2) + a_{3_1}(3x_2^2) &= 0,
\end{aligned}
$$

or in matrix notation,

$$
\begin{bmatrix}
1 & x_1 & x_1^2 & x_1^3 \\
1 & x_2 & x_2^2 & x_2^3 \\
0 & 1 & x_1 & x_1^2 \\
0 & 1 & x_2 & x_2^2
\end{bmatrix}
\begin{Bmatrix}
a_{0_1} \\
a_{1_1} \\
a_{2_1} \\
a_{3_1}
\end{Bmatrix}
=
\begin{Bmatrix}
1 \\
0 \\
0 \\
0
\end{Bmatrix}.
$$

Solving this system of equations, we determine the coefficients for $\phi_1^{(e)}(x)$. Doing the same procedure for the other polynomials (see file hermite.m), we obtain the following expressions for the cubic Hermite polynomials in terms of the nodal coordinates x_1 and x_2:

$$
\begin{aligned}
\phi_1^{(e)}(x) &= \left[\frac{1}{(x_1 - x_2)(x_1^2 - 2x_1x_2 + x_2^2)} \right] \left[-2x^3 + 3(x_1 + x_2)x^2 - 6x_1x_2x + x_2^2(3x_1 - x_2) \right], \\
\phi_2^{(e)}(x) &= \left[\frac{1}{x_1^2 - 2x_1x_2 + x_2^2} \right] \left[x^3 - (x_1 + 2x_2)x^2 + (x_2^2 + 2x_1x_2)x \right] - \frac{x_1 x_2^2}{(x_1 - x_2)^2}, \\
\phi_3^{(e)}(x) &= \left[\frac{1}{((x_1 - x_2)(x_1^2 - 2x_1x_2 + x_2^2))} \right] \left[2x^3 - 3(x_1 + x_2)x^2 + 6x_1x_2x + x_1^2(x_1 - 3x_2) \right], \\
\phi_4^{(e)}(x) &= \left[\frac{1}{x_1^2 - 2x_1x_2 + x_2^2} \right] \left[x^3 - (2x_1 + x_2)x^2 + (x_1^2 + 2x_2x_1)x \right] - \frac{x_1^2 x_2}{(x_1 - x_2)^2}.
\end{aligned}
$$

Substituting $x_1 = 0$ and $x_2 = h^{(e)}$, the Hermite polynomials for the beam element are determined, that is,

$$
\begin{aligned}
\phi_1^{(e)}(x) &= 2\left(\frac{x}{h^{(e)}}\right)^3 - 3\left(\frac{x}{h^{(e)}}\right)^2 + 1, \\
\phi_2^{(e)}(x) &= h^{(e)}\left[\left(\frac{x}{h^{(e)}}\right)^3 - 2\left(\frac{x}{h^{(e)}}\right)^2 + \frac{x}{h^{(e)}}\right], \\
\phi_3^{(e)}(x) &= -2\left(\frac{x}{h^{(e)}}\right)^3 + 3\left(\frac{x}{h^{(e)}}\right)^2, \\
\phi_4^{(e)}(x) &= h^{(e)}\left[\left(\frac{x}{h^{(e)}}\right)^3 - \left(\frac{x}{h^{(e)}}\right)^2\right].
\end{aligned}
\qquad (5.122)
$$

The first-, second- and third-order derivatives of the Hermite polynomials are given in Table 5.1. The polynomials and their first-order derivatives are illustrated in Figure 5.54 for $h^{(e)} = 1$.

i	$\phi_i^{(e)'}(x)$	$\phi_i^{(e)''}(x)$	$\phi_i^{(e)'''}(x)$
1	$\dfrac{6}{h^{(e)}}\left(\dfrac{x}{h^{(e)}}\right)\left(\dfrac{x}{h^{(e)}}-1\right)$	$\dfrac{6}{h^{(e)2}}\left(2\dfrac{x}{h^{(e)}}-1\right)$	$\dfrac{12}{h^{(e)3}}$
2	$3\left(\dfrac{x}{h^{(e)}}\right)^2-4\left(\dfrac{x}{h^{(e)}}\right)+1$	$\dfrac{1}{h^{(e)}}\left(6\dfrac{x}{h^{(e)}}-4\right)$	$\dfrac{6}{h^{(e)2}}$
3	$-\dfrac{6}{h^{(e)}}\left(\dfrac{x}{h^{(e)}}\right)\left(\dfrac{x}{h^{(e)}}-1\right)$	$-\dfrac{6}{h^{(e)2}}\left(2\dfrac{x}{h^{(e)}}-1\right)$	$-\dfrac{12}{h^{(e)3}}$
4	$3\left(\dfrac{x}{h^{(e)}}\right)^2-2\left(\dfrac{x}{h^{(e)}}\right)$	$\dfrac{1}{h^{(e)}}\left(6\dfrac{x}{h^{(e)}}-2\right)$	$\dfrac{6}{h^{(e)2}}$

Table 5.1

First-, second- and third-order derivatives of the cubic Hermite polynomials.

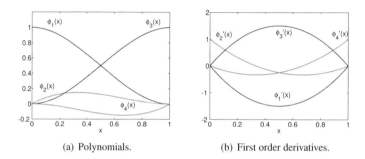

(a) Polynomials.   (b) First order derivatives.

Figure 5.54 Cubic Hermite polynomials and the first-order derivatives.

Calculating expressions (5.118) and (5.119) at coordinates $x=0$ and $x=h^{(e)}$ of beam element ends, the approximation coefficients, due to the collocation property of the Hermite polynomials, have the following interpretation:

$$
\begin{aligned}
u_{y4}^{(e)}(0) &= a_1\phi_1(0)+a_2\phi_2(0)+a_3\phi_3(0)+a_4\phi_4(0) \rightarrow a_1 = \bar{u}_{y1}^{(e)}, \\
u_{y4}^{(e)}(h^{(e)}) &= a_1\phi_1(h^{(e)})+a_2\phi_2(h^{(e)})+a_3\phi_3(h^{(e)})+a_4\phi_4(h^{(e)}) \rightarrow a_3 = \bar{u}_{y2}^{(e)}, \\
\theta_{z4}^{(e)}(0) &= a_1\phi_1'(0)+a_2\phi_2'(0)+a_3\phi_3'(0)+a_4\phi_4'(0) \rightarrow a_2 = \bar{\theta}_{z1}^{(e)}, \\
\theta_{z4}^{(e)}(h^{(e)}) &= a_1\phi_1'(h^{(e)})+a_2\phi_2'(h^{(e)})+a_3\phi_3'(h^{(e)})+a_4\phi_4'(h^{(e)}) \rightarrow a_4 = \bar{\theta}_{z2}^{(e)}.
\end{aligned}
$$

Therefore, the approximation coefficients a_i represent the displacements and rotations at the beam ends. Equations (5.118) and (5.119) can be rewritten as

$$
u_{y4}^{(e)}(x) = \bar{u}_{y1}^{(e)}\phi_1^{(e)}(x)+\bar{\theta}_{z1}^{(e)}\phi_2^{(e)}(x)+\bar{u}_{y2}^{(e)}\phi_3^{(e)}(x)+\bar{\theta}_{z2}^{(e)}\phi_4^{(e)}(x), \tag{5.123}
$$
$$
\theta_{z4}^{(e)}(x) = \bar{u}_{y1}^{(e)}\phi_1^{(e)'}(x)+\bar{\theta}_{z1}^{(e)}\phi_2^{(e)'}(x)+\bar{u}_{y2}^{(e)}\phi_3^{(e)'}(x)+\bar{\theta}_{z2}^{(e)}\phi_4^{(e)'}(x). \tag{5.124}
$$

In matrix form, we have the following approximations for the transversal displacement and rotation for the two-node beam element:

$$
u_{y4}^{(e)}(x) = \left[\begin{array}{cccc} \phi_1^{(e)}(x) & \phi_2^{(e)}(x) & \phi_3^{(e)}(x) & \phi_4^{(e)}(x) \end{array}\right]
\left\{\begin{array}{c} \bar{u}_{y1}^{(e)} \\ \bar{\theta}_{z1}^{(e)} \\ \bar{u}_{y2}^{(e)} \\ \bar{\theta}_{z2}^{(e)} \end{array}\right\} = [\bar{N}_1^{(e)}]\{\bar{u}^{(e)}\}, \tag{5.125}
$$

$$\theta_{z4}^{(e)}(x) = \left[\begin{array}{cccc} \phi_1^{(e)'}(x) & \phi_2^{(e)'}(x) & \phi_3^{(e)'}(x) & \phi_4^{(e)'}(x) \end{array} \right] \left\{ \begin{array}{c} \bar{u}_{y1}^{(e)} \\ \bar{\theta}_{z1}^{(e)} \\ \bar{u}_{y2}^{(e)} \\ \bar{\theta}_{z2}^{(e)} \end{array} \right\} = [\bar{N}_2^{(e)}]\{\bar{u}^{(e)}\}. \tag{5.126}$$

with $[\bar{N}_1^{(e)}]$ and $[\bar{N}_2^{(e)}]$ the element matrices of the shape functions and their first-order derivatives.

The specific normal strain is given by

$$\varepsilon_{xx4}^{(e)}(x) = -y\frac{d^2 u_{y4}(x)}{dx^2} = -y \left[\begin{array}{cccc} \phi_1^{(e)''}(x) & \phi_2^{(e)''}(x) & \phi_3^{(e)''}(x) & \phi_4^{(e)''}(x) \end{array} \right] \left\{ \begin{array}{c} \bar{u}_{y1}^{(e)} \\ \bar{\theta}_{z1}^{(e)} \\ \bar{u}_{y2}^{(e)} \\ \bar{\theta}_{z2}^{(e)} \end{array} \right\}$$

$$= [\bar{B}^{(e)}]\{\bar{u}^{(e)}\}, \tag{5.127}$$

where the element strain matrix $[\bar{B}^{(e)}]$ contains the second-order derivatives of the shape functions.

The stiffness matrix $[\bar{K}^{(e)}]$ of the beam element with constant cross-section, in the local reference system $\bar{x}\bar{y}$, is determined replacing the matrices $[D] = E$ and $[\bar{B}^{(e)}]$ and $dV = dA\,dx$ into expression

$$\left[\bar{K}^{(e)} \right] = \int_0^{h^{(e)}} \int_A [\bar{B}^{(e)}]^T [D][\bar{B}^{(e)}] \, dA dx$$

$$= E^{(e)} \underbrace{\int_A y^2 dA}_{I_z^{(e)}} \left(\int_0^{h^{(e)}} \left[\begin{array}{c} \phi_1^{(e)''}(x) \\ \phi_2^{(e)''}(x) \\ \phi_3^{(e)''}(x) \\ \phi_4^{(e)''}(x) \end{array} \right] \left[\begin{array}{cccc} \phi_1^{(e)''}(x) & \phi_2^{(e)''}(x) & \phi_3^{(e)''}(x) & \phi_4^{(e)''}(x) \end{array} \right] dx \right)$$

$$= E^{(e)} I_z^{(e)} \int_0^{h^{(e)}} \left[\begin{array}{cccc} \phi_1^{(e)''}(x)\phi_1^{(e)''}(x) & \phi_1^{(e)''}(x)\phi_2^{(e)''}(x) & \phi_1^{(e)''}(x)\phi_3^{(e)''}(x) & \phi_1^{(e)''}(x)\phi_4^{(e)''}(x) \\ \phi_1^{(e)''}(x)\phi_2^{(e)''}(x) & \phi_2^{(e)''}(x)\phi_2^{(e)''}(x) & \phi_2^{(e)''}(x)\phi_3^{(e)''}(x) & \phi_2^{(e)''}(x)\phi_4^{(e)''}(x) \\ \phi_1^{(e)''}(x)\phi_3^{(e)''}(x) & \phi_2^{(e)''}(x)\phi_3^{(e)''}(x) & \phi_3^{(e)''}(x)\phi_3^{(e)''}(x) & \phi_3^{(e)''}(x)\phi_4^{(e)''}(x) \\ \phi_1^{(e)''}(x)\phi_4^{(e)''}(x) & \phi_2^{(e)''}(x)\phi_4^{(e)''}(x) & \phi_3^{(e)''}(x)\phi_4^{(e)''}(x) & \phi_4^{(e)''}(x)\phi_4^{(e)''}(x) \end{array} \right] dx.$$

Using the second-order derivatives of the Hermite polynomials given in Table 5.1 and performing the given integrations, we obtain the stiffness matrix of the beam element, represented in the $\bar{x}\bar{y}$ local system, as

$$\left[\bar{K}^{(e)} \right] = \frac{E^{(e)} I_z^{(e)}}{h^{(e)3}} \left[\begin{array}{cccc} 12 & 6h^{(e)} & -12 & 6h^{(e)} \\ 6h^{(e)} & 4h^{(e)2} & -6h^{(e)} & 2h^{(e)2} \\ -12 & -6h^{(e)} & 12 & -6h^{(e)} \\ 6h^{(e)} & 2h^{(e)2} & -6h^{(e)} & 4h^{(e)2} \end{array} \right]. \tag{5.128}$$

The equivalent nodal vector to the distributed load $q_y(x)$ is given by

$$\{\bar{f}^{(e)}\} = \int_0^{h^{(e)}} q_y(x) \left\{ \begin{array}{cccc} \phi_1^{(e)}(x) & \phi_2^{(e)}(x) & \phi_3^{(e)}(x) & \phi_4^{(e)}(x) \end{array} \right\}^T dx. \tag{5.129}$$

For a constant distributed load $q_y(x) = q_0$, we have

$$\{\bar{f}^{(e)}\} = \frac{q_0 h^{(e)}}{2} \left\{ \begin{array}{cccc} 1 & \dfrac{h^{(e)}}{6} & 1 & -\dfrac{h^{(e)}}{6} \end{array} \right\}^T. \tag{5.130}$$

The local system of equations for the two-node beam element with $q_y(x) = q_0$ is

$$\frac{E^{(e)}I_z^{(e)}}{h^{(e)3}} \begin{bmatrix} 12 & 6h^{(e)} & -12 & 6h^{(e)} \\ 6h^{(e)} & 4h^{(e)2} & -6h^{(e)} & 2h^{(e)2} \\ -12 & -6h^{(e)} & 12 & -6h^{(e)} \\ 6h^{(e)} & 2h^{(e)2} & -6h^{(e)} & 4h^{(e)2} \end{bmatrix} \begin{Bmatrix} \bar{u}_{y_1}^{(e)} \\ \bar{\theta}_{z_1}^{(e)} \\ \bar{u}_{y_2}^{(e)} \\ \bar{\theta}_{z_2}^{(e)} \end{Bmatrix} = \frac{q_0 h^{(e)}}{2} \begin{Bmatrix} 1 \\ \dfrac{h^{(e)}}{6} \\ 1 \\ -\dfrac{h^{(e)}}{6} \end{Bmatrix} + \begin{Bmatrix} \bar{F}_{y_1} \\ \bar{M}_{z_1} \\ \bar{F}_{y_2} \\ \bar{M}_{z_2} \end{Bmatrix}.$$

(5.131)

The last vector is for the external concentrated forces and moments applied at the element ends.

In the case of the mass matrix, the matrix of shape functions $[\bar{N}_1^{(e)}]$ is substituted in expression

$$[\bar{M}^{(e)}] = \int_0^{h^{(e)}} \int_A \rho^{(e)} [\bar{N}_1^{(e)}]^T [\bar{N}_1^{(e)}] \, dA dx$$

$$= \rho^{(e)} A^{(e)} \left(\int_0^{h^{(e)}} \begin{bmatrix} \phi_1^{(e)}(x) \\ \phi_2^{(e)}(x) \\ \phi_3^{(e)}(x) \\ \phi_4^{(e)}(x) \end{bmatrix} \begin{bmatrix} \phi_1^{(e)}(x) & \phi_2^{(e)}(x) & \phi_3^{(e)}(x) & \phi_4^{(e)}(x) \end{bmatrix} dx \right)$$

(5.132)

$$= \rho^{(e)} A^{(e)} \int_0^{h^{(e)}} \begin{bmatrix} \phi_1^{(e)}(x)\phi_1^{(e)}(x) & \phi_1^{(e)}(x)\phi_2^{(e)}(x) & \phi_1^{(e)}(x)\phi_3^{(e)}(x) & \phi_1^{(e)}(x)\phi_4^{(e)}(x) \\ \phi_1^{(e)}(x)\phi_2^{(e)}(x) & \phi_2^{(e)}(x)\phi_2^{(e)}(x) & \phi_2^{(e)}(x)\phi_3^{(e)}(x) & \phi_2^{(e)}(x)\phi_4^{(e)}(x) \\ \phi_1^{(e)}(x)\phi_3^{(e)}(x) & \phi_2^{(e)}(x)\phi_3^{(e)}(x) & \phi_3^{(e)}(x)\phi_3^{(e)}(x) & \phi_3^{(e)}(x)\phi_4^{(e)}(x) \\ \phi_1^{(e)}(x)\phi_4^{(e)}(x) & \phi_2^{(e)}(x)\phi_4^{(e)}(x) & \phi_3^{(e)}(x)\phi_4^{(e)}(x) & \phi_4^{(e)}(x)\phi_4^{(e)}(x) \end{bmatrix} dx.$$

Substituting the Hermite polynomials given in (5.123), we obtain, after performing the indicated integrations, the following mass matrix:

$$[\bar{M}_e] = \frac{\rho^{(e)} A^{(e)} h^{(e)}}{420} \begin{bmatrix} 156 & 22h^{(e)} & 54 & -13h^{(e)} \\ 22h^{(e)} & 4h^{(e)2} & 13h^{(e)} & -3h^{(e)2} \\ 54 & 13h^{(e)} & 156 & -22h^{(e)} \\ -13h^{(e)} & -3h^{(e)2} & -22h^{(e)} & 4h^{(e)2} \end{bmatrix}.$$

(5.133)

The element bending moment and shear force can be calculated from the nodal displacements and rotations, respectively, as

$$M_{z_4}^{(e)}(x) = E^{(e)} I_z^{(e)} \begin{bmatrix} \phi_1^{(e)''}(x) & \phi_2^{(e)''}(x) & \phi_3^{(e)''}(x) & \phi_4^{(e)''}(x) \end{bmatrix} \begin{Bmatrix} \bar{u}_{y_1}^{(e)} \\ \bar{\theta}_{z_1}^{(e)} \\ \bar{u}_{y_2}^{(e)} \\ \bar{\theta}_{z_2}^{(e)} \end{Bmatrix},$$

(5.134)

$$V_{y_4}^{(e)}(x) = E^{(e)} I_z^{(e)} \begin{bmatrix} \phi_1^{(e)'''}(x) & \phi_2^{(e)'''}(x) & \phi_3^{(e)'''}(x) & \phi_4^{(e)'''}(x) \end{bmatrix} \begin{Bmatrix} \bar{u}_{y_1}^{(e)} \\ \bar{\theta}_{z_1}^{(e)} \\ \bar{u}_{y_2}^{(e)} \\ \bar{\theta}_{z_2}^{(e)} \end{Bmatrix}.$$

(5.135)

Note that, as the second and third derivatives of the Hermite polynomials are respectively linear and constant functions, the bending moment varies linearly and the shear force is constant in each element. Thus, the approximate solution coincides with the exact solution only when the latter is of third order. This corresponds to the case of beams subjected to concentrated loads, as illustrated in the following example.

Example 5.21 *Consider the steel beam with length $L = 1$ m, rectangular cross-section 15×25 cm, supported and loaded as indicated in Figure 5.55(a). The beam is discretized with three elements of size $h^{(e)} = \frac{L}{3}$, as illustrated in Figure 5.55(b). Determine the nodal displacements and rotations, as well as the support reactions.*

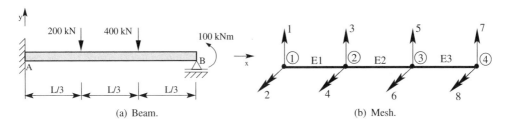

(a) Beam.   (b) Mesh.

Figure 5.55 Example 5.21: Beam and mesh.

For the solution of this example, the same procedure used in Chapter 4 is applied.

- *The global coordinate system is placed at the left end of the beam, as illustrated in Figure 5.55(a).*
- *The elements, nodal incidence, and degrees of freedom are numbered according to Figure 5.55(b) and Table 5.2.*
- *As the elements have the same size, cross-section, and material, the stiffness matrices are also the same. Using equation (5.128), the stiffness matrix is*

$$
[\bar{K}_e] = EI_z
\begin{bmatrix}
324 & 54 & -324 & 54 \\
54 & 12 & -54 & 6 \\
-324 & -54 & 324 & -54 \\
54 & 6 & -54 & 121
\end{bmatrix}.
$$

Element	Incidence	Degrees of freedom
1	1-2	1 2 3 4
2	2-3	3 4 5 6
3	3-4	5 6 7 8

Table 5.2
Example 5.21: nodal incidence and mesh degrees of freedom numbering.

- *Each element stiffness matrix is assembled in the rows and columns of the global stiffness matrix of the beam, corresponding to the degrees of freedom indicated in Table 5.2. The global matrix has rank 8, as there are four nodes with two degrees of freedom at each node.*

Thus,

$$[K] = EI_z \begin{bmatrix} 324 & 54 & -324 & 54 & 0 & 0 & 0 & 0 \\ 54 & 12 & -54 & 6 & 0 & 0 & 0 & 0 \\ -324 & -54 & 324+324 & -54+54 & -324 & 54 & 0 & 0 \\ 54 & 6 & -54+54 & -12+12 & -54 & 6 & 0 & 0 \\ 0 & 0 & -324 & -54 & 324+324 & -54+54 & -324 & 54 \\ 0 & 0 & 54 & 6 & -54+54 & 12+12 & -54 & 6 \\ 0 & 0 & 0 & 0 & -324 & -54 & 324 & -54 \\ 0 & 0 & 0 & 0 & 54 & 6 & -54 & 12 \end{bmatrix}.$$

The global load vector is given by

$$\{F\} = \left\{ \begin{array}{cccccccc} \bar{R}_{y_1} & \bar{M}_{z_2} & -200000 & 0 & -400000 & 0 & \bar{R}_{y_7} & 100000 \end{array} \right\}^T,$$

with $\bar{R}_{y_1}$ and $\bar{M}_{z_2}$ the force and moment reactions in the clamped left end, and $\bar{R}_{y_7}$ the force reaction at the right end.

Hence, the final system of equations for the considered mesh is

$$EI_z \begin{bmatrix} 324 & 54 & -324 & 54 & 0 & 0 & 0 & 0 \\ 54 & 12 & -54 & 6 & 0 & 0 & 0 & 0 \\ -324 & -54 & 628 & 0 & -324 & 54 & 0 & 0 \\ 54 & 6 & 0 & 0 & -54 & 6 & 0 & 0 \\ 0 & 0 & -324 & -54 & 648 & 0 & -324 & 54 \\ 0 & 0 & 54 & 6 & 0 & 24 & -54 & 6 \\ 0 & 0 & 0 & 0 & -324 & -54 & 324 & -54 \\ 0 & 0 & 0 & 0 & 54 & 6 & -54 & 12 \end{bmatrix} \left\{ \begin{array}{c} \bar{u}_{y_1} \\ \bar{\theta}_{z_2} \\ \bar{u}_{y_3} \\ \bar{\theta}_{z_4} \\ \bar{u}_{y_5} \\ \bar{\theta}_{z_6} \\ \bar{u}_{y_7} \\ \bar{\theta}_{z_8} \end{array} \right\} = \left\{ \begin{array}{c} \bar{R}_{y_1} \\ \bar{M}_{z_2} \\ -200000 \\ 0 \\ -400000 \\ 0 \\ \bar{R}_{y_7} \\ 100000 \end{array} \right\}$$

- *Due to the clamped left end, the transversal displacement and rotation of node 1 are zero, i.e., $\bar{u}_{y_1} = \bar{\theta}_{z_2} = 0$. Analogously, the transversal displacement in the simple support at the beam right end is zero and, thus $\bar{u}_{y_7} = 0$. This means that columns 1, 2, and 7 of the global matrix can be eliminated because the coefficients of these columns multiply zero degrees of freedom. The coefficients of rows 1, 2 and 7 may also be eliminated from the global matrix and load vector, because they will be used to calculate the support reactions. Thereafter, the final system of equations to be solved for the beam has five equations and is given by*

$$EI_z \begin{bmatrix} 648 & 0 & -324 & 54 & 0 \\ 0 & 24 & -54 & 6 & 0 \\ -324 & -54 & 648 & 0 & 54 \\ 54 & 6 & 0 & 24 & 6 \\ 0 & 0 & 54 & 6 & 12 \end{bmatrix} \left\{ \begin{array}{c} \bar{u}_{y_3} \\ \bar{\theta}_{z_4} \\ \bar{u}_{y_5} \\ \bar{\theta}_{z_6} \\ \bar{\theta}_{z_8} \end{array} \right\} = \left\{ \begin{array}{c} -200000 \\ 0 \\ -400000 \\ 0 \\ 100000 \end{array} \right\}.$$

The moment of area of the rectangular cross-section is given by

$$I_z = \frac{bh^3}{12} = \frac{(0.15)(0.25)^3}{12} = 1.95 \times 10^{-4} \text{ m}^4$$

and

$$E^{(e)} I_z^{(e)} = \left(210 \times 10^9 \frac{\text{N}}{\text{m}^2} \right) (1.95 \times 10^{-4} \text{ m}^4) = 4.10 \times 10^7 \text{ Nm}^2.$$

The solution of the system of equations results in

$$\begin{array}{rcl} \bar{u}_{y_3} & = & -1.2096 \times 10^{-4} \text{ m,} \\ \bar{\theta}_{z_4} & = & -4.9414 \times 10^{-4} \text{ rad,} \\ \bar{u}_{y_5} & = & -2.0513 \times 10^{-4} \text{ m,} \\ \bar{\theta}_{z_6} & = & 1.3043 \times 10^{-4} \text{ rad,} \\ \bar{\theta}_{z_8} & = & 1.0610 \times 10^{-3} \text{ rad.} \end{array}$$

- The support reactions are obtained by multiplying the coefficients of rows 1, 2, and 7 and columns 3, 4, 5, 6, and 8 of the global matrix by the solution vector, that is,

$$
\left\{ \begin{array}{c} \bar{R}_{y_1} \\ \bar{M}_{z_2} \\ \bar{R}_{y_7} \end{array} \right\} = \left[\begin{array}{ccccc} -324 & 54 & 0 & 0 & 0 \\ -54 & 6 & 0 & 0 & 0 \\ 0 & 0 & -324 & -54 & -54 \end{array} \right] \left\{ \begin{array}{c} \bar{u}_{y_3} \\ \bar{\theta}_{z_4} \\ \bar{u}_{y_5} \\ \bar{\theta}_{z_6} \\ \bar{\theta}_{z_8} \end{array} \right\}.
$$

Hence,

$$
\bar{R}_{y_1} = 512962.96 \text{ N}, \qquad \bar{M}_{z_2} = 146296.29 \text{ Nm}, \qquad \bar{R}_{y_7} = 87037.04 \text{ N}.
$$

- We can calculate the vector of internal forces and moments $\{R_n\}$ multiplying the global stiffness matrix by the solution vector, that is,

$$
\{R_n\} = \frac{EI_z}{10^4} \left[\begin{array}{cccccccc} 324 & 54 & -324 & 54 & 0 & 0 & 0 & 0 \\ 54 & 12 & -54 & 6 & 0 & 0 & 0 & 0 \\ -324 & -54 & 628 & 0 & -324 & 54 & 0 & 0 \\ 54 & 6 & 0 & 0 & -54 & 6 & 0 & 0 \\ 0 & 0 & -324 & -54 & 628 & 0 & -324 & 54 \\ 0 & 0 & 54 & 6 & 0 & 24 & -54 & 6 \\ 0 & 0 & 0 & 0 & -324 & -54 & 324 & -54 \\ 0 & 0 & 0 & 0 & 54 & 6 & -54 & 12 \end{array} \right] \left\{ \begin{array}{c} 0.000 \\ 0.000 \\ -1.2096 \\ -4.9414 \\ -2.0513 \\ 1.3043 \\ 0.000 \\ 10.6102 \end{array} \right\},
$$

resulting in

$$
\{R_n\} = \left\{ \begin{array}{cccccccc} 512.96 & 146.30 & -200.00 & 0.00 & -400.00 & 0.00 & 87.04 & 100.00 \end{array} \right\}^T.
$$

Observe that the nodal internal loads recover the support reactions and the the the external loads applied to the beam.

The analytical solutions for the rotation and transversal displacement are given, respectively, by

$$
\theta_z(x) = \frac{1}{EI_z} \left[-10000 <x - \frac{1}{3}>^2 - 20000 <x - \frac{2}{3}>^2 + \frac{C_1}{2}x^2 + C_2 x \right],
$$

$$
u_y(x) = \frac{1}{EI_z} \left[-\frac{20000}{3} <x - \frac{1}{3}>^3 - \frac{40000}{3} <x - \frac{2}{3}>^3 + \frac{C_1}{6}x^3 + \frac{C_2}{2}x^2 \right],
$$

with $C_1 = 512962.96$ and $C_2 = -146296.30$. As $u_y(x)$ is third order, the obtained approximated solution coincides with the exact solution.

File beamsolapex1.m implements the solution of this example by the FEM with an arbitrary number of elements. The diagrams of shear force, bending moment, and normal stresses for $y^{min} = -7.5$ cm and $y^{max} = 7.5$ cm, obtained by this program with three elements, are illustrated in Figure 5.56. Note that the shear force is constant on each element, while the bending moment varies linearly. Also, as the rectangular section is symmetric with respect to the z axis of the reference system, the minimum and maximum normal stresses in each section have the same absolute value.

□

Example 5.22 *Consider the cantilever beam of length L subjected to a distributed load of constant intensity q_0, as illustrated in Figure 5.57. The expressions for the shear force, bending moment,*

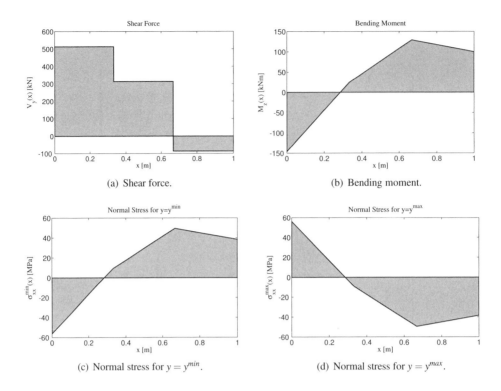

(a) Shear force.

(b) Bending moment.

(c) Normal stress for $y = y^{min}$.

(d) Normal stress for $y = y^{max}$.

Figure 5.56   Example 5.21: diagrams for the shear force, bending moment, and normal stresses.

rotation, and transversal displacement are given, respectively, by

$$
\begin{aligned}
V_y(x) &= -q_0(x - L), \\
M_z(x) &= -\frac{q_0}{2}(x^2 - 2Lx + L^2), \\
\theta_z(x) &= -\frac{q_0}{6EI_z}\left(x^3 - 3Lx^2 + 3L^2x\right), \\
u_y(x) &= -\frac{q_0}{24EI_z}\left(x^4 - 4Lx^3 + 6L^2x^2\right).
\end{aligned}
$$

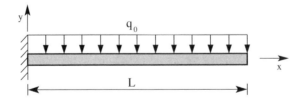

Figure 5.57   Example 5.22: cantilever beam subjected to a distributed load.

As the solution to the transversal displacement is of fourth order and the Hermite polynomials are of third order, the approximated solution by the FEM does not coincide with the exact solution. File beamsolapex2.m implements the solution of this example for an arbitrary number of mesh elements. The error in the energy norm is calculated as the sum of the errors in each mesh element. There-

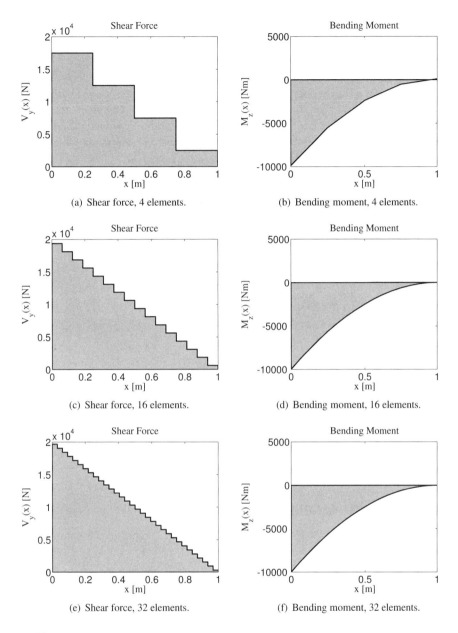

(a) Shear force, 4 elements.

(b) Bending moment, 4 elements.

(c) Shear force, 16 elements.

(d) Bending moment, 16 elements.

(e) Shear force, 32 elements.

(f) Bending moment, 32 elements.

Figure 5.58 Example 5.22: diagrams of the shear force and bending moment for meshes with different numbers of elements.

fore,

$$\|e\|_E^2 = \sum_{e=1}^{Nel} \left[E^{(e)} I_z^{(e)} \int_0^{h^{(e)}} \left(\frac{d\theta_z(x)}{dx} \right)^2 - \left(\frac{d\theta_{z_4}^{(e)}(x)}{dx} \right)^2 \right]^2 dx. \qquad (5.136)$$

Figure 5.58 illustrates diagrams of shear force and bending moment obtained with meshes of 4, 16, and 32 elements. Notice that the shear force is constant and the bending moment varies linearly in each element. When the meshes are refined, the obtained diagrams approach the analytical ones, but with the same constant and linear behaviors on the elements for the shear force and bending moment.

Figure 5.59 presents, in semi-logarithmic scale, the relative error in terms of the number of degrees of freedom for meshes with 2, 4, 8, 16, and 32 elements. The rate of decrease of the relative error is 8 for each successive refinement, as expected by the a priori *error estimate [5].*

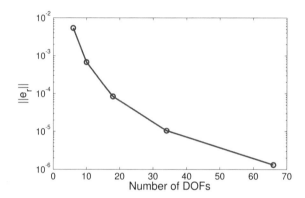

Figure 5.59   Example 5.22: relative error in the energy norm.

□

5.14   HIGH-ORDER BEAM ELEMENT

The term $\dfrac{x}{h^{(e)}}$ in the expressions of Hermite polynomials (5.123) represents a natural coordinate, because it assumes values in the $[0, 1]$ interval. Substituting $L_1 = \dfrac{x}{h^{(e)}}$ in (5.123), we have the representation of the beam shape functions in the normalized coordinate system L_1, that is,

$$\begin{aligned}
\phi_1^{(e)}(L_1) &= 2L_1^3 - 3L_1^2 + 1, \\
\phi_2^{(e)}(L_1) &= h^{(e)} \left(L_1^3 - 2L_1^2 + L_1 \right), \\
\phi_3^{(e)}(L_1) &= -2L_1^3 + 3L_1^2, \\
\phi_4^{(e)}(L_1) &= h^{(e)} \left(L_1^3 - L_1^2 \right).
\end{aligned} \qquad (5.137)$$

Substituting now (4.103) in the previous expressions, we have the Hermite polynomials represented in the normalized local system ξ_1. Hence,

$$
\begin{aligned}
\phi_1^{(e)}(\xi_1) &= \frac{1}{4}(\xi_1^3 - 3\xi_1 + 2) = \frac{1}{4}(1-\xi_1)^2(2+\xi_1), \\
\phi_2^{(e)}(\xi_1) &= \frac{h^{(e)}}{8}(\xi_1^3 - \xi_1^2 - \xi_1 + 1) = \frac{h^{(e)}}{8}(1-\xi_1)^2(1+\xi_1), \\
\phi_3^{(e)}(\xi_1) &= \frac{1}{4}(-\xi_1^3 + 3\xi_1 + 2) = \frac{1}{4}(1+\xi_1)^2(2-\xi_1), \\
\phi_4^{(e)}(\xi_1) &= \frac{h^{(e)}}{8}(\xi_1^3 + \xi_1^2 - \xi_1 - 1) = -\frac{h^{(e)}}{8}(1+\xi_1)^2(1-\xi_1).
\end{aligned}
\tag{5.138}
$$

The transversal displacement of the beam element is expressed in the local system ξ_1 by the following linear combination of the previous basis functions:

$$
u_y^{(e)}(\xi_1) = \bar{u}_{y_1}^{(e)}\phi_1^{(e)}(\xi_1) + \bar{\theta}_{z_1}^{(e)}\phi_2^{(e)}(\xi_1) + \bar{u}_{y_2}^{(e)}\phi_3^{(e)}(\xi_1) + \bar{\theta}_{z_2}^{(e)}\phi_4^{(e)}(\xi_1).
\tag{5.139}
$$

Analogously, the bending rotation is given in terms of ξ_1 as

$$
\theta_z^{(e)}(\xi_1) = \frac{du_y^{(e)}(\xi_1)}{d\xi_1}.
\tag{5.140}
$$

In the case of a straight beam, the x and ξ_1 coordinates are related by the linear shape functions using equation (4.97). Applying the chain rule, we determine the following expression between the bending rotations expressed in the x and ξ_1 coordinates:

$$
\frac{du_y^{(e)}(\xi_1)}{d\xi_1} = \frac{du_y^{(e)}[x(\xi_1)]}{d\xi_1} = \frac{du_y^{(e)}[x(\xi_1)]}{dx}\frac{dx^{(e)}(\xi_1)}{d\xi_1} = [J]\frac{du_y^{(e)}[x(\xi_1)]}{dx}.
\tag{5.141}
$$

In this case, the Jacobian $[J]$ matrix has only one coefficient. From equation (4.138), we have $[J] = \frac{h^{(e)}}{2}$, and the following relations between the derivatives of u_y relative to the x and ξ_1 coordinates are valid:

$$
\frac{du_y^{(e)}[x(\xi_1)]}{dx} = \frac{2}{h^{(e)}}\frac{du_y^{(e)}(\xi_1)}{d\xi_1},
\tag{5.142}
$$

$$
\frac{d^2u_y^{(e)}[x(\xi_1)]}{dx^2} = \left(\frac{2}{h^{(e)}}\right)^2\frac{d^2u_y^{(e)}(\xi_1)}{d\xi_1^2}.
\tag{5.143}
$$

We want to construct high-order basis for the beam model. For this purpose, the continuity of displacements and rotations at the interface of mesh elements must be preserved. The following internal shape functions are defined from $P = 4$ by:

$$
\phi_{P+1}^{(e)}(\xi_1) = (1-\xi_1)(1+\xi_1)\phi_1^{(e)'}(\xi_1)P_{P-4}^{1,1}(\xi_1), \quad P \geq 4.
\tag{5.144}
$$

The main idea is to ensure that the shape function and its first derivative are zero at the ends of the beam element, in order to ensure the continuity of displacements and rotations between elements. This is satisfied by the $(1-\xi_1)(1+\xi_1)$ term and also the derivative $\phi_1^{(e)'}(\xi_1)$. The derivative of $\phi_3^{(e)}(\xi_1)$ could also be used. Jacobi polynomials $P_{P-4}^{1,1}(\xi_1)$ are used to ensure better numerical conditioning of the mass and stiffness matrices calculated with the high-order basis.

The derivative of equation (5.144) is given by

$$
\begin{aligned}
\phi_{P+1}^{(e)'}(\xi_1) &= -2\xi_1\phi_1^{(e)'}(\xi_1)P_{P-4}^{1,1}(\xi_1)+(1-\xi_1^2)\phi_1^{(e)''}(\xi_1)P_{P-4}^{1,1}(\xi_1) \\
&+ (1-\xi_1^2)\phi_1^{(e)'}(\xi_1)P_{P-4}'^{(1,1)}(\xi_1).
\end{aligned}
\tag{5.145}
$$

The first term of the previous expression is zero at the element ends due to $\phi_1^{(e)'}(\xi_1)$; the second term is also zero due to $(1-\xi_1^2)$; and the third one is zero at the ends due to $(1-\xi_1^2)$ and $\phi_1^{(e)'}(\xi_1)$.

Figure 5.60 illustrates the high-order functions and their first derivatives for $P=8$. Observe that they are zero at the local beam element ends. This basis is hierarchical because functions of order $P-1$ are contained in the function set of order P.

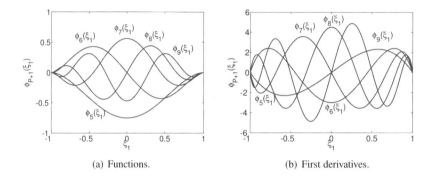

(a) Functions.   (b) First derivatives.

Figure 5.60   High-order shape functions and their first derivatives for $P=8$.

Figure 5.61(a) illustrates the sparsity profile for the mass matrix and $P=8$. The rectangles indicate the blocks of vertex and internal functions. It is noted that there is a coupling between the functions of these blocks. After calculating the Schur complement, we have the profile illustrated in Figure 5.61(b) without the coupling block. The sparsity profile for the stiffness matrix results in a natural decoupling between the vertex and internal blocks, as shown in Figure 5.61(c). Hence, it is not necessary to calculate the Schur complement. Once the solution for a beam with $P=3$ is obtained, we need to solve only the system of equations for each element corresponding to the internal block in a sequence of successive refinements of the approximation order.

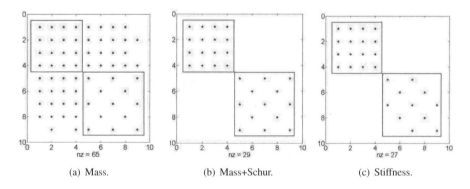

(a) Mass.   (b) Mass+Schur.   (c) Stiffness.

Figure 5.61   Sparsity profiles for the mass and stiffness matrices obtained with the high-order functions for the beam and $P=8$.

The orthogonality between the derivatives of the vertex and internal functions in the coupling matrix can be confirmed taking the local coefficient k_{ij} $(i=1,\ldots,4, j=5,\ldots,P+1)$ and applying integration by parts twice, namely,

$$
\begin{aligned}
k_{ij} &= \int_{-1}^{1} E^{(e)} I_z^{(e)} \frac{d^2 \phi_i^{(e)}(\xi_1)}{d^2 \xi_1} \frac{d^2 \phi_j^{(e)}(\xi_1)}{d^2 \xi_1} d\xi_1 \\
&= -\int_{-1}^{1} E^{(e)} I_z^{(e)} \frac{d^3 \phi_i^{(e)}(\xi_1)}{d^3 \xi_1} \frac{d \phi_j^{(e)}(\xi_1)}{d \xi_1} d\xi_1 + \underbrace{E^{(e)} I_z^{(e)} \frac{d^2 \phi_i^{(e)}(\xi_1)}{d^2 \xi_1} \frac{d \phi_j^{(e)}(\xi_1)}{d \xi_1} \bigg|_{-1}^{1}}_{=0} \\
&= \underbrace{\int_{-1}^{1} E^{(e)} I_z^{(e)} \frac{d^4 \phi_i^{(e)}(\xi_1)}{d^4 \xi_1} \phi_j^{(e)}(\xi_1) d\xi_1}_{=0} + \underbrace{E^{(e)} I_z^{(e)} \frac{d^3 \phi_i^{(e)}(\xi_1)}{d^3 \xi_1} \phi_j^{(e)}(\xi_1) \bigg|_{-1}^{1}}_{=0} \\
&= 0.
\end{aligned}
\tag{5.146}
$$

The boundary terms that arise of the integration by parts are zero because the high-order functions and the respective first derivatives are zero at the element ends. The last integral in the element domain is also zero, because the fourth-order derivative of the cubic Hermite polynomials is zero.

Figure 5.62 illustrates the numerical conditioning of the mass and stiffness matrices, with and without the Schur complement. A moderate increase in the condition numbers is noticed for the mass matrices after the Schur complement and stiffness matrices. In the case of the stiffness matrix, as the vertex and internal blocks are already decoupled, the conditioning with Schur is constant and corresponds to the one calculated with order $P = 3$.

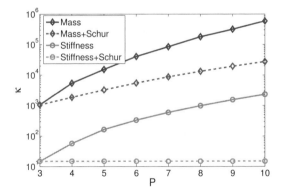

Figure 5.62 Condition number for the mass and stiffness matrices obtained with beam high-order functions and $P = 3$ to $P = 8$.

Example 5.23 *Consider the cantilever beam of Example 5.22. As the solution for the transversal displacement is of fourth order and the Hermite polynomials of third order, the approximated solution with the FEM did not match the exact solution.*

Using higher-order functions with $P = 4$, the error in the energy norm is zero for any number of elements. File beamsolapex3.m implements the solution of this example, for an arbitrary number of elements and approximation order.

Figure 5.63 illustrates the diagrams of shear force and bending moment obtained with a mesh of four elements. The diagrams are exactly the same ones obtained by the analytical solution.

$\square$

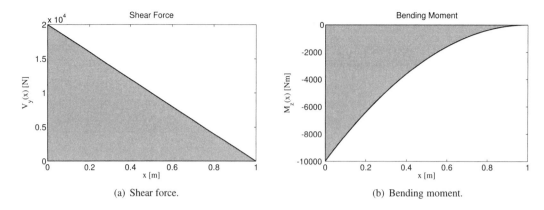

(a) Shear force. (b) Bending moment.

Figure 5.63 Example 5.23: diagrams for the shear force and bending moment for a mesh with four elements and $P = 4$.

Example 5.24 *Consider the simply supported steel beam with length L, rectangular cross-section, and subjected to a distributed senoidal load $q_y(x) = -10000 \sin\left(\dfrac{\pi}{L}x\right)$ N/m as illustrated in Figure 5.64.*

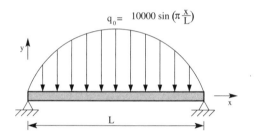

Figure 5.64 Example 5.24: simply supported beam subjected to a distributed senoidal load.

The expressions for the shear force, bending moment, rotation, and transversal displacement are given, respectively, by

$$V_y(x) = 10000\left(\frac{L}{\pi}\right)\cos\left(\frac{\pi}{L}x\right),$$

$$M_z(x) = -10000\left(\frac{L}{\pi}\right)^2\sin\left(\frac{\pi}{L}x\right),$$

$$\theta_z(x) = \frac{10000}{EI_z}\left(\frac{L}{\pi}\right)^3\cos\left(\frac{\pi}{L}x\right),$$

$$u_y(x) = -\frac{10000}{EI_z}\left(\frac{L}{\pi}\right)^4\sin\left(\frac{\pi}{L}x\right).$$

File beamsolapex4.m implements the solution of this example allowing the specification of an arbitrary number of elements and approximation order. The numerical integration with the Gauss-Lobatto-Legendre quadrature are used to make simpler the post-processing of results, due to the coordinates $\xi_1 = \pm 1$ at the local element ends.

Figure 5.65 shows the obtained results with a mesh of four elements and $P = 7$. Note that the shear force is smooth when compared to the results obtained in Example 5.22.

The behavior of the relative error in the energy norm is illustrated in Figure 5.66 for an h re-finement using four meshes with 2, 4, 8, and 16 elements, p refinement using a mesh with two elements and orders from $P = 3, \ldots, 10$, and an hp refinement using a mesh with four elements and $P = 3, \ldots, 10$. A cubic convergence rate is observed for the h refinement and spectral rates for the other refinements. The p version is the most effective in the reduction of the approximation error in terms of the number of degrees of freedom.

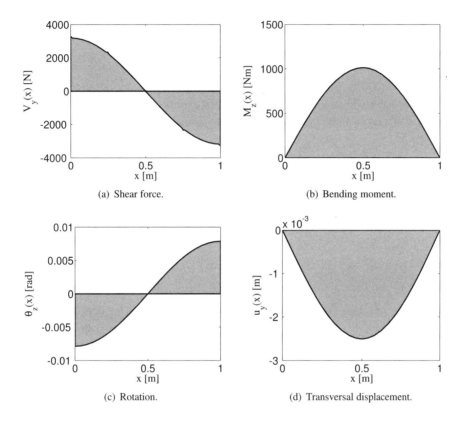

(a) Shear force. (b) Bending moment.

(c) Rotation. (d) Transversal displacement.

Figure 5.65 Example 5.24: diagrams of the shear force, bending moment, rotation, and transversal displacement.

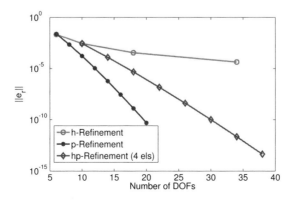

Figure 5.66 Example 5.24: relative error in the energy norm.

□

5.15 MATHEMATICAL ASPECTS OF THE FEM

We can express the strong form of the beam problem with the operator notation given in (4.88) as

$$\mathscr{A} = EI_z \frac{d^4}{dx^4}, \ u = u_y(x), \ f = q_y(x), \tag{5.147}$$

The domain $D_{\mathscr{A}}$ of the solution u of the strong form for beams with a continuous distributed loading function $q_y(x)$ is

$$D_{\mathscr{A}} = \{u \mid u \in C^4(0,L), u \text{ satisfies the boundary conditions}\}, \tag{5.148}$$

with $C^4(0,L)$ the set of functions with continuous derivatives up to the fourth order in the open interval $(0,L)$. The essential boundary conditions are in terms of the transversal displacement $u_y(x)$ and rotation $\theta_z(x)$. The natural boundary conditions are given in terms of the shear force $V_y(x)$ and bending moment $M_z(x)$.

Example 5.25 *Indicate the domain of the operator $\mathscr{A}$ for the cases of cantilever, simply supported, and double-clamped beams.*

As the differential equation is of fourth order, the solution functions must have continuous derivatives up to the fourth order, that is, they must belong to the set $C^4(0,L)$.

For the cantilever, simply supported, and double-clamped beams, we have the following domains, respectively,

$$D_{\mathscr{A}} = \left\{ u_y(x) \mid u_y(x) \in C^4(0,L), u_y(0) = 0, \frac{du_y(0)}{dx} = 0, \frac{d^2u_y(L)}{dx^2} = 0, \frac{d^3u_y(L)}{dx^3} = 0 \right\},$$

$$D_{\mathscr{A}} = \left\{ u_y(x) \mid u_y(x) \in C^4(0,L), u_y(0) = 0, \frac{d^2u_y(0)}{dx^2} = 0, u_y(L) = 0, \frac{d^2u_y(L)}{dx^2} = 0 \right\},$$

$$D_{\mathscr{A}} = \left\{ u_y(x) \mid u_y(x) \in C^4(0,L), u_y(0) = 0, \frac{du_y(0)}{dx} = 0, u_y(L) = 0, \frac{du_y(L)}{dx} = 0 \right\}.$$

□

An operator is a transformation that associates an element u of the vector space X with an element f of another vector space Y and denoted as

$$\begin{array}{rccc} \mathscr{A}: & X & \rightarrow & Y \\ & u & \rightarrow & f = \mathscr{A}u \end{array} . \tag{5.149}$$

For a beam with a continuous distributed load, we have $q_y(x) \in Y = C(0,L)$ and $X = D_{\mathscr{A}}$. Using the previous notation, we have

$$\begin{array}{rccc} \mathscr{A}: & D_{\mathscr{A}} & \rightarrow & C(0,L) \\ & u_y & \rightarrow & q_y(x) = EI_z \frac{d^4u_y(x)}{dx^4} \end{array} . \tag{5.150}$$

The operator $\mathscr{A}: X \rightarrow Y$ is linear if for all $u,v \in X$ and $\alpha \in \mathfrak{R}$, we have

$$\mathscr{A}(u+v) = \mathscr{A}u + \mathscr{A}v, \tag{5.151}$$

$$\mathscr{A}(\alpha u) = \alpha(\mathscr{A}u). \tag{5.152}$$

These two linearity properties can be expressed in a unique expression as

$$\mathscr{A}(\alpha u + \beta v) = \alpha(\mathscr{A}u) + \beta(\mathscr{A}v), \quad \forall u,v \in X \text{ and } \alpha, \beta \in \mathfrak{R}. \tag{5.153}$$

Example 5.26 *Verify that the beam operator $\mathscr{A}$ is linear.*
For any $u(x), v(x) \in D_{\mathscr{A}}$ and $\alpha, \beta \in \Re$, we have

$$EI_z \frac{d^4}{dx^4} (\alpha u(x) + \beta v(x)) = \alpha \left(EI_z \frac{d^4 u(x)}{dx^4} \right) + \beta \left(EI_z \frac{d^4 v(x)}{dx^4} \right).$$

Thus, the beam operator is linear.

$\square$

The linear operator $\mathscr{A} : D_{\mathscr{A}} \to Y$ is symmetric if

$$(\mathscr{A}u, v) = (u, \mathscr{A}v), \tag{5.154}$$

For the beam case, we have

$$(\mathscr{A}u_y, v) = \int_0^L EI_z \frac{d^4 u_y(x)}{dx^4} v(x) dx. \tag{5.155}$$

Example 5.27 *Consider a simply supported beam with length L. Show that the operator $\mathscr{A}$ for the beam is symmetric.*

As the beam is simply supported, the kinematic boundary conditions are $u_y(0) = u_y(L) = 0$ and the natural ones are $M_z(0) = M_z(L) = 0$. The function $v(x)$ also satisfies these boundary conditions. Thus, $v(0) = v(L) = 0$ and $v''(0) = v''(L) = 0$. The domain of the operator $\mathscr{A}$ for a simply supported beam was indicated in the previous example.

Taking $u_y(x)$ and $v(x)$ in $D_{\mathscr{A}}$, integrating equation (5.155) four times by parts and applying the boundary conditions, we have

$$
\begin{aligned}
(\mathscr{A}u_y, v) &= -\int_0^L EI_z \frac{d^3 u_y(x)}{dx^3} \frac{dv(x)}{dx} dx + \underbrace{V_y(x)v(x)\big|_0^L}_{=0} \\
&= \int_0^L EI_z \frac{d^2 u_y(x)}{dx^2} \frac{d^2 v(x)}{dx^2} dx - \underbrace{M_z(x) \frac{dv(x)}{dx}\bigg|_0^L}_{=0} \\
&= -\int_0^L EI_z \frac{du_y(x)}{dx} \frac{d^3 v(x)}{dx^3} dx + \underbrace{EI_z \frac{du_y(x)}{dx} \frac{d^2 v(x)}{dx^2}\bigg|_0^L}_{=0} \\
&= \int_0^L EI_z u_y(x) \frac{d^4 v(x)}{dx^4} dx + \underbrace{EI_z u_y(x) \frac{d^3 v(x)}{dx^3}\bigg|_0^L}_{=0} \\
&= (u_y, \mathscr{A}v).
\end{aligned}
$$

Thus, the operator $\mathscr{A}$ is symmetric.

$\square$

The operator $\mathscr{A}$ is positive-definite if

$$
\begin{aligned}
(Au, u) &> 0, \tag{5.156} \\
(Au, u) &= 0 \text{ if and only if } u = 0. \tag{5.157}
\end{aligned}
$$

Example 5.28 *Consider again the simply supported beam with length L. Show that the beam operator $\mathscr{A}$ is positive-definite.*

Initially, it is demonstrated that $(Au, u) > 0$. From (5.155) and two integration by parts, we have

$$(Au_y, u_y) = \int_0^L EI_z \left(\frac{d^2 u_y}{dx^2} \right)^2 dx.$$

As $EI_z > 0$ and the square of the second derivative of u_y is positive, we have $(Au_y, u_y) > 0$.

To demonstrate (5.157) for the case of beams, note that $u_y = 0 \rightarrow (Au_y, u_y) = 0$. It remains to demonstrate that for $(Au_y, u_y) = 0$, we have $u_y = 0$.

Assuming $(Au_y, u_y) = 0$, it is verified that

$$\frac{d^2 u_y}{dx^2} = \frac{d}{dx} \left(\frac{du_y}{dx} \right) = 0,$$

that is, $\dfrac{du_y}{dx}$ is constant and u_y varies linearly as

$$u_y(x) = ax + b.$$

The coefficients a and b are determined using the boundary conditions of the simply supported beam, that is,

$$u_y(x = 0) = a(0) + b = 0 \rightarrow b = 0,$$
$$u_y(x = L) = a(L) + b = 0 \rightarrow a = 0.$$

Therefore, $u_y = 0$.

Hence, the operator $\mathscr{A}$ for the simply supported beam is positive-definite.

$\square$

It is possible to associate an inner product to a vector space X according to (3.82). The energy inner product for the operator $\mathscr{A} : X \rightarrow Y$ is defined by

$$(u, v)_{\mathscr{A}} = (\mathscr{A}u, v) = \int_\Omega (\mathscr{A}u) v \, d\Omega. \tag{5.158}$$

The inner product induces a norm. Particularly, the energy norm associated to the previous definition is

$$\|u\|_{\mathscr{A}} = \sqrt{(u, u)_{\mathscr{A}}}. \tag{5.159}$$

Consider the symmetric operator $\mathscr{A} : X \rightarrow Y$. It is called a lower bounded positive operator for each $u \in X$ if

$$(u, u)_{\mathscr{A}} \geq C_1^2 (u, u), \tag{5.160}$$

with C_1 a constant strictly positive. In terms of norms, the previous relation is written as

$$\|u\|_{\mathscr{A}} \geq C_1 \|u\|. \tag{5.161}$$

This property plays an important role in the existence of solutions of BVPs.

Example 5.29 *As seen in the previous example, the beam operator is symmetric and positive-definite. We want to demonstrate now that this operator is lower bounded positive.*

Consider the following relation for the derivative of $u_y(x)$:

$$\frac{du_y(x)}{dx} = \int_0^x \frac{du_y^2(t)}{dt^2} dt = \int_0^x (1) \frac{du_y^2(t)}{dt^2} dt.$$

Thus,

$$\left(\frac{du_y(x)}{dx}\right)^2 = \left(\int_0^x (1)\frac{du_y^2(t)}{dt^2}dt\right)^2.$$

Given two elements u and v of a given vector space, the Cauchy-Schwartz inequality states that

$$|(u,v)| \le ||u|| \, ||v||. \tag{5.162}$$

Applying the Cauchy-Schwartz inequality to the previous relation, and as $L > 0$, we have

$$\left(\frac{du_y(x)}{dx}\right)^2 \le \left(\int_0^x (1)^2 dt\right)\int_0^x \left(\frac{du_y^2(t)}{dt^2}\right)^2 dt = x\int_0^L \left(\frac{du_y^2(t)}{dt^2}\right)^2 dt.$$

Integrating the above equation in $(0,L)$ and recalling that $EI_z > 0$, we obtain

$$\underbrace{\int_0^L \left(\frac{du_y(x)}{dx}\right)^2 dx}_{||u_y'||^2} \le \int_0^L x \left(\underbrace{\int_0^L \left(EI_z\frac{du_y^2(t)}{dt^2}\right)^2 dt}_{||u_y||_E^2}\right) dx = \frac{L^2}{2}||u_y||_E^2.$$

Thus,

$$||u_y'|| \le \frac{L\sqrt{2}}{2}||u_y||_E.$$

Taking $C_1 = \dfrac{L\sqrt{2}}{2}$, the $\mathscr{A}$ operator for a simply supported beam is lower bounded positive.
□

A norm allows the generalization of the usual concept of distance between points of the Euclidean space for any vector space with the metric concept. Given two elements u and v of a normed vector space X, the metric d is defined as

$$d = ||u - v||. \tag{5.163}$$

Note that the error given by (5.136) represents a metric between two functions in the energy norm, that is,

$$d = ||u - v||_E. \tag{5.164}$$

Now consider an infinite sequence $\{u_n\}_{n=1}^\infty$ in X. This sequence converges to the element $u \in X$ if

$$\lim_{n\to\infty} u_n = u, \tag{5.165}$$

which in terms of the metric concept results in

$$\lim_{n\to\infty} ||u_n - u||_X. \tag{5.166}$$

The previous sequence is called fundamental or Cauchy if

$$\lim_{m,n\to\infty} ||u_n - u_m||_X = 0. \tag{5.167}$$

Example 5.30 *Consider the following sequence of functions defined in the [-1,1] interval for $n = 1, 2, \ldots$:*

$$u_n(x) = \begin{cases} nx + 1 & x \in [-1/n, 0] \\ -nx + 1 & x \in [0, 1/n] \end{cases}.$$

For the L_2 norm, this sequence of functions converges to the zero function $u = 0$ because

$$
\begin{aligned}
\lim_{n \to \infty} ||u_n - u||_{L_2}^2 &= \lim_{n \to \infty} \left[\int_{-1/n}^{0} (nx+1)^2\, dx + \int_{0}^{1/n} (-nx+1)^2\, dx \right] \\
&= \lim_{n \to \infty} \frac{2}{3n} = 0.
\end{aligned}
$$

Thus, the given sequence of functions converges to the function $u = 0$.

The sequence is also fundamental or Cauchy because

$$
\begin{aligned}
\lim_{m,n \to \infty} ||u_n - u_m||_{L_2}^2 &\leq \lim_{n \to \infty} ||u_n||_{L_2}^2 = \left[\int_{-1/n}^{0} (nx+1)^2\, dx + \int_{0}^{1/n} (-nx+1)^2\, dx \right] \\
&= \lim_{n \to \infty} \frac{2}{3n} = 0.
\end{aligned}
$$

□

A normed space X is complete if every sequence $\{u_n\}_{n=1}^{\infty}$ converges to an element $u \in X$. In this case, X is called Banach space. Every space with a complete inner product is called a Hilbert space. As every inner product induces a norm, it follows that every Hilbert space is also a Banach space. The opposite is not true, because not every norm is induced by an inner product. For example, the maximum norm $||u|| = \max_{x \in [a,b]} |u(x)|$ of function $u(x)$ defined in the interval $[a, b]$.

The norm of the Hilbert space H_1 is defined by

$$||u||_{H_1} = \sqrt{\int_{\Omega} u^2 + (u')^2\, d\Omega}. \tag{5.168}$$

The set of shape functions $\{\phi_n\}_{n=1}^{N}$ constitutes a complete set in the vector space X. Given a number $\varepsilon > 0$, it is possible to find a positive integer number N and coefficients a_i $(i = 1, \ldots, N)$ such that

$$||u - \sum_{i=1}^{N} a_i \phi_i^{(e)}||_X < \varepsilon. \tag{5.169}$$

Thus, it is possible to approximate the function u with any desired precision.

The linear and bilinear forms of the general weak form expression (4.92) for beams are given, respectively, by

$$a(u_y, v) = \int_{0}^{L} EI_z \frac{du_y^2(x)}{dx^2} \frac{dv^2(x)}{dx^2}\, dx, \tag{5.170}$$

$$f(v) = \int_{0}^{L} q_y(x)v(x)\, dx + V_{y_L}v(L) + V_{y_0}v(0) + M_{z_L}v'(L) + M_{z_0}v'(0). \tag{5.171}$$

As seen in the previous examples, to demonstrate that an operator is symmetric and positive-definite, we apply integration by parts and obtain the expression of the bilinear form $a(u, v)$. Hence, the concepts of symmetry and positivity for bilinear forms are extended as

$$a(u, v) = a(v, u), \tag{5.172}$$

$$a(u, u) \geq 0 \text{ and } a(u, u) = 0 \text{ if and only if } u = 0. \tag{5.173}$$

It is said that $a(u,v)$ is lower bounded positive or elliptical in the Hilbert space H_1 if

$$a(u,u) \geq C_1 ||u||_{H_1}. \tag{5.174}$$

Also the bilinear form is continuous or limited if

$$|a(u,v)| \geq C_2 ||u||_{H_1} ||v||_{H_1}, \tag{5.175}$$

with $C_2 > 0$.

The set of functions with infinite energy defines the energy space E for the domain Ω, that is,

$$E(\Omega) = \{u \mid a(u,u) < \infty\}. \tag{5.176}$$

The boundary of Ω is denoted by $\partial\Omega$ and partitioned as

$$\partial\Omega = \partial\Omega_D \cup \partial\Omega_N, \tag{5.177}$$

with $\partial\Omega_D$ and $\partial\Omega_N$, respectively, the boundary regions with kinematical (essential or Dirichlet) and natural (force or Neummann) boundary conditions, in such way that $\partial\Omega_D \cap \partial\Omega_N = \emptyset$.

The functions of the energy space $E(\Omega)$ also belong to the space H_1 and satisfies the condition that their norms in H_1 is limited, i.e., $||u||_{H_1} < \infty$. It is further assumed that the right side f also has finite energy. Generally, the set of trial solutions to weak forms must have finite energy and satisfy the kinematic or essential or Dirichlet boundary conditions, that is,

$$X = \{u \mid u \in H_1, u \text{ satisfies the essential boundary conditions}\}. \tag{5.178}$$

For the case of beams, we have

$$X = \{u_y \mid u_y \in C_{cp}^2(0,L), u_y \text{ satisfies the essential boundary conditions}\}, \tag{5.179}$$

with $C_{cp}^2(0,L)$ the set of functions with piecewise continuous second derivatives. Note that the previous domain is broader than the domain $D_{\mathscr{A}}$ of the strong form. It requires less regularity of functions and needs only to satisfy the Dirichlet boundary conditions. On the other hand, the test function space V is defined by

$$V = \{u_y \mid u_y \in C_{cp}^2(0,L), u_y \text{ satisfies the essential homogeneous boundary conditions}\}. \tag{5.180}$$

Based on that, the solution of the weak form can be denoted in a general way by: *find $u \in X$ such that*

$$a(u,v) = f(v), \quad \forall v \in V. \tag{5.181}$$

When approximating the weak form by the FEM, the subsets $X_N \subset X$ and $V_N \subset V$ are selected with a finite number of shape functions, such that the previous expression is denoted by: *find $u_N \in X_N$ such that*

$$a(u,v) = f(v), \quad \forall v_N \in V. \tag{5.182}$$

When $a(u,v)$ is bilinear, elliptic, and limited, but not necessarily symmetric, and f is limited, the Lax-Milgram theorem guarantees the existence and uniqueness of the Galerkin approximation [15].

5.16 FINAL COMMENTS

In this chapter, the variational formulation of the Euler-Bernoulli model for beams was presented, using the same sequence of steps that obtained the bar and shaft models.

The use of the equilibrium boundary value problems and singularity functions allows solving isostatic and hyperstatic problems using the same solution procedure.

This chapter also presented the approximation of the bending problem using cubic Hermite polynomials and introduced a high-order basis. Several examples were presented with MATLAB solution files. At the end, several formal concepts about approximation were presented, which apply to all considered problems in this book.

5.17 PROBLEMS

1. Plot the diagrams for the shear force and bending moment to the beam AB shown in Figure 5.67. Neglect the weight of the beams and axial forces.

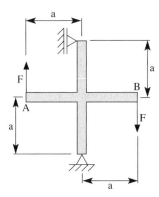

Figure 5.67 Problem 1.

2. For the beam illustrated in Figure 5.68, obtain the free body diagram and plot the shear force and bending moment diagrams.

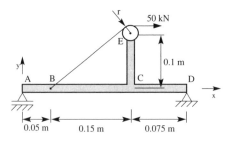

Figure 5.68 Problem 2.

3. Indicate the boundary conditions and use the singularity function notation to the load equations for the beams illustrated in Figure 5.69.
4. A small boat has the load represented in Figure 5.70. Plot the shear force and bending moment diagrams.
5. Use singular functions to indicate the loads and plot the diagrams for the shear force, bending moment, rotation, and transversal displacement for the steel beams ($E = 210$ GPa) illustrated in Figure 5.71.

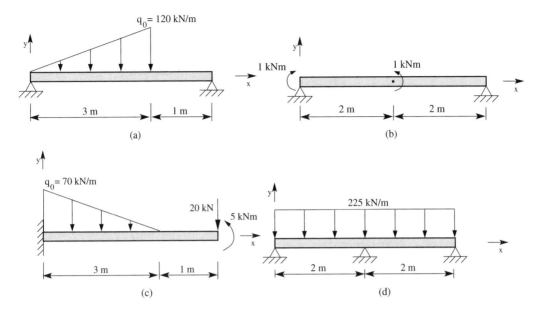

Figure 5.69 Problem 3.

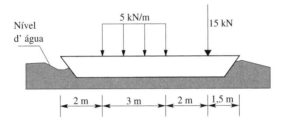

Figure 5.70 Problem 4.

6. For the steel beams of Figure 5.72, calculate the maximum tractive and compressive stresses due to bending. Plot the stress distributions at the beam critical cross section.
7. Design:
   a. The steel beam ($E = 210$ GPa) with circular cross-section and loads illustrated in Figure 5.73(a);
   b. The brass beam ($E = 100$ GPa) with retangular cross-section ($4h/b = 4$) and loads illustrated in Figure 5.73(b).
8. Calculate the expressions of the critical buckling loads given in equation (5.111).
9. Write MATLAB programs to solve the beams in Figure 5.69. Compare the analytical and numerical results using the energy norm.
10. Write a MATLAB program that allows the determination of critical loads in columns.

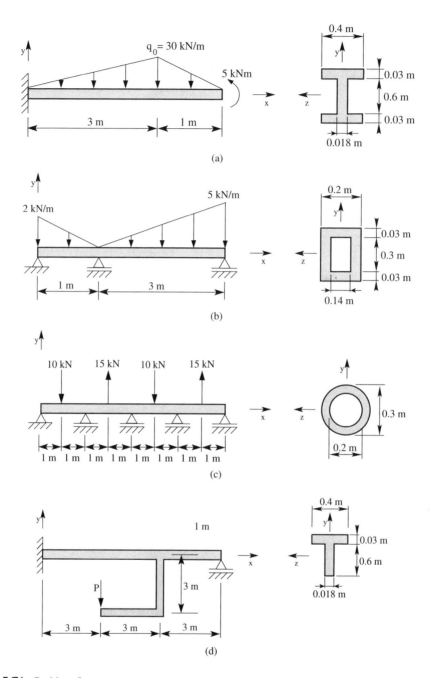

Figure 5.71 Problem 5.

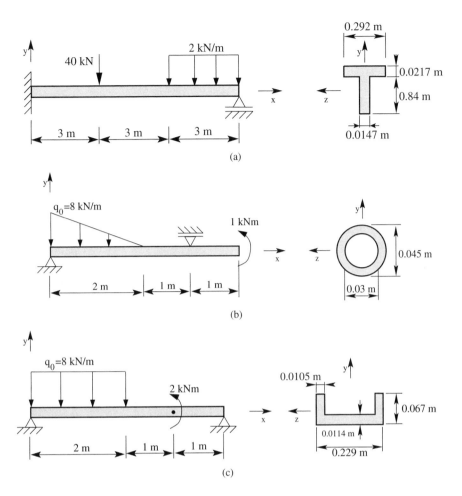

Figure 5.72 Problem 6.

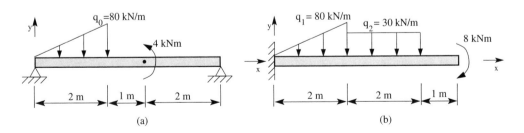

Figure 5.73 Problem 7.

6 FORMULATION AND APPROXIMATION OF BEAM IN SHEAR

6.1 INTRODUCTION

The Euler-Bernoulli beam model does not consider the shear effect present in cross-sections and longitudinal fibers of the beam. For short beams, shear is important and thus the Timoshenko model is employed. Its formulation will be derived below. The steps of the previous variational formulation are employed again. Notice that the same geometrical hypotheses for bars, shafts, and beams in bending only are valid for the Timoshenko beam; that is, we assume the length to be much larger than the cross-section dimensions. Subsequently, we present the approximation with the FEM, using low- and high-order shape functions.

6.2 KINEMATICS

Analogous to the Euler-Bernoulli model, the kinematic bending actions of the Timoshenko beam must be such that the cross-sections remain flat. However, the sections have angular deformations and do not remain orthogonal to the longitudinal axis of the beam anymore. These hypotheses are illustrated in the cantilever beam of Figure 6.1.

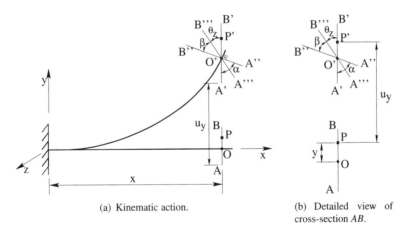

(a) Kinematic action.

(b) Detailed view of cross-section AB.

Figure 6.1 Timoshenko beam kinematics.

The cross-section AB presents a rigid transversal displacement $u_y(x)$, reaching position $A'B'$. Thus, we have a rigid bending rotation $\alpha(x)$ about the z axis, and the section reaches position $A''B''$, orthogonal to the tangent of the deformed longitudinal axis of the beam. As the Timoshenko beam considers the effect of shear, there is a constant angular deformation $\beta(x)$, so that AB reaches final position $A'''B'''$, as shown in Figure 6.1(a). The bending angle $\theta_z(x)$ of section x is given by

$$\theta_z(x) = \alpha(x) - \beta(x), \tag{6.1}$$

with $\theta_z(x)$ the total rotation, $\alpha(x) = \dfrac{du_y(x)}{dx}$ the rotation due to bending only, and $\beta(x)$ the distortion of the section due to the shear effect.

The kinematics of this model is analogous to the Euler-Bernoulli beam, only including the distortion effect of $\beta(x)$ when calculating the axial displacement $u_x(x)$. Thus, from the straight triangle $O'QP'''$ of Figure 6.2, we verify that

$$\sin\theta_z = \frac{\Delta x}{y}.$$

Notice that $O'P' = O'P''' = y$, because the cross-section dimensions do not change, but only distort.

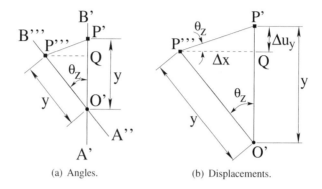

(a) Angles.   (b) Displacements.

Figure 6.2   Kinematics of section AB of the Timoshenko beam.

The axial displacement of section AB is given by the difference between the final and initial positions, that is,

$$u_x = (x - \Delta x) - x = -\Delta x.$$

By combining the two previous equations, we have

$$u_x(x,y) = -y\sin\theta_z(x).$$

The negative sign is because for a positive bending angle θ_z, the axial displacement occurs in the negative direction of the adopted x axis.

Considering small displacements, $\theta_z(x)$ is small and thus $\sin\theta_z(x) \approx \theta_z(x)$. Hence, we have the following expression for the axial displacement in cross-section points of the Timoshenko beam:

$$u_x(x,y) = -y\theta_z(x) = -y[\alpha(x) - \beta(x)]. \tag{6.2}$$

Therefore, the kinematics of a Timoshenko beam is described by the vector field $\mathbf{u}(x,y)$ with the following components:

$$\mathbf{u}(x,y) = \left\{ \begin{array}{c} u_x(x,y) \\ u_y(x) \\ u_z(x) \end{array} \right\} = \left\{ \begin{array}{c} -y\theta_z(x) \\ u_y(x) \\ 0 \end{array} \right\}. \tag{6.3}$$

Taking $\beta(x) = 0$ in (6.2), we recover the same expression for the axial displacement of the Euler-Bernoulli beam.

Example 6.1 *Consider the same cantilever beam with length L, clamped in $x = 0$ and subjected to a transversal displacement δ at the free end $(x = L)$, as illustrated in Figure 5.6.*

The function $u_y(x)$ that gives the transversal displacement in the cross-section points of the beam in this example is

$$u_y(x) = -\left(-\frac{x}{K_cGA} + \frac{(x-3L)x^2}{6EI_z}\right)\frac{\delta}{\dfrac{L^3}{3EI_z} + \dfrac{L}{K_cGA}},$$

with E and G the longitudinal and shear moduli of the material; A, I_z, and K_c are respectively the area, second moment of area about the z axis, and the shear coefficient of the cross-section. This function will be determined later by the solution of the equations to be derived at the end of the Timoshenko beam formulation (see Example 6.3).

□

In fact, the distortion $\beta(x)$ is not constant in the section, and a warping effect occurs, as illustrated in Figure 6.3 for a beam of rectangular section [42]. In this case, the distortion is maximum in the neutral line, and zero at the upper and lower ends of the section. In Figure 6.4(a), the right end of the beam is isolated and two infinitesimal elements are taken and their shapes after the distortion are illustrated in Figure 6.4(b). As the distortion varies in the cross-section, the beam elements, such as the ones illustrated in Figure 6.4(a), will deform distinctly. Reassembling these elements, we have a section which is no longer perpendicular to the neutral line, characterizing the warping, as illustrated in Figure 6.4(b). Subsequently, we include a shear factor K_c in the model to correct the assumed constant distortion of the cross-section.

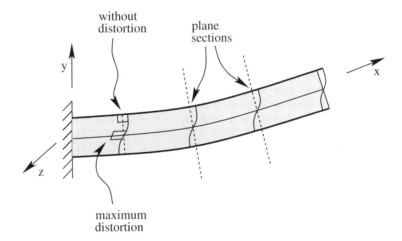

Figure 6.3 Cross-section warping in the Timoshenko beam.

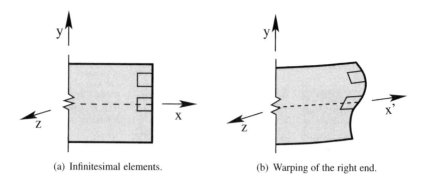

(a) Infinitesimal elements.   (b) Warping of the right end.

Figure 6.4   Warping analysis of the beam right end.

6.3 STRAIN MEASURE

The previously studied mechanical models have only normal or shear strain components. On the other hand, the Timoshenko model presents both normal or longitudinal ε_{xx} and angular γ_{xy} strain components.

In order to determine the expression for the component ε_{xx} on a generic section x of the beam, we should consider Figure 5.8 again, and the specific variation $\dfrac{\Delta u_x}{\Delta x}$ of the axial displacement $u_x(x,y)$ between sections AB and CD. Thus, analogous to equation (5.88), the normal specific strain $\varepsilon_{xx}(x)$ is obtained from

$$\varepsilon_{xx}(x,y) = \lim_{\Delta x \to 0} \frac{u_x(x+\Delta x,y) - u_x(x,y)}{\Delta x}.$$

Substituting (6.2) in the above expression, we have

$$\varepsilon_{xx}(x,y) = \lim_{\Delta x \to 0} \frac{-y\theta_z(x+\Delta x) - [-y\theta_z(x)]}{\Delta x} = -y \lim_{\Delta x \to 0} \overbrace{\frac{\theta_z(x+\Delta x) - \theta_z(x)}{\Delta x}}^{\Delta\theta_z}.$$

By employing the derivative definition for both previous expressions, we obtain the final equation for the normal strain component ε_{xx}, that is,

$$\varepsilon_{xx}(x,y) = \frac{\partial u_x(x,y)}{\partial x} = -y\frac{d\theta_z(x)}{dx}. \tag{6.4}$$

To determine the distortion or angular strain, consider the infinitesimal undeformed $ABCD$ and deformed $A'B'C'D'$ beam elements illustrated in Figure 6.5. Considering only the distortion effect, the initial and final edge lengths are Δx and Δy, respectively. Figure 6.6(a) illustrates the deformed beam element. The distortion in planes x and y are shown separately in Figures 6.6(b) and 6.6(c), respectively. The following geometric relations are valid:

$$\tan\beta_1 = \frac{\Delta u_y}{\Delta x},$$

$$\tan\beta_2 = \frac{\Delta u_x}{\Delta y}.$$

For small values of β_1 and β_2, we have $\tan\beta_1 \approx \beta_1$ and $\tan\beta_2 \approx \beta_2$. The angular strains at point A are obtained by taking $\Delta x \to 0$ and $\Delta y \to 0$. Thus, respectively taking the limits of the previous

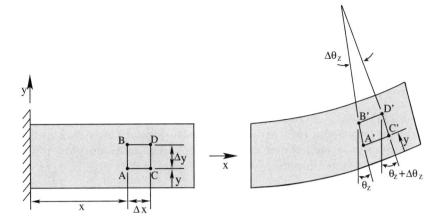

Figure 6.5 Beam element *ABCD* before and after the bending action.

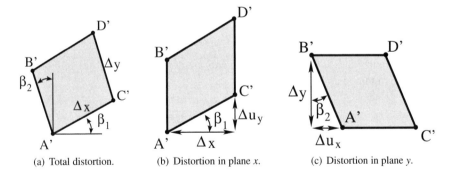

(a) Total distortion.   (b) Distortion in plane *x*.   (c) Distortion in plane *y*.

Figure 6.6 Total distortion and distortion planes *x* and *y* for the *ABCD* beam element.

expressions for $\Delta x \to 0$ and $\Delta y \to 0$ we obtain

$$\beta_1(x) = \lim_{\Delta x \to 0} \frac{\Delta u_y}{\Delta x} = \frac{du_y(x)}{dx}, \tag{6.5}$$

$$\beta_2(x) = \lim_{\Delta y \to 0} \frac{\Delta u_x}{\Delta y} = \frac{\partial u_x(x, y)}{\partial y}. \tag{6.6}$$

We verify that $\beta_1(x)$ and $\beta_2(x)$ represent angular strains which indicate that the initially straight angle BAC is no longer straight. The x axis is orthogonal to edges AB and CD. β_1 is the distortion component γ_{xy}, with x the plane and y the strain direction. Analogously, the y axis is orthogonal to edges AC and BD, and β_2 represents the γ_{yx} shear component. Thus,

$$\beta_1(x) = \gamma_{xy}(x) = \frac{du_y(x)}{dx}, \tag{6.7}$$

$$\beta_2(x) = \gamma_{yx}(x) = \frac{\partial u_x(x, y)}{\partial y}. \tag{6.8}$$

The total distortion at point A according to plane xy is denoted by $\bar{\gamma}_{xy}(x)$. Thus,

$$\bar{\gamma}_{xy}(x) = \beta_1(x) + \beta_2(x) = \frac{\partial u_x(x, y)}{\partial y} + \frac{du_y(x)}{dx}. \tag{6.9}$$

Substituting the kinematics given by (6.3) in the above expression, and performing the indicated derivatives, we have

$$\bar{\gamma}_{xy}(x) = -\theta_z(x) + \frac{du_y(x)}{dx} = \beta(x). \tag{6.10}$$

Thus, the total distortion component is reduced to the characteristic angle β of the shear effect in the Timoshenko beam.

6.4   RIGID ACTIONS

In the case of rigid kinematics, the strain components are zero. Thus, the rigid actions for the Timoshenko beam are obtained as

$$\varepsilon_{xx}(x, y) = -y \frac{d\theta_z(x)}{dx} = 0,$$

$$\bar{\gamma}_{xy}(x) = -\theta_z(x) + \frac{du_y(x)}{dx} = 0.$$

The distortion in the second equation is zero, and thus we recover the definition of bending angle of the Euler-Bernoulli beam, that is,

$$\theta_z(x) = \frac{du_y(x)}{dx},$$

which when substituted in the first expression results in equation (5.6). Consequently, we obtain the same rigid body actions of the beam under bending only, illustrated in Figure 5.8.

6.5   DETERMINATION OF INTERNAL LOADS

In this section, we determine the compatible stress components and internal loads with the kinematics of a Timoshenko beam, using the Newtonian approach. For this, we consider the volume element $dV = dx\,dy\,dz$ illustrated in Figure 6.7(a).

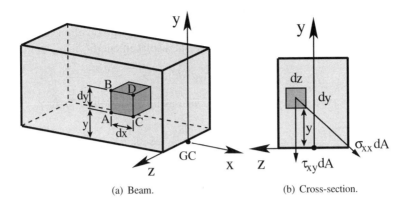

(a) Beam. (b) Cross-section.

Figure 6.7 Volume element indicating the stress components of the Timoshenko beam.

Taking the cross-section of the beam shown in Figure 6.7(b), the normal force in the area element $dA = dydz$ is $dN_x = \sigma_{xx}dA$, with σ_{xx} the normal stress component. The first index represents the plane where the stress acts and the second one represents the direction, both x in this case. Analogous to the case of bending only, the bending moment acting in the area element in direction z is given by

$$dM_z = -ydN_x = -y\sigma_{xx}dA.$$

However, the bending moment in the cross-section is

$$M_z(x) = -\int_A y\sigma_{xx}dA. \tag{6.11}$$

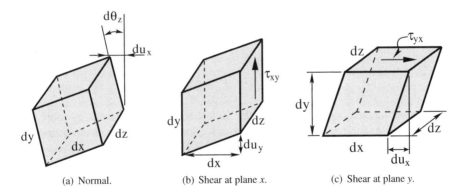

(a) Normal. (b) Shear at plane x. (c) Shear at plane y.

Figure 6.8 Relative displacements and stress components in the Timoshenko beam.

The relative axial displacement between faces CD and AB of the volume element is illustrated in Figure 6.8(a). The strain internal work associated to the normal force dN_x for displacement du_x is

$$dW_{i,\sigma_{xx}} = dN_x du_x = (\sigma_{xx}dydz)du_x.$$

Multiplying and dividing the above equation by dx and substituting the definition of the normal strain component (6.4), we have

$$dW_{i,\sigma_{xx}} = \sigma_{xx}\frac{du_x}{dx}dxdydz = \sigma_{xx}\varepsilon_{xx}dV. \tag{6.12}$$

There is a shear stress component τ_{xy} associated to γ_{xy}, acting on plane x in direction y, as illustrated in Figure 6.8(b). The corresponding force in the $dydz$ differential volume element is $\tau_{xy}dydz$. The work done by this force and the relative transversal displacement du_y is

$$dW_{i,\tau_{xy}} = (\tau_{xy}dydz)du_y.$$

Multiplying and dividing the above equation by dx and using definition (6.7) for γ_{xy}, we have

$$dW_{i,\tau_{xy}} = (\tau_{xy}dxdydz)\frac{du_y}{dx} = \tau_{xy}\gamma_{xy}dV. \tag{6.13}$$

There is a stress component τ_{yx} associated with γ_{yx}. It acts in plane y and direction x, as illustrated in Figure 6.8(c). The internal strain work done by the force $\tau_{yx}dxdz$ with the relative axial displacement du_x is

$$dW_{i,\tau_{yx}} = (\tau_{yx}dxdz)du_x.$$

Multiplying and dividing the above equation by dy and using definition (6.8) for γ_{yx}, we have

$$dW_{i,\tau_{yx}} = (\tau_{yx}dxdydz)\frac{du_x}{dy} = \tau_{yx}\gamma_{yx}dV. \tag{6.14}$$

Consider the representation of the tangent forces obtained from the shear stress components given in Figure 6.8. By calculating the resultant of moments relative to point A, we obtain

$$\sum m_{z_A} = (\tau_{xy}dydz)dx - (\tau_{yx}dxdz)dy = 0.$$

Thus,

$$\tau_{xy} = \tau_{yx}, \tag{6.15}$$

that is, the shear stresses acting on a point according to perpendicular planes are equal.

The shear force acting in the area element illustrated in Figure 6.7(b) is

$$dV_y = -\tau_{xy}dydz.$$

However, the shear force in the cross-section is given by

$$V_y(x) = -\int_A \tau_{xy}dA. \tag{6.16}$$

The shear force acting on a generic section of the Timoshenko beam is shown in Figure 6.9. On the other hand, the bending moment present in the cross-sections of the Timoshenko beam is the same as the Euler-Bernoulli beam, and is illustrated in Figure 5.10.

Taking the summation of the internal work equations (6.12) to (6.14), and using equation (6.9), the internal strain work for the beam volume element dV is

$$dW_i = [\sigma_{xx}\varepsilon_{xx} + \tau_{xy}(\gamma_{xy} + \gamma_{yx})]dV = [\sigma_{xx}\varepsilon_{xx} + \tau_{xy}\bar{\gamma}_{xy})]dV. \tag{6.17}$$

Hence, the strain internal work of the beam is obtained from the following volume integral:

$$W_i = \int_V dW_i = \int_V [\sigma_{xx}(x)\varepsilon_{xx}(x) + \tau_{xy}(x)\bar{\gamma}_{xy}(x)]dV. \tag{6.18}$$

Substituting the normal and shear strain components, respectively given in (6.4) and (6.9), and decomposing the volume integral in the cross-section area A and the length L of the beam, we have

$$W_i = \int_0^L \left[\int_A -\sigma_{xx}(x)ydA\right]\frac{d\theta_z(x)}{dx}dx + \int_0^L \left[\int_A -\tau_{xy}(x)dA\right]\left(\theta_z(x) - \frac{du_y(x)}{dx}\right)dx. \tag{6.19}$$

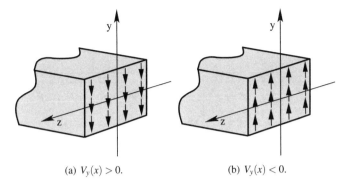

(a) $V_y(x) > 0$.　　　　(b) $V_y(x) < 0$.

Figure 6.9 Shear force in the Timoshenko beam.

The terms inside brackets respectively represent the bending moment $M_z(x)$ in the z axis and the shear force in the y axis, given by (6.11) and (6.16). Thus, the strain internal work is

$$W_i = \int_0^L M_z(x)\frac{d\theta_z(x)}{dx}dx - \int_0^L V_y(x)\frac{du_y(x)}{dx}dx + \int_0^L V_y(x)\theta_z(x)dx. \qquad (6.20)$$

Integrating by parts the two first integrals of the above equation, we have

$$\begin{aligned} W_i &= -\int_0^L \frac{dM_z(x)}{dx}\theta_z(x)dx + M_z(x)\theta_z(x)|_0^L \\ &\quad + \int_0^L \frac{dV_y(x)}{dx}u_y(x)dx - V_y(x)u_y(x)|_0^L \\ &\quad + \int_0^L V_y(x)\theta_z(x)dx. \end{aligned} \qquad (6.21)$$

Rearranging the previous expression, we have

$$\begin{aligned} W_i &= \int_0^L \left[-\frac{dM_z(x)}{dx} + V_y(x) \right]\theta_z(x)dx + \int_0^L \left[\frac{dV_y(x)}{dx} \right]u_y(x)dx \\ &\quad + [M_z(L)\theta_z(L) - M_z(0)\theta_z(0)] + [-V_y(L)v(L) + V_y(0)v(0)]. \end{aligned} \qquad (6.22)$$

Supposing that the force and length measures are in the SI, both terms inside brackets in the above integrals have the following units

$$\begin{aligned} \left[-\frac{dM_z(x)}{dx} + V_y(x) \right] &= \frac{\mathrm{Nm}}{\mathrm{m}} + \mathrm{N} = \frac{\mathrm{Nm}}{\mathrm{m}} + \frac{\mathrm{Nm}}{\mathrm{m}} = \frac{\mathrm{Nm}}{\mathrm{m}}, \\ \left[\frac{dV_y(x)}{dx} \right] &= \frac{\mathrm{N}}{\mathrm{m}}. \end{aligned}$$

Thus, the first term represents an internal bending moment density $m_{z_i}(x)$ in direction z, i.e., we have a distributed bending moment per units of length. The second term represents a transversal force density, that is, a shear force distribution $V_y(x)$ in direction y by units of length. In addition, there are bending moments $M_z(L)$ and $M_z(0)$ and shear forces $V_y(L)$ and $V_y(0)$, which are concentrated loads at the ends of the beam. These internal forces are shown in the free body diagram of Figure 6.10(a), with the compatible directions according to the signs of equation (6.22).

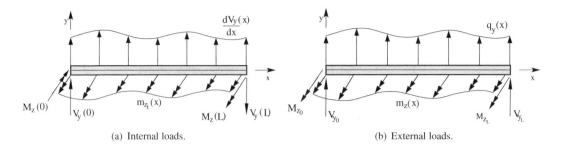

(a) Internal loads.   (b) External loads.

Figure 6.10   Internal and external loads in the Timoshenko beam.

6.6   DETERMINATION OF EXTERNAL LOADS

To characterize the compatible external loads of the Timoshenko beam model, we just analyze equation (6.22). There must exist a corresponding external load for each internal load in (6.22).

Thus, to balance the internal work of the distributed bending moment density $-\dfrac{dM_z(x)}{dx} + V_y(x)$, an external distributed moment density $m_z(x)$ must exist such that the respective external work contribution $\int_0^L m_z(x)\theta_z(x)dx$ balances the internal work term $\int_0^L \left[-\dfrac{dM_z(x)}{dx} + V_y(x) \right] \theta_z(x)dx$ for the rotation $\theta_z(x)$. Analogously, to balance the internal work $\int_0^L \dfrac{dV_y(x)}{dx} u_y(x)dx$, there must be a transversal distributed load $q_y(x)$, such that the external work $\int_0^L q_y(x)u_y(x)dx$ balances the respective internal work. Finally, the external moments in z, M_{z_0} and M_{z_L}, and the external transversal forces in y, V_{y_0} and V_{y_L}, must be present at ends $x = 0$ and $x = L$ to balance the work done by the bending moments and shear forces acting at the ends of the beam. The compatible external loads with the kinematics of the Timoshenko beam are illustrated in Figure 6.10(b).

Hence, the work W_e done by the external loads is given by

$$W_e = \int_0^L m_z(x)\theta_z(x)dx + \int_0^L q_y(x)u_y(x)dx + M_{z_L}\theta_z(L) + M_{z_0}\theta_z(0) + V_{y_L}u_y(L) + V_{y_0}u_y(0). \quad (6.23)$$

6.7   EQUILIBRIUM

The PVW states that the beam is in equilibrium in the deformed configuration if the external δW_e and internal δW_i works for any kinematically compatible virtual action $\delta\mathbf{u}(x,y)$ are equal. Thus,

$$\delta W_e = \delta W_i. \quad (6.24)$$

Substituting (6.22) and (6.23) in the PVW statement (6.24) and rearranging the terms, we have

$$\int_0^L \left[-\frac{dM_z(x)}{dx} + V_y(x) - m_z(x) \right] \delta\theta_z(x)dx + \int_0^L \left[\frac{dV_y(x)}{dx} - q_y(x) \right] \delta u_y(x)dx$$
$$+ \ \left[M_z(L) - M_{z_L} \right] \delta\theta_z(L) + \left[-M_z(0) - M_{z_0} \right] \delta\theta_z(0)$$
$$+ \ \left[-V_y(L) - V_{y_L} \right] \delta u_y(L) + \left[V_y(0) - V_{y_0} \right] \delta u_y(0) = 0,$$

with $\delta\theta_z(x)$ and $\delta u_y(x)$ respectively the rotation in z and the transversal displacement in y, relative to the virtual action $\delta\mathbf{u}(x,y)$.

As the above equation is valid for any arbitrary virtual action $\delta\mathbf{u}(x,y)$, the terms inside brackets must be zero, resulting in the local form of the equilibrium BVP of the Timoshenko beam, that is,

$$
\begin{cases}
\dfrac{dM_z(x)}{dx} - V_y(x) + m_z(x) = 0 & \text{in } x \in (0,L) \\[2mm]
\dfrac{dV_y(x)}{dx} - q_y(x) = 0 & \text{in } x \in (0,L) \\[2mm]
V_y(0) = V_{y_0} & \text{in } x = 0 \\[1mm]
V_y(L) = -V_{y_L} & \text{in } x = L \\[1mm]
M_z(0) = -M_{z_0} & \text{in } x = 0 \\[1mm]
M_z(L) = M_{z_L} & \text{in } x = L
\end{cases}
\tag{6.25}
$$

Therefore, we have two differential equations for the Timoshenko beam, coupled by the shear force $V_y(x)$. However, it is possible to obtain a single differential equation of equilibrium. For this purpose, just take the derivative of the first equation and replace it into the second one, that is,

$$
\frac{d^2 M_z(x)}{dx^2} = q_y(x) - \frac{dm_z(x)}{dx}.
\tag{6.26}
$$

Supposing that the external distributed moment $m_z(x)$ is zero, we obtain the same differential equation of equilibrium of the Euler-Bernoulli model in terms of the bending moment, that is,

$$
\frac{d^2 M_z(x)}{dx^2} = q_y(x).
\tag{6.27}
$$

Besides that, we recover the same definition of the shear force given in (5.19) from the first equation in (6.25), that is,

$$
V_y(x) = \frac{dM_z(x)}{dx}.
\tag{6.28}
$$

In this case, the solution procedure used in the examples of Chapter 5 can be employed, yielding the same shear force and bending moment diagrams of the Euler-Bernoulli beam model.

If $(\delta u_y, \delta\theta_z)$ is a rigid virtual action, the strain measures, and consequently the internal work, are zero. In this case, the PVW states that for any rigid virtual action of a Timoshenko beam in equilibrium, the external work (6.23) is zero, that is,

$$
\begin{aligned}
\delta W_e &= \int_0^L m_z(x)\delta\theta_z(x)dx + \int_0^L q_y(x)\delta u_y(x)dx + M_{z_L}\delta\theta_z(L) + M_{z_0}\delta\theta_z(0) \\
&\quad + V_{y_L}\delta u_y(L) + V_{y_0}\delta u_y(0) = 0.
\end{aligned}
\tag{6.29}
$$

As $\delta u_y(x) = \delta u_y = \text{cte}$ and $\delta\theta_z(x) = \delta\theta_z = \text{cte}$ for a rigid action, the above expression is reduced to

$$
\delta W_e = \left(\int_0^L q_y(x)dx + V_{y_L} + V_{y_0} \right)\delta u_y = 0 + \left(\int_0^L m_z(x)dx + M_{z_L} + M_{z_0} \right)\delta\theta_z.
$$

Thus, the equilibrium conditions for the Timoshenko and Euler-Bernoulli beams are the same and state that the resultants of external forces in y and moments in z must be zero. That is,

$$
\sum f_y = \int_0^L q_y(x)dx + V_{y_L} + V_{y_0} = 0,
\tag{6.30}
$$

$$
\sum m_z = \int_0^L m_z(x)dx + M_{z_L} + M_{z_0} = 0.
\tag{6.31}
$$

The integral of $q_y(x)$ along the length of the beam represents the concentrated force equivalent to the transversal distributed load. Analogously, the integral of $m_z(x)$ along the length results in the concentrated moment in z, equivalent to the distributed moment.

6.8 APPLICATION OF THE CONSTITUTIVE EQUATION

Hooke's law for a homogeneous isotropic linear elastic material states that the normal stress $\sigma_{xx}(x,y)$ is related to the specific longitudinal strain $\varepsilon_{xx}(x,y)$ by the Young's elastic modulus $E(x)$. Similarly, the shear stress $\tau_{xy}(x)$ at section x is related to the angular strain $\tilde{\gamma}_{xy}(x)$ by the shear modulus $G(x)$. Thus,

$$\sigma_{xx}(x,y) = E(x)\varepsilon_{xx}(x,y), \tag{6.32}$$

$$\tau_{xy}(x) = G(x)\tilde{\gamma}_{xy}(x). \tag{6.33}$$

Substituting the strain components expressions (6.4) and (6.9) in the above relations, we have

$$\sigma_{xx}(x.y) = -E(x)\frac{d\theta_z(x)}{dx}y, \tag{6.34}$$

$$\tau_{xy}(x) = G(x)\left(\frac{du_y(x)}{dx} - \theta_z(x)\right). \tag{6.35}$$

On the other hand, substituting (6.34) in the bending moment expression (6.11), we obtain

$$M_z(x) = E(x)\frac{d\theta_z(x)}{dx}\int_A y^2 dA = E(x)I_z(x)\frac{d\theta_z(x)}{dx}, \tag{6.36}$$

with $I_z(x) = \int_A y^2 dA$ the second moment of area of the cross-section relative to axis z of the reference system.

From the above equation, we obtain the following relation for the derivative of $\theta_z(x)$:

$$\frac{d\theta_z(x)}{dx} = \frac{M_z(x)}{E(x)I_z(x)}, \tag{6.37}$$

which, when substituted in (6.36) gives relation (5.30) for the normal stress $\sigma_{xx}(x,y)$ in terms of the bending moment $M_z(x)$ at section x, that is,

$$\sigma_{xx}(x,y) = -\frac{M_z(x)}{I_z(x)}y. \tag{6.38}$$

Thus, the normal bending stress $\sigma_{xx}(x,y)$ has a linear variation along the cross-section of the beam for both Bernoulli and Timoshenko models, as illustrated in Figure 5.21.

Analogously, when substituting (6.35) in expression (6.16) of the shear force, we obtain

$$\begin{aligned} V_y(x) &= -\int_A G(x)\left[\frac{du_y(x)}{dx} - \theta_z(x)\right]dA \\ &= -G(x)\left[\frac{du_y(x)}{dx} - \theta_z(x)\right]\int_A dA \\ &= -\left[\frac{du_y(x)}{dx} - \theta_z(x)\right]G(x)A(x), \end{aligned} \tag{6.39}$$

with $A(x)$ the area of cross-section x of the beam.

Substituting (6.1) in the previous expression, we have

$$V_y(x) = -G(x)A(x)\beta(x). \tag{6.40}$$

From this relation, the distortion $\beta(x)$ in the cross-sections of the beam for a Hookean material is given by

$$\beta(x) = -\frac{V_y(x)}{G(x)A(x)}, \tag{6.41}$$

which, when substituted in (6.35), yields the expression for the shear stress $\tau_{xy}(x)$ in terms of the shear force $V_y(x)$ in the section, that is,

$$\tau_{xy}(x) = -\frac{V_y(x)}{A(x)}. \tag{6.42}$$

Thus, the shear stress $\tau_{xy}(x)$ in the Timoshenko beam model is constant on each section and represents an average stress, because it divides the shear force $V_y(x)$ by the area $A(x)$. This behavior is illustrated in Figure 6.11 and is in accordance with the hypothesis that $\beta(x)$ is constant in the cross-section. The negative sign indicates that the positive shear force is downwards, contrary to axis y, as illustrated in Figure 6.11(b). In addition, the shear stress and shear force have the same direction.

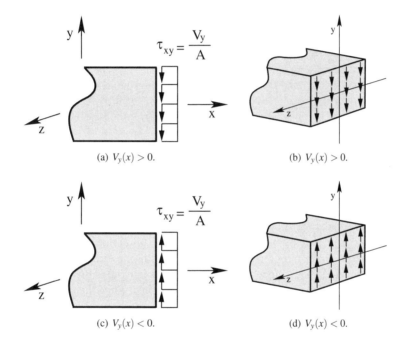

Figure 6.11 Shear stress in the Timoshenko beam.

One of the typical applications of the mean shear stress given in (6.42) is the design of fasteners like rivets and screws, as illustrated in the following example.

Example 6.2 *Figure 6.12(a) illustrates the cross-section of three plates connected by a rivet and subjected to a traction force $F = 5000$ N. Design the rivet knowing that the admissible shear stress is $\bar{\tau} = 20$ MPa.*

As illustrated in Figures 6.12(b) and 6.12(c), the rivet cross-sections are subjected to a shear stress due to the force applied on the plates. This shear stress distribution can be assumed to be uniform and homogeneous in the rivet cross-sections in contact with the plates.

The shear stress is calculated dividing the force F by the area below the shear. In this case, as sections A and B are under shear, the total area is 2A, with A the area of the rivet. Thus,

$$\tau = \frac{F}{2A}.$$

When designing for the maximum stress, the previous expression must be equal to the admissible shear stress of the rivet, that is,

$$\tau = \frac{F}{2A} = \bar{\tau} \rightarrow A = \frac{F}{2\bar{\tau}}.$$

Substituting the values, we obtain $A = 2.50$ cm^2 and the minimum diameter of the rivet to maintain it in the elastic range is

$$d = \sqrt{\frac{4A}{\pi}} = 1.78 \text{ cm}.$$

□

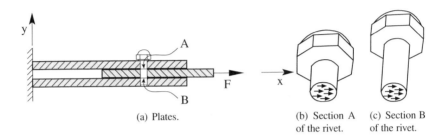

| (a) Plates. | (b) Section A of the rivet. | (c) Section B of the rivet. |

Figure 6.12 Example 6.2: plates connected by a rivet.

All problems considered until here are treated by one-dimensional models. Thus, the internal and external loads illustrated in Figure 6.10 for the Timoshenko beam are applied along the x axis of the adopted reference system, which passes in the geometric centers of the cross-sections. Therefore, the upper and lower ends of the cross-section are free surfaces and no external shear forces are present. Therefore, the shear force $V_y(x)$ and hence the shear stress $\tau_{xy}(x)$ are zero at these ends.

Thus, the constant stress distribution given in equation (6.42) does not properly represent the shear stress in the beam. To correct this limitation, we introduce a dimensionless correction factor called shear factor, denoted by K_c. Since the shear force is the integral of the shear stress in the cross-section, a factor $K_c < 1$ is employed, which yields a lower shear force than (6.40) when multiplied by the shear stress.

Thus, using K_c in (6.40) and (6.42), the shear forces and stresses are respectively given by

$$V_y(x) = -K_c(x)G(x)A(x)\beta(x), \tag{6.43}$$

$$\tau_{xy}(x) = -\frac{V_y(x)}{K_c(x)A(x)}. \tag{6.44}$$

Consequently, the distortion is determined by

$$\beta(x) = -\frac{V_y(x)}{K_c(x)G(x)A(x)}, \tag{6.45}$$

The term K_c is here defined as the ratio between the mean shear strain $\dfrac{V_y(x)}{A(x)G(x)}$ and the effective strain $\bar{\gamma}_{xy}(x)$ [50, 51]. Thus,

$$K_c(x) = \frac{V_y(x)}{A(x)G(x)} \frac{1}{\bar{\gamma}_{xy}(x)}. \tag{6.46}$$

This factor will be determined for rectangular and circular cross-sections in the following section. There are other ways to calculate K_c, according to references [19, 34].

We can now substitute (6.36) and (6.43) in the differential equations indicated in (6.25), obtaining the following equations in terms of rotation $\theta_z(x)$ and transversal displacement $u_y(x)$:

$$\frac{d}{dx}\left(E(x)I_z(x)\frac{d\theta_z(x)}{dx}\right) + K_c(x)G(x)A(x)\left(\frac{du_y(x)}{dx} - \theta_z(x)\right) + m_z(x) = 0, \tag{6.47}$$

$$-\frac{d}{dx}\left[K_c(x)G(x)A(x)\left(\frac{du_y(x)}{dx} - \theta_z(x)\right)\right] - q_y(x) = 0. \tag{6.48}$$

The previous expressions constitute a system of coupled differential equations, because the shear force $V_y(x) = K_cG(x)A(x)\left(\frac{du_y(x)}{dx} - \theta_z(x)\right)$ is present in both equations. Supposing that the distributed moment $m_z(x)$ is zero and the material and cross-section parameters are constant ($E(x) = E$, $I_z(x) = I_z$, $A(x) = A$, $G(x) = G$), we have

$$EI_z\frac{d^2\theta_z(x)}{dx^2} + K_cGA\left(\frac{du_y(x)}{dx} - \theta_z(x)\right) = 0, \tag{6.49}$$

$$-K_cGA\left(\frac{d^2u_y(x)}{dx^2} - \frac{d\theta_z(x)}{dx}\right) - q_y(x) = 0. \tag{6.50}$$

This system of differential equations can be written as two independent equations for $u_y(x)$ and $\theta_z(x)$. For this purpose, the derivative of (6.49) is taken

$$EI_z\frac{d^3\theta_z(x)}{dx^3} + K_cGA\left(\frac{d^2u_y(x)}{dx^2} - \frac{d\theta_z(x)}{dx}\right) = 0$$

and the obtained expression is rewritten as

$$K_cGA\left(\frac{d^2u_y(x)}{dx^2} - \frac{d\theta_z(x)}{dx}\right) = -EI_z\frac{d^3\theta_z(x)}{dx^3}.$$

Substituting in (6.50), we obtain

$$EI_z\frac{d^3\theta_z(x)}{dx^3} - q_y(x) = 0. \tag{6.51}$$

The effect of this procedure is to eliminate the second term of equation (6.49), which refers to the shear force in the beam. Notice that if the shear is zero, that is, $\beta(x) = 0$, then $\theta_z(x) = \alpha(x) = \frac{du_y(x)}{dx}$. Substituting this result in the above expression, we obtain the fourth-order differential equation in terms of $u_y(x)$ given by (5.37) for the bending only model.

The equation in terms of $u_y(x)$ is obtained by differentiating (6.50) twice

$$-K_cGA\left(\frac{d^4u_y(x)}{dx^4} - \frac{d^3\theta_z(x)}{dx^3}\right) - \frac{d^2q_y(x)}{dx^2} = 0,$$

and introducing $\frac{d^3\theta_z(x)}{dx^3} = \frac{1}{EI_z}q_y(x)$, obtained from (6.51), we have

$$-K_cGA\frac{d^4u_y(x)}{dx^4} + \frac{K_cGA}{EI_z}q_y(x) - \frac{d^2q_y(x)}{dx^2} = 0.$$

Multiplying the above equation by factor $-\dfrac{EI_z}{K_cGA}$, the final equation is determined:

$$EI_z\frac{d^4u_y(x)}{dx^4} - \frac{EI_z}{K_cGA}\frac{d^2q_y(x)}{dx^2} + q_y(x) = 0. \tag{6.52}$$

All following examples consider that the distributed bending moment is zero, that is, $m_z(x) = 0$.

Example 6.3 *Consider the beam of Example 6.1. We want to obtain the expression of the shear force, bending moment, rotation, and transversal displacement considering the shear effect.*

The solution procedure is the same used in the previous chapters, but we should integrate both differential equations given in (6.47) and (6.48).

- *Load equation*
  As the distributed transversal load is zero, we have $q_y(x) = 0$.
- *Boundary conditions*
  Because of the clamp, the rotation and transversal displacement are zero in the left end, that is, $\theta_z(x = 0) = 0$ and $u_y(x = 0) = 0$. At the right end, the bending moment is zero and the transversal displacement must be equal to the applied displacement δ, that is, $M_z(x = L) = 0$ and $u_y(x = L) = \delta$.
- *Differential equation integration*
  In this case, differential equations (6.47) and (6.48) are reduced to

$$EI_z\frac{d^2\theta_z(x)}{dx^2} + K_cGA\left(\frac{du_y(x)}{dx} - \theta_z(x)\right) = -m_z(x) = 0,$$

$$-\frac{d}{dx}\left[K_cGA\left(\frac{du_y(x)}{dx} - \theta_z(x)\right)\right] = q_y(x) = 0.$$

Integrating the second equation above, we obtain the shear force expression

$$V_y(x) = -K_cGA\left(\frac{du_y(x)}{dx} - \theta_z(x)\right) = C_1. \tag{6.53}$$

Substituting $V_y(x)$ in the first differential equation, we have

$$EI_z\frac{d^2\theta_z(x)}{dx^2} = C_1.$$

The first integration of this equation yields the bending moment expression

$$M_z(x) = EI_z\frac{d\theta_z(x)}{dx} = C_1x + C_2.$$

The second integration results in the bending rotation, that is,

$$\theta_z(x) = \frac{1}{EI_z}\left(\frac{C_1}{2}x^2 + C_2x + C_3\right).$$

Substituting the rotation in (6.53), we obtain the following expression for the derivative of $u_y(x)$:

$$\frac{du_y(x)}{dx} = -\frac{C_1}{K_cGA} + \frac{1}{EI_z}\left(\frac{C_1}{2}x^2 + C_2x + C_3\right).$$

Integrating the above equation, we have

$$u_y(x) = \left(-\frac{x}{K_cGA} + \frac{x^3}{6EI_z}\right)C_1 + \frac{x^2}{2EI_z}C_2 + \frac{x}{EI_z}C_3 + C_4.$$

- Determination of the integration constants

  The integration constants C_1 to C_4 are determined with the application of boundary conditions at the ends of the beam. Thus,

$$\theta_z(x=0) = \frac{1}{EI_z}\left(\frac{C_1}{2}(0) + C_2(0) + C_3\right) = 0 \rightarrow C_3 = 0,$$

$$u_y(x=0) = (0)C_1 + (0)C_2 + (0)C_3 + C_4 = 0 \rightarrow C_4 = 0,$$

$$M_z(x=L) = C_1(L) + C_2 = 0 \rightarrow C_2 = -C_1 L,$$

$$u_y(x=L) = \left(-\frac{L}{K_c GA} + \frac{L^3}{6EI_z}\right)C_1 + \frac{L^2}{2EI_z}C_2 = \delta.$$

Substituting $C_2 = -C_1 L$ in the above equation, we have

$$C_1 = -\frac{\delta}{\dfrac{L^3}{3EI_z} + \dfrac{L}{K_c GA}}.$$

Thus,

$$C_2 = -C_1 L = \frac{\delta}{\dfrac{L^2}{3EI_z} + \dfrac{1}{K_c GA}}.$$

- Final equations

  Substituting C_1 to C_4, we have the final expressions for the shear force, bending moment, rotation, and transversal displacement, which are respectively given by

$$V_y(x) = -\frac{\delta}{\dfrac{L^3}{3EI_z} + \dfrac{L}{K_c GA}},$$

$$M_z(x) = C_1(x-L) = -\frac{\delta}{\dfrac{L^3}{3EI_z} + \dfrac{L}{K_c GA}}(x-L),$$

$$\theta_z(x) = \frac{C_1}{2EI_z}\left(x^2 - 2Lx\right) = -\frac{\delta}{\dfrac{2L^3}{3} + \dfrac{2LEI_z}{K_c GA}}\left(x^2 - 2Lx\right),$$

$$u_y(x) = \left[-\frac{x}{K_c GA} + \frac{(x-3L)x^2}{6EI_z}\right]C_1 = -\frac{\delta}{\dfrac{L^3}{3EI_z} + \dfrac{L}{K_c GA}}\left[-\frac{x}{K_c GA} + \frac{(x-3L)x^2}{6EI_z}\right].$$

The shear deformation is calculated from (6.45)

$$\beta(x) = -\frac{\delta}{\dfrac{K_c GAL^3}{3EI_z} + L}.$$

Neglecting the shear effect, the integration constant C_1 is reduced to

$$C_1 = -\frac{3EI_z}{L^3}\delta.$$

Substituting this expression for C_1, we recover the same equations of shear force, bending moment, rotation, and transversal displacement for bending only obtained in Example 5.7, that is,

$$V_y(x) = -\frac{3EI_z}{L^3}\delta,$$

$$M_z(x) = -\frac{3EI_z}{L^3}\delta(x-L),$$

$$\theta_z(x) = -\frac{3EI_z}{2L^3}\delta(x-2L)x,$$

$$u_y(x) = -\frac{EI_z}{2L^3}\delta(x-3L)x^2.$$

Notice that in this case, $\theta_z(x) = \dfrac{du_y(x)}{dx}$ and consequently $\beta(x) = 0$.

· *Diagrams*
  The diagrams for the shear force, bending moment, rotation, and transversal displacement considering the shear effect are illustrated in Figure 6.13 for a steel beam ($E = 210$ GPa, $G = 80.8$ GPa) with base $b = 0.1$ m, height $h = 0.2$ m, $K_c = \frac{2}{3}$, and length $L = 1$ m.

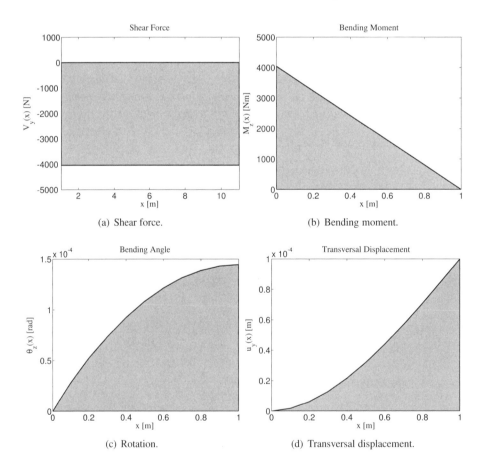

(a) Shear force.

(b) Bending moment.

(c) Rotation.

(d) Transversal displacement.

Figure 6.13 Example 6.3: Shear force, bending moment, rotation, and transversal displacement diagrams.

Figure 6.14 compares the rotation and transversal displacement for the beam considering the effect of shear and bending only. The maximum relative difference for the transversal

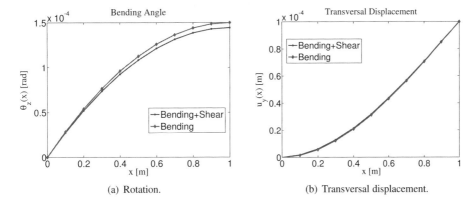

(a) Rotation. (b) Transversal displacement.

Figure 6.14 Example 6.3: rotation and transversal displacement diagrams for the beam with and without shear.

displacement with and without shear is 18.1%. By increasing the length to $L = 2$ m and $L = 10$ m, the relative differences are respectively 10.8% and 2.5%. Thus, the shear effect is less important for longer beams.

- *Support reactions*
  The support reactions are determined from the boundary conditions in terms of the shear force and bending moment as

$$R_{Ay} = V_y(x = 0) = -C_1 = -4042.41 \text{ N} \quad and \quad M_{Az} = -M_z(x = 0) = C_2 = 4042.41 \text{ Nm}.$$

□

Example 6.4 *Consider the simply supported beam of Example 5.4 subjected to a constant distributed load. Draw the diagrams for the shear force, bending moment, rotation, and transversal displacement considering the shear effect.*

The solution procedure is the same used in the previous example, with the integration of differential equations (6.47) and (6.48).

- *Load equation*
  As the distributed transversal load is constant, we have $q_y(x) = -q_0$.
- *Boundary conditions*
  The bending moment and the transversal displacement are zero at both ends of a simply supported beam. Thus, $M_z(x = 0) = 0$, $u_y(x = 0) = 0$, $M_z(x = L) = 0$, and $u_y(x = L) = 0$.
- *Integration of the differential equations*
  Differential equations (6.47) and (6.48) are reduced, in this case, to

$$EI_z \frac{d^2\theta_z(x)}{dx^2} + K_c GA \left(\frac{du_y(x)}{dx} - \theta_z(x) \right) = -m_z(x) = 0,$$

$$-\frac{d}{dx} \left[K_c GA \left(\frac{du_y(x)}{dx} - \theta_z(x) \right) \right] = q_y(x) = -q_0.$$

Integrating the second equation above, we obtain the shear force expression

$$V_y(x) = -K_c GA \left(\frac{du_y(x)}{dx} - \theta_z(x) \right) = -q_0 x + C_1. \tag{6.54}$$

Substituting $V_y(x)$ in the first differential equation, we have

$$EI_z \frac{d^2\theta_z(x)}{dx^2} = -q_0 x + C_1.$$

The first integration yields the bending moment expression

$$M_z(x) = EI_z \frac{d\theta_z(x)}{dx} = -\frac{q_0}{2}x^2 + C_1 x + C_2.$$

The second integration results in the rotation, that is,

$$\theta_z(x) = \frac{1}{EI_z}\left(-\frac{q_0}{6}x^3 + \frac{C_1}{2}x + C_2 x + C_3\right).$$

Substituting the rotation in (6.54), we obtain the following expression for the derivative of $u_y(x)$:

$$\frac{du_y(x)}{dx} = -\frac{-q_0 x + C_1}{K_c GA} + \frac{1}{EI_z}\left(-\frac{q_0}{6}x^3 + \frac{C_1}{2}x^2 + C_2 x + C_3\right).$$

Integrating this equation, we have

$$u_y(x) = \frac{q_0}{2K_c GA}x^2 - \frac{q_0}{24EI_z}x^4 + \left(-\frac{x}{K_c GA} + \frac{x^3}{6EI_z}\right)C_1 + \frac{x^2}{2EI_z}C_2 + \frac{x}{EI_z}C_3 + C_4.$$

- *Determination of the integration constants*
  Applying the boundary conditions at the ends of the beam, we determine the integration constants C_1 to C_4. Thus,

$$\begin{aligned}
M_z(x=0) &= C_1(0) + C_2 = 0 \to C_2 = 0, \\
u_y(x=0) &= (0)C_1 + (0)C_2 + (0)C_3 + C_4 = 0 \to C_4 = 0, \\
M_z(x=L) &= -\frac{q_0}{2}L^2 + C_1(L) = 0 \to C_1 = \frac{q_0 L}{2}, \\
u_y(x=L) &= \frac{q_0 L^2}{2K_c GA} - \frac{q_0 L^4}{24EI_z} + \left(-\frac{L}{K_c GA} + \frac{L^3}{6EI_z}\right)\frac{q_0 L}{2} \\
&+ \frac{L}{EI_z}C_3 = 0 \to C_3 = -\frac{q_0 L^3}{24}.
\end{aligned}$$

- *Final equations*
  Substituting the integration constants, we have the final expressions for the shear force, bending moment, rotation, and transversal displacement, respectively, given by

$$V_y(x) = \frac{q_0}{2}(L - 2x),$$

$$M_z(x) = \frac{q_0}{2}(L - x)x,$$

$$\theta_z(x) = \frac{q_0}{24EI_z}\left(-4x^3 + 6Lx^2 - L^3\right),$$

$$u_y(x) = \frac{q_0}{2K_c GA}(L - x)x + \frac{q_0}{24EI_z}\left(-x^4 + 2Lx^3 - L^3 x\right).$$

Neglecting the shear effect, we recover the same equations for the shear force, bending moment, rotation, and transversal displacement for the beam under bending only, that is,

$$V_y(x) = \frac{q_0}{2}(L - 2x),$$

$$M_z(x) = \frac{q_0}{2}(L-x)x,$$

$$\theta_z(x) = \frac{q_0}{24EI_z}\left(-4x^3 + 6Lx^2 - L^3\right),$$

$$u_y(x) = \frac{q_0}{24EI_z}\left(-x^4 + 2Lx^3 - L^3x\right).$$

- *Diagrams*
  The diagrams for the shear force, bending moment, rotation, and transversal displacement considering the shear effect are illustrated in Figure 6.15 for a steel beam ($E = 210$ GPa, $G = 80.8$ GPa), with rectangular cross-section of base $b = 0.1$ m, height $h = 0.2$ m, $K_c = \frac{2}{3}$, length $L = 2$ m and a distributed load of intensity $q_0 = 1000$ N/m.

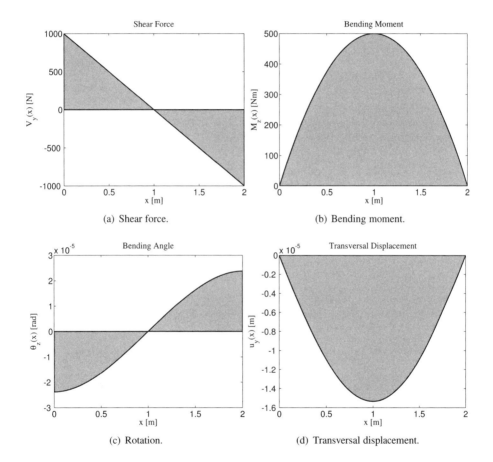

(a) Shear force.

(b) Bending moment.

(c) Rotation.

(d) Transversal displacement.

Figure 6.15 Example 6.4: diagrams for the shear force, bending moment, rotation, and transversal displacement.

- *Support reactions*
  The support reactions are determined from the boundary conditions in terms of shear force as

$$R_{Ay} = V_y(x=0) = 1000 \text{ N} \quad and \quad R_{By} = -V_y(x=L) = 1000 \text{ N}.$$

File beamshearsolver.m solves this example and the previous one employing the symbolic manipulation toolkit available in MATLAB. We should just take care and use the proper load expressions and boundary conditions for each example.

□

Example 6.5 *Consider the cantilever beam of Figure 6.16 subjected to a constant distributed load and a concentrated load in the central section. Draw the diagrams for the shear force, bending moment, rotation, and transversal displacement considering the shear effect.*

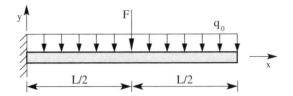

Figure 6.16   Example 6.5: beam with distributed load and concentrated force.

The solution procedure is the same used in the previous examples, with the integration of differential equations (6.47) and (6.48).

- *Load equation*
  As the transversal distributed load is constant, and using the notation in terms of singularity functions for the concentrated force, we have

$$q_y(x) = -q_0 - F < x - \frac{L}{2} >^{-1} .$$

- *Boundary conditions*
  As the beam is of cantilever type, the rotation and transversal displacement are zero at the left end, that is, $\theta_z(x = 0) = 0$ and $u_y(x = 0) = 0$. On the other hand, the shear force and the bending moment are zero at the right end, that is, $V_y(x = L) = 0$ and $M_z(x = L) = 0$.
- *Integration of the differential equations*
  Differential equations (6.47) and (6.48) are reduced in this case to

$$EI_z \frac{d^2 \theta_z(x)}{dx^2} + K_c GA \left(\frac{du_y(x)}{dx} - \theta_z(x) \right) = 0,$$

$$- \frac{d}{dx} \left[K_c GA \left(\frac{du_y(x)}{dx} - \theta_z(x) \right) \right] = -q_0 - F < x - \frac{L}{2} >^{-1} .$$

Integrating the second equation above, we obtain the expression for the shear force

$$V_y(x) = -K_c GA \left(\frac{du_y(x)}{dx} - \theta_z(x) \right) = -q_0 x - F < x - \frac{L}{2} >^0 + C_1. \qquad (6.55)$$

Substituting $V_y(x)$ in the first differential equation, we have

$$EI_z \frac{d^2 \theta_z(x)}{dx^2} = -q_0 x - F < x - \frac{L}{2} >^0 + C_1.$$

The first integration gives the bending moment expression

$$M_z(x) = EI_z \frac{d\theta_z(x)}{dx} = -\frac{q_0}{2} x^2 - F < x - \frac{L}{2} >^1 + C_1 x + C_2.$$

The second integration results in the equations for the bending rotation, that is,

$$\theta_z(x) = \frac{1}{EI_z} \left(-\frac{q_0}{6} x^3 - \frac{F}{2} < x - \frac{L}{2} >^2 + \frac{C_1}{2} x^2 + C_2 x + C_3 \right).$$

Substituting the rotation in (6.55), we obtain the following expression for the derivative of $u_y(x)$:

$$\frac{du_y(x)}{dx} = -\frac{1}{K_c GA}\left(-q_0 x - F < x - \frac{L}{2} >^0 + C_1\right)$$
$$+ \frac{1}{EI_z}\left(-\frac{q_0}{6}x^3 - \frac{F}{2} < x - \frac{L}{2} >^2 + \frac{C_1}{2}x^2 + C_2 x + C_3\right).$$

Integrating the above equation, we obtain

$$u_y(x) = \frac{q_0}{2K_c GA}x^2 + \frac{F}{K_c GA} < x - \frac{L}{2} >^1 - \frac{q_0}{24EI_z}x^4 - \frac{F}{6EI_z} < x - \frac{L}{2} >^3$$
$$+ \left(-\frac{x}{K_c GA} + \frac{x^3}{6EI_z}\right)C_1 + \frac{x^2}{2EI_z}C_2 + \frac{x}{EI_z}C_3 + C_4.$$

- *Determination of the integration constants*
  Applying the boundary conditions at the ends of the beam, we find the integration constants C_1 to C_4. Thus,

$$\theta_z(x=0) = \frac{1}{EI_z}\left(-\frac{q_0}{6}(0) - \frac{F}{2}(0) + \frac{C_1}{2}(0) + C_2(0) + C_3\right) = 0 \rightarrow C_3 = 0,$$

$$u_y(x=0) = \frac{q_0}{2K_c GA}(0) + \frac{F}{K_c GA}(0) + -\frac{q_0}{24EI_z}(0)^4 - \frac{F}{6EI_z}(0)$$
$$+ \left(-\frac{0}{K_c GA} + \frac{0}{6EI_z}\right)C_1 + \frac{0}{2EI_z}C_2 + \frac{0}{EI_z}C_3 + C_4 = 0 \rightarrow C_4 = 0,$$

$$V_y(x=L) = -q_0(L) - F\left(L - \frac{L}{2}\right)^0 + C_1 = 0 \rightarrow C_1 = q_0 L + F,$$

$$M_z(x=L) = -\frac{q_0}{2}L^2 - F\left(L - \frac{L}{2}\right)^1 + C_1 L + C_2 = 0 \rightarrow C_2 = -\frac{q_0 L^2}{2} - \frac{FL}{2},$$

- *Final equations*
  When substituting C_1 to C_4, we obtain the final expressions for the shear force, bending moment, rotation, and transversal displacement, which are respectively given by

$$V_y(x) = -q_0 x - F < x - \frac{L}{2} >^0 + q_0 L + F,$$

$$M_z(x) = -\frac{q_0}{2}x^2 - F < x - \frac{L}{2} >^1 + (q_0 L + F)x - \frac{q_0 L^2}{2} - \frac{FL}{2},$$

$$\theta_z(x) = \frac{1}{EI_z}\left[-\frac{q_0}{6}x^3 - \frac{F}{2} < x - \frac{L}{2} >^2 + \frac{q_0 L + F}{2}x^2 - \left(\frac{q_0 L^2}{2} + \frac{FL}{2}\right)x\right],$$

$$u_y(x) = \frac{q_0}{2K_c GA}x^2 + \frac{F}{K_c GA} < x - \frac{L}{2} >^1 - \frac{q_0 L + F}{K_c GA}x$$
$$- \frac{q_0}{24EI_z}x^4 - \frac{F}{6EI_z} < x - \frac{L}{2} >^3 + \frac{q_0 L + F}{6EI_z}x^3 - \frac{(q_0 L^2 + FL)}{4EI_z}x^2.$$

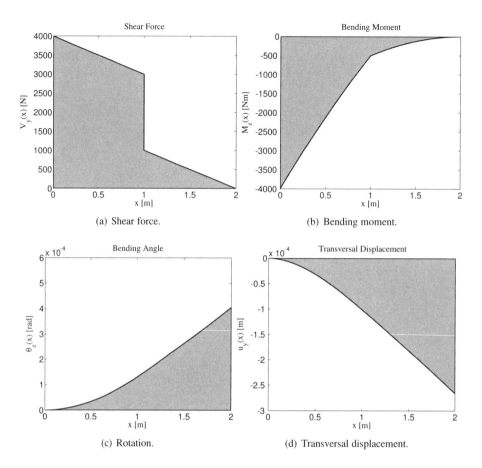

(a) Shear force.

(b) Bending moment.

(c) Rotation.

(d) Transversal displacement.

Figure 6.17 Example 6.5: diagrams for the shear force, bending moment, rotation, and transversal displacement.

- *Diagrams*
  The diagrams for the shear force, bending moment, rotation, and transversal displacement are illustrated in Figure 6.17 for a steel beam ($E = 210$ GPa, $G = 80.8$ GPa) with rectangular cross-section of base $b = 0.1$ m, height $h = 0.2$ m, $K_c = \frac{2}{3}$, length $L = 2$ m, distributed load of intensity $q_0 = 1000$ N/m, and a concentrated load with $F = 2000$ N.
- *Support reactions*
  The support reactions are determined from the boundary conditions in terms of shear force and bending moment as

$$R_{Ay} = V_y(x = 0) = 4000 \text{ N} \quad and \quad M_{Az} = M_z(x = 0) = -4000 \text{ Nm}.$$

□

6.9 SHEAR STRESS DISTRIBUTION

As illustrated in Figure 6.11, the assumed kinematic hypotheses for the Timoshenko beam induce a constant shear stress distribution in the cross-sections. However, this distribution is inconsistent with the warping of sections, as illustrated in Figure 6.3.

To minimize this model limitation, we introduced a shear factor K_c [see equation (6.46)]. The derivation of the exact shear strain, which depends on the warping is described in [51]. Our intention here is to assume a certain variation for the shear stress, in order to obtain a closer distribution to the actual one, when compared to the constant distribution of the Timoshenko model. The cases of rectangular, circular, and I-shaped cross-sections are considered below.

6.9.1 RECTANGULAR CROSS-SECTION

In this section we assume that the shear stress is parallel to the shear force in the vertical direction.

Figure 6.18(a) illustrates the constant shear stress distribution on a beam with rectangular cross-section of base b and height h. As this distribution is incorrect, we initially assume a linear variation for the stress with y at section x of the beam, that is,

$$\tau_{xy}(x,y) = c_1 y + c_2. \tag{6.56}$$

The coefficients c_1 and c_2 are determined knowing that the shear stress is zero at the cross-section upper and lower boundaries. Thus,

$$\begin{cases} \tau_{xy}(x, y = -\frac{h}{2}) = -c_1 \left(\frac{h}{2} \right) + c_2 = 0 \\ \tau_{xy}(x, y = \frac{h}{2}) = c_1 \left(\frac{h}{2} \right) + c_2 = 0 \end{cases} . \tag{6.57}$$

Solving the above system of equations, we obtain $c_1 = c_2 = 0$, and thus $\tau_{xy}(x,y) = 0$. As the shear stress is not necessarily zero at the cross-section, the linear distribution is ruled out for τ_{xy}.

A quadratic variation with the y coordinate is then assumed for the shear stress, that is,

$$\tau_{xy}(x,y) = -c_1 y^2 + c_2 y + c_3. \tag{6.58}$$

The coefficient c_1 is taken negative, because the concavity of the parabola is downwards, as illustrated in Figure 6.18(b). Three conditions are necessary to determine the constants c_1, c_2, and c_3. The first two are the same as the previous ones, that is, the shear stress is zero at the cross-section ends. Thus,

$$\begin{cases} \tau_{xy}(x, y = -\frac{h}{2}) = -c_1 \left(\frac{h}{2} \right)^2 - c_2 \left(\frac{h}{2} \right) + c_3 = 0 \\ \tau_{xy}(x, y = \frac{h}{2}) = -c_1 \left(\frac{h}{2} \right)^2 + c_2 \left(\frac{h}{2} \right) + c_3 = 0 \end{cases} . \tag{6.59}$$

Taking the sum of the above expressions, we have

$$-2c_1 \left(\frac{h}{2} \right)^2 + 2c_3 = 0.$$

Hence, we determine following relation between c_1 and c_3:

$$c_3 = c_1 \left(\frac{h}{2} \right)^2 . \tag{6.60}$$

The third necessary condition is obtained from the symmetry of the rectangular cross-section relative to the z axis of the reference system. As the shear stress is zero at both ends and has a parabolic variation, we have a point of maximum or minimum at the center of the cross-section, which in this case is $y = 0$. This condition implies that the first derivative of $\tau_{xy}(x,y)$, that is,

$$\frac{\partial \tau_{xy}(x,y)}{\partial y} = -2c_1 y + c_2$$

is zero for $y = 0$. Thus,

$$\frac{\partial \tau_{xy}(x,0)}{\partial y} = -2c_1(0) + c_2 = 0 \rightarrow c_2 = 0. \tag{6.61}$$

Substituting (6.60) and (6.61) in (6.58), we have

$$\tau_{xy}(x,y) = c_1\left[\left(\frac{h}{2}\right)^2 - y^2\right]. \tag{6.62}$$

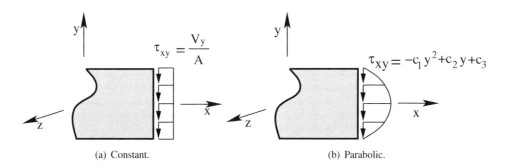

(a) Constant.   (b) Parabolic.

Figure 6.18 Shear stress distributions in a rectangular cross-section of the Timoshenko beam.

The constant c_1 is indeterminate, because the maximum shear stress is still unknown. To solve this, we know that the shear force $V_y(x)$ is obtained from the solution of the BVP (6.25). Hence, we can express c_1 in terms of V_y. Substituting (6.62) in (6.16), we obtain

$$\begin{aligned}
V_y(x) &= -\int_A \tau_{xy}(x,y)dA = -c_1\int_A\left[\left(\frac{h}{2}\right)^2 - y^2\right]dA\\
&= -c_1\left[\left(\frac{h}{2}\right)^2\int_A dA - \int_A y^2 dA\right].
\end{aligned}$$

The integrals in the above equation respectively represent the area $A(x)$ and second moment of area $I_z(x)$ of the cross-section relative to the z axis of the reference system. Thus, the coefficient c_1 can be expressed in terms of V_y as

$$c_1 = -\frac{V_y(x)}{\left(\dfrac{h}{2}\right)^2 A(x) - I_z(x)}. \tag{6.63}$$

Substituting (6.63) in (6.62), we obtain the shear stress $\tau_{xy}(x,y)$ at section x with a parabolic variation with coordinate y, that is,

$$\tau_{xy}(x,y) = -\frac{V_y(x)}{\left(\dfrac{h}{2}\right)^2 A(x) - I_z(x)}\left[\left(\frac{h}{2}\right)^2 - y^2\right]. \tag{6.64}$$

Figure 6.19 illustrates the parabolic shear stress variation in the cross-section. The stress is zero at the ends and assumes a maximum value at the geometrical center. Also notice that the shear stress $\tau_{xy}(x,y)$ has the same direction as the shear force $V_y(x)$.

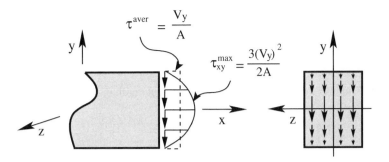

Figure 6.19 Parabolic shear stress distribution in the rectangular cross-section of the Timoshenko beam.

To determine the maximum shear stress, consider the denominator of equation (6.64). Recalling that $A(x) = bh$ and $I_z(x) = \dfrac{bh^3}{12}$ for a rectangular section, we obtain

$$\left(\frac{h}{2}\right)^2 A(x) - I_z(x) = \left(\frac{h}{2}\right)^2 bh - \frac{bh^3}{12} = \frac{bh^3}{4} - \frac{bh^3}{12} = 2I_z(x). \tag{6.65}$$

Thus, using the above relation in (6.64) we have

$$\tau_{xy}(x,y) = -\frac{V_y(x)}{2I_z(x)}\left[\left(\frac{h}{2}\right)^2 - y^2\right]. \tag{6.66}$$

The maximum shear stress $\tau_{xy}^{\max}$ at section x is obtained taking $y = 0$. Thus,

$$\tau_{xy}^{\max}(x,y) = \tau_{xy}(x, y = 0) = -\frac{V_y(x)}{2I_z(x)}\frac{h^2}{4} = -\frac{V_y(x)}{2\dfrac{bh^3}{12}}\frac{h^2}{4} = -\frac{12}{8}\frac{V_y(x)}{A(x)},$$

that is,

$$\tau_{xy}^{\max}(x,y) = -\frac{3}{2}\frac{V_y(x)}{A(x)}. \tag{6.67}$$

Hence, the maximum shear stress on a rectangular section is 50% greater than the initially obtained mean shear stress for the Timoshenko beam model. Notice that the negative sign only indicates the direction of the shear force in the section.

The precision of the parabolic shear stress distribution in the rectangular section depends of the ratio between the height h and base b. For narrower beams, where h is much greater than b $\left(\dfrac{h}{b} > 4\right)$, expression (6.66) is practically exact. When b increases, the precision of the parabolic expression decreases. For square sections, the maximum exact shear stress obtained by the solution of the complete elasticity equations is approximately 13% greater than obtained using (6.66) [27].

Now consider the hatched area in the rectangular cross-section shown in Figure 6.20(a). We want to calculate the static moment $M_{s_z}(x,y)$ of the hatched area relative to the z axis of the reference system. By definition, we have that

$$M_{s_z}(x,y) = \int_A y\,dA. \tag{6.68}$$

Taking $dA = b\,dy$, as illustrated in Figure 6.20(a), the static moment is given by

$$M_{s_z}(x,y) = \int_y^{\frac{h}{2}} by\,dy = b\int_y^{\frac{h}{2}} y\,dy = b\left.\frac{y^2}{2}\right|_y^{\frac{h}{2}} = \frac{b}{2}\left[\left(\frac{h}{2}\right)^2 - y^2\right]. \tag{6.69}$$

Another way to calculate $M_{s_z}(x)$ is from Figure 6.20(b), that is,

$$M_{s_z}(x,y) = A\bar{y}, \tag{6.70}$$

with $A = b\left(\dfrac{h}{2} - y\right)$ the considered area and $\bar{y} = y + \dfrac{1}{2}\left[\dfrac{h}{2} - y\right] = \dfrac{1}{2}\left[\dfrac{h}{2} + y\right]$ the distance between the centroid of the hatched area and the centroid of the rectangular section. Thus,

$$M_{s_z}(x,y) = b\left(\frac{h}{2} - y\right)\frac{1}{2}\left(\frac{h}{2} + y\right) = \frac{b}{2}\left[\left(\frac{h}{2}\right)^2 - y^2\right]. \tag{6.71}$$

As expected, expressions (6.69) and (6.71) are identical and represent the static moment of an area which is y units distant from the centroid, which in turn is x units distant from the origin of the adopted reference system.

From (6.71), we have

$$\left[\left(\frac{h}{2}\right)^2 - y^2\right] = \frac{2M_{s_z}(x,y)}{b}.$$

Substituting the above expression in (6.66), we obtain

$$\tau_{xy}(x,y) = -\frac{V_y(x)M_{s_z}(x,y)}{bI_z(x)}. \tag{6.72}$$

This expression is commonly obtained indirectly from the beam in bending only in the strength of materials literature, as in [42, 27].

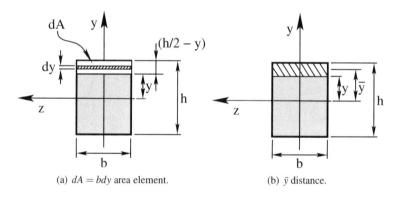

(a) $dA = bdy$ area element. (b) $\bar{y}$ distance.

Figure 6.20 Static moment of area of the rectangular section.

The shear flow $q_c(x,y)$ at section x of an area which is y units distant from the centroid is defined by

$$q_c(x,y) = -\frac{V_y(x)M_s(x,y)}{I_z(x)}. \tag{6.73}$$

By supposing that the force is given in newtons and the length in meters, the shear flow has the following units in the International System:

$$[q_c(x,y)] = \frac{\text{Nm}^3}{\text{m}} = \frac{\text{N}}{\text{m}}.$$

Thus, $q_c(x,y)$ represents the shear force in the vertical direction by units of length that the area above y can transmit at section x. The main use of this concept is in the study of cross-sections built from the union of elements with simpler geometry, as will be illustrated below.

Substituting (6.73) in (6.72), we rewrite the shear flow as

$$\tau_{xy}(x,y) = \frac{q_c(x,y)}{b}. \tag{6.74}$$

The shear coefficient was defined in expression (6.46) as the ratio between the average strain $\bar{\gamma}_{xy}(x) = \tau_{xy}(x)/G(x) = -V_y(x)/A(x)G(x)$ and the strain in the cross-section geometric center. For a rectangular section, the stress is maximum at the centroid. From (6.63) and Hooke's law, the respective strain is

$$\bar{\gamma}_{xy}(x) = \frac{\tau_{xy}(x)}{G(x)} = -\frac{3}{2}\frac{V_y(x)}{A(x)G(x)}. \tag{6.75}$$

To obtain K_c we divide the first above relation by the second one, that is,

$$K_c = \frac{-\dfrac{V_y(x)}{A(x)G(x)}}{-\dfrac{3}{2}\dfrac{V_y(x)}{A(x)G(x)}} = \frac{2}{3}. \tag{6.76}$$

Another frequently used value is $K_c = \frac{5}{6}$. We can also consider the Poisson effect and use $K_c = \dfrac{10(1+v)}{(12+11v)}$ [34, 8].

6.9.2 CIRCULAR CROSS-SECTION

When assuming a quadratic shear stress variation with y for a rectangular cross-section in equation (6.58), we implicitly assumed the stress to be vertically distributed in the cross-section and parallel to the shear force. This assumption is invalid for the circular section. In this case, analogous to the shaft model, the shear stress on each cross-section boundary point has a tangent direction, as illustrated in Figure 6.21(a) for points A and A'.

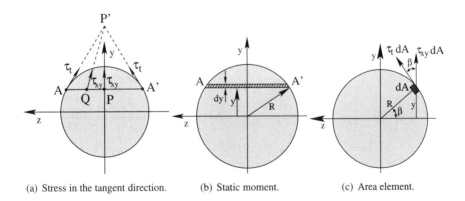

(a) Stress in the tangent direction.   (b) Static moment.   (c) Area element.

Figure 6.21   Shear stress on a circular cross-section.

Due to the symmetry to y and z of the cross-section, the stress at midpoint P of line AA' has the same vertical direction of shear force. Therefore, the directions of the shear stresses at points A and P intersect at point P' along the y axis as shown in Figure 6.21(a). Therefore, we assume that the shear stress direction at any other point Q of line AA' is also directed to P'. Hence, it is possible to determine the direction of the shear stress at any point of line AA', and consequently for any point of the cross-section.

Another possibility is to consider the vertical shear stress components to be equal for all points of line AA'. This is the same assumption used for rectangular cross-sections with a quadratic variation of τ_{xy} in (6.58). This allows us to use equation (6.72) to calculate the vertical shear stress component. In this case, base b in (6.72) is taken as the length of line AA', given according to Figure 6.21(b) by

$$b = \overline{AA'} = 2\sqrt{R^2 - y^2}, \tag{6.77}$$

with R the radius of the cross-section.

We should calculate the static moment of area above line AA' illustrated in Figure 6.21(b), which is y units distant from the z axis. For this purpose, consider the area element dA, given by

$$dA = 2\sqrt{R^2 - y^2}dy.$$

Thus, employing the definition of static moment, we have

$$M_{s_z}(x, y) = \int_A y\, dA = 2\int_y^R y\sqrt{R^2 - y^2}dy.$$

By performing the previous integration, we have

$$M_{s_z}(x, y) = \frac{2}{3}\left(R^2 - y^2\right)^{\frac{3}{2}}. \tag{6.78}$$

Substituting (6.77) and (6.78) in (6.72) we obtain

$$\tau_{xy}(x, y) = -\frac{V_y(x)}{I_z(x)}\frac{\frac{2}{3}\left(R^2 - y^2\right)^{\frac{3}{2}}}{2\sqrt{R^2 - y^2}},$$

that is,

$$\tau_{xy}(x, y) = -\frac{V_y(x)\left(R^2 - y^2\right)}{3I_z(x)}. \tag{6.79}$$

To determine the tangent shear stress τ_t at points A and A', consider the area element dA at the boundary of the cross-section illustrated in Figure 6.21(c). The resultant of forces in y and tangent direction t acting on dA are respectively $\tau_{xy}dA$ and $\tau_t dA$. The following relation is valid for these forces:

$$(\tau_t dA)\cos\beta = \tau_{xy}dA,$$

with $\cos\beta = \dfrac{\sqrt{R^2 - y^2}}{R}$. Thus,

$$\tau_t = \frac{R}{\sqrt{R^2 - y^2}}\tau_{xy}. \tag{6.80}$$

Substituting (6.79) in the above expression, we obtain

$$\tau_t(x, y) = -\frac{V_y(x)}{3I_z(x)}R\sqrt{R^2 - y^2}. \tag{6.81}$$

Notice that from this expression, the maximum shear stress at section x occurs at points located along the z axis, which passes through the geometric center of the cross-section. Thus,

$$\tau_t^{max}(x, y) = \tau_t(x, y = 0) = -\frac{V_y(x)}{3I_z(x)}R^2. \tag{6.82}$$

Substituting $I_z(x) = \dfrac{\pi R^4}{4}$, we have

$$\tau_t^{\max}(x) = -V_y(x)\frac{R^2}{\dfrac{3\pi R^4}{4}} = -\frac{4}{3}\frac{V_y(x)}{\pi R^2}. \tag{6.83}$$

Recalling that $A(x) = \pi R^2$, we obtain the final expression for the maximum shear stress in the circular section

$$\tau_t^{\max}(x) = -\frac{4}{3}\frac{V_y(x)}{A(x)}. \tag{6.84}$$

Hence, the maximum shear stress in the circular section is approximately 33% higher than the mean stress $-\dfrac{V_y(x)}{A(x)}$ determined for the Timoshenko beam model.

Notice that the exact maximum shear stress, obtained with the consideration of warping of the section, is given by [51]

$$\tau_t^{\max}(x) = -1.38\frac{V_y(x)}{A(x)}. \tag{6.85}$$

Thus, the error in expression (6.84) is about 4%, which is reasonable from an engineering standpoint.

Analogous to the rectangular section, the shear coefficient is obtained from (6.84) and is

$$K_c = \frac{3}{4}. \tag{6.86}$$

Considering the Poisson effect, we employ $K_c = \frac{6(1+v)}{(7+6v)}$ [8].

6.9.3 I-SHAPED CROSS-SECTION

The I-shaped cross-section illustrated in Figure 6.22 is part of a standardized set of frequently used cross-sections in structures. In this section we discuss how to determine the shear stress distribution for I-shaped cross-sections. These can be regarded as a combination of three rectangles, neglecting the inner radius. Both horizontal rectangles are called flanges or tables, while the vertical one is the web.

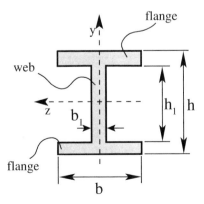

Figure 6.22 I-shaped cross-section.

To determine the shear stress distribution in I-shaped cross-sections, we consider that the stresses are parallel to the shear force and are uniformly distributed along the web thickness, similar to the rectangular section. This hypothesis is more accurate for the shear stress calculation in the web. The actual shear stress distribution in the flanges at the interface area with the web is more complex and the expressions to be obtained are less accurate.

As there are two thickness transitions between the flanges and the web, we consider the three cross-section cuts shown in Figure 6.23 to calculate the static moment of area of the I-shaped cross-section.

The static moment of area above AA', shown in Figure 6.23(a), is given analogously to (6.69), that is

$$M_{S_z}(x,y) = \int_y^h ydA = \frac{b}{2}\left[\left(\frac{h}{2}\right)^2 - y^2\right], \quad \frac{h_1}{2} \leq y \leq \frac{h}{2}. \tag{6.87}$$

Substituting the above expression in (6.72) we have

$$\tau_{xy}(x,y) = -\frac{V_y(x)}{2I_z(x)}\left[\left(\frac{h}{2}\right)^2 - y^2\right]. \tag{6.88}$$

When calculating the shear stress for $y = \frac{h_1}{2}$ and $y = \frac{h}{2}$, we obtain

$$\begin{cases} \tau_{xy}\left(x, y = \frac{h_1}{2}\right) &= -\frac{V_y(x)}{8I_z(x)}[h^2 - h_1^2] \\ \tau_{xy}\left(x, y = \frac{h}{2}\right) &= 0 \end{cases}. \tag{6.89}$$

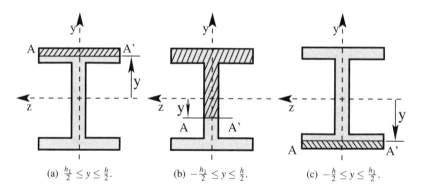

(a) $\frac{h_1}{2} \leq y \leq \frac{h}{2}$.   (b) $-\frac{h_1}{2} \leq y \leq \frac{h}{2}$.   (c) $-\frac{h}{2} \leq y \leq \frac{h_1}{2}$.

Figure 6.23 Cuts of I-shaped cross-sections.

The static moment of area above AA', indicated in Figure 6.23(b), is given by

$$M_{S_z}(x,y) = \int_y^{\frac{h}{2}} ydA = b_1\int_y^{\frac{h_1}{2}} ydy + b\int_{\frac{h_1}{2}}^{\frac{h}{2}} ydy, \quad -\frac{h_1}{2} \leq y \leq \frac{h}{2}.$$

The previous expression was split into two integrals due to the thickness change from b_1 in the web to b in the flange. Performing the indicated integrations, we obtain

$$M_{S_z}(x,y) = \frac{b_1}{2}\left[\left(\frac{h_1}{2}\right)^2 - y^2\right] + \frac{b}{2}\left[\left(\frac{h}{2}\right)^2 - \left(\frac{h_1}{2}\right)^2\right]. \tag{6.90}$$

Thus, the shear stress is given by

$$\tau_{xy}(x,y) = -\frac{V_y(x)}{2b_1I_z(x)}\left\{b_1\left[\left(\frac{h_1}{2}\right)^2 - y^2\right] + b\left[\left(\frac{h}{2}\right)^2 - \left(\frac{h_1}{2}\right)^2\right]\right\}. \tag{6.91}$$

For $y = 0$ and $y = \pm \dfrac{h_1}{2}$, we respectively have the maximum and minimum stresses in the web, that is,

$$\begin{cases} \tau_{xy}(x, y = 0) = -\dfrac{V_y(x)}{8b_1 I_z(x)} \left[bh^2 - h_1^2(b - b_1) \right] \\ \tau_{xy}\left(x, y = \pm \dfrac{h_1}{2}\right) = -\dfrac{V_y(x)}{8b_1 I_z(x)} \left[bh^2 - bh_1^2 \right] \end{cases} \tag{6.92}$$

Notice the difference of the shear stress calculated in (6.89) and (6.92) for $y = \frac{h_1}{2}$. This indicates a stress discontinuity in the flange/web interface.

The last area is obtained with a cut in the inferior flange of the section, as illustrated in Figure 6.23(c). The static moment of area above AA' is given by

$$M_{s_z}(x, y) = \int_y^{\frac{h}{2}} y \, dA.$$

Instead of calculating the static moment of area above the cut, we can take the area below line AA'. Recall that the static moment relative to an axis passing over the geometric center of the section is zero. Thus, the sum of the static moments of the areas above and below line AA' is zero, and the following relation is valid:

$$M_{s_z}(x, y) = \int_y^{\frac{h}{2}} y \, dA = - \int_{-\frac{h}{2}}^{y} y \, dA = \int_y^{-\frac{h}{2}} y \, dA.$$

Thus,

$$M_{s_z}(x, y) = \int_y^{-\frac{h}{2}} y \, dA = \frac{b}{2} \left[\left(\frac{h}{2} \right)^2 - y^2 \right], \quad -\frac{h}{2} \leq y \leq -\frac{h_1}{2}. \tag{6.93}$$

Calculating the static moments for $y = -\frac{h_1}{2}$ and $y = -\frac{h}{2}$, we have the same expressions indicated in (6.89).

Figure 6.24 illustrates the shear stress distribution for an I-shaped cross-section. Notice the stress discontinuity in the points with thickness change at the web/flange interface. Analogous to the rectangular and circular sections, the maximum shear stress in an I-shaped cross-section occurs at the centroid and is given in equation (6.92) for $y = 0$.

For I-shaped cross-sections, we adopt the following shear factor:

$$K_c = \frac{A_a}{A}, \tag{6.94}$$

with A_a the area of the web.

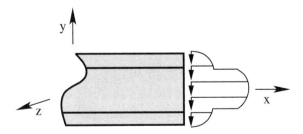

Figure 6.24   Shear stress distribution in the I-shaped cross-section.

We obtain a good approximation for the maximum shear stress when dividing the shear force by the area of the web. This occurs because the shear stress in the web integrated in the area gives a force $V_{y_a}(x)$ which is practically equal to the shear force $V_y(x)$. To show this, consider expressions (6.16) and (6.91) for the area of the web, that is,

$$
\begin{aligned}
V_{y_a}(x) &= -\int_A \tau_{xy}(x,y)dA = -b_1 \int_{-\frac{h_1}{2}}^{\frac{h_1}{2}} \tau_{xy}(x,y)dy \\
&= \frac{V_y(x)}{2I_z(x)} \int_{-\frac{h_1}{2}}^{\frac{h_1}{2}} \left\{ b_1\left[\left(\frac{h_1}{2}\right)^2 - y^2\right] + b\left[\left(\frac{h}{2}\right)^2 - \left(\frac{h_1}{2}\right)^2\right] \right\} dy.
\end{aligned}
$$

Integrating the above expression, we obtain

$$
V_{y_a}(x) = \frac{V_y(x)}{I_z(x)} \left[\frac{b(h-h_1)}{2} \frac{(h+h_1)}{2} \frac{h_1}{2} + \frac{b_1 h_1^3}{12} \right]. \tag{6.95}
$$

The second moment of area $I_z(x)$ of the cross-section is given by the sum of the second moment of area of the web $[I_z(x)]_a$ and flanges $[I_z(x)]_m$ relative to the centroid of the section, that is,

$$
I_z(x) = [I_z(x)]_a + 2[I_z(x)]_m. \tag{6.96}
$$

On the other hand,

$$
\begin{aligned}
[I_z(x)]_a &= \frac{b_1 h_1^3}{12}, \\
[I_z(x)]_m &= \frac{b(h-h_1)^3}{12} + \frac{b(h-h_1)}{2} \frac{(h+h_1)^2}{4}.
\end{aligned}
$$

Notice that the parallel axis theorem was used to calculate $[I_z(x)]_m$. Thus,

$$
I_z(x) = \frac{b_1 h_1^3}{12} + 2\left[\frac{b(h-h_1)^3}{12} + \frac{b(h-h_1)}{2} \frac{(h+h_1)^2}{4} \right]. \tag{6.97}
$$

When the flange thickness is small, that is, h_1 approximates to h, the above expression simplifies to the second moment of area $I_z(x) = \frac{b_1 h_1^3}{12}$ of the web. The same happens for the term inside brackets of expression (6.95). Thus, for flanges with small thicknesses, the web absorbs all the shear force, because the force $V_{y_a}(x)$ given in (6.95) approximates to the shear force $V_y(x)$ in cross-section x.

Example 6.6 *Consider the beam shown in Figure 6.25(a) subjected to a force $F = 150$ kN. The cross-section is I-shaped, and is fastened with two screws with 18 mm of diameter, as illustrated in Figure 6.25(b). The admissible shear stress for the beam and screws is 50 MPa. Calculate the minimum constant spacing between the screws along the length of the beam for both sections. What is the total number of screws?*

The shear force is constant along the entire beam, with intensity equal to F, that is, $V_y = F = 150$ kN.

The second moment of area for the I-shaped cross-section illustrated in Figure 6.25(b) is determined applying the parallel axis theorem as

$$
I_z = \underbrace{\frac{(4)(40)^3}{12}}_{web} + \underbrace{2\left[\frac{(24)(4)^3}{12} + (24)(4)(20+2)^2 \right]}_{flanges} = 115285.33 \text{ cm}^4.
$$

The shear flow indicates the transversal force generated in the beam by units of length. This force density must be compensated by the screws. The static moment of the area above the union is given by

$$
M_{s_z} = A\bar{y} = (4)(24)(20+2) = 2112.00 \text{ cm}^3.
$$

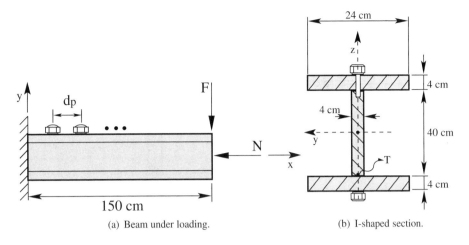

(a) Beam under loading.   (b) I-shaped section.

Figure 6.25 Example 6.6: I-shaped beam connected by screws.

The shear flow is calculated by the expression

$$q_c = \frac{V_y M_{s_z}}{I_z} = \frac{(150000)(2112.00)}{115285.33} = 2747.96 \text{ N/cm}.$$

The maximum force on each screw in the elastic range is

$$F_p = \bar{\tau} A_p = (50)\frac{\pi(18)^2}{4} = 12723.45 \text{ N}.$$

The spacing d_p between the screws is obtained by

$$(d_p)(q_c) = F_p.$$

Substituting the values, we have

$$d_p = \frac{12723.45}{2747.96} = 4.63 \text{ cm}.$$

The amount of screws in the upper part of the beam is given by

$$(n_p)(d_p) = L.$$

Thus,

$$n_p = \frac{150}{4.61} = 32.40.$$

We must employ 66 screws in the upper and lower parts of the beam.
□

Example 6.7 *Consider the beam of the previous example, but with a T-shaped cross-section as illustrated in Figure 6.26(a). The distance between the screws is $d_p = 5$ cm and the force $F = 100$ kN. Adopt an admissible shear stress of 30 MPa. What is the total number of screws necessary to keep the union of both rectangles of the T-shaped section?*

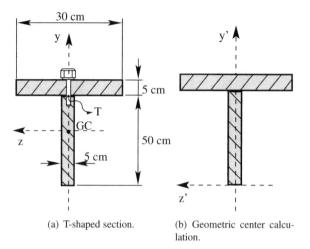

(a) T-shaped section.   (b) Geometric center calculation.

Figure 6.26   Example 6.7: screw-fastened T-shaped beam.

The first step is to calculate the geometric center (GC) coordinates of the T-shaped section. As y is an axis of symmetry, we have $z_{GC} = 0$. The coordinate y_{GC} is calculated using the $y'z'$ coordinate system illustrated in Figure 6.26(b) as

$$y_{GC} = \frac{M_{s'_z}}{A} = \frac{(5)(30)(52.5) + (5)(50)(25)}{(5)(30) + (5)(50)} = 35.31 \text{ cm}.$$

Applying the parallel axis theorem, the second moment of area of the cross-section is calculated as

$$\begin{aligned} I_z &= \frac{(30)(5)^3}{12} + (30)(5)(52.50 - 35.31)^2 + \frac{(5)(50)^3}{12} + (5)(50)(35.31 - 25.00)^2 \\ &= 123294.27 \text{ cm}^4. \end{aligned}$$

The static moment for the area above the screw is

$$M_{s_z} = A\bar{y} = (5)(50)(52.50 - 35.31) = 2578.50 \text{ cm}^3.$$

The shear flow is given by

$$q_c = \frac{V_y M_{s_z}}{I_z} = \frac{(100)(2578.5)}{123294.27} = 2.09 \text{ kN/cm}.$$

The maximum force supported by each screw is

$$F_p = (d_p)(q_c) = (5)(2.09) = 10.45 \text{ kN}.$$

The screw diameter is calculated from $F_p = \bar{\tau} A_p$ as

$$d = \sqrt{\frac{4F_p}{\pi \bar{\tau}}} = \sqrt{\frac{(4)(10.45 \times 10^3)}{(\pi)(30 \times 10^6)}} = 2.11 \text{ cm}.$$

The number of screws is given by

$$(n_p)(d_p) = L \rightarrow n_p = \frac{150}{5} = 30.$$

Thus, 30 screws are required, with a minimum diameter of $d = 2.11$ cm.

□

6.10   DESIGN AND VERIFICATION

The models studied so far have uniaxial normal stress state (bar and beam) and shear stress state (shaft). The Timoshenko beam has both stress components. It is not possible to consider the bending and shear effects separately when designing the beam, just taking the respective maximum normal and shear stresses and their admissible values.

As will be seen in Chapter 8, the normal and shear stress components are used to calculate the equivalent normal stress

$$\sigma_{eqv} = \sqrt{\sigma_{xx}^2 + 3\tau_{xy}^2}. \tag{6.98}$$

The following design procedure of the Timoshenko beam is applied:

1. The functions and respective shear force $V_y(x)$ and bending moment $M_z(x)$ diagrams are determined, integrating the differential equations given in (6.25) for isostatic problems and (6.47) and (6.48) for hyperstatic problems.
2. Based on these diagrams, the critical section is determined, i.e., the section with the largest absolute value for the bending moment, denoted by M_z^{max}. If two or more sections have the same maximum bending moment, the critical section will be the one with the largest absolute value for the shear force, namely V_y^{max}. A more conservative possibility is to choose the maximum absolute values for the shear force and bending moment, even if they occur in distinct sections.
3. The critical points due to bending at the critical section are selected. In the case of bending, the best candidates to be the critical ones are the boundary points of the cross-section, which are most distant from the geometric center.
   The maximum normal stress in the critical point is given by

$$\sigma_{xx}^{max} = \frac{M_z^{max}}{W_z}. \tag{6.99}$$

4. The maximum normal stress value to maintain the beam in the elastic range must be less than or equal to the admissible normal stress $\bar{\sigma} = \min(\bar{\sigma}_c, \bar{\sigma}_t)$ of the material, that is,

$$\sigma_{xx}^{max} \leq \bar{\sigma}. \tag{6.100}$$

In order to obtain the smallest cross-section, we take the equality of the above expression and use (6.99) to obtain

$$W_z = \frac{M_z^{max}}{\bar{\sigma}}. \tag{6.101}$$

From the calculated value for W_z and the shape of the section, we obtain the values of the main cross-section dimensions.
5. After the design based on the normal stress, the obtained values are slightly increased (around 3 to 5%). The beam is then verified, including the shear effect to calculate the maximum equivalent normal stress σ_{eqv}^{max}, using equation (6.98). As the maximum normal stress occurs at the boundaries of the cross-section, and the shear stress is zero at these points, we take the maximum shear stress at the geometrical center of the critical section.

For the verification of a beam, the cross-section dimensions are known and we want to check if it remains in the elastic range when subjected to loading. Thus, we calculate the maximum equivalent normal stress σ_{eqv}^{max} using equation (6.98). Then we check if the maximum value is lower than or equal to the admissible normal stress of the material. In this case, we say that the beam remains in the elastic range. If this condition is violated, the beam must be resized applying the previous procedure.

Generally, we should verify a specific point of the cross-section, as illustrated in the following example.

Example 6.8 *Verify the point T of the cross-section of Figure 6.26(a) for an admissible normal stress of $\bar{\sigma} = 50$ MPa.*
The critical section of the beam is $x = 0$ with loads $M_z = -225$ kNm and $V_y = 150$ kN for point T, $y = 14.69$ cm and normal stress

$$\sigma_{xx} = -\frac{M_z}{I_z}y = -\frac{22500000}{123294.27}14.69 = -26.81 \text{ MPa}.$$

The static moment of area above T is given by

$$M_{s_z} = A\bar{y} = (5)(30)(14.69 + 2.50) = 2578.50 \text{ cm}^3.$$

The shear stress at this point is

$$\tau_{xy} = -\frac{V_y M_{s_z}}{bI_z} = -\frac{(150000)(2578.50)}{(5)(123294.27)} = -6.27 \text{ MPa}.$$

As there is a width discontinuity at point T, we take the smallest value for b in order to obtain a higher shear stress value.
The equivalent normal stress is given by

$$\sigma_{eqv}^{max} = \sqrt{(-26.81)^2 + 3(-6.27)^2} = 28.93 \text{ MPa}.$$

As $\bar{\sigma} = 50$ MPa $\geq \sigma_{eqv}^{max} = 28.93$ MPa, the stress state at point T is in the elastic range.
□

Example 6.9 *Consider the beam of Example 6.6 subjected to a force $F = 1.50 \times 10^6$ N. The cross-section is I-shaped, and is parameterized by dimension a, as shown in Figure 6.27(a). Design the cross-section for an admissible normal stress of $\bar{\sigma} = 40$ MPa and verify the shear effect.*

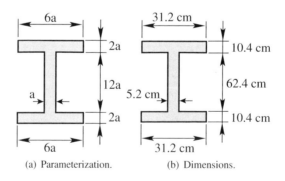

(a) Parameterization.   (b) Dimensions.

Figure 6.27   Example 6.9: beam with a parameterized I-shaped profile.

The only way to design the I-shaped beam is to adopt a parameterization as indicated in Figure 6.27(a). However, such procedure requires the definition of a proportional relation between the dimensions of the flanges and web.
The critical section of the beam is $x = 0$ with $V_y^{max} = 1.50 \times 10^6$ N and $M_z^{max} = (1.50 \times 10^6)(0.15) = 225000$ Nm. The critical points are at the ends of the flanges, that is, $y^{max} = \pm 8a$.

The second moment of area of the cross-section is calculated in terms of a as

$$I_z = \frac{a(12a)^3}{12} + 2\left[\frac{(6a)(2a)^3}{12} + 12a^2(7a)^2\right] = 320a^4.$$

The bending strength modulus is calculated as

$$W_z = \frac{I_z}{y^{max}} = 40a^3.$$

Similarly,

$$W_z = \frac{M_z^{max}}{\bar{\sigma}} = \frac{2.25 \times 10^5}{40 \times 10^6} = 5.63 \times 10^{-3} \text{ m}^3.$$

From the two previous expressions for W_z, we obtain a as

$$a = \left(\frac{5,63 \times 10^{-3}}{40}\right)^{\frac{1}{3}} = 5.20 \text{ cm}.$$

For a = 5.50 cm, we have $W_z = 6.66 \times 10^{-3}$ m^3 and the maximum normal stress is

$$\sigma_{xx}^{max} = \frac{M_z^{max}}{W_z} = 33.81 \text{ MPa}.$$

The maximum shear stress occurs at the geometric center of the cross-section. Equation (6.92) is employed for the calculation with $I_z = 320a^4 = 2.93 \times 10^{-3}$ m^4 and $M_{s_z} = (6a)(2a)(7a) + (a)(6a)(3a) = 102a^3 = 1.70 \times 10^{-2}$ m^3. Thus,

$$\tau_{xy}^{max} = \frac{(1.50 \times 10^6)(1.70 \times 10^{-2})}{(5.50)(2.93 \times 10^{-3})} = 1.58 \text{ MPa}.$$

To verify the shear effect, we calculate the equivalent stress

$$\sigma_{eqv}^{max} = \sqrt{(33.81)^2 + 3(1.58)^2} = 33.91 \text{ MPa}.$$

As $\bar{\sigma} = 40$ MPa $\geq \sigma_{eqv}^{max} = 33.91$ MPa, the beam is in the elastic range. The final dimensions of the I-shaped cross-section are illustrated in Figure 6.27(b).

$\square$

6.11 STANDARDIZED CROSS-SECTION SHAPES

Some types of cross-sections were standardized for large-scale production and reduction of construction costs of metallic structures. Figure 6.28 illustrates some standard cross-sections, such as the I-shaped and C-shaped ones. Several properties such as dimensions, area, second moment of area, and strength modulus are tabulated and used to select the appropriate shape to design the beam. These shapes can be used in the vertical [Figures 6.28(a) and 6.28(b)] and horizontal [Figures 6.28(c) and 6.28(d)] positions. Besides that, we can weld or fasten several shapes to construct the cross-section, as illustrated in Figure 6.28(e) for a double C-shaped cross-section, which forms a rectangular tube.

The design procedure of beams under bending only with these shapes is analogous to the previous section and is given below:

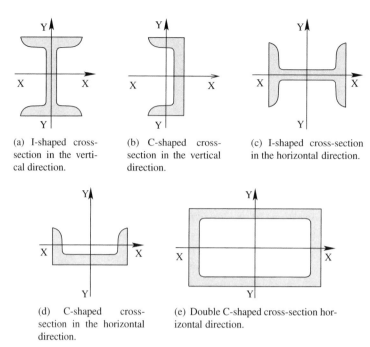

(a) I-shaped cross-section in the vertical direction.

(b) C-shaped cross-section in the vertical direction.

(c) I-shaped cross-section in the horizontal direction.

(d) C-shaped cross-section in the horizontal direction.

(e) Double C-shaped cross-section horizontal direction.

Figure 6.28   Standard cross-section shapes.

1. Find the critical section of the beam solving the equilibrium BVP (6.25) for isostatic problems and (6.47) and (6.48) for hyperstatic problems. This section has the maximum absolute bending moment. If there is more than one section with the same maximum bending moment, the one with highest absolute value for the shear force is selected. The maximum bending moment is called $M_z^{\max}$.

2. The bending strength modulus W_z is calculated from the normal stress $\sigma_{xx}^{\max}$ expression, that is,

$$\sigma_{xx}^{\max} = \frac{M_z^{\max}}{W_z}. \qquad (6.102)$$

In the design for bending only, we impose that $\sigma_{xx}^{\max}$ is equal to the admissible normal stress $\bar{\sigma} = \min(\bar{\sigma}_c, \bar{\sigma}_t)$. From this condition and the above relation, we have the following expression for W_z:

$$W_z = \frac{M_z^{\max}}{\bar{\sigma}}.$$

3. Being n_p the number of shapes, the bending strength modulus W_z^p of each profile is given by

$$W_z^p = \frac{W_z}{n_p}.$$

For instance, for Figures 6.28(a) and 6.28(e), we have $n_p = 1$ and $n_p = 2$, respectively.

4. By knowing the minimum strength modulus of each shape to keep the beam in the elastic range, we select a shape in a given table of standardized profiles with a modulus higher than or equal to W_z^p. Notice that if the profile is in the vertical direction, we should use the $X - X$ axis column. If the shape is horizontal, the $Y - Y$ column is taken (see Figure 6.28).

5. The selected cross-section shape is indicated providing the number of shapes, the type, the specific weight, and height. For example, 2C 6" 35.7 Kg/m indicates that two C profiles with 6" of height and 35.7 Kg/m of specific weight were selected.

6. We should verify if the beam remains in the elastic range, including the specific weight as a distributed constant load in the beam.

The new maximum stress is calculated with (6.102) recalling that M_z^{max} is now obtained including the weight in the original load equation of the beam. If the calculated σ_{xx}^{max} is less than $\bar{\sigma}$, the beam remains in the elastic range. Otherwise, we should resize the beam applying the above procedure. Generally, this process is iterated until a suitable profile is selected.

Example 6.10 *Consider the simply supported beam of Example 6.4 subjected to a $q_0 = 1000$ kN/m distributed load. Indicate the I-shaped cross-section with broad flanges according to the table presented in http://www.structural-drafting-net- expert.com/steel-sections-i-beam-w-shape.html. Adopt $\bar{\sigma} = 50$ MPa.*

The critical section is $x = 1$ with a maximum bending moment of $M_z^{max} = 500$ kNm. The bending strength modulus is

$$W_z = \frac{500000}{50 \times 10^6} = 10000 \text{ cm}^3 = 610.24 \text{ in}^3.$$

The W36" 182 lb/ft standard shape is selected. The specific weight in N/m is determined using the conversions 1 lb = 0.453592 kg and 1 ft = 0.304800 m and multiplying it by the acceleration due to gravity. Thus,

$$q_p = 182\frac{0.453592}{0.304800}9.81 = 2657.0 \text{ N/m}.$$

With the inclusion of the self-weight effect, the new value for the bending strength modulus is

$$W_z = \frac{502657}{50 \times 10^6} = 10053.14 \text{ cm}^3 = 613.48 \text{ in}^3.$$

As the selected shape has a bending strength modulus of $W_{X-X} = 623$ in^3, above the minimum calculated, the beam remains in the elastic range even with the self-weight consideration.

□

6.12 SHEAR CENTER

Consider the cantilever beam with a C-shaped cross-section illustrated in Figure 6.29(a), subjected to a concentrated force P at the free end. The force P is applied along the y axis passing through the geometrical center of the section.

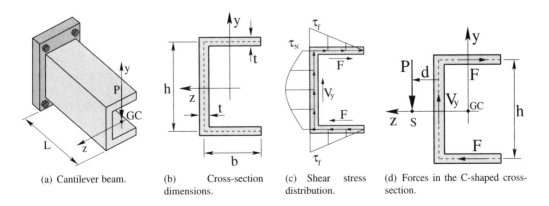

(a) Cantilever beam.   (b)   Cross-section dimensions.   (c) Shear stress distribution.   (d) Forces in the C-shaped cross-section.

Figure 6.29 Shear center in the beam with a C-shaped cross-section.

We assume that the cross-section is thin-walled, that is, $b, h \gg t$, such that the dimensions along the center lines are considered, as illustrated in Figure 6.29(b),

The shear force in the beam is considered constant and equal to P, that is, $V_y(x) = P$ $(0 \leq x \leq L)$. The distributions of stress and shear flow in the flanges and web are illustrated in Figure 6.29(c) (see Example 6.11). When integrated in the flange, the shear stress distribution τ_f gives rise to a horizontal force $F = \dfrac{\tau_f}{2} bt$. Similarly, the integral of τ_w in the web of the profile generates a shear force $V_y(x) = P$.

Notice that the force couple F generates a twist of the cross-section about the longitudinal axis. One way to avoid this effect is applying a force P at a distance d from the centerline of the web, as shown in Figure 6.29(d). Making a balance of forces in the cross-section, we have

$$
\begin{aligned}
i) & \quad \sum f_y = 0: & V_y - P = 0, \\
ii) & \quad \sum f_z = 0: & -F + F = 0, \\
iii) & \quad \sum m_x = 0: & -Pd + Fh = 0.
\end{aligned}
$$

The last expression allows the determination of d as

$$
d = \frac{F}{P} h = \frac{\tau_f bth}{2P}. \tag{6.103}
$$

On the other hand, the shear stress τ_f in the flange is

$$
\tau_f = \frac{V_y M_{s_z}}{I_z b} = \frac{V_y bth}{2 I_z b} = \frac{Pth}{2 I_z}.
$$

Thus,

$$
d = \frac{bh^2 t^2}{4 I_z}. \tag{6.104}
$$

Thus, the distance d is independent of the applied force magnitude. The line of action of P and the z axis intersect at the shear center S of the cross-section. The shear center for any cross-section is located in a line parallel to the longitudinal axis of the beam. Any transversal force applied on S does not cause any twist of the beam.

For cross-sections with only one axis of symmetry, the shear center is always located in the given axis. In the case of cross-sections with two axes of symmetry, the centroid and the shear center are coincident. For nonsymmetric, thick-walled cross-sections, the exact location of S is difficult to determine.

Example 6.11 *We want to determine the stress distributions and shear flow in the flanges and web of the C-shaped section illustrated in Figure 6.29(b).*

Consider the hatched area in the superior flange of the section shown in Figure 6.30(a). The static moment of area is $M_{s_z} = \dfrac{h}{2}(b - x)t$. The shear flow is calculated using equation (6.73) as

$$
q_c = \frac{V_y M_{s_z}}{I_z} = \frac{V_y}{2 I_z} ht(b - x).
$$

Thus, the shear flow has a linear variation in the flange area, with $q_c = 0$ at $x = b$ and $q_c = \dfrac{V_y}{2 I_z} htb$ at $x = 0$. The shear stress in the flange also has a linear variation of the following form:

$$
\tau_c = \frac{q_c}{b} = \frac{V_y}{2 b I_z} ht(b - x).
$$

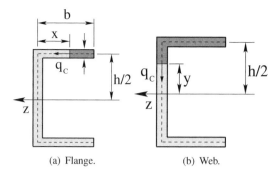

(a) Flange.   (b) Web.

Figure 6.30 Shear flow in the C-shaped cross-section.

The horizontal force F in the flange is

$$F = \int_0^b q_c dx = \int_0^b \frac{V_y}{2I_z} ht(b-x)dx = \frac{V_y ht b^2}{4I_z}.$$

The same expression can be obtained from the area of the triangle representing the linear distribution of q_c, illustrated in Figure 6.29(c). Thus,

$$F = \frac{1}{2}\left(\frac{V_y}{2I_z} htb\right)(b) = \frac{V_y}{4I_z} htb^2.$$

In the case of the web, we consider the area illustrated in Figure 6.30(b), where the static moment is

$$M_{S_z} = (bt)\frac{h}{2} + \left[y + \frac{1}{2}\left(\frac{h}{2} - y\right)\right]t\left(\frac{h}{2} - y\right) = \frac{bth}{2} + \frac{t}{2}\left(\frac{h^2}{4} - y^2\right).$$

Thus, the shear flow in the web is

$$q_c = \frac{V_y M_{S_z}}{I_z} = \frac{V_y t}{2I_z}\left(bh + \frac{h^2}{4} - y^2\right).$$

The parabolic variation of $q_c = \frac{V_y bth}{2I_z}$ in $y = \frac{h}{2}$ has a maximum value of $q_c = \frac{V_y t}{2I_z}\left(bh + \frac{h^2}{4}\right)$ at $y = 0$. The force in the web is obtained integrating the shear flow as

$$F_w = \int_{-h/2}^{h/2} q_c dy = \frac{V_y t h^2}{4I_z}\left(2b + \frac{h}{3}\right).$$

The second moment of area of the cross-section relative to the z axis is

$$I_z = 2\left[\frac{bt^3}{12} + bt\left(\frac{h}{2}\right)^2\right] + \frac{th^3}{12}.$$

As the flange thickness is small, the first term of the above expression is neglected. Thus,

$$I_z = \frac{th^2}{4}\left(2b + \frac{h}{3}\right).$$

Substituting I_z in the force expression for the web, we recover $F_w = V_y$, as expected.

$\square$

A similar analysis allows the identification of the shear centers for open cross-sections, as the ones illustrated in Figure 6.31.

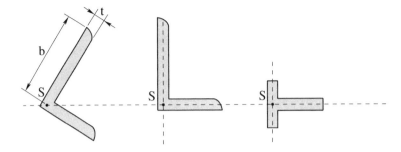

Figure 6.31   Shear centers in open cross-sections.

6.13 SUMMARY OF THE VARIATIONAL FORMULATION OF BEAMS WITH SHEAR

The kinematics of the Timoshenko beam is described by a vector field $\mathbf{u}(x,y)$ with the following components:

$$\mathbf{u}(x,y) = \left\{ \begin{array}{c} u_x(x,y) \\ u_y(x) \\ u_z(x) \end{array} \right\} = \left\{ \begin{array}{c} -y\theta_z(x) \\ u_y(x) \\ 0 \end{array} \right\}. \tag{6.105}$$

Thus, the set $\mathscr{V}$ of kinematically admissible actions is described by

$$\mathscr{V} = \left\{ \mathbf{u}, u_x(x,y) = -y\theta_z(x), u_y(x), u_z(x) = 0 \right\}. \tag{6.106}$$

The subset of rigid actions is

$$\mathscr{N}(\mathscr{D}) = \left\{ \mathbf{u}(x) \in \mathscr{V} \mid \theta_z(x) = \theta_z = \text{cte and } u_y(x) = u_y = \text{cte} \right\}. \tag{6.107}$$

For this case, the strain operator $\mathscr{D}$ is indicated in matrix form as

$$\begin{aligned}
\mathscr{D}: \quad & \mathscr{V} \to \mathscr{W} \\
& \mathbf{u}(x,y) \to \mathscr{D}\mathbf{u}(x,y) \\
& \left\{ \begin{array}{c} -y\theta_z(x) \\ u_y(x) \end{array} \right\} \to \left[\begin{array}{cc} \dfrac{d}{dx} & 0 \\[2mm] \dfrac{1}{y} & \dfrac{d}{dx} \end{array} \right] \left\{ \begin{array}{c} -y\theta_z(x) \\ u_y(x) \end{array} \right\} \\
& = \left\{ \begin{array}{c} -y\dfrac{d\theta_z(x)}{dx} \\[2mm] -\theta_z(x) + \dfrac{du_y(x)}{dx} \end{array} \right\} = \left\{ \begin{array}{c} \varepsilon_{xx}(x,y) \\ \bar{\gamma}_{xy}(x) \end{array} \right\}.
\end{aligned} \tag{6.108}$$

The space of internal loads $\mathscr{W}'$ is constituted of continuous functions $V_y(x)$ and $M_z(x)$, which respectively represent the shear force and bending moment in the cross-sections of the beam. Notice that the terms $m_z(x)$, $q_y(x)$, V_{y0}, V_{yL}, M_{z0}, and M_{zL}, relative to the external loads, define the vector space $\mathscr{V}'$.

From (6.25), the equilibrium differential operator $\mathscr{D}^*$ is defined between the internal and external loads, which can be denoted as

$$\mathscr{D}^*: \quad \mathscr{W}' \to \mathscr{V}'$$

$$\left\{ \begin{array}{c} M_z(x) \\ V_y(x) \end{array} \right\} \to \mathscr{D}^* \left\{ \begin{array}{c} M_z(x) \\ V_y(x) \end{array} \right\}$$

$$\left\{ \begin{array}{c} M_z(x) \\ V_y(x) \end{array} \right\} \to \begin{bmatrix} \dfrac{d}{dx} & -1 \\[2ex] 0 & \dfrac{d}{dx} \\[2ex] 0 & 1|_{x=0} \\[1ex] 0 & -1|_{x=L} \\[1ex] -1|_{x=0} & 0 \\[1ex] 1|_{x=L} & 0 \end{bmatrix} \left\{ \begin{array}{c} M_z(x) \\ V_y(x) \end{array} \right\} = \left\{ \begin{array}{c} \dfrac{dM_z(x)}{dx} - V_y(x) \\[2ex] \dfrac{dV_y(x)}{dx} \\[1ex] V_y(x)|_{x=0} \\[0.5ex] -V_y(x)|_{x=L} \\[0.5ex] -M_z(x)|_{x=0} \\[0.5ex] M_z(x)|_{x=L} \end{array} \right\} . \quad (6.109)$$

Figure 6.32 schematically presents the Timoshenko beam formulation.

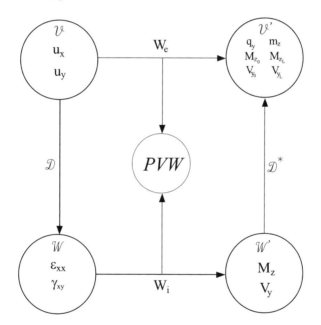

Figure 6.32 Variational formulation of the Timoshenko beam.

6.14 ENERGY METHODS

Until now, we have used the PVW to determine the support reactions and stability conditions for rigid bodies, as shown in Chapter 2. In other chapters, the equilibrium BVPs of the mechanical models were determined equating the work done by internal and external loads for the given virtual kinematics. This variant of the PVW is also referred to in the literature as the method of virtual displacements [42, 44].

We can also calculate the internal and external virtual work considering the virtual loads and the real kinematics of a mechanical model. The virtual loads are a set of internal and/or external loads in equilibrium and not necessarily related to the actual loads in the component. The PVW version that calculates the work using virtual loads is called the principle of complementary virtual work. For instance, it can be applied to determine displacements and rotations at specific points of a component, resulting in the method of virtual forces [42]. Thus, instead of solving the differential equations of equilibrium, we employ this method to determine the kinematics on selected points of a body or set of bodies. This approach is useful when we work with mechanical systems constituted of distinct bodies, and we want to determine the displacements at specific points of these systems.

In the following sections, we present the concepts of strain energy, complementary virtual work, and energy methods derived from the PVW.

6.14.1 STRAIN ENERGY

Generally, the internal energy comes from chemical, thermal, strain, and other phenomena. Only the strain energy due to mechanical loads is considered here, and is called internal strain energy, denoted by U.

For cases of large strains and nonlinear materials, the strain energy is equal to the strain internal work, that is,

$$U = W_i. \tag{6.110}$$

For a body with a Hooke material in the elastic range, the above relation must include the factor $\frac{1}{2}$ due to the linear relation between the stress and strain measures, that is,

$$U = \frac{1}{2}W_i. \tag{6.111}$$

Substituting Hooke's law in expression (3.28) of the work done by the internal loads in a bar, the respective strain energy is given from equation (6.110) as

$$U = \frac{1}{2} \int_V E(x)\varepsilon_{xx}^2(x)dV, \tag{6.112}$$

or,

$$U = \frac{1}{2} \int_V \frac{\sigma_{xx}^2(x)}{E(x)}dV. \tag{6.113}$$

We can denote the strain energy in the bar in terms of the normal force using equation (3.42) and Hooke's law as

$$U = \frac{1}{2} \int_0^L \sigma_{xx}(x)\varepsilon_{xx}(x)A(x)dx = \frac{1}{2} \int_0^L N_x(x)\varepsilon_{xx}(x)dx = \frac{1}{2} \int_0^L \frac{N_x^2(x)}{E(x)A(x)}dx. \tag{6.114}$$

For the case of a bar with an axial force P in the end, constant cross-section, and same material, we respectively have $N_x(x) = P$, $A(x) = A$ and $E(x) = E$. Consequently, the above expression reduces to

$$U = \frac{1}{2}\frac{P^2}{EA} \int_0^L dx = \frac{1}{2}\frac{P^2L}{EA}. \tag{6.115}$$

Recalling that the elongation in the free end of the bar is

$$\delta = \frac{PL}{EA},$$

the strain energy is rewritten as

$$U = \frac{EA\delta^2}{2L}. \tag{6.116}$$

Analogously, the strain energy for a beam under bending only is given by

$$U = \frac{1}{2}\int_V \sigma_{xx}(x,y)\varepsilon_{xx}(x,y)dV = \frac{1}{2}\int_V \frac{\sigma_{xx}^2(x,y)}{E(x)}dV. \tag{6.117}$$

Substituting expression (5.39) for the normal stress in a beam, we have

$$U = \frac{1}{2}\int_V \frac{M_z^2(x)}{E(x)I_z^2(x)}y^2 dV = \frac{1}{2}\int_0^L \underbrace{\left(\int_A y^2 dA\right)}_{I_z(x)}\left(\frac{M_z^2(x)}{E(x)I_z^2(x)}dx\right) = \frac{1}{2}\int_0^L \frac{M_z^2(x)}{E(x)I_z(x)}dx. \tag{6.118}$$

For the case of circular torsion, the strain energy in the elastic range is

$$U = \frac{1}{2}\int_V \tau_t(x,r)\gamma_t(x,r)dV. \tag{6.119}$$

Substituting the expressions for Hooke's law (4.36) and shear stress (4.55), we have

$$U = \frac{1}{2}\int_V \frac{\tau_t^2(x)}{G(x)}dV = \frac{1}{2}\int_V \frac{M_x^2(x)}{G(x)I_p^2(x)}r^2 dV = \frac{1}{2}\int_0^L \underbrace{\left(\int_A r^2 dA\right)}_{I_p(x)}\frac{M_x^2(x)}{G(x)I_p^2(x)}dx.$$

Thus,

$$U = \frac{1}{2}\int_0^L \frac{M_x^2(x)}{G(x)I_p(x)}dx. \tag{6.120}$$

In the case of a beam with shear, the strain energy is

$$U = \frac{1}{2}\int_V \left(\sigma_{xx}(x,y)\varepsilon_{xx}(x,y) + \tau_{xy}(x)\bar{\gamma}_{xy}(x)\right)dV.$$

Using result (6.119) for the bending term and equations (6.33) and (6.45), we have

$$U = \frac{1}{2}\int_0^L \left(\frac{M_z^2(x)}{E(x)I_z(x)} + \frac{V_y^2(x)}{G(x)K_c(x)A(x)}\right)dx. \tag{6.121}$$

Example 6.12 *We can employ the strain energy concept to define the shear coefficient K_c in a beam with rectangular cross-section. The strain energy relative to shear is given by*

$$U = \frac{1}{2}\int_V \tau_{xy}\bar{\gamma}_{xy}dV = \frac{1}{2G}\int_V \tau_{xy}^2 dV.$$

Calculating the strain energy for the constant and parabolic shear stress distributions, we have

$$U_c = \frac{1}{2}L\int_A \frac{V_y^2(x)}{A(x)^2}dA = \frac{V_y^2(x)L}{2Gbh},$$

$$U_p = \frac{1}{2}L\int_A \left[\frac{3V_y(x)}{bh}\left(1 - \left(\frac{2y}{h}\right)^2\right)\right]^2 dA = \frac{6V_y^2(x)L}{10Gbh}.$$

The shear factor can be defined as the ratio of the above expressions, resulting in $K_c = \dfrac{5}{6}$.

□

The term $\sigma_{xx}\varepsilon_{xx}$ represents a strain energy density, as seen in Chapters 3 and 5. The strain energy density U_0 in the elastic range is

$$U_0 = \frac{1}{2}\sigma_{xx}(x)\varepsilon_{xx}(x). \tag{6.122}$$

Thus, the strain energy is expressed by the following volume integral of U_0:

$$U = \frac{1}{2}\int_V U_0 dV. \tag{6.123}$$

The strain energy density can be interpreted as the area below the curve of the tensile strength test, as illustrated in Figure 6.33(a).

The virtual internal energy δU is determined by the internal loads for a virtual kinematic action. Expression (3.11) for the internal work or strain energy in a bar or beam under bending only can be denoted for a virtual action δu_x as

$$\delta U = \int_V \sigma_{xx}\delta\varepsilon_{xx}dV, \tag{6.124}$$

with

$$\delta\varepsilon_{xx} = \frac{d\delta u_x}{dx} = \delta\frac{du_x}{dx}.$$

We can change the order of the variation and differentiation operators, because we assume u_x to be a continuous function. The variation of the strain energy density δU_0 is illustrated in Figure 6.33(b). Notice that equation (6.124) is valid for linear and nonlinear material behaviors.

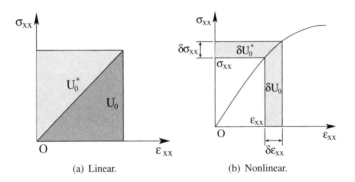

(a) Linear.   (b) Nonlinear.

Figure 6.33   Normal and complementary strain energy densities.

Assuming a linear elastic behavior and substituting Hooke's law for a bar in (6.124) we have

$$\begin{aligned}
\delta U &= \frac{1}{2}\int_0^L \left(\int_A \sigma_{xx}(x)dA\right)\delta\varepsilon_{xx}(x)dx \\
&= \frac{1}{2}\int_0^L N_x(x)\frac{\delta\sigma_{xx}(x)}{E(x)}dx = \frac{1}{2}\int_0^L \frac{N_x(x)\delta N_x(x)}{E(x)A(x)}dx. \tag{6.125}
\end{aligned}$$

The virtual strain energy for a beam in bending with Hookean material is given from (6.124) by

$$\begin{aligned}
\delta U &= -\frac{1}{2}\int_V \frac{M_z(x)}{I_z(x)}\frac{\delta\sigma_{xx}(x)}{E(x)}y\,dV \\
&= \frac{1}{2}\int_0^L \frac{M_z(x)\delta M_z(x)}{E(x)I_z^2(x)}\left(\int_A y^2 dA\right)dx = \frac{1}{2}\int_0^L \frac{M_z(x)\delta M_z(x)}{E(x)I_z(x)}. \tag{6.126}
\end{aligned}$$

Analogously, the following strain energy expressions are valid for the virtual kinematic actions in shafts and beams with shear:

$$\delta U = \int_V \tau_t \delta \gamma_t dV, \tag{6.127}$$

$$\delta U = \int_V (\sigma_{xx} \delta \varepsilon_{xx} + \tau_{xy} \delta \gamma_{xy}) dV. \tag{6.128}$$

For the case of a Hooke material, the above expressions are reduced to

$$\delta U = \frac{1}{2} \int_0^L \frac{M_x(x) \delta M_x(x)}{G(x) I_p(x)} dx, \tag{6.129}$$

$$\delta U = \frac{1}{2} \int_0^L \left[\frac{M_z(x) \delta M_z(x)}{E(x) I_z(x)} + \frac{V_y(x) \delta V_y(x)}{G(x) K_c(x) A(x)} \right] dx. \tag{6.130}$$

In the above expressions, δN_x, δV_y, δM_x, and δM_z are virtual internal loads.

6.14.2 COMPLEMENTARY STRAIN ENERGY

The virtual complementary strain energy δU^* is defined by the internal virtual work from the virtual stress distribution and strain measures that result from a real kinematic action of the mechanical component. The following expressions are valid for bars, shafts, and beams with shear, respectively:

$$\delta U^* = \int_V \delta \sigma_{xx} \varepsilon_{xx} dV, \tag{6.131}$$

$$\delta U^* = \int_V \delta \tau_t \gamma_t dV, \tag{6.132}$$

$$\delta U^* = \int_V (\delta \sigma_{xx} \varepsilon_{xx} + \delta \tau_{xy} \gamma_{xy}) dV. \tag{6.133}$$

The complementary strain energy density δU_0^* is illustrated in Figure 6.33(b), and is valid for linear and nonlinear material behaviors. The complementary strain energy is

$$\delta U^* = \int_V \delta U_0^* dV. \tag{6.134}$$

For the linear elastic range, the above expressions must be multiplied by the factor $\frac{1}{2}$. Moreover, the strain energies δU and δU^* are coincident, as illustrated in Figure 6.33(a).

Expressions (6.125), (6.126), (6.129), and (6.130) are also valid for the virtual complementary strain energy in the elastic range.

6.14.3 COMPLEMENTARY EXTERNAL WORK

The external virtual work of a bar was determined in Chapter 3 by the product of the external axial loads applied in a bar and a virtual displacement action, that is,

$$\delta W_e = \int_0^L q_x(x) \delta u_x(x) dx + P_0 \delta u_x(0) + P_L \delta u_x(L). \tag{6.135}$$

We can also determine the external virtual work using virtual forces and real axial displacements over the bar. In this case, the complementary external virtual work is given by

$$\delta W_e^* = \int_0^L \delta q_x(x) u_x(x) dx + \delta P_0 u_x(0) + \delta P_L u_x(L), \tag{6.136}$$

with $\delta q_x(x)$ the virtual distributed axial force and δP_0 and δP_L the concentrated axial forces applied at the bar ends.

Analogously, the expressions of the complementary external virtual work for a shaft and a Timoshenko beam are respectively given by

$$\delta W_e^* = \int_0^L \delta m_x(x)\theta_x(x)dx + \delta T_0\theta_x(0) + \delta T_L\theta_x(L), \tag{6.137}$$

$$\delta W_e^* = \int_0^L \delta m_z(x)\theta_z(x)dx + \int_0^L \delta q_y(x)u_y(x)dx + \delta V_{y_0}u_y(0) + \delta V_{y_L}u_y(L)$$
$$+ \delta M_{z_0}\theta_z(0) + \delta M_{z_L}\theta_z(L). \tag{6.138}$$

The complementary PVW is defined equating the complementary external work and the complementary strain energy for any set of virtual loads in equilibrium, that is,

$$\delta W_e^* = \delta W_i^*. \tag{6.139}$$

6.14.4 PRINCIPLE OF ENERGY CONSERVATION

A mechanical system is conservative if the work done by the internal and external loads is only dependent on the initial and final states of a deformation process. We assume the mechanical models presented here to be conservative. Consequently, the work done by the external loads is entirely converted into strain energy accumulated in the body. Thus,

$$W_e = U, \tag{6.140}$$

with W_e the work done by the external loads and U the internal strain energy stored in the body.

We can use this principle to determine the displacements and rotations of the points where the forces and moments are applied, as illustrated in the following example.

Example 6.13 *Calculate the rotation at point C in the beam illustrated in Figure 6.34, using the principle of energy conservation.*

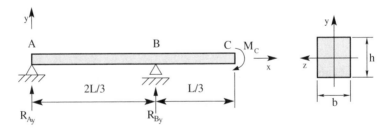

Figure 6.34 Example 6.13: beam with a concentrated moment at the right end.

The support reactions at A and B are calculated as

$$\sum F_y = 0: \quad R_{Ay} + R_{By} = 0,$$
$$\sum M_{z_A} = 0: \quad \tfrac{2}{3}R_{Ay} - M_c = 0.$$

Thus, $R_{Ay} = -\dfrac{3}{2}\dfrac{M_c}{L}$ and $R_{By} = \dfrac{3}{2}\dfrac{M_c}{L}$.

The equations for the shear force and bending moment are respectively given by

$$V_y(x) = \begin{cases} -\dfrac{3M_c}{2L} & 0 < x < \tfrac{2}{3}L \\ 0 & \tfrac{2}{3}L < x < L \end{cases} \quad and \quad M_z(x) = \begin{cases} -\dfrac{3M_c}{2L}x & 0 < x < \tfrac{2}{3}L \\ M_c & \tfrac{2}{3}L < x < L \end{cases}.$$

The diagrams for the shear force and bending moment due to the load applied over the beam are illustrated in Figure 6.35.

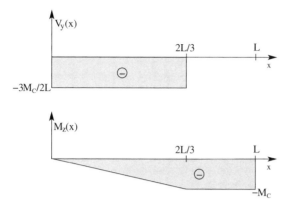

Figure 6.35 Example 6.13: diagrams for the shear force and bending moment for the real load applied to the beam.

The strain energy associated to the real load is given by expression (6.130). Thus,

$$U = \frac{1}{2} \int_0^L \frac{M_z^2(x)}{E(x)I_z(x)} dx + \frac{1}{2} \int_0^L \frac{V_y^2(x)}{K_c(x)G(x)A(x)} dx.$$

Substituting the equations for $V_y(x)$ and $M_z(x)$ on intervals $0 < x < \frac{2}{3}L$ and $\frac{2}{3}L < x < L$, we have

$$
\begin{aligned}
U &= \frac{1}{2E\dfrac{bh^3}{12}} \left(\int_0^{2/3L} \frac{9}{4}\frac{M_c^2}{L^2}x^2 dx + \int_{2/3L}^L M_c^2 dx \right) \\
&+ \frac{1}{2Gbh}\frac{6}{5} \left(\int_0^{2/3L} \frac{9}{4}\frac{M_c^2}{L^2} dx + \int_{2/3L}^L 0^2 dx \right) \\
&= \frac{10M_c^2}{3Ebh^3} + \frac{9M_c^2}{10GbhL}.
\end{aligned}
$$

Applying the principle of energy conservation, $W_e = U$, we have

$$\frac{1}{2}M_c\theta_c = \frac{10M_c^2}{3Ebh^3} + \frac{9M_c^2}{10GbhL}.$$

Thus,

$$\theta_c = \frac{20M_c}{3Ebh^3} + \frac{9M_c}{5GbhL}.$$

The factor $\frac{1}{2}$ was introduced in the external work expression because we assumed the external moment M_c to be gradually applied, and the rotation starts from zero until reaching its nominal value.

□

The limitation of this principle is that displacements and rotations are calculated only for the points where concentrated loads are applied.

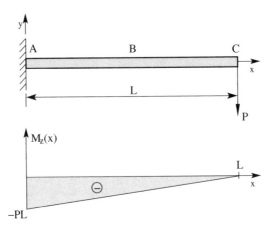

Figure 6.36   Cantilever beam with a concentrated force P at the free end.

6.14.5 METHOD OF VIRTUAL FORCES

The complementary PVW can be used to define the method of virtual force (MVF). This method allows the determination of displacements and rotations in a point of a body or system of bodies for a given direction.

For instance, we want to calculate the transversal displacement u_{y_B} at point B of the cantilever beam illustrated in Figure 6.36, and the bending moment diagrams. For this purpose, a unit virtual force $\delta F = 1$ is applied at point B in the vertical direction, as illustrated in Figure 6.37. Using the complementary PVW given in (6.139) and equation (6.130) for the complementary internal virtual work $(\delta W_i^* = 2\delta U_i^*)$ and neglecting the effect of shear force, we have

$$(\delta F)(u_{y_B}) = \int_0^L \frac{M_z(x)\delta M_z(x)}{E(x)I_z(x)}.$$

In this case, δM_z is the bending moment equation of the beam due to the unit virtual load applied. Thus,

$$u_{y_B} = \int_0^L \frac{M_z(x)\delta M_z(x)}{E(x)I_z(x)}.$$

In order to obtain the rotation θ_{z_B} for the same point, a virtual unit moment is applied at B and the diagram for the virtual bending moment δM_z is obtained, as illustrated in Figure 6.38. Then, the rotation θ_{z_B} is calculated as

$$\theta_{z_B} = \int_0^L \frac{M_z(x)\delta M_z(x)}{E(x)I_z(x)}. \tag{6.141}$$

Example 6.14 *For the beam illustrated in Figure 6.34, calculate the rotation at point B using the method of virtual force.*

Consider the auxiliary system illustrated in Figure 6.39, in which a virtual moment δM is applied at B. The support reactions at A and B respectively are $\delta R_{A_y} = -\dfrac{3\delta M}{2L}$ and $\delta R_{B_y} = \dfrac{3\delta M}{2L}$. The equations of the virtual shear force and virtual bending moment are

$$\delta V_y(x) = \begin{cases} -\dfrac{3\delta M}{2L} & 0 < x < \tfrac{2}{3}L \\ 0 & \tfrac{2}{3}L < x < L \end{cases} \quad and \quad \delta M_z(x) = \begin{cases} -\dfrac{3\delta M}{2L}x & 0 < x < \tfrac{2}{3}L \\ 0 & \tfrac{2}{3}L < x < L \end{cases}.$$

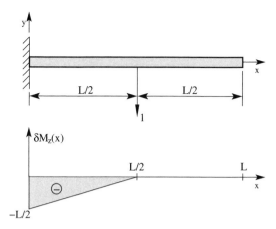

Figure 6.37 Unit virtual force in the displacement direction.

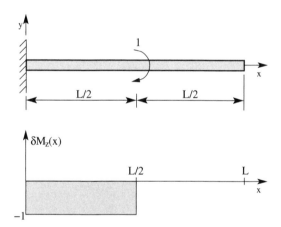

Figure 6.38 Beam with unit virtual moment.

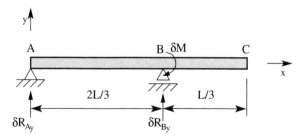

Figure 6.39 Example 6.14: auxiliary system with virtual moment δM.

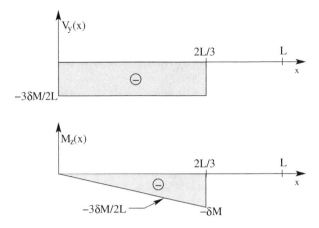

Figure 6.40 Example 6.14: diagrams for the shear force and bending moment due to the virtual unit moment.

The respective diagrams are illustrated in Figure 6.40.
The complementary strain internal virtual work is given by

$$\delta W_i^* = \int_0^L \frac{\delta M_z(x) M_z(x)}{E(x) I_z(x)} dx + \int_0^L \frac{\delta V_y(x) V_y(x)}{K_c(x) G(x) A(x)} dx$$

$$= \frac{12}{Ebh^3} \int_0^{\frac{2}{3L}} \left(-\frac{3}{2} \frac{\delta M}{L} x \right) \left(-\frac{3}{2} \frac{M_c}{L} x \right) dx +$$

$$\frac{1}{Gbh} \frac{3}{2} \int_0^{\frac{2}{3L}} \left(-\frac{3}{2} \frac{\delta M}{L} \right) \left(-\frac{3}{2} \frac{M_c}{L} \right) dx.$$

Thus,

$$\delta W_i^* = \frac{8}{3} \frac{\delta M M_c L}{Ebh^3} + \frac{9}{4} \frac{\delta M M_c}{GbhL}.$$

From the method of virtual force, we have

$$\theta_{zB} = \frac{8}{3} \frac{M_c L}{Ebh^2} + \frac{9}{4} \frac{M_c}{GbhL}.$$

□

Consider a bar fixed in one end and loaded by a force P at the free end. The axial displacement at the free end is given by

$$u_x = \frac{PL}{AE}.$$

On the other hand, the Δ displacement on a given direction for a truss node shared by n bars is given by

$$\Delta = \sum_{i=1}^{n} \delta P_i \frac{P_i L_i}{E_i A_i}, \tag{6.142}$$

with δP_i the normal virtual force on each bar due to the applied unit virtual force in the displacement direction.

One of the advantages of the energy methods is the simple displacement and rotation calculations for points of a mechanical system constituted of several structural elements, as illustrated in the following example.

Example 6.15 *Consider the structure illustrated in Figure 6.41(a), constituted of a beam AB and bars AC and BC, all made of steel ($E = 210$ GPa). The cross-section areas of the bars are 1200 cm^2. The beam has a cross-section area of 4000 cm^2 and second moment of area relative to z of 200000 cm^4. We want to calculate the horizontal displacement of point C and the rotation of point B. Neglect the effect of shear force.*

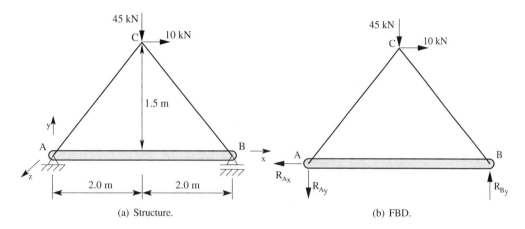

(a) Structure. (b) FBD.

Figure 6.41 Example 6.15: structure with beams and bars.

The support reactions at points A and B, due to the applied load, are indicated in the FBD illustrated in Figure 6.41(b). The values are $R_{A_x} = 10$ kN, $R_{A_y} = 1.5$ kN, and $R_{B_y} = 6$ kN. The diagrams for the normal force, shear force, and bending moment of beam AB are shown in Figure 6.42. The normal forces on bars AC and BC are obtained by the method of nodes, with $N_{x_{AC}} = 2.5$ kN and $N_{x_{BC}} = -10$ kN.

Now consider the FBD of the auxiliary system illustrated in Figure 6.43(a) with a unit virtual force applied in the horizontal direction of node C. The values of the support reactions are $\delta R_{A_x} = 1$ kN, $\delta R_{A_y} = 0.375$ kN, and $\delta R_{B_y} = 0.375$ kN. The diagrams for the normal force, shear force, and bending moment on beam AB are illustrated in Figure 6.44. The normal forces on bars AC and BC are obtained by the method of nodes, with $\delta N_{x_{AC}} = 0.5$ kN and $\delta N_{x_{BC}} = -0.625$ kN.

Applying the method of virtual force (MVF), the internal virtual work is given by the sum of the the works of beam AB and bars AC and BC. For beam AB, we sum the terms of the internal work relative to the normal force and bending moment. Thus,

$$
\begin{aligned}
u_{x_C} &= \delta T_{i_{AB}} + \delta T_{i_{AC}} + \delta T_{i_{BC}} \\
&= \int_0^4 \frac{M_z(x)\delta M_z(x)}{EI_{z_{AB}}}dx + \int_0^4 \frac{N_x(x)\delta N_x(x)}{EA_{AB}}dx \\
&\quad + \int_0^{2.5} \frac{N_x(x)\delta N_x(x)}{EA_{AC}}dx + \int_0^{2.5} \frac{N_x(x)\delta N_x(x)}{EA_{BC}}dx \\
&\quad + \frac{1}{EI_{z_{AB}}}\int_0^2 (-1500x)(-375x)dx + \frac{1}{EI_{z_{AB}}}\int_2^4 (24000 - 6000x)(1500 - 375x)dx \\
&\quad + \frac{1}{EA_{AB}}\int_0^2 (10000)(1000)dx + \frac{1}{EA_{AC}}\int_0^{2.5} (2500)(625)dx \\
&\quad + \frac{1}{EA_{BC}}\int_0^{2.5} (-10000)(-625)dx = 1.89 \text{ cm.}
\end{aligned}
$$

Consider now the FBD of the auxiliary system illustrated in Figure 6.43(b) with the virtual unit moment applied at point B. The support reactions are $\delta R_{A_x} = 0$ kN, $\delta R_{A_y} = 0.25$ kN, and

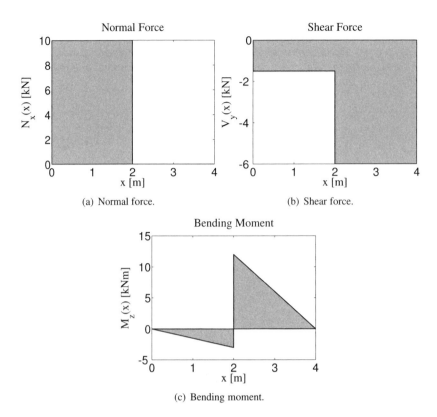

(a) Normal force.

(b) Shear force.

(c) Bending moment.

Figure 6.42 Example 6.15: diagrams due to the real load.

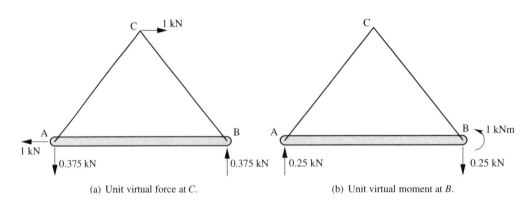

(a) Unit virtual force at C.

(b) Unit virtual moment at B.

Figure 6.43 Example 6.15: FBDs of the auxiliary systems.

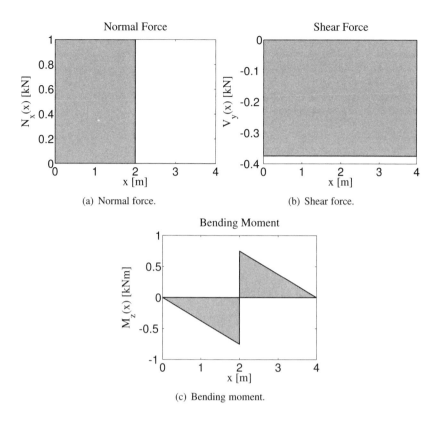

(a) Normal force.

(b) Shear force.

(c) Bending moment.

Figure 6.44 Example 6.15: diagrams due to the horizontal virtual unit force at point C.

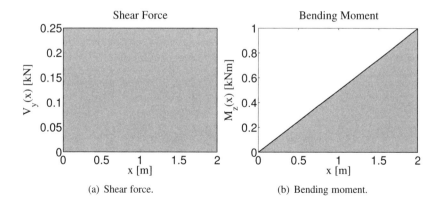

(a) Shear force.

(b) Bending moment.

Figure 6.45 Example 6.15: diagrams due to the virtual unit moment at point B.

$\delta R_{B_y} = 0.25$ kN. *The diagrams for the shear force and bending moment for beam AB are illustrated in Figure 6.45. The normal force on beam AB is zero. The normal forces on bars AC and BC are obtained by the method of nodes, with $\delta N_{x_{AC}} = -0.3125$ kN and $\delta N_{x_{BC}} = 0.3125$ kN.*

The rotation of point B is determined by using the MVF as

$$
\begin{aligned}
\theta_{z_B} &= \delta T_{i_{AB}} + \delta T_{i_{AC}} + \delta T_{i_{BC}} \\
&= \int_0^4 \frac{M_z(x)\delta M_z(x)}{EI_{z_{AB}}}dx + \int_0^{2,5} \frac{N_x(x)\delta N_x(x)}{EA_{AC}}dx + \int_0^{2,5} \frac{N_x(x)\delta N_x(x)}{EA_{BC}}dx \\
&= \frac{1}{EI_{z_{AB}}}\int_0^2 (-1500x)(250x)dx + \frac{1}{EI_{z_{AB}}}\int_2^4 (24000-6000x)(250x)dx \\
&\quad + \frac{1}{EA_{AC}}\int_0^{2,5}(2500)(-312,5)dx + \frac{1}{EA_{BC}}\int_0^{2,5}(-10000)(312,5)dx \\
&= 0.016 \text{ rad.}
\end{aligned}
$$

□

6.15   APPROXIMATED SOLUTION

The BVP of the Timoshenko beam is given in terms of equations (6.47) and (6.48). The former states the balance in terms of internal and external moments in direction z. The latter represents the balance of forces in the transversal direction y. Therefore, the test functions $\alpha(x)$ and $v(x)$, which multiply the respective differential equations to obtain the weak form, physically represent rotations in z and transversal displacements in y.

Performing the multiplications of (6.47) and (6.48), respectively, by the test functions $\alpha(x)$ and $v(x)$ and integrating along the length of the beam, we have

$$
\int_0^L \frac{d}{dx}\left(E(x)I_z(x)\frac{d\theta_z(x)}{dx}\right)\alpha(x)dx + \int_0^L K_c(x)G(x)A(x)\left(\frac{du_y(x)}{dx}-\theta_z(x)\right)\alpha(x)dx
$$
$$
+ \int_0^L m_z(x)\alpha(x)dx = 0,
$$
$$
\int_0^L \frac{d}{dx}\left[K_c(x)G(x)A(x)\left(\frac{du_y(x)}{dx}-\theta_z(x)\right)\right]v(x)dx + \int_0^L q_y(x)v(x)dx = 0.
$$

The above integrals represent the work of the internal and external loads for the rotation $\alpha(x)$ and displacement $v(x)$. We can sum both above equations, resulting in

$$
\int_0^L \frac{d}{dx}\left(E(x)I_z(x)\frac{d\theta_z(x)}{dx}\right)\alpha(x)dx + \int_0^L \frac{d}{dx}\left[K_c(x)G(x)A(x)\left(\frac{du_y(x)}{dx}-\theta_z(x)\right)\right]v(x)dx
$$
$$
+ \int_0^L K_c(x)G(x)A(x)\left(\frac{du_y(x)}{dx}-\theta_z(x)\right)\alpha(x)dx + \int_0^L m_z(x)\alpha(x)dx + \int_0^L q_y(x)v(x)dx = 0.
$$

Integrating by parts the two first terms of the above equation, we have

$$
-\int_0^L E(x)I_z(x)\frac{d\theta_z(x)}{dx}\frac{d\alpha(x)}{dx}dx + E(x)I_z(x)\frac{d\theta_z(x)}{dx}\alpha(x)\Big|_0^L
$$
$$
-\int_0^L K_c(x)G(x)A(x)\left(\frac{du_y(x)}{dx}-\theta_z(x)\right)\frac{dv(x)}{dx}dx + K_c(x)G(x)A(x)\left(\frac{du_y(x)}{dx}-\theta_z(x)\right)v(x)\Big|_0^L
$$
$$
+\int_0^L K_c(x)G(x)A(x)\left(\frac{du_y(x)}{dx}-\theta_z(x)\right)\alpha(x)dx + \int_0^L m_z(x)\alpha(x)dx + \int_0^L q_y(x)v(x)dx = 0.
$$

Using definitions (6.36) and (6.39) for the shear force and bending moment, the above expression is rewritten as

$$\int_0^L E(x)I_z(x)\frac{d\theta_z(x)}{dx}\frac{d\alpha(x)}{dx}dx + \int_0^L K_c(x)G(x)A(x)\left(\frac{du_y(x)}{dx} - \theta_z(x)\right)\left(\frac{dv(x)}{dx} - \alpha(x)\right)dx =$$
$$\int_0^L m_z(x)\alpha(x)dx + \int_0^L q_y(x)v(x)dx + M_z(x)\alpha(x)\big|_0^L - V_y(x)v(x)\big|_0^L.$$

If the test functions $\alpha(x)$ and $v(x)$ are respectively the virtual rotation $\delta\theta_z$ and virtual transversal displacement δu_y, the above expression represents the PVW, stating the equality of the work done by internal and external loads.

Substituting the boundary conditions given in the BVP (6.25), we obtain the final expression of the weak form of the Timoshenko beam

$$\int_0^L E(x)I_z(x)\frac{d\theta_z(x)}{dx}\frac{d\alpha(x)}{dx}dx + \int_0^L K_c(x)G(x)A(x)\left(\frac{du_y(x)}{dx} - \theta_z(x)\right)\left(\frac{dv(x)}{dx} - \alpha(x)\right)dx =$$
$$\int_0^L m_z(x)\alpha(x)dx + \int_0^L q_y(x)v(x)dx + M_{z_L}\alpha(L) + M_{z_0}\alpha(0) + V_{y_L}v(L) + V_{y_0}v(0). \quad (6.143)$$

Consider the two-node beam element illustrated in Figure 5.53. As in the Timoshenko beam, the transversal displacements and rotations are not directly related to each other, they must be interpolated independently. Hence, the two-node element results in the linear interpolation of these variables. In terms of the local coordinate ξ_1, we have

$$u_{y_2}^{(e)}(\xi_1) = \bar{u}_{y_1}^{(e)}\phi_1^{(e)}(\xi_1) + \bar{u}_{y_2}^{(e)}\phi_2^{(e)}(\xi_1), \quad (6.144)$$
$$\theta_{z_2}^{(e)}(\xi_1) = \bar{\theta}_{z_1}^{(e)}\phi_1^{(e)}(\xi_1) + \bar{\theta}_{z_2}^{(e)}\phi_2^{(e)}(\xi_1). \quad (6.145)$$

In matrix form,

$$\left\{\begin{array}{c} u_{y_2}^{(e)}(\xi_1) \\ \theta_{z_2}^{(e)}(\xi_1) \end{array}\right\} = \left[\begin{array}{cccc} \phi_1^{(e)}(\xi_1) & 0 & \phi_2^{(e)}(\xi_1) & 0 \\ 0 & \phi_1^{(e)}(\xi_1) & 0 & \phi_2^{(e)}(\xi_1) \end{array}\right] \left\{\begin{array}{c} \bar{u}_{y_1}^{(e)} \\ \bar{\theta}_{z_1}^{(e)} \\ \bar{u}_{y_2}^{(e)} \\ \bar{\theta}_{z_2}^{(e)} \end{array}\right\} = [\bar{N}^{(e)}]\left\{\begin{array}{c} \bar{u}_{y_1}^{(e)} \\ \bar{\theta}_{z_1}^{(e)} \\ \bar{u}_{y_2}^{(e)} \\ \bar{\theta}_{z_2}^{(e)} \end{array}\right\},$$

$$\quad (6.146)$$

with $[\bar{N}^{(e)}]$ the matrix with shape functions $\phi_1^{(e)}(\xi_1) = \frac{1}{2}(1 - \xi_1)$ and $\phi_2^{(e)}(\xi_1) = \frac{1}{2}(1 + \xi_1)$.

The interpolation of the global derivatives of $u_y^{(e)}[x(\xi_1)]$ and $\theta_z^{(e)}[x(\xi_1)]$ are given, in matrix notation, by

$$\left\{\begin{array}{c} \dfrac{du_{y_2}^{(e)}[x(\xi_1)]}{dx} \\ \dfrac{d\theta_{z_2}^{(e)}[x(\xi_1)]}{dx} \end{array}\right\} = \frac{2}{h^{(e)}}\left[\begin{array}{cccc} -1 & 0 & 1 & 0 \\ 0 & -1 & 0 & 1 \end{array}\right]\left\{\begin{array}{c} \bar{u}_{y_1}^{(e)} \\ \bar{\theta}_{z_1}^{(e)} \\ \bar{u}_{y_2}^{(e)} \\ \bar{\theta}_{z_2}^{(e)} \end{array}\right\}, \quad (6.147)$$

where the term $\dfrac{2}{h^{(e)}}$ is the inverse of the Jacobian.

We interpolate the element test functions $v^{(e)}$ and $\alpha^{(e)}$ and the respective global derivatives using the Galerkin method in an analogous way

$$\left\{\begin{array}{c} v_2^{(e)}(\xi_1) \\ \alpha_2^{(e)}(\xi_1) \end{array}\right\} = \left[\begin{array}{cccc} \phi_1^{(e)}(\xi_1) & 0 & \phi_2^{(e)}(\xi_1) & 0 \\ 0 & \phi_1^{(e)}(\xi_1) & 0 & \phi_2^{(e)}(\xi_1) \end{array}\right]\left\{\begin{array}{c} \bar{v}_1^{(e)} \\ \bar{\alpha}_1^{(e)} \\ \bar{v}_2^{(e)} \\ \bar{\alpha}_2^{(e)} \end{array}\right\}, \quad (6.148)$$

$$
\left\{
\begin{array}{c}
\dfrac{dv_2^{(e)}[x(\xi_1)]}{dx} \\[2mm]
\dfrac{d\alpha_2^{(e)}[x(\xi_1)]}{dx}
\end{array}
\right\}
=
\frac{2}{h^{(e)}}
\left[
\begin{array}{cccc}
-1 & 0 & 1 & 0 \\
0 & -1 & 0 & 1
\end{array}
\right]
\left\{
\begin{array}{c}
\bar{v}_1^{(e)} \\
\bar{\alpha}_1^{(e)} \\
\bar{v}_2^{(e)} \\
\bar{\alpha}_2^{(e)}
\end{array}
\right\},
\tag{6.149}
$$

with $(\bar{v}_1^{(e)}, \bar{v}_2^{(e)})$ and $(\bar{\alpha}_1^{(e)}, \bar{\alpha}_2^{(e)})$ the approximation coefficients of the test functions $v^{(e)}$ and $\alpha^{(e)}$, respectively.

The weak form in the local element system is given by

$$
\int_{-1}^{1} EI_z \frac{d\theta_z^{(e)}[x(\xi_1)]}{dx} \frac{d\alpha^{(e)}[x(\xi_1)]}{dx} |J|d\xi_1 + \int_{-1}^{1} K_c GA \left(\frac{du_y^{(e)}[x(\xi_1)]}{dx} - \theta_z^{(e)}[x(\xi_1)] \right)
$$
$$
\left(\frac{dv^{(e)}[x(\xi_1)]}{dx} - \alpha^{(e)}[x(\xi_1)] \right) |J|d\xi_1 = \int_{-1}^{1} m_z[x(\xi_1)]\alpha^{(e)}[x(\xi_1)]|J|d\xi_1 \tag{6.150}
$$
$$
+ \int_{-1}^{1} q_y[x(\xi_1)]v^{(e)}[x(\xi_1)]|J|d\xi_1 + \bar{M}_{z_2}\alpha^{(e)}(1) + \bar{M}_{z_1}\alpha^{(e)}(-1) + \bar{V}_{y_2}v^{(e)}(1) + \bar{V}_{y_1}v^{(e)}(-1),
$$

with $|J| = \dfrac{h^{(e)}}{2}$, and we assume the material and geometrical properties to be constant within the element.

The first term in the left-hand side is related to bending. Substituting approximations given in (6.147) and (6.149) in this term, we have

$$
E^{(e)}I_z^{(e)} \int_{-1}^{1} \frac{4}{h^{(e)2}} \left\{
\begin{array}{cccc}
\bar{u}_{y_1}^{(e)} & \bar{\theta}_{z_1}^{(e)} & \bar{u}_{y_2}^{(e)} & \bar{\theta}_{z_2}^{(e)}
\end{array}
\right\}
\left[
\begin{array}{c}
0 \\
-1 \\
0 \\
1
\end{array}
\right]
\left[
\begin{array}{cccc}
0 & -1 & 0 & 1
\end{array}
\right]
\left\{
\begin{array}{c}
\bar{v}_1^{(e)} \\
\bar{\alpha}_1^{(e)} \\
\bar{v}_2^{(e)} \\
\bar{\alpha}_2^{(e)}
\end{array}
\right\}
\frac{h^{(e)}}{2}d\xi_1
$$

$$
= \left\{
\begin{array}{cccc}
\bar{v}_1^{(e)} & \bar{\alpha}_1^{(e)} & \bar{v}_2^{(e)} & \bar{\alpha}_2^{(e)}
\end{array}
\right\}
\left(\frac{E^{(e)}I_z^{(e)}}{h^{(e)}}
\left[
\begin{array}{cccc}
0 & 0 & 0 & 0 \\
0 & 1 & 0 & -1 \\
0 & 0 & 0 & 0 \\
0 & -1 & 0 & 1
\end{array}
\right]
\left\{
\begin{array}{c}
\bar{u}_{y_1}^{(e)} \\
\bar{\theta}_{z_1}^{(e)} \\
\bar{u}_{y_2}^{(e)} \\
\bar{\theta}_{z_2}^{(e)}
\end{array}
\right\}
\right)
$$

$$
= \left\{
\begin{array}{cccc}
\bar{v}_1^{(e)} & \bar{\alpha}_1^{(e)} & \bar{v}_2^{(e)} & \bar{\alpha}_2^{(e)}
\end{array}
\right\}
[\bar{K}_f^{(e)}]
\left\{
\begin{array}{c}
\bar{u}_{y_1}^{(e)} \\
\bar{\theta}_{z_1}^{(e)} \\
\bar{u}_{y_2}^{(e)} \\
\bar{\theta}_{z_2}^{(e)}
\end{array}
\right\},
\tag{6.151}
$$

with $[\bar{K}_f^{(e)}]$ the bending stiffness matrix.

The second term of the left-hand side in (6.150) represents the shear effect. Substituting the approximations given in (6.146) and (6.149), we have

$$
K_c^{(e)} G^{(e)} A^{(e)} \int_{-1}^{1} \left\{ \begin{array}{cccc} \bar{u}_{y_1}^{(e)} & \bar{\theta}_{z_1}^{(e)} & \bar{u}_{y_2}^{(e)} & \bar{\theta}_{z_2}^{(e)} \end{array} \right\} \begin{bmatrix} \dfrac{2}{h^{(e)}} \phi_1'(\xi_1) \\ -\phi_1^{(e)}(\xi_1) \\ \dfrac{2}{h^{(e)}} \phi_2'(\xi_1) \\ -\phi_2^{(e)}(\xi_1) \end{bmatrix}
$$

$$
\begin{bmatrix} \dfrac{2}{h^{(e)}} \phi_1'(\xi_1) & -\phi_1^{(e)}(\xi_1) & \dfrac{2}{h^{(e)}} \phi_2'(\xi_1) & -\phi_2^{(e)}(\xi_1) \end{bmatrix} \left\{ \begin{array}{c} \bar{v}_1^{(e)} \\ \bar{\alpha}_1^{(e)} \\ \bar{v}_2^{(e)} \\ \bar{\alpha}_2^{(e)} \end{array} \right\} \dfrac{h^{(e)}}{2} d\xi_1
$$

$$
= \left\{ \begin{array}{cccc} \bar{v}_1^{(e)} & \bar{\alpha}_1^{(e)} & \bar{v}_2^{(e)} & \bar{\alpha}_2^{(e)} \end{array} \right\} \left(\dfrac{K_c^{(e)} G^{(e)} A^{(e)}}{h^{(e)}} \begin{bmatrix} 1 & \dfrac{h^{(e)}}{2} & -1 & \dfrac{h^{(e)}}{2} \\ \dfrac{h^{(e)}}{2} & \dfrac{h^{(e)^2}}{3} & -\dfrac{h^{(e)}}{2} & \dfrac{h^{(e)^2}}{6} \\ -1 & -\dfrac{h^{(e)}}{2} & 1 & -\dfrac{h^{(e)}}{2} \\ \dfrac{h^{(e)}}{2} & \dfrac{h^{(e)^2}}{6} & -\dfrac{h^{(e)}}{2} & \dfrac{h^{(e)^2}}{3} \end{bmatrix} \left\{ \begin{array}{c} \bar{u}_{y_1}^{(e)} \\ \bar{\theta}_{z_1}^{(e)} \\ \bar{u}_{y_2}^{(e)} \\ \bar{\theta}_{z_2}^{(e)} \end{array} \right\} \right)
$$

$$
= \left\{ \begin{array}{cccc} \bar{v}_1^{(e)} & \bar{\alpha}_1^{(e)} & \bar{v}_2^{(e)} & \bar{\alpha}_2^{(e)} \end{array} \right\} [\bar{K}_c^{(e)}] \left\{ \begin{array}{c} \bar{u}_{y_1}^{(e)} \\ \bar{\theta}_{z_1}^{(e)} \\ \bar{u}_{y_2}^{(e)} \\ \bar{\theta}_{z_2}^{(e)} \end{array} \right\}, \tag{6.152}
$$

with $[\bar{K}_c^{(e)}]$ the shear stiffness matrix.

The terms of the right-hand side of the weak form are respectively given by

$$
\int_{-1}^{1} m_z[x(\xi_1)] \alpha^{(e)}(\xi_1) |J| d\xi_1 = \left\{ \begin{array}{cccc} \bar{v}_1^{(e)} & \bar{\alpha}_1^{(e)} & \bar{v}_2^{(e)} & \bar{\alpha}_2^{(e)} \end{array} \right\} \int_{-1}^{1} m_z[x(\xi_1)] \left\{ \begin{array}{c} 0 \\ \phi_1^{(e)}(\xi_1) \\ 0 \\ \phi_2^{(e)}(\xi_1) \end{array} \right\} \dfrac{h^{(e)}}{2} d\xi_1
$$

$$
= \left\{ \begin{array}{cccc} \bar{v}_1^{(e)} & \bar{\alpha}_1^{(e)} & \bar{v}_2^{(e)} & \bar{\alpha}_2^{(e)} \end{array} \right\} \{ \bar{f}_{m_z}^{(e)} \}, \tag{6.153}
$$

$$
\int_{-1}^{1} q_y[x(\xi_1)] v^{(e)}(\xi_1) |J| d\xi_1 = \left\{ \begin{array}{cccc} \bar{v}_1^{(e)} & \bar{\alpha}_1^{(e)} & \bar{v}_2^{(e)} & \bar{\alpha}_2^{(e)} \end{array} \right\} \int_{-1}^{1} q_y[x(\xi_1)] \left\{ \begin{array}{c} \phi_1^{(e)}(\xi_1) \\ 0 \\ \phi_2^{(e)}(\xi_1) \\ 0 \end{array} \right\} \dfrac{h^{(e)}}{2} d\xi_1
$$

$$
= \left\{ \begin{array}{cccc} \bar{v}_1^{(e)} & \bar{\alpha}_1^{(e)} & \bar{v}_2^{(e)} & \bar{\alpha}_2^{(e)} \end{array} \right\} \{ \bar{f}_{q_y}^{(e)} \}, \tag{6.154}
$$

$$
\bar{M}_{z_2} \alpha(1) + \bar{M}_{z_1} \alpha(-1) + \bar{V}_{y_2} v(1) + \bar{V}_{y_1} v(-1) = \left\{ \begin{array}{cccc} \bar{v}_1^{(e)} & \bar{\alpha}_1^{(e)} & \bar{v}_2^{(e)} & \bar{\alpha}_2^{(e)} \end{array} \right\} \left\{ \begin{array}{c} \bar{V}_{y_1} \\ \bar{M}_{z_1} \\ \bar{V}_{y_2} \\ \bar{M}_{z_2} \end{array} \right\}
$$

$$
= \left\{ \begin{array}{cccc} \bar{v}_1^{(e)} & \bar{\alpha}_1^{(e)} & \bar{v}_2^{(e)} & \bar{\alpha}_2^{(e)} \end{array} \right\} \{ \bar{f}_{VM}^{(e)} \}. \tag{6.155}
$$

Substituting (6.151) to (6.155) in the weak form (6.143) and simplifying the test function coefficients, we have the following element system of equations for $K_c^{(e)} G^{(e)} A^{(e)} = K_c GA$ and $E^{(e)} I_z^{(e)} = EI_z$:

$$
\begin{bmatrix}
\dfrac{K_c GA}{h^{(e)}} & \dfrac{K_c GA}{2} & -\dfrac{K_c GA}{h^{(e)}} & \dfrac{K_c GA}{2} \\[2mm]
\dfrac{K_c GA}{2} & \dfrac{K_c GA h^{(e)}}{3} + \dfrac{EI_z}{h^{(e)}} & -\dfrac{K_c GA}{2} & \dfrac{K_c GA h^{(e)}}{6} - \dfrac{EI_z}{h^{(e)}} \\[2mm]
-\dfrac{K_c GA}{h^{(e)}} & -\dfrac{K_c GA}{2} & \dfrac{K_c GA}{h^{(e)}} & -\dfrac{K_c^{(e)} GA}{2} \\[2mm]
\dfrac{K_c GA}{2} & \dfrac{K_c GA h^{(e)}}{6} - \dfrac{EI_z}{h^{(e)}} & -\dfrac{K_c GA}{2} & \dfrac{K_c GA h^{(e)}}{3} + \dfrac{EI_z}{h^{(e)}}
\end{bmatrix}
\begin{Bmatrix}
\bar{u}_{y_1}^{(e)} \\[1mm] \bar{\theta}_{z_1}^{(e)} \\[1mm] \bar{u}_{y_2}^{(e)} \\[1mm] \bar{\theta}_{z_2}^{(e)}
\end{Bmatrix}
=
\begin{Bmatrix}
\bar{V}_{y_1} \\[1mm] \bar{M}_{z_1} \\[1mm] \bar{V}_{y_2} \\[1mm] \bar{M}_{z_2}
\end{Bmatrix}
$$

$$
+ \int_{-1}^{1} m_z[x(\xi_1)]
\begin{Bmatrix}
0 \\ \phi_1^{(e)}(\xi_1) \\ 0 \\ \phi_2^{(e)}(\xi_1)
\end{Bmatrix}
\frac{h^{(e)}}{2} d\xi_1
+ \int_{-1}^{1} q_y[x(\xi_1)]
\begin{Bmatrix}
\phi_1^{(e)}(\xi_1) \\ 0 \\ \phi_2^{(e)}(\xi_1) \\ 0
\end{Bmatrix}
\frac{h^{(e)}}{2} d\xi_1 . \tag{6.156}
$$

In a summarized form,

$$
([\bar{K}_f^{(e)}] + [\bar{K}_c^{(e)}]) \{\bar{u}_e\} = \{\bar{f}_{m_z}^{(e)}\} + \{\bar{f}_{q_y}^{(e)}\} + \{\bar{f}_{VM}^{(e)}\}. \tag{6.157}
$$

Defining [43, 8]

$$
\kappa_c^{(e)} = \frac{12 E^{(e)} I_z^{(e)}}{h^{(e)2} K_c^{(e)} G^{(e)} A^{(e)}}, \tag{6.158}
$$

the element stiffness matrix can be rewritten as

$$
[\bar{K}^{(e)}] = \frac{E^{(e)} I_z^{(e)}}{\kappa_c^{(e)} h^{(e)3}}
\begin{bmatrix}
12 & 6h^{(e)} & -12 & 6h^{(e)} \\[1mm]
6h^{(e)} & (\kappa_c^{(e)} + 4) h^{(e)2} & -6h^{(e)} & -(\kappa_c^{(e)} - 2) h^{(e)2} \\[1mm]
-12 & -6h^{(e)} & 12 & -6h^{(e)} \\[1mm]
6h^{(e)} & -(\kappa_c^{(e)} - 2) h^{(e)2} & -6h^{(e)} & (\kappa_c^{(e)} + 4) h^{(e)2}
\end{bmatrix} . \tag{6.159}
$$

Note that for $\kappa_c^{(e)} = 0$, the coefficients in parentheses are reduced to the bending terms given in (5.128). However, as $\kappa_c^{(e)}$ is present in the denominator of the constant that is multiplying the matrix, the coefficients become very small when $\kappa_c^{(e)} \to 0$. This leads to an ill-conditioned element matrix, making the solution more difficult when the cross-sections are thin. This phenomenon is called shear locking, and is presented below.

Example 6.16 *Consider the cantilever beam of Example 6.5 with the concentrated force F equal to zero. The solution of this beam using a single element in bending leads to the following system of equations:*

$$
\frac{EI_z}{L^3}
\begin{bmatrix}
12 & 6L & -12 & 6L \\
6L & 4L^2 & -6L & 2L^2 \\
-12 & -6L & 12 & -6L \\
6L & 2L^2 & -6L & 4L^2
\end{bmatrix}
\begin{Bmatrix}
\bar{u}_{y_1} \\ \bar{\theta}_{z_1} \\ \bar{u}_{y_2} \\ \bar{\theta}_{z_2}
\end{Bmatrix}
= \frac{q_0 L}{2}
\begin{Bmatrix}
1 \\ 0 \\ 1 \\ 0
\end{Bmatrix} .
$$

After applying the boundary conditions $\bar{u}_{y_1} = \bar{\theta}_{z_1} = 0$, *we have*

$$
\frac{EI_z}{L^3}
\begin{bmatrix}
12 & -6L \\
-L & 4L^2
\end{bmatrix}
\begin{Bmatrix}
\bar{u}_{y_2} \\ \bar{\theta}_{z_2}
\end{Bmatrix}
= \frac{q_0 L}{2}
\begin{Bmatrix}
1 \\ 0
\end{Bmatrix} .
$$

The solution of this system of equations results in the following expressions for the transversal displacement and rotation at the right end of the beam:

$$\bar{\theta}_{z2} = -\frac{q_0 L^3}{4EI_z} \quad and \quad \bar{u}_{y2} = -\frac{q_0 L^4}{6EI_z}.$$

Now using the element with shear, we obtain the following system of equations after applying the boundary conditions:

$$\frac{EI_z}{\kappa_c L^3} \begin{bmatrix} 12 & -6L \\ -L & (\kappa_c + 4)L^2 \end{bmatrix} \begin{Bmatrix} \bar{u}_{y2} \\ \bar{\theta}_{z2} \end{Bmatrix} = \frac{q_0 L}{2} \begin{Bmatrix} 1 \\ 0 \end{Bmatrix},$$

which solution results in

$$\bar{\theta}_{z2} = -\frac{\kappa_c}{\kappa_c + 1} \frac{q_0 L^3}{4EI_z} \quad and \quad \bar{u}_{y2} = -\frac{\kappa_c(\kappa_c + 4)}{\kappa_c + 1} \frac{q_0 L^4}{6EI_z}.$$

For $\kappa_c \to 0$, we recover the solution with bending.

Figure 6.46(a) illustrates the absolute values of transversal displacement at the right end of the beam according to the previous expressions and the analytical solution for the height h of the cross-section in the range $[0.1; 1.0]$ cm, with $\Delta h = 0.05$ cm increments. Notice that as the section becomes thinner, the obtained solution with shear does not match with the analytical and bending ones. This behavior characterizes the shear locking, and is related to the most significant influence of shear in the solution.

For $E = 210.00$ GPa and $G = 80.80$ GPa, we have $\kappa_c = 3.12h^2$. The height corresponding to $\kappa_c = 1$ is $h = 0.56$ m. This is the maximum height at which the shear effect becomes more prevalent and is shown by the vertical line in Figure 6.46(a). On the other hand, Figure 6.46(b) illustrates the behavior of κ_c in terms of the height of the section. Notice that as κ_c goes to zero, the solution has a bending behavior, but the approximate solution with shear does not match this behavior.

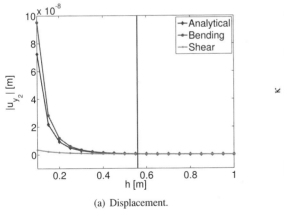

(a) Displacement.

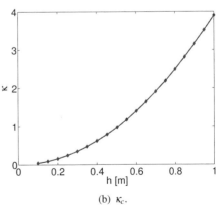

(b) κ_c.

Figure 6.46 Example 6.10: effect of shear locking in the solution.

□

Several techniques can be employed to solve the solution locking problem, such as the selective numerical integration and the mixed formulation [33, 8]. The use of high-order methods to approximate the beam with shear is considered here, which overcomes the locking using of a minimal adequate polynomial order.

In this case, the transversal displacements and rotations are interpolated in the local system ξ_1 using $N+1$ shape functions until order P as

$$u_{y_{N+1}}^{(e)}(\xi_1) = \sum_{i=0}^{N} a_{u_{y_i}} \phi_i^{(e)}(\xi_1), \tag{6.160}$$

$$\theta_{z_{N+1}}^{(e)}(\xi_1) = \sum_{i=0}^{N} a_{\theta_{z_i}} \phi_i^{(e)}(\xi_1), \tag{6.161}$$

with $a_{u_{y_i}}$ and $a_{\theta_{z_i}}$ the approximation coefficients and the shape functions can be given in terms of the Lagrange and Jacobi polynomials. We have, in matrix notation,

$$\left\{ \begin{array}{c} u_{y_{N+1}}^{(e)}(\xi_1) \\ \theta_{z_{N+1}}^{(e)}(\xi_1) \end{array} \right\} = \left[\begin{array}{ccccc} \phi_0^{(e)}(\xi_1) & 0 & \cdots & \phi_N^{(e)}(\xi_1) & 0 \\ 0 & \phi_0^{(e)}(\xi_1) & \cdots & 0 & \phi_N^{(e)}(\xi_1) \end{array} \right] \left\{ \begin{array}{c} a_{u_{y_0}} \\ a_{\theta_{z_0}} \\ \vdots \\ a_{u_{y_N}} \\ a_{\theta_{z_N}} \end{array} \right\}. \tag{6.162}$$

The approximations of the global derivatives are given by

$$\left\{ \begin{array}{c} \dfrac{du_{y_{N+1}}^{(e)}[x(\xi_1)]}{dx} \\ \dfrac{d\theta_{z_{N+1}}^{(e)}[x(\xi_1)]}{dx} \end{array} \right\} = \dfrac{2}{h^{(e)}} \left[\begin{array}{ccccc} \phi_0^{(e)\prime}(\xi_1) & 0 & \cdots & \phi_N^{(e)\prime}(\xi_1) & 0 \\ 0 & \phi_0^{(e)\prime}(\xi_1) & \cdots & 0 & \phi_N^{(e)\prime}(\xi_1) \end{array} \right] \left\{ \begin{array}{c} a_{u_{y_0}} \\ a_{\theta_{z_0}} \\ \vdots \\ a_{u_{y_N}} \\ a_{\theta_{z_N}} \end{array} \right\}. \tag{6.163}$$

By performing the same previous procedure, the bending and shear stiffness matrices are given by

$$[\bar{K}_f^{(e)}] = E^{(e)} I_z^{(e)} \int_{-1}^{1} \left[\begin{array}{ccccccc} 0 & 0 & 0 & 0 & \cdots & 0 & 0 \\ 0 & \phi_0^{(e)\prime}\phi_0^{(e)\prime} & 0 & \phi_0^{(e)\prime}\phi_1^{(e)\prime} & \cdots & 0 & \phi_0^{(e)\prime}\phi_N^{(e)\prime} \\ 0 & 0 & 0 & 0 & \cdots & 0 & 0 \\ 0 & \phi_0^{(e)\prime}\phi_1^{(e)\prime} & 0 & \phi_1^{(e)\prime}\phi_1^{(e)\prime} & \cdots & 0 & \phi_1^{(e)\prime}\phi_N^{(e)\prime} \\ \vdots & \vdots & \vdots & \vdots & \vdots & \vdots & \vdots \\ 0 & 0 & 0 & 0 & \cdots & 0 & 0 \\ 0 & \phi_0^{(e)\prime}\phi_N^{(e)\prime} & 0 & \phi_1^{(e)\prime}\phi_N^{(e)\prime} & \cdots & 0 & \phi_N^{(e)\prime}\phi_N^{(e)\prime} \end{array} \right] \dfrac{h^{(e)}}{2} d\xi_1, \tag{6.164}$$

$$[\bar{K}_c^{(e)}] = K_c^{(e)} G^{(e)} A^{(e)}$$

$$\int_{-1}^{1} \left[\begin{array}{ccccc} \dfrac{4}{h^{(e)2}}\phi_0^{(e)\prime}\phi_0^{(e)\prime} & -\dfrac{2}{h^{(e)}}\phi_0^{(e)}\phi_0^{(e)\prime} & \cdots & \dfrac{4}{h^{(e)2}}\phi_0^{(e)\prime}\phi_N^{(e)\prime} & \dfrac{2}{h^{(e)}}\phi_0^{(e)}\phi_N^{(e)} \\ -\dfrac{2}{h^{(e)}}\phi_0^{(e)}\phi_0^{(e)\prime} & \phi_0^{(e)}\phi_0^{(e)} & \cdots & -\dfrac{2}{h^{(e)}}\phi_0^{(e)}\phi_N^{(e)\prime} & \phi_0^{(e)}\phi_N^{(e)} \\ \vdots & \vdots & \vdots & \vdots & \vdots \\ \dfrac{4}{h^{(e)2}}\phi_0^{(e)\prime}\phi_N^{(e)\prime} & -\dfrac{2}{h^{(e)}}\phi_0^{(e)}\phi_N^{(e)\prime} & \cdots & \dfrac{4}{h^{(e)2}}\phi_N^{(e)\prime}\phi_N^{(e)\prime} & -\dfrac{2}{h^{(e)}}\phi_N^{(e)}\phi_N^{(e)\prime} \\ -\dfrac{2}{h^{(e)}}\phi_0^{(e)\prime}\phi_N^{(e)} & \phi_0^{(e)}\phi_N^{(e)} & \cdots & -\dfrac{2}{h^{(e)}}\phi_N^{(e)}\phi_N^{(e)\prime} & \phi_N^{(e)}\phi_N^{(e)} \end{array} \right] \dfrac{h^{(e)}}{2} d\xi_1. \tag{6.165}$$

The vectors of the equivalent distributed forces (6.156) are given for the high-order elements as

$$\{\bar{f}_{m_z}^{(e)}\} = \int_{-1}^{1} m_z[x(\xi_1)]\left\{ 0 \quad \phi_1^{(e)}(\xi_1) \quad \cdots \quad 0 \quad \phi_N^{(e)}(\xi_1) \right\}^T \frac{h^{(e)}}{2}d\xi_1, \qquad (6.166)$$

$$\{\bar{f}_{q_y}^{(e)}\} = \int_{-1}^{1} q_y[x(\xi_1)]\left\{ \phi_1^{(e)}(\xi_1) \quad 0 \quad \cdots \quad \phi_N^{(e)}(\xi_1) \quad 0 \right\}^T \frac{h^{(e)}}{2}d\xi_1. \qquad (6.167)$$

The weak form (6.143) can be written in terms of the bilinear $a(\cdot,\cdot)$ and linear $f(\cdot)$ operators. The bilinear form is given by the sum of bending and shear contributions

$$a\left((\theta_z, u_y), (\alpha, v)\right) = a_f(\theta_z, \alpha) + a_c((\theta_z, u_y), (\alpha, v)), \qquad (6.168)$$

with

$$a_f(\theta_z, \alpha) = \int_0^L E(x)I_z(x)\frac{d\theta_z(x)}{dx}\frac{d\alpha(x)}{dx}dx, \qquad (6.169)$$

$$a_c((\theta_z, u_y), (\alpha, v)) = \int_0^L K_c(x)G(x)A(x)\left(\frac{du_y(x)}{dx} - \theta_z(x)\right)\left(\frac{dv(x)}{dx} - \alpha(x)\right)dx. \qquad (6.170)$$

The linear form $f(\cdot)$ is given by

$$f((v, \alpha)) = \int_0^L m_z(x)\alpha(x)dx + \int_0^L q_y(x)v(x)dx + M_{z_L}\alpha(L) + M_{z_0}\alpha(0) + V_{y_L}v(L) + V_{y_0}v(0). \qquad (6.171)$$

The energy norm $||u||_E$ of the solution u for the Timoshenko beam is defined as

$$\begin{aligned} ||u||_E^2 &= a((\theta_z, u_y), (\theta_z, u_y)) = \int_0^L E(x)I_z(x)\left(\frac{d\theta_z(x)}{dx}\right)^2 dx \\ &+ \int_0^L K_c(x)G(x)A(x)\left(\frac{du_y(x)}{dx} - \theta_z(x)\right)^2 dx. \end{aligned} \qquad (6.172)$$

The following example illustrates how the use of high-order polynomials overcomes the shear locking phenomenon and at the same time reaches a spectral convergence for the approximate solution.

Example 6.17 *Consider the beam of Example 6.10. Figure 6.47(a) presents the absolute value of the transversal displacement at the right end of the beam using an element with polynomial orders 1, 2, and 3. Notice that for an order $P = 3$, the analytical and approximate solutions are practically coincident, avoiding the observed solution locking when using a linear element.*

Figure 6.47(b) shows the relative error in the energy norm for several height values. Since the analytical solution is of fourth order, the error between the analytical and approximate solutions is zero only when $P = 4$.

Thus, we notice that when using higher-order approximations it is possible to prevent the shear locking and at the same time achieve the exact solution of the beam under consideration. File vcisexemp17.m implements the solution of this example.

□

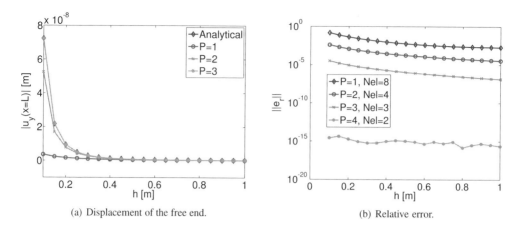

(a) Displacement of the free end.   (b) Relative error.

Figure 6.47   Example 6.17: shear locking effect overcomed by high-order approximations.

6.16   MATHEMATICAL ASPECTS OF THE FEM

The error e between the exact solution u and approximated Galerkin solution u_N of a BVP, that is,

$$e = u - u_N, \tag{6.173}$$

is orthogonal to all functions of the test space V_N in the energy norm. Thus,

$$a(e, v_N) = 0. \tag{6.174}$$

To demonstrate this property, recall that the exact and approximate solutions respectively satisfy equations (5.181) and (5.182). Particularly, as $V_N \subset V$, the exact solution also satisfies

$$a(u, v_N) = f(v_N) \text{ for all } v_N \in V. \tag{6.175}$$

Subtracting the above relation from (5.181), we obtain equation (6.174).

Another property of the approximate solution is that it minimizes the energy norm of the error, that is,

$$\|u - u_N\|_E = \min_{w_N \in X_N} \|u - w_N\|_E. \tag{6.176}$$

6.17   FINAL COMMENTS

This chapter presented the formulation and approximation of the Timoshenko beam model. Regarding the formulation, the same previous steps of the variational formulation were applied and resulted in two differential equations of equilibrium, coupled by the shear force. That was the first considered problem with normal and shear strain and stress components. For this purpose, we defined the equivalent normal stress to design and verify the beam with the shear effect. The constant distortion and shear stress distribution for the Timoshenko beam were partially corrected by setting the shear factor. Subsequently, the shear stress distributions for rectangular, circular, and I-shaped cross-sections were determined. The design of beams for standardized cross-sections was also presented. The concept of shear center and the energy methods were also considered.

For the approximation of the beam model, we presented the shear locking phenomenon for low-order approximations. With the application of high-order methods, we could overcome this problem and also obtain approximations with exponential convergence for smooth solutions.

6.18 PROBLEMS

1. Derive the equations of shear force, bending moment, rotation, and transversal displacement for the beams illustrated in Figure 5.67 considering the shear effect. Use a rectangular section of 20×40 cm.
2. Implement a MATLAB program to solve the beams illustrated in Figure 5.67. Plot the diagrams of shear force, bending moment, rotation, and transversal displacement.
3. Consider the fourth beam of Figure 5.67. Design it for the cross-sections illustrated in Figure 6.48. Use $\bar{\sigma} = 100$ MPa and check the shear effect.

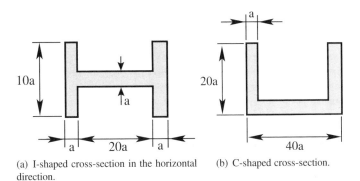

(a) I-shaped cross-section in the horizontal direction.

(b) C-shaped cross-section.

Figure 6.48 Problem 3.

4. Represent the shear stress distribution in the cross-section of Figure 6.49(a).

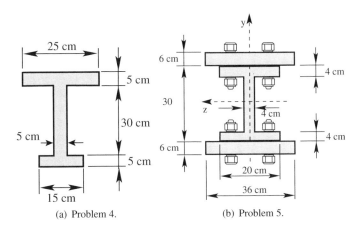

(a) Problem 4.

(b) Problem 5.

Figure 6.49 Problems 4 and 5.

5. Consider a beam with length $L = 4$ m with a maximum shear force of $V_y = 40$ kN. The cross-section is illustrated in Figure 6.49(b) and is fastened by screws with shear resistance of 400 N. Calculate the longitudinal spacing between the screws.
6. Find the shear center for the profile illustrated in Figure 6.31.
7. Calculate the horizontal and vertical displacements of point E of the truss illustrated in Figure 6.50 using the methods of energy.
8. Consider the beam with varying cross-section illustrated in Figure 5.37. Calculate the rotation at point C.
9. Write a MATLAB program to approximate the solution for the beams in Figure 5.67.

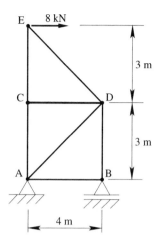

Figure 6.50   Problem 7.

Determine the order of the shape functions to avoid the solution locking for height values of the cross-section within the interval $[0.1; 1.0]$ cm, with increments of 0.1 cm. Plot the approximation errors in the energy norm for polynomial orders from 1 until the solution order of the transversal displacement equation.

7 FORMULATION AND APPROXIMATION OF TWO/THREE-DIMENSIONAL BEAMS

7.1 INTRODUCTION

Until this point, we considered one-dimensional models of bars, shafts, and Euler-Bernoulli and Timoshenko beams for small strains and Hookean material. The main features of these models are illustrated in Table 7.1. In general, the axial, twisting, and bending loads act simultaneously in a one-dimensional element, which from now will be called beam.

In the next sections we present the derivation of equations for the two-dimensional beam model, obtained by the superposition of traction, twist, and bending effects. It will be a good opportunity to review the mechanical models discussed in the previous chapters. Subsequently, we consider the three-dimensional beam model, including bending in y and shear in z. A MATLAB program to solve the differential equations for beams with superposition of loads will be discussed. Finally, the finite element approximations for the two- and three-dimensional beam models are presented.

	Bar	Shaft	Timoshenko beam
Kinematics	$\mathbf{u} = \{\begin{array}{ccc} u_x(x) & 0 & 0 \end{array}\}^T$	$\mathbf{u} = \{\begin{array}{ccc} 0 & -z\theta_x(x) & y\theta_x(x) \end{array}\}^T$	$\mathbf{u} = \{\begin{array}{ccc} -y\theta_z(x) & u_y(x) & 0 \end{array}\}^T$
Strains	$\varepsilon_{xx}(x) = \dfrac{du_x(x)}{dx}$	$\begin{cases} \gamma_{xy}(x,z) = -z\dfrac{d\theta_x(x)}{dx} \\ \gamma_{xz}(x,y) = y\dfrac{d\theta_x(x)}{dx} \end{cases}$	$\begin{cases} \varepsilon_{xx}(x,y) = -y\dfrac{d\theta_z(x)}{dx} \\ \gamma_{xy}(x) = -\theta_z(x) + \dfrac{du_y(x)}{dx} \end{cases}$
Rigid actions	translation in x	rotation in x	translation in y and rotation in z
Internal loads	$N_x(x) = \int_A \sigma_{xx}(x)\,dydz$	$M_x(x) = \int_A r\tau_t(x,r,\theta)dA$	$\begin{cases} M_z(x) = -\int_A \sigma_{xx}(x)ydA \\ V_y(x) = -\int_A \tau_{xy}(x)dA \end{cases}$
External loads	$q_x(x),\ P_0,\ P_L$	$m_x(x),\ T_0,\ T_L$	$q_y(x),\ m_z(x),\ V_{y_0},\ V_{y_L},\ M_{z_0},\ M_{z_L}$
Equilibrium BVP	$\begin{cases} \dfrac{dN_x(x)}{dx} + q_x(x) = 0, \\ N_x(L) = P_L,\ N_x(0) = -P_0 \end{cases}$	$\begin{cases} \dfrac{dM_x(x)}{dx} + m_x(x) = 0 \\ M_x(L) = T_L,\ M_x(0) = -T_0 \end{cases}$	$\begin{cases} \dfrac{dM_z(x)}{dx} - V_y(x) + m_z(x) = 0 \\ \dfrac{dV_y(x)}{dx} - q_y(x) = 0 \\ V_y(0) = V_{y_0},\ V_y(L) = -V_{y_L} \\ M_z(0) = -M_{z_0},\ M_z(L) = M_{z_L} \end{cases}$
Hooke's law	$\begin{cases} \sigma_{xx}(x) = E(x)\varepsilon_{xx}(x), \\ \varepsilon_{yy}(x) = -\nu(x)\varepsilon_{xx}(x), \\ \varepsilon_{zz}(x) = -\nu(x)\varepsilon_{xx}(x). \end{cases}$	$\begin{cases} \tau_{xy}(x,z) = G(x)\gamma_{xy}(x,z), \\ \tau_{xz}(x,y) = G(x)\gamma_{xz}(x,y). \end{cases}$	$\begin{cases} \sigma_{xx}(x,y) = E(x)\varepsilon_{xx}(x,y), \\ \tau_{xy}(x) = G(x)\bar{\gamma}_{xy}(x). \end{cases}$
Kinematic BVP	$\dfrac{d}{dx}\left(E(x)A(x)\dfrac{du_x(x)}{dx}\right) + q_x(x) = 0.$	$\dfrac{d}{dx}\left(G(x)I_p(x)\dfrac{d\theta_x(x)}{dx}\right) + m_x(x) = 0.$	$\begin{cases} EI_z\dfrac{d^3\theta_z(x)}{dx^3} + q_y(x) = 0 \\ K_cGA\left(\dfrac{d^4u_y(x)}{dx^4} - \dfrac{d^3\theta_z(x)}{dx^3}\right) - \\ \dfrac{d^2q_y(x)}{dx^2} = 0 \end{cases}$
Stresses	$\sigma_{xx}(x) = \dfrac{N_x(x)}{A(x)}$	$\tau_t(x,r) = \dfrac{M_x(x)}{I_p(x)}r$	$\begin{cases} \sigma_{xx}(x,y) = -\dfrac{M_z(x)}{I_z(x)}y, \\ \tau_{xy}(x) = -\dfrac{V_y(x)}{K_cA(x)}. \end{cases}$

ures of bar, shaft, and beam mechanical models.

7.2 TWO-DIMENSIONAL BEAM

7.2.1 KINEMATICS

In the case of a beam with traction, circular torsion, and bending, the kinematics is obtained by the superposition of the displacement components given in Figure 7.1 for each case.

Consider a section AB which is x units distant from the origin of the adopted Cartesian reference system, as shown in Figure 7.1(a). Let P be a point of section AB with (x, y, z) coordinates. The x coordinate indicates the section while y and z locate the point in plane yz of the cross-section.

Notice that after an axial displacement $u_x^n(x)$, section AB moves to position $A'B'$ and remains orthogonal to the x axis. Hence, bending occurs, and is characterized by a rigid vertical displacement $u_y(x)$ in direction y and a rigid rotation $\theta_z(x)$ about z, with the final position of the section indicated by $A'''B'''$. In the Timoshenko model, the cross-sections are still flat, but they are no longer orthogonal to the tangent of the deformed x axis. Thus, each section has a constant distortion.

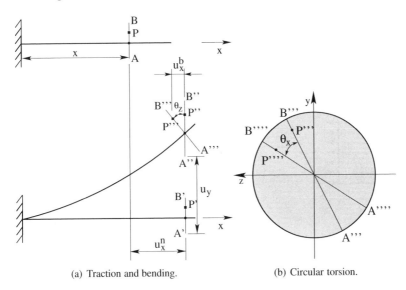

(a) Traction and bending. (b) Circular torsion.

Figure 7.1 Kinematics of the two-dimensional beam.

From Figure 7.1(a), the total axial displacement $u_x(x, y)$ in section x is given by the sum of the axial displacements due to simple traction/compression $u_x^n(x)$ and bending $u_x^b(x)$. Thus,

$$u_x(x, y) = u_x^n(x) + u_x^b(x) = u_x^n(x) - y\theta_z(x). \tag{7.1}$$

Considering a circular cross-section, section AB also has a constant rotation about x given by the angle of twist $\theta_x(x)$, and reach the final position $A''''B''''$ illustrated in Figure 7.1(b). Thus, point P presents the infinitesimal transversal displacements $u_y(x, z) = -z\theta_x(x)$ and $u_z(x, y) = y\theta_x(x)$ due to twist.

The kinematics of the two-dimensional beam with circular cross-section, including the effects of traction, bending, and twisting, is given by the following vector field:

$$\mathbf{u}(x, y, z) = \left\{ \begin{array}{c} u_x(x, y) \\ u_y(x, z) \\ u_z(x, y) \end{array} \right\} = \left\{ \begin{array}{c} u_x^n(x) - y\theta_z(x) \\ u_y(x) - z\theta_x(x) \\ y\theta_x(x) \end{array} \right\}. \tag{7.2}$$

Figure 7.2 illustrates the superposition of axial displacements due to stretching/shortening and bending for several signs. Notice that due to superposition, the neutral line of the cross-section is no longer coincident with z. Figure 7.3 shows the constant distribution of the distortion $\beta(x)$ in the cross-section. Figure 7.4 illustrates the displacement components for positive and negative angles of twist.

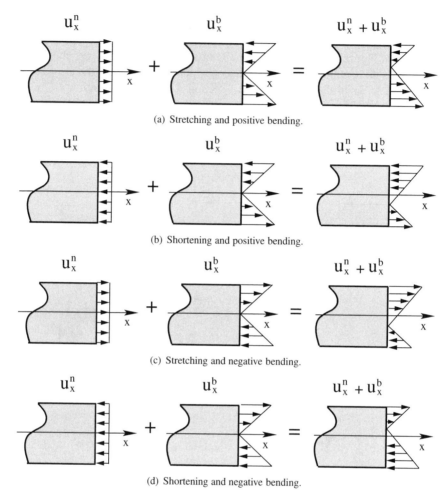

u_x^n u_x^b $u_x^n + u_x^b$

(a) Stretching and positive bending.

u_x^n u_x^b $u_x^n + u_x^b$

(b) Shortening and positive bending.

u_x^n u_x^b $u_x^n + u_x^b$

(c) Stretching and negative bending.

u_x^n u_x^b $u_x^n + u_x^b$

(d) Shortening and negative bending.

Figure 7.2 Superposition of axial displacements in the two-dimensional beam.

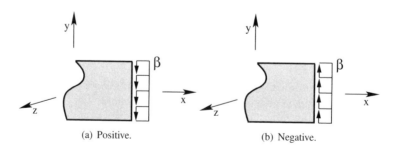

(a) Positive.   (b) Negative.

Figure 7.3 Constant cross-section distortion of the two-dimensional beam.

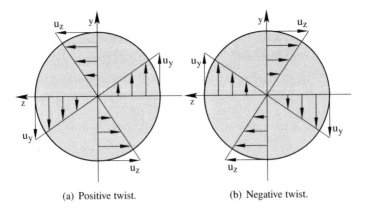

(a) Positive twist.　　　　(b) Negative twist.

Figure 7.4 Displacements in a circular cross-section due to twist of the two-dimensional beam.

7.2.2 STRAIN MEASURE

From the kinematics given in (7.2), the normal strain component ε_{xx} and shear strain components $\bar{\gamma}_{xy}$ and $\bar{\gamma}_{xz}$ are given by the derivatives of the displacement components, that is,

$$\varepsilon_{xx}(x,y) = \frac{\partial u_x(x,y)}{\partial x} = \frac{du_x^n(x)}{dx} - y\frac{d\theta_z(x)}{dx}, \tag{7.3}$$

$$\bar{\gamma}_{xy}(x,z) = \frac{\partial u_x(x,y)}{\partial y} + \frac{\partial u_y(x,z)}{\partial x}$$

$$= -\theta_z(x) - z\frac{d\theta_x(x)}{dx} + \frac{du_y(x)}{dx} = \beta(x) - z\frac{d\theta_x(x)}{dx}, \tag{7.4}$$

$$\bar{\gamma}_{xz}(x,y) = \frac{\partial u_x(x,y)}{\partial z} + \frac{\partial u_z(x,y)}{\partial x} = y\frac{d\theta_x(x)}{dx}. \tag{7.5}$$

The normal strain components due to stretching/shortening $\varepsilon_{xx}^n(x)$ and due to bending $\varepsilon_{xx}^b(x,y)$ are respectively given by

$$\varepsilon_{xx}^n(x) = \frac{du_x^n(x)}{dx}, \tag{7.6}$$

$$\varepsilon_{xx}^b(x,y) = -y\frac{d\theta_z(x)}{dx}. \tag{7.7}$$

In terms of the tangential distortion, the strain measures of the beam are expressed as

$$\varepsilon_{xx}(x,y) = \varepsilon_{xx}^n(x) + \varepsilon_{xx}^b(x,y) = \frac{du_x^n(x)}{dx} - y\frac{d\theta_z(x)}{dx}, \tag{7.8}$$

$$\bar{\gamma}_{xy}(x) = \beta(x) = \frac{du_y(x)}{dx} - \theta_z(x), \tag{7.9}$$

$$\gamma_t(x,r) = r\frac{d\theta_x(x)}{dx}. \tag{7.10}$$

7.2.3 RIGID BODY ACTIONS

The rigid body actions are obtained by imposing the strain components to be zero. Making the longitudinal strain ε_{xx} equal to zero, we have

$$\varepsilon_{xx}(x,y) = \frac{du_x^n(x)}{dx} - y\frac{d\theta_z(x)}{dx} = 0. \tag{7.11}$$

Thus, to satisfy the above expression, we must have $\dfrac{du_x^n(x)}{dx} = 0$ and $\dfrac{d\theta_z(x)}{dx} = 0$, that is, $u_x^n(x) = u_x$ and $\theta_z(x) = \theta_z$ must be constant for all sections of the beam. Thus, a translation along x and a rotation about z are rigid body actions of the two-dimensional beam.

If we substitute the rotation $\theta_z(x) = \dfrac{du_y(x)}{dx} - \beta(x)$ in the second term of the above equation, we obtain

$$y \frac{d}{dx}\left(\frac{du_y(x)}{dx} - \beta(x)\right) = y\frac{d}{dx}\left(\frac{du_y(x)}{dx}\right) = 0, \tag{7.12}$$

recalling that $\beta(x)$ is constant on each section x of the beam and its derivative is zero. For constant $u_y(x) = u_y$, the above expression is satisfied and consequently a translation in y is also a rigid action of the two-dimensional beam.

Imposing the tangential distortional component as zero, we have

$$\gamma_t(x, r) = r\frac{d\theta_x(x)}{dx} = 0. \tag{7.13}$$

The previous expression is satisfied for constant angles of twist $\theta_x(x) = \theta_x$ at all sections of the beam. Thus, a constant rotation around the x axis of the beam is also rigid.

The definition of rotation in terms of the transversal displacement $\theta_z(x) = \dfrac{du_y(x)}{dx}$ is recovered by taking the shear strain component $\bar{\gamma}_{xy}(x, z)$ as zero.

Thus, the rigid body actions for a two-dimensional beam are translations in x and y and rotations about axes x and z.

7.2.4 DETERMINATION OF INTERNAL LOADS

As seen in previous chapters, we have normal $\sigma_{xx}(x, y)$ and shear $\tau_{xy}(x)$ and $\tau_t(x, r)$ stress components associated to the normal $\varepsilon_{xx}(x, y)$ and shear $\bar{\gamma}_{xy}(x, z)$ and $\gamma_t(x, r)$ strain components. These stress components represent the state of internal loads in the two-dimensional beam. Hence, the strain internal work in the beam is given by

$$W_i = \int_V \left(\sigma_{xx}(x, y)\varepsilon_{xx}(x, y) + \tau_{xy}(x)\bar{\gamma}_{xy}(x)\right) dV + \int_V \tau_t(x, r)\gamma_t(x, t)dV. \tag{7.14}$$

Substituting the strain components given by (7.8) to (7.10), we have

$$\begin{aligned} W_i &= \int_V \left[\sigma_{xx}(x)\left(\frac{du_x^n(x)}{dx} - y\frac{d\theta_z(x)}{dx}\right) + \tau_{xy}(x)\left(\frac{du_y(x)}{dx} - \theta_z(x)\right)\right] dV \\ &+ \int_V \tau_t(x, r)\left(r\frac{d\theta_x(x)}{dx}\right) dV. \end{aligned} \tag{7.15}$$

The above integrals can be decomposed into length and area integrals, that is,

$$\begin{aligned} W_i &= \int_0^L \left(\int_A \sigma_{xx}(x, y)dA\right)\frac{du_x^n(x)}{dx} + \int_0^L \left(\int_A -\sigma_{xx}(x, y)ydA\right)\frac{d\theta_z(x)}{dx}dx \\ &+ \int_0^L \left(\int_A r\tau_t(x, r, \theta)dA\right)\frac{d\theta_x(x)}{dx}dx + \int_0^L \left(\int_A -\tau_{xy}(x)dA\right)\left(\theta_z(x) - \frac{du_y(x)}{dx}\right)dx. \end{aligned} \tag{7.16}$$

The area integrals respectively represent the normal force $N_x(x)$ in x, the bending moment $M_z(x)$ in z, the twisting moment $M_x(x)$ in x, and the shear force $V_y(x)$ in y, that is,

$$N_x(x) = \int_A \sigma_{xx}(x,y) dy dz, \tag{7.17}$$

$$M_z(x) = -\int_A \sigma_{xx}(x,y) y dA, \tag{7.18}$$

$$M_x(x) = \int_A r \tau_t(x,r,\theta) dA, \tag{7.19}$$

$$V_y(x) = -\int_A \tau_{xy}(x) dA. \tag{7.20}$$

From these definitions, we can write the strain internal work W_i, given in (7.16), by the following form:

$$
\begin{aligned}
W_i = {} & \int_0^L N_x(x) \frac{du_x^n(x)}{dx} dx + \int_0^L M_z(x) \frac{d\theta_z(x)}{dx} dx + \int_0^L M_x(x) \frac{d\theta_x(x)}{dx} dx \\
& + \int_0^L V_y(x) \left(\theta_z(x) - \frac{du_y(x)}{dx} \right) dx.
\end{aligned}
\tag{7.21}
$$

Integrating by parts the above expressions, we have

$$
\begin{aligned}
W_i = {} & -\int_0^L \frac{dN_x(x)}{dx} u_x^n(x) dx + [N_x(L) u_x^n(L) - N_x(0) u_x^n(0)] \\
& + \int_0^L \left(-\frac{dM_x(x)}{dx} + V_y(x) \right) \theta_z(x) dx + [M_z(L)\theta_z(L) - M_z(0)\theta_z(0)] \\
& - \int_0^L \frac{dM_x(x)}{dx} \theta_x(x) dx + [M_x(L)\theta_x(0) - M_x(L)\theta_x(0)] \\
& + \int_0^L \frac{dV_y(x)}{dx} u_y(x) dx - [V_y(L) u_y(L) - V_y(0) u_y(0)].
\end{aligned}
\tag{7.22}
$$

The compatible internal loads with the kinematics of the two-dimensional beam are illustrated in Figure 7.5.

7.2.5 DETERMINATION OF EXTERNAL LOADS

Inspecting Figure 7.5 we identify the compatible external loads with the kinematic model of the two-dimensional beam. The following loads are present at the beam ends:

- P_0, P_L: axial loads at $x = 0$ and $x = L$;
- T_0, T_L: torques at $x = 0$ and $x = L$;
- M_{z0}, M_{zL}: bending moments in direction z at $x = 0$ and $x = L$;
- V_{y0}, V_{yL}: transversal forces in direction y at $x = 0$ and $x = L$.

The external distributed loads along the beam length are:

- $q_x(x)$: distributed axial force in direction x;
- $m_x(x)$: distributed torque in direction x;
- $q_y(x)$: distributed transversal load in direction y;
- $m_z(x)$: distributed bending moment in direction z.

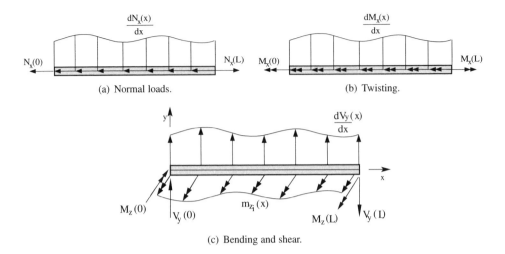

(a) Normal loads.   (b) Twisting.

(c) Bending and shear.

Figure 7.5   Internal loads in the two-dimensional beam.

These external loads are illustrated in Figure 7.6.

The external work expression for any kinematic action $\mathbf{u}(x,y,z)$ given in (7.2) can be written as:

$$
\begin{aligned}
W_e \;=\; & \int_0^L q_x(x)u_x^n(x)\,dx + \left[P_L u_x^n(L) + P_0 u_x^n(0)\right] \\
& + \int_0^L m_z(x)\theta_z(x)\,dx + \left[M_{z_L}\theta_z(L) + M_{z_0}\theta_z(0)\right] \\
& + \int_0^L m_x(x)\theta_x(x)\,dx + \left[T_L\theta_x(0) + T_0\theta_x(0)\right] \\
& + \int_0^L q_y(x)u_y(x)\,dx + \left[V_{y_L}u_y(L) + V_{y_0}u_y(0)\right].
\end{aligned} \tag{7.23}
$$

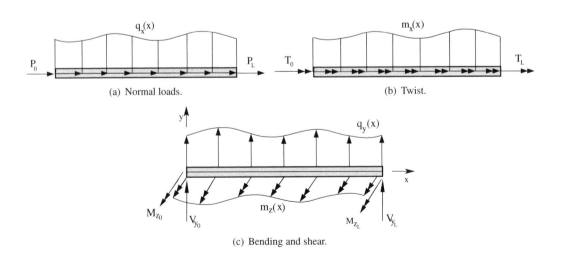

(a) Normal loads.   (b) Twist.

(c) Bending and shear.

Figure 7.6   External loads in the two-dimensional beam.

7.2.6 EQUILIBRIUM

As in previous cases, we apply PVW to find the equilibrium BVPs of the two-dimensional beam. Taking (7.22) equal to (7.23) for any virtual action $\delta\mathbf{u}$ from the deformed configuration, we have

$$\int_0^L \left[-\frac{dN_x(x)}{dx} - q_x(x) \right] \delta u_x^n(x)dx + [N_x(L) - P_L]\,\delta u_x^n(L) + [-N_x(0) - P_0]\,\delta u_x^n(0) +$$

$$\int_0^L \left[-\frac{dM_z(x)}{dx} + V_y(x) + m_z(x) \right] \delta\theta_z(x)dx + [-M_z(L) + M_{z_L}]\,\delta\theta_z(L) + [M_z(0) + M_{z_0}]\,\delta\theta_z(0) +$$

$$\int_0^L \left[-\frac{dM_x(x)}{dx} - m_x(x) \right] \delta\theta_x(x)dx + [-M_x(L) + T_L]\,\delta\theta_x(L) + [M_x(0) + T_0]\,\delta\theta_x(0) +$$

$$\int_0^L \left[-\frac{dV_y(x)}{dx} + q_y(x) \right] \delta u_y(x)dx + [-V_y(L) + V_{y_L}]\,\delta u_y(L) + [V_y(0) + V_{y_0}]\,\delta u_y(0) = 0.$$

The above expression is valid for any arbitrary virtual action if all terms inside brackets are simultaneously zero. This results in the equilibrium BVPs of the two-dimensional beam, that is,

$$\begin{cases} \dfrac{dN_x(x)}{dx} + q_x(x) = 0 & \text{in } x \in (0,L) \\ N_x(L) = P_L & \text{in } x = L \\ N_x(0) = -P_0 & \text{in } x = 0 \end{cases} \tag{7.24}$$

$$\begin{cases} \dfrac{dM_x(x)}{dx} + m_x(x) = 0 & \text{in } x \in (0,L) \\ M_x(L) = T_L & \text{in } x = L \\ M_x(0) = -T_0 & \text{in } x = 0 \end{cases} \tag{7.25}$$

$$\begin{cases} \dfrac{dM_z(x)}{dx} - V_y(x) + m_z(x) = 0 & \text{in } x \in (0,L) \\ \dfrac{dV_y(x)}{dx} - q_y(x) = 0 & \text{in } x \in (0,L) \\ V_y(0) = V_{y_0} & \text{in } x = 0 \\ V_y(L) = -V_{y_L} & \text{in } x = L \\ M_z(0) = -M_{z_0} & \text{in } x = 0 \\ M_z(L) = M_{z_L} & \text{in } x = L \end{cases} \tag{7.26}$$

Thus, the equilibrium BVPs of the beam have four differential equations in terms of the normal and shear forces and twisting and bending moments, along with the boundary conditions in terms of these loads. If the distributed bending moment $m_z(x)$ is zero, the two differential equations of bending in (7.26) reduce to

$$\frac{dV_y(x)}{dx} = q_y(x), \tag{7.27}$$

$$V_y(x) = \frac{dM_z(x)}{dx}. \tag{7.28}$$

Combining the above expressions, we determine the second-order differential equation in terms of the bending moment

$$\frac{dM_z^2(x)}{dx^2} = q_y(x). \tag{7.29}$$

For rigid virtual actions, the strain measures and consequently the internal work are both zero. In this case, the PVW states that the external work (7.23) is zero. Moreover, for a rigid action,

$\delta u_x^n(x) = \delta u_x^n = \text{cte}$, $\delta u_y(x) = \delta u_y = \text{cte}$, $\delta \theta_x(x) = \delta \theta_x = \text{cte}$, and $\delta \theta_z(x) = \delta \theta_z = \text{cte}$. Therefore,

$$
\begin{aligned}
\delta W_e &= \left(\int_0^L q_x(x)dx + P_L + P_0 \right) \delta u_x^n + \left(\int_0^L q_y(x)dx + V_{y_L} + V_{y_0} \right) \delta u_y \\
&+ \left(\int_0^L m_x(x)dx + T_L + T_0 \right) \delta \theta_x + \left(\int_0^L m_z(x)dx + M_{z_L} + M_{z_0} \right) \delta \theta_z = 0.
\end{aligned}
$$

Thus, the equilibrium conditions of the two-dimensional beam state that the resultants of external forces in x and y and resultants of moments in x and z must be zero, i.e.,

$$
\sum f_x = \int_0^L q_x(x)dx + P_L + P_0 = 0, \tag{7.30}
$$

$$
\sum f_y = \int_0^L q_y(x)dx + V_{y_L} + V_{y_0} = 0, \tag{7.31}
$$

$$
\sum m_x = \int_0^L m_x(x)dx + T_L + T_0 = 0, \tag{7.32}
$$

$$
\sum m_z = \int_0^L m_z(x)dx + M_{z_L} + M_{z_0} = 0. \tag{7.33}
$$

7.2.7 APPLICATION OF THE CONSTITUTIVE EQUATION

Hooke's law for a homogeneous isotropic linear elastic material states that the normal stress $\sigma_{xx}(x,y)$ is related to the specific normal strain $\varepsilon_{xx}(x,y)$ by the longitudinal elastic modulus $E(x)$, that is,

$$
\sigma_{xx}(x,y) = E(x)\varepsilon_{xx}(x,y) = \sigma_{xx}^n(x) + \sigma_{xx}^b(x,y), \tag{7.34}
$$

with $\sigma_{xx}^n(x)$ and $\sigma_{xx}^b(x,y)$ the contributions of normal stress in x respectively due to the normal force and bending moment, given by

$$
\sigma_{xx}^n(x) = E(x)\frac{du_x^n(x)}{dx}, \tag{7.35}
$$

$$
\sigma_{xx}^b(x,y) = -E(x)\frac{d\theta_z(x)}{dx}y. \tag{7.36}
$$

Likewise, the shear stresses $\tau_{xy}(x)$ and $\tau_t(x,r)$ are related to the distortions $\bar{\gamma}_{xy}(x)$ and $\gamma_t(x,r)$ by the shear modulus $G(x)$. Thus, the shear stresses are respectively given by

$$
\tau_t(x,r) = G(x)\gamma_t(x,r) = G(x)\frac{d\theta_x(x)}{dx}r, \tag{7.37}
$$

$$
\tau_{xy}(x) = G(x)\bar{\gamma}_{xy}(x) = G(x)\left(\frac{du_y(x)}{dx} - \theta_z(x) \right). \tag{7.38}
$$

Substituting the above relations in the internal load expressions given in (7.17) to (7.20), we respectively obtain the equations of normal force, bending moment, twisting moment, and shear force for a beam with Hooke material

$$
N_x(x) = E(x)\frac{du_x^n(x)}{dx}\int_A dA = E(x)A(x)\frac{du_x^n(x)}{dx}, \tag{7.39}
$$

$$
M_z(x) = E(x)\frac{d\theta_z(x)}{dx}\int_A y^2 dA = E(x)I_z(x)\frac{d\theta_z(x)}{dx}, \tag{7.40}
$$

$$
M_x(x) = G(x)\frac{d\theta_x(x)}{dx}\int_A r^2 dA = G(x)I_p(x)\frac{d\theta_x(x)}{dx}, \tag{7.41}
$$

$$
V_y(x) = -G(x)\left[\frac{du_y(x)}{dx} - \theta_z(x) \right]\int_A dA = -G(x)A(x)\beta(x). \tag{7.42}
$$

Replacing the above expressions in the differential equations of equilibrium and in equations (7.24) to (7.26), we respectively obtain the following differential equations in terms of the kinematics:

$$\frac{d}{dx}\left(E(x)A(x)\frac{du_x^n(x)}{dx}\right) + q_x(x) = 0, \tag{7.43}$$

$$\frac{d}{dx}\left(G(x)I_p(x)\frac{d\theta_x(x)}{dx}\right) + m_x(x) = 0, \tag{7.44}$$

$$\frac{d}{dx}\left[G(x)A(x)\left(\frac{du_y(x)}{dx} - \theta_z(x)\right)\right] + q_y(x) = 0, \tag{7.45}$$

$$\frac{d}{dx}\left(E(x)I_z(x)\frac{d\theta_z(x)}{dx}\right) + G(x)A(x)\left(\frac{du_y(x)}{dx} - \theta_z(x)\right) + m_z(x) = 0. \tag{7.46}$$

To solve a real case using the two-dimensional beam model, we must analyze the types of external loads and integrate the respective differential equations. For instance, for a beam with axial forces and torques, we only integrate the differential equations in terms of the normal force and twisting moment.

Assuming constant cross-section and material properties and $m_z(x) = 0$, the above differential equations of equilibrium simplify to

$$EA\frac{d^2u_x^n(x)}{dx^2} + q_x(x) = 0, \tag{7.47}$$

$$GI_p\frac{d^2\theta_x(x)}{dx^2} + m_x(x) = 0, \tag{7.48}$$

$$GA\left(\frac{d^2u_y(x)}{dx^2} - \frac{d\theta_z(x)}{dx}\right) + q_y(x) = 0, \tag{7.49}$$

$$EI_z\frac{d^2\theta_z(x)}{dx^2} + GA\left(\frac{du_y(x)}{dx} - \theta_z(x)\right) = 0. \tag{7.50}$$

7.2.8 STRESS DISTRIBUTIONS

The derivatives of the kinematic components obtained in equations (7.39)-(7.42) can be respectively substituted in equations (7.35) to (7.38). Thus, we obtain the following normal stress components due to normal force and bending moment and the shear stress components due to twisting moment and shear force:

$$\sigma_{xx}^{N_x}(x) = \frac{N_x(x)}{A(x)}, \tag{7.51}$$

$$\sigma_{xx}^{M_z}(x,y) = -\frac{M_z(x)}{I_z(x)}y, \tag{7.52}$$

$$\tau_t^{M_x}(x,r) = \frac{M_x(x)}{I_p(x)}r, \tag{7.53}$$

$$\tau_{xy}^{V_y}(x) = -\frac{V_y(x)}{K_cA(x)}. \tag{7.54}$$

For the shear stress $\tau_{xy}^{V_y}(x)$ due to the shear force, we consider a shear factor K_c of the cross-section.

The resultant normal stress σ_R is given by the sum of normal stresses from the normal force and bending moment, that is,

$$\sigma_R(x,y) = \sigma_{xx}^{N_x}(x) + \sigma_{xx}^{M_z}(x,y) = \frac{N_x(x)}{A(x)} - \frac{M_z(x)}{I_z(x)}y. \tag{7.55}$$

While the stress due to normal force is constant in the cross-section, the bending stress is maximum at the section extremities. The critical points due to normal stress in a cross-section are those with the highest absolute value of resultant normal stress. Figure 7.7 illustrates the critical points for circular and T-shaped cross-sections.

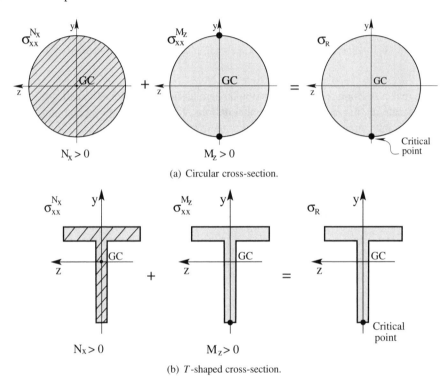

(a) Circular cross-section.

(b) T-shaped cross-section.

Figure 7.7   Critical points due to normal stress.

The resultant normal stress is zero in the neutral line of the section, that is, $\sigma_R(x,y) = 0$. The cross-sections considered in this book are symmetric to the y axis, and thus the $\bar{y}$ neutral line coordinate is calculated as

$$\sigma_R(x,\bar{y}) = \frac{N_x(x)}{A(x)} - \frac{M_z(x)}{I_z(x)}\bar{y} = 0.$$

Thus,

$$\bar{y} = \frac{N_x(x)}{M_z(x)}\frac{I_z(x)}{A(x)}. \tag{7.56}$$

Figure 7.8 illustrates the resultant normal stress and the neutral line position for positive normal forces and bending moments.

For a circular cross-section with diameter d, the resultant shear stress is given by

$$\tau_R(x,r) = \tau_{xy}^{V_y}(x) + \tau_t^{M_x}(x,r). \tag{7.57}$$

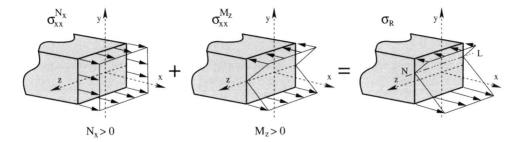

Figure 7.8 Resultant normal stress and neutral line position for positive normal forces and bending moments.

Considering the average shear stress due to shear force, all points of the circular section are equally loaded, as illustrated in Figure 7.9. Including the shear factor, we have

$$\tau_{xy}^{V_y}(x) = -\frac{4V_y(x)}{3A(x)}.$$

The direction of $\tau_{xy}^{V_y}$ is vertical and downwards, with $V_y > 0$.

However, the critical points due to shear stress from the twisting moment are at the extremities of the circular section, as illustrated in Figure 7.9 with

$$\tau_t^{M_x}(x,r) = \frac{16M_x(x)}{\pi d^3}.$$

The component $\tau_t^{M_x}$ is tangent to the boundary points and its direction is counter-clockwise for positive twisting moments.

Points on the boundary of the cross-section are the critical ones to resultant shear stress due to the positive shear force and twisting moment, as indicated in Figure 7.9. When considering only shear stresses due to the twisting moment and normal stresses due to the normal force and bending moment, the critical point is the same, as shown in Figure 7.7, for positive loads. As the shear stress due to shear force is generally lower than from the twisting moment, we can neglect it in the initial design and verify its effect posteriorly.

For a noncircular cross-section, the resultant shear stress is reduced to the stress from the shear force.

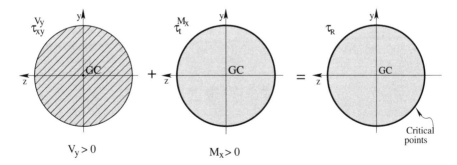

Figure 7.9 Critical cross-section points due to shear stress.

The equivalent normal stress in the cross-section is calculated as

$$\sigma_{eqv} = \sqrt{\sigma_R^2(x,y) + 3\tau_R^2(x,r)}. \tag{7.58}$$

7.2.9 DESIGN AND VERIFICATION

The following design procedure is applied for two-dimensional beams:

1. The functions and respective diagrams for the normal force $N_x(x)$, shear force $V_y(x)$, and bending moment $M_z(x)$ are obtained by the integration of differential equations given in (7.24) and (7.29) for isostatic problems and (7.43), (7.45), and (7.46) for hyperstatic problems. In the case of circular cross-sections and external torques, we must obtain the bending moment $M_x(x)$ integrating (7.25) or (7.44).

2. Based on the diagrams, the critical cross-section is determined, i.e., the section with the highest absolute values in the order: bending moment, twisting moment, normal force, and shear force. These values are denoted respectively $M_z^{\max}$, $M_x^{\max}$, $N_x^{\max}$, and $V_y^{\max}$. The critical point of the critical section is selected, taking into account the load signs in the section.

3. Considering only the bending for noncircular sections, the maximum normal stress at the critical point is given by

$$\sigma_{xx}^{\max} = \frac{M_z^{\max}}{W_z}. \tag{7.59}$$

Using the design criterion of admissible normal stress, the bending strength modulus is given by

$$W_z = \frac{M_z^{\max}}{\bar{\sigma}}. \tag{7.60}$$

Using the calculated value for W_z and the shape of the section, we obtain the principal dimensions. For materials with distinct admissible tensile/compressive stress limits, we use the smaller value between the compressive ($\bar{\sigma}_c$) and tensile ($\bar{\sigma}_t$) stresses, that is, $\bar{\sigma} = \min(\bar{\sigma}_c, \bar{\sigma}_t)$.

After the normal stress criterion design, the calculated values are slightly increased (around 3% to 5%) and the beam is verified, including the effects of normal and shear forces in expression (7.58) of the normal equivalent stress. If the maximum normal stress is smaller than the admissible, the design is complete.

If we use a circular cross-section and initially neglect the shear stress from the shear force $V_y^{\max}$ and the normal stress from the normal force $N_x^{\max}$, we obtain the following maximum equivalent normal stress:

$$\sigma_{eqv}^{\max} = \sqrt{\left(\frac{M_z^{\max}}{W_z}\right)^2 + 3\left(\frac{M_x^{\max}}{W_x}\right)^2}.$$

As $W_x = 2W_z$, the above expression is rewritten as

$$\sigma_{eqv}^{\max} = \frac{1}{W_z}\sqrt{\left(M_z^{\max}\right)^2 + \frac{3}{4}\left(M_x^{\max}\right)^2}.$$

For normal stress criterion design, we use $\sigma_{eqv}^{\max} = \bar{\sigma}$ and we can calculate the cross-section strength modulus as

$$W_z = \frac{1}{\bar{\sigma}}\sqrt{\left(M_z^{\max}\right)^2 + \frac{3}{4}\left(M_x^{\max}\right)^2}. \tag{7.61}$$

Hence, the cross-section diameter d is calculated as

$$d = \sqrt[3]{\frac{32}{\pi}W_z} = \sqrt[3]{\frac{32}{\pi}\left(\frac{1}{\bar{\sigma}}\sqrt{\left(M_z^{\max}\right)^2 + \frac{3}{4}\left(M_x^{\max}\right)^2}\right)}. \tag{7.62}$$

Again, we slightly increase the section diameter around 3% to 5% and calculate the maximum equivalent normal stress including the effect of normal and shear forces. If less than the admissible normal stress, the beam is well designed.

In the verification, the cross-section dimensions are known and we want to check if the beam remains in the elastic range when subjected to loadings. For this, the maximum equivalent normal stress σ_{eqv}^{max} is calculated using (7.58). Then we verify if the calculated value is equal or lower than the admissible normal stress of the material. In this case, we say that the beam remains in the elastic range. If the condition does not hold, we must resize the beam applying the design steps.

The following examples illustrate the use of the above procedure.

Example 7.1 *Design the beam with rectangular cross-section of base b and height h ($h = 4b$) subjected to the indicated loading. Adopt $\bar{\sigma} = 100$ MPa.*

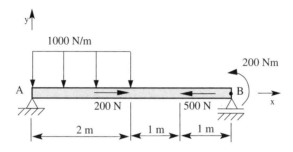

Figure 7.10   Example 7.1: two-dimensional beam with rectangular cross-section.

As the cross-section is rectangular and no torques are applied, we should obtain only the diagrams for the normal force, shear force, and bending moment, integrating the respective differential equations for isostatic beams given in (7.24) and (7.29).

The axial distributed load expression using singular function notation is

$$q_x(x) = 200 <x-2>^{-1} -500 <x-3>^{-1},$$

The beam has a sliding support at the right end, and no axial external forces are applied. Thus, the normal force is zero, that is, $N_x(x = 4) = 0$. The differential equation of the normal force is integrated

$$\frac{dN_x(x)}{dx} = -q_x(x) = -200 <x-2>^{-1} +500 <x-3>^{-1},$$

resulting in

$$N_x(x) = -200 <x-2>^0 +500 <x-3>^0 +C_1.$$

Applying the boundary condition, we obtain the integration constant C_1 as

$$N_x(x = 4) = -200(0)^0 + 500(0)^0 + C_1 = 0 \rightarrow C_1 = -300.$$

Thus, the final normal force equation is

$$N_x(x) = -200 <x-2>^0 +500 <x-3>^0 -300.$$

For the bending problem, the equation for the transversal distributed loading function is

$$q_y(x) = -1000 <x-0>^0 +1000 <x-2>^0 .$$

The bending moment is zero at the left end, and equal to the bending moment applied at the right end. Thus, $M_z(x = 0) = 0$ and $M_z(x = 4) = 200$.

The differential equation of equilibrium is integrated in terms of the bending moment

$$\frac{d^2 M_z(x)}{dx^2} = q_y(x) = -1000 < x - 0 >^0 + 1000 < x - 2 >^0 .$$

The first integration results in the shear force expression

$$V_y(x) = -1000 < x - 0 >^1 + 1000 < x - 2 >^1 + C_2,$$

The second integration results in the bending moment expression

$$M_z(x) = -500 < x - 0 >^2 + 500 < x - 2 >^2 + C_2 x + C_3.$$

We obtain the integration constants C_2 and C_3 applying the boundary conditions. Thus,

$$M_z(x = 0) = -500(0) + 500(0) + (0)C_2 + C_3 = 0 \rightarrow C_3 = 0,$$

$$M_z(x = 4) = -500(4)^2 + 500(4 - 2)^2 + (4)C_2 = 200 \rightarrow C_2 = 1550,$$

The final equations of the shear force and bending moment are respectively given by

$$V_y(x) = -1000 < x - 0 >^1 + 1000 < x - 2 >^1 + 1550,$$

$$M_z(x) = -500 < x - 0 >^2 + 500 < x - 2 >^2 + 1550x.$$

The expressions for the internal loads on each interval of the beam and the values at the ends are:

$0 < x < 2$ m :

$$N_x(x) = -300 = \begin{cases} N_x(0) = -300 \text{ N} \\ N_x(2) = -300 \text{ N} \end{cases}$$

$$V_y(x) = -1000x + 1550 = \begin{cases} V_y(0) = 1550 \text{ N} \\ V_y(2) = -450 \text{ N} \end{cases} ,$$

$$M_z(x) = -500x^2 + 1550x = \begin{cases} M_z(0) = 0 \text{ Nm} \\ M_z(2) = 1100 \text{ Nm} \end{cases}$$

$2 < x < 3$ m :

$$N_x(x) = -500 = \begin{cases} N_x(2) = -500 \text{ N} \\ N_x(3) = -500 \text{ N} \end{cases}$$

$$V_y(x) = -450 = \begin{cases} V_y(2) = -450 \text{ N} \\ V_y(3) = -450 \text{ N} \end{cases} ,$$

$$M_z(x) = -450x + 2000 = \begin{cases} M_z(2) = 1100 \text{ Nm} \\ M_z(3) = 650 \text{ Nm} \end{cases}$$

$3 < x < 4$ m :

$$N_x(x) = 0 = \begin{cases} N_x(3) = 0 \text{ N} \\ N_x(4) = 0 \text{ N} \end{cases}$$

$$V_y(x) = 450 = \begin{cases} V_y(3) = -450 \text{ N} \\ V_y(4) = -450 \text{ N} \end{cases} .$$

$$M_z(x) = -450x + 2000 = \begin{cases} M_z(3) = 650 \text{ Nm} \\ M_z(4) = 200 \text{ Nm} \end{cases}$$

The internal load diagrams are illustrated in Figure 7.11.

The support reactions at sections A and B are obtained as $R_{A_x} = -N_x(x = 0) = 300$ N, $R_{A_y} = V_y(x = 0) = 1550$ N, and $R_{B_y} = -V_y(x = 4) = 450$ N.

As the shear force is zero in the interval $0 < x < 2$ m, the bending moment has a maximum or minimum value, which must be calculated. The section $\bar{x}$ with zero shear force is calculated as

$$V_y(\bar{x}) = -1000x + 1550 = 0 \quad \rightarrow \quad \bar{x} = 1.55 \text{ m.}$$

The bending moment is calculated for $x = 1.55$ m and

$$M_z(1.55) = -500(1.55)^2 + 500(0)^2 + 1550(1.55) = 1201.25 \text{ Nm.}$$

This is the critical section of the beam and the maximum loads are $N_x^{max} = 300$ N and $M_z^{max} = 1201.25$ Nm. The shear force is zero at this section. As the normal force is compressive and the bending is positive, the critical points are in the upper end of the section, with $y = \dfrac{h}{2}$.

For the bending design, the bending strength modulus is obtained as

$$W_z = \frac{M_z^{max}}{\bar{\sigma}} = \frac{1201.25}{1.0 \times 10^8} = 1.20 \times 10^{-5} \text{ m}^3.$$

With $W_z = \dfrac{bh^2}{6}$ for a rectangular section. For $h = 4b$, we have $W_z = \dfrac{h^3}{24}$. Thus,

$$h = \sqrt[3]{24W_z} = 6.60 \text{ cm.}$$

Hence, $b = 1.65$ cm.

For $h = 6.80$ cm and $b = 1.70$ cm, we have $A = 1.16 \times 10^{-3}$ m^2 and $W_z = 1.31 \times 10^{-5}$ m^3. For the normal force verification, the equivalent normal stress is calculated, which in this case reduces to the resultant normal stress σ_R, that is,

$$\sigma_{xx}^{max} = \sigma_R = \frac{N_x^{max}}{A} + \frac{M_z^{max}}{W_z} = \frac{300}{1.16 \times 10^{-3}} + \frac{1201.25}{1.31 \times 10^{-5}} = 91.95 \text{ MPa.}$$

As $\sigma_{xx}^{max} \leq \bar{\sigma}$, the beam remains in the elastic range and is well designed.

The neutral line position $\bar{y}$ in the critical section is obtained as

$$\bar{y} = \frac{N_x \, I_z}{M_z \, A} = \frac{(4.45 \times 10^{-7})(-300)}{(1201.25)(1.16 \times 10^{-3})} = -9.62 \times 10^{-3} \text{ cm.}$$

To calculate the neutral line, we consider the signs of the normal force and bending moment acting in the section.

Figure 7.12 illustrates the normal stress distributions in the critical section. Notice that the normal stress due to bending moment is generally much higher than the normal stress due to normal force.

In this example, the loads at section $x = 2^+$ m are close to the values of the critical section. The absolute values of the normal force and bending moment are respectively higher and lower than the design values. Thus, it is interesting to calculate the maximum resultant normal stress in this section, where $N_x = -500$ N and $M_z = 1100$ Nm. The critical points are located at $y = \dfrac{h}{2}$ and the resultant normal stress is obtained as

$$\sigma_R = \frac{500.0}{1.16 \times 10^{-3}} + \frac{1100.0}{1.31 \times 10^{-5}} = 84.40 \text{ MPa.}$$

The shear stress due to the shear force is given by

$$\tau_{xy}^{V_y} = -\frac{3}{2} \frac{(-450)}{1.16 \times 10^{-3}} = 0.58 \text{ MPa.}$$

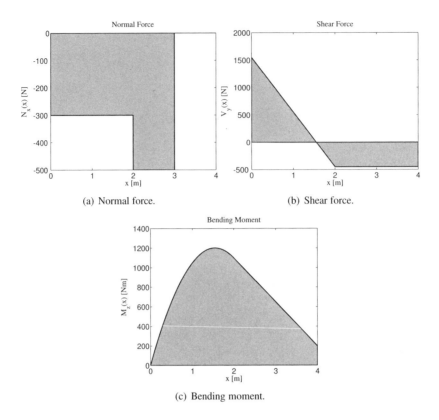

(a) Normal force.

(b) Shear force.

(c) Bending moment.

Figure 7.11   Example 7.1: diagrams for the normal force, shear force, and bending moment.

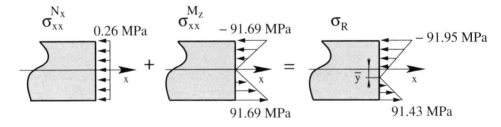

Figure 7.12   Example 7.1: distribution of normal stresses.

Notice that in this case, the obtained value is negligible when compared to the resultant normal stress.

The equivalent normal stress is calculated by

$$\sigma_{eqv} = \sqrt{84.40^2 + 3(0.58^2)} = 84.41 \text{ MPa}.$$

As the obtained stress is lower than the admissible normal stress of the material, the beam remains in the elastic range also considering the loads at section $x = 2^+$ m.

□

Example 7.2 *Consider the beam of circular cross-section with $d = 20$ cm, illustrated in Figure 7.13, subjected to the indicated loading. We want to verify if the beam remains in the elastic range for an admissible normal stress of $\bar{\sigma} = 100$ MPa. Resize the beam if necessary.*

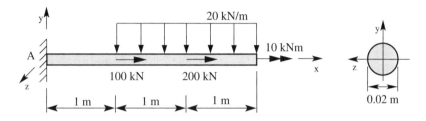

Figure 7.13 Example 7.2: two-dimensional beam with circular section.

The beam is subjected to bending, traction, and twisting loads. Thus, all load diagrams should be obtained. As the problem is isostatic, the equilibrium equations given in (7.24), (7.25), and (7.29) are integrated.

The normal force is initially considered and the same previous integration procedure is applied. The axial distributed load function in singularity notation is

$$q_x(x) = 100 < x - 1 >^{-1} + 200 < x - 2 >^{-1}.$$

The beam is clamped at $x = 0$ with zero axial displacement, that is, $u_x(x = 0) = 0$. The normal force at the right end is also zero, namely, $N_x(x = 3) = 0$, because there is no applied axial force.

The equilibrium equation in terms of the normal force is

$$\frac{dN_x(x)}{dx} = -q_x(x) = -100 < x - 1 >^{-1} - 200 < x - 2 >^{-1}.$$

The integration of the above expression results in

$$N_x(x) = -100 < x - 1 >^0 - 200 < x - 2 >^0 + C_1.$$

Applying the boundary condition in terms of the normal force allows the determination of the integration constant C_1

$$N_x(x = 3) = -100(3 - 1)^0 - 200(3 - 2)^0 + C_1 = 0 \rightarrow C_1 = 300.$$

The final normal force expression is

$$N_x(x) = -100 < x - 1 >^0 - 200 < x - 2 >^0 + 300.$$

The circular torsion problem is now considered. As no external distributed torque is present, we have $m_x(x) = 0$. The beam is clamped in the left end, and thus the angle of twist is zero in this section, namely, $\theta_x(x = 0) = 0$. Due to the applied torque at the right end, we have $M_x(x = 3) = 10$ kNm.

The differential equation in terms of the twist moment is integrated

$$\frac{dM_x(x)}{dx} = m_x(x) = 0,$$

resulting in

$$M_x(x) = C_2.$$

Applying the boundary condition in terms of torque, we obtain C_2

$$M_x(x = 3) = C_2 = 10 \rightarrow C_2 = 10.$$

The final expression for the twisting moment is

$$M_x(x) = 10.$$

Finally, we consider the bending problem. The equation of the transversal distributed load is given by

$$q_y(x) = -20 < x - 1 >^0.$$

As the right end of the beam is free and no bending loads are applied, we have the boundary conditions $V_y(x = 3) = 0$ and $M_z(x = 3) = 0$.

The differential equation of equilibrium in terms of bending moment is

$$\frac{d^2 M_z(x)}{dx^2} = -20 < x - 1 >^0.$$

The first integration yields the shear force, that is,

$$V_y(x) = \frac{dM_z(x)}{dx} = -20 < x - 1 >^1 + C_3.$$

The second integration results in the bending moment equation

$$M_z(x) = -10 < x - 1 >^2 + C_3 x + C_4.$$

We calculate the integration constants C_3 and C_4 with the application of the boundary conditions as

$$V_y(x = 3) = -20(3 - 1) + C_3 = 0 \rightarrow C_3 = 40,$$

$$M_z(x = 3) = -10(3 - 1)^2 + 40(3) + C_4 = 0 \rightarrow C_4 = -80.$$

After substituting the integration constants we obtain the expressions of shear force and bending moment for the beam, namely,

$$V_y(x) = -20 < x - 1 >^1 + 40,$$

$$M_z(x) = -10 < x - 1 >^2 + 40x - 80.$$

The support reactions are obtained as $R_{Ax} = -N_x(x = 0) = -300$ kN, $R_{Ay} = V_y(x = 0) = 40$ kN, $M_{Ax} = -M_x(x = 0) = -10$ kNm, and $M_{Az} = -M_z(x = 0) = -80$ kNm.

Figure 7.14 illustrates the diagrams of normal and shear forces and the twisting and bending moments. The critical section is $x = 0$ and the maximum internal loads in absolute values are $N_x^{max} = 300$ kN, $V_y^{max} = 40$ kN, $M_x^{max} = 10$ kNm, and $M_z^{max} = 80$ kNm. The critical points of the critical section for each internal load are illustrated in Figure 7.15. As the bending is negative and the traction is positive, the critical point is in the upper extremity of the section.

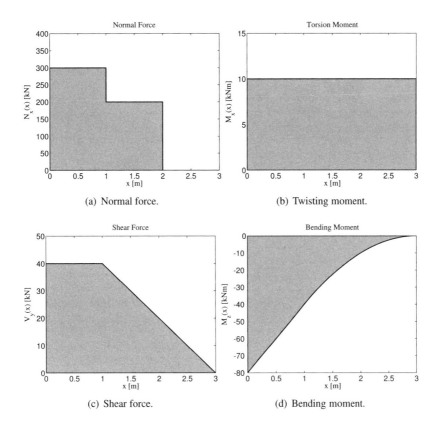

(a) Normal force.

(b) Twisting moment.

(c) Shear force.

(d) Bending moment.

Figure 7.14 Example 7.2: internal load diagrams.

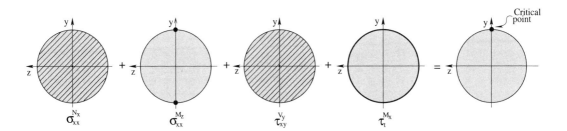

Figure 7.15 Example 7.2: critical point of the beam.

The normal traction and bending stresses at the critical point are:

$$\sigma_{xx}^{N_x} = \frac{N_x^{\mathrm{max}}}{A} = \frac{(300000)(4)}{\pi(0.2)^2} = 9.55 \text{ MPa},$$

$$\sigma_{xx}^{M_z} = \frac{M_z^{\mathrm{max}}}{W_z} = \frac{(80000)(32)}{\pi(0.2)^3} = 101.86 \text{ MPa}.$$

The tangent shear stress due to twisting moment at the critical point is

$$\tau_t^{M_x} = \frac{M_x^{\mathrm{max}}}{W_x} = \frac{(10000)(16)}{\pi(0.2)^3} = 6.37 \text{ MPa}.$$

Assuming a constant shear stress due to shear force, we have for the critical point

$$\tau_{xy}^{V_y} = -\frac{4V_y^{\mathrm{max}}}{3A} = \frac{(16)(40000)}{(3)[\pi(0.2)^2]} = 5.09 \text{ MPa}.$$

The equivalent normal stress at the critical point is calculated as

$$\sigma_{eqv}^{\mathrm{max}} = \sqrt{(\sigma_R^{\mathrm{max}})^2 + 3(\tau_R^{\mathrm{max}})^2} = \sqrt{(9.55+101.6)^2 + 3(6.37+5.09)^2} = 112.91 \text{ MPa}.$$

Because $\sigma_{eqv}^{\mathrm{max}} > \bar{\sigma}$, the beam must be resized.

Equation (7.62) is first employed to calculate the minimum diameter considering the effects of bending and twisting. Thus, the section diameter is given by

$$d = \sqrt[3]{\frac{32}{\pi(1.0 \times 10^8)}} \sqrt{(80000)^2 + \frac{3}{4}(10000)^2} = \sqrt[3]{\frac{32}{\pi} 8.05 \times 10^{-4}} = 20.16 \text{ cm}.$$

The above value is increased to $d = 20.80$ cm and the normal and shear stress components are recalculated for the new diameter. Thus,

$$\sigma_{xx}^{N_x} = \frac{N_x^{\mathrm{max}}}{A} = \frac{(300000)(4)}{\pi(0.208)^2} = 8.66 \text{ MPa},$$

$$\sigma_{xx}^{M_z} = \frac{M_z^{\mathrm{max}}}{W_z} = \frac{(90000)(32)}{\pi(0.208)^3} = 88.00 \text{ MPa},$$

$$\tau_t^{M_x} = \frac{M_x^{\mathrm{max}}}{W_x} = \frac{(10000)(16)}{\pi(0.208)^3} = 5.50 \text{ MPa},$$

$$\tau_{xy}^{V_y} = \frac{4V_y^{\mathrm{max}}}{3A} = \frac{(16)(40000)}{(3)(\pi(0.208)^2)} = 4.71 \text{ MPa}.$$

The equivalent normal stress for the new diameter is

$$\sigma_{eqv}^{\mathrm{max}} = \sqrt{(\sigma_R^{\mathrm{max}})^2 + 3(\tau_R^{\mathrm{max}})^2} = \sqrt{(8.66+88.00)^2 + 3(5.50+4.71)^2} = 98.26 \text{ MPa}.$$

As $\sigma_{eqv}^{\mathrm{max}} < \bar{\sigma}$, the beam is in the elastic range with a diameter of $d = 20.80$ cm.

The normal stress distribution is illustrated in Figure 7.16. The new neutral line position at the critical section is obtained as

$$\bar{y} = \frac{N_x}{M_z} \frac{I_z}{A} = \frac{300000}{-90000} \frac{4\pi(0.216)^4}{64\pi(0.216)^2} = -0.97 \text{ cm}.$$

□

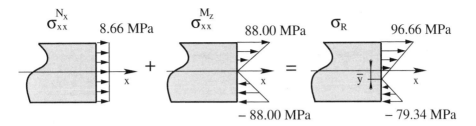

Figure 7.16 Example 7.2: normal stress distribution at the critical section.

Example 7.3 *Consider the beam with T-shaped cross-section illustrated in Figure 6.7. The critical section has a normal force of 300 kN and the bending moment is −240 kNm. The tractive and compressive admissible stresses respectively are $\bar{\sigma}_t = 80$ MPa and $\bar{\sigma}_c = 70$ MPa. Verify the beam, neglecting the shear effect due to shear forces.*

The area and second moment of area of the cross-section are $A = 400$ cm^2 and $I_z = 123294.27$ cm^4, respectively.

The normal stresses due to normal force and bending moment are respectively given as

$$\sigma_{xx}^{N_x} = \frac{N_x}{A} = \frac{300000}{4.0 \times 10^{-2}} = 7.50,$$

$$\sigma_{xx}^{M_z} = -\frac{M_z}{I_z}y = -\frac{-240000}{1.23 \times 10^{-3}} = 1.95 \times 10^8 y.$$

The maximum normal stresses due to bending occur at the section extremities, with values

$$\sigma_{xx}^{M_z} = 1.95 \times 10^8 y = \begin{cases} 38.42 \text{ MPa} & y = 19.69 \text{ cm} \\ -68.90 \text{ MPa} & y = -35.31 \text{ cm.} \end{cases}$$

On the other hand, the resultant normal stress at these points is

$$\sigma_R = \sigma_{xx}^{N_x} + \sigma_{xx}^{M_z} = 7.50 + 1.95 \times 10^8 y = \begin{cases} 45.92 \text{ MPa} & y = 19.69 \text{ cm} \\ -61.40 \text{ MPa} & y = -35.31 \text{ cm.} \end{cases}$$

As the maximum tensile and compressive normal stresses are lower than the admissible ones, namely, $\sigma_t^{\max} = 45.92$ MPa $< \bar{\sigma}_t$ and $\sigma_c^{\max} = 61.40$ MPa $< \bar{\sigma}_c$, the beam remains in the elastic range.

The position of the neutral line is calculated as

$$\bar{y} = \frac{N_x}{M_z}\frac{I_z}{A} = \frac{300000}{-240000}\frac{1.23 \times 10^{-3}}{4.0 \times 10^{-2}} = -3.84 \text{ cm.}$$

Figure 7.17 illustrates the normal stress distributions due to normal force, bending moment, and the resultant.

□

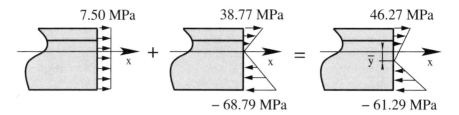

7.50 MPa   38.77 MPa   46.27 MPa

          + x   + x   = x

          − 68.79 MPa   − 61.29 MPa

Figure 7.17 Example 7.3: normal stress distributions at the critical section.

7.3 THREE-DIMENSIONAL BEAM

Now we want to develop a three-dimensional beam model that includes the effects of traction/compression in x, torsion in x for circular sections, and bending and shear in y and z. The combination of bending in y and z is called oblique bending. The variational formulation steps are used again to derive this beam model. We assume that y and z are axes of symmetry of the cross-section, as in rectangular and circular cross-sections.

7.3.1 KINEMATICS

The expressions for the components of axial displacement due to traction and transversal displacements in y and z in the torsion of circular croos-sections are the same indicated in equation (7.2) for the two-dimensional beam.

For the three-dimensional beam, we have bending actions in planes y and z, as illustrated in Figure 7.18. While a positive deflection in y results in a positive bending angle in z, as shown in Figure 7.18(b), this doesn't hold for plane xz. According to Figure 7.18(c), a positive bending angle in y is associated with a negative deflection in z. Similarly, a negative bending angle in y is associated to a positive deflection in z, as shown in Figure 7.18(d).

Notice that for a positive bending angle θ_y, the points with negative z coordinates have negative axial displacements; however, the points with positive z coordinates have positive axial displacements [see Figure 7.18(c)]. Likewise, for a negative bending angle θ_y, we have $u_x < 0$ when $z > 0$ and $u_x > 0$ when $z < 0$ [see Figure 7.18(d)]. Following, the case of positive bending angle in y and negative deflection in z is adopted, as shown in Figure 7.18(d).

Consider section AB in plane xz, which is x units distant from the origin of the reference system, as shown in Figure 7.18(c). Analogous to the bending in z [see Figure 7.18(c)], section AB has a traversal displacement u_z, reaching position $A'B'$. Thereafter, the section has a positive rotation α_y about y, reaching position $A''B''$, orthogonal to the tangent at O'. Due to the inclusion of shear, the section presents distortion β_y, reaching the final position $A'''B'''$.

According to Figure 7.19, due to bending in y, the axial displacement of point P is given by the difference between the final and initial positions, that is,

$$u_x = (x + \Delta x) - x = \Delta x. \tag{7.63}$$

The following trigonometric relation is valid for triangle $QO'P'''$ in Figure 7.19(c):

$$\sin \theta_y = \frac{\Delta x}{z}. \tag{7.64}$$

Substituting (7.64) in (7.63), we have

$$u_x(x, z) = z \sin \theta_y(x). \tag{7.65}$$

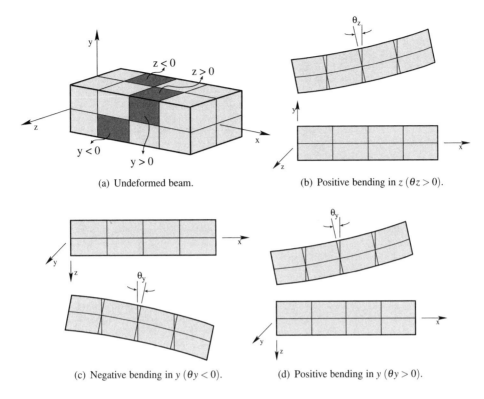

(a) Undeformed beam.

(b) Positive bending in z $(\theta z > 0)$.

(c) Negative bending in y $(\theta y < 0)$.

(d) Positive bending in y $(\theta y > 0)$.

Figure 7.18   Bending actions in planes xy and xz of the three-dimensional beam.

Assuming small bending angle θ_y, we have $\sin\theta_y \approx \theta_y$ and the above expression reduces to

$$u_x(x,z) = z\theta_y(x). \tag{7.66}$$

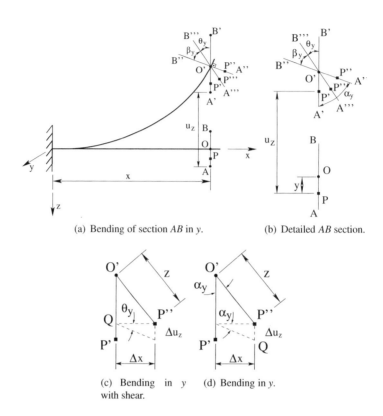

(a) Bending of section AB in y.   (b) Detailed AB section.

(c) Bending in y (d) Bending in y.
with shear.

Figure 7.19   Bending kinematics in plane xz of the three-dimensional beam.

Analogous to bending in z, the axial displacement due to bending in y has a linear variation with the z coordinate of the points, as shown in Figure 7.20.

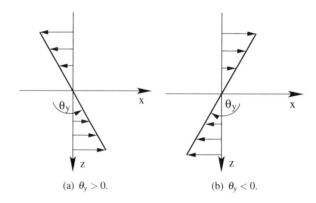

(a) $\theta_y > 0$.   (b) $\theta_y < 0$.

Figure 7.20   Axial displacement due to bending in y for the three-dimensional beam.

The following trigonometric relation is valid for triangle $QP'P'''$ of Figure 7.19(d), only considering the effect of bending,

$$\tan \alpha_y = -\frac{\Delta u_z}{\Delta x}. \tag{7.67}$$

For small rotations, we have $\tan \alpha_y \approx \alpha_y$. Taking the limit for $\Delta x \to 0$, the bending angle for section x is

$$\alpha_y(x) = -\frac{du_z(x)}{dx}. \tag{7.68}$$

Considering the shear effect, the bending angle θ_y of section x is given in terms of bending α_y and distortion β_y angles by

$$\theta_y(x) = \alpha_y(x) - \beta_y(x). \tag{7.69}$$

Substituting (7.69) in (7.68), the final expression for the bending angle in y is

$$\theta_y(x) = -\frac{du_z(x)}{dx} - \beta_y(x). \tag{7.70}$$

The bending in z is the same as the two-dimensional beam, only the distortion due to shear in plane xy will now be denoted by $\beta_z(x)$.

Thus, the resultant axial displacement for oblique bending is given by

$$u_x^b(x)(x,y,z) = u_x^{by}(x,z) + u_x^{bz}(x,y) = z\theta_y(x) - y\theta_z(x). \tag{7.71}$$

The oblique bending angles in y and z, including the shear effect, are respectively given by

$$\theta_y(x) \quad = \quad -\frac{du_z(x)}{dx} - \beta_y(x), \tag{7.72}$$

$$\theta_z(x) \quad = \quad \frac{du_y(x)}{dx} - \beta_z(x). \tag{7.73}$$

Figure 7.21 illustrates the contributions for the axial displacement due to bending in axes y and z, and the resultant axial displacement.

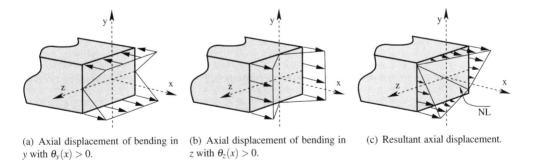

(a) Axial displacement of bending in y with $\theta_y(x) > 0$.

(b) Axial displacement of bending in z with $\theta_z(x) > 0$.

(c) Resultant axial displacement.

Figure 7.21   Axial displacements due to bending in y and z of the three-dimensional beam.

The kinematics of the three-dimensional beam with circular cross-section is given by the following vector field:

$$\mathbf{u}(x,y,z) = \left\{ \begin{array}{c} u_x(x,y,z) \\ u_y(x,z) \\ u_z(x,y) \end{array} \right\} = \left\{ \begin{array}{c} u_x^n(x) + z\theta_y(x) - y\theta_z(x) \\ u_y(x) - z\theta_x(x) \\ u_z(x) + y\theta_x(x) \end{array} \right\}. \tag{7.74}$$

7.3.2 STRAIN MEASURE

The normal and shear strain components of the three-dimensional beam are given by the derivatives of the kinematics given in (7.74).

For the normal strain components in directions x, y, and z, we have

$$\varepsilon_{xx}(x,y,z) = \frac{\partial u_x(x,y,z)}{\partial x} = \frac{du_x^n(x)}{dx} + z\frac{d\theta_y(x)}{dx} - y\frac{d\theta_z(x)}{dx}, \tag{7.75}$$

$$\varepsilon_{yy}(x,z) = \frac{\partial u_y(x,z)}{\partial y} = 0, \tag{7.76}$$

$$\varepsilon_{zz}(x,y) = \frac{\partial u_z(x,y)}{\partial z} = 0. \tag{7.77}$$

Thus, only the normal strain component ε_{xx} has a nonzero value. The other components are zero, because the displacements u_y and u_z are rigid and thus constant for all points of section x.

The normal strain components ε_{xx}^n due to normal force and ε_{xx}^b due to oblique bending are given by

$$\varepsilon_{xx}^n(x) = \frac{du_x^n(x)}{dx}, \tag{7.78}$$

$$\varepsilon_{xx}^b(x,y,z) = z\frac{d\theta_y(x)}{dx} - y\frac{d\theta_z(x)}{dx}. \tag{7.79}$$

The shear strain components are respectively determined according to planes xy, xz, and yz, as

$$\bar{\gamma}_{xy}(x) = \frac{\partial u_x(x,y,z)}{\partial y} + \frac{\partial u_y(x,z)}{\partial x}$$
$$= -\theta_z(x) - z\frac{d\theta_x(x)}{dx} + \frac{du_y(x)}{dx} = \beta_z(x) - z\frac{d\theta_x(x)}{dx}, \tag{7.80}$$

$$\bar{\gamma}_{xz}(x) = \frac{\partial u_x(x,y,z)}{\partial z} + \frac{\partial u_z(x,y)}{\partial x}$$
$$= \theta_y(x) + \frac{du_z(x)}{dx} + y\frac{d\theta_x(x)}{dx} = -\beta_y(x) + y\frac{d\theta_x(x)}{dx}, \tag{7.81}$$

$$\bar{\gamma}_{yz}(x) = \frac{\partial u_y(x,z)}{\partial z} + \frac{\partial u_z(x,y)}{\partial y} = -\theta_x(x) + \theta_x(x) = 0. \tag{7.82}$$

In terms of the tangential distortion, the strain components of the three-dimensional beam are

$$\varepsilon_{xx}(x,y,z) = \varepsilon_{xx}^n(x) + \varepsilon_{xx}^b(x,y,z), \tag{7.83}$$

$$\bar{\gamma}_{xy}(x) = \beta_z(x) = \frac{du_y(x)}{dx} - \theta_z(x), \tag{7.84}$$

$$\bar{\gamma}_{xz}(x) = -\beta_y(x) = \frac{du_z(x)}{dx} + \theta_y(x), \tag{7.85}$$

$$\gamma_t(x,r) = r\frac{d\theta_x(x)}{dx}. \tag{7.86}$$

7.3.3 RIGID BODY ACTIONS

We obtain the rigid body actions imposing the strain components to be zero. Considering the longitudinal strain ε_{xx} equal to zero, we have

$$\varepsilon_{xx}(x,y,z) = \frac{du_x^n(x)}{dx} + z\frac{d\theta_y(x)}{dx} - y\frac{d\theta_z(x)}{dx} = 0.$$

Thus, to satisfy the above expression, we must have $\dfrac{du_x^n(x)}{dx} = 0$, $\dfrac{d\theta_y(x)}{dx} = 0$, $\dfrac{d\theta_z(x)}{dx} = 0$; that is, $u_x^n(x) = u_x$ and $\theta_y(x) = \theta_y$ and $\theta_z(x) = \theta_z$ must be constant for all sections of the beam. Thus, besides the translation in x and rotation about z, there is also a rotation about y as a rigid body action of the three-dimensional beam.

Substituting the definitions of angles $\theta_y(x)$ and $\theta_z(x)$ given in (7.72) and (7.73) for the oblique bending term of the above expression, we have

$$
\begin{aligned}
\varepsilon_{xx}^b(x,y,z) &= z\frac{d}{dx}\left(-\beta_y(x) - \frac{du_z(x)}{dx}\right) + y\frac{d}{dx}\left(-\beta_z(x) + \frac{du_y(x)}{dx}\right)\\
&= y\frac{du_y^2(x)}{dx^2} + z\frac{d^2u_z(x)}{dx^2} = 0,
\end{aligned}
$$

recalling that $\beta_y(x)$ and $\beta_z(x)$ are constant in each section of the beam. For constant values of $u_y(x) = u_y$ and $u_z(x) = u_z$, the previous expression is satisfied, and consequently the translations in y and z are also rigid body actions of the three-dimensional beam.

Analogous to the two-dimensional beam, we enforce the tangential distortional component to be zero and obtain a constant rotation about x, which is also a rigid body action.

Thus, we conclude that the rigid body actions of a three-dimensional beam are translations and rotations in x, y, and z. The translation in z and rotation in y are illustrated in Figure 7.22.

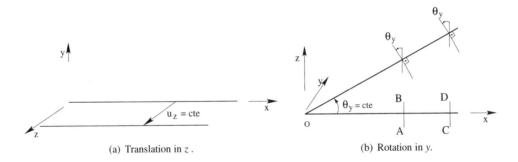

(a) Translation in z. (b) Rotation in y.

Figure 7.22 Rigid body actions for the three-dimensional beam.

7.3.4 DETERMINATION OF INTERNAL LOADS

Analogous to the previous models, associated to the normal $\varepsilon_{xx}(x,y,z)$ and shear $\bar{\gamma}_{xy}(x)$, $\bar{\gamma}_{xz}(x)$, $\gamma_t(x,r)$ strain components, we respectively have the normal $\sigma_{xx}(x,y,z)$ and shear $\tau_{xy}(x)$, $\tau_{xz}(x)$, $\tau_t(x,r)$ stress components, representing the internal loads on the three-dimensional beam. The strain internal work is given by

$$
\begin{aligned}
W_i &= \int_V \left(\sigma_{xx}(x,y,z)\varepsilon_{xx}(x,y,z) + \tau_{xy}(x)\bar{\gamma}_{xy}(x) + \tau_{xz}(x)\bar{\gamma}_{xz}(x)\right)dV\\
&\quad + \int_V \tau_t(x,r)\gamma_t(x,t)dV.
\end{aligned}
\tag{7.87}
$$

We replace the strain components given in (7.83) to (7.86), to obtain

$$
\begin{aligned}
W_i &= \int_V \sigma_{xx}(x,y,z)\left(\frac{du_x^n(x)}{dx} + z\frac{d\theta_y(x)}{dx} - y\frac{d\theta_z(x)}{dx}\right)dV + \int_V \tau_t(x,r)r\frac{d\theta_x(x)}{dx}dV\\
&\quad + \int_V \tau_{xy}(x)\left(\frac{du_y(x)}{dx} - \theta_z(x)\right)dV + \int_V \tau_{xz}(x)\left(\frac{du_z(x)}{dx} + \theta_y(x)\right)dV.
\end{aligned}
\tag{7.88}
$$

The above volume integrals can be decomposed along the length and cross-section area of the beam, that is,

$$
\begin{aligned}
W_i &= \int_0^L \left(\int_A \sigma_{xx}(x,y,z)dA \right) \frac{du_x^n(x)}{dx} + \int_0^L \left(\int_A -\sigma_{xx}(x,y,z)ydA \right) \frac{d\theta_z(x)}{dx} dx \\
&+ \int_0^L \left(\int_A \sigma_{xx}(x,y,z)zdA \right) \frac{d\theta_y(x)}{dx} dx + \int_0^L \left(\int_A r\tau_t(x,r)dA \right) \frac{d\theta_x(x)}{dx} dx \\
&+ \int_0^L \left(\int_A -\tau_{xy}(x)dA \right) \left(\theta_z(x) - \frac{du_y(x)}{dx} \right) dx \\
&+ \int_0^L \left(\int_A -\tau_{xz}(x)dA \right) \left(-\theta_y(x) - \frac{du_z(x)}{dx} \right) dx.
\end{aligned}
\tag{7.89}
$$

The area integrals represent the internal loads on the three-dimensional beam. Besides those indicated in equations (7.39) to (7.42), we have the shear force in z and the bending moment in y, respectively given by

$$
V_z(x) = -\int_A dV_z = -\int_A \tau_{xz}(x)dA,
\tag{7.90}
$$

$$
M_y(x) = \int_A dM_y = \int_A \sigma_{xx}(x,y,z)zdA.
\tag{7.91}
$$

The normal force dN_x, shear forces dV_y and dV_z, and bending moments dM_y and dM_z for an area element of a rectangular cross-section are illustrated in Figure 7.23.

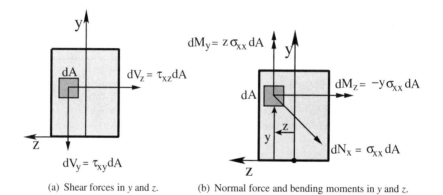

(a) Shear forces in y and z.   (b) Normal force and bending moments in y and z.

Figure 7.23 Normal and shear forces in a cross-section area element of a three-dimensional beam.

From these definitions, we can write the strain internal work (7.89) in the following way:

$$
\begin{aligned}
W_i &= \int_0^L N_x(x) \frac{du_x^n(x)}{dx} dx + \int_0^L V_y(x) \left(\theta_z(x) - \frac{du_y(x)}{dx} \right) dx \\
&+ \int_0^L V_z(x) \left(-\theta_y(x) - \frac{du_z(x)}{dx} \right) dx \int_0^L M_x(x) \frac{d\theta_x(x)}{dx} dx \\
&+ \int_0^L M_z(x) \frac{d\theta_z(x)}{dx} dx + \int_0^L M_y(x) \frac{d\theta_y(x)}{dx} dx.
\end{aligned}
\tag{7.92}
$$

Integrating by parts and rearranging the terms, we have

$$
\begin{aligned}
W_i &= -\int_0^L \frac{dN_x(x)}{dx} u_x^n(x)dx + [N_x(L)u_x^n(L) - N_x(0)u_x^n(0)] \\
&+ \int_0^L \frac{dV_y(x)}{dx} u_y(x)dx - [V_y(L)u_y(L) - V_y(0)u_y(0)] \\
&+ \int_0^L \frac{dV_z(x)}{dx} u_z(x)dx - [V_z(L)u_z(L) - V_z(0)u_z(0)] \\
&- \int_0^L \frac{dM_x(x)}{dx} \theta_x(x)dx + [M_x(L)\theta_x(0) - M_x(L)\theta_x(0)] \\
&+ \int_0^L \left(-\frac{dM_y(x)}{dx} - V_z(x) \right) \theta_y(x)dx + [M_y(L)\theta_y(L) - M_y(0)\theta_y(0)] \\
&+ \int_0^L \left(-\frac{dM_z(x)}{dx} + V_y(x) \right) \theta_z(x)dx + [M_z(L)\theta_z(L) - M_z(0)\theta_z(0)].
\end{aligned}
\tag{7.93}
$$

The compatible internal loads with the kinematics of the three-dimensional beam are the same as in Figure 7.5, and the bending loads in y are shown in Figure 7.24. The internal distributed bending moment in y is denoted as m_{y_i} and defined by

$$
m_{y_i}(x) = -\frac{dM_y(x)}{dx} - V_z(x).
\tag{7.94}
$$

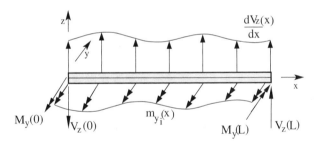

Figure 7.24 Internal y bending and z shear loads for the three-dimensional beam.

7.3.5 DETERMINATION OF EXTERNAL LOADS

Besides the external loads in the two-dimensional beam, we now have the following loads at the ends of the three-dimensional beam:

- M_{y_0}, M_{y_L}: bending moments at $x = 0$ and $x = L$ in direction y
- V_{z_0}, V_{z_L}: transversal forces at $x = 0$ and $x = L$ in direction z

The additional distributed external loads along the length of the beam are:

- $q_z(x)$: transversal distributed force in direction z
- $m_y(x)$: distributed bending moment in direction y

These additional external loads are illustrated in Figure 7.25 according to the positive directions of the adopted Cartesian reference system.

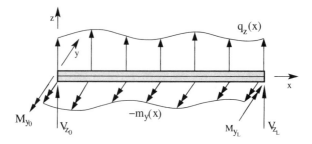

Figure 7.25   External y bending and z shear loads for the three-dimensional beam.

The external work for any kinematic action $\mathbf{u}(x,y,z)$ given in (7.74) can be written as:

$$
\begin{aligned}
W_e &= \int_0^L q_x(x)u_x^n(x)dx + \left[P_L u_x^n(L) + P_0 u_x^n(0)\right] \\
&+ \int_0^L q_y(x)u_y(x)dx + \left[V_{y_L} u_y(L) + V_{y_0} u_y(0)\right] \\
&+ \int_0^L q_z(x)u_z(x)dx + \left[V_{z_L} u_z(L) + V_{z_0} u_z(0)\right] \\
&+ \int_0^L m_x(x)\theta_x(x)dx + \left[T_L \theta_x(0) + T_0 \theta_x(0)\right] \\
&+ \int_0^L m_y(x)\theta_y(x)dx + \left[M_{y_L} \theta_y(L) + M_{y_0} \theta_y(0)\right] \\
&+ \int_0^L m_z(x)\theta_z(x)dx + \left[M_{z_L} \theta_z(L) + M_{z_0} \theta_z(0)\right].
\end{aligned}
\tag{7.95}
$$

7.3.6   EQUILIBRIUM

As in the previous cases, the PVW is applied to determine the equilibrium BVPs of the three-dimensional beam. When equating (7.93) with (7.95) for a virtual action $\delta\mathbf{u}$ from the deformed configuration, we have

$$
\int_0^L \left[-\frac{dN_x(x)}{dx} - q_x(x) \right] \delta u_x^n(x)dx + [N_x(L) - P_L]\,\delta u_x^n(L) + [-N_x(0) - P_0]\,\delta u_x^n(0) +
$$

$$
\int_0^L \left[\frac{dV_y(x)}{dx} - q_y(x) \right] \delta u_y(x)dx + [-V_y(L) - V_{y_L}]\,\delta u_y(L) + [V_y(0) - V_{y_0}]\,\delta u_y(0) +
$$

$$
\int_0^L \left[\frac{dV_z(x)}{dx} - q_z(x) \right] \delta u_z(x)dx + [-V_z(L) - V_{z_L}]\,\delta u_z(L) + [V_z(0) - V_{z_0}]\,\delta u_z(0) +
$$

$$
\int_0^L \left[-\frac{dM_x(x)}{dx} - m_x(x) \right] \delta\theta_x(x)dx + [M_x(L) - T_L]\,\delta\theta_x(L) + [-M_x(0) - T_0]\,\delta\theta_x(0) +
$$

$$
\int_0^L \left[-\frac{dM_y(x)}{dx} - V_z(x) - m_y(x) \right] \delta\theta_y(x)dx + [M_y(L) - M_{y_L}]\,\delta\theta_y(L) + [-M_y(0) - M_{y_0}]\,\delta\theta_y(0) +
$$

$$
\int_0^L \left[-\frac{dM_z(x)}{dx} + V_y(x) - m_z(x) \right] \delta\theta_z(x)dx + [M_z(L) - M_{z_L}]\,\delta\theta_z(L) + [-M_z(0) - M_{z_0}]\,\delta\theta_z(0) = 0.
$$

For the above expression to be valid for any arbitrary virtual action, all terms inside brackets must be simultaneously zero, resulting in the local form in terms of the equilibrium BVPs of the three-dimensional beam, that is,

$$\begin{cases} \dfrac{dN_x(x)}{dx} + q_x(x) = 0 & \text{in } x \in (0,L) \\ N_x(L) = P_L & \text{in } x = L \\ N_x(0) = -P_0 & \text{in } x = 0 \end{cases} \qquad (7.96)$$

$$\begin{cases} \dfrac{dM_x(x)}{dx} + m_x(x) = 0 & \text{in } x \in (0,L) \\ M_x(L) = T_L & \text{in } x = L \\ M_x(0) = -T_0 & \text{in } x = 0 \end{cases} \qquad (7.97)$$

$$\begin{cases} \dfrac{dM_z(x)}{dx} - V_y(x) + m_z(x) = 0 & \text{in } x \in (0,L) \\ \dfrac{dV_y(x)}{dx} - q_y(x) = 0 & \text{in } x \in (0,L) \\ V_y(0) = V_{y_0} & \text{in } x = 0 \\ V_y(L) = -V_{y_L} & \text{in } x = L \\ M_z(0) = -M_{z_0} & \text{in } x = 0 \\ M_z(L) = M_{z_L} & \text{in } x = L \end{cases} \qquad (7.98)$$

$$\begin{cases} \dfrac{dM_y(x)}{dx} + V_z(x) + m_y(x) = 0 & \text{in } x \in (0,L) \\ \dfrac{dV_z(x)}{dx} - q_z(x) = 0 & \text{in } x \in (0,L) \\ V_z(0) = V_{z_0} & \text{in } x = 0 \\ V_z(L) = -V_{z_L} & \text{in } x = L \\ M_y(0) = -M_{y_0} & \text{in } x = 0 \\ M_y(L) = M_{y_L} & \text{in } x = L \end{cases} \qquad (7.99)$$

Thus, the equilibrium BVPs of the three-dimensional beam are given by six differential equations in terms of the normal force in x, shear forces in y and z, twisting moments in x, and bending moments in y and z, along with the boundary conditions in terms of these loads.

Assuming the distributed external moment in y as zero, i.e., $m_y(x) = 0$, the differential equations of equilibrium given in (7.99) reduce to

$$\frac{dV_z(x)}{dx} = q_z(x), \qquad (7.100)$$

$$V_z(x) = -\frac{dM_y(x)}{dx}. \qquad (7.101)$$

Combining the above expressions, we obtain the second-order differential equation of equilibrium in terms of the bending moment in y, that is,

$$\frac{d^2 M_y(x)}{dx^2} = -q_z(x). \qquad (7.102)$$

For a rigid virtual action, the strain measures and consequently the internal work are zero. In this case, the PVW states that the external work (7.95) is zero. Besides that, because of rigid actions, $\delta u_x^n(x) = \delta u_x^n = \text{cte}, \delta u_y(x) = \delta u_y = \text{cte}, \delta u_z(x) = \delta u_z = \text{cte}, \delta \theta_x(x) = \delta \theta_x = \text{cte}, \delta \theta_y(x) = \delta \theta_y =$

cte, and $\delta\theta_z(x) = \delta\theta_z = $ cte, we have

$$
\begin{aligned}
\delta W_e &= \left(\int_0^L q_x(x)dx + P_L + P_0\right)\delta u_x^n + \left(\int_0^L q_y(x)dx + V_{y_L} + V_{y_0}\right)\delta u_y \\
&+ \left(\int_0^L q_z(x)dx + V_{z_L} + V_{z_0}\right)\delta u_y + \left(\int_0^L m_x(x)dx + T_L + T_0\right)\delta\theta_x \\
&+ \left(\int_0^L m_y(x)dx + M_{y_L} + M_{y_0}\right)\delta\theta_y + \left(\int_0^L m_z(x)dx + M_{z_L} + M_{z_0}\right)\delta\theta_z = 0.
\end{aligned}
$$

Hence, the equilibrium conditions for the three-dimensional beam state that the resultant of external forces and moments in directions x, y and z must be zero, that is,

$$\sum f_x = \int_0^L q_x(x)dx + P_L + P_0 = 0, \tag{7.103}$$

$$\sum f_y = \int_0^L q_y(x)dx + V_{y_L} + V_{y_0} = 0, \tag{7.104}$$

$$\sum f_z = \int_0^L q_z(x)dx + V_{z_L} + V_{z_0} = 0, \tag{7.105}$$

$$\sum m_x = \int_0^L m_x(x)dx + T_L + T_0 = 0, \tag{7.106}$$

$$\sum m_y = \int_0^L m_y(x)dx + M_{y_L} + M_{y_0}, \tag{7.107}$$

$$\sum m_z = \int_0^L m_z(x)dx + M_{z_L} + M_{z_0} = 0. \tag{7.108}$$

The previous integrals represent the concentrated forces and moments equivalent to the distributed forces and moments.

7.3.7   APPLICATION OF THE CONSTITUTIVE EQUATION

Hooke's law for an isotropic homogeneous linear elastic material relates the normal stress $\sigma_{xx}(x,y,z)$ to the specific normal strain $\varepsilon_{xx}(x,y,z)$ by the longitudinal elastic modulus $E(x)$, that is,

$$\sigma_{xx}(x,y,z) = E(x)\varepsilon_{xx}(x,y,z) = \sigma_{xx}^n(x) + \sigma_{xx}^{by}(x,z) + \sigma_{xx}^{bz}(x,y). \tag{7.109}$$

with $\sigma_{xx}^n(x)$, $\sigma_{xx}^{by}(x,z)$ and $\sigma_{xx}^{bz}(x,y)$ the contributions of normal stress in x due to the normal force and bending moments in y and z respectively given by

$$\sigma_{xx}^n(x) = E(x)\frac{du_x^n(x)}{dx}, \tag{7.110}$$

$$\sigma_{xx}^{by}(x,z) = E(x)\frac{d\theta_y(x)}{dx}z, \tag{7.111}$$

$$\sigma_{xx}^{bz}(x,y) = -E(x)\frac{d\theta_z(x)}{dx}y. \tag{7.112}$$

Likewise, the shear stresses $\tau_{xy}(x)$, $\tau_{xz}(x)$, and $\tau_t(x,r)$ are respectively related to the shear strain components $\bar{\gamma}_{xy}(x)$, $\bar{\gamma}_{xz}(x)$, and $\gamma_t(x,r)$ by the shear modulus $G(x)$, that is,

$$\tau_t(x,r) = G(x)\gamma_t(x,r) = G(x)\frac{d\theta_x(x)}{dx}r, \tag{7.113}$$

$$\tau_{xy}(x) = G(x)\bar{\gamma}_{xy}(x) = G(x)\left(\frac{du_y(x)}{dx} - \theta_z(x)\right), \tag{7.114}$$

$$\tau_{xz}(x) = G(x)\bar{\gamma}_{xz}(x) = G(x)\left(-\frac{du_z(x)}{dx} - \theta_y(x)\right). \tag{7.115}$$

Substituting the above constitutive relations in the internal load expressions given in (7.17) to (7.20), (7.90), and (7.91), we obtain

$$N_x(x) = E(x)\frac{du_x^n(x)}{dx}\int_A dA = E(x)A(x)\frac{du_x^n(x)}{dx}, \tag{7.116}$$

$$V_y(x) = -G(x)\left[\frac{du_y(x)}{dx} - \theta_z(x)\right]\int_A dA = -G(x)A(x)\beta_z(x), \tag{7.117}$$

$$V_z(x) = -G(x)\left[-\frac{du_z(x)}{dx} - \theta_y(x)\right]\int_A dA = -G(x)A(x)\beta_y(x), \tag{7.118}$$

$$M_x(x) = G(x)\frac{d\theta_x(x)}{dx}\int_A r^2 dA = G(x)I_p(x)\frac{d\theta_x(x)}{dx}, \tag{7.119}$$

$$M_y(x) = E(x)\frac{d\theta_y(x)}{dx}\int_A z^2 dA = E(x)I_y(x)\frac{d\theta_y(x)}{dx}, \tag{7.120}$$

$$M_z(x) = E(x)\frac{d\theta_z(x)}{dx}\int_A y^2 dA = E(x)I_z(x)\frac{d\theta_z(x)}{dx}, \tag{7.121}$$

with $I_y(x)$ the second moment of area of the cross-sections relative to axis y of the reference system.

Now, substituting the above expressions in the differential equations of equilibrium given in (7.96) to (7.99), we obtain the kinematic differential equations of the three-dimensional beam

$$\frac{d}{dx}\left(E(x)A(x)\frac{du_x^n(x)}{dx}\right) + q_x(x) = 0, \tag{7.122}$$

$$\frac{d}{dx}\left(G(x)I_p(x)\frac{d\theta_x(x)}{dx}\right) + m_x(x) = 0, \tag{7.123}$$

$$\frac{d}{dx}\left[G(x)A(x)\left(\frac{du_y(x)}{dx} - \theta_z(x)\right)\right] + q_y(x) = 0, \tag{7.124}$$

$$\frac{d}{dx}\left(E(x)I_z(x)\frac{d\theta_z(x)}{dx}\right) + G(x)A(x)\left(\frac{du_y(x)}{dx} - \theta_z(x)\right) + m_z(x) = 0, \tag{7.125}$$

$$\frac{d}{dx}\left[G(x)A(x)\left(\frac{du_z(x)}{dx} + \theta_y(x)\right)\right] + q_z(x) = 0, \tag{7.126}$$

$$\frac{d}{dx}\left(E(x)I_y(x)\frac{d\theta_y(x)}{dx}\right) - G(x)A(x)\left(\frac{du_z(x)}{dx} + \theta_y(x)\right) + m_y(x) = 0. \tag{7.127}$$

To solve a real problem using the three-dimensional beam model, we should analyze the types of the applied external loads and integrate their respective differential equations.

Neglecting the effect of shear due to the bending in y, the differential equation in terms of the displacement u_z is obtained by substituting (7.120) in (7.102). Thus,

$$\frac{d^2}{dx^2}\left(E(x)I_y(x)\frac{d^2u_z(x)}{dx^2}\right) = -q_z(x). \tag{7.128}$$

7.3.8 STRESS DISTRIBUTIONS

If we respectively replace the derivatives of the kinematic components obtained from equations (7.116) to (7.121), in constitutive equations (7.110) to (7.115), we obtain the following expressions for the normal and shear stress components due to internal loads of the three-dimensional beam:

$$\sigma_{xx}^{N_x}(x) \;=\; \frac{N_x(x)}{A(x)}, \tag{7.129}$$

$$\sigma_{xx}^{M_y}(x,z) \;=\; \frac{M_y(x)}{I_y(x)}z, \tag{7.130}$$

$$\sigma_{xx}^{M_z}(x,y) \;=\; -\frac{M_z(x)}{I_z(x)}y, \tag{7.131}$$

$$\tau_t(x,r) \;=\; \frac{M_x(x)}{I_p(x)}r, \tag{7.132}$$

$$\tau_{xy}^{V_y}(x) \;=\; -\frac{V_y(x)}{K_cA(x)}, \tag{7.133}$$

$$\tau_{xz}^{V_z}(x) \;=\; -\frac{V_z(x)}{K_cA(x)}. \tag{7.134}$$

In the case of shear stresses $\tau_{xy}^{V_y}$ and $\tau_{xz}^{V_z}$, the shear factor K_c was included for the cross-section.

The resultant normal stress is given by the sum of the normal stresses due to normal force and oblique bending, that is,

$$\sigma_R(x,y,z) = \sigma_{xx}^{N_x}(x) + \sigma_{xx}^{M_y}(x,z) + \sigma_{xx}^{M_z}(x,y) = \frac{N_x(x)}{A(x)} + \frac{M_y(x)}{I_y(x)}z - \frac{M_z(x)}{I_z(x)}y. \tag{7.135}$$

While the stress due to normal force is constant at all points of the cross-section, the bending stresses have a linear variation with y and z.

Considering only the oblique bending and neglecting the normal force effect, we obtain the neutral line when the resultant normal stress is zero, i.e.

$$\sigma_R(x,y,z) = \sigma_{xx}^{M_y}(x,z) + \sigma_{xx}^{M_z}(x,y) = 0. \tag{7.136}$$

From the above expression, we have

$$\frac{M_z(x)}{I_z(x)}y = \frac{M_y(x)}{I_y(x)}z \rightarrow \frac{y}{z} = \frac{M_y(x)}{M_z(x)}\frac{I_z(x)}{I_y(x)}. \tag{7.137}$$

Defining the angle β between the neutral line and the z axis of the reference system, as shown in Figure 7.26, the following geometric relationship is valid:

$$\tan\beta = \frac{y}{z}. \tag{7.138}$$

The bending moment resultant is represented in Figure 7.26(b) and is given by

$$M_R = \sqrt{M_y^2 + M_z^2}. \tag{7.139}$$

Besides that,

$$\tan\alpha = \frac{M_y}{M_z}. \tag{7.140}$$

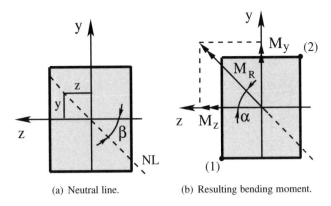

(a) Neutral line.   (b) Resulting bending moment.

Figure 7.26 Neutral line and moment resultant in oblique bending.

Substituting the previous expression in equation (7.137) of the neutral line, we have

$$\tan \beta = \frac{I_z}{I_y} \tan \alpha. \tag{7.141}$$

Thus, the general expression for the angle β, which locates the neutral line in the cross-section, is given by

$$\beta = \arctan \left(\frac{I_z}{I_y} \tan \alpha \right). \tag{7.142}$$

For a circular section, we have $I_z = I_y = I$ and hence $\tan \beta = \tan \alpha$. Thus, the neutral line is in the direction of the bending moment resultant, as shown in Figure 7.27.

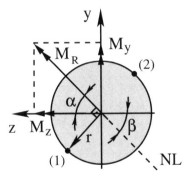

Figure 7.27 Resulting bending moment in the beam with circular cross-section.

The points (1) and (2) which are further away from the neutral line have coordinates (y_1, z_1) and (y_2, z_2), as illustrated in Figures 7.26 and 7.27. These respectively are the critical points due to traction and compression of the cross-section for positive bending moments in y and z. The resulting maximum tractive normal stress in cross-section x occurs at point (1) with coordinates (y_1, z_1). Thus,

$$\sigma_R^{\max,t} = -\frac{M_z(x)}{I_z(x)} y_1 + \frac{M_y(x)}{I_y(x)} z_1. \tag{7.143}$$

The maximum compressive stress resultant occurs at point (2), with coordinates (y_2, z_2), that is,

$$\sigma_R^{\max,c} = -\frac{M_z(x)}{I_z(x)} y_2 + \frac{M_y(x)}{I_y(x)} z_2. \tag{7.144}$$

According to Figure 7.27, the coordinates of point 1 are

$$y_1 = -r\cos\alpha,$$
$$z_1 = r\sin\alpha.$$

Besides that, the following relations are valid for the bending moments:

$$\cos\alpha = \frac{M_z}{M_R},$$
$$\sin\alpha = \frac{M_y}{M_R}.$$

In circular sections, the maximum resultant normal stress due to traction and compression are equal in magnitude. Substituting the above expressions in equation (7.143) for the bending normal stress and making some simplifications, we have

$$\sigma_R^{max} = \frac{\sqrt{M_z^2 + M_y^2}}{W_{NL}} = \frac{M_R}{W_{NL}}, \tag{7.145}$$

with the bending strength modulus relative to the neutral line W_{LN} given by

$$W_{NL} = \frac{I}{\dfrac{d}{2}} = \frac{\pi d^3}{32}. \tag{7.146}$$

Figure 7.28 illustrates the stress distributions due to the positive normal force and oblique bending, which is also positive in the cross-section of the beam.

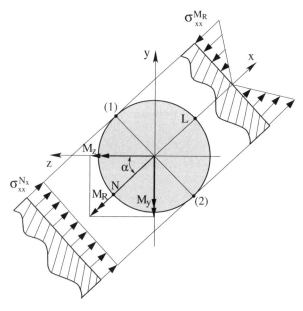

Figure 7.28 Distribution of normal stress for a positive normal force and oblique bending in a beam with circular cross-section.

7.3.9 DESIGN AND VERIFICATION

The following steps are applied to calculate the minimum cross-section dimensions of a three-dimensional beam:

1. Draw the diagrams of the bending moments $M_y(x)$ and $M_z(x)$, twisting moment $M_x(x)$, normal force $N_x(x)$, and shear forces $V_y(x)$ and $V_z(x)$. In the case of noncircular sections, the twisting moment diagram is not considered.
2. Determine the critical cross-section of the beam according to the following criteria:
   a. Section with the maximum absolute resultant moment $M_R = \sqrt{M_y^2 + M_z^2}$
   b. Section with the maximum absolute twisting moment
   c. Section with the maximum absolute normal force
   d. Section with the maximum absolute resultant shear force $V_R = \sqrt{V_y^2 + V_z^2}$

   The loads in the critical section are indicated by $N_x^{\max}$, $V_y^{\max}$, $V_z^{\max}$, $M_x^{\max}$, $M_y^{\max}$, and $M_z^{\max}$.
3. Determine the critical point of the critical cross-section. For this purpose, the normal stress distributions are analyzed in terms of the normal force and oblique bending in the critical section. The critical point considering only the oblique bending will be the most distant from the neutral line.
4. Calculate the maximum equivalent normal stress in the critical points, that is,

$$\sigma_{eqv}^{\max} = \sqrt{(\sigma_R^{\max})^2 + 3(\tau_R^{\max})^2}. \tag{7.147}$$

For a circular section, the maximum resultant normal and shear stresses are given by

$$\sigma_R^{\max} = \frac{N_x^{\max}}{A} + \frac{M_R^{\max}}{W_{NL}}, \tag{7.148}$$

$$\tau_R^{\max} = \frac{M_x^{\max}}{W_x} + \frac{4V_y^{\max}}{3A} + \frac{4V_z^{\max}}{3A}, \tag{7.149}$$

with W_x the torsional strength modulus and W_{NL} the oblique bending strength modulus relative to the neutral line. If the shear stresses due to shear forces are neglected, the shear stress is given only in terms of the twisting moment, that is,

$$\tau_R^{\max} = \frac{M_x^{\max}}{W_x}. \tag{7.150}$$

The maximum equivalent normal stress is now calculated, neglecting the normal and shear forces. Only the normal stresses due to the oblique bending and shear stresses due to twisting moment are considered. Thus, from (7.147), (7.148), and (7.150) we have

$$\sigma_{eqv}^{\max} = \sqrt{\left(\frac{\sqrt{(M_z^{max})^2 + (M_y^{max})^2}}{W_{NL}}\right)^2 + 3\left(\frac{M_x^{max}}{W_x}\right)^2}.$$

Recalling that $W_{NL} = 2W_x$, we obtain

$$\sigma_{eqv}^{\max} = \frac{32}{\pi d^3} \sqrt{(M_z^{max})^2 + (M_y^{max})^2 + \frac{3}{4}(M_x^{max})^2}. \tag{7.151}$$

Using the economic condition $\sigma_{eqv}^{\max} = \bar{\sigma}$, we obtain the expression for the minimum cross-section diameter as

$$d = \sqrt[3]{\frac{32}{\pi\bar{\sigma}} \sqrt{(M_z^{max})^2 + (M_y^{max})^2 + \frac{3}{4}(M_x^{max})^2}}. \tag{7.152}$$

To verify the effect of normal and shear forces, the previous diameter is increased around 5% to 10%, and the beam is verified using equations (7.147), (7.148), and (7.149). If $\sigma_{eqv}^{max} \leq \bar{\sigma}$, the beam is well designed. Otherwise, we increase the diameter again and verify the beam, until we find the diameter that keeps the beam in the elastic range.

For a noncircular cross-section, the maximum resultant normal and shear stresses are given by

$$
\begin{aligned}
\sigma_R^{max} &= \frac{N_x^{max}}{A} - \frac{M_z^{max}}{I_z} y^{max} + \frac{M_y^{max}}{I_y} z^{max} \\
&= \frac{N_x^{max}}{A} - \frac{M_z^{max}}{W_z^{max}} + \frac{M_y^{max}}{W_y^{max}},
\end{aligned}
\tag{7.153}
$$

$$
\tau_R^{max} = \frac{V_y^{max}}{K_c A} + \frac{V_z^{max}}{K_c A},
\tag{7.154}
$$

with W_y and W_z the bending strength modulus in y and z.

Neglecting the normal and shear forces, the maximum equivalent stress is equal to the normal stress due to oblique bending, that is,

$$
\sigma_{eqv}^{max} = -\frac{M_z^{max}}{W_z^{max}} + \frac{M_y^{max}}{W_y^{max}}.
\tag{7.155}
$$

The economic condition is employed, and the obtained equation is solved for the main dimensions in the expressions of W_y and W_z. To verify the effects of normal and shear forces, we apply an analogous procedure for the circular section.

The following examples illustrate the above procedure.

Example 7.4 *Design the beam with circular section of Figure 7.29, subjected to the bending and twisting loads, for $\bar{\sigma} = 70$ MPa. Neglect the shear stresses due to shear forces. The bearing in A is fixed and has three reactions in terms of forces in directions x, y, and z. The bearing at D only has reactions in directions y and z.*

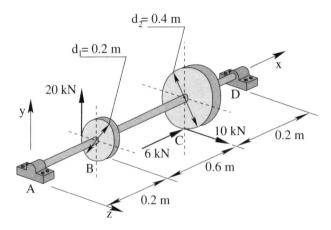

Figure 7.29 Example 7.4: Three-dimensional beam of circular cross-section with bending and twisting loads.

As two mechanical transmissions are assembled in sections B and C, we must first transfer the forces 10000 N and 20000 N to the respective centers of the beam cross-sections. After this, we have the following forces and moments at the centroids of cross-sections B and C:

$$F_{B_y} = 20000 \text{ N}, \quad M_{B_x} = (20000)(0.1) = 2000 \text{ Nm},$$
$$F_{C_x} = 6000 \text{ N}, \quad M_{C_z} = (6000)(0.2) = 1200 \text{ Nm},$$
$$F_{C_y} = 10000 \text{ N}, \quad M_{C_x} = (10000)(0.2) = 2000 \text{ Nm}.$$

These external loads are illustrated in the free body diagram of Figure 7.30.

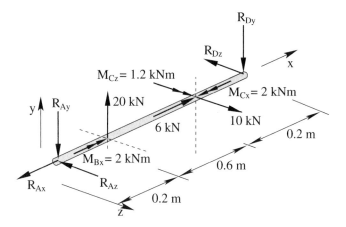

Figure 7.30 Example 7.4: free body diagram.

As the problem is isostatic with only concentrated loads, the method of sections is employed to obtain the diagrams. The following equilibrium conditions are considered to calculate the support reactions:

$$\Sigma F_x = 0: \quad -R_{Ax} + 6000 = 0 \rightarrow R_{Ax} = 6000 \text{ N},$$
$$\Sigma F_y = 0: \quad -R_{Ay} + 20000 - R_{Dy} = 0 \rightarrow R_{Ay} = 20000 - R_{Dy},$$
$$\Sigma F_z = 0: \quad -R_{Az} + 10000 - R_{Dz} = 0 \rightarrow R_{Az} = 10000 - R_{Dz},$$
$$\Sigma M_{z_A} = 0: \quad (20000)(0.2) + 1200 - R_{Dy}(1.0) = 0 \rightarrow R_{Dy} = 5200 \text{ N},$$
$$\Sigma M_{y_A} = 0: \quad -(10000)(0.8) + R_{Dz}(1.0) = 0 \rightarrow R_{Dz} = 8000 \text{ N}.$$

Consequently, $R_{Ay} = 14800 \text{ N}$ and $R_{Az} = 2000 \text{ N}$. The twisting moment must be self-balanced, which can be checked by taking the balance of moments in direction x, that is,

$$\Sigma M_x = -2000 + 2000 = 0.$$

The external loads acting in direction x and planes xy and xz are illustrated in Figure 7.31. The internal load diagrams illustrated in Figure 7.32 for normal and shear forces and twisting moment are obtained following the directions of the applied external loads, as in Chapter 5. The bending moment diagrams in y and z are determined from the areas of the shear force diagrams in z and y, respectively.

The critical cross-section is $x = 0.2^+ m$, with the following loads:

$$\begin{cases} N_x^{max} = 6000 \text{ N}, \\ V_y^{max} = 14800 \text{ N}, \\ V_z^{max} = 2000 \text{ N}, \\ M_x^{max} = 2000 \text{ Nm}, \\ M_y^{max} = 400 \text{ Nm}, \\ M_z^{max} = 2960 \text{ Nm}. \end{cases}$$

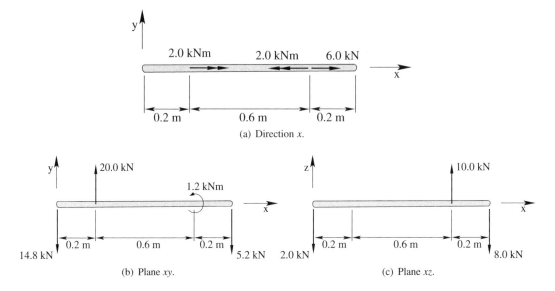

Figure 7.31 Example 7.4: free body diagrams for direction x and planes xy and xz.

The shear force in y at section $x = 0.2^+$ m is $V_y = 5200$ N. The value of $V_y = 14800$ N at section $x = 0.2^-$ m is used because is larger. The maximum resultant bending moment is

$$M_R^{\max} = \sqrt{(400)^2 + (2960)^2} = 2986.9 \text{ Nm}.$$

The minimum diameter of the critical cross-section is first determined considering only the bending and twisting effect from (7.152) and

$$d = \sqrt[3]{\frac{32}{\pi 70 \times 10^6}} \sqrt{(2986.9)^2 + \frac{3}{4}(2000.0)^2} = 7.95 \text{ cm}.$$

The diameter is increased to $d = 8.2$ cm and using equations (7.143) and (7.149), we obtain

$$\sigma_{xx}^{\max} = \frac{6000.0}{\frac{\pi}{4}(0.082)^2} + \frac{2986.9}{\frac{\pi}{32}(0.082)^3} = 45.6 \text{MPa},$$

$$\tau_t^{\max} = \frac{M_x^{\max}}{W_x} = \frac{2000.0}{\frac{\pi}{32}(0.082)^3} = 29.9 \text{ MPa}.$$

Using these values, the equivalent normal stress is $\sigma_{eqv} = 69.0$ MPa. As $\sigma_{eqv} < \bar{\sigma}$, the beam remains in the elastic range.

The neutral line position in the critical cross-section is calculated as

$$\tan \beta = \tan \alpha = \frac{M_y}{M_z} = \frac{-400}{-2960} = 0.9.$$

Thus,

$$\beta = \alpha = 7.7^o.$$

Figure 7.33(a) illustrates the position of the neutral line. On the other hand, Figure 7.33(b) shows the distributions of normal stress due to the normal force ($\sigma_{xx}^{N_x}$), the oblique bending ($\sigma_{xx}^{M_R}$), and resultant (σ_R) in the critical section.

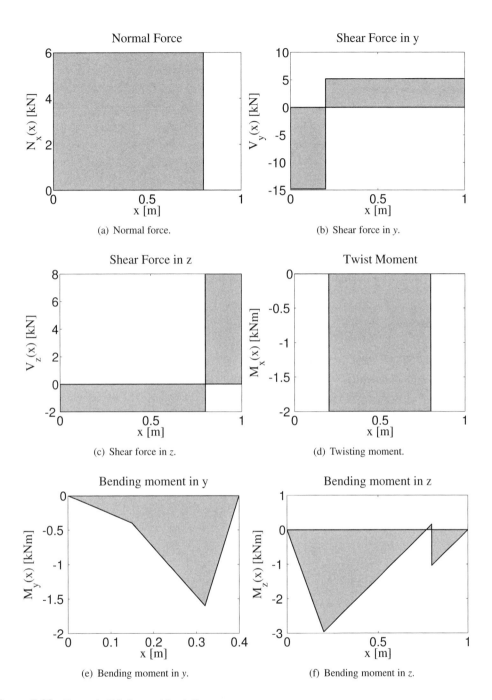

Figure 7.32 Example 7.2: internal load diagrams.

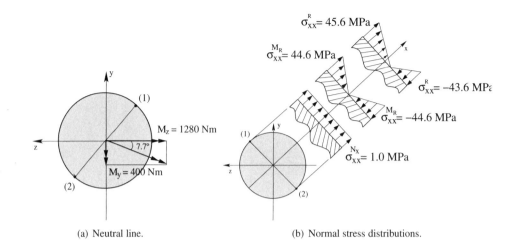

(a) Neutral line.   (b) Normal stress distributions.

Figure 7.33   Example 7.4: neutral line and normal stress distributions.

Example 7.5 *Consider the beam shown in Figure 7.34 mounted on bearings at A and B. The beam supports a gear of primitive diameter $d = 200$ mm. In the left end, an electric engine with power P and angular velocity ω applies a torque T at the beam given by $T = \dfrac{P}{\omega}$. When we transfer the axial (F_a), radial (F_r), and tangential (F_t) gearing forces from the tooth to the center of the circular cross-section, we have a beam model simultaneously subjected to axial and transversal forces, bending and twisting moments, as shown in the free body diagram of Figure 7.35. We assume an anchor-type bearing in A, which prevents translations in directions x, y, and z, giving rise to support reactions R_{Ax}, R_{Ay}, and R_{Az}. The rolling-type bearing in B has constraints in directions y and z, giving rise to support reactions R_{By} and R_{Bz}. We want to verify if the beam remains in the elastic range for $F_a = 3200$ N, $F_r = 3500$ N, $F_r = 8000$ N, and $\bar{\sigma} = 120$ MPa.*

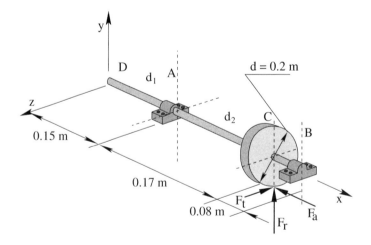

Figure 7.34   Example 7.5: three-dimensional beam with a gear.

The force transfer from the gear to the center of cross-section C gives rise to torque T_{C_x} in x and

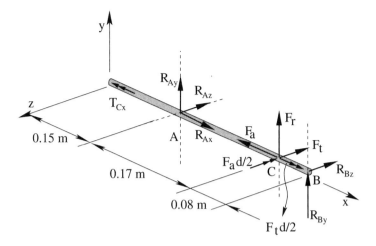

Figure 7.35   Example 7.5: equivalent beam model.

moment M_{C_z} in z respectively given by

$$T_{C_x} = -F_t \frac{d}{2} = -(8000) \frac{(0.2)}{2} = -800 \text{ Nm},$$

$$M_{C_z} = F_a \frac{d}{2} = (3200) \frac{(0.2)}{2} = 320 \text{ Nm}.$$

As the beam of this example is isostatic, the equilibrium BVPs are considered in terms of the loads given in equations (7.96) to (7.99). The method of sections can also be used to obtain the internal load diagrams.

We first consider the traction/compression problem, and the load equation is

$$q_x(x) = R_{Ax} < x - L_1 >^{-1} - F_a < x - L_1 - L_2 >^{-1}$$

with $L_1 = 0.15$ m, $L_2 = 0.17$ m, and $L_1 = 0.08$ m. The boundary conditions at the ends of the beam model are $N_x(x = 0) = 0$ and $N_x(x = L) = 0$, with $L = L_1 + L_2 + L_3 = 0.40$ m. Besides that, we have the auxiliary condition $u_x(x = L_1) = 0$, which will not be used here, because the problem is isostatic.

The integration of the differential equation for the normal force

$$\frac{dN_x(x)}{dx} = -q_x(x) = -R_{Ax} < x - L_1 >^{-1} + F_a < x - L_1 - L_2 >^{-1}$$

results in

$$N_x(x) = -R_{Ax} < x - L_1 >^0 + F_a < x - L_1 - L_2 >^0 + C_1.$$

Using the boundary condition in terms of the normal force, we determine the integration constant C_1 and support reaction R_{Ax} as

$$N_x(x = 0) = -R_{Ax}(0) + F_a(0) + C_1 = 0 \rightarrow C_1 = 0,$$

$$N_x(x = L) = -R_{Ax}(L - L_1)^0 + F_a(L - L_1 - L_2)^0 + 0 = 0 \rightarrow R_{Ax} = F_a = 2400 \text{ N}.$$

Replacing the values, the final equation for the normal force is given by

$$N_x(x) = -2400 < x - 0.15 >^0 + 2400 < x - 0.29 >^0.$$

The equation of the distributed twisting moment is

$$m_x(x) = T_{C_x} <x - L_1 - L_2>^{-1}.$$

The boundary conditions in terms of the twisting moment at the extremities of the beam are $M_x(x = 0) = T_{C_x}$ and $M_x(x = L) = 0$.

The differential equation to be integrated is

$$\frac{dM_x(x)}{dx} = -m_x(x) = -T_{C_x} <x - L_1 - L_2>^{-1}.$$

After integrating the above equation, we obtain the expression of the twisting moment:

$$M_x(x) = -T_{C_x} <x - L_1 - L_2>^0 + C_2.$$

The constant C_2 is determined using the boundary condition at $x = 0$. Thus,

$$M_x(x = 0) = -T_{C_x}(0) + C_2 = T_{C_x} \rightarrow C_2 = T_{C_x}.$$

The final equation for the twisting moment, after the substitution of C_2 and T_{C_x}, is

$$M_x(x) = -800 <x - 0.29>^0 + 800.$$

For bending in z, we have $m_z = 0$ and the expression for the distributed transversal load in y is

$$q_y(x) = R_{Ay} <x - L_1>^{-1} + F_r <x - L_1 - L_2>^{-1} - M_{C_z} <x - L_1 - L_2>^{-2}.$$

In this case, the boundary conditions are $V_y(x = 0) = 0$, $M_z(x = 0) = 0$, and $M_z(x = L) = 0$.

The differential equation in terms of bending moment in z is

$$\frac{d^2 M_z(x)}{dx^2} = q_y(x) = R_{Ay} <x - L_1>^{-1} + F_r <x - L_1 - L_2>^{-1} - M_{C_z} <x - L_1 - L_2>^{-2}.$$

The two indefinite integrations result in the following expressions for the shear force in y and bending moment in z, respectively:

$$V_y(x) = \frac{dM_z(x)}{dx} = R_{Ay} <x - L_1>^0 + F_r <x - L_1 - L_2>^0 - M_{C_z} <x - L_1 - L_2>^{-1} + C_3,$$

$$M_z(x) = R_{Ay} <x - L_1>^1 + F_r <x - L_1 - L_2>^1 - M_{C_z} <x - L_1 - L_2>^0 + C_3 x + C_4.$$

Applying the boundary conditions in $x = 0$, we find C_3 and C_4, that is,

$$V_y(x = 0) = R_{Ay}(0) + F_r(0) + C_3 = 0 \rightarrow C_3 = 0,$$

$$M_z(x = 0) = R_{Ay}(0) + F_r(0) - M_{C_z}(0) + C_1(0) + C_4 = 0 \rightarrow C_4 = 0.$$

Using the boundary condition in terms of bending moment in z at the right end of the beam, we have

$$M_z(x = L) = R_{Ay}(L - L_1) + F_r(L - L_1 - L_2) - M_{C_z}(L - L_1 - L_2)^0 = 0.$$

Hence, the reaction R_{Ay} is determined as

$$R_{Ay} = \frac{-F_r L_3 + M_{C_z}}{L - L_1} = \frac{-(3500)(0.08) + 320}{0.40 - 0.15} = 160 \text{ N}.$$

The final equations for the shear force in y and bending moment in z are respectively given by

$$V_y(x) = 160 <x - 0.15>^0 + 3500 <x - 0.32>^0,$$

$$M_z(x) = 160 < x - 0.15 >^1 + 3500 < x - 0.32 >^1 - 320 < x - 0.32 >^0 .$$

The support reaction at B in direction y is $R_{By} = -V_y(x = L) = 3660$ *N.*

The distributed load equation for bending in y is

$$q_z(x) = R_{Az} < x - L_1 >^{-1} + F_t < x - L_1 - L_2 >^{-1} .$$

The boundary conditions are $V_z(x = 0) = 0$, $M_y(x = 0) = 0$, *and* $M_y(x = L) = 0$.

The differential equation to be integrated is

$$\frac{d^2 M_y(x)}{dx^2} = -q_z(x) = -R_{Az} < x - L_1 >^{-1} - F_t < x - L_1 - L_2 >^{-1} .$$

The two indefinite integrals of the above equation result in shear force in z and bending moment in y, that is,

$$V_z(x) = \frac{dM_y(x)}{dx} = -R_{Az} < x - L_1 >^0 - F_t < x - L_1 - L_2 >^0 + C_5,$$

$$M_y(x) = -R_{Az} < x - L_1 >^1 - F_t < x - L_1 - L_2 >^1 + C_5 x + C_6.$$

Applying the boundary conditions in x = 0, we determine the integration constants C_5 *and* C_6 *as*

$$V_z(x = 0) = -R_{Az}(0) - F_t(0) + C_5 = 0 \rightarrow C_5 = 0,$$

$$M_y(x = 0) = -R_{Az}(0) - F_t(0) + C_5(0) + C_6 = 0 \rightarrow C_6 = 0.$$

From the boundary condition of zero bending moment at x = L, we obtain

$$M_y(x = L) = R_{Az}(L - L_1) + F_t(L - L_1 - L_2) = 0.$$

Hence, the support reaction R_{Az} *is*

$$R_{Az} = -\frac{F_t L_3}{L_2 + L_3} = -\frac{(8000)(0.08)}{0.25} = -2560 \text{ N}.$$

The final equations for shear force in z and bending moment in y are respectively given by

$$V_z(x) = 2560 < x - 0.15 >^0 - 8000 < x - 0.32 >^0,$$

$$M_y(x) = 2560 < x - 0.15 >^1 - 8000 < x - 0.32 >^1 .$$

The support reaction at B in direction z is $R_{Bz} = -V_z(x = L) = 5440$ *N.*

The loads for each beam segment are:

- $0 < x < 0.15$ m:

$$N_x(x) = 0 = \begin{cases} N_x(0) = 0 \text{ N} \\ N_x(0.15) = 0 \text{ N} \end{cases}$$

$$V_y(x) = 0 = \begin{cases} V_y(0) = 0 \text{ N} \\ V_y(0.15) = 0 \text{ N} \end{cases}$$

$$V_z(x) = 0 = \begin{cases} V_z(0) = 0 \text{ N} \\ V_z(0.15) = 0 \text{ N} \end{cases}$$

$$M_x(x) = 911.2 = \begin{cases} M_z(0) = 800 \text{ Nm} \\ M_z(0.15) = 800 \text{ Nm} \end{cases} ;$$

$$M_y(x) = 0 = \begin{cases} M_y(0) = 0 \text{ Nm} \\ M_y(0.15) = 0 \text{ Nm} \end{cases}$$

$$M_z(x) = 0 = \begin{cases} M_z(0) = 0 \text{ Nm} \\ M_z(0.15) = 0 \text{ Nm} \end{cases}$$

- $0.15 \text{ m} < x < 0.32 \text{ m}$:

$$N_x(x) = -2400 = \begin{cases} N_x(0.15) = -2400 \text{ N} \\ N_x(0.32) = -2400 \text{ N} \end{cases}$$

$$V_y(x) = 160 = \begin{cases} V_y(0.15) = 160 \text{ N} \\ V_y(0.32) = 160 \text{ N} \end{cases}$$

$$V_z(x) = 2560 = \begin{cases} V_z(0.15) = 2560 \text{ N} \\ V_z(0.32) = 2560 \text{ N} \end{cases} \qquad ;$$

$$M_x(x) = 800 = \begin{cases} M_x(0.15) = 800 \text{ Nm} \\ M_x(0.32) = 800 \text{ Nm} \end{cases}$$

$$M_y(x) = 2560(x - 0.15) = \begin{cases} M_y(0.15) = 0 \text{ Nm} \\ M_y(0.32) = 435.2 \text{ Nm} \end{cases}$$

$$M_z(x) = 160(x - 0.15) = \begin{cases} M_z(0.15) = 0 \text{ Nm} \\ M_z(0.32) = 27.2 \text{ Nm} \end{cases}$$

- $0.32 \text{ m} < x < 0.40 \text{ m}$:

$$N_x(x) = 0 = \begin{cases} N_x(0.32) = 0 \text{ N} \\ N_x(0.40) = 0 \text{ N} \end{cases}$$

$$V_y(x) = 3660 = \begin{cases} V_y(0.32) = 3660 \text{ N} \\ V_y(0.40) = 3660 \text{ N} \end{cases}$$

$$V_z(x) = 5440 = \begin{cases} V_z(0.32) = 5440 \text{ N} \\ V_z(0.40) = 5440 \text{ N} \end{cases}$$

$$M_x(x) = 0 = \begin{cases} M_x(0.32) = 0 \text{ Nm} \\ M_x(0.40) = 0 \text{ Nm} \end{cases}$$

$$M_y(x) = 2560(x - 0.15) - 8000(x - 0.32) = \begin{cases} M_y(0.32) = 435.2 \text{ Nm} \\ M_y(0.40) = 0 \text{ Nm} \end{cases}$$

$$M_z(x) = 160(x - 0.15) + 3500(x - 0.32) - 320 = \begin{cases} M_z(0.32) = -292.8 \text{ Nm} \\ M_z(0.40) = 0 \text{ Nm} \end{cases}$$

The internal load diagrams are illustrated in Figure 7.36.

According to the proposed criterion for a beam with constant circular section, the critical section in this example is $x = 0.32^+$ m with $M_R = 524.5$ Nm. Since the beam has two distinct diameters, namely d_1 and d_2, and furthermore, the critical cross-section is in the interface of both diameters, we check both sections with d_1 and d_2. The internal loads in these cross-sections are:

$$x = 0.32^- \text{ m} : \begin{cases} N_x = -2400 \text{ N} \\ V_y = 160 \text{ N} \\ V_z = 2560 \text{ N} \\ M_x = 800 \text{ Nm} \\ M_y = 435.2 \text{ Nm} \\ M_z = 27.2 \text{ Nm} \\ \overline{M_R = \sqrt{M_y^2 + M_z^2} = 436.0 \text{ Nm}} \end{cases}$$

and

$$x = 0.32^+ \text{ m} : \begin{cases} N_x = 0 \\ V_y = 3660 \text{ N} \\ V_z = 5440 \text{ N} \\ M_x = 0 \\ M_y = 435.2 \text{ Nm} \\ M_z = -292.8 \text{ Nm} \\ M_R = \sqrt{M_y^2 + M_z^2} = 524.5 \text{ Nm} \end{cases}$$

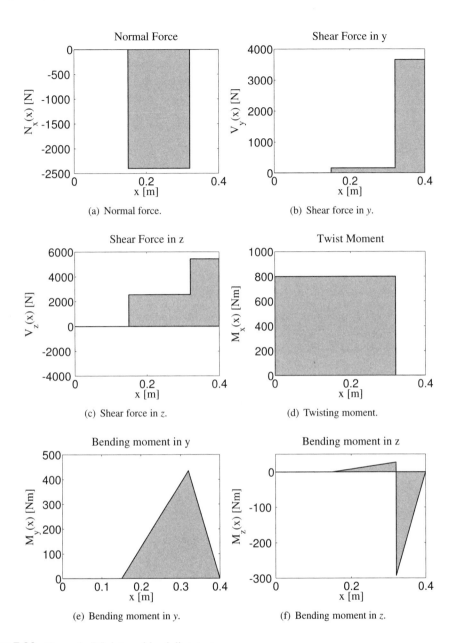

(a) Normal force.

(b) Shear force in y.

(c) Shear force in z.

(d) Twisting moment.

(e) Bending moment in y.

(f) Bending moment in z.

Figure 7.36   Example 7.5: internal load diagrams.

The diameter of section $x = 0.32^-$ m is $d_1 = 50$ mm. Thus, the area and strength modulus to oblique bending relative to the neutral line are obtained as

$$A = \frac{\pi d_1^2}{4} = \frac{\pi}{4}(50 \times 10^{-3})^2 = 1.96 \times 10^{-3} \; m^2,$$

$$W_{NL} = \frac{\pi d_1^3}{32} = \frac{\pi}{32}(50 \times 10^{-3})^3 = 1.23 \times 10^{-5} \; m^3.$$

For verification, the critical point is in section $x = 0.32^-$ m. As the normal force is negative and therefore compressive, point (2) illustrated in Figure 7.37(a) is the critical one for normal stress, which value is given by

$$\sigma_R = \frac{N_x}{A} + \frac{M_R}{W_{NL}} = -\frac{2400}{1.96 \times 10^{-3}} - \frac{436}{1.23 \times 10^{-5}} = -36.7 \; \text{MPa}.$$

Considering the maximum stress based on shear force, the shear stress is

$$\tau_R = \frac{M_x}{2W_{NL}} + \frac{4V_R}{3A} = -\frac{800}{2(1.23 \times 10^{-5})} - \frac{(4)(2565)}{(3)(1.96 \times 10^{-3})} = -34.3 \; \text{MPa}.$$

The maximum normal equivalent stress in the critical point of section $x = 0.32^-$ m is calculated as

$$\sigma_{eqv}^{max} = \sqrt[2]{\sigma_R^2 + 3\tau_R^2} = \sqrt[2]{(-36.7)^2 + 3(-34.3)^2} = 69.8 \; \text{MPa}.$$

As $\sigma_{eqv}^{max} < \bar{\sigma} = 120$ MPa, section $x = 0.32^-$ m remains in the elastic range. The angle α_1 that locates the neutral line in Figure 7.37(a) is

$$\alpha_1 = \left(\tan^{-1} \frac{435.2}{27.2} \right) \frac{180}{\pi} = 86.4^o.$$

The geometric properties of section $x = 0.32^+$ m with diameter $d_2 = 45$ mm are

$$A = \frac{\pi d_2^2}{4} = \frac{\pi(45 \times 10^{-3})^2}{4} = 1.59 \times 10^{-3} \; m^2,$$

$$W_{NL} = \frac{\pi d_2^3}{32} = \frac{\pi(45 \times 10^{-3})^3}{32} = 8.95 \times 10^{-6} \; m^3.$$

Since the normal force is zero at $x = 0.32^+$ m, the corresponding normal stress is also zero. Thus, points (1) and (2) shown in Figure 7.37(b) are the critical ones. The resultant normal and shear stresses in the section are given by

$$\sigma_R = \frac{524.5}{8.95 \times 10^{-6}} = 58.6 \; \text{MPa}.$$

$$\tau_R = \frac{(4)(6556.6)}{(3)(1.6 \times 10^{-3})} = 5.5 \; \text{MPa}.$$

Thus, the maximum equivalent normal stress at section $x = 0.32^+$ m is

$$\sigma_{eqv}^{max} = \sqrt[2]{(\sigma_R)^2 + 3(\tau_R)^2} = \sqrt[2]{(58.6)^2 + 3(5.5)^2} = 59.4 \; \text{MPa}.$$

As $\sigma_{eqv}^{max} < \bar{\sigma}$, section $x = 0.32^+$ m also remains in the elastic range. The angle α_2 that locates the neutral line in Figure 7.37(b) is

$$\alpha_1 = \left(\tan^{-1} \frac{435.2}{-292.8} \right) \frac{180}{\pi} = -56.1^o.$$

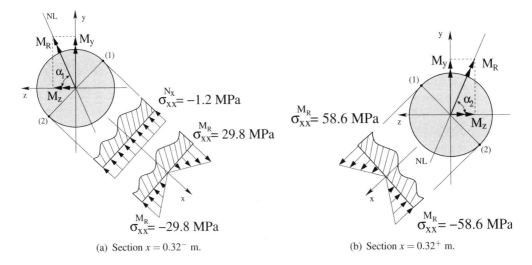

(a) Section $x = 0.32^-$ m. (b) Section $x = 0.32^+$ m.

Figure 7.37 Example 7.5: normal stress distributions at $x = 0.32^-$ m and $x = 0.32^+$ m.

Thus, notice that even for the critical section according to the defined criterion, the equivalent stress at $x = 0.32^+$ m is lower than at $x = 0.32^-$ m. In general, we must check the critical sections for each of the distinct diameters.

Taking the admissible normal stress $\bar{\sigma} = 65$ MPa, the beam does not remain in the elastic range, because $\sigma_{eqv}^{max} = 69.8$ MPa at the critical section $x = 0.32^-$ m. Thus, the diameter d_1 should be resized. For this purpose, expression (7.152) is employed, that is,

$$d_1 = \sqrt[3]{\frac{32}{\pi 60 \times 10^6} \sqrt[2]{(435.2)^2 + (27.2)^2 + \frac{3}{4}(800)^2}} = 51.8 \text{ mm.}$$

Because this design criterion does not account for normal and shear forces, we must verify the beam with a higher diameter value. For $d_1 = 53$ mm, we have the following values of geometrical properties at section $x = 0.32^-$ m:

$$A = \frac{\pi}{4}(53 \times 10^{-3})^2 = 2.21 \times 10^{-3} \text{ m}^2,$$

$$W_{NL} = \frac{\pi}{32}(53 \times 10^{-3})^3 = 1.46 \times 10^{-5} \text{ m}^3.$$

Point (2) is the critical one with the following resultant normal and shear stresses:

$$\sigma_R = -\frac{2400}{2.21 \times 10^{-3}} - \frac{436}{1.46 \times 10^{-5}} = -30.9 \text{ MPa.}$$

$$\tau_R = -\frac{800}{2(1.46 \times 10^{-5})} + \frac{(4)(2565)}{(3)(2.21 \times 10^{-3})} = -28.9 \text{ MPa.}$$

The maximum equivalent normal stress is

$$\sigma_{eqv}^{max} = \sqrt[2]{(-30.9)^2 + 3(-28.9)^2} = 58.9 \text{ MPa.}$$

As $\sigma_{eqv}^{max} < \bar{\sigma}$, the beam with $d_1 = 53$ mm remains in the elastic range.

$\square$

Example 7.6 *The three-dimensional beam model presented may be used in several engineering applications, such as the crankshaft of a four-cylinder internal combustion engine illustrated in Figure 7.38. The main journals and crankpins are the hydrodynamic bearings of the crankshaft. The connecting rods are mounted on the crankpins and the journals are placed in the engine block.*

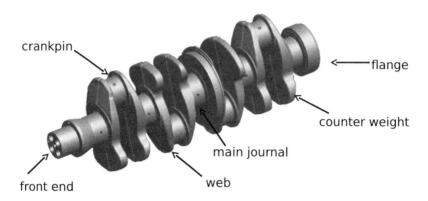

Figure 7.38   Example 7.6: crankshaft of a four-cylinder engine.

The crankshaft is subjected to bending and torsional loads from the explosions in the combustion chambers and from the torsional vibration [31, 16]. Because the crankshaft is not a mechanical component with simple geometry, some considerations must be taken in order to obtain a beam model with equivalent circular cross-section to the crankshaft.

The equivalent diameter and length of the beam may be determined from the formulation given by the BICERA British Association. This formulation is based on empirical curves that provide a series of correction factors to obtain a beam with constant circular section, which is equivalent to the crankshaft. The twisting and bending loads are applied to obtain the internal load diagrams and to determine the critical sections for later design and verification.

The bending loads in directions y and z and torques in x are applied in the equivalent beam model. Figure 7.39(a) shows the forces y and torques in x. Figure 7.39(b) shows the applied forces in direction z. It is possible to derive the equations of the shear forces, bending, and twisting moments by integrating the equilibrium BVPs given in (7.122) to (7.127), because the considered three-dimensional beam model is hyperstatic.

For the bending problem in plane xy, the transversal distributed load equation $q_y(x)$, written in singularity function notation, is given by

$$
\begin{aligned}
q_y(x) \;=\; & F_{y_1} <x-L_1>^{-1} + F_{y_2} <x-L_2>^{-1} + F_{y_3} <x-L_3>^{-1} \\
+\; & F_{y_4} <x-L_4>^{-1} + R_{y_1} <x-L_{a_1}>^{-1} + R_{y_2} <x-L_{a_2}>^{-1} \\
+\; & R_{y_3} <x-L_{a_3}>^{-1} + R_{y_4} <x-L_{a_4}>^{-1} + + R_{y_5} <x-L_{a_5}>^{-1},
\end{aligned}
$$

where L_1, L_2, L_3, and L_4 indicate the sections of the beam, from the left end, where the dynamic loads from the engine are applied [16]; L_{a_1}, L_{a_2}, L_{a_3}, L_{a_4}, and L_{a_5} refer to the positions of the main bearings along the beam with the respective support reactions R_{y_1} to R_{y_5} in y.

The bending differential equation of equilibrium of the beam in plane xy is

$$
\begin{aligned}
EI_z \frac{d^4 u_y(x)}{dx^4} \;=\; & F_{y_1} <x-L_1>^{-1} + F_{y_2} <x-L_2>^{-1} + F_{y_3} <x-L_3>^{-1} \\
+\; & F_{y_4} <x-L_4>^{-1} + R_{y_1} <x-L_{a_1}>^{-1} + R_{y_2} <x-L_{a_2}>^{-1} \\
+\; & R_{y_3} <x-L_{a_3}>^{-1} + R_{y_4} <x-L_{a_4}>^{-1} + R_{y_5} <x-L_{a_5}>^{-1}.
\end{aligned}
$$

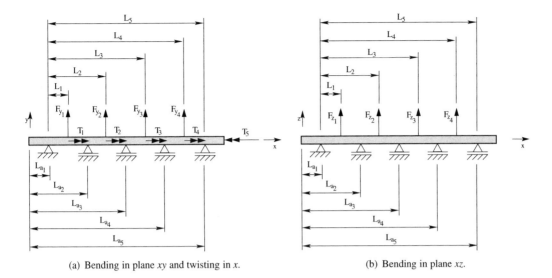

(a) Bending in plane xy and twisting in x. (b) Bending in plane xz.

Figure 7.39 Example 7.6: external loads.

Integrating the previous equation four times, we respectively obtain the expressions for the shear force $V_y(x)$, bending moment $M_z(x)$, bending angle $\theta_z(x)$, and transversal displacement $u_y(x)$, that is,

$$
\begin{aligned}
V_y(x) &= F_{y_1} <x-L_1>^0 + F_{y_2} <x-L_2>^0 + F_{y_3} <x-L_3>^0 \\
&+ F_{y_4} <x-L_4>^0 + R_{y_1} <x-L_{a_1}>^0 + R_{y_2} <x-L_{a_2}>^0 \\
&+ R_{y_3} <x-L_{a_3}>^0 + R_{y_4} <x-L_{a_4}>^0 + R_{y_5} <x-L_{a_5}>^0 + C_1,
\end{aligned}
$$

$$
\begin{aligned}
M_z(x) &= F_{y_1} <x-L_1>^1 + F_{y_2} <x-L_2>^1 + F_{y_3} <x-L_3>^1 \\
&+ F_{y_4} <x-L_4>^1 + R_{y_1} <x-L_{a_1}>^1 + R_{y_2} <x-L_{a_2}>^1 \\
&+ R_{y_3} <x-L_{a_3}>^1 + R_{y_4} <x-L_{a_4}>^1 + R_{y_5} <x-L_{a_5}>^1 + C_1 x + C_2,
\end{aligned}
$$

$$
\begin{aligned}
EI_z\theta_z(x) &= \frac{F_{y_1}}{2} <x-L_1>^2 + \frac{F_{y_2}}{2} <x-L_2>^2 + \frac{F_{y_3}}{2} <x-L_3>^2 \\
&+ \frac{F_{y_4}}{2} <x-L_4>^2 + \frac{R_{y_1}}{2} <x-L_{a_1}>^2 + \frac{R_{y_2}}{2} <x-L_{a_2}>^2 \\
&+ \frac{R_{y_3}}{2} <x-L_{a_3}>^2 + \frac{L_{y4}}{2} <x-L_{a_4}>^2 + \frac{R_{y_5}}{2} <x-L_{a_5}>^2 \\
&+ \frac{C_1}{2} x^2 + C_2 x + C_3,
\end{aligned}
$$

$$
\begin{aligned}
EI_z u_y(x) &= \frac{F_{y_1}}{6} <x-L_1>^3 + \frac{F_{y_2}}{6} <x-L_2>^3 + \frac{F_{y_3}}{6} <x-L_3>^3 \\
&+ \frac{F_{y_4}}{6} <x-L_4>^3 + \frac{R_{y_1}}{6} <x-L_{a_1}>^3 + \frac{R_{y_2}}{6} <x-L_{a_2}>^3 \\
&+ \frac{R_{y_3}}{6} <x-L_{a_3}>^3 + \frac{R_{y_4}}{6} <x-L_{a_4}>^3 + \frac{R_{y_5}}{6} <x-L_{a_5}>^3 \\
&+ \frac{C_1}{6} x^3 + \frac{C_2}{2} x^2 + C_3 x + C_4.
\end{aligned}
$$

There are four boundary conditions of the bending problem in plane xy, that is, $V_y(x = 0) = 0$, $V_y(x = L) = 0$, $M_z(x = 0) = 0$, and $M_z(x = L) = 0$. Besides these conditions, there are five auxiliary conditions in the bearings, which don't allow displacements in direction y. Thus, $u_y(x = L_{a_1}) = 0$, $u_y(x = L_{a_2}) = 0$, $u_y(x = L_{a_3}) = 0$, $u_y(x = L_{a_4}) = 0$, and $u_y(x = L_{a_5}) = 0$. Using the boundary and auxiliary conditions in the obtained expressions for the shear force, bending moment, and transversal displacement, we calculate the four integration constants and five support reactions solving the following system of equations:

$$
\begin{bmatrix}
0 & 0 & 0 & 0 & 0 & 0 & 0 & 6L_{a_1} & 6 \\
(L_{a_2} - L_{a_1})^3 & 0 & 0 & 0 & 0 & 0 & 0 & 6L_{a_2} & 6 \\
(L_{a_3} - L_{a_1})^3 & (L_{a_3} - L_{a_2})^3 & 0 & 0 & 0 & 0 & 0 & 6L_{a_3} & 6 \\
(L_{a_4} - L_{a_1})^3 & (L_{a_4} - L_{a_2})^3 & (L_{a_4} - L_{a_3})^3 & 0 & 0 & 0 & 0 & 6L_{a_4} & 6 \\
(L_{a_5} - L_{a_1})^3 & (L_{a_5} - L_{a_2})^3 & (L_{a_5} - L_{a_3})^3 & (L_{a_5} - L_{a_4})^3 & 0 & 0 & 0 & 6L_{a_5} & 6 \\
0 & 0 & 0 & 0 & 0 & 1 & 0 & 0 & 0 \\
0 & 0 & 0 & 0 & 0 & 0 & 1 & 0 & 0 \\
1 & 1 & 1 & 1 & 1 & 0 & 0 & 0 & 0 \\
(L - L_{a_1}) & (L - L_{a_2}) & (L - L_{a_3}) & (L - L_{a_4}) & (L - L_{a_5}) & 0 & 0 & 0 & 0
\end{bmatrix}
$$

$$
\begin{Bmatrix}
R_{y_1} \\
R_{y_2} \\
R_{y_3} \\
R_{y_4} \\
R_{y_5} \\
C_1 \\
C_2 \\
C_3 \\
C_4
\end{Bmatrix}
=
\begin{Bmatrix}
0 \\
-F_{y_1}(L_{a_2} - L_1)^3 \\
-F_{y_1}(L_{a_3} - L_1)^3 - F_{y_2}(L_{a_3} - L_2)^3 \\
-F_{y_1}(L_{a_4} - L_1)^3 - F_{y_2}(L_{a_4} - L_2)^3 - F_{y_3}(L_{a_4} - L_3)^3 \\
-F_{y_1}(L_{a_5} - L_1)^3 - F_{y_2}(L_{a_5} - L_2)^3 - F_{y_3}(L_{a_5} - L_3)^3 - F_{y_4}(L_{a_5} - L_4)^3 \\
0 \\
0 \\
-F_{y_1} - F_{y_2} - F_{y_3} - F_{y_4} \\
-F_{y_1}(L - L_1) - F_{y_2}(L - L_2) - F_{y_3}(L - L_3) - F_{y_4}(L - L_4)
\end{Bmatrix}.
$$

The final equations of the shear force, bending moment, bending angle, and transversal displacement in plane xy are determined substituting the calculated values for the boundary conditions and support reactions. At the end of the example, the values of the lengths and applied external loads will be replaced to obtain the numerical results.

The expression for the distributed transversal load function in the z axis direction is

$$
\begin{aligned}
q_z(x) \;=\; & F_{z_1} <x - L_1>^{-1} + F_{z_2} <x - L_2>^{-1} + F_{z_3} <x - L_3>^{-1} \\
+ \; & F_{z_4} <x - L_4>^{-1} + R_{z_1} <x - L_{a_1}>^{-1} + R_{z_2} <x - L_{a_2}>^{-1} \\
+ \; & R_{z_3} <x - L_{a_3}>^{-1} + R_{z_4} <x - L_{a_4}>^{-1} + R_{z_5} <x - L_{a_5}>^{-1},
\end{aligned}
$$

with F_{z_1} to F_{z_4} the acting forces in direction z and R_{z_1} to R_{z_5} correspond to the support reactions in the main bearings, analogous to plane xy.

Integrating four times the equilibrium equation

$$
\begin{aligned}
EI_y \frac{d^4 u_z(x)}{dx^4} \;=\; & -F_{z_1} <x - L_1>^{-1} - F_{z_2} <x - L_2>^{-1} - F_{z_3} <x - L_3>^{-1} \\
- \; & F_{z_4} <x - L_4>^{-1} - R_{z_1} <x - L_{a_1}>^{-1} - R_{z_2} <x - L_{a_2}>^{-1} \\
- \; & R_{z_3} <x - L_{a_3}>^{-1} - R_{z_4} <x - L_{a_4}>^{-1} - R_{z_5} <x - L_{a_5}>^{-1},
\end{aligned}
$$

we respectively obtain the shear force $V_z(x)$, bending moment $M_y(x)$, bending angle $\theta_y(x)$, and *transversal displacement $u_z(x)$ equations*

$$
\begin{aligned}
V_z(x) = {} & -F_{z_1}<x-L_1>^0 -F_{z_2}<x-L_2>^0 -F_{z_3}<x-L_3>^0 \\
& -F_{z_4}<x-L_4>^0 -R_{z_1}<x-L_{a_1}>^0 -R_{z_2}<x-L_{a_2}>^0 \\
& -R_{z_3}<x-L_{a_3}>^0 -R_{z_4}<x-L_{a_4}>^0 -R_{z_5}<x-L_{a_5}>^0 \\
& +C_5,
\end{aligned}
$$

$$
\begin{aligned}
M_y(x) = {} & -F_{z_1}<x-L_1>^1 -F_{z_2}<x-L_2>^1 -F_{z_3}<x-L_3>^1 \\
& -F_{z_4}<x-L_4>^1 -R_{z_1}<x-L_{a_1}>^1 -R_{z_2}<x-L_{a_2}>^1 \\
& -R_{z_3}<x-L_{a_3}>^1 -R_{z_4}<x-L_{a_4}>^1 -R_{z_5}<x-L_{a_5}>^1 \\
& +C_5x+C_6,
\end{aligned}
$$

$$
\begin{aligned}
EI_y\theta_y(x) = {} & -\frac{F_{z_1}}{2}<x-L_1>^2 -\frac{F_{z_2}}{2}<x-L_2>^2 -\frac{F_{z_3}}{2}<x-L_3>^2 \\
& -\frac{F_{z_4}}{2}<x-L_4>^2 -\frac{R_{z_1}}{2}<x-L_{a_1}>^2 -\frac{R_{z_2}}{2}<x-L_{a_2}>^2 \\
& -\frac{R_{z_3}}{2}<x-L_{a_3}>^2 -\frac{R_{z_4}}{2}<x-L_{a_4}>^2 -\frac{R_{z_5}}{2}<x-L_{a_5}>^2 \\
& +\frac{C_5}{2}x^2 +\frac{C_6}{x} +C_7,
\end{aligned}
$$

$$
\begin{aligned}
EI_yu_z(x) = {} & -\frac{F_{z_1}}{6}<x-L_1>^3 -\frac{F_{z_2}}{6}<x-L_2>^3 -\frac{F_{z_3}}{6}<x-L_3>^3 \\
& -\frac{F_{z_4}}{6}<x-L_4>^3 -\frac{R_{z_1}}{6}<x-L_{a_1}>^3 -\frac{R_{z_2}}{6}<x-L_{a_2}>^3 \\
& -\frac{R_{z_3}}{6}<x-L_{a_3}>^3 -\frac{R_{z_4}}{6}<x-L_{a_4}>^3 -\frac{R_{z_5}}{6}<x-L_{a_5}>^3 \\
& +\frac{C_5}{6}x^3 +\frac{C_6}{2}x^2 +C_7x+C_8.
\end{aligned}
$$

The four boundary conditions in plane xz are $V_z(x=0)=0$, $V_z(x=L)=0$, $M_y(x=0)=0$, and $M_y(x=L)=0$. The five auxiliary conditions related to the support bearings, which also constrain the displacements in direction z, are $u_z(x=L_{a_1})=0$, $u_z(x=L_{a_2})=0$, $u_z(x=L_{a_3})=0$, $u_z(x=L_{a_4})=0$, and $u_z(x=L_{a_5})=0$.

The four integration constants, as well as support reactions, are found with the solution of the *following system of equations:*

$$
\begin{bmatrix}
0 & 0 & 0 & 0 & 0 & 0 & 0 & 6L_{a_1} & 6 \\
(L_{a_2}-L_{a_1})^3 & 0 & 0 & 0 & 0 & 0 & 0 & 6L_{a_2} & 6 \\
(L_{a_3}-L_{a_1})^3 & (L_{a_3}-L_{a_2})^3 & 0 & 0 & 0 & 0 & 0 & 6L_{a_3} & 6 \\
(L_{a_4}-L_{a_1})^3 & (L_{a_4}-L_{a_2})^3 & (L_{a_4}-L_{a_3})^3 & 0 & 0 & 0 & 0 & 6L_{a_4} & 6 \\
(L_{a_5}-L_{a_1})^3 & (L_{a_5}-L_{a_2})^3 & (L_{a_5}-L_{a_3})^3 & (L_{a_5}-L_{a_4})^3 & 0 & 0 & 0 & 6L_{a_5} & 6 \\
0 & 0 & 0 & 0 & 0 & 1 & 0 & 0 & 0 \\
0 & 0 & 0 & 0 & 0 & 0 & 1 & 0 & 0 \\
1 & 1 & 1 & 1 & 1 & 0 & 0 & & \\
(L-L_{a_1}) & (L-L_{a_2}) & (L-L_{a_3}) & (L-L_{a_4}) & (L-L_{a_5}) & 0 & 0 & 0 & 0
\end{bmatrix}
$$

$$
\left\{
\begin{array}{c}
R_{z_1} \\
R_{z_2} \\
R_{z_3} \\
R_{z_4} \\
R_{z_5} \\
C_5 \\
C_6 \\
C_7 \\
C_8
\end{array}
\right\} =
\left\{
\begin{array}{l}
0 \\
-F_{z_1}(L_{a_2}-L1)^3 \\
-F_{z_1}(L_{a_3}-L1)^3 - F_{z_2}(L_{a_3}-L_2)^3 \\
-F_{z_1}(L_{a_4}-L1)^3 - F_{z_2}(L_{a_4}-L_2)^3 - F_{z_3}(L_{a_4}-L_3)^3 \\
-F_{z_1}(L_{a_5}-L1)^3 - F_{z_2}(L_{a_5}-L_2)^3 - F_{z_3}(L_{a_5}-L_3)^3 - F_{z_4}(L_{a_5}-L_4)^3 \\
0 \\
0 \\
-F_{z_1}-F_{z_2}-F_{z_3}-F_{z_4} \\
-F_{z_1}(L-L_1) - F_{z_2}(L-L_2) - F_{z_3}(L-L_3) - F_{z_4}(L-L_4)
\end{array}
\right\}.
$$

After solving the above linear system, we obtain the final equations for the shear force $V_z(x)$, bending moment $M_y(x)$, bending angle $\theta_y(x)$, and transversal displacement $u_z(x)$.

The distributed torque function is denoted as

$$
m_x(x) = T_1 <x-L_1>^{-1} + T_2 <x-L_2>^{-1} + T_3 <x-L_3>^{-1} + T_4 <x-L_4>^{-1},
$$

with T_1 to T_4 the sum of the engine and torsional vibration torques.

The differential equation of equilibrium for the circular torsion problem is

$$
GI_p \frac{d^2\theta_x(x)}{dx^2} = -T_1 <x-L_1>^{-1} - T_2 <x-L_2>^{-1} - T_3 <x-L_3>^{-1} - T_4 <x-L_4>^{-1}.
$$

Integrating the previous equation twice, we respectively obtain the equations for the twisting moment $M_x(x)$ and angle of twist $\theta_x(x)$ as

$$
\begin{aligned}
M_x(x) = {}& -T_1 <x-L_1>^0 - T_2 <x-L_2>^0 - T_3 <x-L_3>^0 \\
& - T_4 <x-L_4>^0 + C_9,
\end{aligned}
$$

$$
\begin{aligned}
GI_p \theta_x(x) = {}& -T_1 <x-L_1>^1 - T_2 <x-L_2>^1 - T_3 <x-L_3>^1 \\
& - T_4 <x-L_4>^1 + C_9 x + C_{10}.
\end{aligned}
$$

The boundary conditions for the twist problem are $M_x(x=0)=0$ and $\theta_x(x=L)=0$. Substituting the first boundary condition, we obtain $C_9 = 0$ and the following final equation for the twisting moment:

$$
M_x(x) = -T_1 <x-L_1>^0 - T_2 <x-L_2>^0 - T_3 <x-L_3>^0 - T_4 <x-L_4>^0.
$$

The combustion engine has four operation cycles, regarding admission, compression, combustion, and exhaustion. In the case of a four-cylinder engine, the most commonly employed explosion order of cylinders is 1-3-4-2. Considering the explosion of the fourth cylinder and 2100 rpm, the following loads indicated in the previous expressions are determined [16]: $F_{y_1} = 25.0$ kN, $F_{y_2} = -15.0$ kN, $F_{y_3} = -20.0$ kN, $F_{y_4} = 80.0$ kN, $F_{z_1} = -0.5$ kN, $F_{z_2} = 2.0$ kN, $F_{z_3} = 2.0$ kN, $F_{z_4} = -4.5$ kN, $T_1 = 430.0$ Nm, $T_2 = 1200.0$ Nm, $T_3 = 2000.0$ Nm, *and* $T_4 = 2400.0$ Nm.

The total length of the crankshaft is $L = 0.9$ m *and the sections indicated in Figures 7.39(a) and 7.39(b) are* $L_{a_1} = 0.09$ m, $L_{a_2} = 0.27$ m, $L_{a_3} = 0.45$ m, $L_{a_4} = 0.63$ m, $L_{a_5} = 0.81$ m, $L_1 = 0.18$ m, $L_2 = 0.36$ m, $L_3 = 0.54$ m, *and* $L_4 = 0.72$ m. *The equivalent diameter employed in the beam model is* $d = 65$ cm.

Figures 7.40 to 7.42 illustrate the diagrams for circular torsion, bending in xy, and bending in xz. The diagrams were obtained using the BeamLab program, presented in the following section.

The obtained support reactions are $M_{Ax} = -6030.00$ Nm, $R_{y_1} = -10156.25$ N, $R_{y_2} = -15312.50$ N, $R_{y_3} = 38125.00$ N, $R_{y_4} = -49687.50$ N, $R_{y_5} = -32968.75$ N, $R_{z_1} = -276.79$ N, $R_{z_2} = 348.21$ N, $R_{z_3} = 3232.14$ N, $R_{z_4} = -2401.79$ N, *and* $R_{z_5} = -1901.79$ N.

The critical section is $x = 0.72$ m *and* $M_x^{max} = 6030.00$ Nm, $M_y^{max} = 171.16$ Nm, *and* $M_z^{max} = 2967.19$ Nm. *Using expression (7.151), we calculate the maximum equivalent normal stress as*

$$\sigma_{eqv}^{max} = \frac{32}{\pi(0.065)^3}\sqrt{(2967.19)^2 + (171.16)^2 + \frac{3}{4}(6030.00)^2} = 222.86 \text{ MPa}.$$

For a standard steel with admissible normal stress of $\bar{\sigma} = 120$ MPa, *the crankshaft must be resized. However, the manufacturing process of this component employs steels with higher admissible stresses and thermal treatments in the critical regions.*

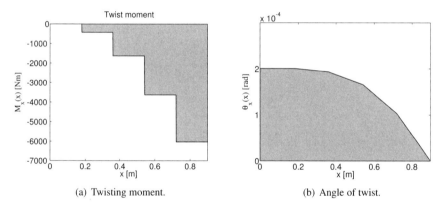

(a) Twisting moment. (b) Angle of twist.

Figure 7.40 Example 7.6: diagrams of twisting moment and angle of twist.

□

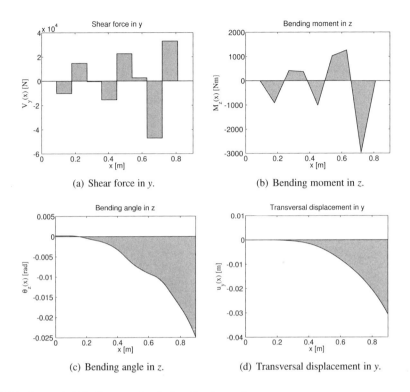

(a) Shear force in y.   (b) Bending moment in z.

(c) Bending angle in z.   (d) Transversal displacement in y.

Figure 7.41 Example 7.6: diagrams for the bending in xy plane.

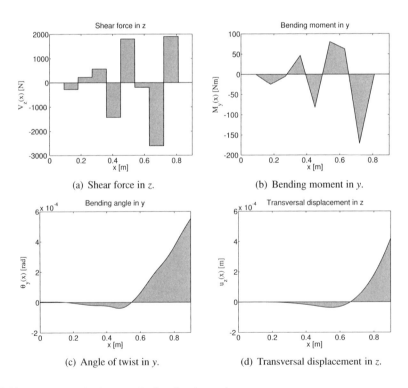

(a) Shear force in z.   (b) Bending moment in y.

(c) Angle of twist in y.   (d) Transversal displacement in z.

Figure 7.42 Example 7.6: diagrams for bending in xz plane.

7.4 BEAMLAB PROGRAM

The BeamLab program is written for MATLAB environment, and solves differential equations for traction, circular torsion, and bending in y and z for isostatic and hyperstatic cases using symbolic manipulation. It considers concentrated and distributed loads of constant intensity. The loading functions are denoted by the notation in terms singularity functions. It also permits the inclusion of supports and hinges in internal sections of the beam.

The program was developed using the paradigm of object-oriented programming. It is organized into four main classes, namely AxialEquations, TorsionalEquations, YBendingEquations, and ZBendingEquations which are respectively responsible by the differential equations of traction/compression, circular torsion, and bending in y and z. All classes have as main methods Read for reading parameters from a text file and Solve to solve the solution of differential equations in terms of internal loads, kinematic, strain, and stress components. The Material, GeometricProperties, and SingularFunctionNotation classes are respectively used to store the material properties, geometrical cross-section properties and length of the beam, and representation, integration, and calculation of terms in singular notation.

The input file with parameters of the problem employs keywords. For instance, consider Example 5.18 for the beam in bending only with the input file given below.

```
*Z_BENDING_LOADS
1
1 0 -1200 0 2

*Z_BENDING_OVERHANGINGS
1
1 1.0 -3

*Z_BENDING_BCS
UY 0
TZ 0
UY 0
MZ 400

*MATERIAL_PROPERTIES
210e9 0.3

*GEOMETRIC_PROPERTIES
2 1E-4 4E-4 2E-4 6.67e-5 -0.1 0.1
```

The keyword *Z_BENDING_LOADS specifies the number of bending loads in z and, for each load, indicates the power of the singularity notation, the load intensity, and initial and final coordinates. In the above example, there is only a distributed load of constant intensity with -1200 Nm between sections $x = 0$ and $x = 2$ m.

The keyword *Z_BENDING_OVERHANGINGS specifies the number of internal supports and hinges. For each one, the coordinate and power of the singularity notation must be specified, with -1 for a support and -3 for a hinge. In this case, there is a joint at $x = 1$ m.

The keyword *Z_BENDING_BCS specifies the boundary conditions of bending in the z plane, with two conditions for each end of the beam. For this purpose, the flag representing the loads or kinematics is specified, followed by the value of the boundary condition. In this case, we have $u_y(x = 0) = 0$ and $\theta_z(x = 0) = 0$ in the left end, and $u_y(x = 2) = 0$ and $M_z(x = 2) = 400$ in the right end. The following flags are available:

- Displacements in x, y, and z: UX, UY, UZ
- Rotations in x, y, and z: TX, TY, TZ
- Normal and shear forces in x, y, and z: NX, VY, VZ
- Moments in x, y, and z: MX, MY, MZ

*MATERIAL_PROPERTIES is used to enter the Young's modulus and Poisson's coefficient of the material. On the other hand, *GEOMETRIC_PROPERTIES allows the specification of element length, area, and second moments of area (I_p, I_y, I_z) and the distances between the geometrical center and the boundaries of the cross-section. Similar keywords are available for other implemented problems. The file manual.doc presents all available keywords.

The program plots the diagrams for loads, kinematics, stress, and strain components and prints the support reactions for each problems in the input file. To use the program, just execute BeamLab from the program's root directory and specify the input file.

7.5 SUMMARY OF THE VARIATIONAL FORMULATION OF BEAMS

The kinematics of the three-dimensional beam with circular cross-section is described by the vector field $\mathbf{u}(x,y,z)$ with the following components:

$$\mathbf{u}(x,y,z) = \left\{ \begin{array}{c} u_x(x,y,z) \\ u_y(x,z) \\ u_z(x,y) \end{array} \right\} = \left\{ \begin{array}{c} u_x^n(x) + z\theta_y(x) - y\theta_z(x) \\ u_y(x) - z\theta_x(x) \\ u_z(x) + y\theta_x(x) \end{array} \right\}. \tag{7.156}$$

The set $\mathcal{V}$ of possible kinematic actions is described by

$$\mathcal{V} = \left\{ \{ \mathbf{u} |\, u_x = u_x^n(x) + z\theta_y(x) - y\theta_z(x), u_y = u_y(x) - z\theta_x(x), u_z = u_z(x) + y\theta_x(x) \} \right\}. \tag{7.157}$$

On the other hand, the subset of rigid actions is given by

$$\mathcal{N}(\mathcal{D}) = \left\{ \mathbf{u}(x,y,z) \in \mathcal{V} \mid u_x = u_y = u_z = \theta_x = \theta_y = \theta_z = \text{cte} \right\}. \tag{7.158}$$

In this case, the strain operator $\mathcal{D}$ is indicated in matrix form as

$$\mathcal{D}: \quad \mathcal{V} \to \mathcal{W}$$

$$\mathbf{u}(x,y,z) \to \mathcal{D}\mathbf{u}(x,y,z) = \begin{bmatrix} \dfrac{d}{dx} & 0 & 0 & 0 & z\dfrac{d}{dx} & -y\dfrac{d}{dx} \\ 0 & \dfrac{d}{dx} & 0 & 0 & 0 & -1 \\ 0 & 0 & \dfrac{d}{dx} & 0 & 1 & 0 \\ 0 & 0 & 0 & r\dfrac{d}{dx} & 0 & 0 \end{bmatrix} \left\{ \begin{array}{c} u_x(x) \\ u_y(x) \\ u_z(x) \\ \theta_x(x) \\ \theta_y(x) \\ \theta_z(x) \end{array} \right\}. \tag{7.159}$$

Thus, the space $\mathcal{W}'$ of internal loads is constituted by continuous functions $N_x(x)$, $V_y(x)$, $V_z(x)$, $M_x(x)$, $M_y(x)$, and $M_z(x)$, which respectively represent the normal force in x, shear forces in y and z, twisting moment in x, and bending moments in y and z in the cross-sections of the beam.

The terms $q_x(x)$, $q_y(x)$, $q_z(x)$, $m_x(x)$, $m_y(x)$, $m_z(x)$, P_0, P_L, V_{y0}, V_{yL}, V_{z0}, V_{zL}, $T_)$, T_L, M_{y0} M_{yL}, M_{z0}, and M_{zL}, relative to the external loads, define the vector space $\mathcal{V}'$.

The equilibrium differential operator $\mathcal{D}^*$, between the internal and external loads, will be denoted here as the composition of the respective operators of the problem of traction $\mathcal{D}_{N_x}^*$, twisting $\mathcal{D}_{M_x}^*$, bending in y, $\mathcal{D}_{M_y}^*$, and bending in z, $\mathcal{D}_{M_z}^*$, that is,

$$\mathcal{D}_{N_x}^*: \quad \mathcal{W}' \to \mathcal{V}'$$

$$N_x(x) \to \mathcal{D}^* N_x(x) = \left\{ \begin{array}{ll} -\dfrac{d}{dx} N_x(x) = q_x(x) & \text{in } x \in (0, L) \\ -N_x(x)|_{x=0} = P_0 & \text{in } x = L \\ N_x(x)|_{x=L} = P_L & \text{in } x = 0 \end{array} \right., \tag{7.160}$$

$$\mathscr{D}_{M_x}^*: \quad \mathscr{W} \to \mathscr{V}'$$

$$M_x(x) \to \mathscr{D}^* M_x(x) = \begin{cases} -\dfrac{d}{dx} M_x(x) = m_x(x) & \text{in } x \in (0,L) \\ - M_x(x)|_{x=0} = T_0 & \text{in } x = L \\ M_x(x)|_{x=L} = T_L & \text{in } x = 0 \end{cases}, \qquad (7.161)$$

$$\mathscr{D}_{M_y}^*: \quad \mathscr{W}' \to \mathscr{V}'$$

$$\left\{ \begin{matrix} M_y(x) \\ V_z(x) \end{matrix} \right\} \to \mathscr{D}^* \left\{ \begin{matrix} M_y(x) \\ V_z(x) \end{matrix} \right\}$$

$$\left\{ \begin{matrix} M_y(x) \\ V_z(x) \end{matrix} \right\} \to \begin{bmatrix} \dfrac{d}{dx} & -1 \\ 0 & \dfrac{d}{dx} \\ 0 & 1|_{x=0} \\ 0 & -1|_{x=L} \\ -1|_{x=0} & 0 \\ 1|_{x=L} & 0 \end{bmatrix} \left\{ \begin{matrix} M_y(x) \\ V_z(x) \end{matrix} \right\} = \begin{cases} \dfrac{dM_y(x)}{dx} - V_z(x) \\ \dfrac{dV_z(x)}{dx} \\ V_z(x)|_{x=0} \\ -V_z(x)|_{x=L} \\ -M_y(x)|_{x=0} \\ M_y(x)|_{x=L} \end{cases}, \qquad (7.162)$$

$$\mathscr{D}_{M_z}^*: \quad \mathscr{W}' \to \mathscr{V}'$$

$$\left\{ \begin{matrix} M_z(x) \\ V_y(x) \end{matrix} \right\} \to \mathscr{D}^* \left\{ \begin{matrix} M_z(x) \\ V_y(x) \end{matrix} \right\}$$

$$\left\{ \begin{matrix} M_z(x) \\ V_y(x) \end{matrix} \right\} \to \begin{bmatrix} \dfrac{d}{dx} & -1 \\ 0 & \dfrac{d}{dx} \\ 0 & 1|_{x=0} \\ 0 & -1|_{x=L} \\ -1|_{x=0} & 0 \\ 1|_{x=L} & 0 \end{bmatrix} \left\{ \begin{matrix} M_z(x) \\ V_y(x) \end{matrix} \right\} = \begin{cases} \dfrac{dM_z(x)}{dx} - V_y(x) \\ \dfrac{dV_y(x)}{dx} \\ V_y(x)|_{x=0} \\ -V_y(x)|_{x=L} \\ -M_z(x)|_{x=0} \\ M_z(x)|_{x=L} \end{cases}. \qquad (7.163)$$

Figure 7.43 schematically presents the formulation of the three-dimensional beam with load superposition.

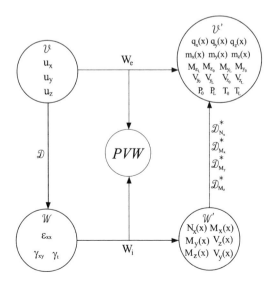

Figure 7.43 Variational formulation of the three-dimensional beam.

7.6   APROXIMATED SOLUTION

The equilibrium BVPs in terms of the kinematics for a three-dimensional beam with circular cross-section are given in equations (7.122) to (7.127). Multiplying these equations by the test functions $\alpha_i(x)$ and $v_i(x)$ $(i = x, y, z)$ and integrating along the length of the beam, we have

$$\int_0^L \frac{d}{dx}\left(E(x)A(x)\frac{du_x(x)}{dx}\right)v_x(x)dx + \int_0^L q_x(x)v_x(x)dx = 0,$$

$$\int_0^L \frac{d}{dx}\left(G(x)I_p(x)\frac{d\theta_x(x)}{dx}\right)\alpha_x(x)dx + \int_0^L m_x(x)\alpha_x(x)dx = 0,$$

$$\int_0^L \frac{d}{dx}\left[K_c(x)G(x)A(x)\left(\frac{du_y(x)}{dx} - \theta_z(x)\right)\right]v_y(x)dx + \int_0^L q_y(x)v_y(x)dx = 0,$$

$$\int_0^L \frac{d}{dx}\left(E(x)I_z(x)\frac{d\theta_z(x)}{dx}\right)\alpha_z(x)dx + \int_0^L K_c(x)G(x)A(x)\left(\frac{du_y(x)}{dx} - \theta_z(x)\right)\alpha_z(x)dx$$

$$+ \int_0^L m_z(x)\alpha_z(x)dx = 0,$$

$$\int_0^L \frac{d}{dx}\left[K_c(x)G(x)A(x)\left(\frac{du_z(x)}{dx} + \theta_y(x)\right)\right]v_z(x)dx + \int_0^L q_z(x)v_z(x)dx = 0,$$

$$\int_0^L \frac{d}{dx}\left(E(x)I_y(x)\frac{d\theta_y(x)}{dx}\right)\alpha_y(x)dx - \int_0^L K_c(x)G(x)A(x)\left(\frac{du_z(x)}{dx} + \theta_y(x)\right)\alpha_y(x)dx$$

$$+ \int_0^L m_y(x)\alpha_y(x)dx = 0.$$

The test functions $v_i(x)$ and $\alpha_i(x)$ respectively represent displacements and rotations in directions x, y and z.

The previous integrals are the work of internal and external loads. We can sum these expressions, resulting in the integral form of equilibrium for the three-dimensional beam

$$\int_0^L \frac{d}{dx}\left(E(x)A(x)\frac{du_x(x)}{dx}\right)v_x(x)dx + \int_0^L \frac{d}{dx}\left(G(x)I_p(x)\frac{d\theta_x(x)}{dx}\right)\alpha_x(x)dx$$

$$+ \int_0^L \frac{d}{dx}\left(E(x)I_z(x)\frac{d\theta_z(x)}{dx}\right)\alpha_z(x)dx + \int_0^L \frac{d}{dx}\left[K_c(x)G(x)A(x)\left(\frac{du_y(x)}{dx} - \theta_z(x)\right)\right]v_y(x)dx$$

$$+ \int_0^L \frac{d}{dx}\left(E(x)I_y(x)\frac{d\theta_y(x)}{dx}\right)\alpha_y(x)dx - \int_0^L \frac{d}{dx}\left[K_c(x)G(x)A(x)\left(\frac{du_z(x)}{dx} + \theta_y(x)\right)\right]v_z(x)dx$$

$$+ \int_0^L K_c(x)G(x)A(x)\left(\frac{du_y(x)}{dx} - \theta_z(x)\right)\alpha_z(x)dx$$

$$+ \int_0^L K_c(x)G(x)A(x)\left(\frac{du_z(x)}{dx} + \theta_y(x)\right)\alpha_y(x)dx$$

$$+ \int_0^L m_x(x)\alpha_x(x)dx + \int_0^L m_y(x)\alpha_y(x)dx + \int_0^L m_z(x)\alpha(x)dx$$

$$+ \int_0^L q_x(x)v_x(x)dx + \int_0^L q_y(x)v_y(x)dx + \int_0^L q_z(x)v_z(x)dx = 0.$$

Integrating by parts the first six terms of the above equation, we have

$$
-\int_0^L E(x)A(x)\frac{du_x(x)}{dx}\frac{dv_x(x)}{dx}dx + E(x)A(x)\frac{du_x(x)}{dx}v_x(x)\Big|_0^L
$$

$$
-\int_0^L G(x)I_p(x)\frac{d\theta_x(x)}{dx}\frac{d\alpha_x(x)}{dx}dx + G(x)I_p(x)\frac{d\theta_x(x)}{dx}\alpha_x(x)\Big|_0^L
$$

$$
-\int_0^L E(x)I_z(x)\frac{d\theta_z(x)}{dx}\frac{d\alpha_z(x)}{dx}dx + E(x)I_z(x)\frac{d\theta_z(x)}{dx}\alpha_z(x)\Big|_0^L
$$

$$
-\int_0^L K_c(x)G(x)A(x)\left(\frac{du_y(x)}{dx}-\theta_z(x)\right)\frac{dv_y(x)}{dx}dx + K_c(x)G(x)A(x)\left(\frac{du_y(x)}{dx}-\theta_z(x)\right)v_y(x)\Big|_0^L
$$

$$
-\int_0^L E(x)I_y(x)\frac{d\theta_y(x)}{dx}\frac{d\alpha_y(x)}{dx}dx + E(x)I_y(x)\frac{d\theta_y(x)}{dx}\alpha_y(x)\Big|_0^L
$$

$$
+\int_0^L K_c(x)G(x)A(x)\left(\frac{du_z(x)}{dx}+\theta_y(x)\right)\frac{dv_z(x)}{dx}dx - K_c(x)G(x)A(x)\left(\frac{du_z(x)}{dx}+\theta_y(x)\right)v_z(x)\Big|_0^L
$$

$$
+\int_0^L K_c(x)G(x)A(x)\left(\frac{du_y(x)}{dx}-\theta_z(x)\right)\alpha_z(x)dx
$$

$$
+\int_0^L K_c(x)G(x)A(x)\left(\frac{du_z(x)}{dx}+\theta_y(x)\right)\alpha_y(x)dx
$$

$$
+\int_0^L q_x(x)v_x(x)dx + \int_0^L q_y(x)v_y(x)dx + \int_0^L q_z(x)v_z(x)dx
$$

$$
+\int_0^L m_x(x)\alpha_x(x)dx + \int_0^L m_y(x)\alpha_y(x)dx + \int_0^L m_z(x)\alpha_z(x)dx = 0.
$$

Using definitions (7.116) to (7.121) for the normal and shear forces and bending and twisting moments, the previous expression is rewritten as

$$
-\int_0^L E(x)A(x)\frac{du_x(x)}{dx}\frac{dv_x(x)}{dx}dx - \int_0^L G(x)I_p(x)\frac{d\theta_x(x)}{dx}\frac{d\alpha_x(x)}{dx}dx
$$

$$
-\int_0^L E(x)I_z(x)\frac{d\theta_z(x)}{dx}\frac{d\alpha_z(x)}{dx}dx - \int_0^L K_c(x)G(x)A(x)\left(\frac{du_y(x)}{dx}-\theta_z(x)\right)\frac{dv_y(x)}{dx}dx
$$

$$
-\int_0^L E(x)I_y(x)\frac{d\theta_y(x)}{dx}\frac{d\alpha_y(x)}{dx}dx + \int_0^L K_c(x)G(x)A(x)\left(\frac{du_z(x)}{dx}+\theta_y(x)\right)\frac{dv_z(x)}{dx}dx
$$

$$
+\int_0^L K_c(x)G(x)A(x)\left(\frac{du_y(x)}{dx}-\theta_z(x)\right)\alpha_z(x)dx
$$

$$
+\int_0^L K_c(x)G(x)A(x)\left(\frac{du_z(x)}{dx}+\theta_y(x)\right)\alpha_y(x)dx
$$

$$
+\int_0^L q_x(x)v_x(x)dx + \int_0^L q_y(x)v_y(x)dx + \int_0^L q_z(x)v_z(x)dx
$$

$$
+\int_0^L m_x(x)\alpha_x(x)dx + \int_0^L m_y(x)\alpha_y(x)dx + \int_0^L m_z(x)\alpha_z(x)dx
$$

$$
+ N_x(x)v_x(x)\big|_0^L - V_y(x)v_y(x)\big|_0^L - V_z(x)v_z(x)\big|_0^L
$$

$$
+ M_x(x)\alpha_x(x)\big|_0^L + M_y(x)\alpha_y(x)\big|_0^L + M_z(x)\alpha_z(x)\big|_0^L = 0.
$$

Substituting the boundary conditions given in BVPs (7.96) to (7.99), we obtain the final expression for the weak form of the three-dimensional beam

$$
\begin{aligned}
&\int_0^L E(x)A(x)\frac{du_x(x)}{dx}\frac{dv_x(x)}{dx}dx + \int_0^L G(x)I_p(x)\frac{d\theta_x(x)}{dx}\frac{d\alpha_x(x)}{dx}dx \\
&+ \int_0^L E(x)I_z(x)\frac{d\theta_z(x)}{dx}\frac{d\alpha_z(x)}{dx}dx + \int_0^L E(x)I_y(x)\frac{d\theta_y(x)}{dx}\frac{d\alpha_y(x)}{dx}dx \\
&+ \int_0^L K_c(x)G(x)A(x)\left(\frac{du_y(x)}{dx}-\theta_z(x)\right)\left(\frac{dv_y(x)}{dx}-\alpha_z(x)\right)dx \\
&- \int_0^L K_c(x)G(x)A(x)\left(\frac{du_z(x)}{dx}+\theta_y(x)\right)\left(\frac{dv_z(x)}{dx}+\alpha_y(x)\right)dx \\
&- \int_0^L K_c(x)G(x)A(x)\left(\frac{du_y(x)}{dx}-\theta_z(x)\right)\alpha_z(x)dx \\
&- \int_0^L K_c(x)G(x)A(x)\left(\frac{du_z(x)}{dx}+\theta_y(x)\right)\alpha_y(x)dx \\
&= \int_0^L q_x(x)v_x(x)dx + \int_0^L q_y(x)v_y(x)dx + \int_0^L q_z(x)v_z(x)dx \\
&+ \int_0^L m_x(x)\alpha_x(x)dx + \int_0^L m_y(x)\alpha_y(x)dx + \int_0^L m_z(x)\alpha_z(x)dx \\
&+ P_0 v_x(0) + P_L v_x(L) + V_{y_0}v_y(0) + V_{y_L}v_y(L) + V_{z_0}v_z(0) + V_{z_L}v_z(L) \\
&+ T_0\alpha_x(0) + T_L\alpha_x(L) + M_{y_0}\alpha_y(0) + M_{y_L}\alpha_y(L) + +M_{z_0}\alpha_z(0) + M_{z_L}\alpha_z(L).
\end{aligned}
\tag{7.164}
$$

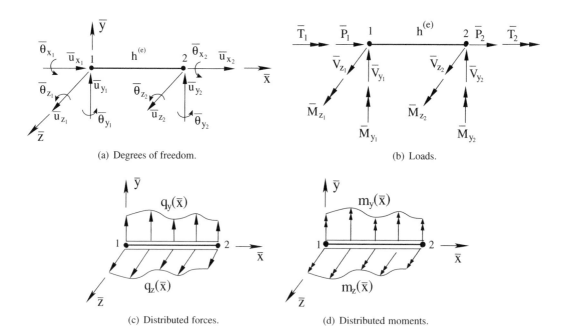

(a) Degrees of freedom.   (b) Loads.

(c) Distributed forces.   (d) Distributed moments.

Figure 7.44   Two-node finite element for the three-dimensional beam.

Consider the two-node beam element illustrated in Figure 7.44 according to the non-normalized local coordinate system $\bar{x}\bar{y}\bar{z}$. Each node has six degrees of freedom corresponding to displacements and rotations according to x, y, and z, as illustrated in Figure 7.44(a). The concentrated nodal loads are illustrated in Figure 7.44(b). The distributed loads in terms of forces and moments in the transversal directions are illustrated in Figures 7.44(c) and 7.44(d).

The linear interpolation of displacements and rotations in the normalized local coordinate system ξ_1 is given by

$$u_{x_2}^{(e)}(\xi_1) = \bar{u}_{x_1}^{(e)}\phi_1^{(e)}(\xi_1) + \bar{u}_{x_2}^{(e)}\phi_2^{(e)}(\xi_1), \tag{7.165}$$

$$u_{y_2}^{(e)}(\xi_1) = \bar{u}_{y_1}^{(e)}\phi_1^{(e)}(\xi_1) + \bar{u}_{y_2}^{(e)}\phi_2^{(e)}(\xi_1), \tag{7.166}$$

$$u_{z_2}^{(e)}(\xi_1) = \bar{u}_{z_1}^{(e)}\phi_1^{(e)}(\xi_1) + \bar{u}_{z_2}^{(e)}\phi_2^{(e)}(\xi_1), \tag{7.167}$$

$$\theta_{x_2}^{(e)}(\xi_1) = \bar{\theta}_{x_1}^{(e)}\phi_1^{(e)}(\xi_1) + \bar{\theta}_{x_2}^{(e)}\phi_2^{(e)}(\xi_1), \tag{7.168}$$

$$\theta_{y_2}^{(e)}(\xi_1) = \bar{\theta}_{y_1}^{(e)}\phi_1^{(e)}(\xi_1) + \bar{\theta}_{y_2}^{(e)}\phi_2^{(e)}(\xi_1), \tag{7.169}$$

$$\theta_{z_2}^{(e)}(\xi_1) = \bar{\theta}_{z_1}^{(e)}\phi_1^{(e)}(\xi_1) + \bar{\theta}_{z_2}^{(e)}\phi_2^{(e)}(\xi_1). \tag{7.170}$$

In matrix notation,

$$\{u^{(e)}\} = [\bar{N}^{(e)}]\{\bar{u}^{(e)}\}, \tag{7.171}$$

where

$$\{u^{(e)}\} = \left\{ \begin{array}{cccccc} u_{x_2}^{(e)}(\xi_1) & u_{y_2}^{(e)}(\xi_1) & u_{z_2}^{(e)}(\xi_1) & \theta_{x_2}^{(e)}(\xi_1) & \theta_{y_2}^{(e)}(\xi_1) & \theta_{z_2}^{(e)}(\xi_1) \end{array} \right\}^T,$$

$$\{\bar{u}^{(e)}\} = \left\{ \begin{array}{cccccccccccc} \bar{u}_{x_1}^{(e)} & \bar{u}_{y_1}^{(e)} & \bar{u}_{z_1}^{(e)} & \bar{\theta}_{x_1}^{(e)} & \bar{\theta}_{y_1}^{(e)} & \bar{\theta}_{z_1}^{(e)} & \bar{u}_{x_2}^{(e)} & \bar{u}_{y_2}^{(e)} & \bar{u}_{z_2}^{(e)} & \bar{\theta}_{x_2}^{(e)} & \bar{\theta}_{y_2}^{(e)} & \bar{\theta}_{z_2}^{(e)} \end{array} \right\}^T.$$

$[\bar{N}^{(e)}]$ is the matrix of the linear shape functions $\phi_1^{(e)}(\xi_1) = \frac{1}{2}(1-\xi_1)$ and $\phi_2^{(e)}(\xi_1) = \frac{1}{2}(1+\xi_1)$

$$[\bar{N}^{(e)}] = \begin{bmatrix} \phi_1^{(e)} & 0 & 0 & 0 & 0 & 0 & \phi_2^{(e)} & 0 & 0 & 0 & 0 & 0 \\ 0 & \phi_1^{(e)} & 0 & 0 & 0 & 0 & 0 & \phi_2^{(e)} & 0 & 0 & 0 & 0 \\ 0 & 0 & \phi_1^{(e)} & 0 & 0 & 0 & 0 & 0 & \phi_2^{(e)} & 0 & 0 & 0 \\ 0 & 0 & 0 & \phi_1^{(e)} & 0 & 0 & 0 & 0 & 0 & \phi_2^{(e)} & 0 & 0 \\ 0 & 0 & 0 & 0 & \phi_1^{(e)} & 0 & 0 & 0 & 0 & 0 & \phi_2^{(e)} & 0 \\ 0 & 0 & 0 & 0 & 0 & \phi_1^{(e)} & 0 & 0 & 0 & 0 & 0 & \phi_2^{(e)} \end{bmatrix} \tag{7.172}$$

The interpolation of the global displacement derivatives and rotations are given, in matrix notation, as

$$\left\{ \begin{array}{c} \frac{du_{x_2}^{(e)}[x(\xi_1)]}{dx} \\ \frac{du_{y_2}^{(e)}[x(\xi_1)]}{dx} \\ \frac{du_{z_2}^{(e)}[x(\xi_1)]}{dx} \\ \frac{d\theta_{x_2}^{(e)}[x(\xi_1)]}{dx} \\ \frac{d\theta_{y_2}^{(e)}[x(\xi_1)]}{dx} \\ \frac{d\theta_{z_2}^{(e)}[x(\xi_1)]}{dx} \end{array} \right\} = \frac{1}{h^{(e)}} \underbrace{\begin{bmatrix} -1 & 0 & 0 & 0 & 0 & 0 & 1 & 0 & 0 & 0 & 0 & 0 \\ 0 & -1 & 0 & 0 & 0 & 0 & 0 & 1 & 0 & 0 & 0 & 0 \\ 0 & 0 & -1 & 0 & 0 & 0 & 0 & 0 & 1 & 0 & 0 & 0 \\ 0 & 0 & 0 & -1 & 0 & 0 & 0 & 0 & 0 & 1 & 0 & 0 \\ 0 & 0 & 0 & 0 & -1 & 0 & 0 & 0 & 0 & 0 & 1 & 0 \\ 0 & 0 & 0 & 0 & 0 & -1 & 0 & 0 & 0 & 0 & 0 & 1 \end{bmatrix}}_{[\bar{B}^{(e)}]} \left\{ \begin{array}{c} \bar{u}_{x_1}^{(e)} \\ \bar{u}_{y_1}^{(e)} \\ \bar{u}_{z_1}^{(e)} \\ \bar{\theta}_{x_1}^{(e)} \\ \bar{\theta}_{y_1}^{(e)} \\ \bar{\theta}_{z_1}^{(e)} \\ \bar{u}_{x_2}^{(e)} \\ \bar{u}_{y_2}^{(e)} \\ \bar{u}_{z_2}^{(e)} \\ \bar{\theta}_{x_2}^{(e)} \\ \bar{\theta}_{y_2}^{(e)} \\ \bar{\theta}_{z_2}^{(e)} \end{array} \right\} \tag{7.173}$$

with $[\bar{B}^{(e)}]$ the deformation matrix and the term $\dfrac{1}{h^{(e)}}$ comes from the inverse of the Jacobian.

Analogously, the test functions $v_i^{(e)}(\xi_1)$ and $\alpha_i^{(e)}(\xi_1)$ are interpolated in the element using the Galerkin method as

$$\{v^{(e)}\} = [\bar{N}^{(e)}]\{\bar{v}^{(e)}\}, \tag{7.174}$$

where

$$\{v^{(e)}\} = \left\{ v_{x_2}^{(e)}(\xi_1) \quad v_{y_2}^{(e)}(\xi_1) \quad v_{z_2}^{(e)}(\xi_1) \quad \alpha_{x_2}^{(e)}(\xi_1) \quad \alpha_{y_2}^{(e)}(\xi_1) \quad \alpha_{z_2}^{(e)}(\xi_1) \right\}^T,$$

$$\{\bar{v}^{(e)}\} = \left\{ \bar{v}_{x_1}^{(e)} \quad \bar{v}_{y_1}^{(e)} \quad \bar{v}_{z_1}^{(e)} \quad \bar{\alpha}_{x_1}^{(e)} \quad \bar{\alpha}_{y_1}^{(e)} \quad \bar{\alpha}_{z_1}^{(e)} \quad \bar{v}_{x_2}^{(e)} \quad \bar{v}_{y_2}^{(e)} \quad \bar{v}_{z_2}^{(e)} \quad \bar{\alpha}_{x_2}^{(e)} \quad \bar{\alpha}_{y_2}^{(e)} \quad \bar{\alpha}_{z_2}^{(e)} \right\}^T.$$

Their derivatives are interpolated as

$$\left\{ \begin{array}{c} \frac{dv_x^{(e)}[x(\xi_1)]}{dx} \\ \frac{dv_y^{(e)}[x(\xi_1)]}{dx} \\ \frac{dv_z^{(e)}[x(\xi_1)]}{dx} \\ \frac{d\alpha_x^{(e)}[x(\xi_1)]}{dx} \\ \frac{d\alpha_y^{(e)}[x(\xi_1)]}{dx} \\ \frac{d\alpha_z^{(e)}[x(\xi_1)]}{dx} \end{array} \right\} = \frac{1}{h^{(e)}} \begin{bmatrix} -1 & 0 & 0 & 0 & 0 & 0 & 1 & 0 & 0 & 0 & 0 & 0 \\ 0 & -1 & 0 & 0 & 0 & 0 & 0 & 1 & 0 & 0 & 0 & 0 \\ 0 & 0 & -1 & 0 & 0 & 0 & 0 & 0 & 1 & 0 & 0 & 0 \\ 0 & 0 & 0 & -1 & 0 & 0 & 0 & 0 & 0 & 1 & 0 & 0 \\ 0 & 0 & 0 & 0 & -1 & 0 & 0 & 0 & 0 & 0 & 1 & 0 \\ 0 & 0 & 0 & 0 & 0 & -1 & 0 & 0 & 0 & 0 & 0 & 1 \end{bmatrix} \left\{ \begin{array}{c} \bar{v}_{x_1}^{(e)} \\ \bar{v}_{y_1}^{(e)} \\ \bar{v}_{z_1}^{(e)} \\ \bar{\alpha}_{x_1}^{(e)} \\ \bar{\alpha}_{y_1}^{(e)} \\ \bar{\alpha}_{z_1}^{(e)} \\ \bar{v}_{x_2}^{(e)} \\ \bar{v}_{y_2}^{(e)} \\ \bar{v}_{z_2}^{(e)} \\ \bar{\alpha}_{x_2}^{(e)} \\ \bar{\alpha}_{y_2}^{(e)} \\ \bar{\alpha}_{z_2}^{(e)} \end{array} \right\},$$

$$(7.175)$$

with $(\bar{v}_{x_j}^{(e)}, \bar{v}_{y_j}^{(e)}, \bar{v}_{z_j}^{(e)})$ and $(\bar{\alpha}_{x_j}^{(e)}, \bar{\alpha}_{y_j}^{(e)}, \bar{\alpha}_{z_j}^{(e)})$ $(j = 1, 2)$ respectively the approximation coefficients of test functions $v_i^{(e)}(\xi_1)$ and $\alpha_i^{(e)}(\xi_1)$.

From (7.165), the weak form in the element local system is given by

$$\int_{-1}^{1} EA \frac{du_x^{(e)}[x(\xi_1)]}{dx} \frac{dv_x^{(e)}[x(\xi_1)]}{dx} |J| d\xi_1 + \int_{-1}^{1} GI_p \frac{d\theta_x^{(e)}[x(\xi_1)]}{dx} \frac{d\alpha_x^{(e)}[x(\xi_1)]}{dx} |J| d\xi_1$$

$$+ \int_{-1}^{1} EI_z \frac{d\theta_z^{(e)}[x(\xi_1)]}{dx} \frac{d\alpha_z^{(e)}[x(\xi_1)]}{dx} |J| d\xi_1 + \int_{-1}^{1} EI_y \frac{d\theta_y^{(e)}[x(\xi_1)]}{dx} \frac{d\alpha_y^{(e)}[x(\xi_1)]}{dx} |J| d\xi_1$$

$$+ \int_{-1}^{1} K_c GA \left(\frac{du_y^{(e)}[x(\xi_1)]}{dx} - \theta_z^{(e)}[x(\xi_1)] \right) \left(\frac{dv_y^{(e)}[x(\xi_1)]}{dx} - \alpha_z^{(e)}(x(\xi_1)) \right) |J| d\xi_1$$

$$- \int_{-1}^{1} K_c GA \left(\frac{du_z^{(e)}[x(\xi_1)]}{dx} + \theta_y^{(e)}[x(\xi_1)] \right) \left(\frac{dv_z^{(e)}[x(\xi_1)]}{dx} + \alpha_y^{(e)}[x(\xi_1)] \right) |J| d\xi_1$$

$$= \int_{-1}^{1} q_x[x(\xi_1)] v_x^{(e)}[x(\xi_1)] |J| d\xi_1 + \int_{-1}^{1} q_y[x(\xi_1)] v_y^{(e)}[x(\xi_1)] |J| d\xi_1 \qquad (7.176)$$

$$+ \int_{-1}^{1} q_z[x(\xi_1)] v_z^{(e)}[x(\xi_1)] |J| d\xi_1 + \int_{-1}^{1} m_x[x(\xi_1)] \alpha_x^{(e)}[x(\xi_1)] |J| d\xi_1$$

$$+ \int_{-1}^{1} m_y[x(\xi_1)] \alpha_y^{(e)}[x(\xi_1)] |J| d\xi_1 + \int_{-1}^{1} m_z[x(\xi_1)] \alpha_z^{(e)}[x(\xi_1)] |J| d\xi_1$$

$$+ \bar{P}_2 v_x^{(e)}(1) + \bar{P}_1 v_x^{(e)}(-1) + \bar{V}_{y_2} v_y^{(e)}(1) + \bar{V}_{y_1} v_y^{(e)}(-1) + \bar{V}_{z_2} v_z^{(e)}(1) + \bar{V}_{z_1} v_z^{(e)}(-1)$$

$$+ \bar{T}_2 \alpha_x^{(e)}(1) + \bar{T}_1 \alpha_x^{(e)}(-1) + \bar{M}_{y_2} \alpha_y^{(e)}(1) + \bar{M}_{y_1} \alpha_y^{(e)}(-1) + + \bar{M}_{z_2} \alpha_z^{(e)}(1) + \bar{M}_{z_1} \alpha_z^{(e)}(-1).$$

with $|J| = \dfrac{h^{(e)}}{2}$ and assuming the material and geometrical properties to be constant in the element.

Substituting approximations (7.173) to (7.177) for the terms of traction and twist in the above weak form, we respectively have the following local systems of equations for each problem:

$$\frac{E^{(e)}A^{(e)}}{h^{(e)}} \begin{bmatrix} 1 & -1 \\ -1 & 1 \end{bmatrix} \left\{ \begin{array}{c} \bar{u}_{x_1}^{(e)} \\ \bar{u}_{x_2}^{(e)} \end{array} \right\} = \int_{-1}^{1} q_x(x(\xi_1)) \left\{ \begin{array}{c} \phi_1^{(e)} \\ \phi_2^{(e)} \end{array} \right\} |J| d\xi_1 + \left\{ \begin{array}{c} \bar{P}_1 \\ \bar{P}_2 \end{array} \right\}, \qquad (7.177)$$

$$\frac{G^{(e)}I_p^{(e)}}{h^{(e)}}\begin{bmatrix} 1 & -1 \\ -1 & 1 \end{bmatrix}\begin{Bmatrix} \bar{\theta}_{x_1}^{(e)} \\ \bar{\theta}_{x_2}^{(e)} \end{Bmatrix} = \int_{-1}^{1} m_x(x(\xi_1))\begin{Bmatrix} \phi_1^{(e)} \\ \phi_2^{(e)} \end{Bmatrix}|J|d\xi_1 + \begin{Bmatrix} \bar{T}_1 \\ \bar{T}_2 \end{Bmatrix}. \tag{7.178}$$

Now considering the bending terms in z in the weak form of the element and the respective approximations given in (7.173) to (7.177), we have the following element system of equations in terms of the stiffness matrix, analogous to equation (6.159):

$$\frac{E^{(e)}I_z^{(e)}}{\kappa_{c_z}^{(e)}h^{(e)3}}\begin{bmatrix} 12 & 6h^{(e)} & -12 & 6h^{(e)} \\ 6h^{(e)} & (\kappa_{c_z}^{(e)}+4)h^{(e)2} & -6h^{(e)} & -(\kappa_{c_z}^{(e)}-2)h^{(e)2} \\ -12 & -6h^{(e)} & 12 & -6h^{(e)} \\ 6h^{(e)} & -(\kappa_{c_z}^{(e)}-2)h^{(e)2} & -6h^{(e)} & (\kappa_{c_z}^{(e)}+4)h^{(e)2} \end{bmatrix}\begin{Bmatrix} \bar{u}_{y_1}^{(e)} \\ \bar{\theta}_{z_1}^{(e)} \\ \bar{u}_{y_2}^{(e)} \\ \bar{\theta}_{z_2}^{(e)} \end{Bmatrix} = \begin{Bmatrix} \bar{V}_{y_1} \\ \bar{M}_{z_1} \\ \bar{V}_{y_2} \\ \bar{M}_{z_2} \end{Bmatrix}$$

$$\int_{-1}^{1} m_z[x(\xi_1)]\begin{Bmatrix} 0 \\ \phi_1^{(e)}(\xi_1) \\ 0 \\ \phi_2^{(e)}(\xi_1) \end{Bmatrix}\frac{h^{(e)}}{2}d\xi_1 + \int_{-1}^{1} q_y[x(\xi_1)]\begin{Bmatrix} \phi_1^{(e)}(\xi_1) \\ 0 \\ \phi_2^{(e)}(\xi_1) \\ 0 \end{Bmatrix}\frac{h^{(e)}}{2}d\xi_1, \tag{7.179}$$

where

$$\kappa_{c_z}^{(e)} = \frac{12E^{(e)}I_z^{(e)}}{h^{(e)2}K_c^{(e)}G^{(e)}A^{(e)}}. \tag{7.180}$$

Analogous to bending in y, we have the following system of equations for the element:

$$\frac{E^{(e)}I_y^{(e)}}{\kappa_{c_y}^{(e)}h^{(e)3}}\begin{bmatrix} 12 & 6h^{(e)} & -12 & 6h^{(e)} \\ 6h^{(e)} & (\kappa_{c_y}^{(e)}+4)h^{(e)2} & -6h^{(e)} & -(\kappa_{c_y}^{(e)}-2)h^{(e)2} \\ -12 & -6h^{(e)} & 12 & -6h^{(e)} \\ 6h^{(e)} & -(\kappa_{c_y}^{(e)}-2)h^{(e)2} & -6h^{(e)} & (\kappa_{c_y}^{(e)}+4)h^{(e)2} \end{bmatrix}\begin{Bmatrix} \bar{u}_{z_1}^{(e)} \\ \bar{\theta}_{y_1}^{(e)} \\ \bar{u}_{z_2}^{(e)} \\ \bar{\theta}_{y_2}^{(e)} \end{Bmatrix} = \begin{Bmatrix} \bar{V}_{z_1} \\ \bar{M}_{y_1} \\ \bar{V}_{z_2} \\ \bar{M}_{y_2} \end{Bmatrix}$$

$$\int_{-1}^{1} m_y[x(\xi_1)]\begin{Bmatrix} 0 \\ \phi_1^{(e)}(\xi_1) \\ 0 \\ \phi_2^{(e)}(\xi_1) \end{Bmatrix}\frac{h^{(e)}}{2}d\xi_1 + \int_{-1}^{1} q_z[x(\xi_1)]\begin{Bmatrix} \phi_1^{(e)}(\xi_1) \\ 0 \\ \phi_2^{(e)}(\xi_1) \\ 0 \end{Bmatrix}\frac{h^{(e)}}{2}d\xi_1, \tag{7.181}$$

where

$$\kappa_{c_y}^{(e)} = \frac{12E^{(e)}I_y^{(e)}}{h^{(e)2}K_c^{(e)}G^{(e)}A^{(e)}}. \tag{7.182}$$

By the superposition of the previous effects, the local stiffness matrix for the two-node three-dimensional beam element is given by

$$[\bar{K}^{(e)}] = \begin{bmatrix} k_1 & 0 & 0 & 0 & 0 & 0 & -k_1 & 0 & 0 & 0 & 0 & 0 \\ 0 & k_2 & 0 & 0 & 0 & k_3 & 0 & -k_2 & 0 & 0 & 0 & k_3 \\ 0 & 0 & k_2 & 0 & k_3 & 0 & 0 & 0 & -k_2 & 0 & k_3 & 0 \\ 0 & 0 & 0 & k_4 & 0 & 0 & 0 & 0 & 0 & -k_4 & 0 & 0 \\ 0 & 0 & k_3 & 0 & k_5 & 0 & 0 & 0 & -k_3 & 0 & -k_6 & 0 \\ 0 & k_3 & 0 & 0 & 0 & k_7 & 0 & -k_3 & 0 & 0 & 0 & -k_8 \\ -k_1 & 0 & 0 & 0 & 0 & 0 & k_1 & 0 & 0 & 0 & 0 & 0 \\ 0 & -k_2 & 0 & 0 & 0 & -k_3 & 0 & k_2 & 0 & 0 & 0 & -k_3 \\ 0 & 0 & -k_2 & 0 & -k_3 & 0 & 0 & 0 & k_2 & 0 & -k_3 & 0 \\ 0 & 0 & 0 & -k_4 & 0 & 0 & 0 & 0 & 0 & k_4 & 0 & 0 \\ 0 & 0 & k_3 & 0 & -k_6 & 0 & 0 & 0 & -k_3 & 0 & k_5 & 0 \\ 0 & k_3 & 0 & 0 & 0 & -k_8 & 0 & -k_3 & 0 & 0 & 0 & k_7 \end{bmatrix}, \tag{7.183}$$

where the nonzero coefficients indicated as k_i are

$$
\begin{aligned}
k_1 &= k_b^{(e)}, \\
k_2 &= 12k_f^{(e)}, \\
k_3 &= 6h^{(e)}k_f^{(e)}, \\
k_4 &= k_t^{(e)}, \\
k_5 &= \varphi_{y_1}^{(e)}h^{(e)^2}k_f^{(e)}, \\
k_6 &= \varphi_{y_2}^{(e)}h^{(e)^2}k_f^{(e)}, \\
k_7 &= \varphi_{z_1}^{(e)}h^{(e)^2}k_f^{(e)}, \\
k_8 &= \varphi_{z_2}^{(e)}h^{(e)^2}k_f^{(e)},
\end{aligned}
\tag{7.184}
$$

and $\varphi_{i_1}^{(e)} = (\kappa_{c_i}^{(e)} + 4)$ and $\varphi_{i_2}^{(e)} = (\kappa_{c_i}^{(e)} - 2)$ $(i = y, z)$. The elastic constants of traction $(k_b^{(e)})$, twisting $(k_t^{(e)})$, and bending in y and z $(k_f^{(e)})$ of the element are respectively given by

$$
k_b^{(e)} = \frac{E^{(e)}A^{(e)}}{h^{(e)}}, \tag{7.185}
$$

$$
k_t^{(e)} = \frac{G^{(e)}I_p^{(e)}}{h^{(e)}}, \tag{7.186}
$$

$$
k_f^{(e)} = \frac{K_c^{(e)}G^{(e)}A^{(e)}}{12h^{(e)}}. \tag{7.187}
$$

Considering that the intensities of the distributed loads are constant, that is, $q_x(x) = q_{x_0}, q_y(x) = q_{y_0}, q_z(x) = q_{z_0}, m_x(x) = m_{x_0}, m_y(x) = m_{y_0}$, and $m_z(x) = m_{z_0}$, the nodal equivalent load vector for a three-dimensional beam element in the non-normalized local coordinate system $\bar{x}\bar{y}\bar{z}$ is given by

$$
\{\bar{f}^{(e)}\} =
\begin{Bmatrix}
q_{x_0}\frac{h^{(e)}}{2} + \bar{P}_1 \\
q_{y_0}\frac{h^{(e)}}{2} + \bar{V}_{y_1} \\
q_{z_0}\frac{h^{(e)}}{2} + \bar{V}_{z_1} \\
m_{x_0}\frac{h^{(e)}}{2} + \bar{T}_1 \\
m_{y_0}\frac{h^{(e)}}{2} + \bar{M}_{y_1} \\
m_{z_0}\frac{h^{(e)}}{2} + \bar{M}_{z_1} \\
q_{x_0}\frac{h^{(e)}}{2} + \bar{P}_2 \\
q_{y_0}\frac{h^{(e)}}{2} + \bar{V}_{y_2} \\
q_{z_0}\frac{h^{(e)}}{2} + \bar{V}_{z_2} \\
m_{x_0}\frac{h^{(e)}}{2} + \bar{T}_2 \\
m_{y_0}\frac{h^{(e)}}{2} + \bar{M}_{y_2} \\
m_{z_0}\frac{h^{(e)}}{2} + \bar{M}_{z_2}
\end{Bmatrix}. \tag{7.188}
$$

From (7.183) and (7.188), the element system of equations in the local coordinate system $\bar{x}\bar{y}\bar{z}$ is

$$
[\bar{K}^{(e)}]\{\bar{u}^{(e)}\} = \{\bar{f}^{(e)}\} \tag{7.189}
$$

Example 7.7 *Consider the beam of Example 7.5. We want to obtain the numerical solution using three beam elements, as illustrated in Figure 7.45. We assume that the beam is made of steel with $E = 210.0$ GPa and $G = 80.8$ GPa. The shear factor $K_c = \frac{3}{4}$ is considered.*

Each node has six degrees of freedom, resulting in a system of equations with rank 24. The obtained solution is given in Table 7.2. The angles of twist are higher than the other degrees of freedom, because the beam is free to rotate about x.

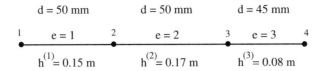

Figure 7.45   Example 7.7: mesh with three beam elements.

Node	$\bar{u}_x$ [m]	$\bar{u}_y$ [m]	$\bar{u}_z$ [m]	$\bar{\theta}_x$ [rad]	$\bar{\theta}_y$ [rad]	$\bar{\theta}_z$ [rad]
1	0.0000	0.0039	0.0146	-7.3319	-0.0975	0.0257
2	0.0000	0.0000	0.0000	-4.5740	-0.0975	0.0257
3	-0.0008	0.0039	-0.0097	-2.0000	0.0035	0.0333
4	-0.0008	0.0000	0.0000	-2.0000	0.1636	-0.0718

Table 7.2
Example 7.7: numerical results.

We employ equation (7.189) to calculate the element internal loads, simply replacing the vector $\{\bar{u}^{(e)}\}$ with the displacement and rotation results for the degrees of freedom in each element. The directions of forces and moments on the element local system are in agreement with those obtained by the equilibrium BVPs. Figure 7.46 illustrates the directions of the nodal internal loads of elements $e-1$ and e. The global internal loads in the respective degrees of freedom of node i shared by elements $e-1$ and e are obtained as

$$\begin{aligned}
N_{x_i} &= N_{x_2}^{e-1} - N_{x_1}^e, \\
V_{y_{i+1}} &= -V_{y_2}^{e-1} + V_{y_1}^e, \\
V_{z_{i+2}} &= V_{z_2}^{e-1} - V_{y_1}^e, \\
M_{x_{i+3}} &= M_{x_2}^{e-1} - M_{x_1}^e, \\
M_{y_{i+4}} &= -M_{y_2}^{e-1} + M_{y_1}^e, \\
M_{z_{i+5}} &= M_{z_2}^{e-1} - M_{z_1}^e.
\end{aligned}$$

Notice that the signs of some local loads should be taken negative in order to make the compatibility with the positive directions of the axes of the employed global coordinate system. The same internal loads diagrams illustrated in Figure 7.35 are also obtained by approximation. File beam3dexample7.m implements this example in MATLAB.

□

Generally, the beam element has an arbitrary orientation with respect to the xyz global coordinate system, as illustrated in Figure 7.47(a). We want to obtain a coordinate transformation matrix $[T]$, such that the vectors of degrees of freedom $\{\bar{u}^{(e)}\}$ and loads $\{\bar{f}^{(e)}\}$, given in the local coordinate system of the element, can be expressed in terms of vectors of degrees of freedom $\{u^{(e)}\}$ and loads $\{f^{(e)}\}$ of the element in the global reference system as follows

$$\{u^{(e)}\} = [T]\{\bar{u}^{(e)}\}, \tag{7.190}$$

$$\{f^{(e)}\} = [T]\{\bar{f}^{(e)}\}. \tag{7.191}$$

As $[T]$ is an orthogonal matrix, relation $[T]^T[T] = [I]$ is valid and the previous equations can be rewritten, expressing the element variables of the local system in terms of the global ones, that is,

$$\{\bar{u}^{(e)}\} = [T]^T \{u^{(e)}\}, \tag{7.192}$$

$$\{\bar{f}^{(e)}\} = [T]^T \{f^{(e)}\}. \tag{7.193}$$

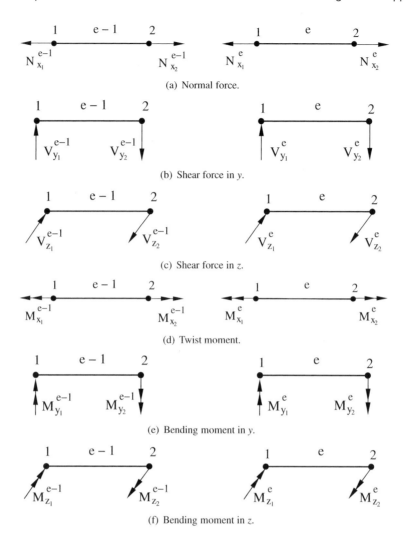

(a) Normal force.

(b) Shear force in y.

(c) Shear force in z.

(d) Twist moment.

(e) Bending moment in y.

(f) Bending moment in z.

Figure 7.46 Example 7.7: internal loads at the interface of two beam elements.

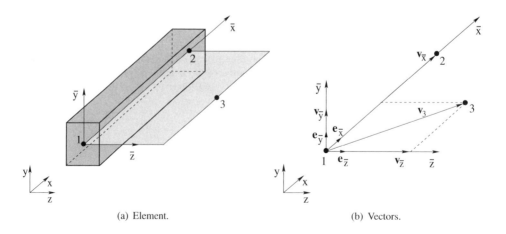

(a) Element.

(b) Vectors.

Figure 7.47 Three-node beam element in the global reference system.

Substituting these expressions in (7.189), we have the following expression:

$$[K^{(e)}][T]^T \{u^{(e)}\} = [T]^T \{f^{(e)}\},$$

Pre-multiplying both sides of the above equation by the coordinate transformation matrix $[T]$, we have the element system of equations in terms of the element variables expressed in the global reference system. Thus,

$$[K^{(e)}]\{u^{(e)}\} = \{f^{(e)}\}, \tag{7.194}$$

with $[K^{(e)}]$ the element stiffness matrix of a three-dimensional beam in the global reference system, given by

$$[K^{(e)}] = [T][\bar{K}^{(e)}][T]^T. \tag{7.195}$$

One way to determine the transformation matrix $[T]$ is considering a third node for the beam element, in such way that nodes 1 to 3 define the plane $\bar{x}\bar{z}$ of the element reference system, as illustrated in Figure 7.47.

The nodal coordinates, according to the global reference system, are denoted by (x_1, y_1, z_1), (x_2, y_2, z_2), and (x_3, y_3, z_3). From Figure 7.47(b), the components of vector $\mathbf{v}_{\bar{x}}$ along the element are obtained by the difference of coordinates of nodes 2 and 1, that is,

$$\{v_{\bar{x}}\} = \left\{ \begin{array}{c} x_2 - x_1 \\ y_2 - y_1 \\ z_2 - z_1 \end{array} \right\}. \tag{7.196}$$

Analogously, the components of vector $\mathbf{v}_3$ are obtained by the difference of coordinates of nodes 3 and 1, that is,

$$\{v_3\} = \left\{ \begin{array}{c} x_3 - x_1 \\ y_3 - y_1 \\ z_3 - z_1 \end{array} \right\}. \tag{7.197}$$

Vector $\mathbf{v}_{\bar{y}}$ along axis $\bar{y}$ of the element coordinate system is determined by the cross product of $\mathbf{v}_{\bar{x}}$ and $\mathbf{v}_3$. Thus,

$$\mathbf{v}_{\bar{y}} = \mathbf{v}_{\bar{x}} \times \mathbf{v}_3. \tag{7.198}$$

On the other hand, vector $\mathbf{v}_{\bar{z}}$, along axis $\bar{z}$ of the element coordinate system, is given by the cross product of $\mathbf{v}_{\bar{y}}$ and $\mathbf{v}_{\bar{x}}$, that is,

$$\mathbf{v}_{\bar{z}} = \mathbf{v}_{\bar{y}} \times \mathbf{v}_{\bar{x}}. \tag{7.199}$$

The unit vectors along axes $\bar{x}$, $\bar{y}$, and $\bar{z}$ are respectively given by

$$\mathbf{e}_{\bar{x}} = \frac{\mathbf{v}_{\bar{x}}}{||\mathbf{v}_{\bar{x}}||}, \tag{7.200}$$

$$\mathbf{e}_{\bar{y}} = \frac{\mathbf{v}_{\bar{y}}}{||\mathbf{v}_{\bar{y}}||}, \tag{7.201}$$

$$\mathbf{e}_{\bar{z}} = \frac{\mathbf{v}_{\bar{z}}}{||\mathbf{v}_{\bar{z}}||}. \tag{7.202}$$

The coordinate transformation matrix $[\bar{T}]$ between the local and global measures is given from the components of the above unit vectors as

$$[\bar{T}] = \left[\ \{e_{\bar{x}}\} \quad \{e_{\bar{y}}\} \quad \{e_{\bar{z}}\} \ \right]. \tag{7.203}$$

Example 7.8 *Consider a beam element with nodes 1, 2, and 3 with the respective global coordinates* $(0.0; 0.5; 1.2)$, $(1.7; 0.6; 1.6)$, *and* $(0.2; 1.5; 0.3)$.

The components of vectors $\{v_{\bar{x}}\}$, $\{v_3\}$, $\{v_{\bar{y}}\}$, *and* $\{v_{\bar{z}}\}$ *are respectively given by*

$$\{v_{\bar{x}}\} = \left\{ \begin{array}{ccc} 1.7 & 0.1 & 0.4 \end{array} \right\}^T,$$

$$\{v_3\} = \left\{ \begin{array}{ccc} 0.2 & 1.0 & -0.9 \end{array} \right\}^T,$$

$$\{v_{\bar{y}}\} = \left\{ \begin{array}{ccc} 1.7 & 0.1 & 0.4 \end{array} \right\}^T,$$

$$\{v_{\bar{z}}\} = \left\{ \begin{array}{ccc} 1.7 & 0.1 & 0.4 \end{array} \right\}^T.$$

On the other hand, the components of the unit vectors are:

$$\{e_{\bar{x}}\} = \{\ 0.97 \quad 0.06 \quad 0.23\ \}^T,$$

$$\{e_{\bar{y}}\} = \{\ -0.21 \quad 0.68 \quad 0.71\ \}^T,$$

$$\{e_{\bar{z}}\} = \{\ 0.11 \quad 0.73 \quad -0.67\ \}^T.$$

The coordinate transformation matrix $[\bar{T}]$ *is given by*

$$[\bar{T}] = \begin{bmatrix} 0.97 & -0.21 & 0.11 \\ 0.06 & 0.68 & 0.73 \\ 0.23 & 0.71 & -0.67 \end{bmatrix}.$$

Considering that the origins of the local and global coordinate systems are coincident, the coordinates of element nodes 1 to 3 in the local reference system are obtained from vectors $\{v_{\bar{x}}\}$ *and* $\{v_3\}$ *as*

$$\begin{bmatrix} 0.97 & -0.21 & 0.11 \\ 0.06 & 0.68 & 0.73 \\ 0.23 & 0.71 & -0.67 \end{bmatrix} \begin{bmatrix} 0.0 & 0.0 & 0.0 \\ 1.7 & 0.1 & 0.4 \\ 0.2 & 1.0 & -0.9 \end{bmatrix} = \begin{bmatrix} 0.00 & 0.00 & 0.00 \\ 1.75 & 0.00 & 0.00 \\ 0.05 & 0.00 & 1.36 \end{bmatrix}.$$

Thus, we have that the $\bar{x}$ *coordinates of nodes 1 to 3 are* $\bar{x}_1 = 0.00$, $\bar{x}_2 = 1.75$, *and* $\bar{x}_3 = 0.05$. *However, the nodal coordinates in the other directions are zero, except for* $\bar{z}_3 = 1.36$ *at node 3. This happens because in the local system only the* $\bar{x}$ *coordinates are necessary.*

In a general way, we can write the following relation between the local and global nodal coordinates:

$$\begin{bmatrix} 0 & 0 & 0 \\ x_2 & y_2 & z_2 \\ x_3 & y_3 & z_3 \end{bmatrix} [\bar{T}] = \begin{bmatrix} 0 & 0 & 0 \\ \bar{x}_2 & 0 & 0 \\ \bar{x}_3 & 0 & \bar{z}_3 \end{bmatrix}. \tag{7.204}$$

File beam3dsolapex8.m implements the solution of this example in MATLAB.
□

The matrix $[\bar{T}]$ can be applied to the set of three displacement and rotation components of each node, allowing the obtention of the transformation matrix $[T]$ in the following way:

$$[T] = \begin{bmatrix} [\bar{T}] & [0] & [0] & [0] \\ [0] & [\bar{T}] & [0] & [0] \\ [0] & [0] & [\bar{T}] & [0] \\ [0] & [0] & [0] & [\bar{T}] \end{bmatrix}. \tag{7.205}$$

Once the displacements and rotations are calculated in the global reference system, we can calculate the forces and moments in the element using equations (7.189) and (7.190). Thus,

$$\{\bar{f}^{(e)}\} = \{\bar{K}^{(e)}\}[T]^T \{u^{(e)}\}. \tag{7.206}$$

The high-order three-dimensional beam element is obtained by interpolating the displacements and rotations in the local system ξ_1 using $N+1$ shape functions until order P as

$$u_{x_{N+1}}^{(e)}(\xi_1) = \sum_{i=0}^{N} a_{u_{x_i}} \phi_i^{(e)}(\xi_1), \tag{7.207}$$

$$u_{y_{N+1}}^{(e)}(\xi_1) = \sum_{i=0}^{N} a_{u_{y_i}} \phi_i^{(e)}(\xi_1), \tag{7.208}$$

$$u_{z_{N+1}}^{(e)}(\xi_1) = \sum_{i=0}^{N} a_{u_{z_i}} \phi_i^{(e)}(\xi_1), \tag{7.209}$$

$$\theta_{x_{N+1}}^{(e)}(\xi_1) = \sum_{i=0}^{N} a_{\theta_{x_i}} \phi_i^{(e)}(\xi_1), \tag{7.210}$$

$$\theta_{y_{N+1}}^{(e)}(\xi_1) = \sum_{i=0}^{N} a_{\theta_{y_i}} \phi_i^{(e)}(\xi_1), \tag{7.211}$$

$$\theta_{z_{N+1}}^{(e)}(\xi_1) = \sum_{i=0}^{N} a_{\theta_{z_i}} \phi_i^{(e)}(\xi_1), \tag{7.212}$$

with $a_{u_{x_i}}$, $a_{u_{y_i}}$, $a_{u_{z_i}}$, $a_{\theta_{x_i}}$, $a_{\theta_{y_i}}$ and $a_{\theta_{z_i}}$ the approximation coefficients. The shape functions can be given in terms of the Lagrange or Jacobi polynomials. As the shape functions starting from second order are zero at the element ends, we can employ distinct orders for each element.

The three-dimensional beam element can be simplified to obtain the two-dimensional counterpart. In this case, each element node has only three degrees of freedom, with the axial displacement $\bar{u}_x^{(e)}$ in the local direction $\bar{x}$, the transversal displacement $\bar{u}_y^{(e)}$ in $\bar{y}$, and the rotation $\bar{\theta}_z^{(e)}$ in $\bar{z}$, as illustrated in Figure 7.48. For a circular section, the rotation $\bar{\theta}_x^{(e)}$ is also included in the local direction $\bar{x}$.

The local element system of equations is obtained from expressions (7.183) and (7.188), that is,

$$\begin{bmatrix} k_1 & 0 & 0 & 0 & -k_1 & 0 & 0 & 0 \\ 0 & k_2 & 0 & k_3 & 0 & -k_2 & 0 & k_3 \\ 0 & 0 & k_4 & 0 & 0 & 0 & -k_4 & 0 \\ 0 & k_3 & 0 & k_7 & 0 & -k_3 & 0 & -k_8 \\ -k_1 & 0 & 0 & 0 & k_1 & 0 & 0 & 0 \\ 0 & -k_2 & 0 & -k_3 & 0 & k_2 & 0 & -k_3 \\ 0 & 0 & -k_4 & 0 & 0 & 0 & k_4 & 0 \\ 0 & k_3 & 0 & -k_8 & 0 & -k_3 & 0 & k_7 \end{bmatrix} \begin{Bmatrix} \bar{u}_{x_1}^{(e)} \\ \bar{u}_{y_1}^{(e)} \\ \bar{\theta}_{x_1}^{(e)} \\ \bar{\theta}_{z_1}^{(e)} \\ \bar{u}_{x_2}^{(e)} \\ \bar{u}_{y_2}^{(e)} \\ \bar{\theta}_{x_2}^{(e)} \\ \bar{\theta}_{z_2}^{(e)} \end{Bmatrix} = \begin{Bmatrix} q_{x0}\frac{h^{(e)}}{2} + \bar{P}_1 \\ q_{y0}\frac{h^{(e)}}{2} + \bar{V}_{y_1} \\ m_{x0}\frac{h^{(e)}}{2} + \bar{T}_1 \\ m_{z0}\frac{h^{(e)}}{2} + \bar{M}_{z_1} \\ q_{x0}\frac{h^{(e)}}{2} + \bar{P}_2 \\ q_{y0}\frac{h^{(e)}}{2} + \bar{V}_{y_2} \\ m_{x0}\frac{h^{(e)}}{2} + \bar{T}_2 \\ m_{z0}\frac{h^{(e)}}{2} + \bar{M}_{z_2} \end{Bmatrix}, \tag{7.213}$$

where the coefficients k_i are given in equation (7.185). In a compact form,

$$[\bar{K}^{(e)}]\{\bar{u}^{(e)}\} = \{\bar{f}^{(e)}\}. \tag{7.214}$$

The global-local transformation is performed analogously to the case of bars under traction. From Figure 7.49, we obtain the following relations between the local ($\bar{u}_x^{(e)}$, $\bar{u}_y^{(e)}$, $\bar{\theta}_x^{(e)}$, $\bar{\theta}_z^{(e)}$) and global ($u_x^{(e)}$, $u_y^{(e)}$, $\theta_x^{(e)}$, $\theta_z^{(e)}$) degrees of freedom for nodes 1 and 2:

$$\begin{aligned} \bar{u}_{x_1}^{(e)} &= u_{x_1}^{(e)}\cos\theta + u_{y_1}^{(e)}\sin\theta, \\ \bar{u}_{x_2}^{(e)} &= u_{x_2}^{(e)}\cos\theta + u_{y_2}^{(e)}\sin\theta, \\ \bar{u}_{y_1}^{(e)} &= -u_{x_1}^{(e)}\sin\theta + u_{y_1}^{(e)}\cos\theta, \\ \bar{u}_{y_2}^{(e)} &= -u_{x_2}^{(e)}\sin\theta + u_{y_2}^{(e)}\cos\theta, \\ \bar{\theta}_{x_1}^{(e)} &= \theta_{x_1}^{(e)}, \\ \bar{\theta}_{z_1}^{(e)} &= \theta_{z_1}^{(e)}, \\ \bar{\theta}_{x_2}^{(e)} &= \theta_{x_2}^{(e)}, \\ \bar{\theta}_{z_2}^{(e)} &= \theta_{z_2}^{(e)}. \end{aligned} \tag{7.215}$$

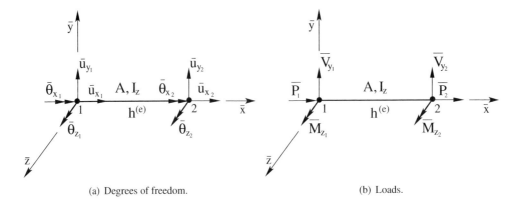

(a) Degrees of freedom.   (b) Loads.

Figure 7.48 Two-dimensional beam element in circular twist.

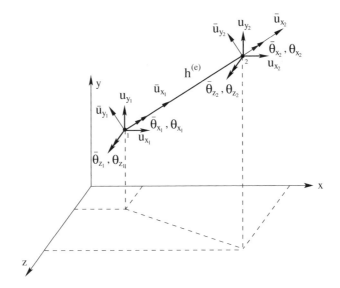

Figure 7.49 Global-local transformation for the two-dimensional beam element with twist.

In matrix form,

$$
\begin{Bmatrix}
\bar{u}_{x_1}^{(e)} \\
\bar{u}_{y_1}^{(e)} \\
\bar{\theta}_{x_1}^{(e)} \\
\bar{\theta}_{z_1}^{(e)} \\
\bar{u}_{x_2}^{(e)} \\
\bar{u}_{y_2}^{(e)} \\
\bar{\theta}_{x_2}^{(e)} \\
\bar{\theta}_{z_2}^{(e)}
\end{Bmatrix}
=
\begin{bmatrix}
\cos\theta & \sin\theta & 0 & 0 & 0 & 0 & 0 & 0 \\
-\sin\theta & \cos\theta & 0 & 0 & 0 & 0 & 0 & 0 \\
0 & 0 & 1 & 0 & 0 & 0 & 0 & 0 \\
0 & 0 & 0 & 1 & 0 & 0 & 0 & 0 \\
0 & 0 & 0 & 0 & \cos\theta & \sin\theta & 0 & 0 \\
0 & 0 & 0 & 0 & -\sin\theta & \cos\theta & 0 & 0 \\
0 & 0 & 0 & 0 & 0 & 0 & 1 & 0 \\
0 & 0 & 0 & 0 & 0 & 0 & 0 & 1
\end{bmatrix}
\begin{Bmatrix}
u_{x_1}^{(e)} \\
u_{y_1}^{(e)} \\
\theta_{x_1}^{(e)} \\
\theta_{z_1}^{(e)} \\
u_{x_2}^{(e)} \\
u_{y_2}^{(e)} \\
\theta_{x_2}^{(e)} \\
\theta_{z_2}^{(e)}
\end{Bmatrix}
\tag{7.216}
$$

In summarized form,

$$
\{\bar{u}^{(e)}\} = [T]\{u^{(e)}\},
\tag{7.217}
$$

with $[T]$ the coordinate transformation matrix between the global and local reference systems given in (7.216).

The high-order two-dimensional beam element is obtained interpolating the displacements $u_x^{(e)}$ and $u_y^{(e)}$ and rotations $\theta_x^{(e)}$ and $\theta_z^{(e)}$ in the local system ξ_1 of the element and using $N+1$ shape functions until order P as

$$u_{x_{N+1}}^{(e)}(\xi_1) = \sum_{i=0}^{N} a_{u_{x_i}} \phi_i^{(e)}(\xi_1), \tag{7.218}$$

$$u_{y_{N+1}}^{(e)}(\xi_1) = \sum_{i=0}^{N} a_{u_{y_i}} \phi_i^{(e)}(\xi_1), \tag{7.219}$$

$$\theta_{x_{N+1}}^{(e)}(\xi_1) = \sum_{i=0}^{N} a_{\theta_{x_i}} \phi_i^{(e)}(\xi_1), \tag{7.220}$$

$$\theta_{z_{N+1}}^{(e)}(\xi_1) = \sum_{i=0}^{N} a_{\theta_{z_i}} \phi_i^{(e)}(\xi_1), \tag{7.221}$$

with $a_{u_{x_i}}$, $a_{u_{y_i}}$, $a_{\theta_{x_i}}$, and $a_{\theta_{z_i}}$ the approximation coefficients.

7.7 FINAL COMMENTS

This chapter presented the formulation of a two-dimensional beam model, including the effects of traction, circular torsion, and bending with shear from the superposition of the models individually studied in the previous chapters.

Subsequently, the three-dimensional beam model was formulated with the inclusion of bending and shear effects in plane xz. Then, the concept of oblique bending and design and verification procedures were presented. A program for solution of differential equations of the three-dimensional beam model was considered, as well as some applications of the studied model. The same steps of the variational formulation were applied again for both beam models.

Finally, the approximation by low- and high-order finite elements of the considered beam models was introduced.

Thus, the introduction of one-dimensional mechanical models of this book is complete. The next chapter presents the variational formulation of linear elastic solids, introducing the concept of second-order tensors, as well as the finite element approximation. Subsequently, the two-dimensional plane stress, plane strain, and noncircular torsion problems are presented.

7.8 PROBLEMS

1. Consider the beam with cross-section of diameter d subjected to the loads indicated in Figure 7.50. Design the beam for an admissible normal stress of 120 MPa. Neglect the shear stress due to shear force. Plot the normal and shear stress distributions in the critical section.

2. Consider the beam of rectangular cross-section with area $b \times 4b$ subjected to the loading indicated in Figure 7.51. Design the beam for an admissible normal stress of 100 MPa, including the shear stress due to the shear force.

3. Verify the steel beam with rectangular cross-section subjected to the loading indicated in Figure 7.52, including the shear stress due to shear force. Plot the normal and shear stress distributions in the critical section. Adopt $\bar{\sigma} = 120$ MPa.

4. Verify the beam with circular cross-section subjected to the loading indicated in Figure 7.53, including the shear stress due to shear force. Plot the normal and shear stresses in the critical cross-section. Adopt $\bar{\sigma} = 120$ MPa.

5. Verify the critical section in the interface of the diameters for the beam subjected to the loading indicated in Figure 7.54, including the shear stress due to shear force. Adopt $\bar{\sigma} = 120$ MPa.

6. Implement a MATLAB program to analyze the steel frames of Figures 7.55 and 7.56. Assume circular cross-sections with diameter $d = 5$ cm.

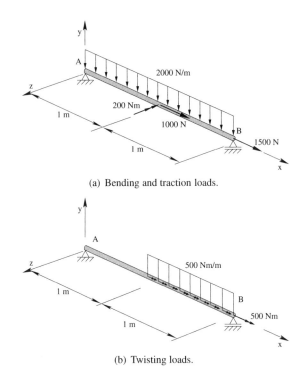

(a) Bending and traction loads.

(b) Twisting loads.

Figure 7.50 Problem 1.

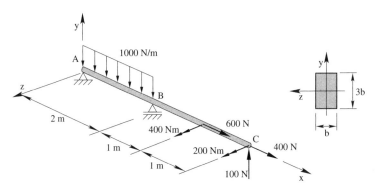

Figure 7.51 Problem 2.

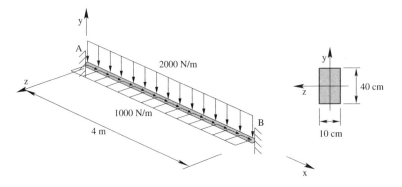

Figure 7.52 Problem 3.

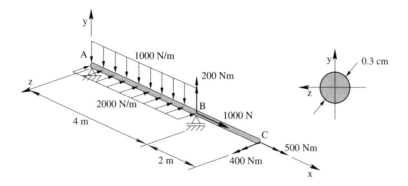

Figure 7.53 Problem 4.

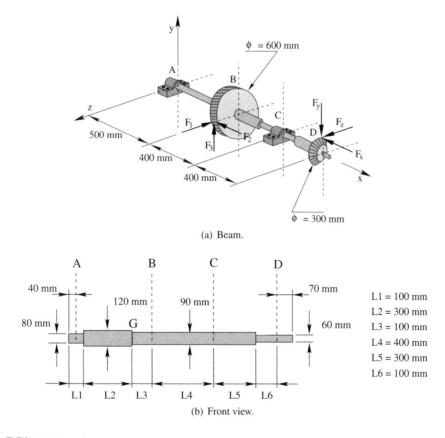

(a) Beam.

(b) Front view.

Figure 7.54 Problem 5.

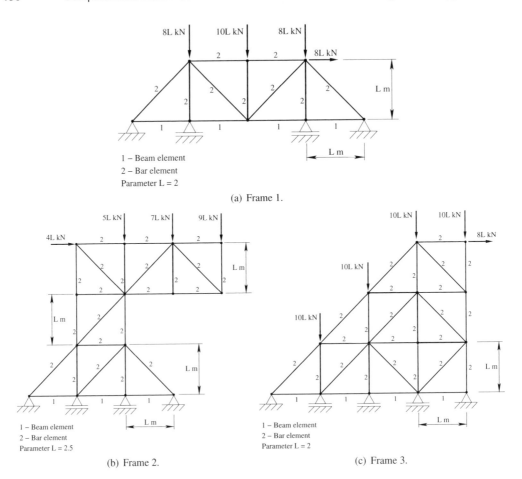

(a) Frame 1.

(b) Frame 2.

(c) Frame 3.

Figure 7.55 Problem 6.

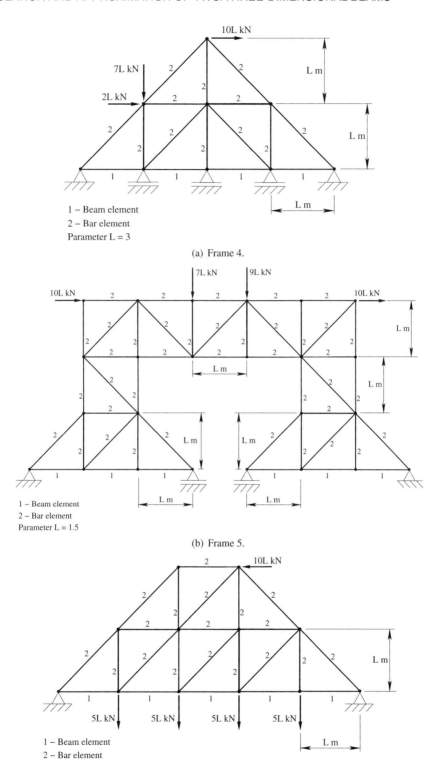

(a) Frame 4.

(b) Frame 5.

(c) Frame 6.

Figure 7.56 Problem 6.

8 FORMULATION AND APPROXIMATION OF SOLIDS

8.1 INTRODUCTION

For the mechanical models studied in the previous chapters, we considered the geometric hypothesis that the length is much greater than the cross-section dimensions. Thus, the obtained mathematical models were one dimensional, with the main variables varying along the x axis of the adopted reference system.

The strain and stress analysis in solids is now considered. In this case, no geometric hypothesis is made for the body and we consider small displacements and strains and as well as isotropic linear elastic material. The same steps of the previous chapters are used for the variational formulation. Initially, we apply scalar and vector functions to describe the kinematics and strain and stress components. To allow a more compact and general notation for the model formulation, the equations for solids will be later rewritten using second-order tensors.

To illustrate the concepts of the general formulation of solids, the previously studied one-dimensional models of bars, shafts, and beams are now derived from the general solid model. For this, the simplifying kinematic hypotheses are considered for each particular case.

Several concepts will be presented throughout this chapter, like the generalized Hooke's law and design criteria for verification of a linear elastic solid. At the end, the approximation for linear elastic solids is presented, introducing shape functions for different elements and mapping between local and global coordinate systems.

In this chapter, we write vectors in bold lowercase letters and their components are given in curlybraces. Second-order tensors are expressed by uppercase bold letters and their components in brackets.

This chapter is mainly based on references [30, 36, 38, 42, 25].

8.2 KINEMATICS

Consider a three-dimensional solid body $\mathscr{B}$ and the Cartesian reference system xyz illustrated in Figure 8.1. Let P_1 be any point of the body $\mathscr{B}$ with coordinates (x, y, z) according to the adopted reference system, denoted as $P_1(x, y, z)$. $\mathscr{B}$ is the initial, reference, or undeformed configuration of the body. If $\{\mathbf{e}_x, \mathbf{e}_y, \mathbf{e}_z\}$ is an orthonormal basis of the adopted reference system, the position vector $\mathbf{r}_{P_1}$ of point P_1 is given by

$$\mathbf{r}_{P_1} = x\mathbf{e}_x + y\mathbf{e}_y + z\mathbf{e}_z.$$

Now suppose that the body $\mathscr{B}$ has a kinematic action and assumes the deformed configuration, indicated by $\mathscr{B}'$ in Figure 8.1. Point P_1 assumes the final position $P_1'(x', y', z')$ in $\mathscr{B}'$, and the respective position vector in the reference system xyz is given by

$$\mathbf{r}_{P_1'} = x'\mathbf{e}_x + y'\mathbf{e}_y + z'\mathbf{e}_z.$$

The displacement vector $\mathbf{u}$ of point P_1 is defined as the difference between the final $\mathbf{r}_{P_1'}$ and initial $\mathbf{r}_{P_1}$ position vectors, as illustrated in Figure 8.1. Thus,

$$\mathbf{u} = \mathbf{r}_{P_1'} - \mathbf{r}_{P_1} = (x' - x)\mathbf{e}_x + (y' - y)\mathbf{e}_y + (z' - z)\mathbf{e}_z. \tag{8.1}$$

Notice that $u_x = (x' - x)$, $u_y = (y' - y)$, and $u_z = (z' - z)$ respectively are the components of the displacement vector $\mathbf{u}$ in directions x, y, and z. Hence, the above expression can be rewritten as

$$\mathbf{u} = u_x\mathbf{e}_x + u_y\mathbf{e}_y + u_z\mathbf{e}_z, \tag{8.2}$$

or, in matrix form,

$$\mathbf{u} = \left\{ \begin{array}{c} u_x \\ u_y \\ u_z \end{array} \right\}. \tag{8.3}$$

439

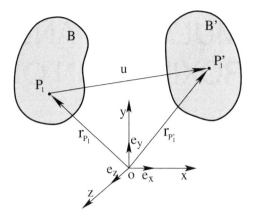

Figure 8.1 Displacement and position vectors of a point in the initial and deformed configurations of a solid body.

Due to the hypothesis of continuum media, the body $\mathscr{B}$ has an infinite number of points. Each point has a displacement vector $\mathbf{u}$ when the body changes its configuration. Thus, the kinematics of a solid body is described by an infinite number of displacement vectors, similar to the vector given in (8.3). This infinite number of vectors defines a three-dimensional displacement vector field $\mathbf{u}(x,y,z)$. Hence, when replacing coordinates (x,y,z) of an arbitrary point P_1, $\mathbf{u}(x,y,z)$ gives the respective displacement vector $\mathbf{u}$ of this point, according to (8.3). Thus, the kinematics of a solid is given by the following displacement vector field:

$$\mathbf{u}(x,y,z) = u_x(x,y,z)\mathbf{e}_x + u_y(x,y,z)\mathbf{e}_y + u_z(x,y,z)\mathbf{e}_z = \left\{ \begin{array}{c} u_x(x,y,z) \\ u_y(x,y,z) \\ u_z(x,y,z) \end{array} \right\}. \tag{8.4}$$

Example 8.1 *The kinematics of the bar model consists of axial displacement actions $u_x(x)$ and can be represented using expression (8.4) as*

$$\mathbf{u}(x) = \left\{ \begin{array}{ccc} u_x(x) & 0 & 0 \end{array} \right\}^T.$$

Analogously, for the shaft and the beam in bending only, we have, respectively,

$$\mathbf{u}(x,y,z) = \left\{ \begin{array}{ccc} 0 & -z\theta_x(x) & y\theta_x(x) \end{array} \right\}^T,$$

$$\mathbf{u}(x,y) = \left\{ \begin{array}{ccc} -y\theta_z(x) & u_y(x) & 0 \end{array} \right\}^T,$$

where $\theta_x(x)$ is the angle of twist of the cross-sections along axis x. The bending rotation $\theta_z(x)$ of the cross-sections is given by

$$\theta_z(x) = \frac{du_y(x)}{dx}.$$

Note that the kinematics of the bar is only dependent on x. However, for shafts and beams, as the displacement components have a linear variation with the coordinates of the cross-section points, the kinematics is respectively given in terms of (x,y,z) and (x,y).

□

8.3 STRAIN MEASURE

Now we want to characterize the distance variation between two arbitrary points of the solid before and after the displacement action. This will allow the definition of strain measures for points of the solid.

Consider the arbitrary points $P_1(x,y,z)$ and $P_2(x+\Delta x,y+\Delta y,z+\Delta z)$, illustrated in Figure 8.2 and the respective position vectors

$$\begin{aligned} \mathbf{r}_{P_1} &= x\mathbf{e}_x + y\mathbf{e}_y + z\mathbf{e}_z, & (8.5) \\ \mathbf{r}_{P_2} &= (x+\Delta x)\mathbf{e}_x + (y+\Delta y)\mathbf{e}_y + (z+\Delta z)\mathbf{e}_z. & (8.6) \end{aligned}$$

According to Figure 8.2, the initial distance $\mathbf{d}$ between points P_2 and P_1 is given by the difference between the position vectors, that is,

$$\mathbf{d} = \mathbf{r}_{P_2} - \mathbf{r}_{P_1} = \Delta x \mathbf{e}_x + \Delta y \mathbf{e}_y + \Delta z \mathbf{e}_z.$$

After the displacement action of the body, points P_1 and P_2 respectively assume the final positions $P_1'(x', y', z')$ and $P_2'(x' + \Delta x', y' + \Delta y', z' + \Delta z')$ with the following position vectors:

$$\mathbf{r}_{P_1'} = x' \mathbf{e}_x + y' \mathbf{e}_y + z' \mathbf{e}_z, \tag{8.7}$$

$$\mathbf{r}_{P_2'} = (x' + \Delta x') \mathbf{e}_x + (y' + \Delta y') \mathbf{e}_y + (z' + \Delta z') \mathbf{e}_z. \tag{8.8}$$

The final distance $\mathbf{d}'$ between points P_2 and P_1 in the deformed configuration is given by the difference of these position vectors. Thus,

$$\mathbf{d}' = \mathbf{r}_{P_2'} - \mathbf{r}_{P_1'} = \Delta x' \mathbf{e}_x + \Delta y' \mathbf{e}_y + \Delta z' \mathbf{e}_z.$$

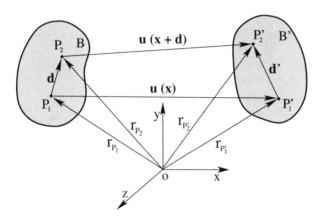

Figure 8.2 Distance variation between points of a solid.

We want to express the distance $\mathbf{d}'$ in terms of the displacement components of points P_1 and P_2. From Figure 8.2, and adopting an analogous procedure as used to obtain equation (8.4), the displacement vectors of points P_1 and P_2 between the final and initial configurations are respectively given by

$$\begin{aligned}
\mathbf{u}(\mathbf{x}) &= \mathbf{r}_{P_1'} - \mathbf{r}_{P_1} = u_x(\mathbf{x}) \mathbf{e}_x + u_y(\mathbf{x}) \mathbf{e}_y + u_z(\mathbf{x}) \mathbf{e}_z, \\
\mathbf{u}(\mathbf{x}+\mathbf{d}) &= \mathbf{r}_{P_2'} - \mathbf{r}_{P_2} = u_x(\mathbf{x}+\mathbf{d}) \mathbf{e}_x + u_y(\mathbf{x}+\mathbf{d}) \mathbf{e}_y + u_z(\mathbf{x}+\mathbf{d}) \mathbf{e}_z,
\end{aligned}$$

where $\mathbf{x} = (x, y, z)$ and $(\mathbf{x} + \mathbf{d}) = (x + \Delta x, y + \Delta y, z + \Delta z)$.

From these expressions, we can write the position vectors of P_1' and P_2' in terms of the respective displacement and initial position vectors as

$$\begin{aligned}
\mathbf{r}_{P_1'} &= \mathbf{r}_{P_1} + \mathbf{u}(\mathbf{x}) \\
&= [x + u_x(\mathbf{x})] \mathbf{e}_x + [y + u_y(\mathbf{x})] \mathbf{e}_y + [z + u_z(\mathbf{x})] \mathbf{e}_z, \\
\mathbf{r}_{P_2'} &= \mathbf{r}_{P_2} + \mathbf{u}(\mathbf{x}+\mathbf{d}) \\
&= [x + \Delta x + u_x(\mathbf{x}+\mathbf{d})] \mathbf{e}_x + [y + \Delta y + u_y(\mathbf{x}+\mathbf{d})] \mathbf{e}_y + [z + \Delta z + u_z(\mathbf{x}+\mathbf{d})] \mathbf{e}_z.
\end{aligned}$$

Thus, $\mathbf{d}'$ is expressed as

$$\mathbf{d}' = \mathbf{r}_{P_2'} - \mathbf{r}_{P_1'} = (\Delta x + \Delta u_x) \mathbf{e}_x + (\Delta y + \Delta u_y) \mathbf{e}_y + (\Delta z + \Delta u_z) \mathbf{e}_z, \tag{8.9}$$

with the difference of the displacement components between points P_2 and P_1 in directions x, y, and z respectively given by

$$\begin{aligned}
\Delta u_x &= u_x(\mathbf{x}+\mathbf{d}) - u_x(\mathbf{x}) = u_x(x+\Delta x, y+\Delta y, z+\Delta z) - u_x(x, y, z), \\
\Delta u_y &= u_y(\mathbf{x}+\mathbf{d}) - u_y(\mathbf{x}) = u_y(x+\Delta x, y+\Delta y, z+\Delta z) - u_y(x, y, z), \\
\Delta u_z &= u_z(\mathbf{x}+\mathbf{d}) - u_z(\mathbf{x}) = u_z(x+\Delta x, y+\Delta y, z+\Delta z) - u_z(x, y, z).
\end{aligned}$$

Finally, the variation of distance $\Delta\mathbf{d}$ is

$$\Delta\mathbf{d} = \mathbf{d}' - \mathbf{d} = \Delta u_x \mathbf{e}_x + \Delta u_z \mathbf{e}_y + \Delta u_z \mathbf{e}_z. \tag{8.10}$$

Following the same procedure of the previous chapters, the idea is to define a strain measure in P_1 dividing the previous expression by $\mathbf{d}$ and taking the limit

$$\lim_{\mathbf{d} \to 0} \frac{\Delta\mathbf{d}}{\mathbf{d}}.$$

However, in standard differential calculus, only the derivatives of the scalar components of vector functions are considered. Thus, the strain measure components will be obtained by taking the components of the relative displacement vector.

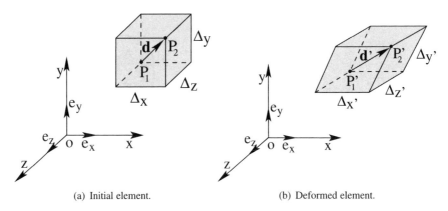

(a) Initial element.
(b) Deformed element.

Figure 8.3 Differential volume elements in the initial and final configurations of a solid body.

Consider the volume elements in the initial and final configurations, respectively, in the neighborhood of points (P_1, P_2) and (P_1', P_2') illustrated in Figure 8.3. The diagonals of these elements are respectively given by $\mathbf{d}$ and $\mathbf{d}'$. The undeformed element is a cube of dimensions Δx, Δy, and Δz and its edges are straight lines forming right angles to each other. After the displacement action, we have a deformed element between points P_1' and P_2' with dimensions $\Delta x'$, $\Delta y'$, and $\Delta z'$. The edges change in length, characterizing the longitudinal strains. The angles between the edges are no longer right angles, characterizing distortions. We want to determine the stretching/shortening and the distortions, defining the strain measures for each point of the solid. We consider planes xy, xz, and yz individually, to make the representation easier.

Figure 8.4 illustrates the projections of the undeformed and deformed elements of Figure 8.3 in plane xy, with the respective displacement components u_x and u_y of points P_1 and P_2, and distortions γ_1 and γ_2. The first analysis is the element stretching in directions x and y, as illustrated in Figure 8.4(a). The element unit elongation in x is given by the length variation $\Delta x' - \Delta x$ divided by the initial length Δx, that is,

$$\frac{\Delta x' - \Delta x}{\Delta x}.$$

From Figure 8.4(a), we have $\Delta x' = \Delta x + \Delta u_x$. Thus,

$$\frac{\Delta x' - \Delta x}{\Delta x} = \frac{\Delta x + \Delta u_x - \Delta x}{\Delta x} = \frac{\Delta u_x}{\Delta x}. \tag{8.11}$$

Taking small values for Δx, point P_2 approximates to P_1. The specific longitudinal strain ε_{xx} is defined for point P_1 in direction x as the limit of the above equation with Δx going to zero, that is,

$$\varepsilon_{xx}(x,y,z) = \lim_{\Delta x \to 0} \frac{\Delta u_x}{\Delta x} = \lim_{\Delta x \to 0} \frac{u_x(x + \Delta x, y + \Delta y, z + \Delta z) - u_x(x,y,z)}{\Delta x}. \tag{8.12}$$

The above limit is the definition of the partial derivative of the continuous scalar function $u_x(x,y,z)$. Thus,

$$\varepsilon_{xx}(x,y,z) = \frac{\partial u_x(x,y,z)}{\partial x}. \tag{8.13}$$

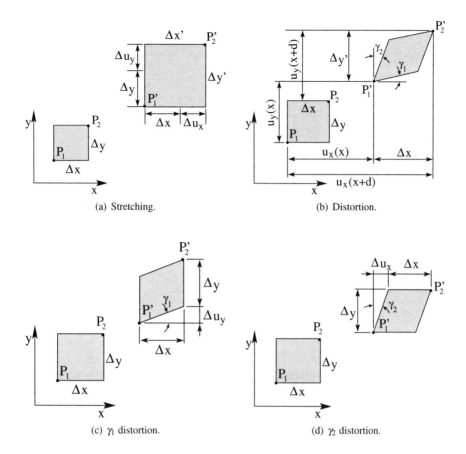

(a) Stretching.

(b) Distortion.

(c) γ_1 distortion.

(d) γ_2 distortion.

Figure 8.4 Strains in plane xy.

The same procedure can be used to obtain the specific longitudinal strain at P_1 in direction y, that is,

$$\varepsilon_{yy}(x,y,z) = \lim_{\Delta y \to 0} \frac{\Delta u_y}{\Delta x} = \lim_{\Delta y \to 0} \frac{u_y(x+\Delta x, y+\Delta y, z+\Delta z) - u_y(x,y,z)}{\Delta y}. \tag{8.14}$$

Thus,

$$\varepsilon_{yy}(x,y,z) = \frac{\partial u_y(x,y,z)}{\partial y}. \tag{8.15}$$

Now consider the distortion of the element illustrated in Figure 8.4(b). In this case, the element dimensions do not change, but the angles are no longer right angles, characterizing a pure element distortion. We initially analyze direction y, which only has a distortion γ_1, as shown in Figure 8.4(c). The following trigonometric relation is valid:

$$\tan \gamma_1 = \frac{\Delta u_y}{\Delta x}. \tag{8.16}$$

For small values of γ_1, the tangent of γ_1 is approximately equal to γ_1, that is, $\tan \gamma_1 \approx \gamma_1$. Thus, taking the limit $\Delta x \to 0$, the following relation is valid for the distortion at point P_1 due to displacement u_y:

$$\gamma_1 = \lim_{\Delta x \to 0} \frac{\Delta u_y}{\Delta x} = \lim_{\Delta x \to 0} \frac{u_y(x+\Delta x, y+\Delta y, z+\Delta z) - u_y(x,y,z)}{\Delta x} = \frac{\partial u_y(x,y,z)}{\partial x}. \tag{8.17}$$

Now, considering only distortion γ_2, according to Figure 8.4(d), we have

$$\tan \gamma_2 = \frac{\Delta u_x}{\Delta y}. $$

For small values of γ_2, we have $\tan \gamma_2 \approx \gamma_2$. Taking the limit to $\Delta y \to 0$, we have the distortion at P_1 due to u_x. Thus,

$$\gamma_2 = \lim_{\Delta y \to 0} \frac{\Delta u_x}{\Delta y} = \lim_{\Delta y \to 0} \frac{u_x(x+\Delta x, y+\Delta y, z+\Delta z) - u_x(x,y,z)}{\Delta y} = \frac{\partial u_x(x,y,z)}{\partial y}. \tag{8.18}$$

The total distortion at point P_1 according to plane xy, denoted as $\bar{\gamma}_{xy}(x,y,z)$, is given by the sum of γ_1 and γ_2, that is,

$$\bar{\gamma}_{xy}(x,y,z) = \gamma_1 + \gamma_2 = \frac{\partial u_y(x,y,z)}{\partial x} + \frac{\partial u_x(x,y,z)}{\partial y}. \tag{8.19}$$

The same procedure is done for plane xz. Taking the displacement components u_x and u_z of points P_1 and P_2, we determine the specific longitudinal strain at point P_1 in direction z, analogous to plane xy, as

$$\varepsilon_{zz}(x,y,z) = \frac{\partial u_z(x,y,z)}{\partial z}. \tag{8.20}$$

The total distortion $\bar{\gamma}_{xz}(x,y,z)$ at P_1 in plane xz is given by

$$\bar{\gamma}_{xz}(x,y,z) = \frac{\partial u_x(x,y,z)}{\partial z} + \frac{\partial u_z(x,y,z)}{\partial x}. \tag{8.21}$$

Finally, the total distortion $\bar{\gamma}_{yz}(x,y,z)$ at P_1 according to plane yz is

$$\bar{\gamma}_{yz}(x,y,z) = \frac{\partial u_y(x,y,z)}{\partial z} + \frac{\partial u_z(x,y,z)}{\partial y}. \tag{8.22}$$

The above strain components can be rearranged in the following matrix notation:

$$\left\{ \begin{array}{c} \varepsilon_{xx}(x,y,z) \\ \varepsilon_{yy}(x,y,z) \\ \varepsilon_{zz}(x,y,z) \\ \bar{\gamma}_{xy}(x,y,z) \\ \bar{\gamma}_{xz}(x,y,z) \\ \bar{\gamma}_{yz}(x,y,z) \end{array} \right\} = \begin{bmatrix} \dfrac{\partial}{\partial x} & 0 & 0 \\ 0 & \dfrac{\partial}{\partial y} & 0 \\ 0 & 0 & \dfrac{\partial}{\partial z} \\ \dfrac{\partial}{\partial y} & \dfrac{\partial}{\partial x} & 0 \\ \dfrac{\partial}{\partial z} & 0 & \dfrac{\partial}{\partial x} \\ 0 & \dfrac{\partial}{\partial z} & \dfrac{\partial}{\partial y} \end{bmatrix} \left\{ \begin{array}{c} u_x(x,y,z) \\ u_y(x,y,z) \\ u_z(x,y,z) \end{array} \right\}, \tag{8.23}$$

or,

$$\{\varepsilon\} = [\mathbf{L}]\{\mathbf{u}\}, \tag{8.24}$$

where $[\mathbf{L}]$ is a differential operator, $\{\varepsilon\}$ and $\{\mathbf{u}\}$ respectively are the vectors with strain and displacement components.

Thus, the state of small strains at each point of a solid body is characterized by six components. Notice that the specific strain components ε_{xx}, ε_{yy}, and ε_{zz} are dimensionless, and establish a specific variation relation of the respective displacement components of points along a given direction. In turn, the total distortions $\bar{\gamma}_{xy}$, $\bar{\gamma}_{xz}$, and $\bar{\gamma}_{yz}$ are also dimensionless and represent the angular strains in radians for each point of the solid body.

Finally, we must highlight that the above derivation, as well as continuum mechanics, are fully based on the idea of differential. From Figure 8.2, a comparison was made for the relative kinematics of two arbitrary points P_1 and P_2 of the solid. The distance $\mathbf{d}$ between these points can be taken as small as we want, in such way that the strain measure of point P_1 is considered.

Example 8.2 *Substituting the kinematics of bars from Example 8.1 in the expressions given in equation (8.23), we obtain the following strain components for a bar:*

$$
\begin{aligned}
\varepsilon_{xx}(x) &= \frac{\partial u_x(x)}{\partial x} = \frac{du_x(x)}{dx}, \\
\varepsilon_{yy}(x) &= \frac{\partial u_y(x)}{\partial y} = 0, \\
\varepsilon_{zz}(x) &= \frac{\partial u_z(x)}{\partial z} = 0, \\
\bar{\gamma}_{xy}(x) &= \frac{\partial u_x(x)}{\partial y} + \frac{\partial u_y(x)}{\partial x} = 0, \\
\bar{\gamma}_{xz}(x) &= \frac{\partial u_x(x)}{\partial z} + \frac{\partial u_z(x)}{\partial x} = 0, \\
\bar{\gamma}_{yz}(x) &= \frac{\partial u_y(x)}{\partial z} + \frac{\partial u_z(x)}{\partial y} = 0.
\end{aligned}
$$

For a beam in bending only, we have

$$
\begin{aligned}
\varepsilon_{xx}(x,y) &= \frac{\partial u_x(x,y)}{\partial x} = -y\frac{d\theta_z(x)}{dx}, \\
\varepsilon_{yy}(x,y) &= \frac{\partial u_y(x,y)}{\partial y} = 0, \\
\varepsilon_{zz}(x,y) &= \frac{\partial u_z(x,y)}{\partial z} = 0, \\
\bar{\gamma}_{xy}(x,y) &= \frac{\partial u_x(x,y)}{\partial y} + \frac{\partial u_y(x,y)}{\partial x} = -\theta_z(x) + \frac{du_y(x)}{dx} = 0, \\
\bar{\gamma}_{xz}(x,y) &= \frac{\partial u_x(x,y)}{\partial z} + \frac{\partial u_z(x,y)}{\partial x} = 0, \\
\bar{\gamma}_{yz}(x,y) &= \frac{\partial u_y(x,y)}{\partial z} + \frac{\partial u_z(x,y)}{\partial y} = 0.
\end{aligned}
$$

Note that as the displacement component u_y is constant for all cross-section points, the derivative in y is zero. Besides that, the distortion components $\bar{\gamma}_{xy}$ is zero, consistent with the bending only hypothesis of the Euler-Bernoulli beam. Thus, we recover that ε_{xx} is the only nonzero strain component for bar and bending only beam models.

We have the following strain components for the shaft:

$$\varepsilon_{xx}(x,y,z) = \frac{\partial u_x(x,y,z)}{\partial x} = 0,$$

$$\varepsilon_{yy}(x,y,z) = \frac{\partial u_y(x,y,z)}{\partial y} = 0,$$

$$\varepsilon_{zz}(x,y,z) = \frac{\partial u_z(x,y,z)}{\partial z} = 0,$$

$$\bar{\gamma}_{xy}(x,y,z) = \frac{\partial u_x(x,y,z)}{\partial y} + \frac{\partial u_y(x,y,z)}{\partial x} = -z\frac{d\theta_x(x)}{dx},$$

$$\bar{\gamma}_{xz}(x,y,z) = \frac{\partial u_x(x,y,z)}{\partial z} + \frac{\partial u_z(x,y,z)}{\partial x} = y\frac{d\theta_x(x)}{dx},$$

$$\bar{\gamma}_{yz}(x,y,z) = \frac{\partial u_y(x,y,z)}{\partial z} + \frac{\partial u_z(x,y,z)}{\partial y} = -\theta_x(x) + \theta_x(x) = 0.$$

There are no normal strains, but only distortions and the shaft has a pure shear strain state.

□

8.4   RIGID ACTIONS

If the norms of vectors **d** and **d'** shown in Figure 8.2 are equal, then the body has a rigid displacement action. Recall that for a rigid body the distance between any two points is constant for any kinematic action. This implies that all strain components for all points of the body are zero, that is,

$$\begin{cases} \varepsilon_{xx}(x,y,z) = \dfrac{\partial u_x(x,y,z)}{\partial x} = 0 \\[2mm] \varepsilon_{yy}(x,y,z) = \dfrac{\partial u_y(x,y,z)}{\partial y} = 0 \\[2mm] \varepsilon_{zz}(x,y,z) = \dfrac{\partial u_z(x,y,z)}{\partial z} = 0 \\[2mm] \bar{\gamma}_{xy}(x,y,z) = \dfrac{\partial u_y(x,y,z)}{\partial x} + \dfrac{\partial u_x(x,y,z)}{\partial y} = 0 \\[2mm] \bar{\gamma}_{xz}(x,y,z) = \dfrac{\partial u_x(x,y,z)}{\partial z} + \dfrac{\partial u_z(x,y,z)}{\partial x} = 0 \\[2mm] \bar{\gamma}_{yz}(x,y,z) = \dfrac{\partial u_y(x,y,z)}{\partial z} + \dfrac{\partial u_z(x,y,z)}{\partial y} = 0 \end{cases} . \tag{8.25}$$

If the kinematics $\mathbf{u} = \mathbf{u}_0 = \{u_{x_0}\ u_{y_0}\ u_{z_0}\}^T$ is such that the displacement components u_{x_0}, u_{y_0}, and u_{z_0} are constant for all points of $\mathcal{B}$, that is, we only have a rigid translation, the previous conditions are satisfied.

Now, if the body has infinitesimal rotations about axes x, y, and z, respectively represented by constant angles θ_x, θ_y, and θ_z, the following displacement vector field

$$\begin{aligned} \mathbf{u}(x,y,z) &= \theta \times \mathbf{r} = \det \begin{bmatrix} \mathbf{e}_x & \mathbf{e}_y & \mathbf{e}_z \\ \theta_x & \theta_y & \theta_z \\ x & y & z \end{bmatrix} \\ &= (z\theta_y - y\theta_z)\mathbf{e}_x + (x\theta_z - z\theta_x)\mathbf{e}_y + (y\theta_x - x\theta_y)\mathbf{e}_z, \end{aligned} \tag{8.26}$$

is also rigid, with $u_x = (\theta_y z - \theta_z y)$, $u_y = (\theta_z x - \theta_x z)$, and $u_z = (\theta_x y - \theta_y z)$. Substituting these displacement components in (8.25), the strain components are zero. Notice that the finite rotations are not commutative. Because of this, only infinitesimal rotations should be considered.

The above expression can be written in matrix notation as

$$\{\mathbf{u}(x,y,z)\} = [\mathbf{W}]\{\mathbf{r}\} = \begin{bmatrix} 0 & -\theta_z & \theta_y \\ \theta_z & 0 & -\theta_x \\ -\theta_y & \theta_x & 0 \end{bmatrix} \begin{Bmatrix} x \\ y \\ z \end{Bmatrix} = \begin{Bmatrix} z\theta_y - y\theta_z \\ x\theta_z - z\theta_x \\ y\theta_x - x\theta_y \end{Bmatrix}.$$

$$\tag{8.27}$$

Notice that $[\mathbf{W}]$ is skew-symmetric and $\{\theta\} = \{\theta_x \ \theta_y \ \theta_z\}^T$ is the axial vector associated to $[\mathbf{W}]$, as will be considered later.

A general rigid displacement action is given by the sum of a translation and a rigid rotation as follows:

$$\mathbf{u} = \mathbf{u}_0 + \theta \times \mathbf{r} = \left\{ \begin{array}{c} u_{x_0} \\ u_{y_0} \\ u_{z_0} \end{array} \right\} + \left\{ \begin{array}{c} z\theta_y - y\theta_z \\ x\theta_z - z\theta_x \\ y\theta_x - x\theta_y \end{array} \right\}. \tag{8.28}$$

Example 8.3 *In the case of a bar, the rigid action corresponds to the axial displacement u_{x_0}. For the beam in bending only, the transversal displacement u_{y_0} and rotation θ_z are constant. For circular torsion, the rigid action is described by a constant axial rotation θ_x.*

□

8.5 DETERMINATION OF INTERNAL LOADS

We employ the concept of strain internal work to determine the internal loads, associated to the strain measures compatible with the kinematic actions of the solid.

Thus, associated to the normal strain components ε_{xx}, ε_{yy}, and ε_{zz} for each point of the body, we have the respective normal stresses σ_{xx}, σ_{yy}, and σ_{zz} in directions x, y, and z. Analogous, associated to distortions $\bar{\gamma}_{xy}$, $\bar{\gamma}_{xz}$, and $\bar{\gamma}_{yz}$, we have the respective shear stress components τ_{xy}, τ_{xz}, and τ_{yz}. The normal stresses are responsible for the stretching/shortening in directions x, y, and z. On the other hand, the shear stresses cause distortions in planes xy, xz, and yz.

The strain internal work for a differential volume element dV in the neighborhood of point P_1, illustrated in Figure 8.3(a), is given by

$$dW_i = \left(\sigma_{xx}\varepsilon_{xx} + \sigma_{yy}\varepsilon_{yy} + \sigma_{zz}\varepsilon_{zz} + \tau_{xy}\bar{\gamma}_{xy} + \tau_{xz}\bar{\gamma}_{xz} + \tau_{yz}\bar{\gamma}_{yz} \right) dV.$$

The strain internal work in the solid is obtained by the work summation for each differential element, that is, by the following volume integral:

$$W_i = \int_V dW_i = \int_V \left(\sigma_{xx}\varepsilon_{xx} + \sigma_{yy}\varepsilon_{yy} + \sigma_{zz}\varepsilon_{zz} + \tau_{xy}\bar{\gamma}_{xy} + \tau_{xz}\bar{\gamma}_{xz} + \tau_{yz}\bar{\gamma}_{yz} \right) dV. \tag{8.29}$$

Making a dimensional analysis of the first integrand term of the above expression in the SI, the resultant unit must be equal to the internal work, that is,

$$[\sigma_{xx}\varepsilon_{xx}dV] = \frac{\text{N}}{\text{m}^2}\frac{\text{m}}{\text{m}}\text{m}^3 = \text{Nm}. \tag{8.30}$$

Thus, there must be a continuous function σ_{xx}, associated to the specific normal strain measure ε_{xx}, which represents the normal internal loads in x. It is dependent on the position (x,y,z) of the points of the body and has units of $\dfrac{\text{N}}{\text{m}^2}$. Thus, when performing the integration on the volume of the body, with units of m³, we obtain units of work or energy, i.e., Nm. The function $\sigma_{xx}(x,y,z)$ is called normal stress component in plane x and direction x. In the following equations, the coordinates (x,y,z) are n ot explicitly shown with the purpose of simplifying the notation.

Substituting the strain components given in (8.23) in the internal work expression (8.29), we have

$$\begin{aligned} W_i = {} & \int_V \left[\sigma_{xx}\frac{\partial u_x}{\partial x} + \sigma_{yy}\frac{\partial u_y}{\partial y} + \sigma_{zz}\frac{\partial u_z}{\partial z} + \tau_{xy}\left(\frac{\partial u_x}{\partial y} + \frac{\partial u_y}{\partial x} \right) \right. \\ & \left. + \tau_{xz}\left(\frac{\partial u_x}{\partial z} + \frac{\partial u_z}{\partial x} \right) + \tau_{yz}\left(\frac{\partial u_y}{\partial z} + \frac{\partial u_z}{\partial y} \right) \right] dV. \end{aligned} \tag{8.31}$$

We want to obtain an internal work expression in terms of the displacement components of the body points instead of their derivatives, as in equation (8.31). Considering that the stress and displacement components of equation (8.31) are continuous functions in the entire domain of the body, we can apply the integration by parts to reduce the derivative order of the displacement components. The integration by parts for continuous scalar functions $f(x,y,z)$ and $g(x,y,z)$ is defined as

$$\int_V f \frac{\partial g}{\partial x} dV = -\int_V \frac{\partial f}{\partial x} g dV + \int_S fg n_x dS,$$

$$\int_V f \frac{\partial g}{\partial y} dV = -\int_V \frac{\partial f}{\partial y} g dV + \int_S fg n_y dS, \qquad (8.32)$$

$$\int_V f \frac{\partial g}{\partial z} dV = -\int_V \frac{\partial f}{\partial z} g dV + \int_S fg n_z dS,$$

where $\mathbf{n} = n_x \mathbf{e}_x + n_y \mathbf{e}_y + n_z \mathbf{e}_z$ is the normal vector at each point of S (boundary of V), as illustrated in Figure 8.5, and (n_x, n_y, n_z) are the associated direction cosines.

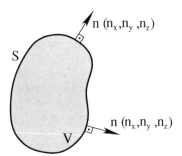

Figure 8.5 Surface S with volume V of a solid and normal vector field.

The integration by parts is applied for each term of the integrand of the internal work expression (8.31). Thus, for instance

$$\int_V \sigma_{xx} \frac{\partial u_x}{\partial x} dV = -\int_V \frac{\partial \sigma_{xx}}{\partial x} u_x dV + \int_S \sigma_{xx} u_x n_x dS,$$

$$\int_V \tau_{yz} \left(\frac{\partial u_y}{\partial z} + \frac{\partial u_z}{\partial y} \right) dV = -\int_V \left(\frac{\partial \tau_{yz}}{\partial z} u_y + \frac{\partial \tau_{yz}}{\partial y} u_z \right) dV + \int_S \left(\tau_{yz} u_y n_z + \tau_{yz} u_z n_y \right) dS.$$

Substituting the obtained expressions after the integration by parts in equation (8.31) and rearranging the terms in volume and surface integrals, we obtain

$$W_i = W_i^V + W_i^S, \qquad (8.33)$$

with

$$W_i^V = -\int_V \left[\left(\frac{\partial \sigma_{xx}}{\partial x} + \frac{\partial \tau_{xy}}{\partial y} + \frac{\partial \tau_{xz}}{\partial z} \right) u_x + \left(\frac{\partial \tau_{xy}}{\partial x} + \frac{\partial \sigma_{yy}}{\partial y} + \frac{\partial \tau_{yz}}{\partial z} \right) u_y \right. $$
$$\left. + \left(\frac{\partial \tau_{xz}}{\partial x} + \frac{\partial \tau_{yz}}{\partial y} + \frac{\partial \sigma_{zz}}{\partial z} \right) u_z \right] dV \qquad (8.34)$$

and

$$W_i^S = \int_S \left[\left(\sigma_{xx} n_x + \tau_{xy} n_y + \tau_{xz} n_z \right) u_x + \left(\tau_{xy} n_x + \sigma_{yy} n_y + \tau_{yz} n_z \right) u_y \right.$$
$$\left. + \left(\tau_{xz} n_x + \tau_{yz} n_y + \sigma_{zz} n_z \right) u_z \right] dS. \qquad (8.35)$$

Making a dimensional analysis in terms of the integrands of W_i^V and W_i^S, we observe that

$$\left[\frac{\partial \sigma_{xx}}{\partial x} \right] = \frac{\text{N}}{\text{m}^2} \frac{1}{\text{m}} = \frac{\text{N}}{\text{m}^3}, \qquad (8.36)$$

$$\left[\sigma_{xx} n_x \right] = \frac{\text{N}}{\text{m}^2}. \qquad (8.37)$$

The term $\dfrac{\partial \sigma_{xx}}{\partial x}$ represents an internal force density by units of volume, also known as internal body force or volume force. Furthermore, $\sigma_{xx}n_x$ represents the distributed internal force in the surface of the solid, also known as internal surface force. Thus, the internal forces compatible with the strain state of a solid are given by a force density by units of volume, present in all points of the body, and a force density by units of area, present in the the surface. Thus, W_i^V and W_i^S respectively represent the strain internal works of the volume and surface forces.

Example 8.4 *As there is only the normal strain component ε_{xx} for the bar, the internal work expression (8.33) is reduced to*

$$W_i = -\int_V \frac{d\sigma_{xx}(x)}{dx}u_x(x)dV + \int_S \sigma_{xx}(x)n_x(x)u_x(x)dS.$$

Making $dV = dSdx$, the term of the volume integral is rewritten as

$$\int_V \frac{d\sigma_{xx}(x)}{dx}u_x(x)dV = \int_0^L \frac{d}{dx}\left(\int_S \sigma_{xx}(x)dS\right)u_x(x)dx = \int_0^L \frac{dN_x(x)}{dx}u_x(x)dx,$$

where $N_x(x) = \int_S \sigma_{xx}(x)dS$ is the normal force in section x.
The surface integral can be written as

$$\int_S \sigma_{xx}(x)n_x(x)u_x(x)dS = \left(\int_S \sigma_{xx}(x)dS\right)n_x(x)u_x(x) = N_x(x)u_x(x)n_x(x).$$

The boundaries of the bar consist of only two sections at ends $x = 0$ and $x = L$, with normal vectors $\mathbf{n}(0) = (-1,0,0)$ and $\mathbf{n}(L) = (1,0,0)$, respectively. Thus, the surface integral is reduced to a sum of the previous expression at these two sections

$$\int_S \sigma_{xx}(x)n_x(x)u_x(x)dS = -N_x(0)u_x(0) + N_x(L)u_x(L).$$

Thus, the final expression for the strain internal work of the bar is

$$W_i = -\int_0^L \frac{dN_x(x)}{dx}u_x(x)dx + N_x(L)u_x(L) - N_x(0)u_x(0).$$

□

Example 8.5 *The strain internal work expression is the same for bars and beams in bending only. The volume integral can be written, after the substitution of $u_x(x,y) = -y\theta_z(x)$, as*

$$\int_V \frac{\partial \sigma_{xx}(x,y)}{\partial x}u_x(x,y)dV = \int_0^L \frac{d}{dx}\left[-\int_S y\sigma_{xx}(x,y)dS\right]\theta_z(x)dx.$$

The term inside brackets represents the bending moment $M_z(x)$ at section x. Thus,

$$\int_V \frac{d\sigma_{xx}(x,y)}{dx}u_x(x,y)dV = \int_0^L \frac{dM_z(x)}{dx}\theta_z(x)dx.$$

Furthermore, the surface integral can be expressed as

$$\int_S \sigma_{xx}(x,y)n_x(x)u_x(x,y)dS = \left(-\int_S y\sigma_{xx}(x)dS\right)n_x(x)\theta_z(x) = M_z(x)n_x(x)\theta_z(x).$$

Substituting $x = 0$ and $x = L$ in the above expression, we have

$$\int_S \sigma_{xx}(x,y)n_x(x)u_x(x,y)dS = M_z(L)\theta_z(L) - M_z(0)\theta_z(0).$$

Hence, the final expression for the strain internal work is

$$W_i = -\int_0^L \frac{dM_z(x)}{dx}\theta_z(x)dx + M_z(L)\theta_z(L) - M_z(0)\theta_z(0).$$

Substituting the definition of $\theta_z(x) = \dfrac{du_y(x)}{dx}$ *and integrating by parts, we obtain*

$$W_i = \int_0^L \frac{d^2 M_z(x)}{dx^2} u_y(x) dx + M_z(L)\theta_z(L) - M_z(0)\theta_z(0) - V_y(L)u_y(L) + V_y(0)u_y(0).$$

☐

Example 8.6 *In the case of circular torsion, the internal work expression is reduced to*

$$W_i = -\int_V \left(\frac{\partial \tau_{xy}}{\partial x} u_y + \frac{\partial \tau_{xz}}{\partial x} u_z \right) dV + \int_S \left(\tau_{xy}u_y + \tau_{xz}u_z \right) n_x dS.$$

Substituting the displacement components of the kinematics of circular torsion, the volume integral is rewritten as

$$\int_V \left(\frac{\partial \tau_{xy}}{\partial x} u_y + \frac{\partial \tau_{xz}}{\partial x} u_z \right) dV = \int_0^L \frac{d}{dx} \left[\int_S \left(-\tau_{xy}z + \tau_{xz}y \right) dS \right] \theta_x(x) dx.$$

The term inside brackets represents the twisting moment $M_x(x)$ *at section x. Thus,*

$$\int_V \left(\frac{\partial \tau_{xy}}{\partial x} u_y + \frac{\partial \tau_{xz}}{\partial x} u_z \right) dV = \int_0^L \frac{dM_x(x)}{dx} \theta_x(x) dx.$$

The surface integral is expressed as

$$\int_S \left(\tau_{xy}u_y + \tau_{xz}u_z \right) n_x dS = \left[\int_S \left(-\tau_{xy}z + \tau_{xz}y \right) dS \right] n_x(x)\theta_x(x) = M_x(x)n_x(x)\theta_x(x).$$

For x = 0 and x = L, we have

$$\int_S \left(\tau_{xy}u_y + \tau_{xz}u_z \right) n_x dS = -M_x(0)\theta_x(0) + M_x(L)\theta_x(L).$$

Hence, the strain internal work expression for circular torsion is

$$W_i = -\int_0^L \frac{dM_x(x)}{dx} \theta_x(x) dx + M_x(L)\theta_x(L) - M_x(0)\theta_x(0).$$

Thus, when introducing the kinematic and strain measure components in the internal work expression (8.33) of a solid body, we recover the previously obtained strain internal work expressions for bars, beams in bending only, and shafts of the previous chapters.

☐

8.6   DETERMINATION OF EXTERNAL LOADS

In this section we determine the external loads, which are compatible with the internal ones and consequently with the kinematics of the solid for small displacements and strains.

From the analysis of expressions (8.34) and (8.35), the vector fields $\mathbf{b} = (b_x, b_y, b_z)$ and $\mathbf{t} = (t_x, t_y, t_z)$ are defined, which respectively characterize the external force densities per unit volume and unit surface area of the solid. Thus, the expression for the external work of a solid is given by

$$
\begin{aligned}
W_e &= W_e^V + W_e^S \\
&= \int_V \mathbf{b}(x,y,z) \cdot \mathbf{u}(x,y,z) dV + \int_S \mathbf{t}(x,y,z) \cdot \mathbf{u}(x,y,z) dS \\
&= \int_V \left(b_x(x,y,z)u_x(x,y,z) + b_y(x,y,z)u_y(x,y,z) + b_z(x,y,z)u_z(x,y,z) \right) dV \\
&\quad + \int_S \left(t_x(x,y,z)u_x(x,y,z) + t_y(x,y,z)u_y(x,y,z) + t_z(x,y,z)u_z(x,y,z) \right) dS.
\end{aligned}
\tag{8.38}
$$

Example 8.7 *Substituting the bar kinematics, the external work expression (8.38) is reduced to*

$$W_e = \int_V b_x(x)u_x(x)dV + \int_S t_x(x)u_x(x)dS.$$

Using $dV = A(x)dx$, where $A(x)$ is the area of cross-section x of the bar, the volume integral is rewritten as the following integral along the length of the bar:

$$\int_V b_x(x)u_x(x)dV = \int_0^L b_x(x)A(x)u_x(x)dx = \int_0^L q_x(x)u_x(x)dx.$$

The term $b_x(x)A(x)$ corresponds to the distributed load $q_x(x)$ along the length of the bar.
  On the other hand, the surface integral is written as

$$\int_S t_x(x)u_x(x)dS = \left(\int_S t_x(x)dS \right) u_x(x).$$

The integration of the force density t_x results in an external axial force P_x in the cross-section, as illustrated in Figure 8.6. Thus, for $x = 0$ and $x = L$, we have

$$\int_S t_x(x)u_x(x)dS = t_x(0)u_x(0)A(0) + t_x(L)u_x(L)A(L) = P_0 u_x(0) + P_L u_x(L).$$

Thus, the final expression for the work done by external forces in the bar is

$$W_e = \int_0^L q_x(x)u_x(x)dx + P_0 u_x(0) + P_L u_x(L).$$

Figure 8.6 Surface traction at the bar ends.

☐

Example 8.8 *The external work expression (8.38) in the case of circular torsion is reduced to*

$$\begin{aligned} W_e &= \int_V \left(b_y(x,y,z)u_y(x,y,z) + b_z(x,y,z)u_z(x,y,z) \right) dV \\ &+ \int_S \left(t_y(x,y,z)u_y(x,y,z) + t_z(x,y,z)u_z(x,y,z) \right) dS. \end{aligned}$$

After substituting the displacement components u_y and u_z of the circular torsion and $dV = dAdx$, the volume integral is rewritten as

$$\begin{aligned} \int_V \left(b_y(x,y,z)u_y(x,y,z) + b_z(x,y,z)u_z(x,y,z) \right) dV &= \int_0^L \left(\int_S (-zb_y + yb_z)dA \right) \theta_x(x)dx \\ &= \int_0^L m_x(x)\theta_x(x)dx, \end{aligned}$$

where $m_x(x)$ is the external distributed torque applied in the shaft, as illustrated in Figure 8.7(a) for an area element at section x.
  The surface integral reduces to

$$\begin{aligned} \int_S \left(t_y(x,y,z)u_y(x,y,z) + t_z(x,y,z)u_z(x,y,z) \right) dS &= \left[\int_S (-zt_y + yt_z)dA \right] \theta_x(x) \Big|_0^L \\ &= T_0 \theta_x(0) + T_L \theta_x(L), \end{aligned}$$

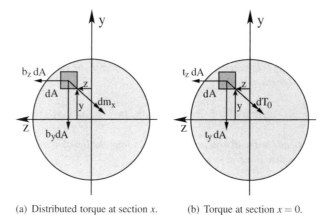

(a) Distributed torque at section x.   (b) Torque at section $x = 0$.

Figure 8.7   Distributed and concentrated torques in area elements of sections x and $x = 0$.

with T_0 and T_L the concentrated torques at the shaft ends, as illustrated in Figure 8.7(b) for an area element at section $x = 0$. Hence, the final expression for the work done by external loads is

$$W_e = \int_0^L m_x(x)\theta_x(x)dx + T_L\theta_x(L) + T_0\theta_x(0).$$

□

Example 8.9 *In the case of a beam in bending only, the external work given in (8.38) reduces to*

$$W_e = \int_V \left(b_x(x,y)u_x(x,y) + b_y(x)u_y(x)\right)dV + \int_S \left(t_x(x,y)u_x(x,y) + t_y(x)u_y(x)\right)dS.$$

The body force component b_x is zero, because there are no forces acting in x for a beam under bending only. The remaining term in the volume integral can be written as

$$\int_V b_y(x)u_y(x)dV = \int_0^L \left(\int_S b_y dA\right)u_y(x)dx = \int_0^L q_y(x)u_y(x)dx.$$

The surface integral inside parentheses represents the distributed load $q_y(x)$ in direction y along the length of the beam, as illustrated in Figure 8.8(a) for the area element at a generic x section.

Furthermore, the surface integral in the external work expression can be rewritten as

$$\begin{aligned}
\int_S \left(t_x(x,y)u_x(x,y) + t_y(x)u_y(x)\right)dS &= \left(-\int_A t_x y\,dS\right)\theta_z(x) + \left(\int_A t_y dA\right)u_y(x) \\
&= M_z(x)\theta_z(x) + V_y(x)u_y(x).
\end{aligned}$$

The surface integrals represent the moment in z generated by the surface force component t_x and the resultant of the transversal force due to t_y, respectively, as illustrated in Figure 8.8(b). Applying the previous expression at $x = 0$ and $x = L$ and substituting, along with the resulting term of the volume integral in the external work expression, we have

$$W_e = \int_0^L q_y(x)u_y(x)dx + M_{z_0}\theta_z(0) + M_{z_L}\theta_z(L) + V_{y_0}(0)u_y(0) + V_{y_L}(L)u_y(L).$$

Analogous to the internal work, when introducing the kinematic components in expression (8.38) of the external work of a solid, we recover the previously obtained external work expressions for bars, shafts, and beams under bending only of the previous chapters.

□

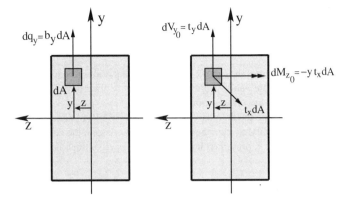

(a) Distributed force at section x.

(b) Transversal force and bending moment at section $x = 0$.

Figure 8.8 Distributed force and bending moment in area elements of sections x and $x = 0$.

8.7 EQUILIBRIUM

The PVW states that the solid is in equilibrium in the deformed configuration if the works done by external and internal loads are equal for any virtual action $\delta\mathbf{u} = \{\delta u_x \ \delta u_y \ \delta u_z\}^T$. Thus,

$$\delta W_e = \delta W_i. \tag{8.39}$$

In terms of the volume and surface integrals given in (8.34), (8.35), and (8.38), we have

$$\delta W_e^V + \delta W_e^S = \delta W_i^V + \delta W_i^S, \tag{8.40}$$

that is,

$$\delta W_e^V = \delta W_i^V, \tag{8.41}$$

$$\delta W_e^S = \delta W_i^S. \tag{8.42}$$

Substituting the expressions of the work done by the internal volume loads (8.34), internal surface loads (8.35), and the work done by the external volume and surface loads given in (8.38), respectively, in (8.41) and (8.42), we have

$$\int_V \left[\left(\frac{\partial \sigma_{xx}}{\partial x} + \frac{\partial \tau_{xy}}{\partial y} + \frac{\partial \tau_{xz}}{\partial z} + b_x \right) \delta u_x + \left(\frac{\partial \tau_{xy}}{\partial x} + \frac{\partial \sigma_{yy}}{\partial y} + \frac{\partial \tau_{yz}}{\partial z} + b_y \right) \delta u_y + \right.$$
$$\left. \left(\frac{\partial \tau_{xz}}{\partial x} + \frac{\partial \tau_{yz}}{\partial y} + \frac{\partial \sigma_{zz}}{\partial z} + b_z \right) \delta u_z \right] dV = 0, \tag{8.43}$$

$$\int_S \left[\left(\sigma_{xx} n_x + \tau_{xy} n_y + \tau_{xz} n_z - t_x \right) \delta u_x + \left(\tau_{xy} n_x + \sigma_{yy} n_y + \tau_{yz} n_z - t_y \right) \delta u_y + \right.$$
$$\left. \left(\tau_{xz} n_x + \tau_{yz} n_y + \sigma_{zz} n_z - t_z \right) \delta u_z \right] dS = 0. \tag{8.44}$$

As $\delta\mathbf{u} = \{\delta u_x \ \delta u_y \ \delta u_z\}^T$ is an arbitrary virtual displacement action, compatible with the kinematics, we conclude that the previous equations will only be satisfied when the terms inside parentheses are simultaneously zero. This results in the following differential equations

$$\begin{cases} \dfrac{\partial \sigma_{xx}}{\partial x} + \dfrac{\partial \tau_{xy}}{\partial y} + \dfrac{\partial \tau_{xz}}{\partial z} + b_x = 0 \\[2mm] \dfrac{\partial \tau_{xy}}{\partial x} + \dfrac{\partial \sigma_{yy}}{\partial y} + \dfrac{\partial \tau_{yz}}{\partial z} + b_y = 0 \\[2mm] \dfrac{\partial \tau_{xz}}{\partial x} + \dfrac{\partial \tau_{yz}}{\partial y} + \dfrac{\partial \sigma_{zz}}{\partial z} + b_z = 0 \end{cases} \tag{8.45}$$

and boundary conditions

$$\begin{cases} \sigma_{xx}n_x + \tau_{xy}n_y + \tau_{xz}n_z - t_x = 0 \\ \tau_{xy}n_x + \sigma_{yy}n_y + \tau_{yz}n_z - t_y = 0 \\ \tau_{xz}n_x + \tau_{yz}n_y + \sigma_{zz}n_z - t_z = 0 \end{cases} . \tag{8.46}$$

The set of equations in (8.45) defines the system of differential equations of equilibrium between the internal and external volume loads, which are valid for the entire domain of the body. The set of equations in (8.46) defines the boundary conditions in the surface of the body.

Equations (8.45) and (8.46) define the boundary value problem (BVP) of equilibrium of solids under small strains and displacements. No simplifying hypothesis is introduced despite the continuity of the possible kinematic actions. Hence, the formulation presented here is valid for any continuous medium, independent of the material.

Example 8.10 *To obtain the equilibrium BVPs of a bar, shaft, and beam under bending only, we can equate the internal and external work expressions obtained in the previous examples for a compatible virtual action with the kinematics of these models. Another possibility is to directly use the differential equation of equilibrium and boundary conditions given in (8.45) and (8.46), as illustrated below.*

The only nonzero stress component of the bar is $\sigma_{xx}(x)$. Thus, equations (8.45) and (8.46) respectively reduce to

$$\frac{d\sigma_{xx}(x)}{dx} = -b_x(x),$$

$$\sigma_{xx}(x)n_x(x) = t_x(x).$$

For a bar with constant cross-section, the normal stress in terms of the normal force is given by

$$\sigma_{xx}(x) = \frac{N_x(x)}{A}.$$

Thus, substituting it in the above expressions, we have

$$\frac{dN_x(x)}{dx} = -Ab_x(x) = -q_x(x),$$

$$N_x(x)n_x(x) = At_x(x) = P_x.$$

Substituting $x = 0$ and $x = L$ in the above expression, we have the boundary conditions given in (3.20). Thus, the equilibrium BVP of the bar is recovered

$$\begin{cases} \dfrac{dN_x(x)}{dx} + q_x(x) = 0 & in\ x \in (0,L) \\ N_x(L) = P_L & in\ x = L \\ N_x(0) = -P_0 & in\ x = 0 \end{cases} .$$

□

Example 8.11 *For the case of circular torsion, the differential equations of equilibrium (8.45) reduce to*

$$\frac{\partial \tau_{xy}}{\partial x} + b_y = 0$$
$$\frac{\partial \tau_{xz}}{\partial x} + b_z = 0$$.

Multiplying the two above expressions, respectively by z and y, and taking the difference between the first and second expressions, we have

$$\frac{\partial}{\partial x}(\tau_{xz}y - \tau_{xy}z) + (b_z y - b_y z) = 0.$$

Integrating the above expression along the length and cross-section area, we obtain

$$\int_0^L \frac{\partial}{\partial x} \underbrace{\left(\int_A (\tau_{xz}y - \tau_{xy}z)dA \right)}_{M_x(x)} dx + \int_0^L \underbrace{\left(\int_A (b_z y - b_y z)dA \right)}_{m_x(x)} dx = 0.$$

The area integrals respectively represent the twisting moment and external distributed torque generated by the body force components b_y and b_z. Thus, the above expression can be rewritten as

$$\int_0^L \left(\frac{dM_x(x)}{dx} + m_x(x) \right) dx = 0,$$

giving rise to the differential equation of equilibrium of the circular torsion problem

$$\frac{dM_x(x)}{dx} = -m_x(x).$$

The boundary conditions for the circular torsion are obtained analogously.

□

Example 8.12 *The differential equations of equilibrium (8.45) for a beam in bending only are reduced to*

$$\frac{\partial \sigma_{xx}}{\partial x} + \frac{\partial \tau_{xy}}{\partial y} = 0$$
$$\frac{\partial \tau_{xy}}{\partial x} + b_y = 0$$.

The body force component b_x is zero, because only the case of bending is considered.

Integrating the second differential equation in the volume of the beam, we have

$$\int_0^L \frac{\partial}{\partial x} \underbrace{\left(\int_A \tau_{xy} dA \right)}_{-V_y(x)} dx + \int_0^L \underbrace{\left(\int_A b_y(x) dA \right)}_{q_y(x)} dx = 0.$$

The first area integral defines the shear force at section x. The second area integral results in the distributed load by units of length $q_y(x)$. These forces are illustrated in Figure 8.9 for a cross-section area element. Thus, the above equation results in

$$\int_0^L \left(-\frac{dV_y(x)}{dx} + q_y(x) \right) dx = 0.$$

To satisfy it, we have

$$\frac{dV_y(x)}{dx} = q_y(x).$$

The shear stress in the region of the cross-section which is y units distant from z, illustrated in Figure 8.9(a), is obtained with the integration of the first above differential equation in direction y as

$$\int_0^y \frac{\partial \tau_{xy}}{\partial y} dy = -\int_0^y \frac{\partial \sigma_{xx}}{\partial x} dy,$$

resulting in

$$\tau_{xy}(x, y) = -\frac{\partial \sigma_{xx}(x)}{\partial x} y.$$

Now, with the integration of the above expression over the cross-section area and length of the beam, we have

$$\int_0^L \underbrace{\left(\int_A \tau_{xy} dA \right)}_{-V_y(x)} dx + \int_0^L \frac{\partial}{\partial x} \underbrace{\left(\int_A \sigma_{xx}(x) y dA \right)}_{M_z(x)} dx = 0.$$

The two area integrals are the shear force $V_y(x)$ and bending moment $M_z(x)$ at cross-section x. Thus, the above expression results in a differential equation relating both forces

$$V_y(x) = \frac{dM_z(x)}{dx}.$$

Combining the obtained differential equations, we have

$$\frac{d^2 M_z(x)}{dx^2} = q_y(x).$$

The boundary conditions for the beam under bending only are obtained in an analogous way.

□

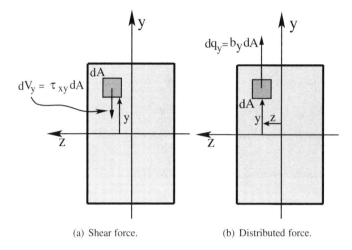

(a) Shear force.   (b) Distributed force.

Figure 8.9 Distributed and shear forces in cross-sectional area elements of a beam.

8.8 GENERALIZED HOOKE'S LAW

Until now we established the concepts regarding stress and strain measures, which are applicable to any material in equilibrium that satisfies the continuum media hypothesis with small displacements and strains.

In this section, we consider a generalization of Hooke's law as seen in the previous chapters, for a solid body of elastic, linear, homogeneous, and isotropic material. The term elastic means that the material returns to its original shape after removing the external applied load, i.e., there are no permanent deformations. Linear means that the relationship between stress and strain measures is given by a linear equation. Thus, an increase in stress causes a proportional increase in strains. Homogeneous indicates that the material properties are equal for all parts of the body. Isotropic means that the mechanical properties of the material are the same regardless of direction. A class of materials which meets this law for the elastic range is the class of metallic materials (steel, aluminum, copper, etc.) at room temperature.

Consider the stress states in an element cube, around a point P of a solid body, illustrated in Figure 8.10. As seen earlier for uniaxial stress states, i.e., normal stresses in one direction, there is a range where the stress-strain relationship has a linear elastic behavior. Thus, for cases of pure traction in x, y, and z, illustrated in Figures 8.10(a) to 8.10(c), Hooke's law establishes the following relationships between the stress and strain measures:

$$\sigma_{xx} = E\varepsilon_{xx} \rightarrow \varepsilon_{xx} = \frac{\sigma_{xx}}{E}, \tag{8.47}$$

$$\sigma_{yy} = E\varepsilon_{yy} \rightarrow \varepsilon_{yy} = \frac{\sigma_{yy}}{E}, \tag{8.48}$$

$$\sigma_{zz} = E\varepsilon_{zz} \rightarrow \varepsilon_{zz} = \frac{\sigma_{zz}}{E}, \tag{8.49}$$

where E is the longitudinal elastic modulus, or Young's modulus, representing the elastic behavior of a material when subjected to uniaxial loading. The behavior of an isotropic material is the same for all directions.

In the case of uniaxial loading, there are strains in the transversal direction to the loading. Considering an elongation ε_{xx} in direction x, there are shortenings in the transversal directions y and z, which are proportional to the stretch along x. For instance, for a bar with circular cross-section being pulled in direction x, there is a diameter reduction and similarly, an increase in compression. Thus, in the case of uniaxial stress state in x, we have the following relations for normal strain components in the transversal directions y and z:

$$\varepsilon_{yy} = \varepsilon_{zz} = -\nu\varepsilon_{xx} \rightarrow \varepsilon_{yy} = \varepsilon_{zz} = -\frac{\nu}{E}\sigma_{xx}. \tag{8.50}$$

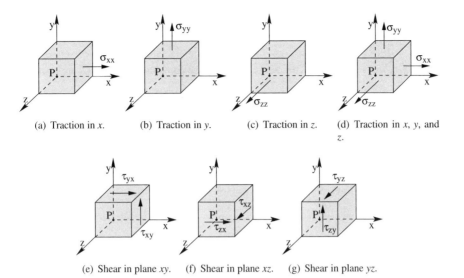

(a) Traction in x.   (b) Traction in y.   (c) Traction in z.   (d) Traction in x, y, and z.

(e) Shear in plane xy.   (f) Shear in plane xz.   (g) Shear in plane yz.

Figure 8.10   Stress states in a point of a solid.

Analogous for the other directions, considering the material isotropy, we have

$$\varepsilon_{xx} = \varepsilon_{zz} = -\nu\varepsilon_{yy} \rightarrow \varepsilon_{xx} = \varepsilon_{zz} = -\frac{\nu}{E}\sigma_{yy}, \tag{8.51}$$

$$\varepsilon_{xx} = \varepsilon_{yy} = -\nu\varepsilon_{zz} \rightarrow \varepsilon_{xx} = \varepsilon_{yy} = -\frac{\nu}{E}\sigma_{zz}, \tag{8.52}$$

where ν is the Poisson ratio. The sign $-$ in equations (8.50) to (8.52) is only employed to represent the observed physical phenomenon, that is, the transversal strains are negative for a positive normal strain.

For three-dimensional stress states (normal stresses simultaneously in directions x, y, and z), as shown in Figure 8.10(d), there is a superposition of loading effects for each direction. Thus, taking the sum of the effects indicated in equations (8.47) to (8.52) we have

$$\varepsilon_{xx} = \frac{\sigma_{xx}}{E} - \frac{\nu}{E}\sigma_{yy} - \frac{\nu}{E}\sigma_{zz} = \frac{1}{E}[\sigma_{xx} - \nu(\sigma_{yy} + \sigma_{zz})], \tag{8.53}$$

$$\varepsilon_{yy} = \frac{\sigma_{yy}}{E} - \frac{\nu}{E}\sigma_{xx} - \frac{\nu}{E}\sigma_{zz} = \frac{1}{E}[\sigma_{yy} - \nu(\sigma_{xx} + \sigma_{zz})], \tag{8.54}$$

$$\varepsilon_{zz} = \frac{\sigma_{zz}}{E} - \frac{\nu}{E}\sigma_{yy} - \frac{\nu}{E}\sigma_{xx} = \frac{1}{E}[\sigma_{zz} - \nu(\sigma_{yy} + \sigma_{xx})]. \tag{8.55}$$

Analogously, considering only the pure shear cases shown in Figures 8.10(e) to 8.10(g), we verify that

$$\tau_{xy} = G\bar{\gamma}_{xy} = \frac{E}{2(1+\nu)}\bar{\gamma}_{xy} \rightarrow \bar{\gamma}_{xy} = \frac{2(1+\nu)}{E}\tau_{xy}, \tag{8.56}$$

$$\tau_{xz} = G\bar{\gamma}_{xz} = \frac{E}{2(1+\nu)}\bar{\gamma}_{xz} \rightarrow \bar{\gamma}_{xz} = \frac{2(1+\nu)}{E}\tau_{xz}, \tag{8.57}$$

$$\tau_{yz} = G\bar{\gamma}_{yz} = \frac{E}{2(1+\nu)}\bar{\gamma}_{yz} \rightarrow \bar{\gamma}_{yz} = \frac{2(1+\nu)}{E}\tau_{yz}, \tag{8.58}$$

where G is the shear modulus, which is related to the longitudinal elastic modulus E and Poisson's ratio ν as

$$G = \frac{E}{2(1+\nu)}. \tag{8.59}$$

Notice from equations (8.56) to (8.58) that the shear effect in a given plane does not give rise to distortions in the others. Thus, τ_{xy}, τ_{xz}, and τ_{yz} are independent stress components.

We can write relations (8.53) to (8.58) in the following matrix notation:

$$\{\varepsilon\} = [\mathbf{C}]\{\sigma\}, \tag{8.60}$$

where the strain $\{\varepsilon\}$ and stress $\{\sigma\}$ vector components and the compliance matrix $[\mathbf{C}]$ are respectively given by

$$\{\varepsilon\} = \left\{\begin{array}{cccccc} \varepsilon_{xx} & \varepsilon_{yy} & \varepsilon_{zz} & \bar{\gamma}_{xy} & \bar{\gamma}_{xz} & \bar{\gamma}_{yz} \end{array}\right\}^T, \tag{8.61}$$

$$\{\sigma\} = \left\{\begin{array}{cccccc} \sigma_{xx} & \sigma_{yy} & \sigma_{zz} & \tau_{xy} & \tau_{xz} & \tau_{yz} \end{array}\right\}^T, \tag{8.62}$$

$$[\mathbf{C}] = \frac{1}{E}\begin{bmatrix} 1 & -v & -v & 0 & 0 & 0 \\ -v & 1 & -v & 0 & 0 & 0 \\ -v & -v & 1 & 0 & 0 & 0 \\ 0 & 0 & 0 & 2(1+v) & 0 & 0 \\ 0 & 0 & 0 & 0 & 2(1+v) & 0 \\ 0 & 0 & 0 & 0 & 0 & 2(1+v) \end{bmatrix}. \tag{8.63}$$

We can express the stress components in terms of the strain components as

$$\{\sigma\} = [\mathbf{D}]\{\varepsilon\}, \tag{8.64}$$

with the elasticity matrix $[\mathbf{D}]$ given by

$$[\mathbf{D}] = [\mathbf{C}]^{-1} = \frac{E}{(1+v)(1-2v)}\begin{bmatrix} 1-v & v & v & 0 & 0 & 0 \\ v & 1-v & v & 0 & 0 & 0 \\ v & v & 1-v & 0 & 0 & 0 \\ 0 & 0 & 0 & \frac{1-2v}{2} & 0 & 0 \\ 0 & 0 & 0 & 0 & \frac{1-2v}{2} & 0 \\ 0 & 0 & 0 & 0 & 0 & \frac{1-2v}{2} \end{bmatrix}. \tag{8.65}$$

Expanding the previous expression for σ_{xx}, we have

$$\sigma_{xx} = \frac{E}{(1+v)(1-2v)}\varepsilon_{xx} + \frac{Ev}{(1+v)(1-2v)}(\varepsilon_{yy} + \varepsilon_{zz}). \tag{8.66}$$

Taking the sum and subtraction of the term $\dfrac{Ev}{(1+v)(1-2v)}\varepsilon_{xx}$ in the right-hand side of equation (8.66) and rearranging, we obtain

$$\sigma_{xx} = \frac{E}{(1+v)}\varepsilon_{xx} + \frac{Ev}{(1+v)(1-2v)}(\varepsilon_{xx} + \varepsilon_{yy} + \varepsilon_{zz}) = 2\mu\varepsilon_{xx} + \lambda e, \tag{8.67}$$

where μ and λ are the Lam coefficients given by

$$\mu = \frac{E}{2(1+v)}, \tag{8.68}$$

$$\lambda = \frac{Ev}{(1+v)(1-2v)}. \tag{8.69}$$

The term e is the dilatation, that is, the sum of the normal strain components

$$e = \varepsilon_{xx} + \varepsilon_{yy} + \varepsilon_{zz}.$$

The dilatation e represents the specific volumetric strain, that is, the volume variation per unit volume for each point of the solid (see Section 3.12 for a uniaxial stress state).

Performing the same procedure for the other normal stress components, we have the expressions for the generalized Hooke's law for an elastic, linear, homogeneous, and isotropic material in terms of the Lam coefficients, that is,

$$
\begin{aligned}
\sigma_{xx} &= 2\mu\varepsilon_{xx} + \lambda e, \\
\sigma_{yy} &= 2\mu\varepsilon_{yy} + \lambda e, \\
\sigma_{zz} &= 2\mu\varepsilon_{zz} + \lambda e, \\
\tau_{xy} &= \mu\bar{\gamma}_{xy}, \\
\tau_{xz} &= \mu\bar{\gamma}_{xz}, \\
\tau_{yz} &= \mu\bar{\gamma}_{yz}.
\end{aligned}
\tag{8.70}
$$

Summing the first three expressions above, we obtain

$$
\bar{\sigma} = (2\mu + 3\lambda)e,
\tag{8.71}
$$

where

$$
\bar{\sigma} = \sigma_{xx} + \sigma_{yy} + \sigma_{zz}.
\tag{8.72}
$$

Equation (8.71) can be used to denote the dilatation in terms of the normal stress components as

$$
e = \frac{\bar{\sigma}}{2\mu + 3\lambda}.
\tag{8.73}
$$

On the other hand, the above relation is employed to express the strain components in terms of the stress components in (8.70) as

$$
\begin{aligned}
\varepsilon_{xx} &= \frac{1}{2\mu}\left(\sigma_{xx} - \frac{\lambda}{2\mu + 3\lambda}\bar{\sigma}\right), \\
\varepsilon_{yy} &= \frac{1}{2\mu}\left(\sigma_{yy} - \frac{\lambda}{2\mu + 3\lambda}\bar{\sigma}\right), \\
\varepsilon_{zz} &= \frac{1}{2\mu}\left(\sigma_{zz} - \frac{\lambda}{2\mu + 3\lambda}\bar{\sigma}\right), \\
\bar{\gamma}_{xy} &= \frac{1}{\mu}\tau_{xy}, \\
\bar{\gamma}_{xz} &= \frac{1}{\mu}\tau_{xz}, \\
\bar{\gamma}_{yz} &= \frac{1}{\mu}\tau_{yz}.
\end{aligned}
\tag{8.74}
$$

The previous equations may be rewritten in terms of the material properties as

$$
\begin{aligned}
\varepsilon_{xx} &= \frac{(1+v)}{E}\sigma_{xx} - \frac{v}{E}\bar{\sigma}, \\
\varepsilon_{yy} &= \frac{(1+v)}{E}\sigma_{yy} - \frac{v}{E}\bar{\sigma}, \\
\varepsilon_{zz} &= \frac{(1+v)}{E}\sigma_{zz} - \frac{v}{E}\bar{\sigma}, \\
\bar{\gamma}_{xy} &= \frac{\tau_{xy}}{G}, \\
\bar{\gamma}_{xz} &= \frac{\tau_{xz}}{G}, \\
\bar{\gamma}_{yz} &= \frac{\tau_{yz}}{G}.
\end{aligned}
\tag{8.75}
$$

The hydrostatic stress state is the one where all shear stress components are zero and the normal stress components are equal, that is, $\sigma_{xx} = \sigma_{yy} = \sigma_{zz} = \sigma$. The bulk modulus of the material is defined as

$$
k = \frac{\sigma}{e},
\tag{8.76}
$$

where e is the dilatation. We can measure the bulk modulus by experiment, and the value for steel is 138 GPa.

For a hydrostatic state, equation (8.71) simplifies to

$$\sigma = \left(\frac{2}{3}\mu + \lambda\right) e.$$

Consequently, we can write the bulk modulus in term of the Lam coefficients as

$$k = \frac{\sigma}{e} = \frac{2}{3}\mu + \lambda. \tag{8.77}$$

A material is incompressible when there is no volume change for any stress state. Taking the sum of the normal strain components given in (8.53) to (8.55), we have

$$e = \frac{1 - 2v}{E}\bar{\sigma}.$$

As e represents the volumetric strain, it must be zero for an incompressible material. The condition for this is $v = \frac{1}{2}$.

Substituting $v = \frac{1}{2}$ in (8.59), we obtain $G = \mu = \frac{E}{3}$ for an incompressible material. Besides that, when the dilatation e tends to zero, the bulk modulus goes to infinity, and so does λ. However, we can observe from (8.77) that the difference $k - \lambda = \frac{2}{3}\mu$ is finite.

8.9   APPLICATION OF THE CONSTITUTIVE EQUATION

The expressions in (8.70) provide the stress components at each point of the solid in terms of the respective strain components for a linear, elastic, homogeneous, and isotropic material. Substituting these relations in the differential equations of equilibrium (8.45), we obtain the equilibrium conditions in terms of displacement components given in (8.4).

For the first equation in (8.45) we have

$$\frac{\partial}{\partial x}(2\mu\varepsilon_{xx} + \lambda e) + \frac{\partial}{\partial y}(\mu\bar{\gamma}_{xy}) + \frac{\partial}{\partial z}(\mu\bar{\gamma}_{xz}) + b_x = 0.$$

Assuming constant Lam coefficients and rearranging the terms, we have

$$\lambda\frac{\partial e}{\partial x} + 2\mu\frac{\partial\varepsilon_{xx}}{\partial x} + \mu\frac{\partial}{\partial y}\left(\frac{\partial u_x}{\partial y} + \frac{\partial u_y}{\partial x}\right) + \mu\frac{\partial}{\partial z}\left(\frac{\partial u_x}{\partial z} + \frac{\partial u_z}{\partial x}\right) + b_x = 0. \tag{8.78}$$

As the displacement components are continuous functions, we can commute the differential operators. Thus,

$$\frac{\partial}{\partial y}\left(\frac{\partial u_x}{\partial y} + \frac{\partial u_y}{\partial x}\right) = \frac{\partial^2 u_x}{\partial y^2} + \frac{\partial}{\partial x}\left(\frac{\partial u_y}{\partial y}\right) = \frac{\partial^2 u_x}{\partial y^2} + \frac{\partial\varepsilon_{yy}}{\partial x}, \tag{8.79}$$

$$\frac{\partial}{\partial z}\left(\frac{\partial u_x}{\partial z} + \frac{\partial u_z}{\partial x}\right) = \frac{\partial^2 u_x}{\partial z^2} + \frac{\partial}{\partial x}\left(\frac{\partial u_z}{\partial z}\right) = \frac{\partial^2 u_x}{\partial z^2} + \frac{\partial\varepsilon_{zz}}{\partial x}. \tag{8.80}$$

Substituting these relations in (8.78), we have

$$\lambda\frac{\partial e}{\partial x} + 2\mu\frac{\partial\varepsilon_{xx}}{\partial x} + \mu\left(\frac{\partial^2 u_x}{\partial y^2} + \frac{\partial\varepsilon_{yy}}{\partial x}\right) + \mu\left(\frac{\partial^2 u_x}{\partial y^2} + \frac{\partial\varepsilon_{zz}}{\partial x}\right) + b_x = 0. \tag{8.81}$$

Recalling that $e = \varepsilon_{xx} + \varepsilon_{yy} + \varepsilon_{zz}$ and $\dfrac{\partial\varepsilon_{xx}}{\partial x} = \dfrac{\partial^2 u_x}{\partial x^2}$ and rearranging the terms, we obtain

$$(\lambda + \mu)\frac{\partial e}{\partial x} + \mu\left(\frac{\partial^2}{\partial x^2} + \frac{\partial^2}{\partial y^2} + \frac{\partial^2}{\partial z^2}\right)u_x + b_x = 0. \tag{8.82}$$

Performing the same procedure for the two other equations in (8.45), we obtain the Navier equations in terms of the displacement components and the dilatation e, that is,

$$\begin{cases} (\lambda + \mu)\dfrac{\partial e}{\partial x} + \mu \left(\dfrac{\partial^2}{\partial x^2} + \dfrac{\partial^2}{\partial y^2} + \dfrac{\partial^2}{\partial z^2} \right) u_x + b_x = 0 \\[2mm] (\lambda + \mu)\dfrac{\partial e}{\partial y} + \mu \left(\dfrac{\partial^2}{\partial x^2} + \dfrac{\partial^2}{\partial y^2} + \dfrac{\partial^2}{\partial z^2} \right) u_y + b_y = 0 \\[2mm] (\lambda + \mu)\dfrac{\partial e}{\partial z} + \mu \left(\dfrac{\partial^2}{\partial x^2} + \dfrac{\partial^2}{\partial y^2} + \dfrac{\partial^2}{\partial z^2} \right) u_z + b_z = 0 \end{cases} . \tag{8.83}$$

Notice that while equilibrium equations (8.45) are valid for any continuous three-dimensional medium with small deformations, the Navier equations provide the equilibrium in terms of the displacement components only for a Hookean material.

We outline that the analytical solution of the system of equations in (8.83) can be obtained only for some particular cases. When there is no closed-form solution for a given problem, numerical solution techniques like the finite element method (FEM) are applied.

Example 8.13 *For the bar, we have only the displacement $u_x(x)$, longitudinal strain $\varepsilon_{xx}(x)$, and normal stress $\sigma_{xx}(x)$, and thus the dilatation $e(x) = \varepsilon_{xx}(x)$. Hence, from the first Navier equation, we have*

$$(\lambda + 2\mu)\frac{d^2 u_x(x)}{dx^2} = -b_x(x).$$

The following relation is valid:

$$E = \lambda + 2\mu.$$

Besides that, for a bar with constant cross-section, we have $b_x(x) = \dfrac{q_x(x)}{A}$. Thus, the previous expression recovers the differential equation of equilibrium for the bar in terms of u_x, that is,

$$EA\frac{d^2 u_x(x)}{dx^2} = -q_x(x).$$

□

8.10 FORMULATION EMPLOYING TENSORS

The employed formulation so far used scalars and vectors as basic mathematical entities, as well as scalar and vector functions. Another important concept in the study of problems in mechanics is tensors. The use of tensors allow us to present the model formulations in a more compact and elegant notation. Another advantage is that the equation presented in a tensor notation is independent of the employed coordinate system. Thus, we can focus only on the concepts involved in the equations, without worrying about unnecessary details, from the formulation point of view. These details are only important when a specific coordinate system is adopted to study the problem.

In fact, the concept of tensor represents a generalization of scalars and vectors, because these can be respectively defined as zero-order and first-order tensors [36]. The second-order tensors are extensively used in mechanics, like the strain, stress, and inertia tensors. Fourth-order tensors are used to represent constitutive equations of materials [36, 12].

Following, we formulate the solid model using the tensor concepts. The same previous steps of the variational formulation are used. Before that, however, it is important to define some basic concepts of continuum mechanics [30, 36].

8.10.1 BODY

As seen in Chapter 1, the geometrical space considered in the study of continuum mechanics is the three-dimensional Euclidean space $\mathscr{E}$. The elements of $\mathscr{E}$ are called points.

The main physical characteristic of a body is that it occupies a region of the three-dimensional Euclidean space. Therefore, any body can occupy different regions in distinct times. Although none of these regions may be associated with the body, it becomes convenient to select one, called reference configuration $\mathscr{B}$, identifying points on the body with positions in $\mathscr{B}$. Thus, a body $\mathscr{B}$ becomes a regular region of $\mathscr{E}$, and the points of $\mathscr{B}$ are called material points. Any limited regular subregion of $\mathscr{B}$ is called part and indicated by $\mathscr{P}$. The boundaries of $\mathscr{B}$ and $\mathscr{P}$ are respectively given by $\partial\mathscr{B}$ and $\partial\mathscr{P}$. These concepts are illustrated in Figure 8.11.

As a body can occupy different regions when moving, it is necessary to introduce a parameter $t \in \left[t_0, t_f\right]$, designating a certain configuration $\mathscr{B}_t$ of the body. Notice that in several problems, such as shape optimization, t will not necessarily represent time.

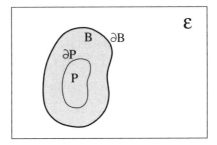

Figure 8.11 Definition of body, part, and boundaries.

8.10.2 VECTORS

Intuitively, we observe that the sum of two points of $\mathscr{E}$ have no physical meaning. However, the difference between points P and Q, as shown in Figure 8.12(a), defines a vector, that is,

$$\mathbf{v} = Q - P, \qquad P, Q \in \mathscr{E}. \tag{8.84}$$

Thus, notice that a vector is formally defined as the difference between points of $\mathscr{E}$.

The set of vectors obtained by the difference between points of $\mathscr{E}$ actually forms a vector space $\mathscr{V}$. Also notice that the sum of a point P and a vector $\mathbf{v}$ defines a new point Q, i.e.,

$$Q = P + \mathbf{v}, \qquad P \in \mathscr{E}, \quad \mathbf{v} \in \mathscr{V}. \tag{8.85}$$

We can only speak of vector components and directions when a coordinate system is adopted. A Cartesian coordinate system consists of an orthonormal basis $\{\mathbf{e}_x, \mathbf{e}_y, \mathbf{e}_z\}$ and an arbitrary point O of $\mathscr{E}$, called origin. Hence, the coordinates of any point P are given by the position vector $\mathbf{r}_P = P - O$ relative to the origin O. These concepts are illustrated in Figure 8.12(b).

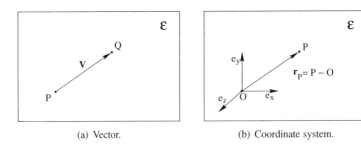

(a) Vector.   (b) Coordinate system.

Figure 8.12 Definition of vector and coordinate system.

Once the coordinate system is defined, we can represent a vector $\mathbf{v}$ in this system with the following linear combination of base vectors $\{\mathbf{e}_x, \mathbf{e}_y, \mathbf{e}_z\}$:

$$\mathbf{v} = v_x \mathbf{e}_x + v_y \mathbf{e}_y + v_z \mathbf{e}_z. \tag{8.86}$$

However, the components (v_x, v_y, v_z) of vector **v** are obtained by the scalar product of **v** with the basis vectors, that is,

$$v_x = \mathbf{v} \cdot \mathbf{e}_x, \qquad v_y = \mathbf{v} \cdot \mathbf{e}_y, \qquad v_z = \mathbf{v} \cdot \mathbf{e}_z. \tag{8.87}$$

The Euclidean norm of vector **v** is given by

$$\|\mathbf{v}\| = \sqrt{\mathbf{v} \cdot \mathbf{v}} = \sqrt{v_x^2 + v_y^2 + v_z^2}. \tag{8.88}$$

Example 8.14 *Figure 8.13 illustrates a vector* **v** *represented in coordinate systems* xy *and* $x'y'$, *respectively, with bases* $\{\mathbf{e}_x, \mathbf{e}_y, \mathbf{e}_z\}$ *and* $\{\mathbf{e}'_x, \mathbf{e}'_y, \mathbf{e}'_z\}$, *rotated by an angle* α *about the* z *axis.*

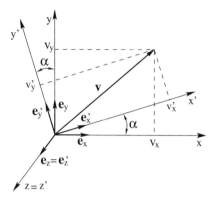

Figure 8.13 Example 8.14: different representations of a vector.

The vector **v** *has the following representations in both coordinate systems*

$$\mathbf{v} = v_x\mathbf{e}_x + v_y\mathbf{e} = v'_x\mathbf{e}'_x + v'_y\mathbf{e}'_y. \tag{8.89}$$

The relations between the base vectors of both reference systems are:

$$\begin{aligned}
\mathbf{e}_x &= \cos\alpha\,\mathbf{e}'_x - \sin\alpha\,\mathbf{e}'_y, \\
\mathbf{e}_y &= \sin\alpha\,\mathbf{e}'_x + \cos\alpha\,\mathbf{e}'_y, \\
\mathbf{e}_z &= \mathbf{e}'_z.
\end{aligned}$$

Hence, vector **v** *can be expressed, after the substitution of the previous relations in (8.89), by*

$$\mathbf{v} = (v_x\cos\alpha + v_y\sin\alpha)\mathbf{e}'_x + (-v_x\sin\alpha + v_y\cos\alpha)\mathbf{e}'_y.$$

Consequently, we have the following relations between the components of **v** *in both coordinate systems:*

$$v'_x = v_x\cos\alpha + v_y\sin\alpha \quad and \quad v'_y = -v_x\sin\alpha + v_y\cos\alpha.$$

The norm of **v** *can be calculated using the components in both bases as*

$$\begin{aligned}
\|\mathbf{v}\|^2 &= v_x^2 + v_y^2, \\
\|\mathbf{v}\|^2 &= \left(v'_x\right)^2 + \left(v'_y\right)^2 = (v_x\cos\alpha + v_y\sin\alpha)^2 + (-v_x\sin\alpha + v_y\cos\alpha)^2 = v_x^2 + v_y^2.
\end{aligned}$$

Thus, notice that while the vector **v** *has only one definition, given by the difference between points O and P, it can have different representations in distinct coordinate systems, but with the same norm.*

$\square$

In the following sections, the solid model previously presented is formulated with the introduction of the concept of tensors. Although one of the advantages of employing tensors is to obtain general expressions for any coordinate system, the Cartesian coordinates (x, y, z) are used below to maintain compatibility with the previous notation.

8.10.3 KINEMATICS

As seen in Section 8.2, the kinematics of a solid is described by a three-dimensional vector field $\mathbf{u}(x,y,z)$, which respectively gives the displacement components u_x, u_y, and u_z in directions $\mathbf{e}_x$, $\mathbf{e}_y$, and $\mathbf{e}_z$, for each point of the body with coordinates (x,y,z). Thus, the kinematics of a solid can be denoted by the following displacement vector field:

$$\mathbf{u}(x,y,z) = \left\{ \begin{array}{c} u_x(x,y,z) \\ u_y(x,y,z) \\ u_z(x,y,z) \end{array} \right\}. \tag{8.90}$$

8.10.4 STRAIN MEASURE

This section starts with the the concept of second-order tensors from the Taylor series expansion of a vector function. Subsequently, we characterize the strain measure for a solid body under small and large strains.

Let $f(x)$ be a scalar function of x. Thus, for each value of x, $f(x)$ gives a real number or scalar. For example, $f(x)$ can represent the axial displacement in bars or transversal displacement in beams. We can expand the function f in the neighborhood of $y = x + d$ using the Taylor series, that is,

$$\begin{aligned} f(y) &= f(x) + \frac{df(x)}{dx}d + \frac{1}{2}\frac{d^2 f(x)}{dx^2}d^2 + \ldots + \frac{1}{n!}\frac{d^{(n)} f(x)}{dx^{(n)}}d^n + \underbrace{\frac{1}{(n+1)!}\frac{d^{(n)} f(\xi)}{dx^{(n)}}d^{n+1}}_{R_n(\xi)} \\ &= f(x) + \frac{df(x)}{dx}d + \mathcal{O}(d^2), \end{aligned} \tag{8.91}$$

with $d = y - x$ and $\mathcal{O}(d^2)$ is a term of order d^2. This means that when y approaches x, that is, when $d = y - x$ goes to zero, d^2 goes faster to zero. Thus,

$$\lim_{y \to x} \frac{d^2}{y - x} = \lim_{y \to x} \frac{(y - x)^2}{y - x} = \lim_{y \to x} (y - x) = 0. \tag{8.92}$$

The symbol $R_n(\xi)$ represents the residual term with $x < \xi < y = x + d$.

Example 8.15 *Consider the function $f(x) = \exp(x)$. Thus, $f^{(n)}(x) = \exp(x)$ for each $n \geq 0$ and the Taylor formula for $x = 0$ reduces to*

$$f(y) = f(0) + f'(0)y + \frac{f''(0)}{2!}y^2 + \ldots + \frac{f^{(n)}(0)}{n!}y^n + \frac{f^{(n+1)}(\xi)}{(n+1)!}y^{n+1},$$

Thus,

$$\exp(y) = 1 + y + \frac{y^2}{2!} + \frac{y^3}{3!} + \ldots + \frac{y^n}{n!} + \frac{y^{n+1}}{(n+1)!}\exp(\xi).$$

As

$$\lim_{n \to \infty} \frac{y^n}{n!} = 0,$$

we have $\lim_{n \to \infty} R_n(y) = 0$ for every y. Consequently, the above Taylor series converge to $\exp(y)$, for any y. Thus, we can write

$$\exp(y) = \sum_{n=0}^{\infty} \frac{y^n}{n!} = 1 + y + \frac{y^2}{2!} + \frac{y^3}{3!} + \frac{y^4}{4!} + \ldots$$

For $y = 1$ and taking 10 terms of the series, we have

$$e = \exp(1) \approx 1 + \frac{1}{1!} + \frac{1}{2!} + \ldots + \frac{1}{10!} = 2.7182818.$$

□

Now suppose that f is a function that yields scalar values but depends on variables x, y, and z. We can say that f depends on the position vector $\mathbf{x} = (x, y, z)$ of a point of the solid, denoted as $f = f(\mathbf{x}) = f(x, y, z)$. Using the Taylor series, we can expand f around $\mathbf{y} = \mathbf{x} + \mathbf{d}$ by the following way

$$f(\mathbf{y}) = f(\mathbf{x}) + \nabla f^T(\mathbf{x})\mathbf{d} + \mathcal{O}(\|\mathbf{d}\|^2), \tag{8.93}$$

where $\mathbf{d} = \mathbf{y} - \mathbf{x}$ is the difference vector between position vectors $\mathbf{y} = (x', y', z')$ and $\mathbf{x} = (x, y, z)$. The Euclidean norm of $\mathbf{d}$ is indicated by $\|\mathbf{d}\|$ and $\|\mathbf{d}\|^2 = (x' - x)^2 + (y' - y)^2 + (z' - z)^2$. Thus, $\mathcal{O}(\|\mathbf{d}\|^2)$ is a term of order $\|\mathbf{d}\|^2$ and it approaches zero faster than the norm $\|\mathbf{d}\|$ when $\mathbf{y}$ goes to $\mathbf{x}$, that is,

$$\lim_{\mathbf{y} \to \mathbf{x}} \frac{\|\mathbf{d}\|^2}{\|\mathbf{y} - \mathbf{x}\|} = \lim_{\mathbf{y} \to \mathbf{x}} \frac{\|\mathbf{y} - \mathbf{x}\|^2}{\|\mathbf{y} - \mathbf{x}\|} = \lim_{\mathbf{y} \to \mathbf{x}} \|\mathbf{y} - \mathbf{x}\| = 0. \tag{8.94}$$

As f is now a function with three variables, the first derivative in (8.91) is given by the gradient vector field of f, which components are the partial derivatives of f relative to the independent variables x, y, and z. Thus,

$$\{\nabla f(\mathbf{x})\} = \left\{ \begin{array}{ccc} \dfrac{\partial f(\mathbf{x})}{\partial x} & \dfrac{\partial f(\mathbf{x})}{\partial y} & \dfrac{\partial f(\mathbf{x})}{\partial z} \end{array} \right\}^T. \tag{8.95}$$

Example 8.16 *Consider the scalar field $f(x, y, z) = 5x^2 y^3 z$. The vector field corresponding to the gradient of f is given by*

$$\{\nabla f(\mathbf{x})\} = \left\{ \begin{array}{ccc} 10xy^3 z & 15x^2 y^2 z & 5x^2 y^3 \end{array} \right\}^T.$$

□

Let $\mathbf{f}$ be a vector function which is dependent on variables x, y, and z, that is, $\mathbf{f} = \mathbf{f}(\mathbf{x}) = \mathbf{f}(x, y, z)$. Hence, the components of $\mathbf{f}$ are scalar functions in x, y, and z, that is,

$$\{\mathbf{f}(\mathbf{x})\} = \left\{ \begin{array}{c} f_x(\mathbf{x}) \\ f_y(\mathbf{x}) \\ f_z(\mathbf{x}) \end{array} \right\}. \tag{8.96}$$

Expanding $\mathbf{f}$ in the neighborhood of $\mathbf{y} = \mathbf{x} + \mathbf{d}$, we have

$$\mathbf{f}(\mathbf{y}) = \mathbf{f}(\mathbf{x}) + \nabla \mathbf{f}(\mathbf{x})\mathbf{d} + \mathcal{O}(\|\mathbf{d}\|^2). \tag{8.97}$$

In this case, the gradient of $\mathbf{f}(\mathbf{x})$ is given by the partial derivatives of $\mathbf{f}$ relative to x, y, and z. Thus,

$$\nabla \mathbf{f}(\mathbf{x}) = \left[\begin{array}{ccc} \dfrac{\partial \mathbf{f}(\mathbf{x})}{\partial x} & \dfrac{\partial \mathbf{f}(\mathbf{x})}{\partial y} & \dfrac{\partial \mathbf{f}(\mathbf{x})}{\partial z} \end{array} \right]^T. \tag{8.98}$$

According to (8.96), $\mathbf{f}$ is a vector function with components f_x, f_y, and f_z. Thus, each term in the right-hand side of equation (8.98) is a vector, analogous to equation (8.95). Expanding each one of these terms, we have

$$[\nabla \mathbf{f}(\mathbf{x})] = \left[\begin{array}{ccc} \dfrac{\partial f_x(\mathbf{x})}{\partial x} & \dfrac{\partial f_x(\mathbf{x})}{\partial y} & \dfrac{\partial f_x(\mathbf{x})}{\partial z} \\[2ex] \dfrac{\partial f_y(\mathbf{x})}{\partial x} & \dfrac{\partial f_y(\mathbf{x})}{\partial y} & \dfrac{\partial f_y(\mathbf{x})}{\partial z} \\[2ex] \dfrac{\partial f_z(\mathbf{x})}{\partial x} & \dfrac{\partial f_z(\mathbf{x})}{\partial y} & \dfrac{\partial f_z(\mathbf{x})}{\partial z} \end{array} \right]. \tag{8.99}$$

Thus, the gradient of a vector function $\mathbf{f}$, which is dependent on position vector $\mathbf{x} = (x, y, z)$, is a matrix of rank 3. Actually, equation (8.99) is the matrix representation of the second-order tensor $\nabla \mathbf{f}(\mathbf{x})$, according to the adopted Cartesian system.

Example 8.17 *Consider the vector field $\mathbf{f}(x, y, z) = xyz\mathbf{e}_x + 10y^2 z\mathbf{e}_y + y^2 z\mathbf{e}_z$. The corresponding tensor field to the gradient of $\mathbf{f}$ has the following matrix notation:*

$$[\nabla \mathbf{f}(\mathbf{x})] = \left[\begin{array}{ccc} yz & xz & xy \\ 0 & 20yz & 10y^2 \\ 0 & 2yz & y^2 \end{array} \right].$$

□

The multiplication of the matrix representation of tensor $\nabla \mathbf{f}$ given in (8.99) by a vector $\mathbf{v}$, with Cartesian components (v_x, v_y, v_z), results in another vector, that is,

$$
\begin{bmatrix}
\dfrac{\partial f_x}{\partial x} & \dfrac{\partial f_x}{\partial y} & \dfrac{\partial f_x}{\partial z} \\[2mm]
\dfrac{\partial f_y}{\partial x} & \dfrac{\partial f_y}{\partial y} & \dfrac{\partial f_y}{\partial z} \\[2mm]
\dfrac{\partial f_z}{\partial x} & \dfrac{\partial f_z}{\partial y} & \dfrac{\partial f_z}{\partial z}
\end{bmatrix}
\left\{
\begin{array}{c}
v_x \\ v_y \\ v_z
\end{array}
\right\}
=
\left\{
\begin{array}{c}
\dfrac{\partial f_x}{\partial x} v_x + \dfrac{\partial f_x}{\partial y} v_y + \dfrac{\partial f_x}{\partial z} v_z \\[2mm]
\dfrac{\partial f_y}{\partial x} v_x + \dfrac{\partial f_y}{\partial y} v_y + \dfrac{\partial f_y}{\partial z} v_z \\[2mm]
\dfrac{\partial f_z}{\partial x} v_x + \dfrac{\partial f_z}{\partial y} v_y + \dfrac{\partial f_z}{\partial z} v_z
\end{array}
\right\}.
$$

It is important to establish the concept of second-order tensor. Similar to vectors, we have the definition of second-order tensor and its representation in a coordinate system. Thus, the formal definition of a second-order tensor $\mathbf{T}$ is a linear transformation of vector space $\mathscr{V}$ in $\mathscr{V}$, denoted as

$$\mathbf{Tu} = \mathbf{v}. \tag{8.100}$$

This means that when the tensor $\mathbf{T}$ is applied to any vector $\mathbf{u}$, it results in another vector $\mathbf{v}$. As the transformation is linear, the following properties are valid:

$$\mathbf{T(u+v)} \;=\; \mathbf{Tu} + \mathbf{Tv}, \tag{8.101}$$

$$\mathbf{T}(\alpha \mathbf{u}) \;=\; \alpha(\mathbf{Tu}), \tag{8.102}$$

where α is a scalar number. Equations (8.100) to (8.102) define a second-order tensor.

Example 8.18 *If $\mathbf{T}$ transforms any vector in a unique vector $\mathbf{a}$, show that $\mathbf{T}$ is not a tensor.*
For any vectors $\mathbf{u}$ and $\mathbf{v}$, we have

$$\mathbf{Tu} = \mathbf{a}, \qquad \mathbf{Tv} = \mathbf{a} \quad and \quad \mathbf{T(u+v)} = \mathbf{a}.$$

Thus, $\mathbf{T}$ is not a tensor, because it is not a linear transformation, as shown below

$$\mathbf{T(u+v)} = \mathbf{a} \neq \mathbf{Tu} + \mathbf{Tv} = \mathbf{a} + \mathbf{a} = 2\mathbf{a}.$$

$\square$

Using a Cartesian coordinate system with basis $\{\mathbf{e}_1, \mathbf{e}_2, \mathbf{e}_3\}$, the components of $\mathbf{T}$ are defined as

$$T_{ij} = \mathbf{e}_i \cdot \mathbf{Te}_j, \quad i, j = 1, 2, 3. \tag{8.103}$$

Hence, the components of the second-order tensor $\mathbf{T}$ can be represented in matrix notation as

$$
[\mathbf{T}] = [T_{ij}] =
\begin{bmatrix}
T_{11} & T_{12} & T_{13} \\
T_{21} & T_{22} & T_{23} \\
T_{31} & T_{32} & T_{33}
\end{bmatrix}.
$$

Example 8.19 *Consider the vector $\mathbf{u}$ given by the linear combination of its components (u_1, u_2, u_3) and the base vectors $\{\mathbf{e}_1, \mathbf{e}_2, \mathbf{e}_3\}$*

$$\mathbf{u} = u_1 \mathbf{e}_1 + u_2 \mathbf{e}_2 + u_3 \mathbf{e}_3.$$

Using the linearity of the second-order tensor $\mathbf{T}$ we obtain

$$\mathbf{v} = \mathbf{Tu} = \mathbf{T}(u_1 \mathbf{e}_1 + u_2 \mathbf{e}_2 + u_3 \mathbf{e}_3) = u_1 \mathbf{Te}_1 + u_2 \mathbf{Te}_2 + u_3 \mathbf{Te}_3.$$

The components of vector $\mathbf{v}$ are obtained by taking the following dot products:

$$
\begin{aligned}
v_1 &= \mathbf{e}_1 \cdot \mathbf{v} = u_1 (\mathbf{e}_1 \cdot \mathbf{Te}_1) + u_2 (\mathbf{e}_1 \cdot \mathbf{Te}_2) + u_3 (\mathbf{e}_1 \cdot \mathbf{Te}_3), \\
v_2 &= \mathbf{e}_2 \cdot \mathbf{v} = u_1 (\mathbf{e}_2 \cdot \mathbf{Te}_1) + u_2 (\mathbf{e}_2 \cdot \mathbf{Te}_2) + u_3 (\mathbf{e}_2 \cdot \mathbf{Te}_3), \\
v_3 &= \mathbf{e}_3 \cdot \mathbf{v} = u_1 (\mathbf{e}_3 \cdot \mathbf{Te}_1) + u_2 (\mathbf{e}_3 \cdot \mathbf{Te}_2) + u_3 (\mathbf{e}_3 \cdot \mathbf{Te}_3).
\end{aligned}
$$

The terms $T_{11} = \mathbf{e}_1 \cdot \mathbf{Te}_1$ and $T_{21} = \mathbf{e}_2 \cdot \mathbf{Te}_1$ are components of second-order tensor $\mathbf{T}$. They are also components of vector $\mathbf{Te}_1$ in directions $\mathbf{e}_1$ and $\mathbf{e}_2$ respectively.

Expression $\mathbf{v} = \mathbf{Tu}$ may be also written in terms of components as

$$v_1 = T_{11}u_1 + T_{12}u_2 + T_{13}u_3,$$
$$v_2 = T_{21}u_1 + T_{22}u_2 + T_{23}u_3,$$
$$v_3 = T_{31}u_1 + T_{32}u_2 + T_{33}u_3.$$

In matrix notation, we have

$$\left\{ \begin{array}{c} v_1 \\ v_2 \\ v_3 \end{array} \right\} = \left[\begin{array}{ccc} T_{11} & T_{12} & T_{13} \\ T_{21} & T_{22} & T_{23} \\ T_{31} & T_{32} & T_{33} \end{array} \right] \left\{ \begin{array}{c} u_1 \\ u_2 \\ u_3 \end{array} \right\}.$$

□

The kinematics of a body is also described by a vector function $\mathbf{u}$ which depends on position vector $\mathbf{x} = (x, y, z)$, as indicated in (8.90). Expanding $\mathbf{u}(\mathbf{x})$ in the neighborhood of $\mathbf{y} = \mathbf{x} + \mathbf{d}$, analogous to equation (8.97), we have

$$\mathbf{u}(\mathbf{y}) = \mathbf{u}(\mathbf{x}) + \nabla\mathbf{u}(\mathbf{x})\mathbf{d} + \mathcal{O}(\|\mathbf{d}\|^2), \tag{8.104}$$

where $\nabla\mathbf{u}(\mathbf{x})$ is the displacement gradient tensor calculated in $\mathbf{x}$. Its matrix representation in the Cartesian coordinate system is given by

$$[\nabla\mathbf{u}] = \left[\begin{array}{ccc} \dfrac{\partial u_x}{\partial x} & \dfrac{\partial u_x}{\partial y} & \dfrac{\partial u_x}{\partial z} \\ \dfrac{\partial u_y}{\partial x} & \dfrac{\partial u_y}{\partial y} & \dfrac{\partial u_y}{\partial z} \\ \dfrac{\partial u_z}{\partial x} & \dfrac{\partial u_z}{\partial y} & \dfrac{\partial u_z}{\partial z} \end{array} \right]. \tag{8.105}$$

As $\mathbf{y} = \mathbf{x} + \mathbf{d}$, expression (8.104) can be rewritten as

$$\mathbf{u}(\mathbf{x} + \mathbf{d}) = \mathbf{u}(\mathbf{x}) + \nabla\mathbf{u}(\mathbf{x})\mathbf{d} + \mathcal{O}(\|\mathbf{d}\|^2). \tag{8.106}$$

Example 8.20 *The displacement vector fields for the models of bars, shafts, and beams under bending only are, respectively,*

$$\mathbf{u}(x, y, z) = \left\{ \begin{array}{ccc} u_x(x) & 0 & 0 \end{array} \right\}^T,$$
$$\mathbf{u}(x, y, z) = \left\{ \begin{array}{ccc} 0 & -z\theta_x(x) & y\theta_x(x) \end{array} \right\}^T,$$
$$\mathbf{u}(x, y, z) = \left\{ \begin{array}{ccc} -y\theta_z(x) & u_y(x) & 0 \end{array} \right\}^T.$$

The respective displacement gradient tensors are:

$$[\nabla\mathbf{u}(x, y, z)] = \left[\begin{array}{ccc} \dfrac{du_x(x)}{dx} & 0 & 0 \\ 0 & 0 & 0 \\ 0 & 0 & 0 \end{array} \right],$$

$$[\nabla\mathbf{u}(x, y, z)] = \left[\begin{array}{ccc} 0 & 0 & 0 \\ -z\dfrac{d\theta_x(x)}{dx} & 0 & -\theta_x(x) \\ y\dfrac{d\theta_x(x)}{dx} & \theta_x(x) & 0 \end{array} \right],$$

$$[\nabla\mathbf{u}(x, y, z)] = \left[\begin{array}{ccc} -y\dfrac{d\theta_z(x)}{dx} & -\theta_z(x) & 0 \\ \dfrac{du_y(x)}{dx} & 0 & 0 \\ 0 & 0 & 0 \end{array} \right].$$

□

The displacement gradient tensor field can be written as

$$
\begin{aligned}
\nabla \mathbf{u} &= \frac{1}{2}\nabla \mathbf{u} + \frac{1}{2}\nabla \mathbf{u} \\
&= \frac{1}{2}\nabla \mathbf{u} + \frac{1}{2}\nabla \mathbf{u}^T + \frac{1}{2}\nabla \mathbf{u} - \frac{1}{2}\nabla \mathbf{u}^T \\
&= \frac{1}{2}[\nabla \mathbf{u} + \nabla \mathbf{u}^T] + \frac{1}{2}[\nabla \mathbf{u} - \nabla \mathbf{u}^T].
\end{aligned}
\tag{8.107}
$$

In this case, $\nabla \mathbf{u}^T$ is the transposed tensor of $\nabla \mathbf{u}$. To obtain the matrix representation of $\nabla \mathbf{u}^T$ in the Cartesian coordinate system, we should swap the rows and columns in (8.105), that is,

$$
[\nabla \mathbf{u}^T] =
\begin{bmatrix}
\dfrac{\partial u_x}{\partial x} & \dfrac{\partial u_y}{\partial x} & \dfrac{\partial u_z}{\partial x} \\[6pt]
\dfrac{\partial u_x}{\partial y} & \dfrac{\partial u_y}{\partial y} & \dfrac{\partial u_z}{\partial y} \\[6pt]
\dfrac{\partial u_x}{\partial z} & \dfrac{\partial u_y}{\partial z} & \dfrac{\partial u_z}{\partial z}
\end{bmatrix}.
\tag{8.108}
$$

Generally, the transposed tensor $\mathbf{A}^T$ of a tensor $\mathbf{A}$ is unique and has the following definition:

$$
\mathbf{u} \cdot \mathbf{A}\mathbf{v} = \mathbf{A}^T\mathbf{u} \cdot \mathbf{v},
\tag{8.109}
$$

for any vectors $\mathbf{u}$ and $\mathbf{v}$.

The infinitesimal or small strain tensor $\mathbf{E}(\mathbf{x})$ and the infinitesimal rotation tensor $\Omega(\mathbf{x})$ are respectively defined as

$$
\mathbf{E} = \frac{1}{2}[\nabla \mathbf{u} + \nabla \mathbf{u}^T],
\tag{8.110}
$$

$$
\Omega = \frac{1}{2}[\nabla \mathbf{u} - \nabla \mathbf{u}^T].
\tag{8.111}
$$

The matrix representation of the infinitesimal strain tensor $\mathbf{E}$ in the Cartesian coordinate system is obtained when substituting (8.105) and (8.108) in (8.110). Performing the indicated operations, we have

$$
[\mathbf{E}] =
\begin{bmatrix}
\dfrac{\partial u_x}{\partial x} & \dfrac{1}{2}\left(\dfrac{\partial u_y}{\partial x} + \dfrac{\partial u_x}{\partial y}\right) & \dfrac{1}{2}\left(\dfrac{\partial u_z}{\partial x} + \dfrac{\partial u_x}{\partial z}\right) \\[10pt]
\dfrac{1}{2}\left(\dfrac{\partial u_x}{\partial y} + \dfrac{\partial u_y}{\partial x}\right) & \dfrac{\partial u_y}{\partial y} & \dfrac{1}{2}\left(\dfrac{\partial u_z}{\partial y} + \dfrac{\partial u_y}{\partial z}\right) \\[10pt]
\dfrac{1}{2}\left(\dfrac{\partial u_x}{\partial z} + \dfrac{\partial u_z}{\partial x}\right) & \dfrac{1}{2}\left(\dfrac{\partial u_y}{\partial z} + \dfrac{\partial u_z}{\partial y}\right) & \dfrac{\partial u_z}{\partial z}
\end{bmatrix}.
\tag{8.112}
$$

Notice that the Cartesian components of $\mathbf{E}$ have a direct relation with the previously derived strain components in Section 8.3. Thus, we can rewrite (8.112) as

$$
[\mathbf{E}] =
\begin{bmatrix}
\varepsilon_{xx} & \frac{1}{2}\bar{\gamma}_{xy} & \frac{1}{2}\bar{\gamma}_{xz} \\[4pt]
\frac{1}{2}\bar{\gamma}_{xy} & \varepsilon_{yy} & \frac{1}{2}\bar{\gamma}_{yz} \\[4pt]
\frac{1}{2}\bar{\gamma}_{xz} & \frac{1}{2}\bar{\gamma}_{yz} & \varepsilon_{zz}
\end{bmatrix}.
\tag{8.113}
$$

It is common to write the infinitesimal strain tensor as follows

$$
[\mathbf{E}] =
\begin{bmatrix}
\varepsilon_{xx} & \gamma_{xy} & \gamma_{xz} \\
\gamma_{yx} & \varepsilon_{yy} & \gamma_{yz} \\
\gamma_{zx} & \gamma_{zy} & \varepsilon_{zz}
\end{bmatrix}.
\tag{8.114}
$$

The main diagonal components ε_{xx}, ε_{yy}, and ε_{zz} represent the specific normal longitudinal strains in directions x, y, and z calculated in the point with coordinates $\mathbf{x} = (x, y, z)$. The other components are the shear strains also called distortions. Tensor $\mathbf{E}$ is symmetric because

$$
\gamma_{xy} = \gamma_{yx}, \qquad \gamma_{xz} = \gamma_{zx}, \qquad \gamma_{yz} = \gamma_{zy}.
\tag{8.115}
$$

The symmetry of a tensor $\mathbf{T}$ is defined as

$$
\mathbf{T} = \mathbf{T}^T.
\tag{8.116}
$$

In terms of components, this implies that

$$T_{12} = T_{21}, \qquad T_{13} = T_{31}, \qquad T_{23} = T_{32}, \tag{8.117}$$

or in a general way

$$T_{ij} = T_{ji}, \qquad i,j = 1,2,3. \tag{8.118}$$

Recall that the first subscript in γ_{xy} indicates the plane x, while the subscript y indicates the strain direction. This is analogous for γ_{xz} and γ_{yz} (see Figure 8.4). Notice that the total distortions $\bar{\gamma}_{xy}$, $\bar{\gamma}_{xz}$, and $\bar{\gamma}_{yz}$ at planes xy, xz, and yz, given in (8.23), are twice the respective distortions γ_{xy}, γ_{xz}, and γ_{yz}, that is,

$$\bar{\gamma}_{xy} = 2\gamma_{xy}, \qquad \bar{\gamma}_{xz} = 2\gamma_{xz}, \qquad \bar{\gamma}_{yz} = 2\gamma_{yz}. \tag{8.119}$$

Example 8.21 *The infinitesimal strain tensor for each cross-section x of the bar is obtained from (8.114) as*

$$[\mathbf{E}] = \begin{bmatrix} \dfrac{du_x(x)}{dx} & 0 & 0 \\ 0 & 0 & 0 \\ 0 & 0 & 0 \end{bmatrix} = \begin{bmatrix} \varepsilon_{xx}(x) & 0 & 0 \\ 0 & 0 & 0 \\ 0 & 0 & 0 \end{bmatrix}.$$

On the other hand, for shafts and beams in bending only, we respectively have

$$[\mathbf{E}] = \begin{bmatrix} 0 & -\dfrac{1}{2}z\dfrac{d\theta_x(x)}{dx} & \dfrac{1}{2}y\dfrac{d\theta_x(x)}{dx} \\ -\dfrac{1}{2}z\dfrac{d\theta_x(x)}{dx} & 0 & 0 \\ \dfrac{1}{2}y\dfrac{d\theta_x(x)}{dx} & 0 & 0 \end{bmatrix} = \begin{bmatrix} 0 & \gamma_{xy}(x,z) & \gamma_{xz}(x,y) \\ \gamma_{xy}(x,z) & 0 & 0 \\ \gamma_{xz}(x,y) & 0 & 0 \end{bmatrix},$$

$$[\mathbf{E}] = \begin{bmatrix} \dfrac{\partial u_x(x,y)}{\partial x} & 0 & 0 \\ 0 & 0 & 0 \\ 0 & 0 & 0 \end{bmatrix} = \begin{bmatrix} \varepsilon_{xx}(x,y) & 0 & 0 \\ 0 & 0 & 0 \\ 0 & 0 & 0 \end{bmatrix}.$$

The above tensors actually represent tensor fields, because when substituting the values of coordinates (x,y,z) of a point, we obtain the respective small strain tensor.
□

Example 8.22 *Consider the displacement vector field $\mathbf{u}(x,y,z) = [2x^2y^3z\mathbf{e}_x + 6x^2yz\mathbf{e}_y + 6yz^2\mathbf{e}_z] \times \alpha$ [cm], with $\alpha = 10^{-4}$. This factor is employed only to enforce a small displacement field.*

The matrix representation of the infinitesimal strain tensor field is obtained from (8.114). Thus,

$$[\mathbf{E}(x,y,z)] = \begin{bmatrix} 4xy^3z & 3x^2y^2z+3x^2z & x^2y^3 \\ 3x^2y^2z+3x^2z & 6x^2z & 3x^2y+3z^2 \\ x^2y^3 & 3x^2y+3z^2 & 12yz \end{bmatrix} \times 10^{-4}.$$

Substituting the values of coordinates (x,y,z) for any point of the body, we have the infinitesimal strain tensor at the given point. For instance, for $(x,y,z) = (1,1,1)$, we have

$$[\mathbf{E}(1,1,1)] = \begin{bmatrix} 4 & 6 & 1 \\ 6 & 6 & 6 \\ 1 & 6 & 12 \end{bmatrix} \times 10^{-4}.$$

□

Analogously, we obtain the components of the infinitesimal rotation tensor Ω substituting (8.105) and (8.108) in (8.111). Thus,

$$[\Omega] = \begin{bmatrix} 0 & \dfrac{1}{2}\left(\dfrac{\partial u_x}{\partial y} - \dfrac{\partial u_y}{\partial x}\right) & \dfrac{1}{2}\left(\dfrac{\partial u_x}{\partial z} - \dfrac{\partial u_z}{\partial x}\right) \\ -\dfrac{1}{2}\left(\dfrac{\partial u_x}{\partial y} - \dfrac{\partial u_y}{\partial x}\right) & 0 & \dfrac{1}{2}\left(\dfrac{\partial u_y}{\partial z} - \dfrac{\partial u_z}{\partial y}\right) \\ -\dfrac{1}{2}\left(\dfrac{\partial u_x}{\partial z} - \dfrac{\partial u_z}{\partial x}\right) & -\dfrac{1}{2}\left(\dfrac{\partial u_y}{\partial z} - \dfrac{\partial u_z}{\partial y}\right) & 0 \end{bmatrix}. \tag{8.120}$$

We can write the tensor Ω in the following way

$$[\Omega] = \begin{bmatrix} 0 & -\Omega_z & \Omega_y \\ \Omega_z & 0 & -\Omega_x \\ -\Omega_y & \Omega_x & 0 \end{bmatrix}, \tag{8.121}$$

where Ω_x, Ω_y, and Ω_z respectively indicate the infinitesimal rotations of each point $\mathbf{x}$ about the Cartesian axes x, y, and z.

To verify this, consider the infinitesimal element around point P of the solid with a distortion γ_1 in plane xy, as shown in Figure 8.14(a). Notice that the diagonal of the element has a rotation Ω_1 about axis z in the counter-clockwise direction. The following expressions are valid for the angles shown in Figure 8.14(a):

$$2\beta = 2\alpha + \gamma_1 \rightarrow \beta = \alpha + \frac{1}{2}\gamma_1, \tag{8.122}$$

$$\beta + \Omega_1 = \alpha + \gamma_1. \tag{8.123}$$

Substituting (8.122) in (8.123), we have

$$a + \frac{1}{2}\gamma_1 + \Omega_1 = \alpha + \gamma_1 \rightarrow \Omega_1 = \frac{1}{2}\gamma_1. \tag{8.124}$$

Now consider a distortion γ_2 for the element, shown in Figure 8.14(b). Its diagonal has a rotation Ω_2 about z in the clockwise direction, and thus it has a negative value. From Figure 8.14(b), we verify that

$$2\beta = 2\alpha + \gamma_2 \rightarrow \beta = \alpha + \frac{1}{2}\gamma_2, \tag{8.125}$$

$$\beta - \Omega_2 = \alpha + \gamma_2,. \tag{8.126}$$

Substituting (8.125) in (8.126), we obtain

$$\Omega_2 = -\frac{1}{2}\gamma_2. \tag{8.127}$$

For the general case, where the elements have a distortion $\gamma_1 + \gamma_2$ [see Figure 8.14(c)], the diagonal presents a local rigid rotation Ω_z, given by

$$\Omega_z = \Omega_1 + \Omega_2. \tag{8.128}$$

Substituting (8.124) and (8.127) in (8.128), and recalling that $\gamma_1 = \dfrac{\partial u_y}{\partial x}$ and $\gamma_2 = \dfrac{\partial u_x}{\partial y}$, we have

$$\Omega_z = \frac{1}{2}\left(\frac{\partial u_y}{\partial x} - \frac{\partial u_x}{\partial y}\right). \tag{8.129}$$

Analogously, for the other planes [see Figures 8.15(a) and 8.15(b)], we have

$$\Omega_x = \frac{1}{2}\left(\frac{\partial u_y}{\partial z} - \frac{\partial u_z}{\partial y}\right), \tag{8.130}$$

$$\Omega_y = \frac{1}{2}\left(\frac{\partial u_x}{\partial z} - \frac{\partial u_z}{\partial x}\right). \tag{8.131}$$

Thus, the components Ω_x, Ω_y, and Ω_z represent infinitesimal rigid rotations about each point of the solid.

Example 8.23 *The infinitesimal rotation tensor for the bar is null, because the kinematics involves only stretching and shortening of cross-sections. However, for shafts and beams in bending only, the respective infinitesimal rotation tensors are:*

$$[\Omega(x,y,z)] = \begin{bmatrix} 0 & \frac{1}{2}z\frac{d\theta_x(x)}{dx} & -\frac{1}{2}y\frac{d\theta_x(x)}{dx} \\ -\frac{1}{2}z\frac{d\theta_x(x)}{dx} & 0 & -\theta_x(x) \\ \frac{1}{2}y\frac{d\theta_x(x)}{dx} & \theta_x(x) & 0 \end{bmatrix},$$

$$[\Omega(x,y,z)] = \begin{bmatrix} 0 & -\theta_z(x) & 0 \\ \theta_z(x) & 0 & 0 \\ 0 & 0 & 0 \end{bmatrix}.$$

$\square$

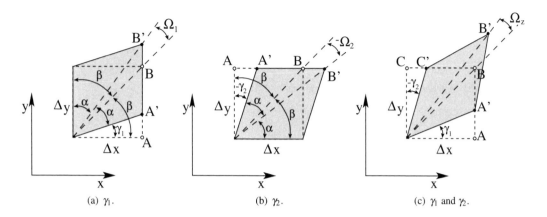

(a) γ_1.   (b) γ_2.   (c) γ_1 and γ_2.

Figure 8.14   Infinitesimal rigid body rotations in plane xy.

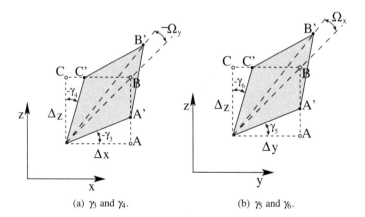

(a) γ_3 and γ_4.   (b) γ_5 and γ_6.

Figure 8.15   Infinitesimal rigid body rotations in planes xz and yz.

Example 8.24 *Consider the displacement vector field of Example 8.22. The matrix representation of the infinitesimal rotation tensorial field is obtained from (8.121) by*

$$[\Omega(x,y,z)] = \begin{bmatrix} 0 & 3x^2y^2z - 6xyz & x^2y^3 \\ -3x^2y^2z + 6xyz & 0 & -3x^2y - 3z^2 \\ -x^2y^3 & 3x^2y + 3z^2 & 0 \end{bmatrix} \times 10^{-4}.$$

The infinitesimal rotation tensor for point $(x,y,z) = (1,1,1)$ is

$$[\Omega(1,1,1)] = \begin{bmatrix} 0 & -3 & 1 \\ 3 & 0 & -6 \\ -1 & 6 & 0 \end{bmatrix} \times 10^{-4}.$$

□

Notice from (8.121) that the tensor Ω is skew-symmetric. Generally, a tensor $\mathbf{T}$ is skew-symmetric if

$$\mathbf{T} = -\mathbf{T}^T. \tag{8.132}$$

In terms of components, this implies that

$$T_{12} = -T_{21}, \qquad T_{13} = -T_{31}, \qquad T_{23} = -T_{32}, \qquad T_{11} = T_{22} = T_{33} = 0, \tag{8.133}$$

or, for $i,j = 1,2,3$,

$$\begin{cases} T_{ij} = -T_{ji}, & i \neq j, \\ T_{ij} = 0 & i = j \end{cases}. \tag{8.134}$$

Substituting (8.110) and (8.111) in (8.107), we have

$$\nabla \mathbf{u} = \mathbf{E} + \Omega, \tag{8.135}$$

that is, the displacement gradient tensor is given by the sum of a symmetric tensor $\mathbf{E}$ and a skew-symmetric tensor Ω. This decomposition is valid for any second-order tensor $\mathbf{A}$. Thus,

$$\mathbf{A} = \mathbf{A}^S + \mathbf{A}^A, \tag{8.136}$$

with the symmetric $\mathbf{A}^S$ and skew-symmetric $\mathbf{A}^A$ parts of $\mathbf{A}$, respectively given by

$$\mathbf{A}^S = \frac{1}{2}(\mathbf{A} + \mathbf{A}^T), \tag{8.137}$$

$$\mathbf{A}^A = \frac{1}{2}(\mathbf{A} - \mathbf{A}^T). \tag{8.138}$$

Example 8.25 *Taking the sum of the infinitesimal strain and rotation tensor fields obtained in Examples 8.21 and 8.23, we obtain the displacement gradient tensorial field given in Example 8.20.*
□

The terms $\mathbf{E}$ and Ω respectively represent the symmetric and skew-symmetric parts of the displacement gradient tensor, denoted by

$$\mathbf{E} = (\nabla \mathbf{u})^S, \tag{8.139}$$

$$\Omega = (\nabla \mathbf{u})^A. \tag{8.140}$$

Now, substituting (8.135) in (8.106) we have

$$\mathbf{u}(\mathbf{x} + \mathbf{d}) = \mathbf{u}(\mathbf{x}) + \mathbf{E}(\mathbf{x})\mathbf{d} + \Omega(\mathbf{x})\mathbf{d} + \mathscr{O}(\|\mathbf{d}\|^2). \tag{8.141}$$

This equation is very important because it shows that the vector field of small displacements of a three-dimensional continuous medium contains a term relative to the infinitesimal strains, given by tensor $\mathbf{E}$, and another one comprising an infinitesimal rotation, given by tensor Ω. Therefore, only the strain components of

E are not sufficient to take the body from its initial to deformed configuration. A rigid infinitesimal rotation occurs in the neighborhood of each point of the body.

To illustrate this fact consider the cantilever beam shown in Figure 8.16(a). Suppose that the beam is built by plates joined by pins. Figure 8.16(b) illustrates the deformed geometry of the beam when subjected to a force at the free end. Removing the pins of the upper part and bending each plate separately, we notice that the obtained deformed geometry is not the same as before [see Figure 8.16(c)] unless there is a rigid rotation $\Omega_z = \theta_z$ around each point (see Example 8.23). Thus, this simple example shows that the term regarding the infinitesimal rotation in (8.141) is generally present when a body has a deformation action.

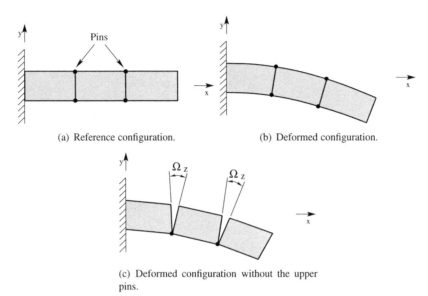

(a) Reference configuration. (b) Deformed configuration.

(c) Deformed configuration without the upper pins.

Figure 8.16 Interpretation of the rigid rotation in a cantilever beam.

Now considering that points $\mathbf{y} = \mathbf{x} + \mathbf{d}$ and $\mathbf{x}$ illustrated in Figure 8.17 are very close to each other, we have an infinitesimal norm for vector $\mathbf{d}$. Thus, the term $\mathcal{O}(\|\mathbf{d}\|^2)$ is neglected in equation (8.141), and we obtain the following expression for the infinitesimal displacement field in the neighborhood of $\mathbf{y} = \mathbf{x} + \mathbf{d}$

$$\mathbf{u}(\mathbf{x}+\mathbf{d}) = \mathbf{u}(\mathbf{x}) + \mathbf{E}(\mathbf{x})\mathbf{d} + \Omega(\mathbf{x})\mathbf{d}, \tag{8.142}$$

or,

$$\mathbf{u}(\mathbf{x}+\mathbf{d}) = \mathbf{u}(\mathbf{x}) + \nabla\mathbf{u}(\mathbf{x})\mathbf{d}. \tag{8.143}$$

Expression (8.143) is rewritten as

$$\mathbf{u}(\mathbf{x}+\mathbf{d}) - \mathbf{u}(\mathbf{x}) = \nabla\mathbf{u}(\mathbf{x})\mathbf{d}. \tag{8.144}$$

The following vector relation is valid from Figure 8.17:

$$\mathbf{d}' = \mathbf{d} + \mathbf{u}(\mathbf{x}+\mathbf{d}) - \mathbf{u}(\mathbf{x}).$$

Substituting (8.144) in the previous expression, we have

$$\mathbf{d}' = \mathbf{d} + \nabla\mathbf{u}(\mathbf{x})\mathbf{d} = [\mathbf{I} + \nabla\mathbf{u}(\mathbf{x})]\mathbf{d}, \tag{8.145}$$

where **I** is the identity tensor with the following matrix representation:

$$[\mathbf{I}] = \begin{bmatrix} 1 & 0 & 0 \\ 0 & 1 & 0 \\ 0 & 0 & 1 \end{bmatrix}. \tag{8.146}$$

Observe that the components of tensor $\mathbf{I}$ are independent of the used coordinate system, and therefore $\mathbf{I}$ is called an isotropic tensor.

The expression

$$\mathbf{F}(\mathbf{x}) = \mathbf{I} + \nabla \mathbf{u}(\mathbf{x}), \tag{8.147}$$

is called deformation gradient tensor. This tensor is fundamental in problems involving large strains [36, 12].

From the previous definition, equation (8.145) assumes the following form:

$$\mathbf{d}' = \mathbf{F}(\mathbf{x})\mathbf{d}. \tag{8.148}$$

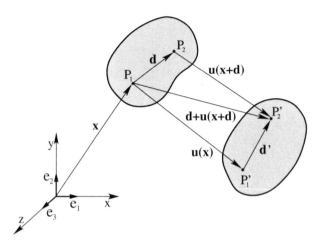

Figure 8.17   Deformation of a solid body.

Equation (8.148) allows the determination of final distance $\mathbf{d}'$ between P_1' and P_2' after the deformation process, through tensor $\mathbf{F}$ and the initial distance $\mathbf{d}$ between points P_1 and P_2, which are sufficiently close to each other. To obtain the strain measure of P_1, we take the difference of the lengths of vectors $\mathbf{d}'$ and $\mathbf{d}$. Thus, using (8.148), we have

$$\Delta \mathbf{d} = \mathbf{d}' \cdot \mathbf{d}' - \mathbf{d} \cdot \mathbf{d} = \mathbf{F}(\mathbf{x})\mathbf{d} \cdot \mathbf{F}(\mathbf{x})\mathbf{d} - \mathbf{d} \cdot \mathbf{d}. \tag{8.149}$$

Based on the definition of transposed tensor given in (8.109), expression (8.149) can be rewritten as

$$\Delta \mathbf{d} = \mathbf{F}^T(\mathbf{x})\mathbf{F}(\mathbf{x})\mathbf{d} \cdot \mathbf{d} - \mathbf{d} \cdot \mathbf{d} = [\mathbf{F}^T(\mathbf{x})\mathbf{F}(\mathbf{x}) - \mathbf{I}]\mathbf{d} \cdot \mathbf{d}. \tag{8.150}$$

Calling

$$\mathbf{E}^*(\mathbf{x}) = \frac{1}{2}[\mathbf{F}^T(\mathbf{x})\mathbf{F}(\mathbf{x}) - \mathbf{I}], \tag{8.151}$$

as the Green-Lagrange strain tensor, equation (8.150) can be rewritten as

$$\Delta \mathbf{d} = 2\mathbf{E}^*(\mathbf{x})\mathbf{d} \cdot \mathbf{d}. \tag{8.152}$$

Substituting (8.147) in (8.151), we have

$$\mathbf{E}^*(\mathbf{x}) = \frac{1}{2}\{[\mathbf{I} + \nabla \mathbf{u}(\mathbf{x})]^T[\mathbf{I} + \nabla \mathbf{u}(\mathbf{x})] - \mathbf{I}\}. \tag{8.153}$$

Given two tensors $\mathbf{A}$ and $\mathbf{B}$, the following relation is valid:

$$(\mathbf{A} + \mathbf{B})^T = \mathbf{A}^T + \mathbf{B}^T. \tag{8.154}$$

As $\mathbf{I}^T = \mathbf{I}$, then

$$
\begin{aligned}
\mathbf{E}^* &= \frac{1}{2}\{[\mathbf{I}+\nabla\mathbf{u}^T][\mathbf{I}+\nabla\mathbf{u}]-\mathbf{I}\} \\
&= \frac{1}{2}[\mathbf{I}+\nabla\mathbf{u}+\nabla\mathbf{u}^T+\nabla\mathbf{u}^T\nabla\mathbf{u}-\mathbf{I}] \\
&= \frac{1}{2}[\nabla\mathbf{u}+\nabla\mathbf{u}^T]+\frac{1}{2}\nabla\mathbf{u}^T\nabla\mathbf{u} \\
&= \mathbf{E}+\frac{1}{2}\nabla\mathbf{u}^T\nabla\mathbf{u}.
\end{aligned}
\tag{8.155}
$$

From equation (8.155) we notice that the Green-Lagrange strain tensor provides a general strain measure, which is applicable for small and large strains. However, for small strains, the norms of $\mathbf{u}$ and $\nabla\mathbf{u}$ are small, i.e., $\|\mathbf{u}\| < \varepsilon$ and $\|\nabla\mathbf{u}\| < \varepsilon$, with ε of order 10^{-4}, for instance. In this case, the nonlinear term $\frac{1}{2}\nabla\mathbf{u}^T\nabla\mathbf{u}$ becomes negligible and the tensor $\mathbf{E}^*$ is reduced to the infinitesimal strain tensor $\mathbf{E}$.

Example 8.26 *Consider the displacement vector field of Example 8.22. The Green-Lagrange tensor field is given by the sum of the term regarding small strains with the nonlinear strain term $\frac{1}{2}\nabla\mathbf{u}^T\nabla\mathbf{u}$. Thus,*

$$
[\mathbf{E}^*(x,y,z)] = \begin{bmatrix} 4xy^3z & 3x^2y^2z+3x^2z & x^2y^3 \\ 3x^2y^2z+3x^2z & 6x^2z & 3x^2y+3z^2 \\ x^2y^3 & 3x^2y+3z^2 & 12yz \end{bmatrix}\alpha +
$$

$$
\begin{bmatrix} 8x^2y^6z^2+72x^2y^2z^2 & 12x^3y^5z^2+36x^3yz^2 & 4zx^3y^6+36zx^3y^2 \\ 12x^3y^5z^2+36x^3yz^2 & 18x^4y^4z^2+18x^4z^2+18z^4 & 6x^4y^5z+18x^4yz+36yz^3 \\ 4zx^3y^6+36zx^3y^2 & 6x^4y^5z+18x^4yz+36yz^3 & 2x^4y^6+18x^4y^2+72y^2z^2 \end{bmatrix}\alpha^2.
$$

For $\alpha = 10^{-2}$ and taking point $(1,1,1)$, the first term of the above tensor is given by

$$
E^*(1,1) = 0.040+0.008 = 0.048.
$$

The ratio between the linear and nonlinear terms is 0.2 or 20%. Thus, we have a case of large strains.

For $\alpha = 10^{-4}$, due to the term $\alpha^2 = 10^{-8}$ multiplying the nonlinear term, its contribution to the strain measure becomes negligible and we have a case of small strains, described only by tensor $\mathbf{E}$.

□

8.10.5 RIGID ACTIONS

It is known that a three-dimensional body has six rigid actions, comprising three translations and three rotations in directions x, y, and z. We want to verify how the rigid actions can be represented, using the concepts presented in the previous section.

The deformation is homogeneous if the displacement gradient tensor $\nabla\mathbf{u}$ is constant for all points $\mathbf{x}$ of the body, indicating $\nabla\mathbf{u} = \nabla\mathbf{u}_0$. In this case, expression (8.106) simplifies to

$$
\mathbf{u}(\mathbf{x}+\mathbf{d}) = \mathbf{u}(\mathbf{x})+\nabla\mathbf{u}_0\mathbf{d}.
\tag{8.156}
$$

Notice that as $\nabla\mathbf{u}_0$ is constant, the other terms of the Taylor series expansion are zero, and thus $\mathcal{O}(\|\mathbf{d}\|^2)$ is zero.

An example of homogeneous deformation is a translation from a given position of the body. As all points of the body have the same displacement, we have

$$
\mathbf{u}(\mathbf{x}+\mathbf{d}) = \mathbf{u}(\mathbf{x}).
\tag{8.157}
$$

Substituting this relation in (8.156), we have

$$
\nabla\mathbf{u}_0\mathbf{d} = \mathbf{0},
\tag{8.158}
$$

As $\mathbf{d}$ is the distance between two arbitrary points of the body, the previous expression is zero if

$$
\nabla\mathbf{u}_0 = \mathbf{0}.
\tag{8.159}
$$

Therefore, as the gradient of the displacement field is zero, the displacement field $\mathbf{u}_0$ for a translation is constant for all points of the body, that is,

$$\mathbf{u}(\mathbf{x}) = \mathbf{u}(\mathbf{x}+\mathbf{d}) = \mathbf{u}_0 = \left\{ \begin{array}{ccc} u_{x_0} & u_{y_0} & u_{z_0} \end{array} \right\}, \tag{8.160}$$

where u_{x_0}, u_{y_0}, and u_{z_0} respectively are the translational components in directions x, y, and z. As u_{x_0}, u_{y_0}, and u_{z_0} are constant, the respective components of the infinitesimal strain tensor $\mathbf{E}$ are zero, which characterizes a rigid rotation. Figure 8.18 illustrates a translation of a solid body.

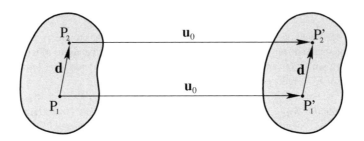

Figure 8.18 Translation of a solid body.

Now consider a rigid rotation of the body, with the origin of the Cartesian reference system at P_1, as illustrated in Figure 8.19. In this case, the displacement $\mathbf{u}(\mathbf{x})$ of point P_1 in equation (8.156) is zero. Thus,

$$\mathbf{u}(\mathbf{x}+\mathbf{d}) = \nabla \mathbf{u}_0 \mathbf{d}. \tag{8.161}$$

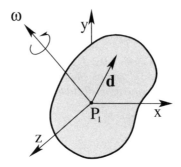

Figure 8.19 Rigid rotation.

As the action is rigid, the symmetric part of $\nabla \mathbf{u}_0$, that is, the infinitesimal strain tensor $\mathbf{E}$ is zero. Thus,

$$\mathbf{u}(\mathbf{x}+\mathbf{d}) = \Omega \mathbf{d}. \tag{8.162}$$

There is an axial vector θ associated with every skew-symmetric tensor Ω, such that

$$\Omega \mathbf{v} = \theta \times \mathbf{v}, \tag{8.163}$$

for every vector $\mathbf{v} = \{v_x \, v_y \, v_z\}^T$. In this case, the components of θ are Ω_x, Ω_y, and Ω_z, that is, the rigid rotations about axes x, y, and z. To verify this relation, we expand both sides of the above expression, that is,

$$\Omega \mathbf{v} = \begin{bmatrix} 0 & -\Omega_z & \Omega_y \\ \Omega_z & 0 & -\Omega_x \\ -\Omega_y & \Omega_x & 0 \end{bmatrix} \left\{ \begin{array}{c} v_x \\ v_y \\ v_z \end{array} \right\} = \left\{ \begin{array}{c} v_z\Omega_y - v_y\Omega_z \\ v_x\Omega_z - v_z\Omega_x \\ v_y\Omega_x - v_x\Omega_y \end{array} \right\}, \tag{8.164}$$

$$\theta \times \mathbf{v} = \begin{bmatrix} \mathbf{e}_x & \mathbf{e}_y & \mathbf{e}_z \\ \theta_x & \theta_y & \theta_z \\ v_x & v_y & v_z \end{bmatrix} \left\{ \begin{array}{c} v_x \\ v_y \\ v_z \end{array} \right\} = \left\{ \begin{array}{c} v_z\theta_y - v_y\theta_z \\ v_x\theta_z - v_z\theta_x \\ v_y\theta_x - v_x\theta_y \end{array} \right\}. \tag{8.165}$$

Thus,

$$\begin{cases} \theta_x = \Omega_x \\ \theta_y = \Omega_y \\ \theta_z = \Omega_z \end{cases}. \tag{8.166}$$

Based on these results, we can write

$$\mathbf{u}(\mathbf{x} + \mathbf{d}) = \theta \times \mathbf{d}. \tag{8.167}$$

Figure 8.20 illustrates a rotation about the z axis of a solid body.

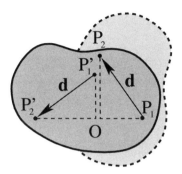

Figure 8.20 Rotation of a solid body about the z axis of the Cartesian reference system.

Hence, a general rigid body action is be given by the superposition of translation and rotation components, respectively expressed by (8.160) and (8.167). Thus, a general rigid body action can be written as

$$\mathbf{u}(\mathbf{x}) = \mathbf{u}_0 + \theta \times \mathbf{d}. \tag{8.168}$$

8.10.6 DETERMINATION OF INTERNAL LOADS

In the general case of infinitesimal strains in a solid, the strain state at each point is given by the components indicated in (8.114). There are normal stress components called σ_{xx}, σ_{yy}, and σ_{zz}, which are respectively associated to the normal strains ε_{xx}, ε_{yy}, and ε_{zz}, and represent the state of internal traction and compression loads at point $\mathbf{x}$ in directions x, y, and z. Likewise, there are shear stresses τ_{xy}, τ_{yx}, τ_{xz}, τ_{zx}, τ_{yz}, and τ_{zy}, which are associated to distortions γ_{xy}, γ_{yx}, γ_{xz}, γ_{zx}, γ_{yz}, and γ_{zy}, where the shear stresses give the state of internal shear loads at point $\mathbf{x}$, according to planes xy, xz, and yz, respectively. Thus, the stress state at each point P of a solid body according to a Cartesian coordinate system is given by nine stress components illustrated in Figure 8.21. Although a cube is employed in the neighborhood of point P to facilitate the stress components visualization, the stress state is effectively represented at point P.

The internal work for the infinitesimal volume element around point P is

$$dW_i = \left[\sigma_{xx}\varepsilon_{xx} + \sigma_{yy}\varepsilon_{yy} + \sigma_{zz}\varepsilon_{zz} + (\tau_{xy} + \tau_{yx})\gamma_{xy} + (\tau_{xz} + \tau_{zx})\gamma_{xz} + (\tau_{yz} + \tau_{zy})\gamma_{yz} \right] dV, \tag{8.169}$$

where the symmetry of the infinitesimal strain tensor $\mathbf{E}$ was used.

The general internal work expression for a three-dimensional body is written as

$$W_i = \int_V dW_i = \int_V [\sigma_{xx}\varepsilon_{xx} + \sigma_{yy}\varepsilon_{yy} + \sigma_{zz}\varepsilon_{zz} + (\tau_{xy} + \tau_{yx})\gamma_{xy} + (\tau_{xz} + \tau_{zx})\gamma_{xz} + (\tau_{yz} + \tau_{zy})\gamma_{yz}]dV. \tag{8.170}$$

The stress components at point $\mathbf{x}$ are denoted by the Cauchy stress tensor $\mathbf{T}(\mathbf{x})$, which has the following matrix representation in the Cartesian system:

$$[\mathbf{T}] = \begin{bmatrix} \sigma_{xx} & \tau_{xy} & \tau_{xz} \\ \tau_{yx} & \sigma_{yy} & \tau_{yz} \\ \tau_{zx} & \tau_{zy} & \sigma_{zz} \end{bmatrix}. \tag{8.171}$$

The Cauchy stress tensor $\mathbf{T}$ represents the internal stress state for each point $\mathbf{x} = (x, y, z)$ of a three-dimensional body.

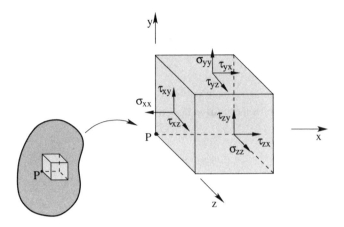

Figure 8.21   Stress state at a point of a solid.

Example 8.27 *In the case of bars and beams in bending only, there is a uniaxial stress state, with the normal stress component σ_{xx} associated to the strain component ε_{xx}. Thus, the Cauchy stress tensor is reduced to*

$$[\mathbf{T}] = \begin{bmatrix} \sigma_{xx} & 0 & 0 \\ 0 & 0 & 0 \\ 0 & 0 & 0 \end{bmatrix}.$$

On the other hand, the stress state for each point of a shaft is given by the following Cauchy stress tensor:

$$[\mathbf{T}] = \begin{bmatrix} 0 & \tau_{xy} & \tau_{xz} \\ \tau_{yx} & 0 & 0 \\ \tau_{zx} & 0 & 0 \end{bmatrix}.$$

It is a pure shear stress state.
□

Recall that the scalar product of two vectors $\mathbf{a} = \{a_1 \; a_2 \; a_3\}^T$ and $\mathbf{b} = \{b_1 \; b_2 \; b_3\}^T$ is calculated as

$$\mathbf{a} \cdot \mathbf{b} = a_1 b_1 + a_2 b_2 + a_3 b_3. \tag{8.172}$$

This scalar product of vectors is a particular case of the general concept of inner product, which can be applied to other mathematical entities, such as functions and tensors. Also notice that the inner product of vectors is commutative, that is,

$$\mathbf{a} \cdot \mathbf{b} = \mathbf{b} \cdot \mathbf{a}. \tag{8.173}$$

Considering the tensors of small strain $\mathbf{E}$ and stress $\mathbf{T}$, the inner product $\mathbf{E} : \mathbf{T}$ is defined as

$$\mathbf{E} : \mathbf{T} = \text{tr}(\mathbf{E}^T \mathbf{T}), \tag{8.174}$$

where $\text{tr}(.)$ is the trace of a tensor, given by the sum of the main diagonal components. When substituting the Cartesian components of $\mathbf{E}$ and $\mathbf{T}$, performing the multiplication, and calculating the trace, we have

$$
\begin{aligned}
\mathbf{E} : \mathbf{T} &= \text{tr}\left(\begin{bmatrix} \varepsilon_{xx} & \gamma_{xy} & \gamma_{xz} \\ \gamma_{yx} & \varepsilon_{yy} & \gamma_{yz} \\ \gamma_{zx} & \gamma_{zy} & \varepsilon_{zz} \end{bmatrix}^T \begin{bmatrix} \sigma_{xx} & \tau_{xy} & \tau_{xz} \\ \tau_{yx} & \sigma_{yy} & \tau_{yz} \\ \tau_{zx} & \tau_{zy} & \sigma_{zz} \end{bmatrix} \right) \\
&= \sigma_{xx}\varepsilon_{xx} + \sigma_{yy}\varepsilon_{yy} + \sigma_{zz}\varepsilon_{zz} + (\tau_{xy} + \tau_{yx})\gamma_{xy} + (\tau_{xz} + \tau_{zx})\gamma_{xz} + (\tau_{yz} + \tau_{zy})\gamma_{yz}. \tag{8.175}
\end{aligned}
$$

Comparing (8.175) and (8.169), we notice that the internal work density dW_i for each point of the body is the inner product $\mathbf{E} : \mathbf{T}$, such that

$$dW_i = \mathbf{T} : \mathbf{E}. \tag{8.176}$$

Thus, we can write the final expression for the strain internal work as follows

$$W_i = \int_V \mathbf{T} : \mathbf{E} dV. \tag{8.177}$$

Example 8.28 *Using the above expression and the tensors given in Examples 8.21 and 8.27, we recover the equation of the strain internal work for the bar and the beam in bending only, that is,*

$$W_i = \int_V \sigma_{xx} \varepsilon_{xx} dV.$$

□

Now consider the inner product of tensors $\mathbf{T}$ and $\nabla \mathbf{u}$, as well as the decomposition of $\nabla \mathbf{u}$ in symmetric $\mathbf{E}$ and skew-symmetric Ω parts, that is,

$$\mathbf{T} : \nabla \mathbf{u} = \mathbf{T} : (\mathbf{E} + \Omega) = \mathbf{T} : \mathbf{E} + \mathbf{T} : \Omega. \tag{8.178}$$

For a rigid action, the components of $\mathbf{E}$ are zero, while the components of Ω are constant. Consequently, the internal work is zero for a rigid action. Thus, from (8.178), we have

$$W_i = \int_V \mathbf{T} : \Omega dV = 0. \tag{8.179}$$

Performing the indicated inner product, we obtain

$$\mathbf{T} : \Omega = \mathrm{tr}\left(\begin{bmatrix} \sigma_{xx} & \tau_{xy} & \tau_{xz} \\ \tau_{yx} & \sigma_{yy} & \tau_{yz} \\ \tau_{zx} & \tau_{zy} & \sigma_{zz} \end{bmatrix}^T \begin{bmatrix} 0 & -\Omega_z & \Omega_y \\ \Omega_z & 0 & \Omega_x \\ -\Omega_y & -\Omega_x & 0 \end{bmatrix}\right)$$

$$= (\tau_{yx} - \tau_{xy})\Omega_z + (\tau_{xz} - \tau_{zx})\Omega_y + (\tau_{yz} - \tau_{zy})\Omega_x. \tag{8.180}$$

Substituting the previous result in (8.179), we have

$$\int_V \left[(\tau_{yx} - \tau_{xy})\Omega_z + (\tau_{xz} - \tau_{zx})\Omega_y + (\tau_{yz} - \tau_{zy})\Omega_x\right] dV = 0. \tag{8.181}$$

As Ω_x, Ω_y, and Ω_z are constant, the above equation is only satisfied when the terms inside brackets are simultaneously zero, that is,

$$\tau_{xy} = \tau_{yx}, \qquad \tau_{xz} = \tau_{zx}, \qquad \tau_{yz} = \tau_{zy}. \tag{8.182}$$

This result implies that the Cauchy stress tensor $\mathbf{T}$ is symmetric, i.e., $\mathbf{T} = \mathbf{T}^T$. This is of main importance in mechanics, and is one of the results of Cauchy's theorem. However, in the variational formulation, the symmetry of the stress tensor is a consequence of the work definition and the zero strain components for rigid actions. Finally, notice that the infinitesimal rotation does not contribute to the strain internal work.

Another important result is that the inner product of a symmetric tensor $\mathbf{A}$ and a skew-symmetric tensor $\mathbf{B}$ is always zero, that is

$$\mathbf{A} : \mathbf{B} = 0.$$

Example 8.29 *Consider the symmetric $\mathbf{A}$ and skew-symmetric $\mathbf{B}$ tensors with the following components:*

$$[\mathbf{A}] = \begin{bmatrix} 1 & 4 & 5 \\ 4 & 2 & 6 \\ 5 & 6 & 3 \end{bmatrix} \quad and \quad [\mathbf{B}] = \begin{bmatrix} 0 & 1 & -2 \\ -1 & 0 & -3 \\ 2 & 3 & 0 \end{bmatrix}.$$

The inner product between these tensors is

$$\mathbf{A} : \mathbf{B} = (4)(1) + (5)(-2) + (4)(-1) + (6)(-3) + (5)(2) + (6)(3) = 0.$$

□

We now integrate by parts the internal work expression. Hence, the concepts of divergence of vector and tensor fields are defined.

Given a vector field $\mathbf{v}$, its divergent is defined as

$$\text{div } \mathbf{v} = \text{tr}(\nabla \mathbf{v}). \tag{8.183}$$

Expanding the above expression in terms of components $[v_x(x,y,z), v_y(x,y,z), v_z(x,y,z)]$ of $\mathbf{v}(x,y,z)$, we have

$$\text{div } \mathbf{v} = \text{tr}\left(\begin{bmatrix} \dfrac{\partial v_x}{\partial x} & \dfrac{\partial v_x}{\partial y} & \dfrac{\partial v_x}{\partial z} \\ \dfrac{\partial v_y}{\partial x} & \dfrac{\partial v_y}{\partial y} & \dfrac{\partial v_y}{\partial z} \\ \dfrac{\partial v_z}{\partial x} & \dfrac{\partial v_z}{\partial y} & \dfrac{\partial v_z}{\partial z} \end{bmatrix}\right) = \dfrac{\partial v_x}{\partial x} + \dfrac{\partial v_y}{\partial y} + \dfrac{\partial v_z}{\partial z}, \tag{8.184}$$

or,

$$\text{div } \mathbf{v} = \left\{\begin{array}{c} \dfrac{\partial}{\partial x} \\ \dfrac{\partial}{\partial y} \\ \dfrac{\partial}{\partial z} \end{array}\right\} \cdot \left\{\begin{array}{c} v_x \\ v_y \\ v_z \end{array}\right\} = \nabla \cdot \mathbf{v}.$$

Example 8.30 *Consider the vector field of Example 8.17. The associated divergent is given by*

$$div \, \mathbf{f} = 21yz + y^2.$$

Thus, notice that the divergent of a vector field yields a scalar field.
□

On the other hand, the divergent of a tensor field $\mathbf{A}$ is defined as

$$(\text{div } \mathbf{A}) \cdot \mathbf{v} = \text{div } (\mathbf{A}^T \mathbf{v}). \tag{8.185}$$

Manipulating the right-hand side of the above expression, we have

$$\begin{aligned} (\text{div } \mathbf{A}) \cdot \mathbf{v} &= \text{div}\left(\begin{bmatrix} A_{xx} & A_{xy} & A_{xz} \\ A_{yx} & A_{yy} & A_{yz} \\ A_{zx} & A_{zy} & A_{zz} \end{bmatrix}^T \left\{\begin{array}{c} v_x \\ v_y \\ v_z \end{array}\right\}\right) \\ &= \text{div}\left\{\begin{array}{c} A_{xx}v_x + A_{yx}v_y + A_{zx}v_z \\ A_{xy}v_x + A_{yy}v_y + A_{zy}v_z \\ A_{xz}v_x + A_{yz}v_y + A_{zz}v_z \end{array}\right\}. \end{aligned} \tag{8.186}$$

Now taking the divergent of a vector field, we have

$$(\text{div } \mathbf{A}) \cdot \mathbf{v} = \left\{\begin{array}{c} \dfrac{\partial}{\partial x} \\ \dfrac{\partial}{\partial y} \\ \dfrac{\partial}{\partial z} \end{array}\right\} \cdot \left\{\begin{array}{c} A_{xx}v_x + A_{yx}v_y + A_{zx}v_z \\ A_{xy}v_x + A_{yy}v_y + A_{zy}v_z \\ A_{xz}v_x + A_{yz}v_y + A_{zz}v_z \end{array}\right\}. \tag{8.187}$$

Taking the scalar product and the factorization of v_x, v_y, and v_z, we have

$$(\text{div } \mathbf{A}) \cdot \mathbf{v} = \left\{\begin{array}{c} \dfrac{\partial A_{xx}}{\partial x} + \dfrac{\partial A_{xy}}{\partial y} + \dfrac{\partial A_{xz}}{\partial z} \\ \dfrac{\partial A_{yx}}{\partial x} + \dfrac{\partial A_{yy}}{\partial y} + \dfrac{\partial A_{yz}}{\partial z} \\ \dfrac{\partial A_{zx}}{\partial x} + \dfrac{\partial A_{zy}}{\partial y} + \dfrac{\partial A_{zz}}{\partial z} \end{array}\right\} \cdot \left\{\begin{array}{c} v_x \\ v_y \\ v_z \end{array}\right\}.$$

Thus,

$$
\operatorname{div} \mathbf{A} = \left\{ \begin{array}{l} \dfrac{\partial A_{xx}}{\partial x} + \dfrac{\partial A_{xy}}{\partial y} + \dfrac{\partial A_{xz}}{\partial z} \\[2mm] \dfrac{\partial A_{yx}}{\partial x} + \dfrac{\partial A_{yy}}{\partial y} + \dfrac{\partial A_{yz}}{\partial z} \\[2mm] \dfrac{\partial A_{zz}}{\partial x} + \dfrac{\partial A_{zy}}{\partial y} + \dfrac{\partial A_{zz}}{\partial z} \end{array} \right\}. \tag{8.188}
$$

Example 8.31 *Consider the tensor field obtained in Example 8.17. The divergent is*

$$
\{div\ \nabla \mathbf{f}\} = \left\{ \begin{array}{c} 0+0+0 \\ 0+20z+0 \\ 0+2z+0 \end{array} \right\} = \left\{ \begin{array}{c} 0 \\ 20z \\ 2z \end{array} \right\}.
$$

Notice that the divergent of a tensor field is a vector field.

□

Given the tensor field $\mathbf{A}$ and vector field $\mathbf{u}$, the following relation is valid:

$$
\mathbf{A} : \nabla \mathbf{u} = \operatorname{div}\left(\mathbf{A}^T \mathbf{u}\right) - (\operatorname{div} \mathbf{A}) \cdot \mathbf{u}. \tag{8.189}
$$

Using this relation in (8.177) and recalling that $\mathbf{E} = (\nabla \mathbf{u})^S$ and $\mathbf{T} = \mathbf{T}^T$, the strain internal work can be denoted by

$$
\begin{aligned}
W_i &= \int_V \mathbf{T} : (\nabla \mathbf{u})^S dV = \int_V \left[\operatorname{div}\left(\mathbf{T}^T \mathbf{u}\right) - (\operatorname{div} \mathbf{T}) \cdot \mathbf{u} \right] dV \\
&= -\int_V (\operatorname{div} \mathbf{T}) \cdot \mathbf{u} dV + \int_V \operatorname{div}(\mathbf{T}\mathbf{u}) dV.
\end{aligned} \tag{8.190}
$$

Example 8.32 *For the bar model and with $dV = A dx$, both integrals of the above expression result in*

$$
\int_V (\operatorname{div} \mathbf{T}) \cdot \mathbf{u} dV = \int_0^L \frac{d\sigma_{xx}(x)}{dx} u_x(x) A dx = \int_0^L \frac{dN_x(x)}{dx} u_x(x) dx,
$$

$$
\int_V \operatorname{div}(\mathbf{T}\mathbf{u}) dV = \int_0^L \frac{d}{dx}(\sigma_{xx}(x) u_x(x)) A dx = \int_0^L \left(\frac{dN_x(x)}{dx} u_x(x) + N_x(x) \frac{du_x(x)}{dx} \right) dx.
$$

Substituting the previous results in the internal work expression (8.190), we have

$$
W_i = \int_0^L N_x(x) \frac{du_x(x)}{dx} dx.
$$

□

The divergence theorem allows the transformation of a volume integral in another integral along the surface S of the body. With the vector field $\mathbf{v}$, this theorem states that

$$
\int_V \operatorname{div} \mathbf{v} dV = \int_S \mathbf{v} \cdot \mathbf{n} dS, \tag{8.191}
$$

where $\mathbf{n}$ is the normal vector field of surface S.

Applying this theorem in the second integral of equation (8.190) we have

$$
\int_V \operatorname{div}(\mathbf{T}\mathbf{u}) dV = \int_S \mathbf{T}\mathbf{u} \cdot \mathbf{n} dS. \tag{8.192}
$$

Substituting this result in (8.190), we obtain

$$
W_i = -\int_V \operatorname{div} \mathbf{T} \cdot \mathbf{u} dV + \int_S \mathbf{T}\mathbf{n} \cdot \mathbf{u} dS, \tag{8.193}
$$

which represents the integration by parts of the internal work expression in tensor form. The term $\operatorname{div} \mathbf{T}$ represents the internal load density by units of volume, which is uniformly distributed in the points of the body. On the other hand, the term $\mathbf{T}\mathbf{n}$ is the internal load density by units of area present at the surface points.

Example 8.33 *The surface integral of equation (8.193) yields for the bar*

$$\int_S \mathbf{Tn} \cdot \mathbf{u} dS = \int_S \sigma_{xx}(x) u_x(x) n_x(x) dS = N_x(L) u_x(L) - N_x(0) u_x(0).$$

Substituting the volume and surface integrals obtained for the bar, we obtain the final expression for the strain internal work, which was derived previously

$$W_i = -\int_0^L \frac{dN_x(x)}{dx} u_x(x) dx + N_x(L) u_x(L) - N_x(0) u_x(0).$$

□

8.10.7 EQUILIBRIUM

As seen in Section 8.6, the compatible external work expression with (8.193) is given by

$$W_e = \int_V \mathbf{b} \cdot \mathbf{u} dV + \int_S \mathbf{t} \cdot \mathbf{u} dS, \tag{8.194}$$

with **b** and **t** respectively being the vector fields of external volume and surface forces.

Applying the PVW for a virtual kinematic action $\delta \mathbf{u}$,

$$\delta W_e = \delta W_i, \tag{8.195}$$

and substituting (8.193) and (8.194), we have

$$\int_V (\text{div } \mathbf{T} + \mathbf{b}) \cdot \delta \mathbf{u} dV + \int_S (\mathbf{Tn} - \mathbf{t}) \cdot \delta \mathbf{u} dS = 0. \tag{8.196}$$

In order for the above expression to be zero, considering that $\delta \mathbf{u}$ is an arbitrary action, the terms inside parentheses must be simultaneously zero, that is,

$$\begin{cases} \text{div } \mathbf{T} + \mathbf{b} = \mathbf{0} \\ \mathbf{Tn} = \mathbf{t} \end{cases}. \tag{8.197}$$

The above equations constitute the same boundary value problem (BVP) obtained in Section 8.6, in terms of stress. However, the expressions in (8.197) are undoubtedly a more compact and elegant representation for the equilibrium of solids. In addition, the above notation is abstract, because it is valid for any adopted coordinate system. As noticed previously, the BVP (8.197) is valid for any continuous medium (solid, liquid, or gas) in small strains. Figure 8.22 illustrates the variational formulation of a solid body in small strain.

8.10.8 APPLICATION OF THE CONSTITUTIVE EQUATION

The generalized Hooke's law given in (8.70) can be written in tensor form as

$$\mathbf{T} = 2\mu \mathbf{E} + \lambda e \mathbf{I} = \mu (\nabla \mathbf{u} + \nabla \mathbf{u}^T) + \lambda (\text{div } \mathbf{u}) \mathbf{I}, \tag{8.198}$$

where **T** is the Cauchy stress tensor, **E** is the infinitesimal strain tensor, **I** the identity tensor, $e = \varepsilon_{xx} + \varepsilon_{yy} + \varepsilon_{zz}$ the dilatation, and μ and λ the Lam coefficients.

To verify that (8.70) and (8.198) are identical, we expand (8.198) according to the Cartesian coordinate system, that is,

$$\begin{bmatrix} \sigma_{xx} & \tau_{xy} & \tau_{xz} \\ \tau_{yx} & \sigma_{yy} & \tau_{yz} \\ \tau_{zx} & \tau_{zy} & \sigma_{zz} \end{bmatrix} = 2\mu \begin{bmatrix} \varepsilon_{xx} & \gamma_{xy} & \gamma_{xz} \\ \gamma_{yx} & \varepsilon_{yy} & \gamma_{yz} \\ \gamma_{zx} & \gamma_{zy} & \varepsilon_{zz} \end{bmatrix} + \lambda e \begin{bmatrix} 1 & 0 & 0 \\ 0 & 1 & 0 \\ 0 & 0 & 1 \end{bmatrix}. \tag{8.199}$$

For instance, $\sigma_{xx} = 2\mu \varepsilon_{xx} + \lambda e$ and $\tau_{xy} = 2\mu \gamma_{xy}$, which is the same expression obtained from (8.70). It is analogous for the other components.

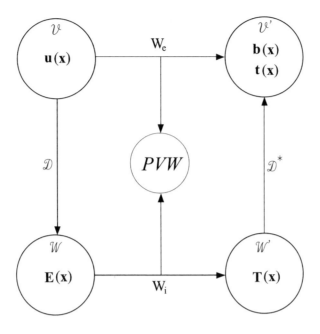

Figure 8.22 Variational formulation of a solid.

The equilibrium BVP in (8.197) is given in terms of the stress components. To obtain the Navier equation in terms of the displacement components, we should substitute (8.198) in (8.197). Thus,

$$\text{div}\,(2\mu\mathbf{E} + \lambda e\mathbf{I}) + \mathbf{b} = \mathbf{0}. \tag{8.200}$$

Assuming constant Lamé coefficients, and because the divergent is a linear operator and $\mathbf{E}$ is given in terms of $\nabla\mathbf{u}$, we can write

$$\mu\,\text{div}\,\nabla\mathbf{u} + \mu\,\text{div}\,\nabla\mathbf{u}^T + \lambda\,\text{div}\,e\mathbf{I} + \mathbf{b} = \mathbf{0}. \tag{8.201}$$

Analyzing the first term of the previous expression, we have

$$\text{div}\,\nabla\mathbf{u} = \text{div}\begin{bmatrix} \dfrac{\partial u_x}{\partial x} & \dfrac{\partial u_x}{\partial y} & \dfrac{\partial u_x}{\partial z} \\[2mm] \dfrac{\partial u_y}{\partial x} & \dfrac{\partial u_y}{\partial y} & \dfrac{\partial u_y}{\partial z} \\[2mm] \dfrac{\partial u_z}{\partial x} & \dfrac{\partial u_z}{\partial y} & \dfrac{\partial u_z}{\partial z} \end{bmatrix} = \left\{ \begin{array}{c} \dfrac{\partial^2 u_x}{\partial x^2} + \dfrac{\partial^2 u_x}{\partial y^2} + \dfrac{\partial^2 u_x}{\partial z^2} \\[2mm] \dfrac{\partial^2 u_y}{\partial x^2} + \dfrac{\partial^2 u_y}{\partial y^2} + \dfrac{\partial^2 u_y}{\partial z^2} \\[2mm] \dfrac{\partial^2 u_z}{\partial x^2} + \dfrac{\partial^2 u_z}{\partial y^2} + \dfrac{\partial^2 u_z}{\partial z^2} \end{array} \right\} = \Delta\mathbf{u}, \tag{8.202}$$

where Δ is the Laplace operator defined for every scalar or vector field ψ, as

$$\Delta\psi = \text{div}\,\nabla\psi. \tag{8.203}$$

For the second term of (8.201), the following expression is valid:

$$\text{div}\,\nabla\mathbf{u}^T = \nabla(\text{div}\,\mathbf{u}). \tag{8.204}$$

On the other hand,

$$\text{div}\,\mathbf{u} = \text{tr}(\nabla\mathbf{u}) = \frac{\partial u_x}{\partial x} + \frac{\partial u_y}{\partial y} + \frac{\partial u_z}{\partial z} = \varepsilon_{xx} + \varepsilon_{yy} + \varepsilon_{zz} = e. \tag{8.205}$$

Thus,

$$\text{div}\,\nabla\mathbf{u}^T = \nabla e. \tag{8.206}$$

For the third term of (8.201), we observe that

$$\text{div } e\mathbf{I} = \text{div } \begin{bmatrix} e & 0 & 0 \\ 0 & e & 0 \\ 0 & 0 & e \end{bmatrix} = \left\{ \begin{array}{c} \dfrac{\partial e}{\partial x} \\[4pt] \dfrac{\partial e}{\partial y} \\[4pt] \dfrac{\partial e}{\partial z} \end{array} \right\} = \nabla e. \tag{8.207}$$

Substituting (8.202), (8.206), and (8.207) in (8.201), we have the Navier equations for a Hookean solid in tensor notation, that is,

$$\mu \Delta \mathbf{u} + (\mu + \lambda)\nabla e + \mathbf{b} = \mathbf{0}. \tag{8.208}$$

Expanding the above equation in the Cartesian coordinate system, we obtain the same three equations given in (8.83).

Example 8.34 *The three terms of the Navier equation for the bar model are* $\mu \dfrac{du_x^2(x)}{dx^2}$, $(\mu + \lambda)\dfrac{du_x^2(x)}{dx^2}$, *and* $b_x = \dfrac{q_x(x)}{A}$. *Substituting these terms in expression (8.208), we obtain*

$$(2\mu + \lambda)A \frac{du_x^2(x)}{dx^2} = -q_x(x).$$

Recalling that $E = 2\mu + \lambda$, *we obtain the differential equation in terms of the axial displacement for the bar*

$$EA\frac{du_x^2(x)}{dx^2} = -q_x(x).$$

□

8.11 VERIFICATION OF LINEAR ELASTIC SOLIDS

The stress state at each point of a solid body is given by the Cauchy stress tensor, which has normal and shear stress components. We want to determine whether the body remains in the linear elastic phase when subjected to a set of external loads. As the normal and shear stress components are of distinct nature, we may wonder about taking the maximum values of each component and comparing them with the respective admissible normal and shear stresses. But the points with maximum values are likely distinct and not necessarily the critical ones.

Thus, it is essential to establish the criteria to determine the critical points for the Hookean solid body. The first step is to determine a coordinate transformation, in such way that the stress state in a point is represented only in terms of normal stresses. Subsequently, we determine the values of these normal stress components. Then, we can establish the resistance criteria for brittle and ductile materials. These steps will be followed in the next sections.

8.11.1 TRANSFORMATION OF VECTORS AND TENSORS

An orthogonal tensor is a linear transformation such that the transformed vectors preserve their lengths and angle between each other. Thus, let $\mathbf{Q}$ be an orthogonal tensor and $\mathbf{u}$ and $\mathbf{v}$ two given vectors. Thus, $||\mathbf{Qu}|| = ||\mathbf{u}||$ and $\cos(\mathbf{u}, \mathbf{v}) = \cos(\mathbf{Qu}, \mathbf{Qv})$. Consequently, $\mathbf{Q}$ preserves the inner or scalar product between vectors, that is,

$$\mathbf{Qu} \cdot \mathbf{Qv} = \mathbf{u} \cdot \mathbf{v}, \qquad \forall \mathbf{u}, \mathbf{v} \in \mathscr{V}. \tag{8.209}$$

From definition (8.109) of the transpose of a tensor, we have

$$\mathbf{Qu} \cdot \mathbf{Qv} = \mathbf{u} \cdot \mathbf{Q}^T \mathbf{Qv}.$$

Thus,

$$\mathbf{u} \cdot \mathbf{v} = \mathbf{u} \cdot \mathbf{Q}^T \mathbf{Qv} \rightarrow \mathbf{u} \cdot \mathbf{Iv} = \mathbf{u} \cdot \mathbf{Q}^T \mathbf{Qv} \rightarrow \mathbf{u} \cdot \left(\mathbf{I} - \mathbf{Q}^T \mathbf{Q} \right) \mathbf{v} = 0.$$

As **u** and **v** are arbitrary vectors, the above relation is zero if and only if

$$\mathbf{Q}^T\mathbf{Q} = \mathbf{I}. \tag{8.210}$$

Thus, we conclude that

$$\mathbf{Q}^T = \mathbf{Q}^{-1}, \tag{8.211}$$

where $\mathbf{Q}^{-1}$ is the inverse tensor of $\mathbf{Q}$.

Hence, the necessary and sufficient condition for $\mathbf{Q}$ to be orthogonal is

$$\mathbf{Q}\mathbf{Q}^T = \mathbf{Q}^T\mathbf{Q} = \mathbf{I}. \tag{8.212}$$

In matrix representation,

$$[\mathbf{Q}][\mathbf{Q}]^T = [\mathbf{Q}]^T[\mathbf{Q}] = [\mathbf{I}]. \tag{8.213}$$

In a general way, from (8.212), we verify that

$$\det\left(\mathbf{Q}\mathbf{Q}^T\right) = \det(\mathbf{I}) \rightarrow \det(\mathbf{Q})\det\left(\mathbf{Q}^T\right) = 1 \rightarrow (\det\mathbf{Q})^2 = 1 \rightarrow \det\mathbf{Q} = \pm 1.$$

If $\det\mathbf{Q} = +1$, then $\mathbf{Q}$ is a rotation. However, if $\det\mathbf{Q} = -1$, $\mathbf{Q}$ is a reflection.

Example 8.35 *Consider the relation between the vectors of the two bases given in Example 8.14*

$$\begin{aligned}
\mathbf{e}'_x &= \cos\theta\,\mathbf{e}_x + \sin\theta\,\mathbf{e}_y, \\
\mathbf{e}'_y &= -\sin\theta\,\mathbf{e}_x + \cos\theta\,\mathbf{e}_y, \\
\mathbf{e}'_z &= \mathbf{e}_z.
\end{aligned}$$

The above relation can be written in matrix form as

$$\left\{ \begin{array}{c} \mathbf{e}'_x \\ \mathbf{e}'_y \\ \mathbf{e}'_z \end{array} \right\} = \left[\begin{array}{ccc} \cos\theta & \sin\theta & 0 \\ -\sin\theta & \cos\theta & 0 \\ 0 & 0 & 1 \end{array} \right] \left\{ \begin{array}{c} \mathbf{e}_x \\ \mathbf{e}_y \\ \mathbf{e}_z \end{array} \right\},$$

or,

$$\{\mathbf{e}\}' = [\mathbf{R}]\{\mathbf{e}\}.$$

Notice that

$$\det[\mathbf{R}] = \cos^2\theta + \sin^2\theta = 1 > 0,$$

which shows that **R** *is, in fact, a rotation.*

□

In a general form, the vectors of bases $\{\mathbf{e}'_1, \mathbf{e}'_2, \mathbf{e}'_3\}$ and $\{\mathbf{e}_1, \mathbf{e}_2, \mathbf{e}_3\}$ of two coordinate systems are related by an orthogonal tensor $\mathbf{Q}$ as follows

$$\mathbf{e}'_i = \mathbf{Q}^T\mathbf{e}_i \rightarrow \left\{ \begin{array}{l} \mathbf{e}'_1 = Q_{11}\mathbf{e}_1 + Q_{21}\mathbf{e}_2 + Q_{31}\mathbf{e}_3 \\ \mathbf{e}'_2 = Q_{12}\mathbf{e}_1 + Q_{22}\mathbf{e}_2 + Q_{32}\mathbf{e}_3 \\ \mathbf{e}'_3 = Q_{13}\mathbf{e}_1 + Q_{23}\mathbf{e}_2 + Q_{33}\mathbf{e}_3 \end{array} \right., \tag{8.214}$$

From the above expression, the components of $\mathbf{Q}$ are given by the direction cosines of the angles between the base vectors, that is,

$$Q_{ji} = Q_{ij}^T = \mathbf{e}_i \cdot \mathbf{Q}^T\mathbf{e}_j = \mathbf{e}_i \cdot \mathbf{e}'_j = \cos\left(\mathbf{e}_i, \mathbf{e}'_j\right). \tag{8.215}$$

Equation (8.214) can be rewritten as

$$\mathbf{e}'_i = \sum_{m=1}^{3} Q_{mi}\mathbf{e}_i, \quad i = 1,2,3. \tag{8.216}$$

Now, taking any vector **a**, its components in both coordinate systems are respectively written as $a_i = \mathbf{e}_i \cdot \mathbf{a}$ and $a'_i = \mathbf{e}'_i \cdot \mathbf{a}$. Once $a'_i = \mathbf{e}'_i \cdot \mathbf{a} = \sum_{m=1}^{3} Q_{mi}\mathbf{e}_m \cdot \mathbf{a}$, we have

$$a'_i = \sum_{m=1}^{3} Q_{mi}a_m, \quad i = 1,2,3. \tag{8.217}$$

In matrix notation,

$$\left\{ \begin{array}{c} a'_1 \\ a'_2 \\ a'_3 \end{array} \right\}_{\mathbf{e}'_i} = \left[\begin{array}{ccc} Q_{11} & Q_{21} & Q_{31} \\ Q_{12} & Q_{22} & Q_{32} \\ Q_{13} & Q_{23} & Q_{33} \end{array} \right]_{\mathbf{e}'_i}^{\mathbf{e}_i} \left\{ \begin{array}{c} a_1 \\ a_2 \\ a_3 \end{array} \right\}_{\mathbf{e}_i} \rightarrow \{\mathbf{a}\}' = [\mathbf{Q}]^T \{\mathbf{a}\}. \tag{8.218}$$

The above expressions constitute the transformation law of the components of a vector with respect to different Cartesian bases. Notice that $\{\mathbf{a}\}' = \{\mathbf{a}\}_{\mathbf{e}'_i}$ and $\{\mathbf{a}\} = \{\mathbf{a}\}_{\mathbf{e}_i}$ are matrix representations of the vector in distinct bases. Thus, expression (8.218) does not correspond to the linear transformation $\mathbf{a}' = \mathbf{Qa}$, which indicates that $\mathbf{a}'$ is the transformed vector of $\mathbf{a}$ by $\mathbf{Q}$ ($\mathbf{a}$ and $\mathbf{a}'$ are two different vectors, while $\{\mathbf{a}\}$ and $\{\mathbf{a}\}'$ are representations of the same vector).

Pre-multiplying (8.218) by $[\mathbf{Q}]$ and using (8.213), the components of $\{\mathbf{a}\}$ are expressed in terms of the components of $\{\mathbf{a}'\}$ as

$$\{\mathbf{a}\} = [\mathbf{Q}]\{\mathbf{a}'\}. \tag{8.219}$$

Now considering a second-order tensor $\mathbf{A}$, the components relative to bases $\{\mathbf{e}_1, \mathbf{e}_2, \mathbf{e}_3\}$ and $\{\mathbf{e}'_1, \mathbf{e}'_2, \mathbf{e}'_3\}$ respectively are $A_{ij} = \mathbf{e}_i \cdot \mathbf{A}\mathbf{e}_j$ and $A'_{ij} = \mathbf{e}'_i \cdot \mathbf{A}\mathbf{e}'_j$. Recalling that $\mathbf{e}'_i = \sum_{m=1}^3 Q_{mi}\mathbf{e}_m$, we have

$$A'_{ij} = \mathbf{e}'_i \cdot \mathbf{A}\mathbf{e}'_j = \sum_{m,n=1}^3 Q_{mi}\mathbf{e}_m \cdot \mathbf{A}Q_{nj}\mathbf{e}_n = Q_{mi}Q_{nj}(\mathbf{e}_m \cdot \mathbf{A}\mathbf{e}_n).$$

Thus,

$$A'_{ij} = \sum_{m,n=1}^3 Q_{mi}Q_{nj}A_{mn}, \quad i,j = 1,2,3. \tag{8.220}$$

In matrix notation,

$$[\mathbf{A}]' = [\mathbf{Q}]^T [\mathbf{A}] [\mathbf{Q}], \tag{8.221}$$

or in expanded form,

$$\left[\begin{array}{ccc} A'_{11} & A'_{12} & A'_{13} \\ A'_{21} & A'_{22} & A'_{23} \\ A'_{31} & A'_{32} & A'_{33} \end{array} \right]_{\mathbf{e}'_i}^{\mathbf{e}'_i} = \left[\begin{array}{ccc} Q_{11} & Q_{21} & Q_{31} \\ Q_{12} & Q_{22} & Q_{32} \\ Q_{13} & Q_{23} & Q_{33} \end{array} \right]_{\mathbf{e}'_i}^{\mathbf{e}_i} \left[\begin{array}{ccc} A_{11} & A_{12} & A_{13} \\ A_{21} & A_{22} & A_{23} \\ A_{31} & A_{32} & A_{33} \end{array} \right]_{\mathbf{e}_i}^{\mathbf{e}_i} \left[\begin{array}{ccc} Q_{11} & Q_{12} & Q_{13} \\ Q_{21} & Q_{22} & Q_{23} \\ Q_{31} & Q_{32} & Q_{33} \end{array} \right]_{\mathbf{e}_i}^{\mathbf{e}'_i}.$$

Analogously,

$$A_{ij} = \sum_{m,n=1}^3 Q_{im}Q_{jn}A'_{mn}, \tag{8.222}$$

or,

$$[\mathbf{A}] = [\mathbf{Q}][\mathbf{A}]'[\mathbf{Q}]^T. \tag{8.223}$$

Equation (8.220) is the transformation law that relates the components of a tensor between different bases. Thus, $[\mathbf{A}]$ and $[\mathbf{A}]'$ are different representations for the same tensor $\mathbf{A}$.

If the components of a vector or tensor relative to the orthonormal basis $\{\mathbf{e}_1, \mathbf{e}_2, \mathbf{e}_3\}$ are known, applying equations (8.217) and (8.220), we find the components relative to any other orthonormal basis $\{\mathbf{e}'_1, \mathbf{e}'_2, \mathbf{e}'_3\}$, obtained with a rotation of $\{\mathbf{e}_1, \mathbf{e}_2, \mathbf{e}_2\}$.

Example 8.36 *Given the representation of a tensor $\mathbf{A}$ in basis $\{\mathbf{e}_1, \mathbf{e}_2, \mathbf{e}_3\}$*

$$[\mathbf{A}] = \left[\begin{array}{ccc} 3 & 2 & 1 \\ 2 & 1 & 0 \\ 0 & 0 & 2 \end{array} \right],$$

calculate the components relative to basis $\{\mathbf{e}'_1, \mathbf{e}'_2, \mathbf{e}'_3\}$, obtained with a rotation of $90°$ about $\mathbf{e}_1$.

For the given rotation, the following relations are valid for vectors of both bases:

$$\mathbf{e}'_1 = \mathbf{e}_1, \qquad \mathbf{e}'_2 = -\mathbf{e}_3, \qquad \mathbf{e}'_3 = -\mathbf{e}_2.$$

Hence, the components of $\mathbf{Q}$ *are determined using (8.215) as*

$$
\begin{aligned}
Q_{11} &= Q_{11}^T = \mathbf{e}_1 \cdot \mathbf{e}_1' = \mathbf{e}_1 \cdot \mathbf{e}_1 = 1, \\
Q_{12} &= Q_{21}^T = \mathbf{e}_2 \cdot \mathbf{e}_1' = \mathbf{e}_2 \cdot \mathbf{e}_1 = 0, \\
Q_{13} &= Q_{31}^T = \mathbf{e}_3 \cdot \mathbf{e}_1' = \mathbf{e}_3 \cdot \mathbf{e}_1 = 0, \\
Q_{21} &= Q_{12}^T = \mathbf{e}_1 \cdot \mathbf{e}_2' = \mathbf{e}_1 \cdot \mathbf{e}_3 = 0, \\
Q_{22} &= Q_{22}^T = \mathbf{e}_2 \cdot \mathbf{e}_2' = \mathbf{e}_2 \cdot \mathbf{e}_3 = 0, \\
Q_{23} &= Q_{32}^T = \mathbf{e}_3 \cdot \mathbf{e}_2' = \mathbf{e}_3 \cdot \mathbf{e}_3 = 1, \\
Q_{31} &= Q_{13}^T = \mathbf{e}_1 \cdot \mathbf{e}_3' = -\mathbf{e}_1 \cdot \mathbf{e}_2 = 0, \\
Q_{32} &= Q_{23}^T = \mathbf{e}_2 \cdot \mathbf{e}_3' = -\mathbf{e}_2 \cdot \mathbf{e}_2 = -1, \\
Q_{33} &= Q_{33}^T = \mathbf{e}_3 \cdot \mathbf{e}_3' = -\mathbf{e}_3 \cdot \mathbf{e}_2 = 0.
\end{aligned}
$$

In matrix notation, we have

$$
[\mathbf{Q}] = \begin{bmatrix} 1 & 0 & 0 \\ 0 & 0 & 1 \\ 0 & -1 & 0 \end{bmatrix}.
$$

The components of $\mathbf{A}$ *in the rotated basis are obtained by*

$$
[\mathbf{A}]' = [\mathbf{Q}]^T [\mathbf{A}] [\mathbf{Q}] = \begin{bmatrix} 1 & 0 & 0 \\ 0 & 0 & -1 \\ 0 & 1 & 0 \end{bmatrix} \begin{bmatrix} 3 & 2 & 1 \\ 2 & 1 & 0 \\ 0 & 0 & 2 \end{bmatrix} \begin{bmatrix} 1 & 0 & 0 \\ 0 & 0 & 1 \\ 0 & -1 & 0 \end{bmatrix} = \begin{bmatrix} 3 & -1 & 2 \\ 0 & 2 & 0 \\ 2 & 0 & 1 \end{bmatrix}.
$$

$\square$

Notice that as we can construct an infinite number of coordinate transformations based on orthogonal tensors, vectors and second-order tensors also have an infinite number of representations.

8.11.2 EIGENVALUE PROBLEM

We want to find an orthogonal tensor, such that the stress state representation at a point contains only normal stresses. This transformation is described by the eigenvectors of the stress tensor. Using the eigenvectors as basis, the stress state at a point is represented as

$$
[\mathbf{T}] = \begin{bmatrix} \sigma_1 & 0 & 0 \\ 0 & \sigma_2 & 0 \\ 0 & 0 & \sigma_3 \end{bmatrix},
$$

where σ_1, σ_2, and σ_3 are the principal stresses given by the eigenvalues of $\mathbf{T}$. The eigenvectors of $\mathbf{T}$ are called principal directions. We adopt the convention $\sigma_1 > \sigma_2 > \sigma_3$.

Generally, given a tensor $\mathbf{A}$, let $\mathbf{e}$ be a unit vector that, when transformed by $\mathbf{A}$, results in a vector parallel to $\mathbf{e}$, that is,

$$
\mathbf{A}\mathbf{e} = \lambda\mathbf{e}. \tag{8.224}
$$

The term $\mathbf{e}$ is called an eigenvector of $\mathbf{A}$ and λ is the correspondent eigenvalue.

As $\mathbf{e}$ is a unit vector, we have

$$
\mathbf{e} \cdot \mathbf{e} = 1. \tag{8.225}
$$

Representing $\mathbf{e}$ as a linear combination of base vectors $\{\mathbf{e}_1, \mathbf{e}_2, \mathbf{e}_3\}$, we obtain

$$
\mathbf{e} = \alpha_1 \mathbf{e}_1 + \alpha_2 \mathbf{e}_2 + \alpha_3 \mathbf{e}_3.
$$

Hence, expression (8.225) reduces to

$$
\alpha_1^2 + \alpha_2^2 + \alpha_3^2 = 1. \tag{8.226}
$$

As $\lambda\mathbf{e} = \lambda\mathbf{I}\mathbf{e}$, the eigenvalue problem (8.224) can be rewritten as

$$
(\mathbf{A} - \lambda\mathbf{I})\mathbf{e} = \mathbf{0},
$$

or in terms of the following components:

$$\begin{cases} (A_{11} - \lambda)\,\alpha_1 + A_{12}\alpha_2 + A_{13}\alpha_3 = 0 \\ A_{21}\alpha_1 + (A_{22} - \lambda)\,\alpha_2 + A_{23}\alpha_3 = 0 \\ A_{31}\alpha_1 + A_{32}\alpha_2 + (A_{33} - \lambda)\,\alpha_3 = 0 \end{cases} . \tag{8.227}$$

In matrix notation, we have

$$([\mathbf{A}] - \lambda[\mathbf{I}])\{\mathbf{e}\} = \{\mathbf{0}\} \rightarrow \begin{bmatrix} A_{11} - \lambda & A_{12} & A_{13} \\ A_{21} & A_{22} - \lambda & A_{23} \\ A_{31} & A_{32} & A_{33} - \lambda \end{bmatrix} \begin{Bmatrix} \alpha_1 \\ \alpha_2 \\ \alpha_3 \end{Bmatrix} = \begin{Bmatrix} 0 \\ 0 \\ 0 \end{Bmatrix} . \tag{8.228}$$

To obtain the solution of this homogeneous system of equations besides the trivial one ($\alpha_1 = \alpha_2 = \alpha_3 = 0$), the matrix determinant must be zero, that is,

$$\det(\mathbf{A} - \lambda\mathbf{I}) = 0. \tag{8.229}$$

In expanded form, we have

$$\begin{vmatrix} A_{11} - \lambda & A_{12} & A_{13} \\ A_{21} & A_{22} - \lambda & A_{23} \\ A_{31} & A_{32} & A_{33} - \lambda \end{vmatrix} = 0. \tag{8.230}$$

The previous determinant results in a cubic equation in λ called the characteristic equation of $\mathbf{A}$. The roots λ_1, λ_2, and λ_3 of this equation are the eigenvalues of $\mathbf{A}$. The corresponding eigenvectors of $\mathbf{A}$ are obtained when substituting each of these eigenvalues in (8.227) and solving the resulting system of equations. To avoid the trivial solution, we must also employ equation (8.226), as illustrated in the following example.

Example 8.37 *Consider the matrix representation of a symmetric tensor* $[\mathbf{A}]$ *relative to basis* $\{\mathbf{e}_1, \mathbf{e}_2, \mathbf{e}_3\}$

$$[\mathbf{A}] = \begin{bmatrix} 2 & 0 & 0 \\ 0 & 3 & 4 \\ 0 & 4 & -3 \end{bmatrix} .$$

Calculate the corresponding eigenvalues and eigenvectors.

The characteristic equation of $\mathbf{A}$ *is*

$$|\mathbf{A} - \lambda\mathbf{I}| = \begin{vmatrix} 2 - \lambda & 0 & 0 \\ 0 & 3 - \lambda & 4 \\ 0 & 4 & -3 - \lambda \end{vmatrix} = (2 - \lambda)(\lambda^2 - 25) = 0.$$

Thus, there are three distinct eigenvalues given by $\lambda_1 = 5$, $\lambda_2 = 2$, *and* $\lambda_3 = -5$.

Substituting $\lambda_1 = 5$ *in the system* $[\mathbf{A} - \lambda\mathbf{I}]\{\mathbf{v}\} = \{\mathbf{0}\}$ *and using (8.226), we have*

$$\begin{cases} -3v_1 = 0 \\ -2v_2 + 4v_3 = 0 \\ 4v_2 - 8v_3 = 0 \\ v_1^2 + v_2^2 + v_3^2 = 1 \end{cases} .$$

Thus, $v_1 = 0$, $v_2 = \dfrac{2}{\sqrt{5}}$, $v_3 = \dfrac{1}{\sqrt{5}}$ *and the correspondent eigenvector is* $\mathbf{v}_1 = \pm\dfrac{1}{\sqrt{5}}(2\mathbf{e}_2 + \mathbf{e}_3)$.

Repeating this procedure for $\lambda_2 = 2$, *we have*

$$\begin{cases} 0v_1 = 0 \\ v_2 + 4v_3 = 0 \\ 4v_2 - 5v_3 = 0 \\ v_1^2 + v_2^2 + v_3^2 = 1 \end{cases} .$$

Thus, $v_2 = v_3 = 0$ *and* $v_1 = \pm1$. *Hence, the associated eigenvector of* $\lambda_2 = 2$ *is* $\mathbf{v}_2 = \pm\mathbf{e}_1$.

Repeating the same procedure for eigenvalue $\lambda_3 = -5$, *the eigenvector is* $\mathbf{v}_3 = \pm\dfrac{1}{\sqrt{5}}(-\mathbf{e}_2 + 2\mathbf{e}_3)$.

□

Expanding the determinant in (8.230), we have

$$
\begin{aligned}
\det\left(\mathbf{A} - \lambda \mathbf{I}\right) &= (A_{11} - \lambda)(A_{22} - \lambda)(A_{33} - \lambda) + A_{12}A_{23}A_{31} + A_{21}A_{32}A_{13} \\
&\quad - A_{13}A_{31}(A_{22} - \lambda) - A_{23}A_{32}(A_{11} - \lambda) - A_{12}A_{21}(A_{33} - \lambda) = 0.
\end{aligned}
$$

The above expression can be rewritten as

$$
\det\left(\mathbf{A} - \lambda \mathbf{I}\right) = -\lambda^3 + I_1\left(\mathbf{A}\right)\lambda^2 - I_2\left(\mathbf{A}\right)\lambda + I_3\left(\mathbf{A}\right), \tag{8.231}
$$

with

$$
\begin{aligned}
I_1\left(\mathbf{A}\right) &= A_{11} + A_{22} + A_{33}, \\
I_2\left(\mathbf{A}\right) &= \begin{vmatrix} A_{11} & A_{12} \\ A_{21} & A_{22} \end{vmatrix} + \begin{vmatrix} A_{22} & A_{23} \\ A_{32} & A_{33} \end{vmatrix} + \begin{vmatrix} A_{11} & A_{13} \\ A_{31} & A_{33} \end{vmatrix}, \\
I_3\left(\mathbf{A}\right) &= \begin{vmatrix} A_{11} & A_{12} & A_{13} \\ A_{21} & A_{22} & A_{23} \\ A_{31} & A_{32} & A_{33} \end{vmatrix}.
\end{aligned} \tag{8.232}
$$

As the eigenvalues of $\mathbf{A}$ are independent of the adopted basis, the coefficients of equation (8.231) must be the same for any used basis to represent the tensor components. Thus, the set

$$
\mathscr{J}_{\mathbf{A}} = (I_1\left(\mathbf{A}\right), I_2\left(\mathbf{A}\right), I_3\left(\mathbf{A}\right))
$$

is called list of principal invariants of the tensor $\mathbf{A}$, because they remain constant for any change of coordinates given by a rotation applied to $\mathbf{A}$. In terms of the trace and determinant operations, the invariants are given by

$$
\begin{aligned}
I_1\left(\mathbf{A}\right) &= \operatorname{tr}\mathbf{A}, \\
I_2\left(\mathbf{A}\right) &= \frac{1}{2}\left[\left(\operatorname{tr}\mathbf{A}\right)^2 - \operatorname{tr}\left(\mathbf{A}\right)^2\right], \\
I_3\left(\mathbf{A}\right) &= \det\mathbf{A}.
\end{aligned} \tag{8.233}
$$

Example 8.38 *For the tensor of Example 8.37, determine the principal invariants and solving (8.231), its eigenvalues.*

The invariants of $\mathbf{A}$ *are*

$$
\begin{aligned}
I_1\left(\mathbf{A}\right) &= 2 + 3 - 3 = 2, \\
I_2\left(\mathbf{A}\right) &= \begin{vmatrix} 2 & 0 \\ 0 & 3 \end{vmatrix} + \begin{vmatrix} 3 & 4 \\ 4 & -3 \end{vmatrix} + \begin{vmatrix} 2 & 0 \\ 0 & -3 \end{vmatrix} = -25, \\
I_3\left(\mathbf{A}\right) &= \begin{vmatrix} 2 & 0 & 0 \\ 0 & 3 & 4 \\ 0 & 4 & -3 \end{vmatrix} = -50.
\end{aligned}
$$

These values give the characteristic equation

$$
\lambda^3 - 2\lambda^2 - 25\lambda + 50 = 0,
$$

or,

$$
(\lambda - 2)(\lambda - 5)(\lambda + 5) = 0.
$$

Hence, we obtain $\lambda_1 = 5$, $\lambda_2 = 2$, *and* $\lambda_3 = -5$.
□

The components of tensor $\mathbf{A}$ relative to its eigenvector basis $\{\mathbf{e}_1, \mathbf{e}_2, \mathbf{e}_3\}$ are given by

$$
\begin{aligned}
A_{11} &= \mathbf{e}_1 \cdot \mathbf{A}\mathbf{e}_1 = \mathbf{e}_1 \cdot (\lambda_1\mathbf{e}_1) = \lambda_1, & A_{12} &= \mathbf{e}_1 \cdot \mathbf{A}\mathbf{e}_2 = \mathbf{e}_1 \cdot (\lambda_2\mathbf{e}_2) = 0 = A_{21}, \\
A_{22} &= \mathbf{e}_2 \cdot \mathbf{A}\mathbf{e}_2 = \mathbf{e}_2 \cdot (\lambda_2\mathbf{e}_1) = \lambda_2, & A_{13} &= \mathbf{e}_1 \cdot \mathbf{A}\mathbf{e}_3 = \mathbf{e}_1 \cdot (\lambda_3\mathbf{e}_3) = 0 = A_{31}, \\
A_{33} &= \mathbf{e}_3 \cdot \mathbf{A}\mathbf{e}_3 = \mathbf{e}_3 \cdot (\lambda_3\mathbf{e}_3) = \lambda_3, & A_{23} &= \mathbf{e}_2 \cdot \mathbf{A}\mathbf{e}_3 = \mathbf{e}_2 \cdot (\lambda_3\mathbf{e}_3) = 0 = A_{32}.
\end{aligned}
$$

Thus,

$$[\mathbf{A}]_{\mathbf{e}_1,\mathbf{e}_2,\mathbf{e}_3} = \begin{bmatrix} \lambda_1 & 0 & 0 \\ 0 & \lambda_2 & 0 \\ 0 & 0 & \lambda_3 \end{bmatrix}, \tag{8.234}$$

that is, the matrix of tensor $\mathbf{A}$ in the eigenvector basis is diagonal and contains the eigenvalues of $\mathbf{A}$.

If tensor $\mathbf{A}$ is symmetric with real components, its respective eigenvalues are always real numbers.

From the infinite number of representations of a second-order tensor obtained from the orthogonal transformations given in (8.221), the largest (λ_1) and lowest (λ_3) eigenvalues are, respectively, the upper and lower limit values for the diagonal components of a tensor $\mathbf{A}$, i.e.,

$$\lambda_3 \leq A_{ii} \leq \lambda_1, \quad i = 1,2,3. \tag{8.235}$$

The previous relation is called the min-max property of the eigenvalue problem of $\mathbf{A}$.

8.11.3 PRINCIPAL STRESSES AND PRINCIPAL DIRECTIONS

The principal stresses at a point of a solid body are obtained from the solution of characteristic equation (8.231), associated to the symmetric Cauchy stress tensor $\mathbf{T}$. We can determine the following analytical expressions to calculate σ_1, σ_2, and σ_3 [20]:

$$\sigma_1 = -2\sqrt{Q}\cos\frac{\theta}{3} + \frac{I_1}{3}, \tag{8.236}$$

$$\sigma_2 = -2\sqrt{Q}\cos\frac{\theta + 2\pi}{3} + \frac{I_1}{3}, \tag{8.237}$$

$$\sigma_3 = -2\sqrt{Q}\cos\frac{\theta - 2\pi}{3} + \frac{I_1}{3}, \tag{8.238}$$

where

$$Q = \frac{I_1^2 - 3I_2}{9}, \tag{8.239}$$

$$\theta = \cos^{-1}\frac{R}{\sqrt{Q^3}}, \tag{8.240}$$

$$R = \frac{-2I_1^3 + 9I_1I_2 - 2 - 27I_3}{54}. \tag{8.241}$$

Example 8.39 *The principal directions of the infinitesimal strain tensor* $\mathbf{E}$ *and Cauchy stress tensor* $\mathbf{T}$ *are the same, as shown below.*

Let $\mathbf{n}_i$ *be an eigenvector of* $\mathbf{E}$. *Using Hooke's law (8.198) we have*

$$\mathbf{T}\mathbf{n}_i = 2\mu\mathbf{E}\mathbf{n}_i + \lambda e\mathbf{I}\mathbf{n}_i = 2\mu E_i\mathbf{n}_i + \lambda e\mathbf{I}\mathbf{n}_i = (2\mu E_i + \lambda e)\mathbf{n}_i = \sigma_i\mathbf{n}_i,$$

where E_i *and* σ_i *are the eigenvalues of tensors* $\mathbf{E}$ *and* $\mathbf{T}$. *Therefore,* $\mathbf{n}_i$ *is an eigenvector of* $\mathbf{E}$ *and* $\mathbf{T}$. *The relation between the eigenvalues is given by*

$$\sigma_i = 2\mu E_i + \lambda e.$$

□

8.11.4 PRINCIPAL STRESSES FOR A PLANE STRESS PROBLEM

The stress tensor at a point of a body in a plane stress state is represented in the Cartesian coordinate system xyz as (see Chapter 9)

$$[\mathbf{T}] = \begin{bmatrix} \sigma_{xx} & \tau_{xy} & 0 \\ \tau_{xy} & \sigma_{yy} & 0 \\ 0 & 0 & 0 \end{bmatrix}. \tag{8.242}$$

In this case, the characteristic equation (8.231) reduces to

$$\lambda^2 - I_1(\mathbf{T})\lambda + I_3(\mathbf{T}) = 0, \tag{8.243}$$

with

$$I_1(\mathbf{T}) = \sigma_{xx} + \sigma_{yy}, \tag{8.244}$$

$$I_3(\mathbf{T}) = \sigma_{xx}\sigma_{yy} - \tau_{xy}^2. \tag{8.245}$$

The solution of the second-order characteristic equation (8.243) results in the following expressions for the principal stresses:

$$\sigma_1 = \frac{\sigma_{xx} + \sigma_{yy}}{2} + \sqrt{\left(\frac{\sigma_{xx} - \sigma_{yy}}{2}\right)^2 + \tau_{xy}^2}, \tag{8.246}$$

$$\sigma_3 = \frac{\sigma_{xx} + \sigma_{yy}}{2} - \sqrt{\left(\frac{\sigma_{xx} - \sigma_{yy}}{2}\right)^2 + \tau_{xy}^2}, \tag{8.247}$$

and $\sigma_2 = 0$.

To verify the min-max property of the eigenvalue problem given by (8.235) in the plane state problem, consider the following coordinate transformation of the stress tensor (8.242) by an orthogonal tensor corresponding to the rotation of an angle θ about z. Thus, the stress tensor in the rotated coordinate system $x'y'z'$ is determined as

$$\begin{bmatrix} \sigma'_{xx} & \tau'_{xy} & 0 \\ \tau'_{xy} & \sigma'_{yy} & 0 \\ 0 & 0 & 0 \end{bmatrix} = \begin{bmatrix} \cos\theta & \sin\theta & 0 \\ -\sin\theta & \cos\theta & 0 \\ 0 & 0 & 1 \end{bmatrix} \begin{bmatrix} \sigma_{xx} & \tau_{xy} & 0 \\ \tau_{xy} & \sigma_{yy} & 0 \\ 0 & 0 & 0 \end{bmatrix} \begin{bmatrix} \cos\theta & -\sin\theta & 0 \\ \sin\theta & \cos\theta & 0 \\ 0 & 0 & 1 \end{bmatrix}. \tag{8.248}$$

Performing the indicated multiplications and employing the trigonometric relations

$$\sin^2\theta = \frac{1+\cos 2\theta}{2},$$

$$\cos^2\theta = \frac{1-\cos 2\theta}{2},$$

$$\sin\theta\cos\theta = \frac{\sin 2\theta}{2},$$

we obtain

$$\sigma'_{xx} = \frac{\sigma_{xx} + \sigma_{yy}}{2} + \frac{\sigma_{xx} - \sigma_{yy}}{2}\cos 2\theta + \tau_{xy}\sin 2\theta, \tag{8.249}$$

$$\sigma'_{yy} = \frac{\sigma_{xx} + \sigma_{yy}}{2} - \frac{\sigma_{xx} - \sigma_{yy}}{2}\cos 2\theta - \tau_{xy}\sin 2\theta, \tag{8.250}$$

$$\tau'_{xy} = -\frac{\sigma_{xx} - \sigma_{yy}}{2}\sin 2\theta + \tau_{xy}\cos 2\theta. \tag{8.251}$$

Now we want to determine the angle θ for which the normal stress σ'_{xx} assumes maximum and minimum values. For this, the derivative of σ'_{xx} relative to θ is taken as zero, that is,

$$\frac{d\sigma'_{xx}}{dx} = -2\frac{\sigma_{xx} - \sigma_{yy}}{2}\sin 2\theta + 2\cos 2\theta\sin 2\theta = 0.$$

Thus,

$$\tan 2\theta_1 = \frac{\tau_{xy}}{(\sigma_{xx} - \sigma_{yy})/2}. \tag{8.252}$$

As indicated in Figure 8.23(a) for plane $\sigma \times \tau$, there are two values of θ_1 that satisfy the previous equation, indicated by angles $2\theta_{1_1}$ and $2\theta_{1_2}$. One of the angles refers to the principal stress σ_1 and the other one to principal stress σ_3.

Notice that the difference between angles $2\theta_{1_1}$ and $2\theta_{1_2}$ is 180 degrees, that is, θ_{1_1} and θ_{1_2} have a phase shift of 90 degrees. The following trigonometric relations are obtained from Figure 8.23(a):

$$\begin{matrix} \cos 2\theta_{1_1} = \dfrac{\tau_{xy}}{OA}, & \sin 2\theta_{1_1} = \dfrac{(\sigma_{xx} - \sigma_{yy})/2}{OA}, \\[2ex] \cos 2\theta_{1_2} = -\dfrac{\tau_{xy}}{OB}, & \sin 2\theta_{1_2} = -\dfrac{(\sigma_{xx} - \sigma_{yy})/2}{OB}, \end{matrix} \tag{8.253}$$

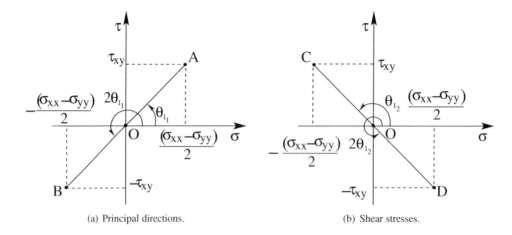

(a) Principal directions.   (b) Shear stresses.

Figure 8.23 Angles defining the principal directions and maximum and minimum shear stresses.

with the length of segments $\overline{OA}$ and $\overline{OB}$ given by

$$\overline{OA} = \overline{OB} = \sqrt{\left(\frac{\sigma_{xx} - \sigma_{yy}}{2}\right)^2 + \tau_{xy}^2}.$$

Substituting the trigonometric relations (8.253) in (8.249), we obtain the same equations for principal stresses σ_1 and σ_3 given in (8.246) and (8.247). We find that σ_1 and σ_3 constitute the maximum and minimum values of normal stresses that the point can assume among the infinite representations of a second-order tensor. Thus, the min-max relationship of the eigenvalue problem given in (8.235) is verified.

The directions where the shear stress is zero are obtained from (8.251) as

$$\tau'_{xy} = -\frac{\sigma_{xx} - \sigma_{yy}}{2} \sin 2\theta + \tau_{xy} \cos 2\theta = 0.$$

Thus,

$$\tan 2\theta = \frac{\sin 2\theta}{\cos 2\theta} = \frac{\tau_{xy}}{(\sigma_{xx} - \sigma_{yy})/2}. \tag{8.254}$$

Therefore, the angles that provide the extreme values of σ'_{xx} also imply in zero shear stresses at the point. This shows that for plane stress, the representation of the stress state at a point according to the principal directions is given by a diagonal matrix with the values of the principal stresses.

Similarly, we can determine the maximum and minimum shear stresses and the associated directions. For this, the derivative of equation (8.251) relative to θ is taken as zero, that is

$$\frac{d\tau'_{xy}}{d\theta} = -2\frac{\sigma_{xx} - \sigma_{yy}}{2} \cos 2\theta - 2\tau_{xy} \sin 2\theta = 0.$$

Hence, we have

$$\tan 2\theta_2 = -\frac{(\sigma_{xx} - \sigma_{yy})/2}{\tau_{xy}}. \tag{8.255}$$

Again, there are two angles (directions) corresponding to the maximum and minimum shear stresses. These angles are shown in Figure 8.23(b) as θ_{2_1} and θ_{2_2}, and are phase-shifted 90 degrees. Notice from Figures 8.23(a) and 8.23(b) that the angles defining the principal normal stresses and maximum shear stresses are phase-shifted 45 degrees.

The following trigonometric relations are observed from Figure 8.23(b):

$$
\begin{aligned}
\cos 2\theta_{2_1} &= \frac{\tau_{xy}}{OC}, & \sin 2\theta_{2_1} &= -\frac{(\sigma_{xx} - \sigma_{yy})/2}{OC}, \\
\cos 2\theta_{2_2} &= -\frac{\tau_{xy}}{OD}, & \cos 2\theta_{2_2} &= \frac{(\sigma_{xx} - \sigma_{yy})/2}{OD},
\end{aligned}
\tag{8.256}
$$

with the length of segments $\overline{OC}$ and $\overline{OD}$ given by

$$\overline{OC} = \overline{OD} = \sqrt{\left(\frac{\sigma_{xx} - \sigma_{yy}}{2}\right)^2 + \tau_{xy}^2}.$$

Substituting the trigonometric relations given by (8.256) in (8.251), we obtain the equations for the maximum τ_{max} and minimum τ_{min} shear stresses, that is,

$$\tau_{max} = \sqrt{\left(\frac{\sigma_{xx} - \sigma_{yy}}{2}\right)^2 + \tau_{xy}^2}, \tag{8.257}$$

$$\tau_{min} = -\sqrt{\left(\frac{\sigma_{xx} - \sigma_{yy}}{2}\right)^2 + \tau_{xy}^2}. \tag{8.258}$$

Notice that the maximum and minimum shear stresses are equal in magnitude but have opposite signs. From the physical point of view, the signs have no meaning. For this reason, the largest stress will always be the maximum, regardless of the sign.

As these stresses are located in planes which are shifted 90 degrees to each other, the shear stresses acting in mutually perpendicular planes are equal. This conclusion is consistent with the symmetry of the Cauchy stress tensor.

The normal stress on planes of maximum and minimum shear stresses is not zero. Just replace relations (8.256) in equation (8.249) and determine that

$$\sigma' = \frac{\sigma_{xx} + \sigma_{yy}}{2}. \tag{8.259}$$

When working with the principal stresses, we have $\tau_{xy} = 0$ and from (8.257) and (8.258) we have

$$\tau_{max} = \frac{(\sigma_1 - \sigma_2)}{2}, \tag{8.260}$$

$$\tau_{min} = -\frac{(\sigma_1 - \sigma_2)}{2}. \tag{8.261}$$

If $\sigma_1 = -\sigma_2$, we have $\tau_{max} = \frac{\sigma_1 + \sigma_2}{2}$. If $\sigma_1 = \sigma_2$, we have $\tau_{max} = 0$.

Example 8.40 *Consider the plane stress representation at point O, according to the Cartesian coordinate system xyz, given by*

$$[\mathbf{T}_{xyz}] = \begin{bmatrix} 3 & 1 & 0 \\ 1 & 2 & 0 \\ 0 & 0 & 0 \end{bmatrix}.$$

We want to determine the stress components in x'y'z', rotated −40 degress about the z axis. We also find the principal and maximum and minimum shear stresses, as well as the angles of these respective directions.

The normal σ'_{xx} and shear τ'_{xy} stresses in a plane inclined −40 degrees are respectively calculated using equations (8.249) and (8.250)

$$\begin{aligned} \sigma'_{xx} &= \frac{\sigma_{xx} + \sigma_{yy}}{2} + \frac{\sigma_{xx} - \sigma_{yy}}{2}\cos 2\theta + \tau_{xy}\sin 2\theta \\ &= \frac{3+2}{2} + \frac{3-2}{2}\cos(-80°) + 1\sin(-80°) = 1.60 \text{ MPa}, \\ \tau'_{xy} &= -\frac{\sigma_{xx} - \sigma_{yy}}{2}\sin 2\theta + \tau_{xy}\cos 2\theta \\ &= -\frac{3+2}{2}\sin(-80°) + 1\cos(-80°) = 0.67 \text{ MPa}. \end{aligned}$$

The stress σ'_{yy} is determined by the first stress invariant I_1 given in (8.244), that is,

$$\sigma'_{yy} = \sigma_{xx} + \sigma_{yy} - \sigma'_{xx} = 3.40 \text{ MPa}.$$

The principal stresses are determined from (8.246) and (8.247). Thus,

$$\sigma_{1,3} = \frac{\sigma_{xx} + \sigma_{yy}}{2} \pm \sqrt{\left(\frac{\sigma_{xx} - \sigma_{yy}}{2}\right)^2 + \tau_{xy}^2}$$

$$= \frac{3+2}{2} \pm \sqrt{\left(\frac{3-2}{2}\right)^2 + 1^2} \rightarrow \begin{cases} \sigma_1 = 3.62 \text{ MPa} \\ \sigma_3 = 1.38 \text{ MPa} \end{cases}.$$

The angles of the principal stress planes are given by (8.252)

$$\tan 2\theta_1 = \frac{\tau_{xy}}{\dfrac{\sigma_{xx} - \sigma_{yy}}{2}} = \frac{1}{\dfrac{3-2}{2}} = 2 \rightarrow \begin{cases} \theta_{1_1} = 31.72° \\ \theta_{1_2} = 121.72° \end{cases}.$$

The maximum and minimum shear stresses are obtained using (8.257) and (8.258), that is,

$$\tau = \pm \sqrt{\left(\frac{\sigma_{xx} - \sigma_{yy}}{2}\right)^2 + \tau_{xy}^2} = \pm \sqrt{\left(\frac{3-2}{2}\right)^2 + 1^2} = \pm 1.12.$$

Thus,

$$\tau_{\max} = 1.12 \text{ MPa} \quad and \quad \tau_{\min} = -1.12 \text{ MPa}$$

The angles of the planes with maximum and minimum shear stresses are calculated from (8.255)

$$\tan 2\theta_2 = -\frac{\dfrac{\sigma_{xx} - \sigma_{yy}}{2}}{\tau_{xy}} = -\frac{\dfrac{3-2}{2}}{1} = -0.5 \rightarrow \begin{cases} \theta_{2_1} = -13.28° \\ \theta_{2_2} = 76.72° \end{cases}.$$

The normal stresses in these planes are determined substituting θ_{2_1} and θ_{2_2} in equations (8.249) and (8.250)

$$\sigma'_{xx} = \frac{3+2}{2} + \frac{3-2}{2} \cos(-26.56°) + 1 \sin(-26.56°) = 2.50 \text{ MPa},$$

$$\sigma'_{yy} = \frac{3+2}{2} + \frac{3-2}{2} \cos(153.44°) + 1 \sin(153.44°) = 2.50 \text{ MPa}.$$

Figure 8.24 illustrates the stress state representation at point O according to reference systems xyz and x'y'z', principal directions, and maximum and minimum shear stresses. The respective stress tensors are given by:

$$[\mathbf{T}_{xyz}] = \begin{bmatrix} 3 & 1 & 0 \\ 1 & 2 & 0 \\ 0 & 0 & 0 \end{bmatrix}, \quad [\mathbf{T}_{x'y'z'}] = \begin{bmatrix} 1.60 & 0.67 & 0 \\ 0.67 & 3.40 & 0 \\ 0 & 0 & 0 \end{bmatrix},$$

$$[\mathbf{T}_{123}] = \begin{bmatrix} 3.62 & 0 & 0 \\ 0 & 1.38 & 0 \\ 0 & 0 & 0 \end{bmatrix}, \quad [\mathbf{T}_\tau] = \begin{bmatrix} 2.50 & 1.12 & 0 \\ 1.12 & 2.50 & 0 \\ 0 & 0 & 0 \end{bmatrix}.$$

MATLAB script stresstransf.m implements the solution of this example.

□

8.11.5 MOHR'S CIRCLE

The plane stress state at a point of the body can be graphically represented using the Mohr's circle. To verify this, we sum the squares of equations (8.249) and (8.251) and after some simplifications, we obtain

$$\left(\sigma'_{xx} - \frac{\sigma_{xx} + \sigma_{yy}}{2}\right)^2 + (\tau'_{xy})^2 = \left(\frac{\sigma_{xx} - \sigma_{yy}}{2}\right)^2 + \tau_{xy}^2, \tag{8.262}$$

or,

$$(\sigma'_{xx} - a)^2 + (\tau'_{xy})^2 = b^2, \tag{8.263}$$

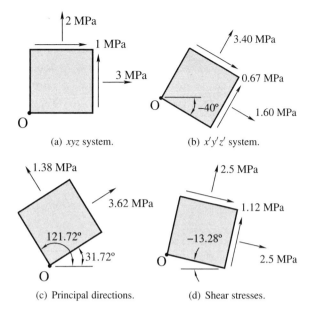

(a) *xyz* system.

(b) *x'y'z'* system.

(c) Principal directions.

(d) Shear stresses.

Figure 8.24   Example 8.40: stress state representations at point O.

with

$$\begin{cases} a = \dfrac{\sigma_{xx} + \sigma_{yy}}{2} \\ b^2 = \left(\dfrac{\sigma_{xx} - \sigma_{yy}}{2}\right)^2 + \tau_{xy}^2 \end{cases}.$$

Comparing expression (8.263) with the general equation of a circle with center (c,d) and radius R

$$(x-c)^2 + (y-d)^2 = R^2,$$

we observe that (8.263) represents the equation of a circle with radius $R = b$ and center $(c,d) = (a,0)$, called Mohr's circle. The radius of the circle is given by

$$R = \sqrt{\left(\frac{\sigma_{xx} - \sigma_{yy}}{2}\right)^2 + \tau_{xy}^2}. \tag{8.264}$$

The following steps are employed to construct the Mohr's circle for a plane stress state, as illustrated in Figure 8.25:

1. From the given stress state, point A (σ_{xx}, τ_{xy}) is plotted at plane $\sigma \times \tau$;

2. The center $\left(a = \dfrac{\sigma_{xx} + \sigma_{yy}}{2}, 0\right)$ of the circle is determined, which must be in axis σ.

3. With both points, the circle is drawn.

4. To determine the stresses for any other plane, we draw a line relative to the plane under consideration in the stress element.

5. A parallel line is copied to the desired plane, passing at point A.

6. Starting from the intersecting point between this line and the circle, another line is drawn perpendicular to axis σ. The point of the circle which is cut by this line is the desired point.

File mohr.m plots the Mohr's circle for a plane stress state at a point.

Some observations can be made regarding the Mohr's circle (see Figure 8.26):

1. The largest value for stress in the considered point is σ_1 and the lowest is σ_3, which are the principal stresses. The shear stress is zero in these points.

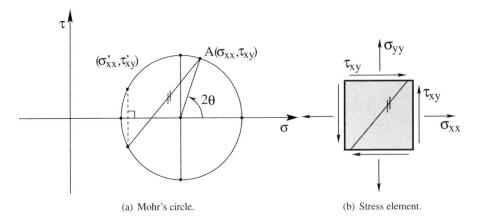

(a) Mohr's circle.   (b) Stress element.

Figure 8.25   Mohr's circle at a point for a plane stress state.

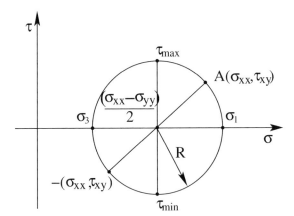

Figure 8.26   Interpretation of the Mohr's circle.

2. The largest shear stress is equal to the radius of the circle, given by

$$\tau_{max} = \frac{\sigma_1 - \sigma_3}{2}.$$

The respective normal stress is

$$\sigma' = \frac{\sigma_1 + \sigma_3}{2}.$$

3. If $\sigma_1 = -\sigma_3$, the center of the Mohr's circle is coincident with the origin of plane $\sigma \times \tau$.
4. If $\sigma_1 = \sigma_3$, the Mohr's circle degenerates to a point, and thus there are no shear stresses.
5. The sum of the stresses of two perpendicular planes represents the first stress invariant, that is,

$$I_1 = \sigma_{xx} + \sigma_{yy} = \sigma'_{xx} + \sigma'_{yy} = \sigma_1 + \sigma_3. \tag{8.265}$$

For a three-dimensional stress state, Mohr's circles are plotted for each principal plane. Thus, in the general stress state, there are three circles, as shown in Figure 8.27. The stresses at any point in any spatial plane lie over one of the three circles or in the shaded area shown in Figure 8.27.

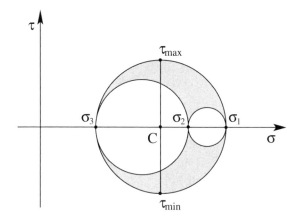

Figure 8.27 Mohr circles for a three-dimensional stress state.

We can also represent a plane stress state using three circles. In this case, we generally take $\sigma_2 = 0$, as illustrated in the following example.

Example 8.41 *Consider the stress element illustrated in Figure 8.28(a) representing a plane stress state at point O according to the Cartesian coordinate system xyz.*

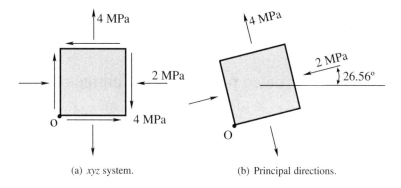

(a) *xyz* system. (b) Principal directions.

Figure 8.28 Example 8.41: stress state representations at point *O*.

The stress state representation at point O is

$$[\mathbf{T}_{xyz}] = \begin{bmatrix} -2 & -4 & 0 \\ -4 & 4 & 0 \\ 0 & 0 & 0 \end{bmatrix}.$$

The principal stresses are calculated using (8.246) and (8.247), that is,

$$\sigma_{1,3} = \frac{\sigma_1 + \sigma_2}{2} \pm \sqrt{\left(\frac{\sigma_1 - \sigma_3}{2}\right)^2 + \tau_{xy}^2} = 1 \pm \sqrt{(3)^2 + (-4)^2} \rightarrow \begin{cases} \sigma_1 = 6 \text{ MPa} \\ \sigma_3 = -4 \text{ MPa} \end{cases}.$$

The angles corresponding to the principal directions are calculated using (8.252)

$$\tan 2\theta_1 = \frac{\tau_{xy}}{\dfrac{\sigma_{xx} - \sigma_{yy}}{2}} = \frac{-4}{\dfrac{-2-4}{2}} = \frac{4}{3} \rightarrow 2\theta_1 = 53.13° \rightarrow \begin{cases} \theta_{1_1} = 26.56° \\ \theta_{1_2} = 116.56° \end{cases}.$$

The stress state at point O can be represented in the principal directions using $\sigma_1 = 6$ MPa, $\sigma_2 = 0$ MPa, and $\sigma_3 = -4$ MPa. Thus, the stress tensor is given by

$$[\mathbf{T}_{123}] = \begin{bmatrix} 6 & 0 & 0 \\ 0 & 0 & 0 \\ 0 & 0 & -4 \end{bmatrix}.$$

The corresponding stress element is illustrated in Figure 8.28(b) and the Mohr's circles are illustrated in Figure 8.29.

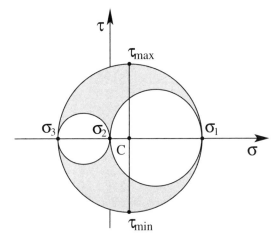

Figure 8.29   Example 8.41: Mohr's circles for the plane stress state at point O.

□

8.11.6   MAXIMUM SHEAR STRESS THEORY (TRESCA CRITERION)

Once an orthogonal tensor is determined, allowing the stress state representation only in terms of the principal normal stresses at a point, three strength criteria are presented to determine whether a solid body remains in the elastic phase when subjected to any stress state. For this purpose, each criterion defines an equivalent normal stress, which should be compared with the admissible normal stress of the material.

The maximum shear stress theory results from the observation that yield occurs in critically oriented planes for ductile material specimens. This indicates that the maximum shear stress plays a key role. Therefore, it is assumed that yielding only depends on the maximum shear stress at the point. When a certain critical value τ_{cr} is reached, the material starts to yield.

For a uniaxial normal tensile or compressive stress state, that is $\sigma_{xx} = \pm\sigma_1$ and $\sigma_{yy} = \tau_{xy} = 0$, the maximum shear stress or critical stress is given from (8.257) by

$$\tau_{\max} = \tau_{cr} = \left|\pm\frac{\sigma_1}{2}\right| \leq \frac{\sigma_y}{2}, \tag{8.266}$$

where σ_y is the normal yield stress of the material, obtained from the uniaxial tensile test. This case is illustrated in Figure 8.30.

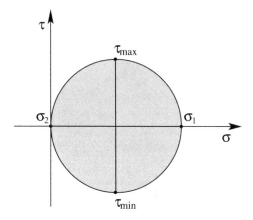

Figure 8.30 Mohr's circle for $\sigma_{xx} = \sigma_1$ and $\sigma_{yy} = 0$.

There are two situations for the plane stress state. Figure 8.31(a) illustrates the Mohr's circle for σ_1 and σ_3, which are both positive and $\sigma_2 = 0$. In this case, the maximum shear stress is the same as the onedimensional case, that is,

$$\tau_{\max} = \left|\pm\frac{\sigma_1}{2}\right| \leq \frac{\sigma_y}{2}.$$

Thus, if $|\sigma_1| > |\sigma_3|$, we have that $|\sigma_1|$ can't exceed σ_y. If $|\sigma_3| > |\sigma_1|$, $|\sigma_3|$ can't exceed σ_y. Both previous conditions can be summarized as

$$|\sigma_1| \leq \sigma_y \quad \text{and} \quad |\sigma_3| \leq \sigma_y,$$

or,

$$\frac{\sigma_1}{\sigma_y} \leq \pm1 \quad \frac{\sigma_3}{\sigma_y} \leq \pm1. \tag{8.267}$$

When the stresses σ_1 and σ_3 have opposite signs and $\sigma_2 = 0$, the correspondent Mohr's circle is illustrated in Figure 8.31(b) for $\sigma_1 > 0$ and $\sigma_3 < 0$. The largest circle passes at σ_1 and σ_3, and the maximum stress is

$$\tau_{\max} = \frac{(|\sigma_1| - |\sigma_3|)}{2}. \tag{8.268}$$

This maximum stress cannot exceed the yield criterion in the simple traction case, that is, ($\tau_{\max} \leq \frac{\sigma_y}{2}$). Thus,

$$\left|\pm\frac{\sigma_1 - \sigma_3}{2}\right| \leq \frac{\sigma_y}{2}. \tag{8.269}$$

or,

$$\frac{\sigma_1}{\sigma_y} - \frac{\sigma_3}{\sigma_y} \leq \pm1. \tag{8.270}$$

Plotting both previous conditions [equations (8.267) and (8.270)] at plane $\frac{\sigma_1}{\sigma_y} \times \frac{\sigma_3}{\sigma_y}$, as shown in Figure 8.32, we obtain the Tresca polygon.

If the stress state at a point is located inside the Tresca polygon, the material remains in the elastic phase. If the stress state corresponds to a point at the boundary of the polygon, the material will yield indefinitely.

The Tresca criterion is insensitive to the superposition of a hydrostatic stress state ($\sigma_1 = \sigma_2 = \sigma_3$). Only a translation of the Mohr's circles occurs.

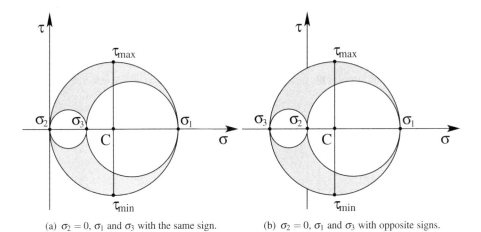

(a) $\sigma_2 = 0$, σ_1 and σ_3 with the same sign.   (b) $\sigma_2 = 0$, σ_1 and σ_3 with opposite signs.

Figure 8.31   Mohr circles for the Tresca strength criterion.

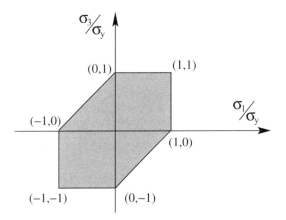

Figure 8.32   Maximum shear stress criterion representation.

In the general stress case, we must take the largest difference between the principal stresses and verify if it exceeds the yield stress of the material.

The Tresca equivalent stress is then defined as follows

$$\sigma_{eqv} = \max\left(|\sigma_1 - \sigma_2|, |\sigma_1 - \sigma_3|, |\sigma_2 - \sigma_3|\right). \tag{8.271}$$

The material remains in the elastic phase if

$$\sigma_{eqv} < \sigma_y. \tag{8.272}$$

The stress state at each point of a Timoshenko beam is given by (σ_{xx}, τ_{xy}) with

$$\sigma_{xx}(x, y) = -\frac{M_z(x)}{I_z(x)}y,$$

$$\tau_{xy}(x, y) = -\frac{V_y(x)M_{sz}(x, y)}{b(y)I_z(x)}.$$

The corresponding principal stresses are given by

$$\sigma_1 = \frac{\sigma_{xx}}{2} + \sqrt{\left(\frac{\sigma_{xx}}{2}\right)^2 + \tau_{xy}^2},$$

$$\sigma_2 = 0,$$

$$\sigma_3 = \frac{\sigma_{xx}}{2} - \sqrt{\left(\frac{\sigma_{xx}}{2}\right)^2 + \tau_{xy}^2}.$$

The corresponding Tresca equivalent stress is

$$\sigma_{eqv} = \sigma_1 - \sigma_3 = \frac{\sigma_{xx}}{2} + \sqrt{\left(\frac{\sigma_{xx}}{2}\right)^2 + \tau_{xy}^2}.$$

8.11.7 MAXIMUM DISTORTION ENERGY THEORY (VON MISES CRITERION)

The theory of maximum distortion strain energy or von Mises criterion is also applied to ductile materials. But it uses the strain energy associated the shear stresses and strains, rather than the critical shear stress of the Tresca criterion. For this, the failure criterion is defined when the distortion energy at a point of a solid body is equal to the energy of the specimen of same material in a tensile test.

Consider the stress state at a point of a body according to the principal directions

$$[\mathbf{T}] = \begin{bmatrix} \sigma_1 & 0 & 0 \\ 0 & \sigma_2 & 0 \\ 0 & 0 & \sigma_3 \end{bmatrix}.$$

The above tensor can be rewritten in the following way (see Figure 8.33):

$$[\mathbf{T}] = \begin{bmatrix} \sigma_p & 0 & 0 \\ 0 & \sigma_p & 0 \\ 0 & 0 & \sigma_p \end{bmatrix} + \begin{bmatrix} \sigma_1 - \sigma_p & 0 & 0 \\ 0 & \sigma_2 - \sigma_p & 0 \\ 0 & 0 & \sigma_3 - \sigma_p \end{bmatrix}, \tag{8.273}$$

with the mean or hydrostatic stress σ_p defined as

$$\sigma_p = \frac{1}{3}\left(\sigma_1 + \sigma_2 + \sigma_3\right). \tag{8.274}$$

The tensor described by σ_p is called a spherical, hydrostatic, or dilatation tensor. The dilatation in terms of the principal strains ε_1, ε_2, and ε_3 is given by

$$e = \varepsilon_1 + \varepsilon_2 + \varepsilon_3.$$

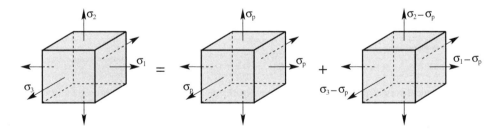

Figure 8.33 Decomposition of the general stress state at a point with hydrostatic and deviatoric components.

The generalized Hooke's law can be expressed in terms of the principal stress and strain components as

$$\varepsilon_1 = \frac{1}{E}\left[\sigma_1 - v\left(\sigma_2 + \sigma_3\right)\right],$$

$$\varepsilon_2 = \frac{1}{E}\left[\sigma_2 - v\left(\sigma_1 + \sigma_3\right)\right], \tag{8.275}$$

$$\varepsilon_3 = \frac{1}{E}\left[\sigma_3 - v\left(\sigma_1 + \sigma_2\right)\right],$$

Hence, the dilatation is written in terms of the principal stresses as

$$e = \frac{1-2v}{E}(\sigma_1 + \sigma_2 + \sigma_3) = \frac{3(1-2v)}{E}\sigma_p. \tag{8.276}$$

The tensor given by the difference between the principal and hydrostatic stresses is called deviatoric or distortion tensor. The tensor decompositions for the general and uniaxial stress states in hydrostatic and deviatoric contributions are respectively illustrated in Figures 8.33 and 8.34.

The strain energy density at each point of a solid is given by

$$U = \frac{1}{2}(\sigma_{xx}\varepsilon_{xx} + \sigma_{yy}\varepsilon_{yy} + \sigma_{zz}\varepsilon_{zz} + \tau_{xy}\gamma_{xy} + \tau_{yz}\gamma_{yz} + \tau_{xz}\gamma_{xz}). \tag{8.277}$$

In terms of principal directions, we have

$$U = \frac{1}{2}(\sigma_1\varepsilon_1 + \sigma_2\varepsilon_2 + \sigma_3\varepsilon_3). \tag{8.278}$$

Substituting the generalized Hooke's law (8.276) in this equation and making some simplifications, we obtain

$$U = \frac{1}{2E}(\sigma_1^2 + \sigma_2^2 + \sigma_3^2) - \frac{v}{E}(\sigma_1\sigma_2 + \sigma_2\sigma_3 + \sigma_3\sigma_1). \tag{8.279}$$

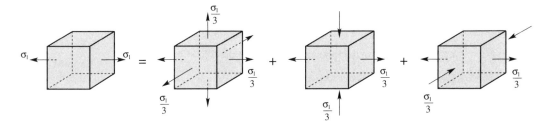

Figure 8.34   Decomposition of a uniaxial stress state at a point in hydrostatic and deviatoric terms.

The dilatational strain energy density of the stress tensor is obtained substituting $\sigma_1 = \sigma_2 = \sigma_3 = \sigma_p$ in the previous expression. Thus,

$$U_{dilat} = \frac{3(1-2v)}{E}\sigma_p^2 = \frac{(1-2v)}{6E}(\sigma_1 + \sigma_2 + \sigma_3)^2. \tag{8.280}$$

Subtracting (8.280) for the strain energy given in equation (8.279), we obtain the distortional energy U_{dist}. Recalling that $G = \dfrac{E}{2(1+v)}$, we have

$$U_{dist} = \frac{1}{12G}\left[(\sigma_1 - \sigma_2)^2 + (\sigma_2 - \sigma_3)^2 + (\sigma_3 - \sigma_2)^2\right]. \tag{8.281}$$

According to the theory of maximum distortion energy, the material in a general stress state will yield when the above distortion energy is equal to the maximum distortion energy obtained from a tensile test.

For the uniaxial stress state, we have $\sigma_2 = \sigma_3 = 0$ and yield begins when $\sigma_1 = \sigma_y$. Thus, the maximum distortion energy in the tensile test is

$$U_{1d} = \frac{2\sigma_y^2}{12G}. \tag{8.282}$$

Making (8.281) equal to (8.282), we have

$$(\sigma_1 - \sigma_2)^2 + (\sigma_2 - \sigma_3)^2 + (\sigma_3 - \sigma_2)^2 = 2\sigma_y^2. \tag{8.283}$$

For a plane stress state, we have $\sigma_2 = 0$ and the above expression simplifies to

$$\left(\frac{\sigma_1}{\sigma_y}\right)^2 - \left(\frac{\sigma_1}{\sigma_y}\frac{\sigma_3}{\sigma_y}\right) + \left(\frac{\sigma_3}{\sigma_y}\right)^2 = 1. \tag{8.284}$$

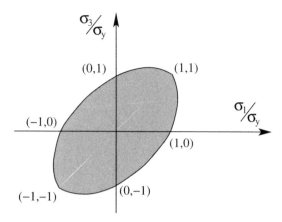

Figure 8.35 Maximum distortion energy criterion representation.

The above expression represents an ellipse equation at plane $\dfrac{\sigma_1}{\sigma_y} \times \dfrac{\sigma_3}{\sigma_y}$, as illustrated in Figure 8.35. Any stress state with representation that lies within the ellipse indicates that the material behaves elastically. The boundary points imply that yielding occurs.

This theory does not predict the response of the material when hydrostatic stress states are added, because only stress differences are involved in equation (8.281). Thus, adding constant stresses to the principal stress components do not change the yield condition.

From (8.283), the von Mises equivalent stress is defined as

$$\sigma_{eqv} = \sqrt{\frac{1}{2}\left[(\sigma_1 - \sigma_2)^2 + (\sigma_2 - \sigma_3)^2 + (\sigma_3 - \sigma_2)^2\right]}. \tag{8.285}$$

The material remains in the elastic phase if

$$\sigma_{eqv} < \sigma_y. \tag{8.286}$$

For a Timoshenko beam, substituting the previously calculated principal stresses in expression (8.285) and taking some simplifications, we obtain

$$\sigma_{eqv} = \sqrt{\sigma_{xx}^2 + 3\tau_{xy}^2}. \tag{8.287}$$

This is the same expression used in Chapters 6 and 7.

8.11.8 RANKINE CRITERION FOR BRITTLE MATERIALS

As observed in Chapter 3, the brittle materials do not present a characteristic yield behavior in the tensile test of a specimen. It is assumed that failure occurs when the maximum normal stress at a point reaches a certain critical value.

This criterion uses only the largest principal stress, which is compared with the ultimate stress (σ_{ult}) determined by the tensile test. Thus, failure occurs when

$$\sigma_1 = \sigma_{ult}. \tag{8.288}$$

The material remains in the elastic range if

$$\sigma_1 < \sigma_{ult}. \tag{8.289}$$

Figure 8.36 represents the Rankine criterion.

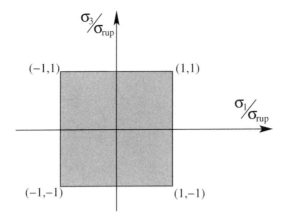

Figure 8.36 Maximum normal stress theory representation.

8.11.9 COMPARISON OF TRESCA, VON MISES, AND RANKINE CRITERIA

The Tresca criterion is based on the maximum shear stress at the critical point of the body. On the other hand, the von Mises criterion considers the energy from shear strains in a three-dimensional body. As the shear stresses are the main parameters for both criteria, we notice that there is great similarity between them, and Tresca's theory is more conservative.

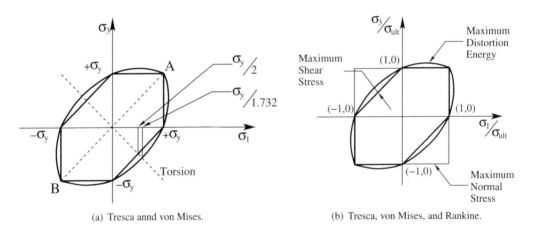

(a) Tresca annd von Mises.   (b) Tresca, von Mises, and Rankine.

Figure 8.37 Comparison of the Tresca, von Mises, and Rankine criteria.

Figure 8.37(a) illustrates the Tresca hexagon and the von Mises ellipse for the plane stress case. The yield of the material occurs when the stress state is in the boundaries of the hexagon or the ellipse.

The uniaxial stresses given by both theories are the same for simple compression or traction. The yield criteria in the second and fourth quadrants indicate less yield resistance than for uniaxial stresses.

The largest difference occurs when two of the principal stresses are equal but with opposite signs, that is, for $\pm\sigma_1 = \mp\sigma_3$. This stress state occurs, for example, in the twist of thin-walled tubes. By the Tresca criterion, these stresses can reach a maximum value of $\dfrac{\sigma_y}{2}$. The von Mises criterion limit these stresses to $\dfrac{\sigma_y}{\sqrt{3}} = 0.577\sigma_y$. On the other hand, the Rankine criterion is used only for brittle materials. Figure 8.37(b) shows a comparison between the three criteria.

8.12 APPROXIMATION OF LINEAR ELASTIC SOLIDS

8.12.1 WEAK FORM

Consider the equilibrium BVP or strong form (8.197) of a solid, expressed in terms of the Cauchy stress tensor. To obtain the weak form, we take the inner product of the differential equation given in (8.197) by the vectorial test function $\mathbf{v} = \mathbf{v}(x, y, z)$, and integrate it in the volume of the body, that is,

$$\int_V \operatorname{div} \mathbf{T} \cdot \mathbf{v} dV + \int_V \mathbf{b} \cdot \mathbf{v} dV = 0.$$

Using (8.189) and the divergence theorem (8.192), we find

$$\int_V \mathbf{T} : \nabla \mathbf{v} dV = \int_V \mathbf{b} \cdot \mathbf{v} dV + \int_{S_N} \mathbf{T} \mathbf{n} \cdot \mathbf{v} dS.$$

We assume that the surface of the solid is divided in regions S_D and S_N where the Dirichlet ($\mathbf{u} = \hat{\mathbf{u}}$) and Neumann ($\mathbf{T}\mathbf{n} = \mathbf{t}$) boundary conditions are applied, respectively. These regions are disjoint, that is, $S_D \cap S_N = \emptyset$ and $S = S_D \cup S_N$.

Using the boundary condition in terms of the stress vector given in (8.197), we obtain the weak form of a solid under small strain

$$\int_V \mathbf{T} : \nabla \mathbf{v} dV = \int_V \mathbf{b} \cdot \mathbf{v} dV + \int_{S_N} \mathbf{t} \cdot \mathbf{v} dS. \qquad (8.290)$$

The weak form for a linear elastic solid in small strain is obtained by substituting Hooke's law (8.198) in the previous expression, that is,

$$\int_V \left(\mu(\nabla \mathbf{u} + \nabla \mathbf{u}^T) + \lambda(\operatorname{div} \mathbf{u})\mathbf{I} \right) : \nabla \mathbf{v} dV = \int_V \mathbf{b} \cdot \mathbf{v} dV + \int_{S_N} \mathbf{t} \cdot \mathbf{v} dS. \qquad (8.291)$$

The terms in left-hand side of the previous equations simplifies to

$$
\begin{aligned}
\mu(\nabla \mathbf{u} + \nabla \mathbf{u}^T) : \nabla \mathbf{v} &= \frac{\mu}{2}(\nabla \mathbf{u} + \nabla \mathbf{u}^T) : \left[(\nabla \mathbf{v} + \nabla \mathbf{v}^T) + (\nabla \mathbf{v} - \nabla \mathbf{v}^T) \right] \\
&= \frac{\mu}{2}(\nabla \mathbf{u} + \nabla \mathbf{u}^T) : (\nabla \mathbf{v} + \nabla \mathbf{v}^T), \\
\mathbf{I} : \nabla \mathbf{v} &= \operatorname{div} \mathbf{v}.
\end{aligned}
$$

Therefore, the weak form (8.291) is rewritten as

$$\int_V \left\{ \frac{\mu}{2}[(\nabla \mathbf{u} + \nabla \mathbf{u}^T) : (\nabla \mathbf{v} + \nabla \mathbf{v}^T)] + \lambda(\operatorname{div} \mathbf{u})(\operatorname{div} \mathbf{v}) \right\} dV = \int_V \mathbf{b} \cdot \mathbf{v} dV + \int_{S_N} \mathbf{t} \cdot \mathbf{v} dS. \qquad (8.292)$$

The above weak form can also be expressed in matrix notation using equations (8.24) and (8.64). Thus,

$$\int_V ([\mathbf{D}][\mathbf{L}]\{\mathbf{u}\})([\mathbf{L}]\{\mathbf{v}\}) dV = \int_V \{\mathbf{b}\}^T \{\mathbf{v}\} dV + \int_{S_N} \{\mathbf{t}\}^T \{\mathbf{v}\} dS. \qquad (8.293)$$

The displacement vector field $\mathbf{u}$ can be approximated by the following linear combination of N global shape functions $\{\Phi_i(x, y, z)\}_{i=1}^N$ as follows:

$$\mathbf{u}(x, y, z) \approx \mathbf{u}_N(x, y, z) = \sum_{i=1}^N \Phi_i(x, y, z)\mathbf{u}_i, \qquad (8.294)$$

where $\mathbf{u}_i = \{u_{x_i}\ u_{y_i}\ u_{z_i}\}^T$ are the approximation coefficients associated to function Φ_i. Using the Galerkin method, the test function is approximated as

$$\mathbf{v}(x, y, z) \approx \mathbf{v}_N(x, y, z) = \sum_{i=1}^N \Phi_i(x, y, z)\mathbf{v}_i, \qquad (8.295)$$

with $\mathbf{v}_i = \{v_{x_i}\ v_{y_i}\ v_{z_i}\}^T$.

Both previous equations can be expressed in matrix notation as

$$
\mathbf{u}_N(x,y,z) =
\begin{bmatrix}
\Phi_1 & 0 & 0 & \dots & \Phi_N & 0 & 0 \\
0 & \Phi_1 & 0 & \dots & 0 & \Phi_N & 0 \\
0 & 0 & \Phi_1 & \dots & 0 & 0 & \Phi_N
\end{bmatrix}
\begin{Bmatrix}
u_{x_1} \\
u_{y_1} \\
u_{z_1} \\
\vdots \\
u_{x_N} \\
u_{y_N} \\
u_{z_N}
\end{Bmatrix}
= [\mathbf{N}]\{\mathbf{u}_N\},
\tag{8.296}
$$

$$
\mathbf{v}_N(x,y,z) =
\begin{bmatrix}
\Phi_1 & 0 & 0 & \dots & \Phi_N & 0 & 0 \\
0 & \Phi_1 & 0 & \dots & 0 & \Phi_N & 0 \\
0 & 0 & \Phi_1 & \dots & 0 & 0 & \Phi_N
\end{bmatrix}
\begin{Bmatrix}
v_{x_1} \\
v_{y_1} \\
v_{z_1} \\
\vdots \\
v_{x_N} \\
v_{y_N} \\
v_{z_N}
\end{Bmatrix}
= [\mathbf{N}]\{\mathbf{v}_N\},
\tag{8.297}
$$

where $\{\mathbf{u}_N\}$ and $\{\mathbf{v}_N\}$ are the approximation coefficient vectors for the displacement and test functions; $[\mathbf{N}]$ is the matrix with the global shape functions.

Substituting the previous expressions in (8.293) and after some manipulations, we obtain

$$
\{\mathbf{v}_N\}^T \left[\left(\int_V [\mathbf{B}]^T [\mathbf{D}][\mathbf{B}] dV \right) \{\mathbf{u}_N\} - \int_V [\mathbf{N}]^T \{\mathbf{b}\} dV - \int_{S_N} [\mathbf{N}]^T \{\mathbf{t}\} dS \right] = 0.
$$

As the approximation coefficients of the test function are arbitrary, to satisfy the previous equation, the term inside brackets must be zero, resulting in the approximated weak form for a linear elastic solid

$$
\left(\int_V [\mathbf{B}]^T [\mathbf{D}][\mathbf{B}] dV \right) \{\mathbf{u}_N\} = \int_V [\mathbf{N}]^T \{\mathbf{b}\} dV + \int_{S_N} [\mathbf{N}]^T \{\mathbf{t}\} dS.
\tag{8.298}
$$

In a summarized way,

$$
[\mathbf{K}]\{\mathbf{u}_N\} = \{\mathbf{f}\},
\tag{8.299}
$$

where $[\mathbf{K}]$ and $\{\mathbf{f}\}$ respectively are the global stiffness matrix and the global equivalent body and surface force vector, given by

$$
[\mathbf{K}] = \int_V [\mathbf{B}]^T [\mathbf{D}][\mathbf{B}] dV,
\tag{8.300}
$$

$$
\{\mathbf{f}\} = \int_V [\mathbf{N}]^T \{\mathbf{b}\} dV + \int_S [\mathbf{N}]^T \{\mathbf{t}\} dS.
\tag{8.301}
$$

The strain-displacement matrix $[\mathbf{B}]$ is given by

$$
[\mathbf{B}] = [\mathbf{L}][\mathbf{N}] = [[\mathbf{B}_1] \mid [\mathbf{B}_2] \mid \dots \mid [\mathbf{B}_N]],
\tag{8.302}
$$

where $[\mathbf{B}_i]$ is the associated matrix to each shape function and

$$
[\mathbf{B}_i] =
\begin{bmatrix}
\Phi_{i,x} & 0 & 0 \\
0 & \Phi_{i,y} & 0 \\
0 & 0 & \Phi_{i,z} \\
\Phi_{i,y} & \Phi_{i,x} & 0 \\
\Phi_{i,z} & 0 & \Phi_{i,x} \\
0 & \Phi_{i,z} & \Phi_{i,y}
\end{bmatrix}.
\tag{8.303}
$$

We can split the solid body in a finite element mesh. The test and displacement field approximations for each element e are expressed as the following linear combinations of N_e element shape functions $\{\phi_i^{(e)}(x,y,z)\}_{i=1}^{N_e}$:

$$
\mathbf{u}_{N_e}^{(e)}(x,y,z) = \sum_{i=1}^{N_e} \phi_i^{(e)}(x,y,z)\{\mathbf{u}_i^{(e)}\} = [\mathbf{N}^{(e)}]\{\mathbf{u}^{(e)}\},
\tag{8.304}
$$

$$\mathbf{v}_{N_e}^{(e)}(x,y,z) = \sum_{i=1}^{N_e} \phi_i^{(e)}(x,y,z)\{\mathbf{v}_i^{(e)}\} = [\mathbf{N}^{(e)}]\{\mathbf{v}^{(e)}\}, \tag{8.305}$$

with the approximation coefficients $\{\mathbf{u}_i^{(e)}\} = \{u_{x_i}^e \ u_{y_i}^e \ u_{z_i}^e\}^T$ and $\{\mathbf{v}_i^{(e)}\} = \{v_{x_i}^e \ v_{y_i}^e \ v_{z_i}^e\}^T$; $[\mathbf{N}^{(e)}]$ is the matrix of the element shape function given by

$$[\mathbf{N}^{(e)}] = \begin{bmatrix} \Phi_1^{(e)} & 0 & 0 & \cdots & \Phi_{N_e}^{(e)} & 0 & 0 \\ 0 & \Phi_1^{(e)} & 0 & \cdots & 0 & \Phi_{N_e}^{(e)} & 0 \\ 0 & 0 & \Phi_1^{(e)} & \cdots & 0 & 0 & \Phi_{N_e}^{(e)} \end{bmatrix}. \tag{8.306}$$

The weak form in element e is obtained analogous to (8.298) as

$$\left(\int_{V^{(e)}} [\mathbf{B}^{(e)}]^T [\mathbf{D}][\mathbf{B}^{(e)}]dV \right)\{\mathbf{u}^{(e)}\} = \int_{V^{(e)}} [\mathbf{N}^{(e)}]^T\{\mathbf{b}\}dV + \int_{S^{(e)}} [\mathbf{N}^{(e)}]^T\{\mathbf{t}\}dS, \tag{8.307}$$

where $V^{(e)}$ and $S^{(e)}$ respectively represent the volume and surface of the element. In a summarized form

$$[\mathbf{K}^{(e)}]\{\mathbf{u}^{(e)}\} = \{\mathbf{f}^{(e)}\}, \tag{8.308}$$

where $[\mathbf{K}^{(e)}]$ and $\{\mathbf{f}^{(e)}\}$ respectively are the element stiffness matrix and equivalent body and surface vectors, given by

$$[\mathbf{K}^{(e)}] = \int_{V^{(e)}} [\mathbf{B}^{(e)}]^T [\mathbf{D}][\mathbf{B}^{(e)}]dV, \tag{8.309}$$

$$\{\mathbf{f}^{(e)}\} = \int_{V^{(e)}} [\mathbf{N}^{(e)}]^T\{\mathbf{b}\}dV + \int_{S^{(e)}} [\mathbf{N}^{(e)}]^T\{\mathbf{t}\}dS. \tag{8.310}$$

On the other hand, the element strain-displacement matrix $[\mathbf{B}^{(e)}]$ is given by

$$[\mathbf{B}^{(e)}] = [\mathbf{L}][\mathbf{N}^{(e)}] = \left[[\mathbf{B}_1^{(e)}] \mid [\mathbf{B}_2^{(e)}] \mid \cdots \mid [\mathbf{B}_{N_e}^{(e)}] \right], \tag{8.311}$$

where $[\mathbf{B}_i^{(e)}]$ is the matrix associated to each element shape function, and is expressed as

$$[\mathbf{B}_i^{(e)}] = \begin{bmatrix} \phi_{i,x}^{(e)}(x,y,z) & 0 & 0 \\ 0 & \phi_{i,y}^{(e)}(x,y,z) & 0 \\ 0 & 0 & \phi_{i,z}^{(e)}(x,y,z) \\ \phi_{i,y}^{(e)}(x,y,z) & \phi_{i,x}^{(e)}(x,y,z) & 0 \\ \phi_{i,z}^{(e)}(x,y,z) & 0 & \phi_{i,x}^{(e)}(x,y,z) \\ 0 & \phi_{i,z}^{(e)}(x,y,z) & \phi_{i,y}^{(e)}(x,y,z) \end{bmatrix}. \tag{8.312}$$

All previous expressions are written in terms of global Cartesian coordinates (x,y,z). The use of these coordinates makes difficult not only the definition of shape functions for distorted global elements, but also the calculation of volume and surface integrals present in the expressions of the element stiffness matrix and load vector, as indicated in equations (8.309) and (8.310).

Analogous to the case of one-dimensional elements presented in the previous chapters, it becomes interesting to work with standard elements defined in local coordinate systems, as shown in Figure 8.38 for squares and hexahedra. Each coordinate ξ_i varies in the closed interval $[-1,1]$.

The approximations for the test and displacement vector fields for a hexahedrical element, expressed in terms of the local coordinates (ξ_1, ξ_2, ξ_3), are respectively given by

$$\mathbf{u}_{N_e}^{(e)}(x,y,z) = \sum_{i=1}^{N_e} \phi_i^{(e)}(\xi_1, \xi_2, \xi_3)\{\mathbf{u}_i^{(e)}\} = [\mathbf{N}^{(e)}]\{\mathbf{u}^{(e)}\}, \tag{8.313}$$

$$\mathbf{v}_{N_e}^{(e)}(x,y,z) = \sum_{i=1}^{N_e} \phi_i^{(e)}(\xi_1, \xi_2, \xi_3)\{\mathbf{v}_i^{(e)}\} = [\mathbf{N}^{(e)}]\{\mathbf{v}^{(e)}\}, \tag{8.314}$$

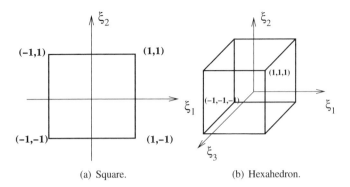

(a) Square.   (b) Hexahedron.

Figure 8.38 Local coordinate systems for squares and hexahedra (adapted from [10]).

where $\{\phi_i^{(e)}\}_{i=1}^{N_e}$ are the element shape functions, defined in terms of the local coordinates. The construction of functions for structured and nonstructured elements will be presented in the next sections based on references [35, 9, 10].

The weak form in element e in terms of the local coordinates is given by

$$\left(\int_{-1}^{1} \int_{-1}^{1} \int_{-1}^{1} [\mathbf{B}^{(e)}]^T [\mathbf{D}][\mathbf{B}^{(e)}] |\det[\mathbf{J}]| d\xi_1 d\xi_2 d\xi_3 \right) \{\mathbf{u}^{(e)}\} =$$
$$\int_{-1}^{1} \int_{-1}^{1} \int_{-1}^{1} [\mathbf{N}^{(e)}]^T \{\mathbf{b}\} |\det[\mathbf{J}]| d\xi_1 d\xi_2 d\xi_3 + \int_{-1}^{1} \int_{-1}^{1} [\mathbf{N}^{(e)}]^T \{\mathbf{t}\} |\det[\mathbf{J}_b]| d\Gamma^{(e)},$$

(8.315)

where $[\mathbf{J}]$ and $[\mathbf{J}_b]$ are the Jacobian matrices, which are respectively associated to the volume $V^{(e)}$ and surface $\Gamma^{(e)}$ of the standard element. These matrices are obtained in Section 8.12.4, along with the mapping of global coordinates (x, y, z) used in approximations (8.313) and (8.314).

The element strain-displacement matrix is composed of local shape functions derivatives with respect to the global variables. For instance, the matrix for the ith element shape functions is given by

$$[\mathbf{B}_i^{(e)}] = \begin{bmatrix} \phi_{i,x}^{(e)}(\xi_1, \xi_2, \xi_3) & 0 & 0 \\ 0 & \phi_{i,y}^{(e)}(\xi_1, \xi_2, \xi_3) & 0 \\ 0 & 0 & \phi_{i,z}^{(e)}(\xi_1, \xi_2, \xi_3) \\ \phi_{i,y}^{(e)}(\xi_1, \xi_2, \xi_3) & \phi_{i,x}^{(e)}(\xi_1, \xi_2, \xi_3) & 0 \\ \phi_{i,z}^{(e)}(\xi_1, \xi_2, \xi_3) & 0 & \phi_{i,x}^{(e)}(\xi_1, \xi_2, \xi_3) \\ 0 & \phi_{i,z}^{(e)}(\xi_1, \xi_2, \xi_3) & \phi_{i,x}^{(e)}(\xi_1, \xi_2, \xi_3) \end{bmatrix}.$$

(8.316)

Obtaining the shape function derivatives, expressed in the local system $\xi_1 \xi_2 \xi_3$, with respect to variables x, y, and z, will also be presented in Section 8.12.4.

The element stiffness matrix and the element body and surface equivalent force vectors can be calculated in the local reference system $\xi_1 \xi_2 \xi_3$ as

$$[\mathbf{K}^{(e)}] = \int_{-1}^{1} \int_{-1}^{1} \int_{-1}^{1} [\mathbf{B}^{(e)}]^T [\mathbf{D}][\mathbf{B}^{(e)}] |\det[\mathbf{J}]| d\xi_1 d\xi_2 d\xi_3,$$

(8.317)

$$\{\mathbf{f}^{(e)}\} = \int_{-1}^{1} \int_{-1}^{1} \int_{-1}^{1} [\mathbf{N}^{(e)}]^T \{\mathbf{b}\} |\det[\mathbf{J}]| d\xi_1 d\xi_2 d\xi_3 + \int_{-1}^{1} \int_{-1}^{1} [\mathbf{N}^{(e)}]^T \{\mathbf{t}\} |\det[\mathbf{J}_b]| d\Gamma^{(e)}.$$

(8.318)

The advantage of employing the above expressions lies with the fixed limits of integration. This favors the use of numerical integration procedures. The Jacobians are responsible for passing the distortion information from the global to the local reference system.

8.12.2 SHAPE FUNCTIONS FOR STRUCTURED ELEMENTS

Squares and hexahedra are considered as structured elements. The modal and nodal shape functions for squares are constructed using the tensor product of onedimensional functions in directions ξ_1 and ξ_2, as illustrated in Figure 8.39 [49, 35, 10].

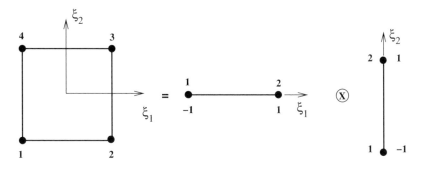

Figure 8.39 Tensor construction of square shape functions (adapted from [10]).

From Figure 8.39, the expressions of the shape functions for squares can be written by the following product of functions in the ξ_1 and ξ_2 directions:

$$\phi_{pq}^{(e)}(\xi_1,\xi_2) = \phi_p^{(e)}(\xi_1)\phi_q^{(e)}(\xi_2), \quad 0 \le p \le P_1 \quad \text{and} \quad 0 \le q \le P_2, \tag{8.319}$$

where P_1 and P_2 are the polynomial orders in directions ξ_1 and ξ_2, as illustrated in Figure 8.40(a).

Analogous to the onedimensional case, we can associate the shape functions to the element topological entities, which in the case of a square are the four vertices (V_1, V_2, V_3, V_4), four edges (E_1, E_2, E_3, E_4), and one face (F_1), illustrated in Figure 8.40(b). The indices p and q of equation (8.319) are associated to the topological entities of the square, according to Figure 8.40(c).

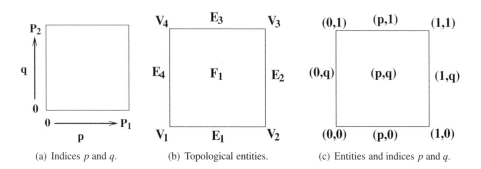

(a) Indices p and q.   (b) Topological entities.   (c) Entities and indices p and q.

Figure 8.40 Association between the topological entities and tensorization indices p and q in the square (adapted from [10]).

Hence, the expressions for the shape functions of vertices V_1, V_2, V_3, V_4 are respectively given by

$$\begin{aligned}
\phi_{00}^{(e)}(\xi_1,\xi_2) &= \phi_0^{(e)}(\xi_1)\phi_0^{(e)}(\xi_2), \\
\phi_{P_1 0}^{(e)}(\xi_1,\xi_2) &= \phi_{P_1}^{(e)}(\xi_1)\phi_0^{(e)}(\xi_2), \\
\phi_{P_1 P_2}^{(e)}(\xi_1,\xi_2) &= \phi_{P_1}^{(e)}(\xi_1)\phi_{P_2}^{(e)}(\xi_2), \\
\phi_{0 P_2}^{(e)}(\xi_1,\xi_2) &= \phi_0^{(e)}(\xi_1)\phi_{P_2}^{(e)}(\xi_2).
\end{aligned} \tag{8.320}$$

Analogously, the functions of edges E_1, E_2, E_3, E_4 are, respectively,

$$
\begin{aligned}
\phi_{p0}^{(e)}(\xi_1,\xi_2) &= \phi_p^{(e)}(\xi_1)\phi_0^{(e)}(\xi_2), && 0 < p < P_1, \\
\phi_{P_1 q}^{(e)}(\xi_1,\xi_2) &= \phi_{P_1}^{(e)}(\xi_1)\phi_q^{(e)}(\xi_2), && 0 < q < P_2, \\
\phi_{p P_2}^{(e)}(\xi_1,\xi_2) &= \phi_p^{(e)}(\xi_1)\phi_{P_2}^{(e)}(\xi_2), && 0 < p < P_1, \\
\phi_{0q}^{(e)}(\xi_1,\xi_2) &= \phi_0^{(e)}(\xi_1)\phi_q^{(e)}(\xi_2), && 0 < q < P_2.
\end{aligned}
\tag{8.321}
$$

Finally, the functions of face F_1 or internal modes are

$$
\phi_{pq}^{(e)}(\xi_1,\xi_2) = \phi_p^{(e)}(\xi_1)\phi_q^{(e)}(\xi_2), \quad 0 < p < P_1 \quad \text{and} \quad 0 < q < P_2.
\tag{8.322}
$$

From the previous procedure, the nodal and modal bases for the square are defined.

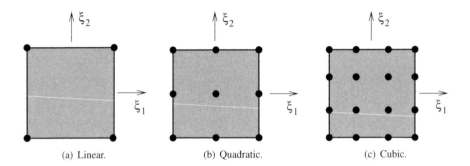

(a) Linear.   (b) Quadratic.   (c) Cubic.

Figure 8.41   Lagrangian square elements.

The standard nodal Lagrangian basis [54, 18, 35] for squares is obtained substituting the one-dimensional basis definition (4.111) in (8.319). The Lagrange elements of first, second, and third orders are illustrated in Figure 8.41.

The linear or first-order element has only vertex functions, and expressions for $P_1 = P_2 = 1$ are obtained substituting (4.111) in equation (8.320). Thus,

$$
\begin{aligned}
\phi_{00}^{(e)}(\xi_1,\xi_2) &= \frac{1}{4}(1-\xi_1)(1-\xi_2), \\
\phi_{10}^{(e)}(\xi_1,\xi_2) &= \frac{1}{4}(1+\xi_1)(1-\xi_2), \\
\phi_{11}^{(e)}(\xi_1,\xi_2) &= \frac{1}{4}(1+\xi_1)(1+\xi_2), \\
\phi_{01}^{(e)}(\xi_1,\xi_2) &= \frac{1}{4}(1-\xi_1)(1+\xi_2).
\end{aligned}
\tag{8.323}
$$

The previous shape functions are illustrated in Figure 8.42.

The vertex functions of the quadratic or second-order element are obtained analogously. Hence,

$$
\begin{aligned}
\phi_{00}^{(e)}(\xi_1,\xi_2) &= \frac{1}{4}\xi_1(1-\xi_1)\xi_2(1-\xi_2), \\
\phi_{20}^{(e)}(\xi_1,\xi_2) &= -\frac{1}{4}\xi_1(1+\xi_1)\xi_2(1-\xi_2), \\
\phi_{22}^{(e)}(\xi_1,\xi_2) &= \frac{1}{4}\xi_1(1+\xi_1)\xi_2(1+\xi_2), \\
\phi_{02}^{(e)}(\xi_1,\xi_2) &= -\frac{1}{4}\xi_1(1-\xi_1)\xi_2(1+\xi_2).
\end{aligned}
\tag{8.324}
$$

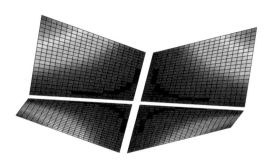

Figure 8.42 Lagrangian shape functions for the linear square.

The edge functions are determined from (4.111) and (8.321), that is,

$$
\begin{aligned}
\phi_{10}^{(e)}(\xi_1, \xi_2) &= -\frac{1}{2}\xi_2(1 - \xi_2)(1 - \xi_1^2), \\
\phi_{21}^{(e)}(\xi_1, \xi_2) &= \frac{1}{2}\xi_1(1 + \xi_1)(1 - \xi_2^2), \\
\phi_{12}^{(e)}(\xi_1, \xi_2) &= \frac{1}{2}\xi_2(1 + \xi_2)(1 - \xi_1^2), \\
\phi_{01}^{(e)}(\xi_1, \xi_2) &= -\frac{1}{2}\xi_1(1 - \xi_1)(1 - \xi_2^2).
\end{aligned}
\tag{8.325}
$$

Finally, the face function is obtained from (8.322), that is,

$$
\phi_{11}^{(e)}(\xi_1, \xi_2) = (1 - \xi_1^2)(1 - \xi_2^2).
\tag{8.326}
$$

The shape functions for the third-order or cubic element are obtained analogously. Figures 8.43 and 8.44 illustrate the functions of second- and third-order elements, respectively.

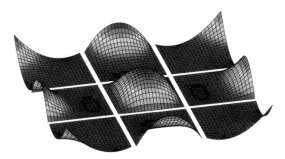

Figure 8.43 Lagrangian shape functions for the second-order square.

As known [54, 18, 35], the Lagrange nodal basis generates additional unnecessary terms to ensure the completeness of the local expansion. To avoid this, we work with the serendipity nodal basis, which requires internal nodes in the square just beyond the fourth order, as shown in Figure 8.45. The linear elements of serendipity and Lagrange families are identical, as well as the vertex shape functions of all elements.

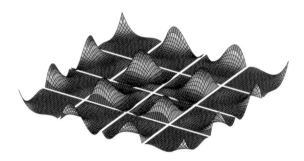

Figure 8.44   Lagrangian shape functions for the third-order square.

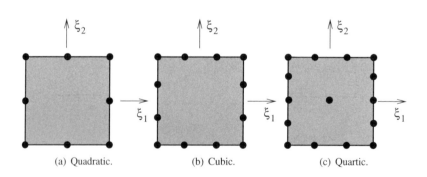

(a) Quadratic.   (b) Cubic.   (c) Quartic.

Figure 8.45   Square elements of the serendipity family.

The edge shape functions of serendipity elements are determined from (8.319), but replacing definition (4.111) by

$$\phi_p^{(e)}(\xi_1) = \begin{cases} \frac{1}{2}(1-\xi_1) & p = 0 \\ \frac{1}{2}(1+\xi_1) & p = P_1 \\ \frac{1}{4}(1-\xi_1)(1+\xi_1)L_p^{(P_1-2)} & 0 < p < P_1 \end{cases}. \tag{8.327}$$

The expressions for the edge functions in the quadratic element are

$$\begin{aligned}
\phi_{10}^{(e)}(\xi_1,\xi_2) &= \phi_1^{(e)}(\xi_1)\phi_0^{(e)}(\xi_2) = \frac{1}{2}(1-\xi_1)(1+\xi_1)(1-\xi_2), \\
\phi_{21}^{(e)}(\xi_1,\xi_2) &= \phi_2^{(e)}(\xi_1)\phi_1^{(e)}(\xi_2) = \frac{1}{2}(1-\xi_1)(1+\xi_1)(1+\xi_2), \\
\phi_{12}^{(e)}(\xi_1,\xi_2) &= \phi_1^{(e)}(\xi_1)\phi_2^{(e)}(\xi_2) = \frac{1}{2}(1+\xi_1)(1+\xi_2)(1-\xi_2), \\
\phi_{01}^{(e)}(\xi_1,\xi_2) &= \phi_0^{(e)}(\xi_1)\phi_1^{(e)}(\xi_2) = \frac{1}{2}(1-\xi_1)(1+\xi_2)(1-\xi_2).
\end{aligned} \tag{8.328}$$

For the cubic element, we determine the edge functions analogously, that is,

$$\begin{aligned}
\phi_{10}^{(e)}(\xi_1,\xi_2) &= \frac{9}{32}(1-3\xi_1)(1-\xi_1^2)(1-\xi_2), \\
\phi_{20}^{(e)}(\xi_1,\xi_2) &= \frac{9}{32}(1+3\xi_1)(1-\xi_1^2)(1-\xi_2), \\
\phi_{31}^{(e)}(\xi_1,\xi_2) &= \frac{9}{32}(1+\xi_1)((1-3\xi_2)(1-\xi_2^2), \\
\phi_{32}^{(e)}(\xi_1,\xi_2) &= \frac{9}{32}(1+\xi_1)(1+3\xi_2)(1-\xi_2^2), \\
\phi_{13}^{(e)}(\xi_1,\xi_2) &= \frac{9}{32}(1-3\xi_1)(1-\xi_1^2)(1+\xi_2), \\
\phi_{23}^{(e)}(\xi_1,\xi_2) &= \frac{9}{32}(1+3\xi_1)(1-\xi_1^2)(1+\xi_2), \\
\phi_{01}^{(e)}(\xi_1,\xi_2) &= \frac{9}{32}(1-\xi_1)(1-3\xi_2)(1-\xi_2^2), \\
\phi_{02}^{(e)}(\xi_1,\xi_2) &= \frac{9}{32}(1-\xi_1)(1+3\xi_2)(1-\xi_2^2).
\end{aligned} \tag{8.329}$$

From the fourth order, the face shape functions are obtained from (8.319), but using definition

$$\phi_p^{(e)}(\xi_1) = \frac{1}{4}(1-\xi_1)(1+\xi_1)L_p^{(P_1-4)}(\xi_1) \tag{8.330}$$

for $P_1 = P_2 = P > 3$ and $1 < p, q < P_1$. The fourth-order element shape functions of the serendipity family are illustrated in Figure 8.46. An alternative family of serendipity elements can be obtained imposing the restriction $p+q \geq P-2$ for the internal modes [35].

The simplest modal basis for square elements is obtained by substituting definition (4.123), in terms of the Jacobi polynomials, in equation (8.319) [35]. The vertex functions are always linear and given by the expressions in (8.323). However, substituting (4.123) in expressions (8.321), we have the edge shape functions

$$\begin{aligned}
\phi_{p0}^{(e)}(\xi_1,\xi_2) &= \frac{1}{8}(1-\xi_1)(1+\xi_1)(1-\xi_2)P_{p-1}^{\alpha_1,\beta_1}(\xi_1), & 1 < p < P_1, \\
\phi_{P_1 q}^{(e)}(\xi_1,\xi_2) &= \frac{1}{8}(1+\xi_1)(1-\xi_2)(1+\xi_2)P_{q-1}^{\alpha_2,\beta_2}(\xi_2), & 1 < q < P_2, \\
\phi_{pP_2}^{(e)}(\xi_1,\xi_2) &= \frac{1}{8}(1-\xi_1)(1+\xi_1)(1+\xi_2)P_{p-1}^{\alpha_1,\beta_1}(\xi_1), & 1 < p < P_1, \\
\phi_{0q}^{(e)}(\xi_1,\xi_2) &= \frac{1}{8}(1-\xi_1)(1+\xi_2)(1-\xi_2)P_{q-1}^{\alpha_2,\beta_2}(\xi_2), & 1 < q < P_2,
\end{aligned} \tag{8.331}$$

where (α_1, β_1) and (α_2, β_2) respectively are the Jacobi weights in directions ξ_1 and ξ_2. These coefficients are selected in a convenient way to obtain a better sparsity of the element matrices. From (4.123) and (8.324), we

Figure 8.46   Serendipity shape functions for the fourth-order square.

obtain the face functions as

$$\phi_{pq}^{(e)}(\xi_1,\xi_2) = \frac{1}{16}(1-\xi_1^2)(1-\xi_2^2)P_{p-1}^{\alpha_1,\beta_1}(\xi_1)P_{q-1}^{\alpha_2,\beta_2}(\xi_2), \quad 0 < p < P_1 \ \text{ and } \ 0 < q < P_2. \tag{8.332}$$

Figures 8.47 and 8.48 illustrate the modal shape functions for the second- and third-order squares with $P_1 = P_2 = P$, with $P = 2$ and $P = 3$, respectively, and $\alpha = \beta = 1$. We can obtain a serendipity modal basis employing condition $p + q \geq P - 2$ [35].

Figure 8.47   Modal shape functions for the second-order square (edge and face functions respectively multiplied by 4 and 16).

Similar to the case of squares, the shape functions for hexahedra are constructed by the tensor product of onedimensional polynomials in directions ξ_1, ξ_2, and ξ_3, as shown in Figure 8.49 [49, 35]. The general expression of the hexahedra shape functions, obtained by tensor product, is given by

$$\phi_{pqr}^{(e)}(\xi_1,\xi_2,\xi_2) = \phi_p^{(e)}(\xi_1)\phi_q^{(e)}(\xi_2)\phi_r^{(e)}(\xi_3), \tag{8.333}$$

with $0 \leq p \leq P_1$, $0 \leq q \leq P_2$, and $0 \leq r \leq P_3$, where P_1, P_2, and P_3 are the polynomial orders in directions ξ_1, ξ_2, and ξ_3, respectively, as illustrated in Figure 8.50(a).

Figure 8.48 Modal shape functions for the third-order square (edge and face functions respectively multiplied by 4 and 16).

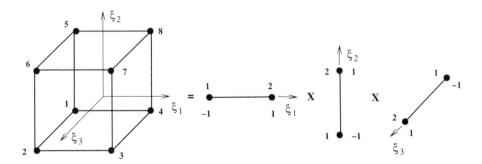

Figure 8.49 Tensorial construction of shape functions for hexahedra (adapted from [10]).

Figures 8.50(b) and 8.50(c) illustrate the topological entities of the hexahedron, which are constituted of eight vertices (V_1 to V_8), twelve edges (E_1 to E_{12}), six faces (F_1 to F_6), and one volume (B_1). Figure 8.51 presents the relation between indices p, q, and r and the topological entities of the hexahedron.

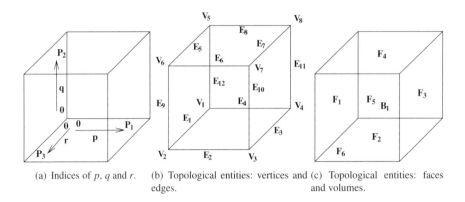

(a) Indices of p, q and r.   (b) Topological entities: vertices and (c) Topological entities: faces
                                      edges.   and volumes.

Figure 8.50 Indices p, q, and r and topological entities of the hexahedron (adapted from [10]).

From Figure 8.51 and equation (8.333), the expressions of the shape functions of vertices (V_1 to V_8) are, respectively,

$$\phi_{000}^{(e)}(\xi_1,\xi_2,\xi_3) = \phi_0^{(e)}(\xi_1)\phi_0^{(e)}(\xi_2)\phi_0^{(e)}(\xi_3),$$
$$\phi_{P_100}^{(e)}(\xi_1,\xi_2,\xi_3) = \phi_{P_1}^{(e)}(\xi_1)\phi_0^{(e)}(\xi_2)\phi_0^{(e)}(\xi_3),$$
$$\phi_{P_1P_20}^{(e)}(\xi_1,\xi_2,\xi_3) = \phi_{P_1}^{(e)}(\xi_1)\phi_{P_2}^{(e)}(\xi_2)\phi_0^{(e)}(\xi_3),$$
$$\phi_{0P_20}^{(e)}(\xi_1,\xi_2,\xi_3) = \phi_0^{(e)}(\xi_1)\phi_{P_2}^{(e)}(\xi_2)\phi_0^{(e)}(\xi_3), \qquad (8.334)$$
$$\phi_{00P_3}^{(e)}(\xi_1,\xi_2,\xi_3) = \phi_0^{(e)}(\xi_1)\phi_0^{(e)}(\xi_2)\phi_{P_3}^{(e)}(\xi_3),$$
$$\phi_{P_10P_3}^{(e)}(\xi_1,\xi_2,\xi_3) = \phi_{P_1}^{(e)}(\xi_1)\phi_0^{(e)}(\xi_2)\phi_{P_3}^{(e)}(\xi_3),$$
$$\phi_{0P_2P_3}^{(e)}(\xi_1,\xi_2,\xi_3) = \phi_0^{(e)}(\xi_1)\phi_{P_2}^{(e)}(\xi_2)\phi_{P_3}^{(e)}(\xi_3),$$
$$\phi_{P_1P_2P_3}^{(e)}(\xi_1,\xi_2,\xi_3) = \phi_{P_1}^{(e)}(\xi_1)\phi_{P_2}^{(e)}(\xi_2)\phi_{P_3}^{(e)}(\xi_3).$$

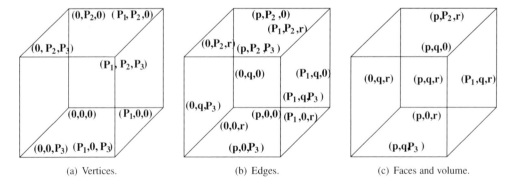

(a) Vertices.   (b) Edges.   (c) Faces and volume.

Figure 8.51 Association between indices p, q, and r and the topological entities of the hexahedron (adapted from [10]).

Analogously, the edge functions are

$$
\begin{aligned}
\phi_{p00}^{(e)}(\xi_1,\xi_2,\xi_3) &= \phi_p^{(e)}(\xi_1)\phi_0^{(e)}(\xi_2)\phi_0^{(e)}(\xi_3), \\
\phi_{P_1q0}^{(e)}(\xi_1,\xi_2,\xi_3) &= \phi_{P_1}^{(e)}(\xi_1)\phi_q^{(e)}(\xi_2)\phi_0^{(e)}(\xi_3), \\
\phi_{pP_20}^{(e)}(\xi_1,\xi_2,\xi_3) &= \phi_p^{(e)}(\xi_1)\phi_{P_2}^{(e)}(\xi_2)\phi_0^{(e)}(\xi_3), \\
\phi_{0q0}^{(e)}(\xi_1,\xi_2,\xi_3) &= \phi_0^{(e)}(\xi_1)\phi_q^{(e)}(\xi_2)\phi_0^{(e)}(\xi_3) \\
\phi_{0P_2r}^{(e)}(\xi_1,\xi_2,\xi_3) &= \phi_0^{(e)}(\xi_1)\phi_{P_2}^{(e)}(\xi_2)\phi_r^{(e)}(\xi_3), \\
\phi_{P_1qP_3}^{(e)}(\xi_1,\xi_2,\xi_3) &= \phi_{P_1}^{(e)}(\xi_1)\phi_q^{(e)}(\xi_2)\phi_{P_3}^{(e)}(\xi_3), \\
\phi_{pP_2P_3}^{(e)}(\xi_1,\xi_2,\xi_3) &= \phi_p^{(e)}(\xi_1)\phi_{P_2}^{(e)}(\xi_2)\phi_{P_3}^{(e)}(\xi_3), \\
\phi_{0qP_3}^{(e)}(\xi_1,\xi_2,\xi_3) &= \phi_0^{(e)}(\xi_1)\phi_q^{(e)}(\xi_2)\phi_{P_3}^{(e)}(\xi_3), \\
\phi_{P_10r}^{(e)}(\xi_1,\xi_2,\xi_3) &= \phi_{P_1}^{(e)}(\xi_1)\phi_0^{(e)}(\xi_2)\phi_r^{(e)}(\xi_3), \\
\phi_{P_1P_2r}^{(e)}(\xi_1,\xi_2,\xi_3) &= \phi_{P_1}^{(e)}(\xi_1)\phi_{P_2}^{(e)}(\xi_2)\phi_r^{(e)}(\xi_3), \\
\phi_{p0P_3}^{(e)}(\xi_1,\xi_2,\xi_3) &= \phi_p^{(e)}(\xi_1)\phi_0^{(e)}(\xi_2)\phi_{P_3}^{(e)}(\xi_3), \\
\phi_{00r}^{(e)}(\xi_1,\xi_2,\xi_3) &= \phi_0^{(e)}(\xi_1)\phi_0^{(e)}(\xi_2)\phi_r^{(e)}(\xi_3).
\end{aligned}
\tag{8.335}
$$

with $0 < p < P_1$, $0 < q < P_2$, and $0 < r < P_3$. The expressions for the face functions are respectively obtained by

$$
\begin{aligned}
\phi_{pq0}^{(e)}(\xi_1,\xi_2,\xi_3) &= \phi_p^{(e)}(\xi_1)\phi_q^{(e)}(\xi_2)\phi_0^{(e)}(\xi_3) \\
\phi_{P_1qr}^{(e)}(\xi_1,\xi_2,\xi_3) &= \phi_{P_1}^{(e)}(\xi_1)\phi_q^{(e)}(\xi_2)\phi_r^{(e)}(\xi_3), \\
\phi_{pqP_3}^{(e)}(\xi_1,\xi_2,\xi_3) &= \phi_p^{(e)}(\xi_1)\phi_q^{(e)}(\xi_2)\phi_{P_3}^{(e)}(\xi_3), \\
\phi_{0qr}^{(e)}(\xi_1,\xi_2,\xi_3) &= \phi_0^{(e)}(\xi_1)\phi_q^{(e)}(\xi_2)\phi_r^{(e)}(\xi_3), \\
\phi_{p0r}^{(e)}(\xi_1,\xi_2,\xi_3) &= \phi_p^{(e)}(\xi_1)\phi_0^{(e)}(\xi_2)\phi_r^{(e)}(\xi_3), \\
\phi_{pP_2r}^{(e)}(\xi_1,\xi_2,\xi_3) &= \phi_p^{(e)}(\xi_1)\phi_{P_2}^{(e)}(\xi_2)\phi_r^{(e)}(\xi_3).
\end{aligned}
\tag{8.336}
$$

Finally, the general expression of the volume function is

$$
\phi_{pqr}^{(e)}(\xi_1,\xi_2,\xi_3) = \phi_p^{(e)}(\xi_1)\phi_q^{(e)}(\xi_2)\phi_r^{(e)}(\xi_3), \quad 0 < p < P_1, \ 0 < q < P_2, \ 0 < r < P_3.
\tag{8.337}
$$

The same nodal basis obtained for the square can be constructed for the hexahedron. The Lagrangian family of elements is determined substituting (4.111) in expressions (8.334) to (8.337). The linear element has only eight vertex functions; the quadratic element still has twelve quadratic edge functions, six face functions, and one volume function; on the other hand, the cubic element has eight vertex functions, twenty-four edge functions, twenty-four face functions, and eight volume functions. The total number of shape functions is given by the product $(P_1 + 1)(P_2 + 1)(P_3 + 1)$. The serendipity family is obtained using definition (8.327) in expressions (8.334) to (8.337).

The modal basis presented in [35] is obtained substituting definition (4.123) in equations (8.334) to (8.337).

Hence, the vertex functions are given by

$$
\begin{aligned}
\phi_{000}^{(e)}(\xi_1,\xi_2,\xi_3) &= \frac{1}{8}(1-\xi_1)(1-\xi_2)(1-\xi_3), \\
\phi_{100}^{(e)}(\xi_1,\xi_2,\xi_3) &= \frac{1}{8}(1+\xi_1)(1-\xi_2)(1-\xi_3), \\
\phi_{110}^{(e)}(\xi_1,\xi_2,\xi_3) &= \frac{1}{8}(1+\xi_1)(1+\xi_2)(1-\xi_3), \\
\phi_{010}^{(e)}(\xi_1,\xi_2,\xi_3) &= \frac{1}{8}(1-\xi_1)(1+\xi_2)(1-\xi_3), \\
\phi_{001}^{(e)}(\xi_1,\xi_2,\xi_3) &= \frac{1}{8}(1-\xi_1)(1-\xi_2)(1+\xi_3), \\
\phi_{101}^{(e)}(\xi_1,\xi_2,\xi_3) &= \frac{1}{8}(1+\xi_1)(1-\xi_2)(1+\xi_3), \\
\phi_{011}^{(e)}(\xi_1,\xi_2,\xi_3) &= \frac{1}{8}(1-\xi_1)(1+\xi_2)(1+\xi_3), \\
\phi_{111}^{(e)}(\xi_1,\xi_2,\xi_3) &= \frac{1}{8}(1+\xi_1)(1+\xi_2)(1+\xi_3).
\end{aligned}
\tag{8.338}
$$

The edge functions have the following expressions:

$$
\begin{aligned}
\phi_{p00}^{(e)}(\xi_1,\xi_2,\xi_3) &= \frac{1}{16}(1-\xi_1)(1+\xi_1)(1-\xi_2)(1-\xi_3)P_{p-1}^{1,1}(\xi_1), \\
\phi_{P_1q0}^{(e)}(\xi_1,\xi_2,\xi_3) &= \frac{1}{16}(1+\xi_1)(1-\xi_2)(1+\xi_2)(1-\xi_3)P_{q-1}^{1,1}(\xi_2), \\
\phi_{pP_20}^{(e)}(\xi_1,\xi_2,\xi_3) &= \frac{1}{16}(1-\xi_1)(1+\xi_1)(1+\xi_2)(1-\xi_3)P_{p-1}^{1,1}(\xi_1), \\
\phi_{0q0}^{(e)}(\xi_1,\xi_2,\xi_3) &= \frac{1}{16}(1-\xi_1)(1-\xi_2)(1+\xi_2)(1-\xi_3)P_{q-1}^{1,1}(\xi_2), \\
\phi_{0P_2r}^{(e)}(\xi_1,\xi_2,\xi_3) &= \frac{1}{16}(1+\xi_1)(1+\xi_2)(1-\xi_3)(1+\xi_3)P_{r-1}^{1,1}(\xi_3), \\
\phi_{P_1qP_3}^{(e)}(\xi_1,\xi_2,\xi_3) &= \frac{1}{16}(1+\xi_1)(1-\xi_2)(1+\xi_2)(1+\xi_3)P_{q-1}^{1,1}(\xi_2), \\
\phi_{pP_2P_3}^{(e)}(\xi_1,\xi_2,\xi_3) &= \frac{1}{16}(1-\xi_1)(1+\xi_1)(1+\xi_2)(1+\xi_3)P_{p-1}^{1,1}(\xi_1), \\
\phi_{0qP_3}^{(e)}(\xi_1,\xi_2,\xi_3) &= \frac{1}{16}(1-\xi_1)(1-\xi_2)(1+\xi_2)(1+\xi_3)P_{q-1}^{1,1}(\xi_2), \\
\phi_{P_10r}^{(e)}(\xi_1,\xi_2,\xi_3) &= \frac{1}{16}(1+\xi_1)(1-\xi_2)(1-\xi_3)(1+\xi_3)P_{r-1}^{1,1}(\xi_3), \\
\phi_{P_1P_2r}^{(e)}(\xi_1,\xi_2,\xi_3) &= \frac{1}{16}(1+\xi_1)(1+\xi_2)(1-\xi_3)(1+\xi_3)P_{r-1}^{1,1}(\xi_3), \\
\phi_{p0P_3}^{(e)}(\xi_1,\xi_2,\xi_3) &= \frac{1}{16}(1-\xi_1)(1+\xi_1)(1-\xi_2)(1+\xi_3)P_{p-1}^{1,1}(\xi_1), \\
\phi_{00r}^{(e)}(\xi_1,\xi_2,\xi_3) &= \frac{1}{16}(1-\xi_1)(1-\xi_2)(1-\xi_3)(1+\xi_3)P_{r-1}^{1,1}(\xi_3).
\end{aligned}
\tag{8.339}
$$

On the other hand, the face functions are

$$\phi_{pq0}^{(e)}(\xi_1,\xi_2,\xi_3) = \frac{1}{32}(1-\xi_1)(1+\xi_1)(1-\xi_2)(1+\xi_2)(1-\xi_3)P_{p-1}^{1,1}(\xi_1)P_{q-1}^{1,1}(\xi_2),$$

$$\phi_{P_1qr}^{(e)}(\xi_1,\xi_2,\xi_3) = \frac{1}{32}(1+\xi_1)(1-\xi_2)(1+\xi_2)(1-\xi_3)(1+\xi_3)P_{q-1}^{1,1}(\xi_2)P_{r-1}^{1,1}(\xi_3),$$

$$\phi_{pqP_3}^{(e)}(\xi_1,\xi_2,\xi_3) = \frac{1}{32}(1-\xi_1)(1+\xi_1)(1+\xi_2)(1-\xi_3)(1+\xi_3)P_{p-1}^{1,1}(\xi_1)P_{r-1}^{1,1}(\xi_3),$$

$$\phi_{0qr}^{(e)}(\xi_1,\xi_2,\xi_3) = \frac{1}{32}(1-\xi_1)(1-\xi_2)(1+\xi_2)(1-\xi_3)(1+\xi_3)P_{q-1}^{1,1}(\xi_2)P_{r-1}^{1,1}(\xi_3),$$

$$\phi_{p0r}^{(e)}(\xi_1,\xi_2,\xi_3) = \frac{1}{32}(1-\xi_1)(1+\xi_1)(1-\xi_2)(1-\xi_3)(1+\xi_3)P_{p-1}^{1,1}(\xi_1)P_{r-1}^{1,1}(\xi_3),$$

$$\phi_{pP_2r}^{(e)}(\xi_1,\xi_2,\xi_3) = \frac{1}{32}(1-\xi_1)(1+\xi_1)(1-\xi_2)(1+\xi_2)(1+\xi_3)P_{p-1}^{1,1}(\xi_1)P_{q-1}^{1,1}(\xi_2),$$

$$(8.340)$$

with $0 < p < P_1$, $0 < q < P_2$, and $0 < r < P_3$. Finally, the volume functions are

$$\phi_{pqr}^{(e)}(\xi_1,\xi_2,\xi_3) = \frac{1}{16}(1-\xi_1^2)(1-\xi_2^2)(1-\xi_3^2)P_{p-1}^{\alpha_1,\beta_1}(\xi_1)P_{q-1}^{\alpha_2,\beta_2}(\xi_2)P_{r-1}^{\alpha_3,\beta_3}(\xi_3). \qquad (8.341)$$

The above modal basis is hierarchical.

Figures 8.52(a) and 8.52(b) present the sparsity profiles of the mass and stiffness matrices for the Poisson problem, obtained by the Jacobi basis with $P = 10$ and weights $\alpha_1 = \beta_1 = \alpha_2 = \beta_2 = 1$.

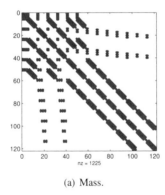
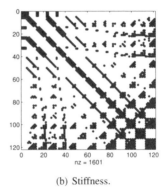

(a) Mass.　　　　　　　　　　　　　　　　(b) Stiffness.

Figure 8.52　Sparsity profiles for the mass and stiffness matrices of squares with Jacobi basis and $P = 10$.

Figures 8.53(a) and 8.53(b) present the behavior of the condition numbers of the mass and stiffness matrices for squares, calculated using Lagrange and Jacobi bases with and without the Schur complement. A better condition number is observed in the mass matrix when using Lagrange polynomials. For the stiffness matrix, and up to order 5, the condition number is better with the use of Lagrange polynomials. Beyond order 5, there is an effective improvement using Jacobi polynomials [see Figure 8.53(b)]. In turn, Figures 8.54 and 8.55 illustrate the sparsity profiles of mass and stiffness matrices for hexahedra with $P = 5$ and the respective condition numbers using Lagrange and Jacobi shape functions.

The solution of a boundary value problem (BVP) with the FEM is approximated by a linear combination of shape functions, as indicated in equation (8.294) for the three-dimensional linear elasticity problem. A fundamental aspect is that, given the desired order, the obtained approximation generates a complete polynomial.

As seen in Chapter 4, a complete polynomial $p(\xi_1)$ of order n in the local variable ξ_1 has all terms of orders from zero until n, that is,

$$p_n(\xi_1) = a_0 + a_1\xi_1 + a_2\xi_1^2 + \ldots + a_{n-1}\xi_1^{n-1} + a_n\xi_1^n.$$

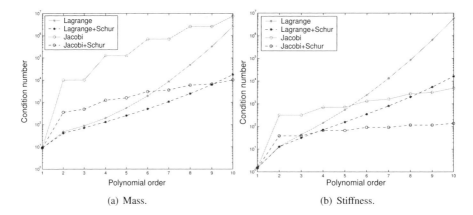

(a) Mass.   (b) Stiffness.

Figure 8.53 Numerical conditioning of mass and stiffness matrices for squares with Lagrange and Jacobi bases.

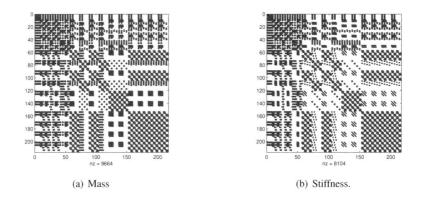

(a) Mass   (b) Stiffness.

Figure 8.54 Sparsity profiles of mass and stiffness matrices of hexahedra with Jacobi basis and $P = 5$.

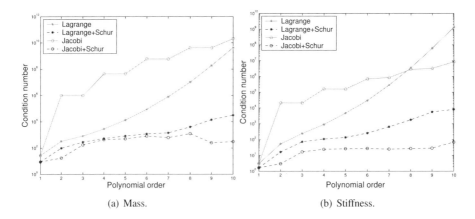

(a) Mass.   (b) Stiffness.

Figure 8.55 Numerical conditioning of mass and stiffness matrices of hexahedra with Lagrange and Jacobi bases.

Example 8.42 *Consider the linear approximation of the axial displacement in a bar under traction, given by*

$$u_x(\xi_1) = \frac{1}{2}(1-\xi_1)\bar{u}_{x_0} + \frac{1}{2}(1+\xi_1)\bar{u}_{x_1}$$

$$= \frac{1}{2}(\bar{u}_{x_1} + \bar{u}_{x_0}) + \frac{1}{2}(-\bar{u}_{x_0})\xi_1$$

$$= a_0 + a_1\xi_1,$$

with $a_0 = \frac{1}{2}(\bar{u}_{x_1} + \bar{u}_{x_0})$ and $a_1 = \frac{1}{2}(\bar{u}_{x_1} - \bar{u}_{x_0})$. Thus, the approximation of the axial displacement is given by a complete first-order polynomial.

Now using a quadratic three-node element, the axial displacement is approximated as

$$u_x(\xi_1) = -\frac{1}{2}\xi_1(1-\xi_1)\bar{u}_{x_0} + \frac{1}{2}\xi_1(1+\xi_1)\bar{u}_{x_1} + (1-\xi_1^2)\bar{u}_{x_2}$$

$$= \bar{u}_{x_2} + \frac{1}{2}(\bar{u}_{x_1} - \bar{u}_{x_0})\xi_1 + \frac{1}{2}(\bar{u}_{x_0} + \bar{u}_{x_1} - 2\bar{u}_{x_1})\xi_1^2$$

$$= a_0 + a_1\xi_1 + a_2\xi_1^2.$$

In this case, we have a complete second-order expansion.

□

The same concept is applied for two- and three-dimensional cases. Complete polynomials of first and second-order in the local variables (ξ_1, ξ_2) are respectively expressed by

$$p_1(\xi_1, \xi_2) = a_0 + a_1\xi_1 + a_2\xi_2 + a_3\xi_1\xi_2,$$
$$p_2(\xi_1, \xi_2) = a_0 + a_1\xi_1 + a_2\xi_2 + a_3\xi_1\xi_2 + a_4\xi_1^2 + a_5\xi_2^2.$$

Example 8.43 *Consider the displacement $u_x(\xi_1, \xi_2)$ in direction x for a plane state problem, interpolated in the four-node square. Thus,*

$$u_x(\xi_1, \xi_2) = \frac{1}{4}(1-\xi_1)(1-\xi_2)\bar{u}_{x_0} + \frac{1}{4}(1+\xi_1)(1-\xi_2)\bar{u}_{x_1} +$$

$$\frac{1}{4}(1+\xi_1)(1+\xi_2)\bar{u}_{x_2} + \frac{1}{4}(1-\xi_1)(1+\xi_2)\bar{u}_{x_3}$$

$$= \frac{1}{4}(\bar{u}_{x_0} + \bar{u}_{x_1} + \bar{u}_{x_2} + \bar{u}_{x_3}) + \frac{1}{4}(-\bar{u}_{x_0} + \bar{u}_{x_1} + \bar{u}_{x_2} - \bar{u}_{x_3})\xi_1 +$$

$$\frac{1}{4}(-\bar{u}_{x_0} - \bar{u}_{x_1} + \bar{u}_{x_2} + \bar{u}_{x_3})\xi_2 + \frac{1}{4}(\bar{u}_{x_0} - \bar{u}_{x_1} + \bar{u}_{x_2} - \bar{u}_{x_3})\xi_1\xi_2$$

$$= a_0 + a_1\xi_1 + a_2\xi_2 + a_3\xi_1\xi_2.$$

Hence, notice that the above expansion generates a complete first-order polynomial for variables ξ_1 and ξ_2.

□

To facilitate the determination of the minimum number of terms necessary to generate a complete expansion for a certain order (ξ_1, ξ_2), we employ the Pascal triangle shown in Figure 8.56. For squares and hexahedra, as the shape functions are constructed by the tensor product of onedimensional functions, it generates a greater number of terms than the minimum required for a given expansion of P order, as illustrated in the following example.

Example 8.44 *Consider the displacement $u_x(\xi_1, \xi_2)$ in the direction of x for a plane state problem interpolated*

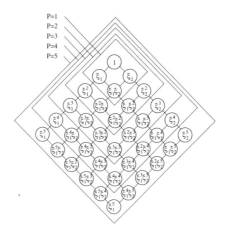

Figure 8.56   Pascal triangle for squares.

by the Lagrangian quadratic square of nine nodes. Thus,

$$
\begin{aligned}
u_x(\xi_1,\xi_2) &= \frac{1}{4}\xi_1\xi_2\,(1-\xi_1)(1-\xi_2)\,\bar{u}_{x_0} - \frac{1}{2}\xi_2\left(1-\xi_1^2\right)(1-\xi_2)\,\bar{u}_{x_1} - \\
&\quad \frac{1}{4}\xi_1\xi_2\,(1+\xi_1)(1-\xi_2)\,\bar{u}_{x_2} - \frac{1}{2}\xi_1\,(1-\xi_1)\left(1-\xi_2^2\right)\bar{u}_{x_3} + \\
&\quad \left(1-\xi_1^2\right)\left(1-\xi_2^2\right)\bar{u}_{x_4} + \frac{1}{2}\xi_1\,(1+\xi_1)\left(1-\xi_2^2\right)\bar{u}_{x_5} + \\
&\quad \frac{1}{4}\xi_1\xi_2\,(1+\xi_1)(1+\xi_2)\,\bar{u}_{x_6} + \frac{1}{2}\xi_2\left(1-\xi_1^2\right)(1+\xi_2)\,\bar{u}_{x_7} - \\
&\quad \frac{1}{4}\xi_1\xi_2\,(1-\xi_1)(1+\xi_2)\,\bar{u}_{x_8}.
\end{aligned}
$$

Grouping up the terms, we obtain

$$
\begin{aligned}
u_x(\xi_1,\xi_2) &= \bar{u}_{x_8} + \frac{1}{2}\left(\bar{u}_{x_7} + \bar{u}_{x_5}\right)\xi_1 + \frac{1}{2}\left(-\bar{u}_{x_4} + \bar{u}_{x_6}\right)\xi_2 + \\
&\quad \frac{1}{4}\left(-\bar{u}_{x_1} + \bar{u}_{x_0} - \bar{u}_{x_3} + \bar{u}_{x_2}\right)\xi_1\xi_2 + \\
&\quad \frac{1}{2}\left(\bar{u}_{x_7} + \bar{u}_{x_5} - 2\bar{u}_{x_8}\right)\xi_1^2 - \frac{1}{2}\left(\bar{u}_{x_4} - 2\bar{u}_{x_8} + \bar{u}_{x_6}\right)\xi_2^2 + \\
&\quad \frac{1}{4}\left(-\bar{u}_{x_1} + \bar{u}_{x_2} + \bar{u}_{x_3} + 2\bar{u}_{x_4} - \bar{u}_{x_0} - 2\bar{u}_{x_6}\right)\xi_1^2\xi_2 + \\
&\quad \frac{1}{4}\left(2\bar{u}_{x_7} + \bar{u}_{x_1} - \bar{u}_{x_3} - \bar{u}_{x_0} - 2\bar{u}_{x_5} + \bar{u}_{x_2}\right)\xi_1\xi_2^2 + \\
&\quad \frac{1}{4}\left(-2\bar{u}_{x_4} - 2\bar{u}_{x_5} - 2\bar{u}_{x_6} + \bar{u}_{x_0} - 2\bar{u}_{x_7} + \bar{u}_{x_2} + \bar{u}_{x_1} + \bar{u}_{x_3} + 4\bar{u}_{x_8}\right)\xi_1^2\xi_2^2.
\end{aligned}
$$

   Notice that the terms in $\xi_1^2\xi_2$, $\xi_1\xi_2^2$, and $\xi_1^2\xi_2^2$, obtained from the tensorization process of one-dimensional second-order functions, are not necessary for a complete second-order expansion, but only for orders 3 and 4. Therefore, these additional terms are called parasites. However, these terms can achieve a better approximation and a faster convergence rate of the approximate solution. The serendipity family removes some of these parasite terms of the expansion, which allows the reduction of the computational cost to obtain an approximate solution.

   $\square$

8.12.3 SHAPE FUNCTIONS FOR NONSTRUCTURED ELEMENTS

Triangles and tetrahedra elements are called unstructured. The natural or barycentric coordinates have been employed for the construction of shape functions for triangles, mainly due to the property of rotational symmetry.

The barycentric or area coordinates for triangles are illustrated in Figure 8.57. Given a triangle of area A and a given point P, three subtriangles with areas A_1, A_2, and A_3 are defined, with $A = A_1 + A_2 + A_3$. The area coordinates $0 \leq L_i \leq 1$ $(i = 1,2,3)$ are defined by the ratio between the areas of the subtriangles and the original triangle as follows [54, 18]:

$$\frac{A_1}{A} + \frac{A_2}{A} + \frac{A_3}{A} = 1 \rightarrow L_1 + L_2 + L_3 = 1. \tag{8.342}$$

Hence, each area coordinate is defined by

$$L_i = \frac{A_i}{A}, \quad i = 1,2,3. \tag{8.343}$$

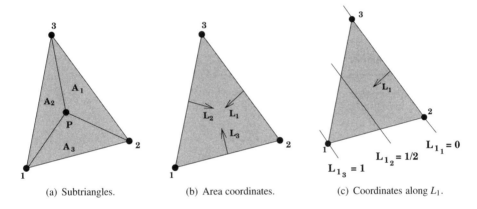

(a) Subtriangles. (b) Area coordinates. (c) Coordinates along L_1.

Figure 8.57 Barycentric coordinates for triangles (adapted from [10]).

From (8.342), we notice that one of the area coordinates depend on the other two. For instance,

$$L_3 = 1 - L_1 - L_2. \tag{8.344}$$

In [9], a procedure for the definition of shape functions for triangles was presented, based on the tensor product of one-dimensional polynomials expressed in natural coordinates. This procedure is presented below, considering the standard Lagrangian nodal basis for triangles presented in the literature [54, 18]. Subsequently, the expressions are rewritten in the same notation used for squares and hexahedra.

The nodal shape functions for triangles can be written as the tensor product of Lagrange polynomials in the area coordinates L_1, L_2, and L_3 as follows:

$$\phi_a^{(e)}(L_1,L_2,L_3) = l_b^{(b-1)}(L_1)l_c^{(c-1)}(L_2)l_d^{(d-1)}(L_3), \tag{8.345}$$

where a is the node number; b, c, and d are the nodal coordinate indices, as illustrated in Figure 8.57(c) for direction L_1. The values of b, c, and d are in the closed interval $[1, P_i + 1]$, and P_i $(i = 1,2,3)$ denotes the polynomial orders in direction L_i.

The Lagrange polynomials indicated in the previous equation are given by

$$
\begin{aligned}
l_b^{(b-1)}(L_1) &= \frac{(L_1 - L_{1_1})(L_1 - L_{1_2})\ldots(L_1 - L_{1_{b-1}})}{(L_{1_b} - L_{1_1})(L_{1_b} - L_{1_2})\ldots(L_{1_b} - L_{1_{b-1}})}, \quad b \geq 2, \\[2mm]
l_c^{(c-1)}(L_2) &= \frac{(L_2 - L_{2_1})(L_2 - L_{2_2})\ldots(L_2 - L_{2_{c-1}})}{(L_{2_c} - L_{2_1})(L_{2_c} - L_{2_2})\ldots(L_{2_c} - L_{2_{c-1}})}, \quad c \geq 2, \\[2mm]
l_d^{(d-1)}(L_3) &= \frac{(L_3 - L_{3_1})(L_3 - L_{3_2})\ldots(L_3 - L_{3_{d-1}})}{(L_{3_d} - L_{3_1})(L_{3_d} - L_{3_2})\ldots(L_{3_d} - L_{3_{d-1}})}, \quad d \geq 2,
\end{aligned}
\tag{8.346}
$$

with $l^{(0)} = 1$.

When compared to equation (4.110), we notice that the Lagrange polynomials in (8.346) are truncated in the nodal coordinates indicated by b, c, and d, respectively, instead of considering all $P_i + 1$ coordinates in each direction L_i [see Figure 8.57(c)]. The polynomial orders in directions L_i are not P_i as would be expected from (4.110), but respectively $b - 1$, $c - 1$, and $d - 1$.

For instance, consider the linear triangle illustrated in Figure 8.58. Along each direction L_i there are two coordinates denoted by $L_{i_1} = 0$ and $L_{i_2} = 1$. The indices a, b, c, and d are given in Table 8.1, and b, c, and d can assume values 1 or 2 for the linear triangle. The shape functions are obtained from expressions (8.345) and (8.346) and Table 8.1 as

$$
\begin{aligned}
\phi_1^{(e)}(L_1, L_2, L_3) &= l_2^{(1)}(L_1) l_1^{(0)}(L_2) l_1^{(0)}(L_3) = \frac{L_1 - L_{1_1}}{L_{1_2} - L_{1_1}} = \frac{L_1 - 0}{1 - 0} = L_1, \\
\phi_2^{(e)}(L_1, L_2, L_3) &= l_1^{(0)}(L_1) l_2^{(1)}(L_2) l_1^{(0)}(L_3) = \frac{L_2 - L_{2_1}}{L_{2_2} - L_{2_1}} = \frac{L_2 - 0}{1 - 0} = L_2, \qquad (8.347) \\
\phi_3^{(e)}(L_1, L_2, L_3) &= l_1^{(0)}(L_1) l_1^{(0)}(L_2) l_2^{(1)}(L_3) = \frac{L_3 - L_{3_1}}{L_{3_2} - L_{3_1}} = \frac{L_3 - 0}{1 - 0} = L_3.
\end{aligned}
$$

These are the nodal Lagrange functions for the linear triangle found in the literature [54, 18].

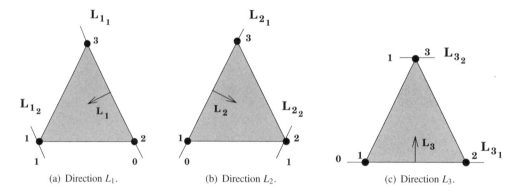

| | (a) Direction L_1. | (b) Direction L_2. | (c) Direction L_3. |

Figure 8.58   Linear triangle (adapted from [10]).

a	1	2	3
b	2	1	1
c	1	2	1
d	1	1	2

Table 8.1
Linear triangle: indices a, b, c, and d [10].

The same procedure can be applied to the quadratic triangle shown in Figure 8.59. In this case, there are three coordinates for each direction L_i. Assuming equally spaced coordinates, we have $L_{i_1} = 0$, $L_{i_2} = \frac{1}{2}$, and $L_{i_3} = 1$. Table 8.2 presents indices a to d. Similarly, the Lagrange shape functions for the second-order triangle are obtained from equations (8.345) and (8.346) and Table 8.2, as

$$\phi_1^{(e)}(L_1,L_2,L_3) = l_3^{(2)}(L_1)l_1^{(0)}(L_2)l_1^{(0)}(L_3) = \frac{(L_1-L_{1_1})(L_2-L_{1_2})}{(L_{1_3}-L_{1_1})(L_{1_3}-L_{1_2})} = L_1(2L_1-1),$$

$$\phi_2^{(e)}(L_1,L_2,L_3) = l_1^{(0)}(L_1)l_3^{(2)}(L_2)l_1^{(0)}(L_3) = L_2(2L_2-1),$$

$$\phi_3^{(e)}(L_1,L_2,L_3) = l_1^{(0)}(L_1)l_1^{(0)}(L_2)l_3^{(2)}(L_3) = L_3(2L_3-1),$$

$$\phi_4^{(e)}(L_1,L_2,L_3) = l_2^{(1)}(L_1)l_2^{(1)}(L_2)l_1^{(0)}(L_3) = 4L_1L_2, \qquad (8.348)$$

$$\phi_5^{(e)}(L_1,L_2,L_3) = l_1^{(0)}(L_1)l_2^{(1)}(L_2)l_2^{(1)}(L_3) = 4L_2L_3,$$

$$\phi_6^{(e)}(L_1,L_2,L_3) = l_1^{(1)}(L_1)l_1^{(0)}(L_2)l_2^{(1)}(L_3) = 4L_1L_3.$$

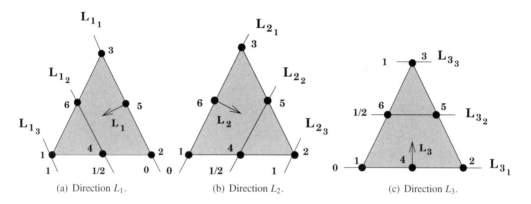

(a) Direction L_1.   (b) Direction L_2.   (c) Direction L_3.

Figure 8.59 Quadratic triangle (adapted from [10]).

a	1	2	3	4	5	6
b	3	1	1	2	1	2
c	1	3	1	2	2	1
d	1	1	3	1	2	2

Table 8.2
Quadratic triangle: indices a, b, c, and d [10].

The same notation used for squares and hexahedra will be employed for triangles. For this purpose, consider the triangle topological entities illustrated in Figure 8.60(a), that is, three vertices (V_1, V_2, V_3), three edges (E_1, E_2, E_3), and one face (F_1). The general expression for the triangle shape functions is

$$\phi_{pqr}^{(e)}(L_1,L_2,L_3) = \phi_p^{(e)}(L_1)\phi_q^{(e)}(L_2)\phi_r^{(e)}(L_3), \qquad (8.349)$$

with $0 \le p \le P_1$, $0 \le q \le P_2$, and $0 \le r \le P_3$, as illustrated in Figure 8.60(b).

The vertex functions are obtained from equation (8.349) and Figure 8.60 as

$$\phi_{P_100}^{(e)}(L_1,L_2,L_3) = \phi_{P_1}^{(e)}(L_1)\phi_0^{(e)}(L_2)\phi_0^{(e)}(L_3) = \phi_{P_1}^{(e)}(L_1),$$

$$\phi_{0P_20}^{(e)}(L_1,L_2,L_3) = \phi_0^{(e)}(L_1)\phi_{P_2}^{(e)}(L_2)\phi_0^{(e)}(L_3) = \phi_{P_2}^{(e)}(L_2), \qquad (8.350)$$

$$\phi_{00P_3}^{(e)}(L_1,L_2,L_3) = \phi_0^{(e)}(L_1)\phi_0^{(e)}(L_2)\phi_{P_3}^{(e)}(L_3) = \phi_{P_3}^{(e)}(L_3).$$

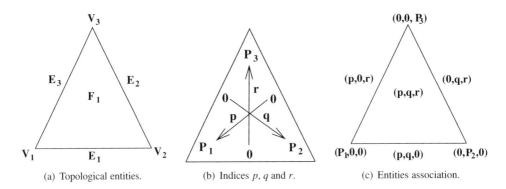

Figure 8.60 Topological entities, tensor indices p, q, r, and association between entities and indices for a triangle (adapted from [10]).

Analogously, the edge functions for $0 < p, q, r < P$ and $P_1 = P_2 = P$ are given by

$$
\begin{aligned}
\phi_{pq0}^{(e)}(L_1, L_2, L_3) &= \phi_p^{(e)}(L_1)\phi_q^{(e)}(L_2)\phi_0^{(e)}(L_3) = \phi_p^{(e)}(L_1)\phi_q^{(e)}(L_2), \\
\phi_{p0r}^{(e)}(L_1, L_2, L_3) &= \phi_p^{(e)}(L_1)\phi_0^{(e)}(L_2)\phi_r^{(e)}(L_3) = \phi_p^{(e)}(L_1)\phi_r^{(e)}(L_3), \\
\phi_{0qr}^{(e)}(L_1, L_2, L_3) &= \phi_0^{(e)}(L_1)\phi_q^{(e)}(L_2)\phi_r^{(e)}(L_3) = \phi_q^{(e)}(L_2)\phi_r^{(e)}(L_3).
\end{aligned}
\tag{8.351}
$$

Finally, the face functions for $0 < p, q, r < P - 1$ are

$$
\phi_{pqr}^{(e)}(L_1, L_2, L_3) = \phi_p^{(e)}(L_1)\phi_q^{(e)}(L_2)\phi_r^{(e)}(L_3).
\tag{8.352}
$$

Following, the nodal and modal bases are described for triangles employing the above notation. For both cases, we assume that $P_1 = P_2 = P_3 = P$ for all functions discussed in this section. In this case, we have three vertex functions, $3(P-1)$ edge functions, and $\frac{1}{2}(P-1)(P-2)$ face functions. Besides that, the bases are constructed for $p + q + r = P$.

Consider the following definition

$$
\phi_p^{(e)}(L_1) = \begin{cases} 1 & p = 0 \\ l_{p+1}^{(p)}(L_1) & 0 < p \le P_1 \end{cases},
\tag{8.353}
$$

with the truncated Lagrange polynomial $l_p^{(p+1)}(L_1)$ given by

$$
l_{p-1}^{(p)}(L_1) = \frac{\prod_{i=1}^{p-1}(L_1 - L_{1_i})}{\prod_{i=1}^{p-1}(L_{1_p} - L_{1_i})}.
\tag{8.354}
$$

Analogously, for $\phi_q^{(e)}(L_2)$ and $\phi_r^{(e)}(L_3)$.

Employing equation (8.353) in expressions (8.350) to (8.352), the nodal Lagrange basis is defined for triangles. The linear, quadratic, and cubic elements are illustrated in Figure 8.61 for the equally spaced coordinates in directions L_1, L_2, and L_3.

The vertex functions of the linear element are given from (8.350) and (8.353) as

$$
\begin{aligned}
\phi_{100}^{(e)}(L_1, L_2, L_3) &= L_1, \\
\phi_{010}^{(e)}(L_1, L_2, L_3) &= L_2, \\
\phi_{001}^{(e)}(L_1, L_2, L_3) &= L_3.
\end{aligned}
\tag{8.355}
$$

Analogously, the vertex functions of the quadratic element are

$$
\begin{aligned}
\phi_{200}^{(e)}(L_1, L_2, L_3) &= l_3^{(2)}(L_1) = L_1(2L_1 - 1), \\
\phi_{020}^{(e)}(L_1, L_2, L_3) &= l_3^{(2)}(L_2) = L_2(2L_2 - 1), \\
\phi_{002}^{(e)}(L_1, L_2, L_3) &= l_3^{(2)}(L_3) = L_3(2L_3 - 1).
\end{aligned}
\tag{8.356}
$$

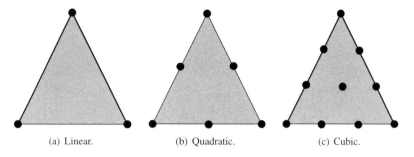

(a) Linear. (b) Quadratic. (c) Cubic.

Figure 8.61 Triangles of the nodal Lagrange family.

The edge functions are

$$\phi_{110}^{(e)}(L_1,L_2,L_3) = 4L_1L_2,$$
$$\phi_{011}^{(e)}(L_1,L_2,L_3) = 4L_2L_3, \qquad (8.357)$$
$$\phi_{101}^{(e)}(L_1,L_2,L_3) = 4L_1L_3.$$

The shape functions for linear, quadratic, and cubic triangles are illustrated in Figures 8.62 to 8.64.

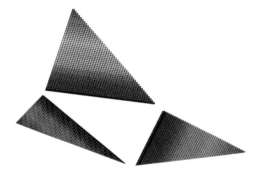

Figure 8.62 Shape functions for the first-order Lagrange triangle.

Consider now the following definition using Jacobi polynomials:

$$\phi_p^{(e)}(L_1) = \begin{cases} 1 & p = 0 \\ L_1 & p = P_1 \\ L_1 P_{p-1}^{\alpha_1,\beta_1}(2L_1 - 1) & 0 < p < P_1 \end{cases} . \qquad (8.358)$$

Analogous to the Lagrange bases, the modal bases for triangles can be defined using equations (8.350) to (8.352). The vertex modes are given by

$$\phi_{P_100}^{(e)}(L_1,L_2,L_3) = \phi_{P_1}^{(e)}(L_1)\phi_0^{(e)}(L_2)\phi_0^{(e)}(L_3) = L_1,$$
$$\phi_{0P_20}^{(e)}(L_1,L_2,L_3) = \phi_0^{(e)}(L_1)\phi_{P_2}^{(e)}(L_2)\phi_0^{(e)}(L_3) = L_2, \qquad (8.359)$$
$$\phi_{00P_3}^{(e)}(L_1,L_2,L_3) = \phi_0^{(e)}(L_1)\phi_0^{(e)}(L_2)\phi_{P_3}^{(e)}(L_3) = L_3.$$

Figure 8.63   Shape functions for the second-order Lagrange triangle.

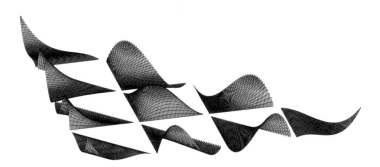

Figure 8.64   Shape functions for the third-order Lagrange triangle.

The edge modes for $P \geq 2$ and $0 < p,q,r < P$ are

$$
\begin{aligned}
\phi_{pq0}^{(e)}(L_1,L_2,L_3) &= \phi_p^{(e)}(L_1)\phi_q^{(e)}(L_2)\phi_0^{(e)}(L_3) \\
&= L_1 L_2 P_{p-1}^{\alpha_1,\beta_1}(2L_1-1)P_{q-1}^{\alpha_2,\beta_2}(2L_2-1), \quad (p+q=P), \\
\phi_{p0r}^{(e)}(L_1,L_2,L_3) &= \phi_p^{(e)}(L_1)\phi_0^{(e)}(L_2)\phi_r^{(e)}(L_3) \\
&= L_1 L_3 P_{p-1}^{\alpha_1,\beta_1}(2L_1-1)P_{r-1}^{\alpha_3,\beta_3}(2L_3-1), \quad (p+r=P), \\
\phi_{0qr}^{(e)}(L_1,L_2,L_3) &= \phi_0^{(e)}(L_1)\phi_q^{(e)}(L_2)\phi_r^{(e)}(L_3) \\
&= L_2 L_3 P_{q-1}^{\alpha_2,\beta_2}(2L_2-1)P_{r-1}^{\alpha_3,\beta_3}(2L_3-1), \quad (q+r=P).
\end{aligned}
\tag{8.360}
$$

Finally, the face modes for $P \geq 3$, $p+q+r=P$, and $0 < p,q,r < P-1$ are

$$
\begin{aligned}
\phi_{pqr}^{(e)}(L_1,L_2,L_3) &= \phi_p^{(e)}(L_1)\phi_q^{(e)}(L_2)\phi_r^{(e)}(L_3) \\
&= L_1 L_2 L_3 P_{p-1}^{\alpha_1,\beta_1}(2L_1-1)P_{q-1}^{\alpha_2,\beta_2}(2L_2-1)P_{r-1}^{\alpha_3,\beta_3}(2L_3-1).
\end{aligned}
\tag{8.361}
$$

Figure 8.65   Modal shape functions for the second-order triangle.

The modal shape functions for the quadratic and cubic elements are illustrated in Figures 8.65 and 8.66.

A hierarchical basis for triangles is generated making $p+q+r \leq P$. For the edge functions, we consider the following relations for indices p, q, r in (8.360), respectively: $r=0$ and $q \geq p$; $q=0$ and $p \geq r$; $p=0$ and $r \geq q$. In the case of the face functions, we use $p \geq q$ and $r \geq q$ in (8.361). Figure 8.67 illustrates the hierarchical shape functions for a third-order element. Notice that we need to multiply the odd functions by -1 to ensure the C_0 continuity at the interface between elements.

The same tensorization procedure of triangles is applied to tetrahedra. For this purpose, the barycentric or volume coordinates presented in Figure 8.68 are used. Given an interior point P, we have four tetrahedra with volumes V_1, V_2, V_3, and V_4. We verify that $V = V_1 + V_2 + V_3 + V_4$ and the coordinates L_i ($i=1,2,3,4$) are defined, in this case, by the ratio between volumes V_i and V, that is, [54, 18]

$$
\frac{V_1}{V} + \frac{V_2}{V} + \frac{V_3}{V} + \frac{V_4}{V} = 1 \rightarrow L_1 + L_2 + L_3 + L_4 = 1.
\tag{8.362}
$$

Analogously, the nodal shape functions for the tetrahedron are given by the following tensor product:

$$
\phi_a^{(e)}(L_1,L_2,L_3,L_4) = l_b^{(b-1)}(L_1)l_c^{(c-1)}(L_2)l_d^{(d-1)}(L_3)l_e^{(e-1)}(L_4),
\tag{8.363}
$$

Figure 8.66   Modal shape functions for the third-order triangle.

Figure 8.67   Hierarchical shape functions for the third-order triangle.

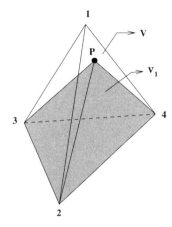

Figure 8.68 Volume coordinates.

where a is the node number; b, c, d, and e respectively are the coordinate indices in directions L_1, L_2, L_3, and L_4. The Lagrange polynomial in direction L_4 is given by

$$l_e^{(e-1)}(L_4) = \frac{(L_4 - L_{4_1})(L_4 - L_{4_2})\ldots(L_4 - L_{4_{e-1}})}{(L_{4_e} - L_{4_1})(L_{4_e} - L_{4_2})\ldots(L_{4_e} - L_{4_{e-1}})}. \tag{8.364}$$

The linear tetrahedron is illustrated in Figure 8.69. Again, there are two coordinates along each L_i coordinate, denoted as $L_{i_1} = 0$ and $L_{i_2} = 1$. Table 8.3 presents indices a, b, c, d, and e. The shape functions are obtained using equations (8.346), (8.363), and (8.364) and Table 8.3. Thus,

$$
\begin{aligned}
\phi_1^{(e)}(L_1, L_2, L_3, L_4) &= l_2^{(1)}(L_1)l_1^{(0)}(L_2)l_1^{(0)}(L_3)l_1^{(0)}(L_4) = \frac{L_1 - L_{1_1}}{L_{1_2} - L_{1_1}} = L_1, \\
\phi_2^{(e)}(L_1, L_2, L_3, L_4) &= l_1^{(0)}(L_1)l_2^{(1)}(L_2)l_1^{(0)}(L_3)l_1^{(0)}(L_4) = L_2, \\
\phi_3^{(e)}(L_1, L_2, L_3, L_4) &= l_1^{(0)}(L_1)l_1^{(0)}(L_2)l_2^{(1)}(L_3)l_1^{(0)}(L_4) = L_3, \\
\phi_4^{(e)}(L_1, L_2, L_3, L_4) &= l_1^{(0)}(L_1)l_1^{(0)}(L_2)l_1^{(0)}(L_3)l_2^{(1)}(L_4) = L_4.
\end{aligned}
\tag{8.365}
$$

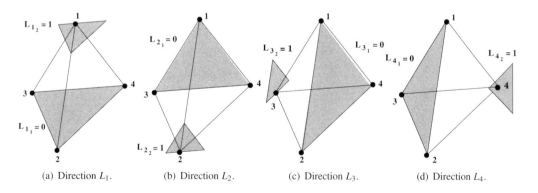

(a) Direction L_1.   (b) Direction L_2.   (c) Direction L_3.   (d) Direction L_4.

Figure 8.69 Barycentric directions for the linear tetrahedron (adapted from [10]).

These are the standard shape functions for the linear tetrahedron [54, 18], but constructed using the tensor product of onedimensional polynomials expressed in barycentric coordinates. The procedure can be applied to tetrahedra of any order.

a	1	2	3	4
b	2	1	1	1
c	1	2	1	1
d	1	1	2	1
e	1	1	1	2

Table 8.3
Linear tetrahedron: indices a, b, c, d, and e.

Similar to the previous elements, the shape functions are associated to the corresponding topological entities illustrated in Figure 8.70(a), which are four vertices (V_1 to V_4), six edges (E_1 to E_6), four faces (F_1 to F_4), and one volume (B_1).

The modal and nodal shape functions for tetrahedra can be written in terms of the topological entities as

$$\phi_{pqrs}^{(e)}(L_1,L_2,L_3,L_4) = \phi_p^{(e)}(L_1)\phi_q^{(e)}(L_2)\phi_r^{(e)}(L_3)\phi_s^{(e)}(L_4), \tag{8.366}$$

with $0 \le p \le P_1$, $0 \le q \le P_2$, $0 \le r \le P_3$, and $0 \le s \le P_4$ where P_1, P_2, P_3, and P_4 are the polynomial orders in directions L_1, L_2, L_3, and L_4, respectively, as illustrated in Figure 8.70(b).

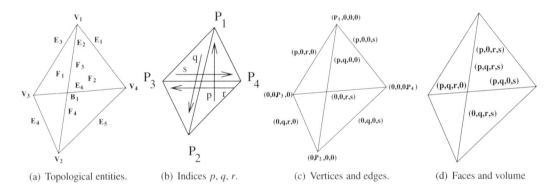

(a) Topological entities.   (b) Indices p, q, r.   (c) Vertices and edges.   (d) Faces and volume

Figure 8.70 Topological entities for tetrahedra, indices p, q, r, and s, and association between entities and indices (adapted from [10]).

The case with $P_1 = P_2 = P_3 = P_4 = P$ is considered, which have four vertex functions, $6(P - 1)$ edge, $\frac{1}{2}(P - 1)(P - 2)$ face, and $\frac{1}{6}(P - 1)(P - 2)(P - 3)$ volume functions.

Based on equation (8.366) and in Figure 8.70, the vertex functions are obtained as

$$
\begin{aligned}
\phi_{P_1000}^{(e)}(L_1,L_2,L_3,L_4) &= \phi_{P_1}^{(e)}(L_1)\phi_0^{(e)}(L_2)\phi_0^{(e)}(L_3)\phi_0^{(e)}(L_4) = \phi_{P_1}^{(e)}(L_1), \\
\phi_{0P_200}^{(e)}(L_1,L_2,L_3,L_4) &= \phi_0^{(e)}(L_1)\phi_{P_2}^{(e)}(L_2)\phi_0^{(e)}(L_3)\phi_0^{(e)}(L_4) = \phi_{P_2}^{(e)}(L_2), \\
\phi_{00P_30}^{(e)}(L_1,L_2,L_3,L_4) &= \phi_0^{(e)}(L_1)\phi_0^{(e)}(L_2)\phi_{P_3}^{(e)}(L_3)\phi_0^{(e)}(L_4) = \phi_{P_3}^{(e)}(L_3), \\
\phi_{000P_4}^{(e)}(L_1,L_2,L_3,L_4) &= \phi_0^{(e)}(L_1)\phi_0^{(e)}(L_2)\phi_0^{(e)}(L_3)\phi_{P_4}^{(e)}(L_4) = \phi_{P_4}^{(e)}(L_4).
\end{aligned}
\tag{8.367}
$$

The edge functions for $P \geq 2$ and $0 < p,q,r,s < P$ are

$$
\begin{aligned}
\phi_{pq00}^{(e)}(\cdot) &= \phi_p^{(e)}(L_1)\phi_q^{(e)}(L_2)\phi_0^{(e)}(L_3)\phi_0^{(e)}(L_4), & (p+q=P),\\
\phi_{p0r0}^{(e)}(\cdot) &= \phi_p^{(e)}(L_1)\phi_0^{(e)}(L_2)\phi_r^{(e)}(L_3)\phi_0^{(e)}(L_4), & (p+r=P),\\
\phi_{p00s}^{(e)}(\cdot) &= \phi_p^{(e)}(L_1)\phi_0^{(e)}(L_2)\phi_0^{(e)}(L_3)\phi_s^{(e)}(L_4), & (p+s=P),\\
\phi_{0qr0}^{(e)}(\cdot) &= \phi_0^{(e)}(L_1)\phi_q^{(e)}(L_2)\phi_r^{(e)}(L_3)\phi_0^{(e)}(L_4), & (q+r=P),\\
\phi_{0q0s}^{(e)}(\cdot) &= \phi_0^{(e)}(L_1)\phi_q^{(e)}(L_2)\phi_0^{(e)}(L_3)\phi_s^{(e)}(L_4), & (q+s=P),\\
\phi_{00rs}^{(e)}(\cdot) &= \phi_0^{(e)}(L_1)\phi_0^{(e)}(L_2)\phi_r^{(e)}(L_3)\phi_s^{(e)}(L_4), & (r+s=P).
\end{aligned}
\tag{8.368}
$$

The face functions for $P \geq 3$ and $0 < p,q,r,s < P-1$ are given by

$$
\begin{aligned}
\phi_{pqr0}^{(e)}(\cdot) &= \phi_p^{(e)}(L_1)\phi_q^{(e)}(L_2)\phi_r^{(e)}(L_3)\phi_0^{(e)}(L_4), & (p+q+r=P),\\
\phi_{pq0s}^{(e)}(\cdot) &= \phi_p^{(e)}(L_1)\phi_q^{(e)}(L_2)\phi_0^{(e)}(L_3)\phi_s^{(e)}(L_4), & (p+q+s=P),\\
\phi_{p0rs}^{(e)}(\cdot) &= \phi_p^{(e)}(L_1)\phi_0^{(e)}(L_2)\phi_r^{(e)}(L_3)\phi_s^{(e)}(L_4), & (p+r+s=P),\\
\phi_{0qrs}^{(e)}(\cdot) &= \phi_0^{(e)}(L_1)\phi_q^{(e)}(L_2)\phi_r^{(e)}(L_3)\phi_s^{(e)}(L_4), & (q+r+s=P).
\end{aligned}
\tag{8.369}
$$

Finally, the volume functions for $P \geq 4$, $p+q+r+s=P$, and $0 < p,q,r,s < P-2$ are

$$
\phi_{pqrs}^{(e)}(\cdot) = \phi_p^{(e)}(L_1)\phi_q^{(e)}(L_2)\phi_r^{(e)}(L_3)\phi_s^{(e)}(L_4).
\tag{8.370}
$$

Employing definition (8.353) in expressions (8.367) to (8.370), we obtain the standard Lagrangian nodal basis found in the literature. Figure 8.71 illustrates the linear, quadratic, and cubic elements.

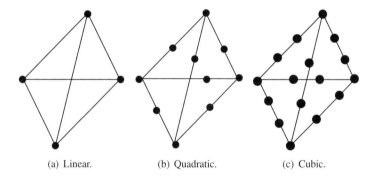

(a) Linear. (b) Quadratic. (c) Cubic.

Figure 8.71 Tetrahedra of the Lagrange nodal family.

The vertex functions of the linear element are

$$
\begin{aligned}
\phi_{1000}^{(e)}(L_1,L_2,L_3,L_4) &= L_1,\\
\phi_{0100}^{(e)}(L_1,L_2,L_3,L_4) &= L_2,\\
\phi_{0010}^{(e)}(L_1,L_2,L_3,L_4) &= L_3,\\
\phi_{0001}^{(e)}(L_1,L_2,L_3,L_4) &= L_4.
\end{aligned}
\tag{8.371}
$$

The vertex functions of the quadratic element are

$$
\begin{aligned}
\phi_{2000}^{(e)}(L_1,L_2,L_3,L_4) &= L_1(2L_1-1),\\
\phi_{0200}^{(e)}(L_1,L_2,L_3,L_4) &= L_2(2L_2-1),\\
\phi_{0020}^{(e)}(L_1,L_2,L_3,L_4) &= L_3(2L_3-1),\\
\phi_{0002}^{(e)}(L_1,L_2,L_3,L_4) &= L_4(2L_4-1).
\end{aligned}
\tag{8.372}
$$

Analogously, the edge functions are

$$
\phi_{1100}^{(e)}(L_1, L_2, L_3, L_4) = 4L_1 L_2,
$$
$$
\phi_{1010}^{(e)}(L_1, L_2, L_3, L_4) = 4L_1 L_3,
$$
$$
\phi_{1001}^{(e)}(L_1, L_2, L_3, L_4) = 4L_1 L_4, \tag{8.373}
$$
$$
\phi_{0110}^{(e)}(L_1, L_2, L_3, L_4) = 4L_2 L_3,
$$
$$
\phi_{0101}^{(e)}(L_1, L_2, L_3, L_4) = 4L_2 L_4,
$$
$$
\phi_{0011}^{(e)}(L_1, L_2, L_3, L_4) = 4L_3 L_4.
$$

To obtain the modal basis for tetrahedra, we substitute (8.358) in expressions (8.367) to (8.370). The vertex functions are

$$
\begin{aligned}
\phi_{1000}^{(e)}(L_1, L_2, L_3, L_4) &= \phi_1^{(e)}(L_1)\phi_0^{(e)}(L_2)\phi_0^{(e)}(L_3)\phi_0^{(e)}(L_4) = L_1, \\
\phi_{0100}^{(e)}(L_1, L_2, L_3, L_4) &= \phi_0^{(e)}(L_1)\phi_1^{(e)}(L_2)\phi_0^{(e)}(L_3)\phi_0^{(e)}(L_4) = L_2, \\
\phi_{0010}^{(e)}(L_1, L_2, L_3, L_4) &= \phi_0^{(e)}(L_1)\phi_0^{(e)}(L_2)\phi_1^{(e)}(L_3)\phi_0^{(e)}(L_4) = L_3, \\
\phi_{0001}^{(e)}(L_1, L_2, L_3, L_4) &= \phi_0^{(e)}(L_1)\phi_0^{(e)}(L_2)\phi_0^{(e)}(L_3)\phi_1^{(e)}(L_4) = L_4.
\end{aligned} \tag{8.374}
$$

The edge modes are also obtained for $P \geq 2$ and $0 < p, q, r, s < P$ as

$$
\begin{aligned}
\phi_{pq00}^{(e)}(\cdot) &= \phi_p^{(e)}(L_1)\phi_q^{(e)}(L_2)\phi_0^{(e)}(L_3)\phi_0^{(e)}(L_4) \\
&= L_1 L_2 P_{p-1}^{\alpha_1,\beta_1}(2L_1-1) P_{q-1}^{\alpha_2,\beta_2}(2L_2-1), \quad (p+q=P), \\
\phi_{p0r0}^{(e)}(\cdot) &= \phi_p^{(e)}(L_1)\phi_0^{(e)}(L_2)\phi_r^{(e)}(L_3)\phi_0^{(e)}(L_4) \\
&= L_1 L_3 P_{p-1}^{\alpha_1,\beta_1}(2L_1-1) P_{r-1}^{\alpha_3,\beta_3}(2L_3-1), \quad (p+r=P), \\
\phi_{p00s}^{(e)}(\cdot) &= \phi_p^{(e)}(L_1)\phi_0^{(e)}(L_2)\phi_0^{(e)}(L_3)\phi_s^{(e)}(L_4) \\
&= L_1 L_4 P_{p-1}^{\alpha_1,\beta_1}(2L_1-1) P_{s-1}^{\alpha_4,\beta_4}(2L_4-1), \quad (p+s=P), \\
\phi_{0pq0}^{(e)}(\cdot) &= \phi_0^{(e)}(L_1)\phi_q^{(e)}(L_2)\phi_r^{(e)}(L_3)\phi_0^{(e)}(L_4) \\
&= L_2 L_3 P_{q-1}^{\alpha_2,\beta_2}(2L_2-1) P_{r-1}^{\alpha_3,\beta_3}(2L_3-1), \quad (q+r=P), \\
\phi_{0q0s}^{(e)}(\cdot) &= \phi_0^{(e)}(L_1)\phi_q^{(e)}(L_2)\phi_0^{(e)}(L_3)\phi_s^{(e)}(L_4) \\
&= L_2 L_4 P_{q-1}^{\alpha_2,\beta_2}(2L_2-1) P_{s-1}^{\alpha_4,\beta_4}(2L_4-1), \quad (q+s=P), \\
\phi_{00rs}^{(e)}(\cdot) &= \phi_0^{(e)}(L_1)\phi_0^{(e)}(L_2)\phi_r^{(e)}(L_3)\phi_s^{(e)}(L_4) \\
&= L_3 L_4 P_{r-1}^{\alpha_3,\beta_3}(2L_3-1) P_{s-1}^{\alpha_4,\beta_4}(2L_4-1), \quad (r+s=P).
\end{aligned} \tag{8.375}
$$

The face modes are defined for $P \geq 3$, $0 < p, q, r, s < P-1$ and given by the following expressions:

$$
\begin{aligned}
\phi_{pqr0}^{(e)}(\cdot) &= \phi_p^{(e)}(L_1)\phi_q^{(e)}(L_2)\phi_r^{(e)}(L_3)\phi_0^{(e)}(L_4) \\
&= L_1 L_2 L_3 P_{p-1}^{\alpha_1,\beta_1}(2L_1-1) P_{q-1}^{\alpha_2,\beta_2}(2L_2-1) P_{r-1}^{\alpha_3,\beta_3}(2L_3-1), \\
&\quad (p+q+r=P), \\
\phi_{pq0s}^{(e)}(\cdot) &= \phi_p^{(e)}(L_1)\phi_q^{(e)}(L_2)\phi_0^{(e)}(L_3)\phi_s^{(e)}(L_4) \\
&= L_1 L_2 L_4 P_{p-1}^{\alpha_1,\beta_1}(2L_1-1) P_{q-1}^{\alpha_2,\beta_2}(2L_2-1) P_{s-1}^{\alpha_4,\beta_4}(2L_4-1), \\
&\quad (p+q+s=P), \\
\phi_{p0rs}^{(e)}(\cdot) &= \phi_p^{(e)}(L_1)\phi_0^{(e)}(L_2)\phi_r^{(e)}(L_3)\phi_s^{(e)}(L_4) \\
&= L_1 L_3 L_4 P_{p-1}^{\alpha_1,\beta_1}(2L_1-1) P_{r-1}^{\alpha_3,\beta_3}(2L_3-1) P_{s-1}^{\alpha_4,\beta_4}(2L_4-1), \\
&\quad (p+r+s=P), \\
\phi_{0qrs}^{(e)}(\cdot) &= \phi_0^{(e)}(L_1)\phi_q^{(e)}(L_2)\phi_r^{(e)}(L_3)\phi_s^{(e)}(L_4) \\
&= L_2 L_3 L_4 P_{q-1}^{\alpha_2,\beta_2}(2L_2-1) P_{r-1}^{\alpha_3,\beta_3}(2L_3-1) P_{s-1}^{\alpha_4,\beta_4}(2L_4-1), \\
&\quad (q+r+s=P).
\end{aligned} \tag{8.376}
$$

Finally, the body modes for $P \geq 4$, $p+q+r+s=P$, and $0 < p, q, r, s < P-2$ are given by the general expression

$$
\begin{aligned}
\phi_{pqrs}^{(e)}(\cdot) &= \phi_p^{(e)}(L_1)\phi_q^{(e)}(L_2)\phi_r^{(e)}(L_3)\phi_s^{(e)}(L_4) \\
&= L_1 L_2 L_3 L_4 P_{p-1}^{\alpha_1,\beta_1}(2L_1-1) P_{q-1}^{\alpha_2,\beta_2}(2L_2-1) P_{r-1}^{\alpha_3,\beta_3}(2L_3-1) P_{s-1}^{\alpha_4,\beta_4}(2L_4-1).
\end{aligned} \tag{8.377}
$$

We can obtain a hierarchical basis substituting the equal symbol in the relations between the indices of the edge and face functions by the sign $\leq$ (that is, $p + q = P$ by $p + q \leq P$ and $p \geq q$), analogous to the case of triangles.

Except for the hierarchy, all nodal Lagrange bases for triangles and tetrahedra have all the properties of any presented modal basis, that is, the vertex modes have unit magnitude at the vertex and are zero at all other vertices; the edge modes have nonzero magnitude in one edge and are zero over all other edges and vertices; face modes have nonzero magnitude along one face, but are all zero over the other faces, edges, and vertices; similar to body modes.

As shown in [9], the modal functions for triangles and tetrahedra (excluding the hierarchical ones) have natural C_0 continuity in the edges and faces of the elements.

Seeking a better sparsity of local matrices, the selection of weights for the Jacobi polynomials in the expressions of the modal shape functions for triangles is not trivial, due to the dependence on barycentric coordinates and the variable integration limits, as shown in [10].

8.12.4 MAPPING

When applying the FEM in the analysis of a linear elastic solid, we should interpolate, besides the displacement field, the geometry of the body by means of the coordinates of the boundary points. For this purpose, we can apply the same set of shape functions used for the displacement field, defining the class of isoparametric finite elements.

Denoting x, y, and z as the coordinates of points of a finite element relative to a global reference system, we can write the following relations to interpolate the element geometry, from the nodal coordinates and interpolation functions expressed in the local system, that is,

$$
\begin{aligned}
x(\xi_1, \xi_2, \xi_3) &= \sum_{i=1}^{N_e} \phi_i^{(e)}(\xi_1, \xi_2, \xi_3) x_i^{(e)}, \\
y(\xi_1, \xi_2, \xi_3) &= \sum_{i=1}^{N_e} \phi_i^{(e)}(\xi_1, \xi_2, \xi_3) y_i^{(e)}, \qquad (8.378) \\
z(\xi_1, \xi_2, \xi_3) &= \sum_{i=1}^{N_e} \phi_i^{(e)}(\xi_1, \xi_2, \xi_3) z_i^{(e)},
\end{aligned}
$$

where N_e is the number of nodes/modes of the element and $(x_i^{(e)}, y_i^{(e)}, z_i^{(e)})$ are the global Cartesian coordinates of node i of the element.

Another possibility is to employ a subset of shape functions used for the displacement field or even a lower order set, generating a family of subparametric elements. For example, for elements with straight edges and plane faces, we can employ linear shape functions to interpolate the element geometry, because the use of higher-order functions doesn't give any advantage in this case. We may also adopt a larger number of shape functions to interpolate the element geometry, defining the class of superparametric elements. Notice that the subparametric elements are more often used than the superparametric ones, because in general we want more accuracy in the displacement field interpolation.

Example 8.45 *Consider the linear square element illustrated in Figure 8.72. Applying (8.378), we obtain the coordinates x and y of points in the global system, that is,*

$$
x(\xi_1, \xi_2) = \phi_1^{(e)}(\xi_1, \xi_2) x_1^e + \phi_2^{(e)}(\xi_1, \xi_2) x_2^e + \phi_3^{(e)}(\xi_1, \xi_2) x_3^e + \phi_4^{(e)}(\xi_1, \xi_2) x_4^e,
$$

$$
y(\xi_1, \xi_2) = \phi_1^{(e)}(\xi_1, \xi_2) y_1^e + \phi_2^{(e)}(\xi_1, \xi_2) y_2^e + \phi_3^{(e)}(\xi_1, \xi_2) y_3^e + \phi_4^{(e)}(\xi_1, \xi_2) y_4^e.
$$

Substituting the expressions of the shape functions given in (8.323) in the previous equations and calculating them in the coordinates of point i ($\xi_1 = -1; \xi_2 = 0$), we have

$$
x_i = x(-1, 0) = \left(\frac{1}{2}\right)(5) + (0)(30) + (0)(20) + \left(\frac{1}{2}\right)(10) = 7.5,
$$

$$
y_i = y(-1, 0) = \left(\frac{1}{2}\right)(15) + (0)(10) + (0)(30) + \left(\frac{1}{2}\right)(25) = 20.
$$

Notice that the calculated coordinates of point i are in accordance to Figure 8.72, since the midpoint of the local edge was mapped in the middle of the correspondent global edge. The constant lines of ξ_1 and ξ_2 are identified in the global system, as shown in Figure 8.72. Therefore, the element shape functions can be used to perform the mapping of a distorted element in the global system to a nondistorted one at the local system.

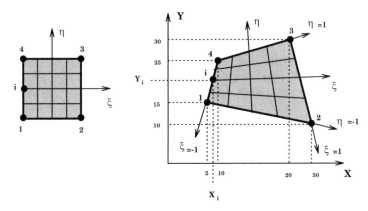

Figure 8.72 Example of transformation between the local and global reference systems using the shape functions.

□

Thus, the shape functions can be used not only to interpolate the geometry and the variables of interest, but they also define a transformation between the global and local reference systems, making the calculation of the finite element matrices easier.

The substitution of coordinate interpolations (8.378) in (8.313) allows us to denote the displacement vector field approximation for a linear elastic solid as

$$\mathbf{u}_{N_e}^{(e)}\left(x(\xi_1,\xi_2,\xi_3),y(\xi_1,\xi_2,\xi_3),z(\xi_1,\xi_2,\xi_3)\right) = \sum_{i=1}^{N_e} \phi_i^{(e)}(\xi_1,\xi_2,\xi_3)\mathbf{u}_i^{(e)} = [\mathbf{N}^{(e)}]\{\mathbf{u}^{(e)}\}. \tag{8.379}$$

Consider the component $u_x(x,y,z)$ of the displacement vector field $\mathbf{u}(x,y,z)$ expressed in terms of the local coordinates by expressions (8.378), that is,

$$u_x(x,y,z) = u_x\left(x(\xi_1,\xi_2,\xi_3),y(\xi_1,\xi_2,\xi_3),z(\xi_1,\xi_2,\xi_3)\right).$$

Using the chain rule, we can calculate the partial derivatives of the scalar function u_x relative to the local coordinates ξ_1, ξ_2 and ξ_3 as

$$\begin{aligned}
u_{x,\xi_1} &= u_{x,x}x_{,\xi_1} + u_{x,y}y_{,\xi_1} + u_{x,z}z_{,\xi_1}, \\
u_{x,\xi_2} &= u_{x,x}x_{,\xi_2} + u_{x,y}y_{,\xi_2} + u_{x,z}z_{,\xi_2}, \\
u_{x,\xi_3} &= u_{x,x}x_{,\xi_3} + u_{x,y}y_{,\xi_3} + u_{x,z}z_{,\xi_3}.
\end{aligned} \tag{8.380}$$

We have, in matrix notation,

$$\left\{\begin{array}{c} u_{x,\xi_1} \\ u_{x,\xi_2} \\ u_{x,\xi_3} \end{array}\right\} = \left[\begin{array}{ccc} x_{,\xi_1} & y_{,\xi_1} & z_{,\xi_1} \\ x_{,\xi_2} & y_{,\xi_2} & z_{,\xi_2} \\ x_{,\xi_3} & y_{,\xi_3} & z_{,\xi_3} \end{array}\right] \left\{\begin{array}{c} u_{x,x} \\ u_{x,y} \\ u_{x,z} \end{array}\right\} = [\mathbf{J}]\left\{\begin{array}{c} u_{x,x} \\ u_{x,y} \\ u_{x,z} \end{array}\right\}, \tag{8.381}$$

where $[\mathbf{J}]$ is the Jacobian matrix of the transformation between the local and global reference systems. Inverting the Jacobian matrix, we obtain the partial derivatives of u_x relative to the global coordinates x, y, and z, that is,

$$\left\{\begin{array}{c} u_{x,x} \\ u_{x,y} \\ u_{x,z} \end{array}\right\} = [\mathbf{J}]^{-1}\left\{\begin{array}{c} u_{x,\xi_1} \\ u_{x,\xi_2} \\ u_{x,\xi_3} \end{array}\right\}. \tag{8.382}$$

This same procedure can be employed to obtain the partial derivatives of the element shape function $\phi_i^{(e)}$ relative to the global variables x, y, and z, as required in the expression of the strain-displacement matrix given in (8.316). Thus,

$$
\left\{
\begin{array}{c}
\phi_{i,x}^{(e)} \\
\phi_{i,y}^{(e)} \\
\phi_{i,z}^{(e)}
\end{array}
\right\}
= [\mathbf{J}]^{-1}
\left\{
\begin{array}{c}
\phi_{i,\xi_1}^{(e)} \\
\phi_{i,\xi_2}^{(e)} \\
\phi_{i,\xi_3}^{(e)}
\end{array}
\right\}.
\tag{8.383}
$$

Using relations (8.378), we obtain the following expression for the Jacobian matrix:

$$
[\mathbf{J}] =
\begin{bmatrix}
\sum_{i=1}^{N_e} \phi_{i,\xi_1}^{(e)} x_i^{(e)} & \sum_{i=1}^{N_e} \phi_{i,\xi_1}^{(e)} y_i^{(e)} & \sum_{i=1}^{N_e} \phi_{i,\xi_1}^{(e)} z_i^{(e)} \\
\sum_{i=1}^{N_e} \phi_{i,\xi_2}^{(e)} x_i^{(e)} & \sum_{i=1}^{N_e} \phi_{i,\xi_2}^{(e)} y_i^{(e)} & \sum_{i=1}^{N_e} \phi_{i,\xi_2}^{(e)} z_i^{(e)} \\
\sum_{i=1}^{N_e} \phi_{i,\xi_3}^{(e)} x_i^{(e)} & \sum_{i=1}^{N_e} \phi_{i,\xi_3}^{(e)} y_i^{(e)} & \sum_{i=1}^{N_e} \phi_{i,\xi_3}^{(e)} z_i^{(e)}
\end{bmatrix}
$$

$$
=
\begin{bmatrix}
\phi_{1,\xi_1}^{(e)} & \phi_{2,\xi_1}^{(e)} & \cdots & \phi_{N_e,\xi_1}^{(e)} \\
\phi_{1,\xi_2}^{(e)} & \phi_{2,\xi_2}^{(e)} & \cdots & \phi_{N_e,\xi_2}^{(e)} \\
\phi_{1,\xi_3}^{(e)} & \phi_{2,\xi_3}^{(e)} & \cdots & \phi_{N_e,\xi_3}^{(e)}
\end{bmatrix}
\begin{bmatrix}
x_1^{(e)} & y_1^{(e)} & z_1^{(e)} \\
x_2^{(e)} & y_2^{(e)} & z_2^{(e)} \\
\vdots & \vdots & \vdots \\
x_{N_e}^{(e)} & y_{N_e}^{(e)} & z_{N_e}^{(e)}
\end{bmatrix}.
\tag{8.384}
$$

The inverse of the Jacobian matrix is calculated as

$$
[\mathbf{J}]^{-1} = \frac{1}{\det[\mathbf{J}]} (\operatorname{cof} [\mathbf{J}])^T,
\tag{8.385}
$$

where cof $[\mathbf{J}]$ is the co-factor matrix of $[\mathbf{J}]$. Its coefficients are calculated as

$$
(\operatorname{cof} [\mathbf{J}])_{ij} = (-1)^{i+j} \det[\mathbf{J}_{ij}],
\tag{8.386}
$$

where $[\mathbf{J}_{ij}]$ is the resultant matrix when we eliminate the row i and column j of the Jacobian matrix.

For squares, the Jacobian matrix is defined by

$$
[\mathbf{J}] =
\begin{bmatrix}
x_{,\xi_1} & y_{,\xi_1} \\
x_{,\xi_2} & y_{,\xi_2}
\end{bmatrix}
=
\begin{bmatrix}
\phi_{1,\xi_1}^{(e)} & \phi_{2,\xi_1}^{(e)} & \cdots & \phi_{N_e,\xi_1}^{(e)} \\
\phi_{1,\xi_2}^{(e)} & \phi_{2,\xi_2}^{(e)} & \cdots & \phi_{N_e,\xi_2}^{(e)}
\end{bmatrix}
\begin{bmatrix}
x_1^{(e)} & y_1^{(e)} \\
x_2^{(e)} & y_2^{(e)} \\
\vdots & \vdots \\
x_{N_e}^{(e)} & y_{N_e}^{(e)}
\end{bmatrix}.
\tag{8.387}
$$

On the other hand, the inverse matrix is

$$
[\mathbf{J}]^{-1} = \frac{1}{\det[\mathbf{J}]}
\begin{bmatrix}
y_{,\xi_2} & -y_{,\xi_1} \\
-x_{,\xi_2} & x_{,\xi_1}
\end{bmatrix}.
\tag{8.388}
$$

Hence, the partial derivatives of the shape functions relative to variables x and y are calculated as

$$
\left\{
\begin{array}{c}
\phi_{i,x}^{(e)} \\
\phi_{i,y}^{(e)}
\end{array}
\right\}
= [\mathbf{J}]^{-1}
\left\{
\begin{array}{c}
\phi_{i,\xi_1}^{(e)} \\
\phi_{i,\xi_2}^{(e)}
\end{array}
\right\}.
\tag{8.389}
$$

Notice that the global-local coordinate mapping cannot be unique in the cases of higly distorted finite elements. To ensure a unique mapping, the sign of the Jacobian determinant must remain the same for all points of the considered domain. The use of high-order elements allows to work with elements of higher aspect ratio.

Example 8.46 *We can use the coordinate transformation given in (8.378) to calculate the element properties. Consider the square element illustrated in Figure 8.72. We want to calculate its area.*
The area $A^{(e)}$ of the element is given by

$$A^{(e)} = \int_{A^{(e)}} dA.$$

The area integral in the global system can be performed in the local coordinate system as

$$A^{(e)} = \int_{A^{(e)}} dA = \int_{-1}^{1} \int_{-1}^{1} |\det[\mathbf{J}]| d\xi_1 d\xi_2.$$

To calculate the Jacobian determinant, the global coordinates (x, y) are initially interpolated using the linear element shape functions. Thus,

$$
\begin{aligned}
x(\xi_1, \xi_2) &= \phi_1^{(e)}(\xi_1, \xi_2) x_1^{(e)} + \phi_2^{(e)}(\xi_1, \xi_2) x_2^{(e)} + \phi_3^{(e)}(\xi_1, \xi_2) x_3^{(e)} + \phi_4^{(e)}(\xi_1, \xi_2) x_4^{(e)}, \\
y(\xi_1, \xi_2) &= \phi_1^{(e)}(\xi_1, \xi_2) y_1^{(e)} + \phi_2^{(e)}(\xi_1, \xi_2) y_2^{(e)} + \phi_3^{(e)}(\xi_1, \xi_2) y_3^{(e)} + \phi_4^{(e)}(\xi_1, \xi_2) y_4^{(e)}.
\end{aligned}
$$

Substituting the expressions for the shape functions and the nodal coordinates, and performing some simplifications, we obtain

$$
\begin{aligned}
x(\xi_1, \xi_2) &= 16.25 - 1.25\,\xi_2 + 8.75\,\xi_1 - 3.75\,\xi_1\,\xi_2, \\
y(\xi_1, \xi_2) &= 20.00 + 7.50\,\xi_2 + 2.50\,\xi_1\,\xi_2.
\end{aligned}
$$

Hence, the Jacobian matrix is given by

$$
[\mathbf{J}] = \begin{bmatrix} 8.75 - 3.75\xi_2 & 2.50\xi_2 \\ -1.25 - 3.75\xi_1 & 7.50 + 2.50\xi_1 \end{bmatrix}.
$$

Consequently, the determinant is

$$\det[\mathbf{J}] = 65.625 + 21.875\xi_1 - 25.000\xi_2.$$

Thus, the area of the element is determined as

$$
\begin{aligned}
A^{(e)} &= \int_{-1}^{1} \int_{-1}^{1} |\det[\mathbf{J}]| d\xi_1 d\xi_2 \\
&= \int_{-1}^{1} \int_{-1}^{1} (65.625 + 21.875\xi_1 - 25.000\xi_2) d\xi_1 d\xi_2 = 262.50 \text{ cm}^2.
\end{aligned}
$$

□

Considering the triangular elements as isoparametric, the global coordinates x and y of the points are interpolated by

$$x(L_1, L_2) = \sum_{i=1}^{N_e} \phi_i^{(e)}(L_1, L_2) x_i^{(e)}, \tag{8.390}$$

$$y(L_1, L_2) = \sum_{i=1}^{N_e} \phi_i^{(e)}(L_1, L_2) y_i^{(e)}, \tag{8.391}$$

where N_e is the number of nodes/modes, $(x_i^{(e)}, y_i^{(e)})$ are the global coordinates of the nodes, and $\phi_i^{(e)}(L_1, L_2)$ indicates the associated shape functions to element node/mode i in terms of the area coordinates (L_1, L_2).

To determine the element stiffness matrices, we calculate the deformation matrix $[B^{(e)}]$, which contains the partial derivatives of the shape functions relative to the global variables x and y. In a general way, the global shape function i is expressed in terms of the global variables x and y, as $\phi_i^{(e)} = \phi_i^{(e)}(x, y)$. Considering the transformation given by equations (8.390) and (8.391) for the local system defined by the area coordinates, we have $\phi_i^{(e)} = \phi_i^{(e)}[x(L_1, L_2), y(L_1, L_2)]$.

There are three area coordinates, where one is dependent, as indicated in (8.344). Considering L_1 and L_2 as independent variables and denoting the shape functions only in terms of L_1 and L_2, we can use the chain rule and write, for the ith shape function,

$$\phi_{i,L_1}^{(e)} = \phi_{i,x}^{(e)} x_{,L_1} + \phi_{i,y}^{(e)} y_{,L_1},$$

$$\phi_{i,L_2}^{(e)} = \phi_{i,x}^{(e)} x_{,L_2} + \phi_{i,y}^{(e)} y_{,L_2}.$$

These relations can be denoted in matrix notation as

$$\left\{ \begin{array}{c} \phi_{i,L_1}^{(e)} \\ \phi_{i,L_2}^{(e)} \end{array} \right\} = \left[\begin{array}{cc} x_{,L_1} & y_{,L_1} \\ x_{,L_2} & y_{,L_2} \end{array} \right] \left\{ \begin{array}{c} \phi_{i,x}^{(e)} \\ \phi_{i,y}^{(e)} \end{array} \right\} = [\mathbf{J}] \left\{ \begin{array}{c} \phi_{i,x}^{(e)} \\ \phi_{i,y}^{(e)} \end{array} \right\}. \tag{8.392}$$

Inverting the Jacobian matrix $[\mathbf{J}]$, the global derivatives of the shape functions are determined relative to x and y. Thus,

$$\left\{ \begin{array}{c} \phi_{i,x}^{(e)} \\ \phi_{i,y}^{(e)} \end{array} \right\} = [\mathbf{J}]^{-1} \left\{ \begin{array}{c} \phi_{i,L_1}^{(e)} \\ \phi_{i,L_2}^{(e)} \end{array} \right\}. \tag{8.393}$$

Denoting the shape function of node i in terms of the three local variables, we have $\phi_i^{(e)} = \phi_i^{(e)}(L_1, L_2, L_3)$. Using the chain rule, we observe that

$$\phi_{i,L_1}^{(e)} \Big|_I = \phi_{i,L_1}^{(e)} L_{1,L_1} + \phi_{i,L_2}^{(e)} L_{2,L_1} + \phi_{i,L_3}^{(e)} L_{3,L_1},$$

where the I subindex indicates that L_1 is an independent coordinate. From (8.344), we verify that

$$L_{1,L_1} = 1, \quad L_{2,L_1} = 0 \quad \text{and} \quad L_{3,L_1} = -1.$$

Thus,

$$\phi_{i,L_1}^{(e)} \Big|_I = \phi_{i,L_1}^{(e)} - \phi_{i,L_3}^{(e)}. \tag{8.394}$$

We obtain an analogous expression for L_{2_i}, that is,

$$\phi_{i,L_2}^{(e)} \Big|_I = \phi_{i,L_2}^{(e)} - \phi_{i,L_3}^{(e)}. \tag{8.395}$$

Example 8.47 *Consider a linear triangle with global nodal coordinates* $(x_1^{(e)}, y_1^{(e)})$, $(x_2^{(e)}, y_2^{(e)})$, *and* $(x_3^{(e)}, y_3^{(e)})$. *We want to determine the matrix and the Jacobian determinant.*

For this purpose, coordinates (x, y) *of any point of the element are interpolated as*

$$\begin{aligned} x(L_1, L_2) &= L_1 x_1^{(e)} + L_2 x_2^{(e)} + (1 - L_1 - L_2) x_3^{(e)}, \\ y(L_1, L_2) &= L_1 y_1^{(e)} + L_2 y_2^{(e)} + (1 - L_1 - L_2) y_3^{(e)}. \end{aligned}$$

Hence, the Jacobian matrix is given by

$$[\mathbf{J}] = \left[\begin{array}{cc} x_1^{(e)} - x_3^{(e)} & y_1^{(e)} - y_3^{(e)} \\ x_2^{(e)} - x_3^{(e)} & y_2^{(e)} - y_3^{(e)} \end{array} \right].$$

The Jacobian determinant is $|J| = \det[\mathbf{J}] = (x_1^{(e)} - x_3^{(e)})(y_2^{(e)} - y_3^{(e)}) - (x_2^{(e)} - x_3^{(e)})(y_1^{(e)} - y_3^{(e)})$.
The area of a triangle can be calculated by the following determinant:

$$A^{(e)} = \frac{1}{2} \left| \begin{array}{ccc} x_1^{(e)} & y_1^{(e)} & 1 \\ x_2^{(e)} & y_2^{(e)} & 1 \\ x_3^{(e)} & y_3^{(e)} & 1 \end{array} \right|.$$

Evaluating the previous expression, we obtain

$$A^{(e)} = \frac{1}{2} [(x_1^{(e)} - x_3^{(e)})(y_2^{(e)} - y_3^{(e)}) - (x_2^{(e)} - x_3^{(e)})(y_1^{(e)} - y_3^{(e)})].$$

Hence, the area of the linear triangle is half of the Jacobian determinant, that is, $A^{(e)} = \det[\mathbf{J}]/2$.

□

Supposing that the tetrahedron shape functions are expressed only in terms of the three barycentric coordinates (L_1, L_2, L_3), the interpolation of any point (x, y, z) of the element is given by

$$
\begin{aligned}
x(L_1, L_2, L_3) &= \sum_{i=1}^{N_e} \phi_i^{(e)}(L_1, L_2, L_3) x_i^{(e)}, \\
y(L_1, L_2, L_3) &= \sum_{i=1}^{N_e} \phi_i^{(e)}(L_1, L_2, L_3) y_i^{(e)}, \\
z(L_1, L_2, L_3) &= \sum_{i=1}^{N_e} \phi_i^{(e)}(L_1, L_2, L_3) z_i^{(e)}.
\end{aligned}
\tag{8.396}
$$

Hence, the global shape function i is expressed as

$$
\phi_i^{(e)}(x, y, z) = \phi_i^{(e)}\left(x(L_1, L_2, L_3), y(L_1, L_2, L_3), z(L_1, L_2, L_3) \right).
$$

Using the chain rule, we obtain

$$
\begin{aligned}
\phi_{i,L_1}^{(e)} &= \phi_{i,x}^{(e)} x_{,L_1} + \phi_{i,y}^{(e)} y_{,L_1} + \phi_{i,z}^{(e)} z_{,L_1}, \\
\phi_{i,L_2}^{(e)} &= \phi_{i,x}^{(e)} x_{,L_2} + \phi_{i,y}^{(e)} y_{,L_2} + \phi_{i,z}^{(e)} z_{,L_2}, \\
\phi_{i,L_3}^{(e)} &= \phi_{i,x}^{(e)} x_{,L_3} + \phi_{i,y}^{(e)} y_{,L_3} + \phi_{i,z}^{(e)} z_{,L_3}.
\end{aligned}
$$

In matrix form,

$$
\left\{ \begin{array}{c} \phi_{i,L_1}^{(e)} \\ \phi_{i,L_2}^{(e)} \\ \phi_{i,L_3}^{(e)} \end{array} \right\}
=
\left[\begin{array}{ccc} x_{,L_1} & y_{,L_1} & z_{,L_1} \\ x_{,L_2} & y_{,L_2} & z_{,L_2} \\ x_{,L_3} & y_{,L_3} & z_{,L_3} \end{array} \right]
\left\{ \begin{array}{c} \phi_{i,x}^{(e)} \\ \phi_{i,y}^{(e)} \\ \phi_{i,z}^{(e)} \end{array} \right\}
= [\mathbf{J}] \left\{ \begin{array}{c} \phi_{i,x}^{(e)} \\ \phi_{i,y}^{(e)} \\ \phi_{i,z}^{(e)} \end{array} \right\},
\tag{8.397}
$$

where $[\mathbf{J}]$ is the Jacobian matrix. Thus, the shape function derivatives relative to the global variables are obtained by inverting the Jacobian matrix, that is,

$$
\left\{ \begin{array}{c} \phi_{i,x}^{(e)} \\ \phi_{i,y}^{(e)} \\ \phi_{i,z}^{(e)} \end{array} \right\}
= [\mathbf{J}]^{-1} \left\{ \begin{array}{c} \phi_{i,L_1}^{(e)} \\ \phi_{i,L_2}^{(e)} \\ \phi_{i,L_3}^{(e)} \end{array} \right\}.
\tag{8.398}
$$

To employ a modal basis in the element coordinate mapping with curved edges and faces, we determine the associated coordinates to the modes of these topological entities, similar to the performed procedure in Chapter 4.

8.12.5 SURFACE JACOBIAN

As shown in equation (8.315), due to the surface traction, the equivalent load vector calculation involves the surface Jacobian $\mathbf{J}_b$. This Jacobian maps the element surface between the local and global reference systems.

In two dimensions, the surface integral is reduced to a line integral, which can be generally expressed as

$$
I = \int_a^b f(x, y) ds,
\tag{8.399}
$$

where ds is the length of the differential arc along the line, expressed in the global coordinate system, as illustrated for the square element in Figure 8.73.

The following relation is valid for ds in terms of the Cartesian differentials dx and dy:

$$
ds = \sqrt{dx^2 + dy^2}.
\tag{8.400}
$$

The mapping between the local and global reference systems is given by

$$
x = x(\xi_1, \xi_2) \quad \text{and} \quad y = y(\xi_1, \xi_2).
\tag{8.401}
$$

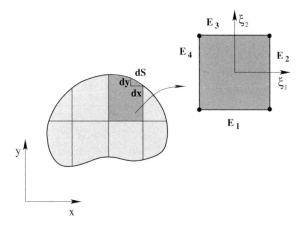

Figure 8.73 Surface integral for square element.

The relation between the global dx and dy and local $d\xi_1$ and $d\xi_2$ differentials is obtained with the chain rule, that is,

$$dx = x,_{\xi_1} d\xi_1 + x,_{\xi_2} d\xi_2, \tag{8.402}$$
$$dy = y,_{\xi_1} d\xi_1 + y,_{\xi_2} d\xi_2. \tag{8.403}$$

When performing the mapping to the local system, each global edge of the element will be in the horizontal or vertical edge of the standard element, being a parameter of only one of the local coordinates ξ_1 and ξ_2. Thus, $d\xi_1$ or $d\xi_2$ will be zero along a standard edge. For instance, consider the element illustrated in Figure 8.73. After substituting (8.402) and (8.403) in (8.401), the arc length ds is respectively denoted for each edge (E_1, E_2, E_3, E_4) as

$$ds = \sqrt{x,_{\xi_1}^2 \Big|_{\xi_2=-1} + y,_{\xi_1}^2 \Big|_{\xi_2=-1}} \, d\xi_1 = |\det[\mathbf{J}_b]| d\xi_1, \tag{8.404}$$

$$ds = \sqrt{x,_{\xi_2}^2 \Big|_{\xi_1=1} + y,_{\xi_2}^2 \Big|_{\xi_1=1}} \, d\xi_2 = |\det[\mathbf{J}_b]| d\xi_2, \tag{8.405}$$

$$ds = \sqrt{x,_{\xi_1}^2 \Big|_{\xi_2=1} + y,_{\xi_1}^2 \Big|_{\xi_2=1}} \, d\xi_1 = |\det[\mathbf{J}_b]| d\xi_1, \tag{8.406}$$

$$ds = \sqrt{x,_{\xi_2}^2 \Big|_{\xi_1=-1} + y,_{\xi_2}^2 \Big|_{\xi_1=-1}} \, d\xi_2 = |\det[\mathbf{J}_b]| d\xi_2. \tag{8.407}$$

In three dimensions, the surface integral is given in a general form as

$$I = \int_S f(x,y,z) dS. \tag{8.408}$$

The element surface is now described by two local coordinates, as illustrated in Figure 8.74, with independent coordinates ξ_1 and ξ_3. In this case, the mapping between the local and global reference systems is

$$x = x(\xi_1, 1, \xi_3), \quad y = y(\xi_1, 1, \xi_3) \quad \text{and} \quad z = z(\xi_1, 1, \xi_3). \tag{8.409}$$

The relation between the global differentials dx, dy, and dz in terms of the local differentials $d\xi_1$ and $d\xi_3$ is obtained using the chain rule, that is,

$$dx = x,_{\xi_1} d\xi_1 + x,_{\xi_3} d\xi_3, \tag{8.410}$$
$$dy = y,_{\xi_1} d\xi_1 + y,_{\xi_3} d\xi_3, \tag{8.411}$$
$$dz = z,_{\xi_1} d\xi_1 + z,_{\xi_3} d\xi_3. \tag{8.412}$$

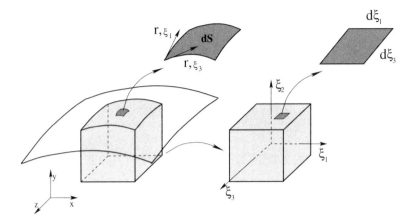

Figure 8.74 Surface integral hexahedral element.

In matrix form,

$$\left\{ \begin{array}{c} dx \\ dy \\ dz \end{array} \right\} = \left[\begin{array}{cc} x_{,\xi_1} & x_{,\xi_3} \\ y_{,\xi_1} & y_{,\xi_3} \\ z_{,\xi_1} & z_{,\xi_3} \end{array} \right] \left\{ \begin{array}{c} d\xi_1 \\ d\xi_3 \end{array} \right\}. \tag{8.413}$$

The columns of the above matrix represent the coordinates of the tangent vectors to the lines with constant ξ_1 and ξ_3, respectively. The previous relation can be rewritten in compact form as

$$d\mathbf{r} = \frac{d\mathbf{r}}{d\xi_1} d\xi_1 + \frac{d\mathbf{r}}{d\xi_3} d\xi_3, \tag{8.414}$$

with $\mathbf{r} = \{x \, y \, z\}^T$.

The surface differential dS is obtained as

$$dS = \left\| \frac{d\mathbf{r}}{d\xi_1} \times \frac{d\mathbf{r}}{d\xi_3} \right\| d\xi_1 d\xi_3 = |\det[\mathbf{J}_b]| d\xi_1 d\xi_3. \tag{8.415}$$

Hence, the surface integral given in (8.408) is rewritten as

$$I = \int_S f(x,y,z) dS = \int_{-1}^{1} \int_{-1}^{1} f(x(\xi_1,1,\xi_3), y(\xi_1,1,\xi_3), z(\xi_1,1,\xi_3)) |\det[\mathbf{J}_b]| d\xi_1 d\xi_3. \tag{8.416}$$

Example 8.48 *Consider the element illustrated in Figure 8.75 where one of the edges is curved. The global-local mapping (8.401) is given, in this case, by*

$$\begin{aligned} x(\xi_1,\xi_2) &= \phi_1(\xi_1,\xi_2)x_1 + \phi_2(\xi_1,\xi_2)x_2 + \phi_3(\xi_1,\xi_2)x_3 + \phi_4(\xi_1,\xi_2)x_4 + \phi_5(\xi_1,\xi_2)x_5, \\ y(\xi_1,\xi_2) &= \phi_1(\xi_1,\xi_2)y_1 + \phi_2(\xi_1,\xi_2)y_2 + \phi_3(\xi_1,\xi_2)y_3 + \phi_4(\xi_1,\xi_2)y_4 + \phi_5(\xi_1,\xi_2)y_5. \end{aligned}$$

The element shape functions for the serendipity nodal basis are given by

$$\begin{aligned} \phi_1(\xi_1,\xi_2) &= \frac{1}{4}(1-\xi_1)(1-\xi_2), \\ \phi_2(\xi_1,\xi_2) &= \frac{1}{4}(1+\xi_1)(1-\xi_2), \\ \phi_3(\xi_1,\xi_2) &= \frac{1}{4}(1+\xi_1)(1+\xi_2)(\xi_1+\xi_2-1), \\ \phi_4(\xi_1,\xi_2) &= -\frac{1}{4}(1-\xi_1)(1+\xi_2)(\xi_1-\xi_2+1), \\ \phi_5(\xi_1,\xi_2) &= \frac{1}{2}(1-\xi_1^2)(1+\xi_2). \end{aligned}$$

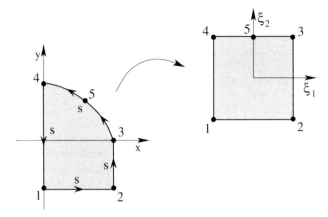

Figure 8.75   Example 8.48: element with a curved edge.

The curved edge in the global reference system corresponds to the third local edge, as illustrated in Figure 8.75. In this case, $\xi_2 = 1$ and the coordinate s is such that $s = -\xi_1$. Substituting these relations and the shape functions given in the mapping expressions, we obtain

$$x(s) \;=\; -\frac{1}{2}s(1-s)x_3 + \frac{1}{2}s(1+s)x_4 + (1-s^2)x_5,$$

$$y(s) \;=\; -\frac{1}{2}s(1-s)y_3 + \frac{1}{2}s(1+s)y_4 + (1-s^2)y_5.$$

The partial derivatives of the previous expression in relation to the curve parameter s are

$$x_{,s}(s) \;=\; (x_3 + x_4 - 2x_5)s + \frac{1}{2}(x_4 - x_3),$$

$$y_{,s}(s) \;=\; (y_3 + y_4 - 2y_5)s + \frac{1}{2}(y_4 - y_3).$$

For the nodal coordinates $(x_3, y_3) = (2,0)$, $(x_4, y_4) = (0,2)$, and $(x_5, y_5) = (\sqrt{2}, \sqrt{2})$, the above expressions reduce to

$$x_{,s}(s) \;=\; -0.8284s - 1,$$

$$y_{,s}(s) \;=\; -0.8284s + 1.$$

Finally, the edge Jacobian is given by

$$\det[\mathbf{J}_b] = \sqrt{x_{,s}^2 + y_{,s}^2} = \sqrt{1.3726s^2 + 2}.$$

For the fourth element edge, the relations $\xi_1 = -1$ and $\xi_2 = -s$ are valid. As the edge is straight, we can use the linear shape functions to interpolate the coordinates (x,y), and using the coordinate relationships, we obtain

$$x(s) \;=\; 0,$$

$$y(s) \;=\; \frac{1}{2}(1+s)y_1 + \frac{1}{2}(1-s)y_4 = \frac{1}{2}(y_1 + y_4) + \frac{1}{2}(y_1 - y_4)s.$$

Hence, the edge Jacobian is obtained as

$$\det[\mathbf{J}_b] = \sqrt{y_{,s}^2} = \sqrt{\frac{1}{4}(y_1 - y_4)^2} = 2.$$

$\square$

8.13 FINAL COMMENTS

This chapter presented the variational formulation of a solid body under small strain and material given by the generalized Hooke's law. Initially, the same systematic as the previous chapters was used, employing continuous functions of three independent variables. Subsequently, the concept of second-order tensors was introduced, with the Taylor series expansion of a vector function with three variables. Hence, the solid model was presented using tensors, allowing us to employ a more compact notation. With the introduction of the kinematics of the mechanical models of bars, shafts, and beams in the general solid formulation, we recovered the same expressions of the previous chapters.

The verification of a solid body required the introduction of a stress transformation at a point, using an orthogonal tensor and the eigenvector basis of the Cauchy stress tensor. Thereafter, three strength criteria were defined for the verification of elastic linear solids of ductile and brittle materials.

At the end, we presented the approximation of the linear elastic solid model, the construction of shape functions for structured and unstructured finite elements, and the global/local mapping between the element coordinate systems. In the next chapter, the concepts of differentiation by collocation and numerical integration will be extended for two- and three-dimensional elements. Examples of computational models using the solid approximation will also be considered.

8.14 PROBLEMS

1. Consider the function $f(x) = x^5 + 3x^3$. Expand this function around $x = 1$ using the four first terms of a Taylor series expansion. Calculate the value of the function for $y = 1.01$, $y = 1.001$, and $y = 1.00001$ and determine the relative contribution of each term for the final value. Determine the error between the calculated and the exact value of the function.

2. Given the vector field $\mathbf{v}(x, y, z) = (x^2 + yz)\mathbf{e}_x + (x + yx^2)\mathbf{e}_y + (z + \sin 2xy)\mathbf{e}_z$, calculate:

    a. $\nabla \mathbf{v}$
    b. $(\nabla \mathbf{v})\mathbf{v}$
    c. div $\mathbf{v}$

3. Let $\phi(x, y, z)$ and $\theta(x, y, z)$ be scalar fields and $\mathbf{v}(x, y, z)$ and $\mathbf{w}(x, y, z)$ vector fields. Using the component notation, verify the following relations:

    a. $\nabla(\phi + \theta) = \nabla\phi + \nabla\theta$
    b. $\text{div}(\mathbf{v} + \mathbf{w}) = \text{div}\mathbf{v} + \text{div}\mathbf{w}$

4. Let $\mathbf{T}$ be a transformation that operates over a vector $\mathbf{a}$ and gives the following relation $\mathbf{Ta} = \mathbf{a}/\|\mathbf{a}\|$, where $\|\mathbf{a}\|$ is the magnitude of $\mathbf{a}$. Show that $\mathbf{T}$ is not a linear transformation.

5. A tensor $\mathbf{T}$ transforms the vector basis $\mathbf{e}_1$ and $\mathbf{e}_2$ in

$$\begin{aligned} \mathbf{Te}_1 &= 2\mathbf{e}_1 + 5\mathbf{e}_2 \\ \mathbf{Te}_2 &= \mathbf{e}_1 - 2\mathbf{e}_2 \end{aligned}.$$

If $\mathbf{a} = \mathbf{e}_1 + 2\mathbf{e}_2$ and $\mathbf{b} = 3\mathbf{e}_1 + \mathbf{e}_2$, use the linear property of $\mathbf{T}$ to find

    a. $\mathbf{Ta}$
    b. $\mathbf{Tb}$
    c. $\mathbf{T}(\mathbf{a} + \mathbf{b})$

6. Tensor $\mathbf{T}$ in the orthonormal basis $\mathbf{e}_i$ is represented as

$$[\mathbf{T}] = \begin{bmatrix} 1 & 3 & -2 \\ 3 & 0 & 1 \\ -2 & 1 & 1 \end{bmatrix}.$$

Find T'_{11}, T'_{12} and T'_{13} in basis $\mathbf{e}'_i$, such that $\mathbf{e}'_1$ is in direction $\mathbf{e}_2 + \mathbf{e}_3$ and $\mathbf{e}'_2$ is in direction $\mathbf{e}_1$.

7. Given the matrix representation of the tensor $\mathbf{T}$

$$[\mathbf{T}] = \begin{bmatrix} 1 & -2 & 5 \\ 4 & 6 & 7 \\ 2 & 3 & 1 \end{bmatrix},$$

    find:

a. the symmetric and skew-symmetric parts of $\mathbf{T}$
b. the axial vector of the skew-symmetric part of $\mathbf{T}$

8. The basis $\mathbf{e}'_i$ is rotated 60 degrees in the clockwise direction relative to $\mathbf{e}_3$. Find the orthogonal tensor and the components of vector $\mathbf{v} = -2\mathbf{e}_1 + 3\mathbf{e}_2$ in the new basis.

9. Given an arbitrary tensor $\mathbf{T}$ and an arbitrary vector $\mathbf{a}$, demonstrate the following relations:

a. $\mathbf{a} \cdot \mathbf{T}^A \mathbf{a} = 0$
b. $\mathbf{a} \cdot \mathbf{T}\mathbf{a} = \mathbf{a} \cdot \mathbf{T}^S \mathbf{a}$
c. $\text{tr}(\mathbf{T}^S \mathbf{T}^A) = 0$

10. For the tensor given in Problem 6, calculate the values and principal directions solving the eigenvalue problem and using the characteristic equation in terms of the tensor invariants.

11. Given the small displacement field

$$\mathbf{u} = [(3x^2 + y)\mathbf{e}_x + 10(3y + z^2)\mathbf{e}_y + 2z^2\mathbf{e}_z] \times 10^{-3} \ [\text{cm}]$$

- Determine the infinitesimal strain and rotation tensors, as well as the axial vector. Particularize them for point $P(2, 1, 3)$.
- If a body is subjected to a small rotation given by vector

$$\boldsymbol{\theta} = 0.002\mathbf{e}_x + 0.005\mathbf{e}_y - 0.002\mathbf{e}_z \ [\text{rad}],$$

what is the corresponding infinitesimal rotation tensor $\boldsymbol{\Omega}$?

12. Given the small displacement field

$$\mathbf{u} = [(6y + 5z)\mathbf{e}_x + (-6x + 3z)\mathbf{e}_y + (-5x - 3y)\mathbf{e}_z] \times 10^{-3} \ [\text{cm}],$$

show that this field induces only a rigid body rotation. It is also asked:
- determine the rotation vector $\boldsymbol{\theta}$ of the body
- calculate the infinitesimal strain tensor $\mathbf{E}$ and the cubic dilatation e

13. Given the small displacement field

$$\mathbf{u} = [(x^3 + 10)\mathbf{e}_x + 3yz\mathbf{e}_y + (z^2 - yx)\mathbf{e}_z] \times 10^{-3} \ [\text{cm}]$$

determine:
- the rigid body translation, taking the origin as reference
- the infinitesimal strain tensor $\mathbf{E}$
- the infinitesimal rotation tensor $\boldsymbol{\Omega}$
- the cubic dilatation e and the strain deviator tensor $\mathbf{E}^D = \mathbf{E} - \dfrac{e}{3}\mathbf{I}$
- particularize the results for point $P(2, 1, 0)$

14. Consider the Timoshenko beam kinematics. Determine the small strain, infinitesimal rotation, and Cauchy stress tensors. Hence, obtain the differential equations of the beam using tensor notation.

15. Determine the expression of the volume of the tetrahedron with straight faces illustrated in Figure 8.71(a).

9 FORMULATION AND APPROXIMATION OF PLANE PROBLEMS

This chapter presents the mechanical models and approximations by the FEM for problems of plane stress, plane strain, and twist of noncircular cross-sections. We also discuss the compatibility conditions, the Airy function for the construction of solutions to linear elastic plane problems, analytical solution of three-dimensional problems, the $(hp)^2$FEM software for high-order approximation, and examples of finite element simulation of three-dimensional models. We consider the generalization of numerical integration rules for two- and three-dimensional domains. At the end, we present a summary of the variational formulation of mechanical models considered in this book. This chapter is mainly based in references [51, 30, 36, 38, 22, 25, 13].

9.1 PLANE STRESS STATE

The plane stress model is usually employed for the analysis of structures such as beams and hookes. These cases may be modeled as plates, where the dimensions of the midplane are much larger than the thickness, as illustrated in Figure 9.1 and $a, b \gg t$. Moreover, the loads must be applied only parallel to the midplane which characterize the membrane effect. Plates under bending loads will be considered in Chapter 10.

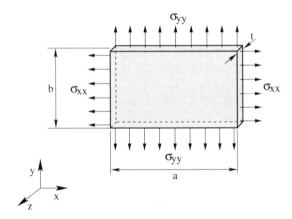

Figure 9.1 Example of a problem with plane stress state.

The basic hypothesis for the plane stress problem according to Figure 9.1 are:

- The thickness t is small when compared to the dimensions in the xy plane. It should be typically 10% or less than the shortest dimension of the xy plane.

- There are no forces applied in the normal direction to the xy plane (z axis of the reference system).

- The components of the volume forces act only in the xy plane and are independent of z, i.e., $b_x = b_x(x,y)$, $b_y = b_y(x,y)$, and $b_z = 0$.

- The traction forces on the surface are independent of z, that is, $t_x = t_x(x,y)$, $t_y = t_y(x,y)$, and $t_z = 0$.

From these hypotheses, it is assumed that the stress components in the z plane (σ_{zz}, τ_{zy}, and τ_{zx}) are small when compared to σ_{xx}, σ_{yy}, and τ_{xy}. Besides that, the variation of these stresses along the z axis is negligible and they depend on x and y coordinates only. Hence, the components of the Cauchy stress tensor are simplified to

$$\sigma_{xx} = \sigma_{xx}(x,y), \qquad \sigma_{yy} = \sigma_{yy}(x,y), \qquad \tau_{xy} = \tau_{xy}(x,y), \quad \text{and} \quad \sigma_{zz} = \tau_{zy} = \tau_{zx} = 0. \tag{9.1}$$

From (9.1), the equilibrium BVP, given in (8.45) and (8.46), simplifies to

$$\begin{cases} \dfrac{\partial \sigma_{xx}(x,y)}{\partial x} + \dfrac{\partial \tau_{xy}(x,y)}{\partial y} + b_x(x,y) = 0 \\ \dfrac{\partial \tau_{xy}(x,y)}{\partial x} + \dfrac{\partial \sigma_{yy}(x,y)}{\partial y} + b_y(x,y) = 0 \end{cases}, \tag{9.2}$$

with the following boundary conditions:

$$\begin{cases} \sigma_{xx} n_x + \tau_{xy} n_y = t_x \\ \tau_{xy} n_x + \sigma_{yy} n_y = t_y \end{cases}. \tag{9.3}$$

The strain component ε_{zz} can be determined in terms of σ_{xx} and σ_{yy}. To check this, consider the constitutive equation of Hooke's material for a plane stress state

$$\begin{bmatrix} \sigma_{xx} & \tau_{xy} & 0 \\ \tau_{xy} & \sigma_{yy} & 0 \\ 0 & 0 & 0 \end{bmatrix} = 2\mu \begin{bmatrix} \varepsilon_{xx} & \gamma_{xy} & 0 \\ \gamma_{xy} & \varepsilon_{yy} & 0 \\ 0 & 0 & \varepsilon_{zz} \end{bmatrix} + \lambda \begin{bmatrix} e & 0 & 0 \\ 0 & e & 0 \\ 0 & 0 & e \end{bmatrix} \tag{9.4}$$

with $e = \dfrac{\partial u_x}{\partial x} + \dfrac{\partial u_y}{\partial y} + \dfrac{\partial u_z}{\partial z}$.

From (9.4), the strain ε_{zz} is determined as

$$2\mu \frac{\partial u_z}{\partial z} + \lambda \left(\frac{\partial u_x}{\partial x} + \frac{\partial u_y}{\partial y} + \frac{\partial u_z}{\partial z} \right) = 0 \rightarrow \varepsilon_{zz} = \frac{\partial u_z}{\partial z} = -\frac{\lambda}{(\mu + \lambda)} \left(\frac{\partial u_x}{\partial x} + \frac{\partial u_y}{\partial y} \right). \tag{9.5}$$

Hence, the Navier equations can be rewritten in the following way:

$$\mu \begin{Bmatrix} \dfrac{\partial^2 u_x}{\partial x^2} + \dfrac{\partial^2 u_x}{\partial y^2} \\ \dfrac{\partial^2 u_y}{\partial x^2} + \dfrac{\partial^2 u_y}{\partial y^2} \end{Bmatrix} + (\mu + \lambda) \begin{Bmatrix} \dfrac{\partial^2 u_x}{\partial x^2} + \dfrac{\partial^2 u_y}{\partial x \partial y} \\ \dfrac{\partial^2 u_x}{\partial x \partial y} + \dfrac{\partial^2 u_y}{\partial y^2} \end{Bmatrix} + \begin{Bmatrix} b_x(x,y) \\ b_y(x,y) \end{Bmatrix} = 0. \tag{9.6}$$

9.2   PLANE STRAIN STATE

This model is generally used to represent the behavior of long domains, such as dams, tubes, and tunnels (see Figure 9.2), where the displacements in the longitudinal direction can be regarded as zero. The plane strain hypotheses are:

- The displacement in the z direction is considered zero, because the length of the body is much larger when compared to the representative dimensions in the x and y directions.

- The volume forces and those applied on the surfaces of the body, normal to the x and y directions, are independent of z.

Based on these hypotheses, the displacement components are such that

$$u_x = u_x(x,y), \qquad u_y = u_y(x,y), \quad \text{and} \quad u_z = 0. \tag{9.7}$$

Consequently, the strain components in the z plane are also zero and

$$\varepsilon_{zz} = \gamma_{yz} = \gamma_{xz} = 0. \tag{9.8}$$

The other components are independent of z, that is, $\varepsilon_{xx} = \varepsilon_{xx}(x,y)$, $\varepsilon_{yy} = \varepsilon_{yy}(x,y)$, and $\gamma_{xy} = \gamma_{xy}(x,y)$. In this case, the stress component σ_{zz} is not zero for a Hookean material and can be determined from the other components.

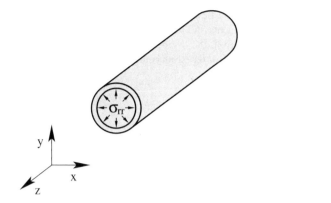

Figure 9.2 Example of plane strain state.

Considering hypotheses (9.7) and (9.8), the equilibrium BVP in terms of the stress components, given in (8.45) and (8.46), simplifies to

$$\begin{cases} \dfrac{\partial \sigma_{xx}(x,y)}{\partial x} + \dfrac{\partial \tau_{xy}(x,y)}{\partial y} + b_x(x,y) = 0 \\ \dfrac{\partial \tau_{xy}(x,y)}{\partial x} + \dfrac{\partial \sigma_{yy}(x,y)}{\partial y} + b_y(x,y) = 0 \quad , \\ \dfrac{\partial \sigma_{zz}(x,y)}{\partial z} = 0 \end{cases} \tag{9.9}$$

with the following boundary conditions:

$$\begin{cases} \sigma_{xx} n_x + \tau_{xy} n_y = t_x \\ \tau_{xy} n_x + \sigma_{yy} n_y = t_y \end{cases} . \tag{9.10}$$

The last equilibrium equation in (9.9) is automatically satisfied, because the normal stress component σ_{zz} depends on (x,y) only.

The constitutive equation for a Hookean material is

$$\begin{bmatrix} \sigma_{xx} & \tau_{xy} & 0 \\ \tau_{xy} & \sigma_{yy} & 0 \\ 0 & 0 & \sigma_{zz} \end{bmatrix} = 2\mu \begin{bmatrix} \varepsilon_{xx} & \gamma_{xy} & 0 \\ \gamma_{xy} & \varepsilon_{yy} & 0 \\ 0 & 0 & 0 \end{bmatrix} + \lambda \begin{bmatrix} e & 0 & 0 \\ 0 & e & 0 \\ 0 & 0 & e \end{bmatrix} . \tag{9.11}$$

From this equation, the stress component σ_{zz} is determined as

$$\sigma_{zz} = 2\lambda \left(\varepsilon_{xx} + \varepsilon_{yy} \right) . \tag{9.12}$$

The Navier equations are exactly the same ones obtained for the plane stress state given in (9.6). l

One of the main difficulties when solving problems of linear elasticity is that they require the determination of the three components of the displacement vectorial field or the six stress components of the stress tensorial field. An alternative is to transform the partial differential equations in terms of a scalar parameter, which is used later to obtain the stress and strain components. This can be achieved with the use of the Airy stress function. However, this procedure requires the use of compatibility equations. Both topics are discussed in the following sections.

9.3 COMPATIBILITY EQUATIONS

If the three components of the displacement vectorial field $\mathbf{u} = (u_x, u_y, u_z)$ are known, it is possible to determine the components of the infinitesimal strain tensorial field $\mathbf{E}$. Then, using the material's constitutive law, we can also obtain the corresponding stress state.

In the analytical treatment of elastic solids, it is very common to know the components of the infinitesimal strain tensor. From these components, the respective displacement field should be determined. The displacement components are obtained from the overdetermined system of equations with six equations and three unknowns given by the expressions of the strain components, that is,

$$
\begin{cases}
\varepsilon_{xx} = \dfrac{\partial u_x}{\partial x} \\[4pt]
\varepsilon_{yy} = \dfrac{\partial u_y}{\partial y} \\[4pt]
\varepsilon_{zz} = \dfrac{\partial u_z}{\partial z} \\[4pt]
\gamma_{xy} = \dfrac{1}{2}\left(\dfrac{\partial u_y}{\partial x} + \dfrac{\partial u_x}{\partial y} \right) \\[4pt]
\gamma_{xz} = \dfrac{1}{2}\left(\dfrac{\partial u_x}{\partial z} + \dfrac{\partial u_z}{\partial x} \right) \\[4pt]
\gamma_{yz} = \dfrac{1}{2}\left(\dfrac{\partial u_y}{\partial z} + \dfrac{\partial u_z}{\partial y} \right)
\end{cases}
\tag{9.13}
$$

The solution of this system of equations is possible only when some constraint conditions are established, which are known as compatibility conditions.

Given an infinitesimal strain field, it is not always possible to find a compatible displacement field, as illustrated in the following example.

Example 9.1 *Consider an infinitesimal strain field with components $\varepsilon_{xx} = \varepsilon_{zz} = \gamma_{xy} = \gamma_{xz} = \gamma_{yz} = 0$ and $\varepsilon_{yy} = kx^2$, and k a constant to enforce the case of infinitesimal strain.*

From the first expression in (9.13), we have

$$
\varepsilon_{xx} = \frac{\partial u_x}{\partial x} = 0.
$$

The indefinite integration of this equation results in

$$
u_x = g(y,z).
$$

From the second expression in (9.13), we have

$$
\varepsilon_{yy} = \frac{\partial u_y}{\partial y} = kx^2.
$$

The indefinite integral of the above equation results in

$$
u_y = kx^2 y + h(x,z),
$$

with $g(y,z)$ and $h(x,z)$ arbitrary integration functions.

From the fourth expression in (9.13), we have

$$
\bar{\gamma}_{xy} = \frac{\partial u_y}{\partial x} + \frac{\partial u_x}{\partial y} = 2kxy + \frac{\partial g(y,z)}{\partial y} + \frac{\partial h(x,z)}{\partial x} = 0.
$$

As g and h are only functions of (y,z) and (x,z), respectively, they do not generate any term like $-2kxy$ that allows us to zero the above expression. Thus, there is no compatible displacement field with the given strain field. Consequently, it is said that the strain components are not kinematically compatible.

□

There are six compatibility equations that are obtained as illustrated below. Taking the second derivative of the equations for ε_{xx} and ε_{yy} in (9.13), respectively, relative to y and x, we obtain

$$
\frac{\partial^2 \varepsilon_{xx}}{\partial y^2} = \frac{\partial^3 u_x}{\partial y^2 \partial x},
\tag{9.14}
$$

$$
\frac{\partial^2 \varepsilon_{yy}}{\partial x^2} = \frac{\partial^3 u_y}{\partial x^2 \partial y}.
\tag{9.15}
$$

Assuming that the strain components are continuous, we can change the differentiation order resulting in

$$\frac{\partial^2 \varepsilon_{xx}}{\partial y^2} = \frac{\partial^3 u_x}{\partial x \partial y^2} = \frac{\partial^2}{\partial x \partial y}\left(\frac{\partial u_x}{\partial y}\right), \tag{9.16}$$

$$\frac{\partial^2 \varepsilon_{yy}}{\partial x^2} = \frac{\partial^3 u_y}{\partial y \partial x^2} = \frac{\partial^2}{\partial x \partial y}\left(\frac{\partial u_y}{\partial x}\right). \tag{9.17}$$

Adding the two previous expressions, we obtain the following equation:

$$\frac{\partial^2 \varepsilon_{xx}}{\partial y^2} + \frac{\partial^2 \varepsilon_{yy}}{\partial x^2} = \frac{\partial^2}{\partial x \partial y}\left(\frac{\partial u_x}{\partial y} + \frac{\partial u_y}{\partial x}\right) = 2\frac{\partial^2 \gamma_{xy}}{\partial x \partial y}. \tag{9.18}$$

Analogously, the following expressions are obtained by respectively taking the pairs of normal strains $(\varepsilon_{yy}, \varepsilon_{zz})$ and $(\varepsilon_{xx}, \varepsilon_{yy})$ given in (9.13):

$$\frac{\partial^2 \varepsilon_{zz}}{\partial y^2} + \frac{\partial^2 \varepsilon_{yy}}{\partial z^2} = 2\frac{\partial^2 \gamma_{yz}}{\partial y \partial z}, \tag{9.19}$$

$$\frac{\partial^2 \varepsilon_{xx}}{\partial z^2} + \frac{\partial^2 \varepsilon_{zz}}{\partial x^2} = 2\frac{\partial^2 \gamma_{xz}}{\partial x \partial z}. \tag{9.20}$$

Consider now the derivatives of the strain components

$$\frac{\partial^2 \varepsilon_{xx}}{\partial y \partial z} = \frac{\partial^3 u_x}{\partial x \partial y \partial z}, \tag{9.21}$$

$$\frac{\partial \gamma_{yz}}{\partial x} = \frac{\partial^2 u_y}{\partial x \partial z} + \frac{\partial^2 u_z}{\partial x \partial y}, \tag{9.22}$$

$$\frac{\partial \gamma_{xz}}{\partial y} = \frac{\partial^2 u_x}{\partial y \partial z} + \frac{\partial^2 u_z}{\partial x \partial y}, \tag{9.23}$$

$$\frac{\partial \gamma_{xy}}{\partial z} = \frac{\partial^2 u_x}{\partial y \partial z} + \frac{\partial^2 u_y}{\partial x \partial z}. \tag{9.24}$$

The following relation is valid:

$$\frac{\partial^2 \varepsilon_{xx}}{\partial y \partial z} = \frac{\partial}{\partial x}\left(-\frac{\partial \gamma_{yz}}{\partial x} + \frac{\partial \gamma_{xz}}{\partial y} + \frac{\partial \gamma_{xy}}{\partial z}\right). \tag{9.25}$$

Analogously,

$$\frac{\partial^2 \varepsilon_{yy}}{\partial x \partial z} = \frac{\partial}{\partial y}\left(\frac{\partial \gamma_{yz}}{\partial x} - \frac{\partial \gamma_{xz}}{\partial y} + \frac{\partial \gamma_{xy}}{\partial z}\right), \tag{9.26}$$

$$\frac{\partial^2 \varepsilon_{zz}}{\partial x \partial y} = \frac{\partial}{\partial z}\left(\frac{\partial \gamma_{yz}}{\partial x} + \frac{\partial \gamma_{xz}}{\partial y} - \frac{\partial \gamma_{xy}}{\partial z}\right). \tag{9.27}$$

In this way, we obtain the compatibility differential equations given by

$$\begin{aligned}
\frac{\partial^2 \varepsilon_{xx}}{\partial y^2} + \frac{\partial^2 \varepsilon_{yy}}{\partial x^2} &= 2\frac{\partial^2 \gamma_{xy}}{\partial x \partial y}, \\
\frac{\partial^2 \varepsilon_{zz}}{\partial y^2} + \frac{\partial^2 \varepsilon_{yy}}{\partial z^2} &= 2\frac{\partial^2 \gamma_{yz}}{\partial y \partial z}, \\
\frac{\partial^2 \varepsilon_{xx}}{\partial z^2} + \frac{\partial^2 \varepsilon_{zz}}{\partial x^2} &= 2\frac{\partial^2 \gamma_{xz}}{\partial x \partial z}, \\
\frac{\partial^2 \varepsilon_{xx}}{\partial y \partial z} &= \frac{\partial}{\partial x}\left(-\frac{\partial \gamma_{yz}}{\partial x} + \frac{\partial \gamma_{xz}}{\partial y} + \frac{\partial \gamma_{xy}}{\partial z}\right), \\
\frac{\partial^2 \varepsilon_{yy}}{\partial x \partial z} &= \frac{\partial}{\partial y}\left(\frac{\partial \gamma_{yz}}{\partial x} - \frac{\partial \gamma_{xz}}{\partial y} + \frac{\partial \gamma_{xy}}{\partial z}\right), \\
\frac{\partial^2 \varepsilon_{zz}}{\partial x \partial y} &= \frac{\partial}{\partial z}\left(\frac{\partial \gamma_{yz}}{\partial x} + \frac{\partial \gamma_{xz}}{\partial y} - \frac{\partial \gamma_{xy}}{\partial z}\right).
\end{aligned} \tag{9.28}$$

The above compatibility conditions assume that the strain components are smooth functions, to ensure the existence of second-order derivatives. Note that if the strain components are linear in the Cartesian coordinates (x, y, z), they satisfy the above conditions. Moreover, these conditions are valid for bodies with simply connected domains. This means that any closed curve contained in the domain can be continuously deformed to a point without leaving the domain. Figure 9.3 illustrates simply and not simply connected domains. In a practical way, domains with holes are not simply connected.

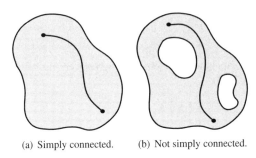

(a) Simply connected.   (b) Not simply connected.

Figure 9.3 Domains simply and not simply connected.

Example 9.2 *Consider a body in plane strain state with strain tensorial field given by*

$$[\mathbf{E}] = \begin{bmatrix} ky^2 & 2kxy & 0 \\ 2kxy & kx^2 & 0 \\ 0 & 0 & 0 \end{bmatrix},$$

and $k = 10^{-4}$.

Only the first compatibility condition in (9.13) has to be verified for this plane case. Thus,

$$\frac{\partial^2 \varepsilon_{xx}}{\partial y^2} + \frac{\partial^2 \varepsilon_{yy}}{\partial x^2} = 2k + 2k = 4k = 2\frac{\partial^2 \gamma_{xy}}{\partial x \partial y}.$$

Hence, the given infinitesimal strain tensorial field is compatible.

□

A set of compatibility equations can be determined in terms of the stress components for a body with homogeneous isotropic linear elastic material, applying Hooke's law to equations (9.28). This set is known as Beltrami-Michell compatibility equations.

Introducing the strain components given in (8.76) in the first compatibility equation given in (9.28), we obtain

$$(1+v)\left(\frac{\partial^2 \sigma_{xx}}{\partial y^2} + \frac{\partial^2 \sigma_{yy}}{\partial x^2}\right) - v\left(\frac{\partial^2 \bar{\sigma}}{\partial x^2} + \frac{\partial^2 \bar{\sigma}}{\partial y^2}\right) = 2(1+v)\frac{\partial^2 \tau_{xy}}{\partial x \partial y}. \tag{9.29}$$

Taking the derivatives of the equilibrium equations (8.45) in terms of the stress components, respectively, relative to x, y, and z, we obtain

$$\frac{\partial^2 \sigma_{xx}}{\partial x^2} + \frac{\partial^2 \tau_{xy}}{\partial x \partial y} + \frac{\partial^2 \tau_{xz}}{\partial x \partial z} + \frac{\partial b_x}{\partial x} = 0,$$
$$\frac{\partial^2 \tau_{xy}}{\partial x \partial y} + \frac{\partial^2 \sigma_{yy}}{\partial y^2} + \frac{\partial^2 \tau_{yz}}{\partial x \partial z} + \frac{\partial b_y}{\partial y} = 0, \tag{9.30}$$
$$\frac{\partial^2 \tau_{xz}}{\partial x \partial z} + \frac{\partial^2 \tau_{yz}}{\partial y \partial z} + \frac{\partial^2 \sigma_{zz}}{\partial z^2} + \frac{\partial b_z}{\partial z} = 0.$$

Subtracting the first and second equations from the third equation, we have

$$2\frac{\partial^2 \tau_{xy}}{\partial x \partial y} = \frac{\partial^2 \sigma_{zz}}{\partial z^2} - \frac{\partial^2 \sigma_{xx}}{\partial x^2} - \frac{\partial^2 \sigma_{yy}}{\partial y^2} + \frac{\partial b_z}{\partial z} - \frac{\partial b_x}{\partial x} - \frac{\partial b_y}{\partial y}.$$

Substituting this relation in (9.29), we have

$$(1+v)\left(\frac{\partial^2\sigma_{xx}}{\partial y^2}+\frac{\partial^2\sigma_{yy}}{\partial x^2}+\frac{\partial^2\sigma_{xx}}{\partial x^2}+\frac{\partial^2\sigma_{yy}}{\partial y^2}-\frac{\partial^2\sigma_{zz}}{\partial z^2}\right)-v\left(\frac{\partial^2\bar{\sigma}}{\partial x^2}+\frac{\partial^2\bar{\sigma}}{\partial y^2}\right)=$$
$$(1+v)\left(\frac{\partial b_z}{\partial z}-\frac{\partial b_x}{\partial x}-\frac{\partial b_y}{\partial y}\right). \tag{9.31}$$

Summing and subtracting $\dfrac{\partial^2\sigma_{xx}}{\partial z^2}$ and $\dfrac{\partial^2\sigma_{yy}}{\partial z^2}$ in the first term of the previous equation and using the Laplace operator Δ, we obtain

$$(1+v)\left(\Delta\sigma_{xx}+\Delta\sigma_{yy}-\frac{\partial^2\bar{\sigma}}{\partial z^2}\right)-v\left(\Delta\bar{\sigma}-\frac{\partial^2\bar{\sigma}}{\partial z^2}\right)=(1+v)\left(\frac{\partial b_z}{\partial z}-\frac{\partial b_x}{\partial x}-\frac{\partial b_y}{\partial y}\right), \tag{9.32}$$

where

$$\begin{aligned}
\Delta\sigma_{xx} &= \frac{\partial^2\sigma_{xx}}{\partial x^2}+\frac{\partial^2\sigma_{xx}}{\partial y^2}+\frac{\partial^2\sigma_{xx}}{\partial z^2}, \\
\Delta\sigma_{yy} &= \frac{\partial^2\sigma_{yy}}{\partial x^2}+\frac{\partial^2\sigma_{yy}}{\partial y^2}+\frac{\partial^2\sigma_{yy}}{\partial z^2}, \\
\Delta\sigma_{zz} &= \frac{\partial^2\sigma_{zz}}{\partial x^2}+\frac{\partial^2\sigma_{zz}}{\partial y^2}+\frac{\partial^2\sigma_{zz}}{\partial z^2}, \\
\Delta\bar{\sigma} &= \Delta\sigma_{xx}+\Delta\sigma_{yy}+\Delta\sigma_{zz}.
\end{aligned}$$

Therefore,

$$(1+v)\left(\Delta\bar{\sigma}-\Delta\sigma_{zz}-\frac{\partial^2\bar{\sigma}}{\partial z^2}\right)-v\left(\Delta\bar{\sigma}-\frac{\partial^2\bar{\sigma}}{\partial z^2}\right)=(1+v)\left(\frac{\partial b_z}{\partial z}-\frac{\partial b_x}{\partial x}-\frac{\partial b_y}{\partial y}\right)$$

or

$$\Delta\bar{\sigma}-(1+v)\Delta\sigma_{zz}-\frac{\partial^2\bar{\sigma}}{\partial z^2}=(1+v)\left(\frac{\partial b_z}{\partial z}-\frac{\partial b_x}{\partial x}-\frac{\partial b_y}{\partial y}\right) \tag{9.33}$$

Doing a similar procedure for the second and third compatibility equations in (9.28), we obtain

$$\Delta\bar{\sigma}-(1+v)\Delta\sigma_{yy}-\frac{\partial^2\bar{\sigma}}{\partial y^2}=(1+v)\left(\frac{\partial b_y}{\partial y}-\frac{\partial b_x}{\partial x}-\frac{\partial b_z}{\partial z}\right), \tag{9.34}$$

$$\Delta\bar{\sigma}-(1+v)\Delta\sigma_{xx}-\frac{\partial^2\bar{\sigma}}{\partial x^2}=(1+v)\left(\frac{\partial b_x}{\partial x}-\frac{\partial b_y}{\partial y}-\frac{\partial b_z}{\partial y}\right). \tag{9.35}$$

Summing the three previous equations, we obtain the following expression for $\Delta\bar{\sigma}$:

$$\Delta\bar{\sigma}=-\frac{(1+v)}{(1-v)}\left(\frac{\partial b_x}{\partial x}+\frac{\partial b_y}{\partial y}+\frac{\partial b_z}{\partial y}\right). \tag{9.36}$$

Substituting this expression in (9.32), we determine

$$\Delta\sigma_{zz}+\frac{1}{(1+v)}\frac{\partial^2\bar{\sigma}}{\partial z^2}=-\frac{1}{(1-v)}\left[2\frac{\partial b_z}{\partial z}-v\left(\frac{\partial b_z}{\partial z}-\frac{\partial b_x}{\partial x}-\frac{\partial b_y}{\partial y}\right)\right]. \tag{9.37}$$

Summing and subtracting $2v\dfrac{\partial b_z}{\partial z}$ in the right-hand side of the previous equation and after some simplification, we obtain

$$\Delta\sigma_{zz}+\frac{1}{(1+v)}\frac{\partial^2\bar{\sigma}}{\partial z^2}=-\frac{v}{(1-v)}\left(\frac{\partial b_x}{\partial x}+\frac{\partial b_y}{\partial y}+\frac{\partial b_z}{\partial z}\right)-2\frac{\partial b_z}{\partial z}. \tag{9.38}$$

Analogously, doing the same procedure for expressions (9.34) and (9.35), we obtain the following equations:

$$\Delta\sigma_{yy} + \frac{1}{(1+\nu)}\frac{\partial^2\bar{\sigma}}{\partial y^2} = -\frac{\nu}{(1-\nu)}\left(\frac{\partial b_x}{\partial x} - \frac{\partial b_y}{\partial y} - \frac{\partial b_z}{\partial z}\right) - 2\frac{\partial b_y}{\partial y}, \tag{9.39}$$

$$\Delta\sigma_{xx} + \frac{1}{(1+\nu)}\frac{\partial^2\bar{\sigma}}{\partial x^2} = -\frac{\nu}{(1-\nu)}\left(\frac{\partial b_x}{\partial x} + \frac{\partial b_y}{\partial y} + \frac{\partial b_z}{\partial z}\right) - 2\frac{\partial b_x}{\partial x}. \tag{9.40}$$

The following expressions are determined from the three remaining compatibility equations:

$$\Delta\tau_{xy} + \frac{1}{(1+\nu)}\frac{\partial^2\bar{\sigma}}{\partial x\partial y} = -\left(\frac{\partial b_x}{\partial x} + \frac{\partial b_y}{\partial y}\right), \tag{9.41}$$

$$\Delta\tau_{xz} + \frac{1}{(1+\nu)}\frac{\partial^2\bar{\sigma}}{\partial x\partial z} = -\left(\frac{\partial b_x}{\partial x} + \frac{\partial b_z}{\partial z}\right), \tag{9.42}$$

$$\Delta\tau_{yz} + \frac{1}{(1+\nu)}\frac{\partial^2\bar{\sigma}}{\partial y\partial z} = -\left(\frac{\partial b_y}{\partial y} + \frac{\partial b_z}{\partial z}\right). \tag{9.43}$$

Finally, if the body forces are neglected, we have the system of compatibility equations for a homogeneous isotropic linear elastic solid, that is,

$$(1+\nu)\Delta\sigma_{xx} + \frac{\partial^2\bar{\sigma}}{\partial x^2} = 0,$$

$$(1+\nu)\Delta\tau_{yz} + \frac{\partial^2\bar{\sigma}}{\partial y\partial z} = 0,$$

$$(1+\nu)\Delta\sigma_{yy} + \frac{\partial^2\bar{\sigma}}{\partial y^2} = 0, \tag{9.44}$$

$$(1+\nu)\Delta\tau_{xz} + \frac{\partial^2\bar{\sigma}}{\partial x\partial z} = 0,$$

$$(1+\nu)\Delta\sigma_{zz} + \frac{\partial^2\bar{\sigma}}{\partial z^2} = 0,$$

$$(1+\nu)\Delta\tau_{xy} + \frac{\partial^2\bar{\sigma}}{\partial x\partial y} = 0.$$

From equations obtained in (9.28) or (9.44), along with the equilibrium equations and the boundary conditions, it is possible to determine the displacement field components in terms of the strain components.

9.4 ANALYTICAL SOLUTIONS FOR PLANE PROBLEMS IN LINEAR ELASTICITY

The differential equations of equilibrium in terms of the stress components for plane stress and plane strain problems are similar, and, respectively given in (9.2) and (9.9). The analytical solution of these equations requires the determination of continuous functions $\sigma_{xx}(x,y)$, $\sigma_{yy}(x,y)$, and $\tau_{xy}(x,y)$, describing the stress components on the considered domain, and satisfying the boundary conditions. This task is not trivial. We wish to determine a simpler procedure involving only the scalar field $\phi = \phi(x,y)$, called potential or Airy stress function.

Note that equilibrium equations (9.2) and (9.9) are automatically satisfied in the absence of body loads, if the following definitions of the stress components in terms of the Airy functions are used:

$$\sigma_{xx}(x,y) = \frac{\partial^2\phi(x,y)}{\partial^2 y},$$

$$\sigma_{yy}(x,y) = \frac{\partial^2\phi(x,y)}{\partial^2 x}, \tag{9.45}$$

$$\tau_{xy}(x,y) = -\frac{\partial^2\phi(x,y)}{\partial x\partial y}.$$

A more general way to define the previous expressions is

$$
\begin{aligned}
\sigma_{xx}(x,y) &= \frac{\partial^2 \phi(x,y)}{\partial^2 y} - \Omega(x,y), \\
\sigma_{yy}(x,y) &= \frac{\partial^2 \phi(x,y)}{\partial^2 x} - \Omega(x,y), \\
\tau_{xy}(x,y) &= -\frac{\partial^2 \phi(x,y)}{\partial x \partial y}.
\end{aligned}
\tag{9.46}
$$

with $\Omega = \Omega(x,y)$ a scalar function of coordinates (x,y). In this case, the body force components must satisfy

$$
b_x(x,y) = \frac{\partial \Omega(x,y)}{\partial x} \quad \text{and} \quad b_y(x,y) = \frac{\partial \Omega(x,y)}{\partial y}.
\tag{9.47}
$$

It is possible to express several types of body loads, such as the body self-weight, using this definition.

Let $\phi(x,y)$ be a given stress function. Thereafter, the stress components are calculated as above and the infinitesimal strain components are determined using Hooke's law. However, we must ensure that the calculated strain components satisfy the compatibility conditions.

For a plane strain case, the nonzero strain components are obtained using Hooke's law (8.60) and equations given in (9.45) by

$$
\begin{aligned}
\varepsilon_{xx} &= \frac{1}{E}\{\sigma_{xx} - \nu[\sigma_{yy} + \nu(\sigma_{xx} + \sigma_{yy})]\} = \frac{1}{E}\left[(1-\nu^2)\frac{\partial^2 \phi}{\partial^2 y} - \nu(1+\nu)\frac{\partial^2 \phi}{\partial^2 x}\right], \\
\varepsilon_{yy} &= \frac{1}{E}\{\sigma_{yy} - \nu[\sigma_{xx} + \nu(\sigma_{xx} + \sigma_{yy})]\} = \frac{1}{E}\left[(1-\nu^2)\frac{\partial^2 \phi}{\partial^2 x} - \nu(1+\nu)\frac{\partial^2 \phi}{\partial^2 y}\right], \\
\gamma_{xy} &= \frac{1}{E}(1+\nu)\tau_{xy} = -\frac{1}{E}(1+\nu)\frac{\partial^2 \phi}{\partial x \partial y}.
\end{aligned}
\tag{9.48}
$$

Only the first compatibility condition in (9.28) is not automatically satisfied for the plane strain case. Substituting the previous strain components in the first equation in (9.28) we have

$$
\nabla^4 \phi = \frac{\partial^4 \phi}{\partial^4 x} + 2\frac{\partial^4 \phi}{\partial^2 x \partial^2 y} + \frac{\partial^4 \phi}{\partial^4 y} = 0.
\tag{9.49}
$$

Hence, any stress function $\phi(x,y)$ that satisfies the previous biharmonic equation generates a possible static solution for a plane strain problem in the absence of body loads.

Example 9.3 *Consider a body in plane strain state with stress function $\phi(x,y) = \alpha x^2 y + 2\beta y^5$. We want to determine the relation between the coefficients α and β in such way that the given function generates a possible stress state in the body.*

For this purpose, the biharmonic equation (9.49) must be satisfied. The fourth-order derivatives of $\phi(x,y)$ are given by

$$
\frac{\partial^4 \phi}{\partial x^4} = 0, \quad \frac{\partial^4 \phi}{\partial x^4} = 240\beta y, \quad \frac{\partial^4 \phi}{\partial x^2 y^2} = \frac{\partial^4 \phi}{\partial y^2 x^2} = 12\alpha y,
$$

which when substituted in (9.49) result in

$$
(24\alpha + 240\beta)y = 0.
$$

Thus, for $\alpha = -10\beta$, we have a possible plane strain state in the body.

$\square$

The nonzero strain components for a body under plane stress state are also obtained from Hooke's law

(8.60), and substituting (9.45) we have

$$
\begin{aligned}
\varepsilon_{xx} &= \frac{1}{E}(\sigma_{xx} - v\sigma_{yy}) = \frac{1}{E}\frac{\partial^2 \phi}{\partial^2 y} - v\frac{\partial^2 \phi}{\partial^2 x}, \\
\varepsilon_{yy} &= \frac{1}{E}(\sigma_{yy} - v\sigma_{xx}) = \frac{1}{E}\frac{\partial^2 \phi}{\partial^2 x} - v\frac{\partial^2 \phi}{\partial^2 y}, \\
\varepsilon_{zz} &= -\frac{v}{E}(\sigma_{xx} + \sigma_{yy}) = -\frac{v}{E}\left(\frac{\partial^2 \phi}{\partial^2 y} + \frac{\partial^2 \phi}{\partial^2 x}\right), \\
\gamma_{xy} &= \frac{1}{E}(1+v)\tau_{xy} = -\frac{1}{E}(1+v)\frac{\partial^2 \phi}{\partial x \partial y}.
\end{aligned}
\tag{9.50}
$$

The first compatibility condition in (9.28) gives the same biharmonic equation (9.49). Substituting the above strain components in the second, third, and sixth compatibility equations, we have

$$
\frac{\partial^2 \varepsilon_{zz}}{\partial^2 x} = \frac{\partial^2 \varepsilon_{zz}}{\partial^2 y} = \frac{\partial^2 \varepsilon_{zz}}{\partial x \partial y} = 0.
\tag{9.51}
$$

This implies that the strain component ε_{zz} varies linearly with (x,y). Due to equation (9.5), there is also a linear variation of $\sigma_{xx} + \sigma_{yy}$. The other compatibility conditions are automatically satisfied. Thus, if $\sigma_{xx} + \sigma_{yy}$ is a linear function of coordinates (x,y), then the plane stress model is a possible stress state for a body with any thickness. However, if $\sigma_{xx} + \sigma_{yy}$ is a nonlinear function of (x,y), then the plane stress model is only an approximation of the real stress state, and more accurate for bodies with small thickness.

The Airy functions must satisfy the following boundary conditions:

$$
\frac{\partial^2 \phi}{\partial^2 y} n_x - \frac{\partial^2 \phi}{\partial x \partial y} n_y = t_x,
\tag{9.52}
$$

$$
\frac{\partial^2 \phi}{\partial^2 x} n_y - \frac{\partial^2 \phi}{\partial x \partial y} n_x = t_y.
\tag{9.53}
$$

Using definition (9.47), the biharmonic equation is then given by

$$
\nabla^4 \phi = \frac{\partial^4 \phi}{\partial^4 x} + 2\frac{\partial^4 \phi}{\partial^2 x \partial^2 y} + \frac{\partial^4 \phi}{\partial^4 y} = C(v)\left(\frac{\partial b_x}{\partial x} + \frac{\partial b_y}{\partial y}\right),
\tag{9.54}
$$

with

$$
C(v) = \begin{cases} \dfrac{1-v}{1-2v} & \text{plane strain state} \\[2mm] \dfrac{1}{1-v} & \text{plane stress state} \end{cases}.
\tag{9.55}
$$

The Airy function is used only for plane stress and strain problems with isotropic materials, Besides that, it is simpler to use for bodies subjected to boundary conditions in terms of prescribed tractions than prescribed displacements.

Example 9.4 *Consider a cantilever beam with rectangular cross-section subjected to a concentrated load P in the left end $x = 0$ and clamped in the right end $x = L$, as illustrated in Figure 9.4.*

Consider the stress function

$$
\phi(x,y) = \alpha x y^3 + \beta x y.
$$

The respective stress components are

$$
\begin{aligned}
\sigma_{xx}(x,y) &= \frac{\partial^2 \phi(x,y)}{\partial^2 y} = 6\alpha x y, \\
\sigma_{yy}(x,y) &= \frac{\partial^2 \phi(x,y)}{\partial^2 x} = 0, \\
\tau_{xy}(x,y) &= -\frac{\partial^2 \phi(x,y)}{\partial x \partial y} = -(3\alpha y^2 + \beta).
\end{aligned}
\tag{9.56}
$$

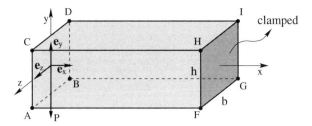

Figure 9.4 Example 9.4: cantilever beam.

Constants α and β can be determined from the beam boundary conditions. The stress vectors are zero on the faces ABFG and CDHI. The vertical coordinates and normal vectors are, respectively, $y = \pm\frac{h}{2}$ and $\mathbf{n} = \pm\mathbf{e}_y$. Thus, using equation (9.10), the stress vector is determined as

$$
\begin{bmatrix} \sigma_{xx} & \tau_{xy} & 0 \\ \tau_{xy} & \sigma_{yy} & 0 \\ 0 & 0 & \sigma_{zz} \end{bmatrix} \begin{Bmatrix} 0 \\ \pm 1 \\ 0 \end{Bmatrix} = \begin{Bmatrix} \pm\tau_{xy} \\ \pm\sigma_{yy} \\ 0 \end{Bmatrix}.
$$

In vector notation,

$$
\mathbf{t} = \pm\left(\tau_{xy}\mathbf{e}_x + \sigma_{yy}\mathbf{e}_y\right)\big|_{y=\pm\frac{h}{2}} = \pm\left(3\alpha\frac{h^2}{4} + \beta\right)\mathbf{e}_x = \mathbf{0}.
$$

Thus,

$$
\beta = -\frac{3h^2}{4}\alpha.
$$

In the left end, we have $x = 0$ and the normal vector is $\mathbf{n} = -\mathbf{e}_x$. The respective stress vector is obtained as

$$
\begin{bmatrix} \sigma_{xx} & \tau_{xy} & 0 \\ \tau_{xy} & \sigma_{yy} & 0 \\ 0 & 0 & \sigma_{zz} \end{bmatrix} \begin{Bmatrix} -1 \\ 0 \\ 0 \end{Bmatrix} = \begin{Bmatrix} -\sigma_{xx} \\ -\tau_{xy} \\ 0 \end{Bmatrix},
$$

or,

$$
\mathbf{t} = -\left(\sigma_{xx}\mathbf{e}_x + \tau_{xy}\mathbf{e}_y\right)\big|_{x=0} = -(3\alpha y^2 + \beta)\mathbf{e}_y.
$$

We have a parabolic distribution for the shear stress on the face ABCD and its resultant must be equal to the applied force P. Thus,

$$
P = \int_A (3\alpha y^2 + \beta)\, dA = 3\alpha b \int_{-h/2}^{h/2} y^2\, dy + \beta \int_A dA = \alpha\frac{bh^3}{4} + \beta bh.
$$

Substituting the expression for β in the above equation and simplifying, we obtain

$$
\alpha = \frac{2P}{bh^3}.
$$

Consequently,

$$
\beta = -\frac{3P}{2bh}.
$$

Based on that, the expressions for the stress components given in (9.56) are rewritten as

$$
\begin{aligned}
\sigma_{xx}(x,y) &= \frac{12P}{bh^3}xy = \frac{P}{I_z}xy, \\
\sigma_{yy}(x,y) &= 0, \\
\tau_{xy}(x,y) &= -3\alpha y^2 - \beta = \frac{P}{I_z}\left(\frac{h^2}{4} - y^2\right).
\end{aligned}
$$

The plane strain components are calculated using (9.48). Hence,

$$\varepsilon_{xx}(x,y) = \frac{(1-v^2)}{E}\frac{P}{I_z}xy,$$

$$\varepsilon_{yy}(x,y) = -\frac{(1+v)v}{E}\frac{P}{I_z}xy,$$

$$\gamma_{xy}(x,y) = \frac{(1+v)}{E}\frac{P}{I_z}\left(\frac{h^2}{4}-y^2\right).$$

The previous equations are valid for any thickness b of the beam cross-section. In this case, the normal stresses on faces ACFG ($z = \frac{b}{2}$) and BDGI ($z = -\frac{b}{2}$) are

$$\sigma_{zz} = v(\sigma_{xx}+\sigma_{yy}) = \frac{Pv}{I_z}xy.$$

As the expression for σ_{zz} is nonlinear, the given stress functions and the previous equations are only valid when the beam thickness b is very small compared with the other dimensions. In this case, the strain components are calculated from (9.50) and

$$\varepsilon_{xx}(x,y) = \frac{P}{EI_z}xy,$$

$$\varepsilon_{yy}(x,y) = \frac{Pv}{EI_z}xy,$$

$$\varepsilon_{zz}(x,y) = \frac{Pv}{EI_z}xy,$$

$$\gamma_{xy}(x,y) = \frac{P(1+v)}{EI_z}\left(\frac{h^2}{4}-y^2\right).$$

The strain component ε_{zz} is not of interest, because the beam is very thin and the compatibility conditions involving ε_{zz} are not satisfied.

$\square$

9.5   ANALYTICAL SOLUTIONS FOR PROBLEMS IN THREE-DIMENSIONAL ELASTICITY

In this section, we present the solution of three problems of three-dimensional linear elasticity and the comparison with the one-dimensional models of bars, shafts, and beams.

Example 9.5 *Consider the cylindrical body with length L and constant cross-section area A placed in the vertical direction, as illustrated in Figure 9.5.*

The body is subjected only to its self-weight. Thus, the volume load vectorial field is

$$\mathbf{b} = \left\{\begin{array}{ccc} b_x & b_y & b_z \end{array}\right\}^T = \left\{\begin{array}{ccc} -\rho g & 0 & 0 \end{array}\right\}^T, \tag{9.57}$$

with ρ the material density, g the acceleration of gravity, and ρg the density of body weight per unit of volume.

It is assumed that each cross-section is subjected to an uniform normal stress induced by the weight of the immediately inferior portion of the bar. Thus, the equilibrium equations for the solid, given in (8.45), are satisfied if

$$\left\{\begin{array}{l} \sigma_{xx} = \rho g x \\ \sigma_{yy} = \sigma_{zz} = \tau_{xy} = \tau_{xz} = \tau_{yz} = 0 \end{array}\right. . \tag{9.58}$$

Note that $\rho g x$ is a linear load distribution in the axial direction due to the body weight.

The normal unit vectors $\mathbf{n}$ to the lateral surface are always perpendicular to the x axis, that is, they can be represented in a general way as $\mathbf{n} = \{0\ n_y\ n_z\}^T$. The surface traction vectors are obtained from (8.197) resulting in

$$\begin{bmatrix} \rho g x & 0 & 0 \\ 0 & 0 & 0 \\ 0 & 0 & 0 \end{bmatrix} \left\{\begin{array}{c} 0 \\ n_y \\ n_z \end{array}\right\} = \left\{\begin{array}{c} 0 \\ 0 \\ 0 \end{array}\right\}. \tag{9.59}$$

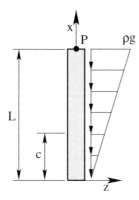

Figure 9.5 Example 9.5: cylindrical body subjected to its self-weight.

Thus, the surface tractions are zero. This is compatible with the fact that there is not any external load applied to the body surface. Analogously, the normal vectors to the inferior $(x = 0)$ and superior $(x = L)$ ends are respectively given by $\mathbf{n} = \{-1\,0\,0\}^T$ and $\mathbf{n} = \{1\,0\,0\}^T$. The respective stress vectors are also zero. Therefore, the body surfaces are free.

The compatibility equations (9.38) to (9.43) for a Hookean material are satisfied because $\sigma_{yy} = \sigma_{zz} = \tau_{xy} = \tau_{xz} = \tau_{yz} = 0$. Besides that, as $\sigma_{xx} = \rho g x$, the partial second-order derivatives are zero, that is, $\dfrac{\partial^2 \sigma_{xx}}{\partial x^2} = \dfrac{\partial^2 \sigma_{xx}}{\partial y^2} = \dfrac{\partial^2 \sigma_{xx}}{\partial z^2} = 0$.

Now using Hooke's law, we have that the distortion components are zero, that is, $\gamma_{xy} = \gamma_{xz} = \gamma_{yz} = 0$ and the normal strain components are $\varepsilon_{xx} = \dfrac{\rho g}{E} x$ and $\varepsilon_{yy} = \varepsilon_{zz} = -\dfrac{\nu \rho g}{E} x$. Hence, the second-order partial derivatives for the stress and strain components are zero, and the first-order partial derivatives are constant in this case.

The next step in the solution procedure is the characterization of the displacement field for this problem. From the expressions of the strain components, we have

$$
\begin{aligned}
\varepsilon_{xx} &= \frac{\partial u_x}{\partial x} = \frac{\rho g}{E} x, \\
\varepsilon_{yy} &= \frac{\partial u_y}{\partial y} = -\frac{\nu \rho g}{E} x, \\
\varepsilon_{zz} &= \frac{\partial u_z}{\partial z} = -\frac{\nu \rho g}{E} x, \\
\gamma_{xy} &= \frac{1}{2}\left(\frac{\partial u_y}{\partial x} + \frac{\partial u_x}{\partial y} \right) = 0, \\
\gamma_{xz} &= \frac{1}{2}\left(\frac{\partial u_x}{\partial z} + \frac{\partial u_z}{\partial x} \right) = 0, \\
\gamma_{yz} &= \frac{1}{2}\left(\frac{\partial u_y}{\partial z} + \frac{\partial u_z}{\partial y} \right) = 0.
\end{aligned}
\tag{9.60}
$$

Integrating the first expression in x, we have

$$
\int \frac{\partial u_x}{\partial x}\, dx = \int \frac{\rho g}{E} x\, dx.
$$

Thus,

$$
u_x(x,y,z) = \frac{\rho g}{2E} x^2 + u_{x_0}(y,z),
\tag{9.61}
$$

with $u_{x_0}(y,z)$ a function to be determined later using the boundary conditions of the problem.

Now substituting this result in the expressions for distortions γ_{xy} and γ_{xz} given in (9.60), we have

$$
\gamma_{xy} = \frac{\partial u_{x_0}(y,z)}{\partial y} + \frac{\partial u_y}{\partial x} = 0 \quad and \quad \gamma_{xz} = \frac{\partial u_{x_0}(y,z)}{\partial z} + \frac{\partial u_z}{\partial x} = 0,
$$

that is,

$$\frac{\partial u_y}{\partial x} = -\frac{\partial u_{x_0}(y,z)}{\partial y} \quad and \quad \frac{\partial u_z}{\partial x} = -\frac{\partial u_{x_0}(y,z)}{\partial z}.$$

Integrating the previous expressions in x, we have

$$\int \frac{\partial u_y}{\partial x} dx = -\int \frac{\partial u_{x_0}(y,z)}{\partial y} dx \quad and \quad \int \frac{\partial u_z}{\partial x} dx = -\int \frac{\partial u_{x_0}(y,z)}{\partial z} dx,$$

resulting in

$$u_y(x,y,z) = -\frac{\partial u_{x_0}(y,z)}{\partial y} x + u_{y_0}(y,z) \quad and \quad u_z(x,y,z) = -\frac{\partial u_{x_0}(y,z)}{\partial z} x + u_{z_0}(y,z), \tag{9.62}$$

with $u_{y_0}(y,z)$ and $u_{z_0}(y,z)$ functions of y and z to be determined later.

Now substituting u_y and u_z in the expression for the normal strain components ε_{yy} and ε_{zz} given in (9.60) we have

$$-x\frac{\partial^2 u_{x_0}(y,z)}{\partial y^2} + \frac{\partial u_{y_0}(y,z)}{\partial y} = -\frac{\nu\rho g}{E}x,$$

$$-x\frac{\partial^2 u_{x_0}(y,z)}{\partial z^2} + \frac{\partial u_{z_0}(y,z)}{\partial z} = -\frac{\nu\rho g}{E}x.$$

Recalling that u_{y_0} and u_{z_0} are independent of x, we conclude that these equations are valid only if

$$\frac{\partial u_{y_0}(y,z)}{\partial y} = \frac{\partial u_{z_0}(y,z)}{\partial z} = 0 \quad and \quad \frac{\partial^2 u_{x_0}(y,z)}{\partial y^2} = \frac{\partial^2 u_{x_0}(y,z)}{\partial z^2} = \frac{\nu\rho g}{E}. \tag{9.63}$$

Substituting expressions for u_y and u_z in the equation for distortion γ_{yz} given in (9.60), taking the derivatives, and recalling that $\gamma_{yz} = 0$, we have

$$-2x\frac{\partial^2 u_{x_0}(y,z)}{\partial y \partial z} + \frac{\partial u_{y_0}(y,z)}{\partial z} + \frac{\partial u_{z_0}(y,z)}{\partial y} = 0.$$

Again, as u_{y_0} and u_{z_0} are independent of x we obtain

$$\frac{\partial u_{y_0}(y,z)}{\partial z} = \frac{\partial u_{z_0}(y,z)}{\partial y} = 0 \quad and \quad \frac{\partial^2 u_{x_0}(y,z)}{\partial y \partial z} = 0. \tag{9.64}$$

Analyzing equations (9.63) and (9.64), we obtain the functions u_{x_0}, u_{y_0}, and u_{z_0}, that is,

$$u_{x_0} = \frac{\nu\rho g}{2E}\left(y^2 + z^2\right) + C_1 y + C_2 z + C_3,$$

$$u_{y_0} = C_4 z + C_5,$$

$$u_{z_0} = C_6 y + C_7,$$

with C_1 to C_7 arbitrary constants to be determined using the boundary conditions of the problem.

Substituting the previous expressions in (9.61) and (9.62), we determine the following displacement components:

$$u_x(x,y,z) = \frac{\rho g}{2E}x^2 + \frac{\nu\rho g}{2E}\left(y^2 + z^2\right) + C_1 y + C_2 z + C_3,$$

$$u_y(x,y,z) = -\frac{\nu\rho g}{E}xy - C_1 x + C_4 x + C_5,$$

$$u_z(x,y,z) = -\frac{\nu\rho g}{E}xz - C_2 x + C_6 y + C_7.$$

The fixing conditions on the surface of the cylindrical body must avoid rigid body movements. In terms of translations, a fixed support is placed at the centroid of the section in the upper end, indicated by point P (see

Figure 9.5) with coordinates $(L,0,0)$, *giving rise to the zero displacement conditions* $u_x(L,0,0) = u_y(L,0,0) = u_z(L,0,0) = 0$. *These boundary conditions, when applied to the previous expressions, result in*

$$
\begin{aligned}
u_x(L,0,0) &= \frac{\rho g}{2E}L^2 + C_3 = 0 \rightarrow C_3 = -\frac{\rho g}{2E}L^2, \\
u_y(L,0,0) &= -C_1 L + C_5 = 0 \rightarrow C_5 = C_1 L, \\
u_z(L,0,0) &= -C_2 L + C_7 = 0 \rightarrow C_7 = C_2 L.
\end{aligned}
\tag{9.65}
$$

To avoid rigid rotations, the following conditions are considered at point P: $\dfrac{\partial u_y}{\partial x} = \dfrac{\partial u_z}{\partial x} = \dfrac{\partial u_y}{\partial z} = \dfrac{\partial u_z}{\partial y} = 0$. *These derivatives of the displacement components are*

$$
\begin{aligned}
u_{y,x}(x,y,z) &= -\frac{\nu \rho g}{E}y - C_1, \\
u_{z,x}(x,y,z) &= -\frac{\nu \rho g}{E}z - C_2, \\
u_{y,z}(x,y,z) &= C_4, \\
u_{z,y}(x,y,z) &= C_6.
\end{aligned}
$$

Particularizing for point $P(L,0,0)$, *we have* $C_1 = C_2 = C_4 = C_6 = 0$. *Substituting these values in expressions* (9.65), *we obtain* $C_5 = C_7 = 0$.

Thus, the final expressions for the displacement components are:

$$
\begin{aligned}
u_x(x,y,z) &= \frac{\rho g}{2E}[x^2 - L^2 + \nu(y^2 + z^2)], \\
u_y(x,y,z) &= -\frac{\nu \rho g}{E}xy, \\
u_z(x,y,z) &= -\frac{\nu \rho g}{E}xz.
\end{aligned}
$$

The behavior of the axial displacement u_x *for a steel bar of length* $L = 1$ m *and square cross-section of edge* 10 cm *is illustrated in Figure 9.6(a).*

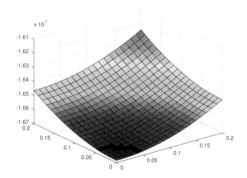

(a) Axial displacement.　　　　　　　(b) Axial displacement in $x = 0.3$ m.

Figure 9.6　Example 9.5: three-dimensional bar solution.

Points located along the x axis $(y = z = 0)$ *have axial displacements only given by*

$$
u_x(x) = -\frac{\rho g}{2E}\left(L^2 - x^2\right).
$$

The other points out of the x axis, due to the lateral contraction that comes from the Poisson's ratio, also present transversal displacements. Consequently, the lines, initially parallel to the x axis, become inclined. The cross-sections, which are initially perpendicular to the x axis, are then curved with the shape of a paraboloid, that is, the initially planar sections do not remain planar in the deformed configuration.

For a generic cross-section $x = c$, the deformed position x' is

$$x' = c + u_x(c, y, z) = c + \frac{\rho g}{2E}[c^2 - L^2 + \nu(y^2 + z^2)].$$

This is the expression of a paraboloid perpendicular to the bar's longitudinal fibers. They become inclined after deformation, in such way that there is no angular change and consequently the shear strains are zero, as illustrated in Figure 9.6(b). File planexemp5.m plots the axial displacements distributions of Figure 9.6.

Solving now the same example using the one-dimensional bar model of Chapter 3, with constant distributed load $q_x(x) = -\rho g A$ and boundary conditions $N_x(0) = 0$ and $u_x(L) = 0$, we obtain the following axial displacement function:

$$u_x(x) = -\frac{\rho g}{2E}\left(L^2 - x^2\right).$$

It is verified that the solution for the x axis in the three-dimensional case is identical to the one obtained by the one-dimensional bar model. However, the one-dimensional model considers that the cross-sections remain plane after deformation. The difference in terms of the axial displacement for both solutions is

$$e_x(x, y, z) = u_x(x, y, z) - u_x(x) = \frac{\nu \rho g}{2E}\left(y^2 + z^2\right).$$

For bars with length much larger than the cross-section dimensions, i.e., $L \gg y_{\max}, z_{\max}$, the difference between the two solutions is negligible. This reveals that the geometric hypothesis of the one-dimensional bar model is quite reasonable.

□

Example 9.6 *Consider the cylindrical body of Figure 9.5 subjected to a concentrated torque. We want to show that the three-dimensional solution and the one obtained using the one-dimensional model for the twist of circular prismatic shafts, discussed in Chapter 4, are coincident.*

The nonzero shear stress components of the one-dimensional model are parallel to the y and z axes and given by

$$\tau_{xy}(x, y, z) = -G\frac{d\theta_x(x)}{dx}z \quad and \quad \tau_{xz}(x, y, z) = G\frac{d\theta_x(x)}{dx}y.$$

The other stress components are zero, that is, $\sigma_{xx} = \sigma_{yy} = \sigma_{zz} = \tau_{yz} = 0$.

As the stress components are zero or linear functions of the coordinates, the compatibility equations (9.44) are satisfied, since the second-order partial derivatives and volume forces are zero for this problem. The equilibrium equations (8.45) are also satisfied because

$$\frac{\partial 0}{\partial x} + \frac{\partial}{\partial y}\left(-Gz\frac{d\theta_x(x)}{dx}\right) + \frac{\partial}{\partial z}\left(Gy\frac{d\theta_x(x)}{dx}\right) + 0 = 0,$$

$$\frac{\partial}{\partial x}\left(-Gz\frac{d\theta_x(x)}{dx}\right) + \frac{\partial 0}{\partial y} + \frac{\partial 0}{\partial z} + 0 = 0,$$

$$\frac{\partial}{\partial x}\left(Gy\frac{d\theta_x(x)}{dx}\right) + \frac{\partial 0}{\partial y} + \frac{\partial 0}{\partial z} + 0 = 0.$$

There are no external loads applied to the surface of the body. In addition, the normal vectors are always perpendicular to the longitudinal axis and parallel to the yz plane. Therefore, the boundary conditions in terms of stress are also satisfied, which can be checked, analogously to equation (9.59) of the previous example, i.e.,

$$\begin{bmatrix} 0 & \tau_{xy} & \tau_{xz} \\ \tau_{xy} & 0 & 0 \\ \tau_{xz} & 0 & 0 \end{bmatrix}\left\{\begin{array}{c} 0 \\ n_y \\ n_z \end{array}\right\} = \left\{\begin{array}{c} 0 \\ 0 \\ 0 \end{array}\right\}.$$

The determination of the three-dimensional displacement field follows analogously from the previous exam-

ple. For this purpose, consider the expressions of strain components

$$\varepsilon_{xx} = \frac{\partial u_x}{\partial x} = 0,$$

$$\varepsilon_{yy} = \frac{\partial u_y}{\partial y} = 0,$$

$$\varepsilon_{zz} = \frac{\partial u_z}{\partial z} = 0,$$

$$\bar{\gamma}_{xy} = \frac{\partial u_y}{\partial x} + \frac{\partial u_x}{\partial y} = \frac{\tau_{xy}}{G} = -z\frac{d\theta_x(x)}{dx},$$

$$\bar{\gamma}_{xz} = \frac{\partial u_x}{\partial z} + \frac{\partial u_z}{\partial x} = \frac{\tau_{xz}}{G} = y\frac{d\theta_x(x)}{dx},$$

$$\bar{\gamma}_{yz} = \frac{\partial u_y}{\partial z} + \frac{\partial u_z}{\partial y} = 0.$$

From the first expression, we have that u_x is independent of x. Considering that the shaft is fixed at point P, we have $u_x(x,y,z) = 0$ for all points of section $x = L$ and constant for other sections. Moreover, sections should not present rigid rotations but θ_x. Hence,

$$\frac{\partial u_x}{\partial y} = \frac{\partial u_x}{\partial z} = 0.$$

The distortion components $\bar{\gamma}_{xy}$ and $\bar{\gamma}_{xz}$ are reduced to

$$\bar{\gamma}_{xy} = \frac{\partial u_y}{\partial x} = -z\frac{d\theta_x(x)}{dx} \quad \text{and} \quad \bar{\gamma}_{xz} = \frac{\partial u_z}{\partial x} = y\frac{d\theta_x(x)}{dx}.$$

Integrating the previous expressions in x, we obtain

$$u_y = -z\theta_x(x) + u_{y_0} \quad \text{and} \quad u_z = y\theta_x(x) + u_{z_0},$$

with u_{y_0} and u_{z_0} functions of coordinates y and z and determined using the boundary conditions of the problem. Substituting these results in the expressions for ε_{zz}, ε_{yy}, and $\bar{\gamma}_{yz}$, we have that

$$\varepsilon_{yy} = \frac{\partial u_y}{\partial y} = \frac{\partial}{\partial y}\left(-z\theta_x(x) + u_{y_0}\right) = 0 \rightarrow \frac{\partial u_{y_0}}{\partial y} = 0, \tag{9.66}$$

$$\varepsilon_{zz} = \frac{\partial u_z}{\partial z} = \frac{\partial}{\partial z}\left(y\theta_x(x) + u_{z_0}\right) = 0 \rightarrow \frac{\partial u_{z_0}}{\partial z} = 0, \tag{9.67}$$

$$\bar{\gamma}_{yz} = \frac{\partial u_y}{\partial z} + \frac{\partial u_z}{\partial y} = \frac{\partial}{\partial z}\left(-z\theta(x) + u_{y_0}\right) + \frac{\partial}{\partial y}\left(y\theta(x) + u_{z_0}\right) = 0$$

$$\rightarrow \frac{\partial u_{z_0}}{\partial y} = -\frac{\partial u_{y_0}}{\partial z}. \tag{9.68}$$

Analyzing the expressions, we conclude that u_{y_0} and u_{z_0} depend only on z and y, respectively. Besides that, u_{y_0} and u_{z_0} are linear functions given by

$$u_{y_0} = C_1 z + C_2 \quad \text{and} \quad u_{z_0} = C_3 y + C_4,$$

with C_1 to C_4 arbitrary constants to be determined from the boundary conditions.
Thus, the transversal displacements are expressed as

$$u_y(x,y,z) = -z\theta_x(x) + C_1 z + C_2, \tag{9.69}$$

$$u_z(x,y,z) = y\theta_x(x) + C_3 y + C_4. \tag{9.70}$$

Analyzing the support conditions, $u_x(L,y,z) = u_y(L,y,z) = u_z(L,y,z) = 0$ and $\dfrac{\partial u_x(L,y,z)}{\partial z} = \dfrac{\partial u_z(L,y,z)}{\partial y} =$

$\dfrac{\partial u_y(L,y,z)}{\partial z} = 0$, we have

$$
\begin{aligned}
u_y(L,y,z) &= -z\theta(L) + C_1 L + C_2 = 0 \rightarrow C_2 = C_1 L, \\
u_z(L,y,z) &= y\theta(L) + C_3 L + C_4 = 0 \rightarrow C_4 = C_3 L, \\
\frac{\partial u_y(L,y,z)}{\partial z} &= C_1 = 0 \rightarrow C_1 = 0, \\
\frac{\partial u_z(L,y,z)}{\partial y} &= C_3 = 0 \rightarrow C_3 = 0.
\end{aligned}
$$

Thus, $C_2 = C_4 = 0$.

Hence, the displacement vectorial field is

$$
\mathbf{u}(x,y,z) = \left\{
\begin{array}{c}
0 \\
-z\theta_x(x) \\
y\theta_x(x)
\end{array}
\right\}.
$$

This is the same result obtained by the one-dimensional model of Chapter 4. Hence, the three-dimensional and one-dimensional solutions for a circular cross-section are coincident.

□

Example 9.7 *Consider again the cylindrical body of Figure 9.5 subjected to a pure moment M at the inferior end and clamped at point P. From the one-dimensional beam theory, the bending moment is $M_z(x) = M$ and the stress components are given by:*

$$
\begin{aligned}
\sigma_{xx} &= -\frac{M}{I_z} y, \\
\sigma_{yy} &= \sigma_{xx} = \tau_{xy} = \tau_{xz} = \tau_{yz} = 0,
\end{aligned}
$$

with I_z the second moment of area of the cross-section relative to the z axis of the reference system. It is assumed that the beam is prismatic with constant cross-section and consequently $I_z(x) = I_z$.

Using the same steps of the previous examples, we want to verify if, in the absence of body forces, this stress distribution satisfies the equilibrium conditions and compatibility equations for a three-dimensional solid. The boundary conditions in terms of stress on the beam lateral surface are satisfied because there are no external loads applied on the body surface.

To characterize the three-dimensional displacement field, consider the following expressions for the strain components:

$$
\begin{aligned}
\varepsilon_{xx} &= \frac{\partial u_x}{\partial x} = \frac{\sigma_{xx}}{E} = -\frac{M}{EI_z} y, \\
\varepsilon_{yy} &= \frac{\partial u_y}{\partial y} = -\nu \varepsilon_{xx} = \frac{M\nu}{EI_z} y, \\
\varepsilon_{zz} &= \frac{\partial u_z}{\partial z} = -\nu \varepsilon_{xx} = \frac{M\nu}{EI_z} y, \\
\gamma_{xy} &= \frac{1}{2}\left(\frac{\partial u_y}{\partial x} + \frac{\partial u_x}{\partial y} \right) = 0, \\
\gamma_{xz} &= \frac{1}{2}\left(\frac{\partial u_x}{\partial z} + \frac{\partial u_z}{\partial x} \right) = 0, \\
\gamma_{yz} &= \frac{1}{2}\left(\frac{\partial u_y}{\partial z} + \frac{\partial u_z}{\partial y} \right) = 0,
\end{aligned}
\tag{9.71}
$$

where E is the Young's modulus and ν is the Poisson's ratio of the material, both assumed constant.

Integrating the previous expression for ε_{xx} in x, we obtain

$$
u_x(x,y,z) = -\frac{M}{EI_z} xy + u_{x_0}(y,z),
$$

with u_{x_0} a function of y and z to be obtained.

From the expressions for the distortion components γ_{xy} and γ_{xz}, we have

$$\frac{\partial u_y}{\partial x} = -\frac{\partial u_x}{\partial y} = \frac{M}{EI_z}x - \frac{\partial u_{x_0}(y,z)}{\partial y},$$

$$\frac{\partial u_z}{\partial x} = -\frac{\partial u_x}{\partial z} = -\frac{\partial u_{x_0}(y,z)}{\partial z}.$$

Integrating the previous equations in x, we have

$$u_y(x,y,z) = \frac{M}{2EI_z}x^2 - \frac{\partial u_{x_0}(y,z)}{\partial y}x + u_{y_0}(y,z), \tag{9.72}$$

$$u_z(x,y,z) = -\frac{\partial u_{x_0}(y,z)}{\partial z}x + u_{z_0}(y,z), \tag{9.73}$$

with u_{y_0} and u_{z_0} functions of y and z to be also determined below.

Substituting these functions in the expressions for ε_{yy} and ε_{zz} given in (9.71), we have

$$-\frac{\partial^2 u_{x_0}}{\partial y^2}x + \frac{\partial u_{y_0}}{\partial y} = \frac{M\nu}{EI_z}y,$$

$$-\frac{\partial^2 u_{x_0}}{\partial z^2}x + \frac{\partial u_{z_0}}{\partial z} = \frac{M\nu}{EI_z}y.$$

As these equations must be satisfied for any x, then

$$\frac{\partial^2 u_{x_0}}{\partial y^2} = \frac{\partial^2 u_{x_0}}{\partial z^2} = 0 \quad and \quad \frac{\partial u_{y_0}}{\partial y} = \frac{\partial u_{z_0}}{\partial z} = \frac{M\nu}{EI_z}.$$

Integrating in y and z the terms in the previous conditions results in

$$u_{y_0}(y,z) = \frac{M\nu}{2EI_z}y^2 + f(z), \tag{9.74}$$

$$u_{z_0}(y,z) = \frac{M\nu}{EI_z}yz + g(y), \tag{9.75}$$

with $f(z)$ and $g(y)$ functions to be determined.

From the expression of γ_{yz} in (9.71) and equations (9.72) and (9.73), we have

$$\frac{\partial u_y}{\partial z} + \frac{\partial u_z}{\partial y} = \frac{\partial u_{y_0}}{\partial z} + \frac{\partial u_{z_0}}{\partial y} = 0.$$

Substituting u_{y_0} and u_{z_0}, given in (9.74) and (9.75), we obtain

$$-2\frac{\partial^2 u_{x_0}}{\partial y\partial z}x + \frac{\partial f(z)}{\partial z} + \frac{\partial g(y)}{\partial y} + \frac{M\nu}{EI_z}z = 0.$$

In order to satisfy this expression, we have

$$\frac{\partial^2 u_{x_0}}{\partial y\partial z} = 0 \quad and \quad \frac{\partial f(z)}{\partial z} + \frac{\partial g(y)}{\partial y} + \frac{M\nu}{EI_z}z = 0.$$

This requires functions $u_{x_0}(y,z)$, $f(z)$, and $g(y)$ to have respectively the following general forms:

$$u_{x_0}(y,z) = C_1 y + C_2 z + C_3,$$

$$f(z) = -\frac{M\nu}{2EI_z}z^2 + C_4 z + C_5,$$

$$g(y) = -C_4 y + C_6,$$

with C_1 to C_6 the arbitrary constants to be determined from the boundary conditions of the problem.

The displacement components are written as

$$u_x(x,y,z) = -\frac{M}{EI_z}xy + C_1 y + C_2 z + C_3,$$

$$u_y(x,y,z) = \frac{M}{2EI_z}[x^2 - v(y^2 - z^2)] - C_1 x + C_4 z + C_5,$$

$$u_z(x,y,z) = \frac{Mv}{EI_z}yz - C_2 x - C_4 y + C_6.$$

Constants C_1 to C_6 are determined from the boundary conditions as

$$u_x(L,0,0) = C_3 = 0 \rightarrow C_3 = 0,$$

$$u_y(L,0,0) = \frac{ML^2}{2EI_z} - C_1 L + C_5 = 0,$$

$$u_z(L,0,0) = -C_2 L + C_6 = 0,$$

$$u_{y,x}(L,0,0) = \frac{ML}{EI_z} - C_1 = 0 \rightarrow C_1 = \frac{ML}{EI_z} \quad and \quad C_5 = \frac{ML^2}{EI_z},$$

$$u_{z,x}(L,0,0) = -C_2 = 0 \rightarrow C_2 = 0 \quad and \quad C_6 = 0,$$

$$u_{y,z}(L,0,0) = C_4 = 0 \rightarrow C_4 = 0.$$

Substituting the constants, the final displacement vectorial field is

$$u_x(x,y,z) = \frac{M}{EI_z}(L-x)y,$$

$$u_y(x,y,z) = \frac{M}{2EI_z}[x^2 - v(y^2 - z^2) - L(2x - L)],$$

$$u_z(x,y,z) = \frac{Mv}{EI_z}yz.$$

The deformed position of the cross-section $x = c$ is determined as

$$x' = c + u_x(c,y,z) = c + \frac{M}{EI_z}(L-c)y.$$

Thus, a planar section remains planar after deformation, according to the one-dimensional model hypotheses. For a rectangular cross-section with $2b \times 2h$ dimensions, edges $z = \pm b$ of section $x = c$ have the following deformed position:

$$z' = \pm b + u_z(c,y,\pm b) = \pm \frac{Mv}{EI_z}yb,$$

which represent two inclined lines. On the other hand, for the edges $y = \pm h$, we have

$$y' = \pm h + u_y(c,\pm h,z) = \frac{M}{2EI_z}[c^2 - v(h^2 - z^2) - L(2c - L)].$$

The above expression constitutes a parabolic curve, in such a way that the concavity is opposite to the deformed configuration. Thus, while the beam concavity is facing upwards, the concavity of the parabola of the upper side is facing downwards.

The solution of the one-dimensional bending problem is determined with zero distributed load $q_y(x) = 0$ and the boundary conditions $V_y(0) = 0$, $M_z(0) = M$, $\theta_z(L) = 0$ and $u_y(L) = 0$. The equations for the shear force, bending moment, rotation, and transversal displacement are given, respectively, by

$$V_y(x) = 0,$$

$$M_z(x) = M,$$

$$\theta_z(x) = \frac{M}{EI_z}(x - L),$$

$$u_y(x) = \frac{M}{2EI_z}(x^2 - 2Lx + L^2).$$

Besides that,

$$u_x(x) = -y\theta_z(x) = \frac{M}{EI_z}(L-x)y.$$

Taking $y = z = 0$ in the expressions of the three-dimensional solution, we obtain the displacements for the neutral line of the beam, that is,

$$u_y(x, y = 0, z = 0) = \frac{M}{2EI_z}(x^2 - 2Lx + L^2) \quad \text{and} \quad u_y = u_z = 0.$$

which coincide with the solution found for the one-dimensional model.

Analogous to the bar example, considering the beam length to be much larger than the cross-section dimensions, the differences between both models can be neglected.

□

9.6 PLANE STATE APPROXIMATION

Analogous to Section 8.12.1, the weak form of element e for a plane state problem is written in matrix notation as

$$\left(t \int_{\Omega^{(e)}} [\mathbf{B}^{(e)}]^T [\mathbf{D}][\mathbf{B}^{(e)}] d\Omega^{(e)}\right) \{\mathbf{u}_i^{(e)}\} = t \int_{\Omega^{(e)}} [\mathbf{N}^{(e)}]^T \{\mathbf{b}\} d\Omega^{(e)} + t \int_{\partial\Omega^{(e)}} [\mathbf{N}^{(e)}]^T \{\mathbf{t}\} d\partial\Omega^{(e)}, \tag{9.76}$$

where $\Omega^{(e)}$ and $\partial\Omega^{(e)}$ are the area and perimeter of the element, respectively. In the plane stress case, t is the element thickness, assumed here to be constant; for the plane strain case, $t = 1$ in the previous expression.

The weak form for the square element e in terms of the local coordinates is given by

$$\left(t \int_{-1}^{1}\int_{-1}^{1} [\mathbf{B}^{(e)}]^T [\mathbf{D}][\mathbf{B}^{(e)}] \det[\mathbf{J}] d\xi_1 d\xi_2\right) \{\mathbf{u}_i^{(e)}\} = t \int_{-1}^{1}\int_{-1}^{1} [\mathbf{N}^{(e)}]^T \{\mathbf{b}\} \det[\mathbf{J}] d\xi_1 d\xi_2$$

$$+ t \int_{-1}^{1} [\mathbf{N}^{(e)}]^T \{\mathbf{t}\} \det[\mathbf{J}_b] d\Gamma^{(e)}, \tag{9.77}$$

with $[\mathbf{J}]$ and $[\mathbf{J}_b]$ the Jacobian matrices associated to the area and perimeter of the element.

The strain-displacement matrix $[\mathbf{B}^{(e)}]$ for an element with N_e nodes or modes is

$$[\mathbf{B}^{(e)}] = [\mathbf{L}][\mathbf{N}^{(e)}] = \left[[\mathbf{B}_1^{(e)}] \mid [\mathbf{B}_2^{(e)}] \mid \cdots \mid [\mathbf{B}_{N_e}^{(e)}]\right], \tag{9.78}$$

with $[\mathbf{B}_i^{(e)}]$ the associated matrix to each element shape functions, expressed as

$$[\mathbf{B}_i^{(e)}] = \begin{bmatrix} \phi_{i,x}^{(e)} & 0 \\ 0 & \phi_{i,y}^{(e)} \\ \phi_{i,y}^{(e)} & \phi_{i,x}^{(e)} \end{bmatrix}. \tag{9.79}$$

The matrix of local shape functions in the plane element is given by

$$[\mathbf{N}_e] = \begin{bmatrix} \phi_1^{(e)} & 0 & \cdots & \phi_{N_e}^{(e)} & 0 \\ 0 & \phi_1^{(e)} & \cdots & 0 & \phi_{N_e}^{(e)} \end{bmatrix}. \tag{9.80}$$

The stiffness matrix and body and surface equivalent force vectors for a square element can be calculated in the local reference system $\xi_1 \times \xi_2$ as

$$[\mathbf{K}^{(e)}] = t \int_{-1}^{1}\int_{-1}^{1} [\mathbf{B}^{(e)}]^T [\mathbf{D}][\mathbf{B}^{(e)}] \det[\mathbf{J}] d\xi_1 d\xi_2, \tag{9.81}$$

$$\{\mathbf{f}^{(e)}\} = t \int_{-1}^{1}\int_{-1}^{1} [\mathbf{N}^{(e)}]^T \{\mathbf{b}\} \det[\mathbf{J}] d\xi_1 d\xi_2 + t \int_{-1}^{1} [\mathbf{N}^{(e)}]^T \{\mathbf{t}\} \det[\mathbf{J}_b] d\Gamma^{(e)}. \tag{9.82}$$

For a triangular element, the previous expressions are

$$[\mathbf{K}^{(e)}] = t \int_0^1 \int_0^{1-L_2} [\mathbf{B}^{(e)}]^T [\mathbf{D}][\mathbf{B}^{(e)}] \det[\mathbf{J}] dL_1 dL_2, \tag{9.83}$$

$$\{\mathbf{f}^{(e)}\} = t \int_0^1 \int_0^{1-L_2} [\mathbf{N}^{(e)}]^T \{\mathbf{b}\} \det[\mathbf{J}] dL_1 dL_2 + t \int_0^1 [\mathbf{N}^{(e)}]^T \{\mathbf{t}\} \det[\mathbf{J}_b] d\Gamma^{(e)}. \tag{9.84}$$

The elasticity matrix for the plane stress and strain states are respectively given by

$$[\mathbf{D}] = \frac{E}{1-\nu^2} \begin{bmatrix} 1 & \nu & 0 \\ \nu & 1 & 0 \\ 0 & 0 & \frac{1-\nu}{2} \end{bmatrix}, \tag{9.85}$$

$$[\mathbf{D}] = \frac{E}{(1+\nu)(1-2\nu)} \begin{bmatrix} 1-\nu & \nu & 0 \\ \nu & 1-\nu & 0 \\ 0 & 0 & 1-2\nu \end{bmatrix}. \tag{9.86}$$

Example 9.8 *File planestress.m implements procedures to calculate the mass and stiffness matrices of quadrangular elements using Lagrange basis and the symbolic manipukation toolkit available in MATLAB.*
□

9.7 $(HP)^2$ FEM PROGRAM

In the context of intense use of computational simulations in the solution of engineering problems, software development is of fundamental importance. Easy inclusion of new models and approximation and solution techniques are some of the desired features of current software engineering. Besides that, with the dissemination of clusters with several processors, the use of high-performance resources becomes fundamental.

The $(hp)^2$ FEM software has been implemented using the object-oriented paradigm with versions for MATLAB and C++. The basic features of the software are [6]:

- Use of tensorization procedures for low- and high-order shape functions, as well as quadrature rules, using tensor products as presented in [9, 10].

- Implementation of the local element operators independently of the employed nodal and modal shape functions.

- Implementation of high-performance extensions.

The program implements the solution of Poisson, linear elasticity, large deformations, contact, and Reynolds equation problems and global and element by element solvers. The program reads two ASCII files. The first file with .fem extension contains definitions of the finite element model, such as dimension, number of nodes, nodal coordinates, type of elements, number of elements, groups of elements and incidence. The second file with .def extension contains parameters that define the mechanical model, boundary conditions, loads, integration rules, shape functions, solution parameters, among others. The program allows the use of symbolic expressions for loads, boundary conditions, and analytical solutions for the calculation of approximation errors.

Example 9.9 *Figure 9.7 illustrates a mesh with four elements used for the study of a plane strain problem in the rectangular domain with 4×1 dimensions. The fabricated solution $u_x(x,y) = xy(y-1)$ and $u_y = 0$ is considered. We adopt $E = 1000$ GPa and $\nu = 0.3$.*

The respective strain components are obtained substituting u_x and u_y given in equation (9.13). Using Hooke's law (9.11), we calculate the stress components. From these components, we can obtain the body force components using the equilibrium equation (9.9) and $b_x = -2\mu x$ and $b_y = -\mu(2y-1)$.

The horizontal edges and the left vertical edge have zero displacements. The components of the traction vector in the right vertical edge are obtained from (9.10), resulting in $t_x = 2\lambda\mu y(y-1)$ and $t_y = \mu x(2y-1)$.

Using second-order approximation with Lagrange polynomials, the error of the approximated solution is zero. Fabricated solutions are a convenient way to verify a finite element program.

Files planestrain.fem and planestrain.def specify the necessary parameters for the analysis of this example with the $(hp)^2$ FEM program.
□

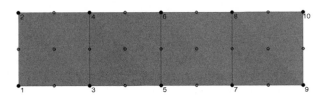

Figure 9.7 Example 9.9: second-order mesh for the analysis of plane strain problem.

Example 9.10 *Consider the three-dimensional meshes illustrated in Figure 9.8 to be analysed using the* $(hp)^2$*FEM software. The files conrod.fem and conrod.def have the input parameters for the mesh and attributes for the conrod showed in Figure 9.8(a). Files piston.fem and piston.def have the attributes for the piston illustrated in Figure 9.8(b). The mesh showed in Figure 9.8(c) for one crankshaft segment is given in files crankshaft.fem and crankshaft.def.*

The results of the finite element analysis may be plotted as illustrated in Figure 9.8(d) for the resultant displacement in the conrod.

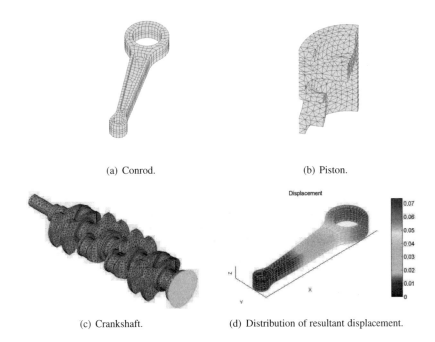

(a) Conrod.

(b) Piston.

(c) Crankshaft.

(d) Distribution of resultant displacement.

Figure 9.8 Example 9.10: three-dimensional meshes solved using the $(hp)^2$FEM software.

□

9.8   TORSION OF GENERIC SECTIONS

In Chapter 4, we considered the torsion model in shafts with a circular cross-section. The kinematics of this model was characterized by a rigid rotation of each cross-section, which remained orthogonal to the x axis of the reference system.

Twist of mechanical components with noncircular cross-sections is very common in engineering applications. Analogous to the circular section, each section has a rigid rotation about x. But now there is warping of the cross-sections, that is, they do not remain orthogonal to the x axis, as illustrated in Figure 9.9 for a retangular section. The formulation of torsion in generic cross-sections was developed by the French mathematician Saint-Venant and is therefore known as the Saint-Venant model. In the following sections, we present the formulation of this model considering the steps of the variational formulation, as well as the tensorial notation presented earlier. The structural element with noncircular cross-section subjected to twist is also denominated here as beam.

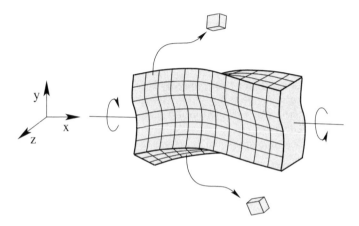

Figure 9.9   Warping of a beam with square cross-section submitted to twist load.

9.8.1   KINEMATICS

The Saint-Venant hypotheses for the problem of twist in generic sections under small strains are:

- As in the case of circular twist, each cross-section has a rigid rotation $\theta_x = \theta_x(x)$ about the x axis of the adopted reference system.

- The longitudinal displacement u_x is independent of x. This means that all sections have the same displacement $u_x(x) = \varphi(y, z)$, with $\varphi(y, z)$ a function to be determined.

As the angle of twist is constant for each cross-section, its rate of variation along the beam length is constant and can be denoted as

$$\alpha = \frac{d\theta_x(x)}{dx} = \text{cte.} \tag{9.87}$$

Then we can rewrite function $\varphi(y, z)$ as

$$\varphi(y, z) = \alpha w(y, z) = \frac{d\theta_x(x)}{dx} w(y, z), \tag{9.88}$$

with $w(y, z)$ called warping function, which is related to the loss of planicity of the cross-section.

The formulation presented here is valid for any cross-section shape, including the circular case. Thus, consider the square cross-section illustrated in Figure 9.10(a). Due to the rotation $\theta_x = \theta_x(x)$, any point P in the cross-section x with initial transversal coordinates (y, z) assumes the final position P' with coordinates (y', z'). The transversal displacements of point P in directions y and z are, respectively, $u_y(x) = -z\theta_x(x)$ and $u_z(x) = y\theta_x(x)$. Due to the cross-section warping, point P' has a longitudinal displacement u_x, assuming the

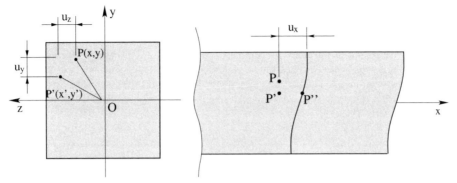

(a) Tranversal displacement components of P in the cross-section.

(b) Warping in the longitudinal direction.

Figure 9.10 Twist kinematics in generic cross-sections.

final position P'' illustrated in Figure 9.10(b). This displacement is given by function $\varphi(y,z)$ according to the Saint-Venant hypotheses, that is, $u_x(x,y,z) = \varphi(y,z)$.

Thus, the twist kinematics in generic cross-sections is given by the three displacement components

$$
\begin{aligned}
u_x(x,y,z) &= \varphi(y,z), \\
u_y(x,y,z) &= -z\theta_x(x), \\
u_z(x,y,z) &= y\theta_x(x).
\end{aligned}
\tag{9.89}
$$

The coordinate x locates the cross-section and the coordinates y and z indicate the point P in section x. The above displacement components constitute the vectorial field $\mathbf{u}(x,y,z)$ given by

$$
\mathbf{u}(x,y,z) = \left\{ \begin{array}{c} u_x(x,y,z) \\ u_y(x,y,z) \\ u_z(x,y,z) \end{array} \right\} = \left\{ \begin{array}{c} \varphi(y,z) \\ -z\theta_x(x) \\ y\theta_x(x) \end{array} \right\}.
\tag{9.90}
$$

9.8.2 STRAIN MEASURES

Once the general expression for the infinitesimal strain tensor given in (8.114) is known, we can particularize it for the case of generic twist. For this purpose, we substitute the kinematics (9.90) in (8.114) and perform the indicated derivatives. Hence, the infinitesimal strain tensor for the twist of generic sections is given by

$$
[\mathbf{E}] = \begin{bmatrix} 0 & \frac{1}{2}\left(-z\frac{d\theta_x(x)}{dx} + \frac{\partial\varphi(y,z)}{\partial y} \right) & \frac{1}{2}\left(y\frac{d\theta_x(x)}{dx} + \frac{\partial\varphi(y,z)}{\partial z} \right) \\ \frac{1}{2}\left(-z\frac{d\theta_x(x)}{dx} + \frac{\partial\varphi(y,z)}{\partial y} \right) & 0 & 0 \\ \frac{1}{2}\left(y\frac{d\theta_x(x)}{dx} + \frac{\partial\varphi(y,z)}{\partial z} \right) & 0 & 0 \end{bmatrix}.
\tag{9.91}
$$

Using (9.88), the previous tensor is expressed as

$$
[\mathbf{E}] = \begin{bmatrix} 0 & \frac{1}{2}\left(-z + \frac{\partial w(y,z)}{\partial y} \right)\frac{d\theta_x(x)}{dx} & \frac{1}{2}\left(y + \frac{\partial w(y,z)}{\partial z} \right)\frac{d\theta_x(x)}{dx} \\ \frac{1}{2}\left(-z + \frac{\partial w(y,z)}{\partial y} \right)\frac{d\theta_x(x)}{dx} & 0 & 0 \\ \frac{1}{2}\left(y + \frac{\partial w(y,z)}{\partial z} \right)\frac{d\theta_x(x)}{dx} & 0 & 0 \end{bmatrix}.
\tag{9.92}
$$

We can also use the same methodology employed in the previous chapters to obtain the strain components for the generic twist model, as considered below.

Analogous to the case of circular twist, we consider two cross-sections located at x and $x + \Delta x$ from the origin of the reference system, as illustrated in Figure 9.11. Due to the axial displacement, we can imagine that the longitudinal strain component $\varepsilon_{xx}(x)$ is not zero. However, the Saint-Venant hypotheses state that $\varphi(y,z)$ is the same for all cross-sections and consequently $\varepsilon_{xx}(x) = 0$. To verify this fact, we just divide the variation of the axial displacement $\Delta u_x = u_x(x + \Delta x, y, z) - u_x(x, y, z)$ by the distance Δx between the sections and take the limit for Δx going to zero. Thus,

$$\varepsilon_{xx}(x) = \lim_{\Delta x \to 0} \frac{u_x(x + \Delta x, y, z) - u_x(x, y, z)}{\Delta x} = \lim_{\Delta x \to 0} \frac{\varphi(y,z) - \varphi(y,z)}{\Delta x} = 0. \tag{9.93}$$

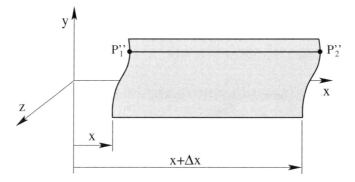

Figure 9.11   Determination of the normal strain component ε_{xx} for the twist model of generic cross-sections.

In order to obtain the total angular strain components $\bar{\gamma}_{xy}$ and $\bar{\gamma}_{xz}$, consider points P_1 and P_2 with coordinates (x,y,z) and $(x + \Delta x, y, z)$ illustrated in Figure 9.12. Due to the rigid rotations θ_{x_1} and θ_{x_2} of sections x and $x + \Delta x$, these points assume positions P_1' and P_2' shown in Figure 9.13(a). To determine the effect of angular variation $\Delta\theta_x = \theta_{x_2} - \theta_{x_1}$ in the transversal displacements u_y and u_z, we just take the difference of the displacements in both cross-sections. However, as illustrated in Figures 9.12 and 9.13(b), the section warping also gives rise to a variation of the transversal displacements u_y and u_z, making points P_1' and P_2' assume the final positions P_1'' and P_2''.

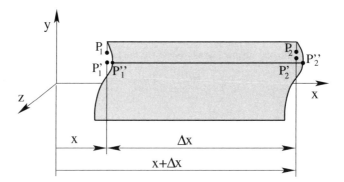

Figure 9.12   Relative displacements of points of two cross-sections.

Hence, the angular strain component $\bar{\gamma}_{xy}$ is the sum of the variation $u_y(x + \Delta x, y, z) - u_y(x, y, z)$, associated to $\Delta\theta_x$, with that one associated to the warping, given by $\varphi(y + \Delta y, z) - \varphi(y, z)$, taking, respectively, the limits for Δx and Δy going to zero. Thus,

$$\bar{\gamma}_{xy} = \lim_{\Delta x \to 0} \frac{u_y(x + \Delta x, y, z) - u_y(x, y, z)}{\Delta x} + \lim_{\Delta y \to 0} \frac{\varphi(y + \Delta y, z) - \varphi(y, z)}{\Delta y}.$$

Using the definition of partial derivative, the expression for the angular strain component $\bar{\gamma}_{xy}$ is

$$\bar{\gamma}_{xy}(x,y,z) = \frac{\partial u_y(x,y,z)}{\partial x} + \frac{\partial \varphi(y,z)}{\partial y}. \tag{9.94}$$

Substituting the displacement component u_y given in (9.90) and the expression (9.88), we have

$$\bar{\gamma}_{xy}(x,y,z) = -z\frac{d\theta_x(x)}{dx} + \frac{\partial \varphi(y,z)}{\partial y} = \left(-z + \frac{\partial w(y,z)}{\partial y}\right)\frac{d\theta_x(x)}{dx}. \tag{9.95}$$

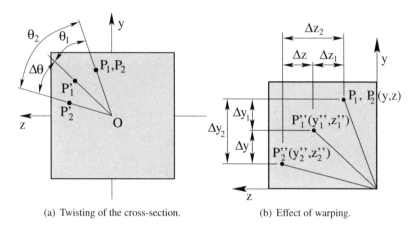

(a) Twisting of the cross-section. (b) Effect of warping.

Figure 9.13 Angular strain components in the twist of generic cross-sections.

Analogously, the strain component $\bar{\gamma}_{xz}$ is the sum of the limits of $u_z(x+\Delta x,y,z) - u_z(x,y,z)$ and $\varphi(y,z+\Delta z) - \varphi(y,z)$, respectively associated to $\Delta\theta_x$ and the warping, for Δx and Δz going to zero, that is,

$$\bar{\gamma}_{xz} = \lim_{\Delta x \to 0} \frac{u_z(x+\Delta x,y,z) - u_z(x,y,z)}{\Delta x} + \lim_{\Delta z \to 0} \frac{\varphi(y,z+\Delta z) - \varphi(y,z)}{\Delta z}.$$

Thus,

$$\bar{\gamma}_{xz}(x,y,z) = \frac{\partial u_z(x,y,z)}{\partial x} + \frac{\partial \varphi(y,z)}{\partial z}. \tag{9.96}$$

Substituting the displacement component u_z given in (9.90) and using (9.88), we have

$$\bar{\gamma}_{xz}(x,y,z) = y\frac{d\theta_x(x,y)}{dx} + \frac{\partial \varphi(y,z)}{\partial y} = \left(y + \frac{\partial w(y,z)}{\partial z}\right)\frac{d\theta_x(x)}{dx}. \tag{9.97}$$

9.8.3 RIGID ACTIONS

To determine the rigid body actions, we should just impose that the strain components $\gamma_{xy}(x)$ and $\gamma_{xz}(x)$ are simultaneously zero, that is,

$$\gamma_{xy}(x,y,z) = \frac{1}{2}\left(-z\frac{d\theta_x(x)}{dx} + \frac{\partial \varphi(y,z)}{\partial y}\right) = 0,$$

$$\gamma_{xz}(x,y,z) = \frac{1}{2}\left(y\frac{d\theta_x(x)}{dx} + \frac{\partial \varphi(y,z)}{\partial z}\right) = 0.$$

Integrating respectively the previous expressions in y and z, we obtain

$$\varphi(y,z) = yz\frac{d\theta_x(x)}{dx} + f(z),$$

$$\varphi(y,z) = -yz\frac{d\theta_x(x)}{dx} + g(y),$$

with $f(z)$ and $g(y)$ functions obtained by the integration procedure. Equating the above expressions, we have

$$yz\frac{d\theta_x(x)}{dx} + f(z) = -yz\frac{d\theta_x(x)}{dx} + g(y),$$

or,

$$2yz\frac{d\theta_x(x)}{dx} + [f(z) - g(y)] = 0.$$

In order to satisfy the above relation, we must have $\dfrac{d\theta_x(x)}{dx} = 0$, implying that $\theta_x(x) = \theta_x$ is constant for all cross-sections. Besides that, functions $f(z)$ and $g(y)$ must be constant, that is, $f(z) = g(y) = C$ with C a constant. Thus, $f(z) - g(y) = 0$, implying that the warping function, and consequently the axial displacement $u_x(x)$, must be equal to this constant, that is, $\varphi(y,z) = C$. Thus, the rigid body actions for the twist in generic sections is composed of a rigid rotation about the x axis and a translation along x. Figure 9.14 illustrates a 90° rotation about x and a constant translation C, that is, $u_x(x) = C$.

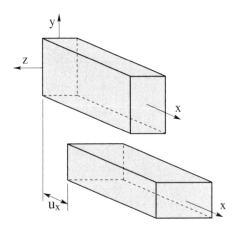

Figure 9.14   Example of a rigid action in generic twist (90° rotation and translation in x).

9.8.4   DETERMINATION OF INTERNAL LOADS

The stress tensor at each point of the beam with generic cross-section subjected to twist is

$$[\mathbf{T}] = \begin{bmatrix} 0 & \tau_{xy} & \tau_{xz} \\ \tau_{xy} & 0 & 0 \\ \tau_{xz} & 0 & 0 \end{bmatrix}. \tag{9.98}$$

Substituting (9.91) and (9.98) in (8.169), we obtain the following expression for the internal work in the volume element dV around any point P:

$$dW_i = \left(\tau_{xy}\gamma_{xy} + \tau_{xz}\gamma_{xz}\right)dV. \tag{9.99}$$

The strain internal work for the whole beam is given by the following volume integral:

$$W_i = \int_V dW_i = \int_V \left[\tau_{xy}(x,y,z)\gamma_{xy}(x,y,z) + \tau_{xz}(x,y,z)\gamma_{xz}(x,y,z)\right]dV. \tag{9.100}$$

Substituting the strain components given in (9.95) and (9.97), we obtain

$$
\begin{aligned}
W_i &= \int_V \left[\tau_{xy}(x,y,z)\left(-z\frac{d\theta_x(x)}{dx} + \frac{\partial\varphi(y,z)}{\partial y}\right) + \tau_{xz}(x,y,z)\left(y\frac{d\theta_x(x)}{dx} + \frac{\partial\varphi(y,z)}{\partial y}\right)\right]dV \\
&= \int_V [-z\tau_{xy}(x,y,z) + y\tau_{xz}(xy,z)]\frac{d\theta_x(x)}{dx}dV + \int_V \left[\tau_{xy}(xy,z)\frac{\partial\varphi(y,z)}{\partial y} + \tau_{xz}(x,y,z)\frac{\partial\varphi(y,z)}{\partial z}\right]dV.
\end{aligned}
$$

The above volume integrals can be rewritten as the product of integrals along the length L and the cross-section area A as

$$W_i = \int_0^L \left[\int_A \left(-z\tau_{xy} + y\tau_{xz} \right) dA \right] \frac{d\theta_x(x)}{dx} dx + \int_0^L \left[\int_A \left(\tau_{xy} \frac{\partial\varphi(y,z)}{\partial y} + \tau_{xz} \frac{\partial\varphi(y,z)}{\partial z} \right) dA \right] dx. \quad (9.101)$$

The first area integral represents the cross-section twisting moment, analogous to the case of circular twist [see equation (4.23)]. The second integral along the length is equal to L, that is, $\int_0^L dx = L$. Thus, the strain internal work is given by

$$W_i = \int_0^L M_x(x) \frac{d\theta_x(x)}{dx} dx + L \int_A \left(\tau_{xy}(x) \frac{\partial\varphi(y,z)}{\partial y} + \tau_{xz}(x) \frac{\partial\varphi(y,z)}{\partial z} \right) dA. \quad (9.102)$$

Note that if no warping of the section occurs, the same expression (4.24) is obtained for the case of circular twist.

It is necessary to analyze the physical meaning of the integrand in the cross-section area of equation (9.102). For a dimensional analysis in the SI, we have

$$\left[\tau_{xy}(x) \frac{\partial\varphi(y,z)}{\partial y} + \tau_{xz}(x) \frac{\partial\varphi(y,z)}{\partial z} \right] = \frac{N}{m^2} \frac{m}{m} + \frac{N}{m^2} \frac{m}{m} = \frac{N}{m^2}.$$

Therefore, the integrand results in a stress, which when integrated in the area gives a resultant of forces in the cross-section. It is noted that this resultant force is inconsistent with the twist model, because due to the kinematics, the internal loads in the section are given by twisting moments. Therefore, the area integral (9.102) must be zero, that is,

$$\int_A \left(\tau_{xy}(x) \frac{\partial\varphi(y,z)}{\partial y} + \tau_{xz}(x) \frac{\partial\varphi(y,z)}{\partial z} \right) dA = 0. \quad (9.103)$$

Hence, the expression for the internal work is reduced to the same one for the circular twist, that is,

$$W_i = \int_0^L M_x(x) \frac{d\theta_x(x)}{dx} dx. \quad (9.104)$$

Integrating the previous expression by parts, we have

$$W_i = -\int_0^L \frac{dM_x(x)}{dx} \theta_x(x) \, dx + M_x(L)\theta_x(L) - M_x(0)\theta_x(0). \quad (9.105)$$

Thus, the compatible internal loads with the kinematics of generic twist are characterized by concentrated twisting moments $M_x(L)$ and $M_x(0)$ at the beam ends, as well as distributed twisting moment $\dfrac{dM_x(x)}{dx}$ along the length. These loads are illustrated in Figure 4.12(a).

However, in the twist of generic sections, we should recall that the area integral (9.103) must be zero. Integrating this equation by parts, we have

$$-\int_A \left[\left(\frac{\partial\tau_{xy}(x,y,z)}{\partial y} + \frac{\partial\tau_{xz}(x,y,z)}{\partial z} \right) \varphi(y,z) \right] dA + \int_{\partial A} \left[\left(\tau_{xy}(x,y,z)n_y + \tau_{xz}(x,y,z)n_z \right) \varphi(y,z) \right] \partial A = 0, \quad (9.106)$$

with ∂A the boundary of the cross-section and n_y and n_z the direction cosines of the normal vector $\mathbf{n}$ in any point P at the section boundary, as illustrated in Figure 9.15(a).

Performing a dimensional analysis of the integrands inside brackets of the previous expression, we have

$$\left(\frac{\partial\tau_{xy}(x,y,z)}{\partial y} + \frac{\partial\tau_{xz}(x,y,z)}{\partial z} \right) \varphi(y,z) = \frac{1}{m} \frac{N}{m^2} m = \frac{N}{m^2},$$

$$\left(\tau_{xy}(x,y,z)n_y + \tau_{xz}(x,y,z)n_z \right) \varphi(y,z) = \frac{N}{m^2} m = \frac{N}{m}.$$

Thus, the first integrand represents a stress and the second represents a force by units of length. Integrating respectively these terms along the area A and the boundary ∂A of the section, we have the resultants of forces in the area and perimeter of the cross-section, which must be zero.

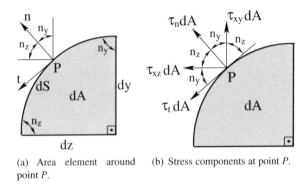

(a) Area element around   (b) Stress components at point P.
point P.

Figure 9.15   Area element around point P on the beam surface.

9.8.5   DETERMINATION OF EXTERNAL LOADS AND EQUILIBRIUM

As done before, we apply the PVW to obtain the compatible external loads with the defined kinematics for the twist of generic sections. The PVW states that the equilibrium of a body in the deformed position can be evaluated by introducing a virtual displacement action.

Thus, suppose that the beam is twisted and is in equilibrium in the deformed configuration. To assess the equilibrium state, we introduce a virtual twist action given by the virtual rotation $\delta\theta_x(x)$ and axial displacement $\delta\varphi(y,z)$.

Following the same procedure considered in Section 4.6, we verify that the twisting moment distribution along the beam is obtained solving the BVP (4.31). Besides that, the compatible external loads with the kinematics of generic sections are illustrated in Figure 4.12(b) and the equilibrium operator $\mathscr{D}^*$ is given in (4.72).

The main difference for the solution of the generic twist problem compared to the circular twist comes from the section warping. In terms of equilibrium, we have that expression (9.106) must be zero for any virtual function $\delta\varphi(y,z)$. Thus,

$$-\int_A \left[\frac{\partial \tau_{xy}(x,y,z)}{\partial y} + \frac{\partial \tau_{xz}(x,y,z)}{\partial z} \right] \delta\varphi(y,z)dA \tag{9.107}$$

$$+\int_{\partial A} [\tau_{xy}(x,y,z)n_y + \tau_{xz}(x,y,z)n_z]\delta\varphi(y,z)\partial A = 0. \tag{9.108}$$

As the virtual function $\delta\varphi(y,z)$ is arbitrary, the above expression is zero when both terms inside brackets are simultaneously zero. This results in the following two-dimensional BVP for the stress components

$$\begin{cases} \dfrac{\partial \tau_{xy}(x,y,z)}{\partial y} + \dfrac{\partial \tau_{xz}(x,y,z)}{\partial z} = 0 \\ \tau_{xy}(x,y,z)n_y + \tau_{xz}(x,y,z)n_z = 0 \end{cases}, \tag{9.109}$$

which must be solved for the posterior determination of the warping function. This BVP is of first order, but it has two unknown functions, that is, $\tau_{xy}(x,y,z)$ and $\tau_{xz}(x,y,z)$.

Figure 9.15(a) illustrates an area element dA of the cross-section and the normal vector $\mathbf{n}$ at the boundary point P; n_y and n_z are the direction cosines with axes y and z, respectively. From Figure 9.15(a), we determine the following relations for n_y and n_z:

$$\begin{aligned} n_y &= -\frac{dy}{ds}, \\ n_z &= \frac{dz}{ds}, \end{aligned} \tag{9.110}$$

with ds a differential element along the boundary of the cross-section.

To interpret the meaning of the boundary condition in (9.109), consider the area element dA of Figure 9.15(b) around point P at the section boundary. The stress components at this point are τ_{xy}, τ_{xz}, τ_t, and τ_n, respectively, in directions y, z, tangential t, and normal n. On the other hand, $\tau_{xy}dA$, $\tau_{xz}dA$, $\tau_t dA$, and $\tau_n dA$ represent the internal forces at point P in the same directions. The decomposition of forces employs the direction cosines n_y and n_z of the normal vector $\mathbf{n}$ at point P. Thus, the following relations are valid:

$$\tau_n dA = (\tau_{xy}dA)n_y + (\tau_{xz}dA)n_z,$$
$$\tau_t dA = (\tau_{xz}dA)n_y - (\tau_{xy}dA)n_z.$$

Simplifying the common dA in the previous expressions, we determine the stress components in the normal and tangent directions, that is,

$$\tau_n = \tau_{xy}n_y + \tau_{xz}n_z, \tag{9.111}$$
$$\tau_t = \tau_{xz}n_y - \tau_{xy}n_z. \tag{9.112}$$

Hence, comparing these equations with the boundary conditions in (9.109), we come to the conclusion that the shear stress on the section boundary is in the tangent direction, as the normal stress τ_n must be zero.

9.8.6 APPLICATION OF THE CONSTITUTIVE EQUATION

Hooke's law for an isotropic linear elastic material states that the shear stress components $\tau_{xy}(x,y,z)$ and $\tau_{xz}(x,y,z)$ are related to the respective angular strain components $\bar{\gamma}_{xy}(x,y,z)$ and $\bar{\gamma}_{xz}(x,y,z)$ by the shear modulus $G(x)$ of section x, that is,

$$\begin{aligned}\tau_{xy}(x,y,z) &= G(x)\bar{\gamma}_{xy}(x,y,z),\\ \tau_{xz}(x,y,z) &= G(x)\bar{\gamma}_{xz}(x,y,z).\end{aligned} \tag{9.113}$$

Applying this constitutive relation in the differential equation of equilibrium in terms of the twisting moment (4.31), we obtain the same differential equation (4.52) in terms of the angle of twist.

The central point here is to solve the BVP (9.109). We observe that the solution gives the functions $\tau_{xy}(x,y,z)$ and $\tau_{xz}(x,y,z)$ describing the stress state at the beam points. To simplify the solution of this BVP, we introduce the stress function $\phi(y,z)$ and write the stress components $\tau_{xy}(x,y,z)$ and $\tau_{xz}(x,y,z)$ in the following way:

$$\tau_{xy}(x,y.z) = \frac{\partial \phi(y,z)}{\partial z}, \tag{9.114}$$

$$\tau_{xz}(x,y,z) = -\frac{\partial \phi(y,z)}{\partial y}. \tag{9.115}$$

Substituting these expressions in the differential equation given in (9.109), we have

$$\frac{\partial \tau_{xy}}{\partial y} + \frac{\partial \tau_{xz}}{\partial z} = \frac{\partial}{\partial y}\left[\frac{\partial \phi(y,z)}{\partial z}\right] + \frac{\partial}{\partial z}\left[-\frac{\partial \phi(y,z)}{\partial y}\right] = 0,$$

that is, the stress components defined in (9.114) and (9.115) in terms of the stress function $\phi(y,z)$ satisfy the differential equation in (9.109).

On the other hand, substituting (9.114) and (9.115) and the angular strain components (9.95) and (9.97) in (9.113), we have

$$\frac{\partial \phi(y,z)}{\partial z} = G(x)\left(\frac{\partial \varphi(y,z)}{\partial y} - z\frac{d\theta_x(x)}{dx}\right),$$
$$-\frac{\partial \phi(y,z)}{\partial y} = G(x)\left(\frac{\partial \varphi(y,z)}{\partial z} + y\frac{d\theta_x(x)}{dx}\right).$$

Taking the derivatives of the above expressions, respectively, in z and y, we obtain

$$\frac{\partial^2 \phi(y,z)}{\partial z^2} = G(x)\left(\frac{\partial \varphi(y,z)}{\partial y \partial z} - \frac{d\theta_x(x)}{dx}\right),$$
$$-\frac{\partial^2 \phi(y,z)}{\partial y^2} = G(x)\left(\frac{\partial \varphi(y,z)}{\partial y \partial y} - \frac{d\theta_x(x)}{dx}\right).$$

Subtracting these equations, we eliminate function $\varphi(y,z)$. Thus,

$$\frac{\partial^2 \phi(y,z)}{\partial y^2} + \frac{\partial^2 \phi(y,z)}{\partial z^2} = -2G(x)\frac{d\theta_x(x)}{dx}. \tag{9.116}$$

Denoting

$$F(x) = -2G(x)\frac{d\theta_x(x)}{dx}, \tag{9.117}$$

we have the second-order differential equation in terms of the stress function

$$\frac{\partial^2 \phi(y,z)}{\partial y^2} + \frac{\partial^2 \phi(y,z)}{\partial z^2} = F(x). \tag{9.118}$$

Substituting (9.114) and (9.115) in the boundary condition given in (9.109), we have

$$\tau_{xy}(x,y,z)n_y + \tau_{xz}(x,y,z)n_z = \frac{\partial \phi(y,z)}{\partial z}n_y - \frac{\partial \phi(y,z)}{\partial y}n_z = 0. \tag{9.119}$$

Using the chain rule, we can take the derivative of the stress function $\phi(y,z)$ along the section boundary, employing the differential element ds, as

$$\frac{\partial \phi(y(s),z(s))}{\partial s} = \frac{\partial \phi(y,z)}{\partial y}\frac{dy(s)}{ds} + \frac{\partial \phi(y,z)}{\partial z}\frac{dz(s)}{ds}. \tag{9.120}$$

Substituting (9.110), the above derivative is rewritten as

$$\frac{\partial \phi(y,z)}{\partial s} = \frac{\partial \phi(y,z)}{\partial y}n_y - \frac{\partial \phi(y,z)}{\partial z}n_z. \tag{9.121}$$

From boundary condition (9.119), the variation of $\phi(y,z)$ along the section boundary must be zero, that is,

$$\frac{\partial \phi(y,z)}{\partial s} = 0 \tag{9.122}$$

Therefore, the stress function $\phi(y,z)$ is constant along the boundary of the cross-section. For solid shafts, this constant can be arbitrarily chosen and taken here as zero [51].

Hence, the stress distribution for an arbitrary section subjected to twisting consists in the determination of the stress function $\phi(y,z)$ which satisfies the differential equation (9.118) and is zero on the section boundary.

Again the introduction of the stress function transformed the first-order BVP with unknowns $\tau_{xy}(x,y,z)$ and $\tau_{xz}(x,y,z)$, in the second-order BVP in terms of the scalar stress function $\phi(y,z)$. The BVP in terms of $\phi(y,z)$ can be resumed to

$$\begin{cases} \dfrac{\partial^2 \phi(y,z)}{\partial y^2} + \dfrac{\partial^2 \phi(y,z)}{\partial z^2} = F(x) \\ \phi(y,z)|_{\partial A} = 0 \end{cases}. \tag{9.123}$$

We can also express the twisting moment $M_x(x)$ in terms of the stress function $\phi(y,z)$. For this purpose, we just substitute (9.114) and (9.115) in (4.23). Thus,

$$M_x(x) = -\int_A \left(z\frac{\partial \phi(y,z)}{\partial z} + y\frac{\partial \phi(y,z)}{\partial y}\right)dydz. \tag{9.124}$$

Integrating the previous expression by parts, we have:

$$M_x(x) = \int_A \left[\phi(y,z)\frac{dz}{dz} + \phi(y,z)\frac{dy}{dy}\right]dA + \int_{\partial A}(z+y)\,\phi(y,z)\partial A.$$

As $\phi(y,z)$ is zero along the section boundary ∂A, the above expression reduces to

$$M_x(x) = 2\int_A \phi(y,z)dA. \tag{9.125}$$

The integral of $\phi(y,z)$ along the area A represents the volume delimited by the stress function and the cross-section. Hence, the twisting moment in each cross-section is proportional to the volume delimited by the stress function.

Substituting the distortion components given in (9.95) and (9.97) in Hooke's law (9.113), we have

$$\tau_{xy}(x,y,z) = G(x)\frac{d\theta_x(x)}{dx}\left(-z+\frac{\partial w(y,z)}{\partial y}\right), \tag{9.126}$$

$$\tau_{xz}(x,y,z) = G(x)\frac{d\theta_x(x)}{dx}\left(y+\frac{\partial w(y,z)}{\partial z}\right). \tag{9.127}$$

From these expressions, the twisting moment is written as

$$M_x(x) = G(x)\frac{d\theta_x(x)}{dx}\int_A\left(\frac{\partial w(y,z)}{\partial y}z - \frac{\partial w(y,z)}{\partial z}y + y^2 + z^2\right)dA. \tag{9.128}$$

The above area integral depends only on the cross-section geometry and is called twist moment of area. If the warping function is zero, the integral reduces to the polar moment of area of the cross-section. Thus, the above expression is rewritten as

$$M_x(x) = G(x)I_t(x)\frac{d\theta_x(x)}{dx} \tag{9.129}$$

with

$$I_t(x) = \int_A\left(\frac{\partial w(y,z)}{\partial y}z - \frac{\partial w(y,z)}{\partial z}y + y^2 + z^2\right)dA. \tag{9.130}$$

Consequently,

$$\frac{d\theta_x(x)}{dx} = \frac{M_x(x)}{G(x)I_t(x)}. \tag{9.131}$$

9.8.7 SHEAR STRESS DISTRIBUTION IN ELLIPTICAL CROSS-SECTION

Figure 9.16(a) illustrates an elliptical cross-section with major and minor radii a and b, respectively. The boundary of the cross-section is described by the ellipse equation

$$\frac{y^2}{b^2} + \frac{z^2}{a^2} = 1. \tag{9.132}$$

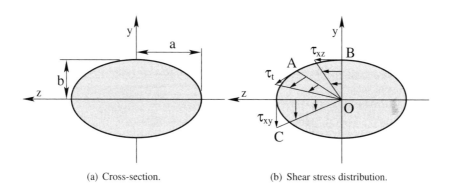

(a) Cross-section. (b) Shear stress distribution.

Figure 9.16 Shear stress distribution in elliptical cross-section.

The differential equation and boundary condition given in (9.109) are satisfied for the following stress function $\phi(y,z)$:

$$\phi(y,z) = m\left(\frac{y^2}{b^2} + \frac{z^2}{a^2} - 1\right), \tag{9.133}$$

with m a constant. Substituting the previous expression in the differential equation given in (9.123) and developing the indicated derivatives, we obtain the following relation for the constant m

$$m = \frac{a^2 b^2}{2(a^2 + b^2)} F(x). \tag{9.134}$$

Substituting this expression in (9.133), we have

$$\phi(y,z) = \frac{a^2 b^2}{2(a^2 + b^2)} \left(\frac{y^2}{b^2} + \frac{z^2}{a^2} - 1 \right) F(x). \tag{9.135}$$

We can obtain an expression for $F(x)$ in terms of the twisting moment $M_x(x)$. For this purpose, we just substitute (9.135) in (9.125). Thus,

$$\begin{aligned}
M_x(x) &= 2 \int_A \phi(y,z)\, dA \\
&= \frac{a^2 b^2}{(a^2 + b^2)} F(x) \left[\frac{1}{b^2} \int_A y^2\, dA + \frac{1}{a^2} \int_A z^2\, dA - \int_A dA \right].
\end{aligned} \tag{9.136}$$

The two first integrals are the second moments of area $I_z(x) = \int_A y^2\, dA$ and $I_y(x) = \int_A z^2\, dA$ of the cross-section, respectively, in relation to the z and y axes of the adopted reference system. The last integral gives the area A of the cross-section, that is, $A(x) = \int_A dA$. For an elliptical section, these geometrical properties are

$$\begin{aligned}
I_y(x) &= \frac{\pi a^3 b}{4}, \\
I_z(x) &= \frac{\pi a b^3}{4}, \\
A(x) &= \pi a b.
\end{aligned} \tag{9.137}$$

The polar moment of area for an elliptical section is given by

$$I_p = I_y + I_z = \frac{\pi}{4} a b (a^2 + b^2). \tag{9.138}$$

Substituting the above relations in (9.136), we have

$$M_x(x) = -\frac{\pi a^3 b^3}{2(a^2 + b^2)} F(x). \tag{9.139}$$

We determine the expression for $F(x)$ in terms of the twisting moment as

$$F(x) = -\frac{2(a^2 + b^2)}{\pi a^3 b^3} M_x(x). \tag{9.140}$$

On the other hand, substituting $F(x)$ in (9.135), we have the following expression of the stress function in terms of the twisting moment $M_x(x)$:

$$\phi(y,z) = -\frac{M_x(x)}{\pi a b} \left(\frac{y^2}{b^2} + \frac{z^2}{a^2} - 1 \right). \tag{9.141}$$

The expression of the twisting moment $M_x(x)$ is obtained integrating the differential equation (4.31). Once the stress function $\phi(y,z)$ is determined, we obtain the stress components τ_{xy} and τ_{xz}, substituting (9.141) in (9.114) and (9.115) and doing the indicated derivatives. Thus,

$$\tau_{xy}(x,y,z) = \frac{2M_x(x)}{\pi a^3 b} z = \frac{M_x(x)}{2 I_y(x)} z, \tag{9.142}$$

$$\tau_{xz}(x,y,z) = -\frac{2M_x(x)}{\pi a b^3} y = -\frac{M_x(x)}{2 I_z(x)} y. \tag{9.143}$$

The ratio $\dfrac{\tau_{xy}}{\tau_{xz}}$ between the stress components is proportional to $\dfrac{z}{y}$ and constant along a radial line OA as indicated in Figure 9.16(b). Hence, the resulting shear stress $\tau_t = \tau_{xy} + \tau_{xz}$ is along the tangent direction to

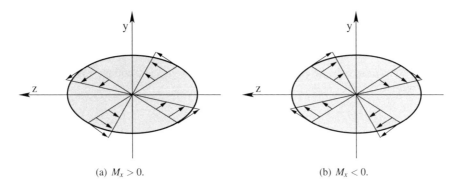

(a) $M_x > 0$.   (b) $M_x < 0$.

Figure 9.17 Shear stress distribution for an elliptical cross-section.

the boundary on point A. Along line OB, the component τ_{xy} is zero and the tangent stress τ_t is equal to τ_{xz}. Analogously, along line OC, we have $\tau_{xz} = 0$ and $\tau_t = \tau_{xy}$. Figure 9.17 illustrates the resulting shear stress distribution τ_t for positive and negative twisting moments.

From Figures 9.16(b) and 9.17, we observe that the maximum shear stress τ_t^{max} occurs at the cross-section boundary and at the minor axis end of the ellipse, for which $\tau_t = \tau_{xz}$ and $y = b$. Thus, from (9.143)

$$\tau_t^{max}(x) = \frac{2M_x(x)}{\pi ab^3}b = \frac{2M_x(x)}{\pi ab^2}. \tag{9.144}$$

Note that for $a = b = \dfrac{d}{2}$, the previous expression reduces to equation (4.58) for the maximum shear stress at the circular cross-section.

We can rewrite (9.117) as

$$\frac{d\theta_x(x)}{dx} = -\frac{F(x)}{2G(x)}.$$

Substituting (9.140), we have

$$\frac{d\theta_x(x)}{dx} = \frac{(a^2 + b^2)}{\pi a^3 b^3}\frac{M_x(x)}{G(x)}. \tag{9.145}$$

Comparing the previous expression with equation (9.131), we conclude that the twist moment of area for an elliptical cross-section is given by

$$I_t = \frac{\pi a^3 b^3}{a^2 + b^2}. \tag{9.146}$$

Integrating expression (9.145) along the beam length, we have the variation $\Delta\theta_x$ of the angle of twist between the beam ends, that is,

$$\Delta\theta_x = \frac{(a^2 + b^2)}{\pi a^3 b^3}\int_0^L \frac{M_x(x)}{G(x)}dx \tag{9.147}$$

If the beam is subjected to a concentrated torque T at the ends, the twisting moment is equal to the applied torque T. Assuming that the shear modulus $G(x)$ is constant, that is, $G(x) = G$, the above expression reduces to

$$\Delta\theta_x = \frac{(a^2 + b^2)}{\pi a^3 b^3 G}TL. \tag{9.148}$$

We can rewrite (9.148) as

$$T = k_t\Delta\theta_x, \tag{9.149}$$

with the torsional stiffness given by

$$k_t = \frac{\pi a^3 b^3 G}{(a^2 + b^2)L} = \frac{GI_t}{L}. \tag{9.150}$$

In order to determine the warping function $w(y, z)$, we substitute (9.142) and (9.145) in (9.126) and integrate in y, obtaining

$$w(y, z) = -\frac{a^2 - b^2}{a^2 + b^2}yz + f(z).$$

Analogously, substituting (9.143) and (9.145) in (9.127) and integrating in z we have

$$w(y,z) = -\frac{a^2 - b^2}{a^2 + b^2} yz + g(y).$$

As the previous expressions must be equal, we have that $f(z) = g(y) = 0$ and the warping function for the elliptical cross-sections is

$$w(y,z) = -\frac{(a^2 - b^2)}{a^2 + b^2} yz. \tag{9.151}$$

9.8.8 ANALOGY WITH MEMBRANES

For other types of cross-sections, the analytical solution of the BVP (9.109) is difficult. In these cases, we can employ the membrane analogy introduced by Prandtl. Hence, instead of solving the BVP (9.109) of the twist problem, we consider the solution of the thin membrane model.

To illustrate this model, consider a plate with a hole fixed at the boundaries. We introduce liquid soap over the hole and continuously inflate air to form a bubble or a thin membrane, as illustrated in Figure 9.18(a). The membrane will then be subjected to a constant distributed load of intensity q and a uniform surface traction per unit of boundary length denoted by S, as illustrated in Figure 9.18(b). The displacement $u_z(y,z)$ of the membrane is given by [51]

$$\frac{\partial^2 u_z(y,z)}{\partial y^2} + \frac{\partial^2 u_z(y,z)}{\partial z^2} = -\frac{q}{S}. \tag{9.152}$$

As the membrane is fixed at the ends, the displacement $u_z(y,z)$ must be zero on the boundary, that is,

$$u_z(y,z) = 0. \tag{9.153}$$

Expressions (9.152) and (9.153) represent the BVP for a thin membrane fixed on the boundary.

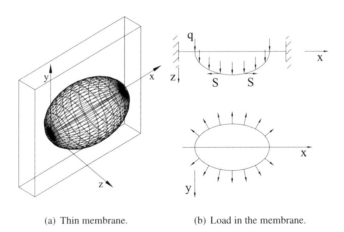

(a) Thin membrane.   (b) Load in the membrane.

Figure 9.18 Membrane analogy.

We observe that the BVP of the thin membrane model is identical to the problem of twist (9.109). Hence, we can study the twist of generic sections making an analogy with a thin membrane, whose boundary is identical to the cross-section under consideration.

From this analogy, the following conclusions are valid [51, 42]:

- The shear stress at any point is proportional to the slope of the deformed membrane at the same point.
- The direction of the shear stress at a point forms a straight angle with the slope of the membrane at the same point.
- Two times the volume defined by the membrane and the x axis is proportional to the twisting moment at the section [see equation (9.125)].

Employing the membrane analogy, we determine the expressions for the maximum shear stress $\tau^{max}(x)$ and the angle of twist $\theta_x(x)$ at section x of a beam with length L and rectangular section of base b and height h (see Figure 9.19) as in [51]

$$\tau^{max}(x) = \frac{M_x(x)}{C_1 hb^2}, \tag{9.154}$$

$$\theta_x(x) = \frac{M_x(x)L}{C_2 hb^3 G(x)}. \tag{9.155}$$

The constants C_1 and C_2 depend on relation $\dfrac{h}{b}$ and are given in Table 9.1.

$\dfrac{h}{b}$	C_1	C_2
1.0	0.208	0.141
1.2	0.219	0.166
1.5	0.231	0.196
2.0	0.246	0.229
2.5	0.258	0.249
3.0	0.267	0.263
4.0	0.282	0.281
5.0	0.291	0.291
10.0	0.312	0.312
∞	0.333	0.333

Table 9.1
Coefficients for the twist of rectangular sections (adapted from [51]).

Example 9.11 *Consider a steel beam with 2 m of length and rectangular cross-section with base 10 cm and height 20 cm. The beam is fixed in the left end and is subjected to the torque 1000 Nm at the right end. Determine the maximum shear stress and the angle of twist of the beam.*

From Table 9.1, $C_1 = 0.246$ and $C_2 = 0.229$ for $\dfrac{h}{b} = 2$. Thus, the maximum shear stress and angle of twist are obtained from (9.154) and (9.155) as

$$\tau^{max} = \frac{1000}{(0.246)(0.2)(0.1^2)} = 20.33 \text{ MPa},$$

$$\theta_x = \frac{(1000)(2)}{(0.229)(0.2)(0.1^3)(80 \times 10^9)} = 5.46 \times 10^{-4} \text{ rad}.$$

$\square$

For slender cross-sections, we have $\dfrac{h}{b} \to \infty$. This case is used to apply expressions (9.154) and (9.155) in cross-sections with thin walls as the ones illustrated in Figure 9.20. For this purpose, we just take the height h in equation (9.154) and (9.155) as the length of the cross-section under consideration. We observe that for sections with reentrant angles, there are stress concentrations that depend on the fillet radius. In these cases, the previous equations cannot be applied to calculate the stresses at the radii. A discussion of these cases is presented in [51].

9.8.9 SUMMARY OF THE VARIATIONAL FORMULATION OF GENERIC TORSION

The set $\mathscr{V}$ of the possible kinematic actions for the twist of generic cross-sections consists of the displacement field $\mathbf{u}(x,y,z)$ given in (9.90), with $\theta_x(x)$ a smooth function. Thus,

$$\mathscr{V} = \{\mathbf{u}, u_x = \varphi(y,z), \ u_y = -z\theta_x(x), \ u_z = y\theta_x(x) \text{ and } \theta_x(x) \text{ is a smooth function}\}. \tag{9.156}$$

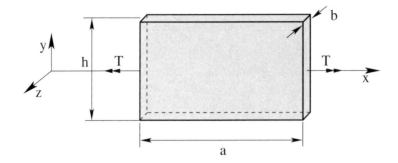

Figure 9.19 Beam with rectangular cross-section subjected to twist.

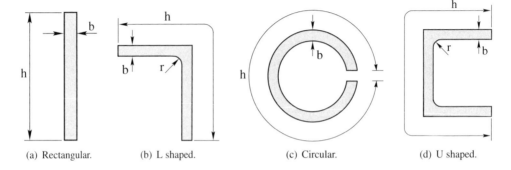

| (a) Rectangular. | (b) L shaped. | (c) Circular. | (d) U shaped. |

Figure 9.20 Membrane analogy for the twist of generic cross-sections.

If no kinematic restrictions are present, all elements $\mathbf{u} \in \mathcal{V}$ are also admissible actions, because there are no physical constraints preventing the twist of the beam ends. When a constraint is present, only the subset $Kin_\mathbf{v}$ of $\mathcal{V}$, given by the functions that respect the kinematic constraints, constitutes the admissible displacement actions.

The space $\mathcal{W}$ of the compatible strain components with the twist of generic sections is constituted by the continuous functions γ_{xy} and γ_{xz}, representing the distortions. We observe that the strain operator $\mathcal{D}$, relating the space $\mathcal{V}$ of the possible displacement actions with the space $\mathcal{W}$ of the compatible strain components with the kinematics defined in $\mathcal{V}$, is indicated in matrix form as

$$\mathcal{D} = \begin{bmatrix} 0 & 0 & 0 \\ \dfrac{\partial}{\partial y} & \dfrac{d}{dx} & 0 \\ \dfrac{\partial}{\partial y} & 0 & \dfrac{d}{dx} \end{bmatrix}, \tag{9.157}$$

in such way that

$$\mathcal{D}: \quad \mathcal{V} \to \mathcal{W}$$

$$\begin{bmatrix} \varphi(y,z) \\ u_y(x,y,z) \\ u_z(x,y,z) \end{bmatrix} \to \frac{1}{2} \begin{bmatrix} 0 & 0 & 0 \\ \dfrac{\partial}{\partial y} & \dfrac{d}{dx} & 0 \\ \dfrac{\partial}{\partial y} & 0 & \dfrac{d}{dx} \end{bmatrix} \begin{bmatrix} \varphi(y,z) \\ u_y(x,y,z) \\ u_z(x,y,z) \end{bmatrix} = \begin{bmatrix} 0 \\ \gamma_{xy}(x,y,z) \\ \gamma_{xz}(x,y,z) \end{bmatrix}. \tag{9.158}$$

Hence, the set $\mathcal{N}(\mathcal{D})$ is composed of the actions given in (9.90), with θ_x and u_x constants. Then we define the set $\mathcal{N}(\mathcal{D})$ as

$$\mathcal{N}(\mathcal{D}) = \{\mathbf{u}; \mathbf{u} \in \mathcal{V} \mid \theta_x \text{ and } u_x \text{ constants}\}. \tag{9.159}$$

The space $\mathscr{W}'$ of internal loads is constituted by the continuous scalar functions $M_x(x)$, characterizing the twisting moment at each cross-section x. The space $\mathscr{V}'$ of the external loads is described by the distributed $m_x(x)$ and concentrated T_0 and T_L torques at the ends.

Figure 9.21 illustrates the variational formulation of twist of generic sections. We notice that if warping is zero, that is, $w(y,z) = 0$, we have the same kinematics of circular twist. Thus, the formulation presented here is valid for generic sections, including the circular section as a particular case.

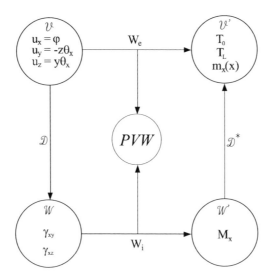

Figure 9.21 Variational formulation of the problem of twist of generic cross-sections.

9.8.10 APPROXIMATED SOLUTION

The BVP in terms of the stress function given in (9.123) is a particular case of the Poisson problem given by

$$\left\{ \begin{array}{l} c\left(\dfrac{\partial^2 u(x,y)}{\partial x^2} + \dfrac{\partial^2 u(x,y)}{\partial y^2}\right) = -f(x,y) \\ + \text{ boundary conditions} \end{array} \right. , \qquad (9.160)$$

with c a coefficient which is assumed here constant.

The previous differential equation can be written in terms of the Laplace operator $\triangle$

$$\triangle = \frac{\partial^2}{\partial x^2} + \frac{\partial^2}{\partial y^2}, \qquad (9.161)$$

as

$$c\triangle u(x,y) = -f(x,y). \qquad (9.162)$$

For the generic twist problem, $c = 1$ and $f(x,y) = -F(x)$; for heat conduction $c = k$ and $f(x,y) = Q_c(x,y)$, with k the thermal conductivity and Q_c the generated heat.

In order to obtain the associated weak form, we multiply the strong form (9.162) by the test function $v(x,y)$ and integrate over the domain Ω, that is,

$$\int_\Omega c\triangle u(x,y)v(x,y)d\Omega = -\int_\Omega f(x,y)v(x,y)d\Omega. \qquad (9.163)$$

Integrating the previous expression by parts, we have

$$\int_\Omega c\nabla u(x,y)\cdot\nabla v(x,y)d\Omega = \int_\Omega f(x,y)v(x,y)d\Omega + c\int_\Gamma \left(\nabla u(x,y)\cdot\mathbf{n}(x,y)\right)v(x,y)d\Gamma, \qquad (9.164)$$

with $\mathbf{n}(x,y)$ the field of normal vectors to the boundary of the domain. Taking

$$q(x,y) = c\nabla u(x,y) \cdot \mathbf{n}(x,y), \tag{9.165}$$

the weak form is given by

$$\int_{\Omega} c\nabla u(x,y) \cdot \nabla v(x,y) d\Omega = \int_{\Omega} f(x,y)v(x,y)d\Omega + \int_{\Gamma} q(x,y)v(x,y)d\Gamma. \tag{9.166}$$

The fields $u(x,y)$ and $v(x,y)$ can be approximated using the Galerkin method by the following linear combinations of N global shape functions $\{\Phi_i(x,y)\}_{i=1}^{N}$ as

$$u(x,y) \approx u_N(x,y) = \sum_{i=1}^{N} \Phi_i(x,y)a_i, \tag{9.167}$$

$$v(x,y) \approx v_N(x,y) = \sum_{i=1}^{N} \Phi_i(x,y)b_i. \tag{9.168}$$

The two previous equations can be expressed in matrix notation as

$$u_N(x,y) = \begin{bmatrix} \Phi_1 & \cdots & \Phi_N \end{bmatrix} \begin{Bmatrix} a_1 \\ \vdots \\ a_N \end{Bmatrix} = [\mathbf{N}]\{\mathbf{a}\}, \tag{9.169}$$

$$v_N(x,y) = \begin{bmatrix} \Phi_1 & \cdots & \Phi_N \end{bmatrix} \begin{Bmatrix} b_1 \\ \vdots \\ b_N \end{Bmatrix} = [\mathbf{N}]\{\mathbf{b}\}, \tag{9.170}$$

with $\{\mathbf{a}\}$ and $\{\mathbf{b}\}$ the vectors with the approximation coefficients for the solution and test functions; $[\mathbf{N}]$ is the matrix with the global shape functions.

Substituting the previous expressions in (9.166) and after some manipulations, we obtain

$$\{\mathbf{b}\}^T \left[\left(\int_{\Omega} [\mathbf{B}]^T [\mathbf{D}][\mathbf{B}]d\Omega \right) \{\mathbf{a}\} - \int_{\Omega} [\mathbf{N}]^T f d\Omega - \int_{\Gamma} [\mathbf{N}]^T q d\Gamma \right] = 0.$$

As the approximation coefficients of the test function are arbitrary, this equation is satisfied if the term inside brackets is zero, resulting in the approximated weak form for the Poisson problem

$$\left(\int_{\Omega} [\mathbf{B}]^T [\mathbf{D}][\mathbf{B}]d\Omega \right) \{\mathbf{a}\} = \int_{\Omega} [\mathbf{N}]^T f d\Omega + \int_{\Gamma} [\mathbf{N}]^T q d\Gamma. \tag{9.171}$$

In compact notation,

$$[\mathbf{K}]\{\mathbf{a}\} = \{\mathbf{f}\}, \tag{9.172}$$

with $[\mathbf{K}]$ and $\{\mathbf{f}\}$, respectively, the global stiffness matrix and global vector of equivalent body and surface loads, given by

$$[\mathbf{K}] = \int_{\Omega} [\mathbf{B}]^T [\mathbf{D}][\mathbf{B}]d\Omega, \tag{9.173}$$

$$\{\mathbf{f}\} = \int_{\Omega} [\mathbf{N}]^T f d\Omega + \int_{\Gamma} [\mathbf{N}]^T q d\Gamma. \tag{9.174}$$

Matrix $[\mathbf{B}]$ is given by

$$[\mathbf{B}] = [\mathbf{L}][\mathbf{N}] = [[\mathbf{B}_1] \mid [\mathbf{B}_2] \mid \cdots \mid [\mathbf{B}_N]], \tag{9.175}$$

with $[\mathbf{B}_i]$ the matrix associated to each global shape function i and

$$[\mathbf{B}_i] = \begin{bmatrix} \Phi_{i,x} & 0 \\ 0 & \Phi_{i,y} \end{bmatrix}. \tag{9.176}$$

Matrix $[\mathbf{D}]$ contains the c coefficient only, that is,

$$[\mathbf{D}] = c. \tag{9.177}$$

We can partition the body in a finite element mesh. The approximations of the solution and test fields for each element e are expressed as the following linear combinations of the N_e element shape functions $\{\Phi_i^{(e)}(x,y)\}_{i=1}^{N_e}$:

$$u_{N_e}(x,y) = \sum_{i=1}^{N_e} \Phi_i^{(e)}(x,y) a_i^{(e)} = [\mathbf{N}^{(e)}]\{\mathbf{a}^{(e)}\}, \tag{9.178}$$

$$v_{N_e}(x,y) = \sum_{i=1}^{N_e} \Phi_i^{(e)}(x,y) b_i^{(e)} = [\mathbf{N}^{(e)}]\{\mathbf{b}^{(e)}\}, \tag{9.179}$$

with $\{\mathbf{a}^{(e)}\}$ and $\{\mathbf{b}^{(e)}\}$ the vectors of approximation coefficients of element e; $[\mathbf{N}^{(e)}]$ is the matrix of element shape functions given by

$$[\mathbf{N}^{(e)}] = \left[\; \Phi_1^{(e)} \quad \cdots \quad \Phi_{N_e}^{(e)} \;\right]. \tag{9.180}$$

The weak form for element e is obtained in an analogous way as (9.171)

$$\left(\int_{\Omega^{(e)}} [\mathbf{B}^{(e)}]^T [\mathbf{D}][\mathbf{B}^{(e)}] d\Omega^{(e)}\right)\{\mathbf{a}^{(e)}\} = \int_{\Omega^{(e)}} [\mathbf{N}^{(e)}]^T f d\Omega^{(e)} + \int_{\Gamma^{(e)}} [\mathbf{N}^{(e)}]^T q d\Gamma^{(e)}, \tag{9.181}$$

In a summarized form,

$$[\mathbf{K}^{(e)}]\{\mathbf{a}^{(e)}\} = \{\mathbf{f}^{(e)}\}, \tag{9.182}$$

with $[\mathbf{K}^{(e)}]$ and $\{\mathbf{f}^{(e)}\}$, respectively, the element stiffness matrix and the equivalent element body and surface load vector, given by

$$[\mathbf{K}^{(e)}] = \int_{\Omega^{(e)}} [\mathbf{B}^{(e)}]^T [\mathbf{D}][\mathbf{B}^{(e)}] d\Omega^{(e)}, \tag{9.183}$$

$$\{\mathbf{f}^{(e)}\} = \int_{\Omega^{(e)}} [\mathbf{N}^{(e)}]^T f d\Omega^{(e)} + \int_{\Gamma^{(e)}} [\mathbf{N}^{(e)}]^T q d\Gamma^{(e)}. \tag{9.184}$$

The element strain-displacement matrix $[\mathbf{B}^{(e)}]$ is

$$[\mathbf{B}^{(e)}] = [\mathbf{L}][\mathbf{N}^{(e)}] = \left[[\mathbf{B}_1^{(e)}] \mid [\mathbf{B}_2^{(e)}] \mid \ldots \mid [\mathbf{B}_{N_e}^{(e)}]\right], \tag{9.185}$$

with $[\mathbf{B}_i^{(e)}]$ the matrix associated to each shape function i of the element, expressed by

$$[\mathbf{B}_i^{(e)}] = \begin{bmatrix} \phi_{i,x}^{(e)} & 0 \\ 0 & \phi_{i,y}^{(e)} \end{bmatrix}. \tag{9.186}$$

The weak form for element e in terms of the local reference system $\xi_1 \times \xi_2$ is given by

$$\left(\int_{-1}^{1}\int_{-1}^{1} [\mathbf{B}^{(e)}]^T [\mathbf{D}][\mathbf{B}^{(e)}] \det[\mathbf{J}] d\xi_1 d\xi_2\right)\{\mathbf{a}^{(e)}\} = \int_{-1}^{1}\int_{-1}^{1} [\mathbf{N}^{(e)}]^T f \det[\mathbf{J}] d\xi_1 d\xi_2$$
$$+ \int_{-1}^{1} [\mathbf{N}^{(e)}]^T q \det[\mathbf{J}_b] d\Gamma^{(e)}, \tag{9.187}$$

with $[\mathbf{J}]$ and $[\mathbf{J}_b]$ the Jacobian matrices associated to the area and perimeter of the element.

The element stiffness matrix and equivalent body and surface load vector in the local system are

$$[\mathbf{K}^{(e)}] = \int_{-1}^{1}\int_{-1}^{1} [\mathbf{B}^{(e)}]^T [\mathbf{D}][\mathbf{B}^{(e)}] \det[\mathbf{J}] d\xi_1 d\xi_2, \tag{9.188}$$

$$\{\mathbf{f}^{(e)}\} = \int_{-1}^{1}\int_{-1}^{1} [\mathbf{N}^{(e)}]^T f \det[\mathbf{J}] d\xi_1 d\xi_2 + \int_{-1}^{1} [\mathbf{N}^{(e)}]^T q \det[\mathbf{J}_b] d\Gamma^{(e)}. \tag{9.189}$$

Example 9.12 *Consider the same mesh of Figure 9.7 to study the Poisson problem with the fabricated solution* $u(x,y) = xy(x-4)(y-1)$. *We adopt* $c = 1$.

Substituting the solution given in (9.162), we obtain $f = -2x(x-4) - 2y(y-1)$, *which is applied as a body load in the domain. The solution is zero at all domain edges.*

Using a second-order approximation with Lagrange polynomials, the error of the approximated solution is zero.

Files poisson.fem and poisson.def specify the necessary parameters to analyze this example in the $(hp)^2$ *FEM program.*

□

9.9 MULTIDIMENSIONAL NUMERICAL INTEGRATION

The coefficients of the element matrices and vectors obtained from the approximations of the two- and three-dimensional models considered in this book are calculated conveniently using numerical integration procedures. In this sense, the Gauss quadrature presented in Chapter 4 can be extended to integrate functions with two and three variables.

Initially consider the function $f(\xi_1,\xi_2)$ defined in a square element in the local system of coordinates $\xi_1 \times \xi_2$. We want to calculate the integral of $f(\xi_1,\xi_2)$, that is,

$$I = \int_{-1}^{1} \int_{-1}^{1} f(\xi_1,\xi_2)\, d\xi_1\, d\xi_2. \tag{9.190}$$

This integral can be calculated numerically considering first only the integral in ξ_2 and keeping ξ_1 constant. Taking n_2 points in the ξ_2 direction and applying (4.160), we have

$$I_{\xi_2} = \int_{-1}^{1} f(\xi_1,\xi_2)\, d\xi_2 = \sum_{j=1}^{n_2} w_{2_j} f(\xi_1,\xi_{2_j}) = g(\xi_1), \tag{9.191}$$

where $g(\xi_1)$ is a function of ξ_1. For n_1 integration points in the ξ_1 direction, the integral of $g(\xi_1)$ is given by

$$I = \int_{-1}^{1} g(\xi_1)\, d\xi_1 = \sum_{i=1}^{n_1} w_{1_i} g(\xi_{1_i}). \tag{9.192}$$

Substituting (9.191) in (9.192) we have

$$I = \int_{-1}^{1} \int_{-1}^{1} g(\xi_1)\, d\xi_1 = \sum_{i=1}^{n_1} w_{1_i} \left[\sum_{j=1}^{n_2} w_{2_j} f(\xi_{1_i},\xi_{2_j}) \right],$$

or,

$$I = \sum_{i=1}^{n_1} \sum_{j=1}^{n_2} w_{1_i} w_{2_j} f(\xi_{1_i},\xi_{2_j}). \tag{9.193}$$

The double summation in (9.193) can be written as a simple one in the following way:

$$I = \sum_{l=1}^{N_{int}} w_l f(\xi_{1_l},\xi_{2_l}), \tag{9.194}$$

where $N_{int} = n_1 n_2$ and $w_l = w_{1_i} w_{2_j}$ $(i = 1,\ldots,n_1; j = 1,\ldots,n_2)$.

The integration points for squares are obtained by the tensor product of coordinates in directions ξ_1 and ξ_2. The weights are determined by the product of the respective weights of the one-dimensional quadrature. Figure 9.22 illustrates the distribution of the Gauss-Legendre integration points for squares.

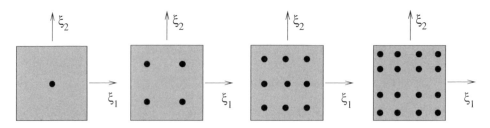

Figure 9.22 Gauss-Legendre integration points for square.

Example 9.13 *We want to calculate numerically the following integral:*

$$I = \int_{-1}^{1} \int_{-1}^{1} (\xi_1^2 + \xi_2^2)\, d\xi_1 d\xi_2.$$

The number of Gauss-Legendre quadrature points in each direction is calculated as

$$n_1 = n_2 = \frac{1}{2}(P+1).$$

As the order for each direction is $P = 2$, we obtain $n_1 = n_2 = 2$.

For the Gauss-Radau-Legendre and Gauss-Lobatto-Legendre quadratures, the number of points in each direction are calculated, respectively, as

$$n_1 = n_2 = \frac{1}{2}(P+2) = 3,$$

$$n_1 = n_2 = \frac{1}{2}(P+3) = 3.$$

The squareni.m program implements the solution of this example for different types of quadratures and plots the coordinates of the integration points for the local square. The analytical and numerical results are coincident and equal to 2.67.

□

Consider now the function $f(\xi_1, \xi_2, \xi_3)$ defined in the hexahedron according to the local reference system. We wish to calculate the following integral:

$$I = \int_{-1}^{1} \int_{-1}^{1} \int_{-1}^{1} f(\xi_1, \xi_2, \xi_3) \, d\xi_1 \, d\xi_2 \, d\xi_3. \tag{9.195}$$

The procedure is analogous to the previous case. Thus, taking n_1, n_2, and n_3 points in directions ξ_1, ξ_2, and ξ_3, respectively, we can integrate (9.195) as

$$I = \sum_{i=1}^{n_1} \sum_{j=1}^{n_2} \sum_{k=1}^{n_3} w_{1_i} w_{2_j} w_{3_k} f(\xi_{1_i}, \xi_{2_j}, \xi_{3_k}). \tag{9.196}$$

The previous expression can be written as a simple summation in the following way:

$$I = \sum_{l=1}^{N_{int}} w_l f(\xi_{1_l}, \xi_{2_l}, \xi_{3_l}), \tag{9.197}$$

where $N_{int} = n_1 n_2 n_3$ and $w_l = w_{1_i} w_{2_j} w_{3_k}$ $(i = 1, \ldots, n_1; j = 1, \ldots, n_2; k = 1, \ldots, n_3)$. The coordinates of the integration points are obtained by the tensor product of the one-dimensional coordinates, analogous to squares.

The integration of the function $h(L_1, L_2, L_3)$ over the area A of the triangle with straight edges is given by [54, 18]

$$I = \int_A h(L_1, L_2, L_3) dA = 2A \int_0^1 \int_0^{1-L_2} \tilde{h}(L_2, L_3) dA, \tag{9.198}$$

where $dA = dL_2 dL_3$ and $L_1 = 1 - L_2 - L_3$.

The numerical integration of the previous equation using the Gauss-Legendre quadrature is expressed by [54, 18]

$$I = 2A \int_A \tilde{h}(L_2, L_3) dA = 2A \sum_{l=1}^{N_{int}} w_l \tilde{h}(L_{2_l}, L_{3_l}), \tag{9.199}$$

where N_{int}, (L_{2_l}, L_{3_l}), and w_l are, respectively, the number, coordinates, and weights of the integration points.

The mapping of the triangle to the square, illustrated in Figure 9.23, is given by [21, 35]

$$\begin{cases} L_2 = \frac{1}{2}(1 + \xi_1), \\ L_3 = \frac{1}{4}(1 - \xi_1)(1 + \xi_2), \\ L_1 = 1 - L_2 - L_3 = \frac{1}{4}(1 - \xi_1)(1 - \xi_2). \end{cases} \tag{9.200}$$

The Jacobian of the mapping is $\frac{1}{8}(1 - \xi_1)$.

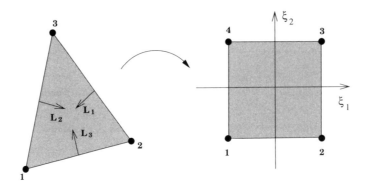

Figure 9.23 Mapping of local square to local triangle.

Using this mapping, the integral in equation (9.198) can be rewritten as

$$I = 2A \int_0^1 \int_0^{1-L_2} \tilde{h}(L_2,L_3)dL_3dL_2 = \frac{A}{4} \int_{-1}^1 \int_{-1}^1 \hat{h}(\xi_1,\xi_2)(1-\xi_1)d\xi_1d\xi_2$$

and the numerical integration is

$$
\begin{aligned}
I &= \frac{A}{4} \int_{-1}^1 \int_{-1}^1 \hat{h}(\xi_1,\xi_2)(1-\xi_1)d\xi_1d\xi_2 \\
&= \frac{A}{4} \sum_{i=1}^{n_1} \sum_{j=1}^{n_2} w_{1_i} w_{2_j} (1-\xi_{1_i}) \hat{h}(\xi_{1_i},\xi_{2_j}),
\end{aligned}
\tag{9.201}
$$

where n_1 and n_2 are the number of integration points in the directions ξ_1 and ξ_2, (ξ_{1_i},ξ_{2_j}) are the coordinates and w_{1_i} and w_{2_j} are the respective weights.

For $\alpha = 1$ and $\beta = 0$, it is possible to consider the Jacobian $(1-\xi_1)$ of the bidirectional mapping in the weight function of the Gauss-Jacobi quadrature. Hence, the Gauss-Jacobi and Gauss-Legendre quadratures can be used, respectively, in the directions ξ_1 and ξ_2 as

$$I = \frac{A}{4} \sum_{i=1}^{n_1} \sum_{j=1}^{n_2} w_{1_i}^{1,0} w_{2_j}^{0,0} \hat{h}(\xi_{1_i},\xi_{2_j}). \tag{9.202}$$

Based on the previous relation, it is not necessary to consider the Jacobian $(1-\xi_1)$ to determine the integrand order in the ξ_1 direction, because it is already included in the weight $w_{1_i}^{1,0}$.

Example 9.14 *We want to calculate the following integral numerically:*

$$I = 1000 \int_0^1 \int_0^{1-L_2} L_1 L_2 L_3 dL_1 dL_2.$$

The number of Gauss-Legendre quadrature points for each direction is calculated as

$$n_1 = n_2 = \frac{1}{2}(P+1) = 2.$$

Program triangleni.m implements the solution of this example for different types of quadratures and plots the coordinates of the integration points for the local triangle. The analytical and numerical results are coincident and equal to 8.33.

□

The integration of function $h(L_1,L_2,L_3,L_4)$ over the volume V of the tetrahedron with straight edges is given by [54, 18]

$$I = \int_V h(L_1,L_2,L_3,L_4)dV.$$

where $dV = (8V)dL_2dL_3dL_4$ and $L_1 = 1 - L_2 - L_3 - L_4$. Thus,

$$I = 8V \int_0^1 \int_0^{1-L_4} \int_0^{1-L_3-L_4} \tilde{h}(L_2, L_3, L_4)dL_2dL_3dL_4. \tag{9.203}$$

When using the Gauss-Legendre quadrature, the previous equation is expressed by [54, 18]

$$I = 8V \int_V \tilde{h}(L_2, L_3, L_4)dV = 8V \sum_{l=1}^{N_{int}} w_l \tilde{h}(L_{2_l}, L_{3_l}, L_{4_l}), \tag{9.204}$$

where N_{int}, $(L_{2_l}, L_{3_l}, L_{4_l})$, and w_l are respectively the number, coordinates and weights of the integration points.

In order to defining a tensorial integration rule for the tetrahedron, we employ the tridirectional mapping and the Gauss-Jacobi quadrature. The tetrahedron to hexahedron mapping is given by [35]

$$\begin{cases} L_4 = \dfrac{1}{2}(1 + \xi_3), \\ L_3 = \dfrac{1}{4}(1 + \xi_2)(1 - \xi_3), \\ L_2 = \dfrac{1}{8}(1 - \xi_1)(1 - \xi_2)(1 - \xi_3), \\ L_1 = 1 - L_2 - L_3 - L_4 = \dfrac{1}{8}(1 + \xi_1)(1 - \xi_2)(1 - \xi_3). \end{cases} \tag{9.205}$$

The Jacobian of this transformation is $\dfrac{1}{64}(1 - \xi_2)(1 - \xi_3)^2$.

Using this mapping, equation (9.203) can be rewritten as

$$\begin{aligned} I &= 8V \int_0^1 \int_0^{1-L_4} \int_0^{1-L_3-L_4} \tilde{h}(L_2, L_3, L_4)dL_2dL_3dL_4 \\ &= 8V \int_{-1}^1 \int_{-1}^1 \int_{-1}^1 \hat{h}(\xi_1, \xi_2, \xi_3)(1 - \xi_2)(1 - \xi_3)^2 d\xi_1 d\xi_2 d\xi_3. \end{aligned} \tag{9.206}$$

Its numerical integration is

$$I = 8V \sum_{i=1}^{n_1} \sum_{j=1}^{n_2} \sum_{k=1}^{n_3} w_{1_i} w_{2_j} w_{3_k} (1 - \xi_{2_j})(1 - \xi_{3_k})^2 \hat{h}(\xi_{1_i}, \xi_{2_j}, \xi_{3_k}), \tag{9.207}$$

where n_1, n_2, and n_3 are the number of integration points in directions ξ_1, ξ_2, and ξ_3, respectively; $(\xi_{1_i}, \xi_{2_j}, \xi_{3_k})$ are the coordinates; and w_{1_i}, w_{2_j}, and w_{3_k} the respective weights.

For $\alpha_2 = 1$, $\beta_2 = 0$, $\alpha_3 = 2$, and $\beta_3 = 0$, the Jacobian $(1 - \xi_2)(1 - \xi_3)^2$ of the tridirectional mapping is the prodduct of weight functions of the Gauss-Jacobi quadrature in directions ξ_2 and ξ_3. Thus, the Gauss-Legendre ($\alpha = 0, \beta = 0$), Gauss-Jacobi ($\alpha = 1, \beta = 0$), and Gauss-Jacobi ($\alpha = 2, \beta = 0$) quadratures can be used respectively in the ξ_1, ξ_2, and ξ_3 directions, that is

$$I = 8V \sum_{i=1}^{n_1} \sum_{j=1}^{n_2} \sum_{k=1}^{n_3} w_{1_i}^{0,0} w_{2_j}^{1,0} w_{3_k}^{2,0} \hat{h}(\xi_{1_i}, \xi_{2_j}, \xi_{3_k}). \tag{9.208}$$

Based on the previous relation, it is not necessary to consider the Jacobian $(1 - \xi_2)(1 - \xi_3)^2$ to determine the integrand order in the ξ_2 and ξ_3 directions, because the Jacobian is represented with weights $w_{2_j}^{1,0}$ and $w_{3_k}^{2,0}$.

9.10 SUMMARY OF THE VARIATIONAL FORMULATION OF MECHANICAL MODELS

The variational formulation was used throughout this book to present the mechanical models under infinitesimal strain with a Hookean material. In this section, we summarize the main aspects of this formulation.

As employed in the previous chapters, we adopt the following steps, illustrated in Figures 9.24 and 9.25, to formulate the mechanical models using the variational approach:

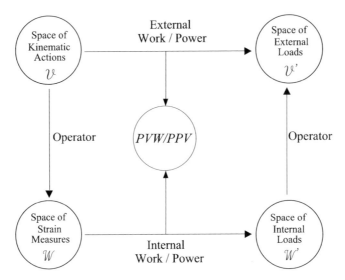

Figure 9.24   Variational formulation of mechanical problems.

1. **Definition of kinematic hypotheses**: In this case, we define the set $\mathcal{V}$ of possible kinematic actions for the body. In general, these actions must satisfy certain kinematic constraints represented by the boundary conditions of the problem. Thus, the subset $Kin_\mathbf{v}$ of $\mathcal{V}$ defines the kinematically admissible actions, i.e., the possible actions that satisfy the physical suports of the problem. The kinematic actions are described by the displacement ($\mathbf{u}$) and velocity ($\mathbf{v}$) vectorial fields.

2. **Strain measures**: From the kinematics, we obtain the strain measure compatible with the adopted kinematic model. We define operator $\mathcal{D}$ which is applied to the kinematic actions to obtain the strain or strain rate measures. Their components constitute the space $\mathcal{W}$.

3. **Characterization of rigid actions**: From the kinematic actions and strain measures, we obtain the subset of rigid actions, i.e., the actions with zero strain or strain rate measures. This subset of $\mathcal{V}$ is denoted by $\mathcal{N}(\mathcal{D})$.

4. **Determination of internal loads**: In the case of deformable bodies, we use the concept of internal work or power to determine the state of internal loads. The internal work or power relates the spaces of strain or strain rate measures $\mathcal{W}$ and internal loads $\mathcal{W}'$.

5. **Determination of external loads**: From the principle of virtual work (PVW) or the principle of virtual power (PVP) and the concept of external work or power, we relate the spaces of kinematic actions $\mathcal{V}$ and external loads $\mathcal{V}'$. In this way, it is possible to characterize the jexternal loads which are present in the considered model.

6. **Application of the PVW or PPV for equilibrium**: With this principle, we relate the internal and external works for a virtual displacement kinematic action, determining an integral expression of equilibrium for the problem. When the kinematic actions are expressed by velocity vectorial fields, we employ the PVP. Using one of these principles, we determine the local equations which constitute the solution of the integral statement of the model, characterizing the operator $\mathcal{D}^*$ and the equilibrium conditions for rigid actions.

7. **Application of constitutive equations**: From the constitutive equations, we have a relation between the stress (strres rate) and strain (strain rate) measures. This allows us to obtain, in the case of linear elastic material, the BVPs in terms of the model kinematics.

9.10.1   EXTERNAL POWER

As illustrated in Chapter 1, in order to evaluate the weight of an object, we lift it slightly and estimate its weight by the work or power done to this movement action (see Figure 1.2). Hence, it is necessary to use a virtual

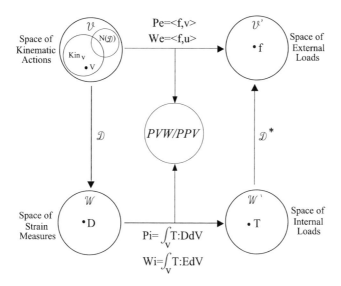

Figure 9.25 $\mathscr{V}$, $\mathscr{V}'$, $\mathscr{W}$, and $\mathscr{W}'$ spaces and the associated internal and external work/power.

movement, removing the object from its natural rest state.

As seen in Chapter 2, the particle movement actions are described by velocity vectors $\mathbf{v}$ (see Figure 2.7). Using the concept of external power P_e, we determined that the external loads, compatible with the kinematics described by $\mathbf{v}$, are the resultant force vectors $\mathbf{F}$ (see Section 2.3). Analogously, the rigid body actions are also given by velocity vectors $\mathbf{v}$, according to equation (2.11). The compatible external loads are resultants of forces $\mathbf{F}$ and moments $\mathbf{M}$ (see Section 2.4). In cases of particles and rigid bodies, the external power is given by the scalar product of vectors of external loads and the kinematic actions.

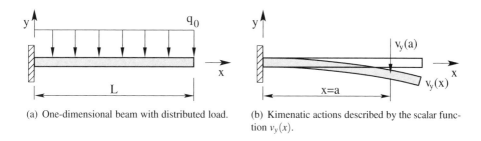

(a) One-dimensional beam with distributed load.

(b) Kimenatic actions described by the scalar function $v_y(x)$.

Figure 9.26 One-dimensional beam model.

Consider now the beam of Figure 9.26(a) subjected to the uniform distributed load q_0. Due to the clamp at the $x = 0$ end, the beam will only deform, because the rigid body movements (in this case, translations in x and y and rotation in z) are constrained. Figure 9.26(b) illustrates the beam-deformed configuration. This configuration is described by function $v_y(x)$, which gives the vertical velocity (rate of the displacement) for each beam cross-section x. Thus, in this example, the movement action is described by a continuous scalar function $v_y(x)$.

As discussed in Chapter 2 for particles and rigid bodies, the concept of external power allows us to associate external loads with kinematic actions. Moreover, the external power varies linearly with the movement action (see Section 2.6).

The same properties of external power are valid to deformable bodies, such as the beam illustrated in Figure 9.26. However, in this case, the movement action is described by continuous functions $v_y(x)$. Since the external power associates the kinematics with the compatible external loads, these loads should be also given by continuous functions $q_y(x)$. As the power is a linear functional of the movement actions, a linear operation associated to the product of functions $q_y(x)$ and $v_y(x)$ that results in a scalar may be an integral along the beam length L. Therefore, the external power is given by

$$P_e = \int_0^L q_y(x)v_y(x)dx. \qquad (9.209)$$

As previously mentioned, the scalar product of vectors is a particular case of the more general concept of inner products. Taking this general concept, it is totally natural to refer to the inner product of continuous functions, such as $q_y(x)$ and $v_y(x)$. Thus, the external power P_e given in (9.209) is the inner product of function $v_y(x)$, describing the beam kinematics, and function $q_y(x)$, representing the compatible external loads with $v_y(x)$. The result of the integral (9.209) is a scalar corresponding to the external power P_e associated to the movement action $v_y(x)$.

For the beam of Figure 9.26(a), the applied load is the constant distributed load q_0. Therefore, $q_y(x) = q_0$ for all $x \in (0, L)$ and expression (9.209) reduces to

$$P_e = \int_0^L q_0 v_y(x)dx.$$

Consider a dimensional analysis of expression (9.209) and suppose that the units of $v_y(x)$ is $\frac{m}{s}$. Thus, power P_e has as units Watts $= N\frac{m}{s}$ and $q_y(x)$ must be expressed in $\frac{N}{m}$. The integrand $q_y(x)v_y(x)$ is given in $\frac{1}{m}\frac{Nm}{s} = \frac{Watts}{m}$ and represents a density of external power, that is, power by units of length. After the integration over the beam length L, we obtain as units $N\frac{m}{s} =$ Watts, indicating that the result represents power. Thus, $q_y(x)$ is a density of force per length unit and indicates exactly the distributed load on the beam. Notice that the integral in (9.209) is calculated along the length, because the beam is formulated by the one-dimensional model. Similarly, the movement action is given by function $v_y(x)$ of one variable x.

The beam of Figure 9.26 can be modeled as a plane problem. Assuming that the variation of internal loads along the thickness of the cross-section of the beam is negligible, we can employ the plane stress model illustrated in Figure 9.27. In this case, the position of each point P of the beam is described by the pair of Cartesian coordinates (x, y) and the velocity by the vector function $\mathbf{v}(x, y)$ of two variables. We say that $\mathbf{v}(x, y)$ is a vectorial function, because for each point P with coordinates (x, y), there is the velocity vector $\mathbf{v}$ with the compnents $v_x(x, y)$ and $v_y(x, y)$ along the x and y directions, respectively. Thus, $\mathbf{v}(x, y)$ can be denoted as

$$\mathbf{v}(x, y) = \left\{ \begin{array}{c} v_x(x, y) \\ v_y(x, y) \end{array} \right\}. \qquad (9.210)$$

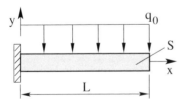

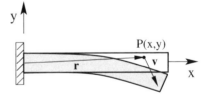

(a) Two-dimensional beam with distributed load.

(b) Position vector $\mathbf{r}$ and movement action described by the vector function $\mathbf{v}(x, y)$.

Figure 9.27   Two-dimensional beam model.

In this two-dimensional model, the external loads compatible with the kinematics $\mathbf{v}(x, y)$ are body loads on the beam surface. They are denoted by the vectorial function $\mathbf{b}(x, y)$

$$\mathbf{b}(x, y) = \left\{ \begin{array}{c} b_x(x, y) \\ b_y(x, y) \end{array} \right\}. \qquad (9.211)$$

The external power is given by the following integral on the beam surface S:

$$P_e = \int_S \mathbf{b}(x,y) \cdot \mathbf{v}(x,y)t\,dS = \int_S \left[b_x(x,y)v_x(x,y) + b_y(x,y)v_y(x,y) \right] t\,dS, \qquad (9.212)$$

with t the beam thickness. The previous expression is the inner product between the vector fields $\mathbf{b}(x,y)$ and $\mathbf{v}(x,y)$, which results the scalar representing the external power. The units of $\mathbf{b}(x,y)t$ are $\dfrac{\mathrm{N}}{\mathrm{m}^2}$, that is, it represents a density of force per surface unit. Thus, the integrand $\mathbf{b}^T(x,y)\mathbf{v}(x,y)t$ has $\dfrac{\mathrm{N}}{\mathrm{m}^2}\dfrac{\mathrm{m}}{\mathrm{s}} = \dfrac{\mathrm{Watts}}{\mathrm{m}^2}$ units, that is, power per units of surface area. Thus, after the integration over the surface S, we have the external power P_e. We observe that the distributed load q_0 illustrated in Figure 9.27(a) is treated as a boundary condition along the upper edge of the beam.

Consider now the same beam of Figures 9.26 and 9.27, but modeled as the three-dimensional body illustrated in Figure 9.28(a). In this model, each point P is described by the Cartesian coordinates (x,y,z). Consequently, the movement action of each point is given by the velocity vector $\mathbf{v}$. The kinematics of the body is described by the vectorial field $\mathbf{v} = \mathbf{v}(x,y,z)$. When we substitute the coordinates (x,y,z) of a point, we obtain its velocity vector $\mathbf{v}$ for a movement action of the beam. The components of $\mathbf{v}(x,y,z)$ in directions x, y, and z are denoted, respectively, by the scalar functions $v_x(x,y,z)$, $v_y(x,y,z)$, and $v_z(x,y,z)$ and as

$$\mathbf{v}(x,y,z) = \left\{ \begin{array}{c} v_x(x,y,z) \\ v_y(x,y,z) \\ v_z(x,y,z) \end{array} \right\}. \qquad (9.213)$$

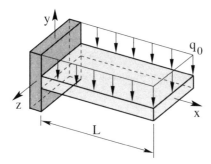

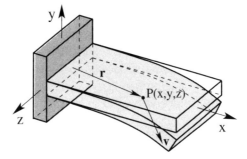

(a) Three-dimensional beam with a distributed load.

(b) Position vector $\mathbf{r}$ and movement action described by the vector function $\mathbf{v}(x,y,z)$.

Figure 9.28 Three-dimensional beam model.

Analogous to the two-dimensional beam model, the compatible external loads with the kinematics $\mathbf{v}(x,y,z)$ are given by body loads, which are no longer applied over the surface or length but over the volume, because the beam is now regarded as a solid. These volume or body forces are indicated by the vectorial function $\mathbf{b}(x,y,z)$ and have units of $\dfrac{\mathrm{N}}{\mathrm{m}^3}$, that is, they indicate a density of force per unit of volume. In terms of components, the vectorial field $\mathbf{b}(x,y,z)$ is given by

$$\mathbf{b}(x,y,z) = \left\{ \begin{array}{c} b_x(x,y,z) \\ b_y(x,y,z) \\ b_z(x,y,z) \end{array} \right\}. \qquad (9.214)$$

The external power P_e is an integral on the volume V of the body, that is,

$$\begin{aligned} P_e &= \int_V \mathbf{b}(x,y,z) \cdot \mathbf{v}(x,y,z)\,dV \\ &= \int_V \left[b_x(x,y,z)v_x(x,y,z) + b_y(x,y,z)v_y(x,y,z)b_z(x,y,z)v_z(x,y,z) \right] dV. \end{aligned} \qquad (9.215)$$

Taking the examples of particles, rigid bodies and one-, two- and three-dimensional beam models, we observe that the movement actions can be described by algebraic vectors $\mathbf{v}$, scalar functions $v_y(x)$, and vectorial functions $\mathbf{v}(x,y)$ and $\mathbf{v}(x,y,z)$. Thus, the nature of the kinematic actions depends on the model under consideration. The set of all movement actions of a certain model constitutes the vector space $\mathscr{V}$, called vector space of the possible movement actions.

Analogously, the compatible external loads with the kinematics of particles, rigid bodies, and the one-, two- and three-dimensional beam models are respectively given by vectors (force $\mathbf{F}$ and moment $\mathbf{M}$), scalar functions [distributed load $q_y(x)$] and vectorial functions [body forces $\mathbf{b}(x,y)$ and $\mathbf{b}(x,y,z)$]. Again, the nature of the external loads depends on the model under consideration, or more specifically, of the adopted kinematics. The set of kinematically compatible external loads for a given model is the vector space of external loads, denoted by $\mathscr{V}'$.

For the considered beam models, we observe that the external loads $q_y(x)$, $\mathbf{b}(x,y)$, and $\mathbf{b}(x,y,z)$ represent, respectively, densities of forces per units of length, surface, and volume. Thus, the external power can be d enoted as an integral of a density of external power p_e over the body $\mathscr{B}_t$ at time t, that is,

$$P_e = \int_{\mathscr{B}_t} p_e \, d\mathscr{B}_t. \tag{9.216}$$

The elements of space $\mathscr{V}'$ must be compatible with the movement actions in $\mathscr{V}$. Besides that, these elements are characterized from the concept of external power P_e. Thus, we say that $\mathscr{V}'$ is the dual space of $\mathscr{V}$. As explained in Section 2.6, power is a linear functional of the elements of $\mathscr{V}$. Formally, $\mathscr{V}'$ is defined as the set of all functionals which are linear and continuous in $\mathscr{V}$.

As the nature of the elements in $\mathscr{V}$ and $\mathscr{V}'$ depends on the model under consideration, we denote the external power P_e by the following general way:

$$P_e = \langle \mathbf{f}, \mathbf{v} \rangle, \tag{9.217}$$

with $\mathbf{f} \in \mathscr{V}'$ the system of loads on body $\mathscr{B}_t$ at time t and $\mathbf{v} \in \mathscr{V}$ the movement action. We observe that $\mathbf{f}$ is characterized by the external power P_e for each movement action $\mathbf{v} \in \mathscr{V}$.

Finally, the movement actions must satisfy certain kinematic constraints of the problem. For the beam of Figure 9.26, the velocity in the y direction and the rate of rotation in z must be zero, that is, $v_y(0) = 0$ and $\dfrac{dv_y(0)}{dx} = 0$. These movement actions belong to the space $\mathscr{V}$ and are called kinematically admissible movement actions. The set containing all these actions defines the subset $Kin_{\mathbf{v}}$ of $\mathscr{V}$. Spaces $\mathscr{V}$ and $\mathscr{V}'$ and subset $Kin_{\mathbf{v}}$ are illustrated in the superior part of Figure 9.25.

9.10.2 INTERNAL POWER

As shown in the previous section, the external power depends only on the movement actions rather than the strain measures. Therefore, for a rigid action, i.e., an action with zero strain measure, no response is obtained about the internal state, given by the bonding forces among the particles of the body. An example is the engine belt illustrated in Figure 1.3. We should perform an action that deforms the belt to evaluate whether it is tractioned. A rigid displacement action does not allow us to evaluate the traction state of the belt.

Considering the one-dimensional beam model shown in Figure 9.26, there is a deformation or configuration change due to the movement action $v_y(x)$. We can imagine that the external power P_e, associated with the distributed load q_0 and the movement action $v_y(x)$, was entirely used to deform the beam. However, the beam does not deform indefinitely. Therefore, when applying load q_0, the beam deforms continuously until reaching the equilibrium configuration. Analogously to the external power P_e, there is an internal power P_i in the beam, such that the internal and external powers in equilibrium are equal, that is, $P_e = P_i$.

The movement action $v_y(x)$ represents an external beam kinematics which, with the distributed load $q_y(x)$, leads to the external power. The strain rate, denoted by $\dot{\varepsilon}_{xx}(x)$, indicates the internal beam kinematics. Strain rate is used because we now consider the concept of power associated to the movement actions described by velocities. The kinematics $v_y(x)$ and strain rate $\dot{\varepsilon}_{xx}(x)$ of the beam are related by $\dot{\varepsilon}_{xx}(x) = -y\dfrac{d^2 v_y(x)}{dx^2}$.

Generally, the strain components are obtained by taking the derivatives of the kinematic components. This is indicated by the strain operator, which is generically denoted by $\mathscr{D}$. In the particular for beams, $\mathscr{D} = y\dfrac{d^2}{dx^2}$.

The main aspect here is to observe that the concept of external power makes it possible to identify the external loads compatible with the adopted kinematics. Similarly, the internal power associates the set of compatible internal loads to the strain measures of the model under consideration. These internal loads are also consistent with the model kinematics, because the strain measures are obtained by derivatives of the movement actions. The internal loads characterize the internal state of a body. Analogously to external power, the internal power is a linear functional of the strain measures, associating the compatible internal loads with the strain measures.

Taking the beam model of Chapter 5, the strain rate measure is given by the continuous functions $\dot{\varepsilon}_{xx}(x)$. There must exist a continuous function $\sigma_{xx}(x)$ associated to $\dot{\varepsilon}_{xx}(x)$, representing the state of internal loads for each beam cross-section x. Recall that in the case of particles, the magnitude of the external power is given in terms of the product of the norms of the resultant force $\mathbf{F}$ and velocity $\mathbf{v}$ vectors. As the beam is a continuous body, there are infinite points. The density of internal power of each point is given, analogous to particles, by the product of strain measure $\dot{\varepsilon}_{xx}(x)$ by continuous function $\sigma_{xx}(x)$ representing the internal loads. This product must be summed for each point of the beam, that is, we have an integration. Thus, the internal power P_i is given as an integral over the volume V of the beam

$$P_i = \int_V \sigma_{xx}(x)\dot{\varepsilon}_{xx}(x)dV. \tag{9.218}$$

Considering a dimensional analysis of expression (9.218), the unit of internal power P_i is Watts $= \dfrac{\text{Nm}}{\text{s}}$. Supposing that $v_y(x)$ unit is $\dfrac{\text{m}}{\text{s}}$, we have $\dot{\varepsilon}_{xx}(x) = -y\dfrac{d^2 u_y(x)}{dx^2}$ given in $\dfrac{1}{\text{s}}$. For the beam volume V in m^3, the previous integral result Watts and function $\sigma_{xx}(x)$ must be necessarily given in $\dfrac{\text{N}}{\text{m}^2}$. Thus, function $\sigma_{xx}(x)$ is actually a density of force per unit of area, called stress. In this particular case, there is a normal stress in the x direction. Observe that the stress concept came from the definition of strain measure and the fact that the internal power is a linear functional of the strain measures.

The previous expression can be rewritten as an integral over the beam length L, that is,

$$P_i = \int_0^L M_z(x)\dfrac{d^2 v_y(x)}{dx^2}dx, \tag{9.219}$$

with $M_z(x)$ the function for the beam bending moment.

The integrand $p_i = M_z(x)\dfrac{d^2 v_y(x)}{dx^2}$ of the previous expression represents a power density, that is, internal power per unit of length. Supposing that $v_y(x)$ is given in $\dfrac{\text{m}}{\text{s}}$ and the bending moment in Nm, p_i has the following units

$$[p_i] = (\text{Nm})\left(\dfrac{\frac{\text{m}}{\text{s}}}{\text{m}^2}\right) = \dfrac{1}{\text{m}}\text{N}\dfrac{\text{m}}{\text{s}} = \dfrac{\text{Watts}}{\text{m}},$$

that is, power per units of length.

Now consider the two-dimensional beam model illustrated in Figure 9.27. There are two longitudinal strain rate components in the x and y directions, denoted by $\dot{\varepsilon}_{xx}(x,y)$ and $\dot{\varepsilon}_{yy}(x,y)$, and two angular strain rate components, denoted by $\dot{\gamma}_{xy}(x,y)$ and $\dot{\gamma}_{yx}(x,y)$. There are four functions associated to the strain components, describing the state of internal loads. These functions represent the normal stress components $\sigma_{xx}(x,y)$ and $\sigma_{yy}(x,y)$ in directions x and y and shear stress components $\tau_{xy}(x,y)$ and $\tau_{yx}(x,y)$, which are respectively associated to $\dot{\varepsilon}_{xx}(x,y)$, $\dot{\varepsilon}_{yy}(x,y)$, $\dot{\gamma}_{xy}(x,y)$, and $\dot{\gamma}_{yx}(x,y)$. Analogous to equation (9.218), the internal power for the plane model is given by the following surface integral:

$$P_i = \int_V \left[\sigma_{xx}(x,y)\dot{\varepsilon}_{xx}(x,y) + \sigma_{yy}(x,y)\dot{\varepsilon}_{yy}(x,y) + \tau_{xy}(x,y)\dot{\gamma}_{xy}(x,y) + \tau_{yx}(x,y)\dot{\gamma}_{yx}(x,y)\right]t\,dS. \tag{9.220}$$

The solid beam model of Figure 9.28 represents the most general strain state case with three components of normal strain rate in directions x, y, and z, denoted by $\dot{\varepsilon}_{xx}(x,y,z)$, $\dot{\varepsilon}_{yy}(x,y,z)$, and $\dot{\varepsilon}_{zz}(x,y,z)$, plus six angular strain rate components, denoted by $\dot{\gamma}_{xy}(x,y,z)$, $\dot{\gamma}_{yx}(x,y,z)$, $\dot{\gamma}_{xz}(x,y,z)$, $\dot{\gamma}_{zx}(x,y,z)$, $\dot{\gamma}_{yz}(x,y,z)$, and $\dot{\gamma}_{zy}(x,y,z)$. Corresponding to the strain rate components, there are the normal stress components $\sigma_{xx}(x,y,z)$, $\sigma_{yy}(x,y,z)$ and $\sigma_{zz}(x,y,z)$ plus six shear strain components, namely $\tau_{xy}(x,y,z)$, $\tau_{yx}(x,y,z)$, $\tau_{xz}(x,y,z)$, $\tau_{zx}(x,y,z)$, $\tau_{yz}(x,y,z)$, and $\tau_{zy}(x,y,z)$. Analogous to (9.220), the internal power is given by

$$P_i = \int_V \left[\begin{array}{l} \sigma_{xx}(x,y,z)\dot{\varepsilon}_{xx}(x,y,z) + \tau_{xy}(x,y,z)\dot{\gamma}_{xy}(x,y,z) + \tau_{xz}(x,y,z)\dot{\gamma}_{xz}(x,y,z) + \\ \tau_{yx}(x,y,z)\dot{\gamma}_{yx}(x,y,z) + \sigma_{yy}(x,y,z)\dot{\varepsilon}_{yy}(x,y,z) + \tau_{yz}(x,y,z)\dot{\gamma}_{yz}(x,y,z) + \\ \tau_{zx}(x,y,z)\dot{\gamma}_{zx}(x,y,z) + \tau_{zy}(x,y,z)\dot{\gamma}_{zy}(x,y,z) + \sigma_{zz}(x,y,z)\dot{\varepsilon}_{zz}(x,y,z) \end{array} \right] dV. \tag{9.221}$$

As can be seen in equations (9.220) and (9.221), the integrands represent the power density p_i, that is, internal power per surface and volume units. To verify this fact, we observe that the strain components are given by the derivative of the kinematic components [for instance, $\dot{\varepsilon}_{xx}(x,y,z) = \dfrac{\partial v_x(x,y,z)}{\partial x}$]. Supposing that the component v_x is given in $\dfrac{\text{m}}{\text{s}}$ and the stress σ_{xx} in $\dfrac{\text{N}^2}{\text{m}}$, the product $\sigma_{xx}\dot{\varepsilon}_{xx}$ has the following units:

$$[\sigma_{xx}\dot{\varepsilon}_{xx}] = \frac{\text{N}}{\text{m}^2}\frac{\text{m}}{\text{s}}\frac{1}{\text{m}} = \frac{\text{Watts}}{\text{m}^3}.$$

From expression (9.221), we verify that in the general case of a three-dimensional body, the infinitesimal strain state at each point will be characterized by nine strain components ($\dot{\varepsilon}_{xx}, \dot{\varepsilon}_{yy}, \dot{\varepsilon}_{zz}, \dot{\gamma}_{xy}, \dot{\gamma}_{xz}, \dot{\gamma}_{yx}, \dot{\gamma}_{yz}, \dot{\gamma}_{zx}, \dot{\gamma}_{zy}$). Similarly, the stress state at each point is characterized by nine stress components ($\sigma_{xx}, \sigma_{yy}, \sigma_{zz}, \tau_{xy}, \tau_{xz}, \tau_{yx},$ $\tau_{yz}, \tau_{zx}, \tau_{zy}$). These stress and strain components can be written in matrix form as

$$[\mathbf{D}] = \begin{bmatrix} \dot{\varepsilon}_{xx} & \dot{\gamma}_{xy} & \dot{\gamma}_{xz} \\ \dot{\gamma}_{yx} & \dot{\varepsilon}_{yy} & \dot{\gamma}_{yz} \\ \dot{\gamma}_{zx} & \dot{\gamma}_{yz} & \dot{\varepsilon}_{zz} \end{bmatrix}, \quad [\mathbf{T}] = \begin{bmatrix} \sigma_{xx} & \tau_{xy} & \tau_{xz} \\ \tau_{yx} & \sigma_{yy} & \tau_{yz} \\ \tau_{zx} & \tau_{yz} & \sigma_{zz} \end{bmatrix}. \tag{9.222}$$

The coordinates (x,y,z) were not explicitly included in the strain and stress components to simplifying notation only. The above matrices contain the Cartesian components of the strain rate tensor $\mathbf{D}$ and Cauchy stress tensor $\mathbf{T}$. Multiplying $[\mathbf{D}]^T$ by $[\mathbf{T}]$, we obtain

$$[\mathbf{D}]^T[\mathbf{T}] = \begin{bmatrix} \dot{\varepsilon}_{xx} & \dot{\gamma}_{yx} & \dot{\gamma}_{zx} \\ \dot{\gamma}_{xy} & \dot{\varepsilon}_{yy} & \dot{\gamma}_{yz} \\ \dot{\gamma}_{xz} & \dot{\gamma}_{yz} & \dot{\varepsilon}_{zz} \end{bmatrix} \begin{bmatrix} \sigma_{xx} & \tau_{xy} & \tau_{xz} \\ \tau_{yx} & \sigma_{yy} & \tau_{yz} \\ \tau_{zx} & \tau_{yz} & \sigma_{zz} \end{bmatrix}$$

$$= \begin{bmatrix} \dot{\varepsilon}_{xx}\sigma_{xx} + \dot{\gamma}_{yx}\tau_{yx} + \dot{\gamma}_{zx}\tau_{zx} & \dot{\varepsilon}_{xx}\tau_{xy} + \dot{\gamma}_{yx}\sigma_{yy} + \dot{\gamma}_{zx}\tau_{yz} & \dot{\varepsilon}_{xx}\tau_{xz} + \dot{\gamma}_{yx}\tau_{yz} + \dot{\gamma}_{zx}\sigma_{zz} \\ \dot{\gamma}_{xy}\sigma_{xx} + \dot{\varepsilon}_{yy}\tau_{yx} + \dot{\gamma}_{yz}\tau_{zx} & \dot{\gamma}_{xy}\tau_{xy} + \dot{\varepsilon}_{yy}\sigma_{yy} + \dot{\gamma}_{yz}\tau_{yz} & \dot{\gamma}_{xy}\tau_{xz} + \dot{\varepsilon}_{yy}\tau_{yz} + \dot{\gamma}_{yz}\sigma_{zz} \\ \dot{\gamma}_{xz}\sigma_{xx} + \dot{\gamma}_{yz}\tau_{yx} + \dot{\varepsilon}_{zz}\tau_{zx} & \dot{\gamma}_{xz}\tau_{yx} + \dot{\gamma}_{yz}\sigma_{yy} + \dot{\varepsilon}_{zz}\tau_{yz} & \dot{\gamma}_{xz}\tau_{xz} + \dot{\gamma}_{yz}\tau_{yz} + \dot{\varepsilon}_{zz}\sigma_{zz} \end{bmatrix}.$$

Taking the sum of the main diagonal coefficients of the product $[\mathbf{D}]^T[\mathbf{T}]$, we obtain a scalar number, called trace and denoted as tr, that is,

$$\text{tr}([\mathbf{D}]^T[\mathbf{T}]) = \dot{\varepsilon}_{xx}\sigma_{xx} + \dot{\gamma}_{yx}\tau_{yx} + \dot{\gamma}_{zx}\tau_{zx} + \dot{\gamma}_{xy}\tau_{xy} + \dot{\varepsilon}_{yy}\sigma_{yy} + \dot{\gamma}_{yz}\tau_{yz} + \dot{\gamma}_{xz}\tau_{xz} + \dot{\gamma}_{yz}\tau_{yz} + \dot{\varepsilon}_{zz}\sigma_{zz}.$$

We observe that the trace is exactly the integrand of expression (9.221). Thus, we can express the power density as $p_i = \text{tr}([\mathbf{D}]^T[\mathbf{T}])$. The inner product $\mathbf{T} : \mathbf{D}$ of tensors $\mathbf{D}$ and $\mathbf{T}$ is exactly the trace of the $[\mathbf{D}]^T[\mathbf{T}]$ product, that is,

$$\mathbf{T} : \mathbf{D} = \text{tr}\left([\mathbf{D}]^T[\mathbf{T}]\right). \tag{9.223}$$

Thus, equation (9.221) can be rewritten as

$$P_i = \int_V \mathbf{T} : \mathbf{D}\,dV = \int_V \text{tr}\left([\mathbf{D}]^T[\mathbf{T}]\right)dV. \tag{9.224}$$

It is observed that from the general case of solid, it is possible to obtain the expressions of the internal power for the one- and two-dimensional models [equations (9.218) and (9.220)], taking only the sum of product of the nonzero components in the tensors $\mathbf{T}$ and $\mathbf{D}$.

Analogous to the case of the external power, the specific form of the internal power depends on the kinematics for the model, as can be seen from expressions (9.218), (9.220), and (9.221). Hence, the internal power P_i is indicated, in a general way, as

$$P_i = (\mathscr{T}, \mathscr{D}) = (\mathscr{T}, \mathscr{D}\mathbf{v}), \tag{9.225}$$

with $\mathscr{D}$ a differential operator applied to the kinematic action $\mathbf{v} \in \mathscr{V}$. For instance, in the case of the one-dimensional beam model, we have $\mathscr{D} = \dfrac{d^2}{dx^2}$ and $\mathscr{D}v_y = \dfrac{d^2 v_y(x)}{dx^2}$. For the case of a solid, we have

$$P_i = (\mathbf{T}, \mathbf{D}) = (\mathscr{T}, \mathscr{D}\mathbf{v}) = \int_V \mathbf{T} : \mathbf{D}\,dV. \tag{9.226}$$

In this case, the kinematics is given by the vectorial field $\mathbf{v}(x,y,z)$ and the operator $\mathscr{D}$ applied to $\mathbf{v}(x,y,z)$ is proportional to the gradient of $\mathbf{v}(x,y,z)$, that is,

$$\mathscr{D}\mathbf{v} = \mathbf{D}(x,y,z) = \frac{1}{2}\left(\operatorname{grad}\mathbf{v}(x,y,z) + \operatorname{grad}\mathbf{v}^T(x,y,z)\right). \tag{9.227}$$

We call $\mathbf{D}$ the strain rate tensor. The gradient operator, denoted by grad, is calculated relative to the spatial coordinates, because we are describing the kinematics using the velocity field, which is a spatial measure.

Taking equations (9.218), (9.220), and (9.221), the integrands represent respectively the densities of internal power by units of length, surface, and volume. The internal power is defined by a functional in terms of a density of internal power p_i by units of volume, surface in a plane case, or length for a one-dimensional model. Thus, the internal power P_i is the integral of a density and thus a scalar function. Hence, in a general way,

$$P_i = \int_{\mathscr{B}_t} p_i\, d\mathscr{B}_t. \tag{9.228}$$

For the cases of rigid actions of a particle or body, the internal power is zero. This is one of the conditions of the principle of virtual power which is discussed in the next section. Thus, for any rigid action $\mathbf{v} \in \mathscr{V}$, we have from (9.225) that the internal power must be zero, that is,

$$P_i = (\mathscr{T}, \mathscr{D}\mathbf{v}) = 0, \tag{9.229}$$

implying that $\mathscr{D}\mathbf{v} = \mathbf{0}$, that is, the strain measure is zero and $\mathbf{v}$ a rigid action. The set of all movement actions $\mathbf{v} \in \mathscr{V}$, for which $\mathscr{D}\mathbf{v} = \mathbf{0}$, defines the subset $\mathscr{N}(\mathscr{D})$ of $\mathscr{V}$ for the rigid movement actions. Symbol $\mathscr{N}$ indicates the null space of $\mathscr{D}$, that is, the subset of rigid movement actions in the space $\mathscr{V}$ of possible movement actions.

On the other hand, the set $\mathscr{D}\mathbf{v}$ defines the space $\mathscr{W}$ of compatible strain measures with the kinematic actions $\mathbf{v} \in \mathscr{V}$. The dual space of $\mathscr{W}$, containing the internal load components, is designated by $\mathscr{W}'$. Figure 9.25 illustrate these spaces, internal power P_i, and subset $\mathscr{N}(\mathscr{D})$.

9.10.3 PRINCIPLE OF VIRTUAL POWER (PVP)

Consider the beam illustrated in Figure 9.26(a). As we know, due to the action of the distributed load q_0, the beam deforms until reaching the equilibrium configuration shown in Figure 9.26(b). In this configuration, the external power P_e, given from the movement action $v_y(x)$ and distributed load q_0, is equal to the internal power P_i, given from the longitudinal strain rate $\dot{\varepsilon}_{xx}(x)$ and normal stress $\sigma_{xx}(x)$.

To check if the beam is indeed in equilibrium, we proceed analogously to the examples of the weight of an object or the traction in the belt. Thus, from the deformed configuration given in Figure 9.26(b), we introduce an arbitrary virtual movement action $\delta v_y(x)$, as illustrated in Figure 9.29. If the external and internal powers originated after the virtual action $\delta v_y(x)$ are equal, the beam is really in equilibrium in the deformed configuration of Figure 9.26(b). Notice in Figure 9.29 that the virtual action $\delta v_y(x)$ is arbitrary and satisfies the kinematic constraints, in this case the clamp at $x = 0$. Thus, the effect of the virtual action $\delta v_y(x)$ is to introduce a perturbation from the deformed position, seeking the evaluation of the beam equilibrium.

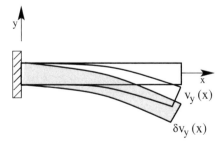

Figure 9.29 Virtual movement action $\delta v_y(x)$ for the beam.

To state formally the PVP, consider a given body in the deformed equilibrium configuration. Imposing a virtual movement action $\delta\mathbf{v}$ from this configuration, the body is in equilibrium if the external and internal powers are equal, that is,

$$\delta P_e = \delta P_i. \tag{9.230}$$

Substituting (9.217) and (9.229), we have

$$\langle \mathbf{f}, \delta \mathbf{v} \rangle = (\mathscr{T}, \mathscr{D} \delta \mathbf{v}), \tag{9.231}$$

for any virtual action $\delta \mathbf{v} \in \mathscr{V}$.

For a rigid virtual action $\delta \mathbf{v} \in \mathscr{N}(\mathscr{D})$, the internal power is zero and expression (9.231) reduces to

$$P_e = \langle \mathbf{f}, \delta \mathbf{v} \rangle = 0,$$

that is, we obtain the equilibrium conditions derived in the previous chapters for the considered mechanical models.

When applying the PVP, it is possible to characterize the external loads compatible with the model kinematic. Thus, the set $\mathscr{V}'$ of the external loads is defined. Besides that, we have the integral equilibrium form, from which we obtain the local form given in terms of the differential equations and boundary conditions. The set formed by the differential equation and boundary conditions is called boundary value problem (BVP). This local form is characterized by the operator $\mathscr{D}^*$ mapping elements of space $\mathscr{W}'$ of internal loads into the space $\mathscr{V}'$ of external loads.

We can generalize the PVP for the case of deformable bodies in movement. For this purpose, we include the inertia forces to the expression of external loads. In expression (9.215), $\mathbf{b}(x, y, z)$ represents the density of force per unit of volume. Thus, we introduce the inertia force as $-\rho \mathbf{a}(x, y, z)$, with ρ the density, given in units of $\frac{Kg}{m^3}$, and $\mathbf{a}(x, y, z)$ the acceleration vectorial field expressed in $\frac{m}{s^2}$. The product $\rho \mathbf{a}$ has units

$$[\rho a] = \frac{Kg\ m}{m^3\ s^2} = \frac{N}{m^3},$$

that is, we have again a density of force per unit of volume.

Thus, we rewrite equation (9.231) as

$$\langle \mathbf{f}^*, \delta \mathbf{v} \rangle - (\mathscr{T}, \mathscr{D} \delta \mathbf{v}) = 0, \tag{9.232}$$

with $\mathbf{f}^* = \mathbf{f} - \rho \mathbf{a}$. The above expression is called D'Alambert's principle. The main advantage of this principle is to allow the study of dynamic problems analogously to static equilibrium problems, because the inertia force component is introduced as an external force in $\mathbf{f}^*$.

The entire formulation presented in this section is based on the idea of power, implying that the movement actions are described by velocity fields $\mathbf{v} \in \mathscr{V}$. As we known, the velocity is the rate of variation of the displacement over time. The virtual actions have the objective of introducing a perturbation that allows the evaluation of the equilibrium of a body. Thus, we can employ virtual actions that are as small as we want, that is, an infinitesimal displacement can be considered in a time interval which is also infinitesimal. Hence, we can characterize the movement action by a displacement. As the problems considered in this book are static, we used movement actions given by displacements in a small time interval. This simplifies the notation, because instead of refering to a strain rate component $\dot{\varepsilon}_{xx}$, we only refer to the strain measure ε_{xx}.

9.11   FINAL COMMENTS

This chapter presented the formulation and approximation of the mechanical models for plane stress, plane strain, and generic twist. The numerical integration quadratures were extended to two- and three-dimensional elements. At the end, a summary of the variational formulation were also presented. Software $(hp)^2$FEM was introduced and will be used in the problems in the next section.

9.12   PROBLEMS

1. Consider a square steel plate with $1\ m \times 1\ m$ and $1\ cm$ of thickness. Let $u_x = k(x - 1)$ and $u_y = ky$ be the displacement field with k a constant to represent an infinitesimal displacement. Does it represent a possible solution for a plane stress model? Represent the boundary conditions in terms of prescribed displacements and tractions at the edges of the plate.

2. Check if the compatibility conditions are satisfied for the following tensorial field of infinitesimal strains:

$$[\mathbf{E}] = \begin{bmatrix} x+4y & x & y \\ x & 2y+z & z \\ y & z & x+2z \end{bmatrix}.$$

3. Let $\phi(x,y) = ax_1^2 + 2bxy + cy^2$ and $\phi(x,y) = 2axy^2$ be the stress functions. We ask: a) check if these functions satisfy the biharmonic equation; b) determine the plane stress components σ_{xx}, σ_{yy}, and τ_{xy}; c) plot the tractions in the faces of a unit square; d) determine the stress components τ_{xz}, τ_{yz}, and σ_{zz} and strain components for the plane stress and plane strain cases.

4. Generate a mesh with hexahedra for a steel bar of length $L = 2$ m and solve Example 9.5 with the $(hp)^2$FEM program.

5. Consider the hook illustrated in Figure 1.10 and the hook.fem file with the finite element mesh. Considering a plane stress model with 5 cm of thickness, analyze the hook with program $(hp)^2$FEM. Use the hook.def definition file.

6. Consider the finite element mesh used in Example 9.12. Plot the error in the L_2 norm with $P = 1,\dots,8$ for solution $u(x,y) = \sin(xy)$ using the $(hp)^2$FEM program. The mesh (poisson.fem) and definition (poisson.def) files are given. The loads, boundary conditions, and theoretical solution must be included in the poisson.def file.

7. Files conrod.fem and conrod.def have the discretization of the conrod illustrated in Figure 9.8(a). Solve it using the $(hp)^2$FEM program and print or plot the Tresca and von Mises stress distributions.

10 FORMULATION AND APPROXIMATION OF PLATES

10.1 INTRODUCTION

Plates are structural elements submitted to bending loads and for which the thickness is much smaller than the other dimensions. Consequently, the plate mathematical models are described by two-variable functions. Therefore, the mean surface along the thickness is considered for the study of bending actions.

The Kirchhoff-Love and Reissner-Mindlin are the most well known plate models. They consider, respectively, the effects of bending only and bending with shear, analogously to the Euler-Bernoulli and Timoshenko beam models. In this chapter, we present the variational formulation and FEM approximation for the Kirchhoff plate model, including the membrane effect. We apply the same steps of the variational formulation and present the variables of interest using second-order tensors. The Kirchhoff-Love model is also known as the plate classical theory and applied to the analysis of thin plates.

The adopted Cartesian reference system for the plate model formulation and the mean surface are illustrated in Figure 10.1. We assume a constant thickness for the plate.

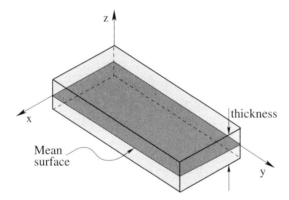

Figure 10.1 Mean surface and adopted reference system for the Kirchhoff plate.

The main objective of this chapter is to show that the previous steps of the formulation and approximation of mechanical models are quite general and robust, even in the case of more complex models, such as plates. This chapter is based mainly on references [23, 3, 41, 44].

10.2 KINEMATICS

The kinematics of plates consists of displacement actions that cause bending. In the Kirchhoff model, the kinematic actions are such that the normal vectors to the undeformed reference surface remain orthogonal to the deformed surface and do not change in length. Thus, the transversal shear strains are zero. The kinematics of this plate model is analogous to the Euler-Bernoulli beam model.

To illustrate the Kirchhoff bending kinematics, consider the plate of Figure 10.2. Let AB be the normal to the reference surface which is far x and y units from the origin of the adopted Cartesian reference system, as shown in Figure 10.2(a). After the plate bending, normal AB rotates in such way to remain perpendicular to the deformed reference surface, as illustrated in Figure 10.2(b). Specifically, points of the normal AB have a rigid displacement or translation u_z in direction z. Thereafter, normal AB rotates about x and y in such way to remain perpendicular to the deformed reference surface.

To characterize the displacements of points on the normal AB for a plate bending action, Figure 10.3 illustrates normal AB according to planes xz and yz. We observe that due to the transversal displacement $u_z(x, y)$,

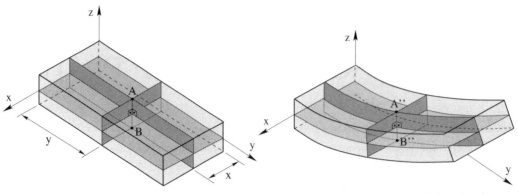

(a) Normal *AB* perpendicular to the undeformed surface.   (b) Normal *AB* perpendicular to the deformed surface.

Figure 10.2   Bending kinematics of the Kirchhoff model (adapted from [41]).

normal *AB* assumes position $A'B'$. Rigid rotations then occur with angles θ_y and θ_x, respectively, about axes y and x in such way that the final position $A''B''$ is perpendicular to the deformed reference surface.

Now consider point P with initial coordinates (x, y, z). Notice that coordinates x and y locate normal AB, while the coordinate z indicates point P on the normal AB. The final position P'' after bending is indicated by $(x - \Delta x, y - \Delta y, z + u_z)$, as can be seen in Figure 10.3. Thus, in addition to the transversal displacement u_z, point P has the displacement components u_x and u_y, respectively, in directions x and y. These displacements are given by the difference between the final and initial positions of point P, that is,

$$u_x(x, y, z) \quad = \quad x - \Delta x - x = -\Delta x, \tag{10.1}$$

$$u_y(x, y, z) \quad = \quad y - \Delta y - y = -\Delta y. \tag{10.2}$$

As the length of normal AB does not change, lines OP' and OP'' indicated in Figure 10.3(b) have the same length z. Thus, the following trigonometric relations are valid:

$$\sin \theta_y \quad = \quad \frac{\Delta x}{z}, \tag{10.3}$$

$$\tan \theta_y \quad = \quad \frac{\Delta u_z}{\Delta x}. \tag{10.4}$$

Assuming small displacements, we have $\sin \theta_y \approx \theta_y$, $\tan \theta_y \approx \theta_y$, and Δx small. Hence, angle θ_y given in equation (10.4) is expressed as

$$\theta_y(x, y) = \lim_{\Delta x \to 0} \frac{\Delta u_z}{\Delta x} = \frac{\partial u_z(x, y)}{\partial x}. \tag{10.5}$$

Substituting (10.1) and (10.5) in (10.3), and recalling that $\sin \theta_y \approx \theta_y$, we have

$$u_x(x, y, z) = -z\theta_y(x, y) = -z\frac{\partial u_z(x, y)}{\partial x}. \tag{10.6}$$

Analogously, we have from Figure 10.3(d)

$$\sin \theta_x \quad = \quad \frac{\Delta y}{z}, \tag{10.7}$$

$$\tan \theta_x \quad = \quad \frac{\Delta u_z}{\Delta y}. \tag{10.8}$$

Assuming small displacements, we have

$$\theta_x(x, y) = \lim_{\Delta y \to 0} \frac{\Delta u_z}{\Delta y} = \frac{\partial u_z(x, y)}{\partial y}. \tag{10.9}$$

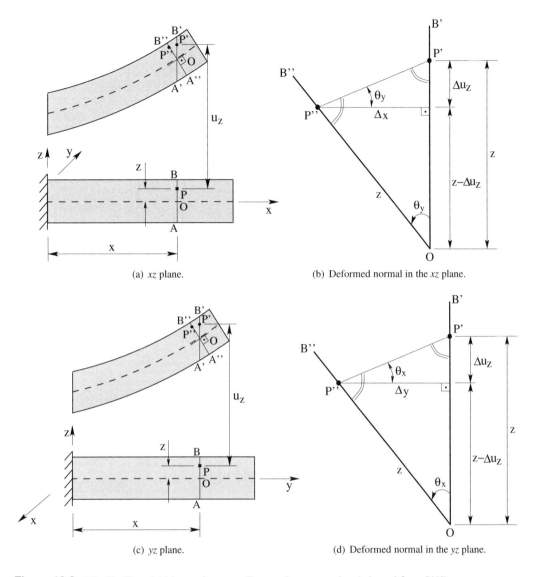

(a) *xz* plane.

(b) Deformed normal in the *xz* plane.

(c) *yz* plane.

(d) Deformed normal in the *yz* plane.

Figure 10.3 Kirchhoff model kinematics according to planes *xz* and *yz* (adapted from [41]).

Thus, from (10.2), (10.7), and (10.9), displacement u_y of point P is given by

$$u_y(x,y,z) = -z\theta_x(x,y) = -z\frac{\partial u_z(x,y)}{\partial y}. \tag{10.10}$$

We have considered the bending kinematics supposing the mean reference surface has only the transversal displacement u_z. We may include the longitudinal or membrane displacements $u_{x_0}(x,y)$ and $u_{y_0}(x,y)$ of points on the reference surface. For the case of infinitesimal strain, the longitudinal displacements on plane xy are decoupled of the transversal displacement, that is, the membrane effect is independent of u_z [44].

Finally, the Kirchhoff plate kinematics, including bending and membrane effects, is given by the following displacement vectorial field:

$$\mathbf{u}(x,y,z) = \left\{ \begin{array}{c} u_x(x,y,z) \\ u_y(x,y,z) \\ u_z(x,y) \end{array} \right\} = \left\{ \begin{array}{c} u_{x_0}(x,y) - z\dfrac{\partial u_z(x,y)}{\partial x} \\ u_{y_0}(x,y) - z\dfrac{\partial u_z(x,y)}{\partial y} \\ u_z(x,y) \end{array} \right\}. \tag{10.11}$$

We note that displacements u_x and u_y due to the plate bending have linear variations with the z coordinate, analogous to the beam in bending.

Example 10.1 *Consider a square plate with lentgh 1 m and transversal displacement given by*

$$u_z(x,y) = kx^2y^2(x-1)^2(y-1)^2\sin(\pi x)\sin(\pi y),$$

where k is a constant to enforce small displacements.

The bending rotations are given by

$$\theta_x(x,y) = u_{z,y} = kx^2(x-1)^2\sin(\pi x)y(y-1)\left(2\sin(\pi y)(2y-1) + \pi y(y-1)\cos(\pi y)\right),$$

$$\theta_y(x,y) = u_{z,x} = kx(x-1)\left(2\sin(\pi x)(2x-1) + \pi x(x-1)\cos(\pi x)\right)y^2(y-1)^2\sin(\pi y).$$

These functions are illustrated in Figure 10.4 for $k = 10^{-3}$. Notice that these functions are zero at the boundary, characterizing a clamped plate.

$\square$

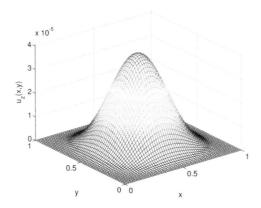

(a) Transversal displacement.

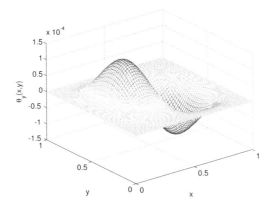

(b) Rotation in x.

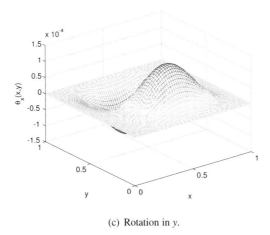

(c) Rotation in y.

Figure 10.4 Example 10.1: transversal displacement and rotations in a Kirchhoff plate.

10.3   STRAIN MEASURE

Using the definition of the displacement gradient tensorial field given in (8.105) for the Kirchhoff plate kinematics (10.11), we have

$$[\nabla \mathbf{u}] = \begin{bmatrix} \dfrac{\partial u_{x_0}(x,y)}{\partial x} - z\dfrac{\partial^2 u_z(x,y)}{\partial x^2} & \dfrac{\partial u_{x_0}(x,y)}{\partial y} - z\dfrac{\partial^2 u_z(x,y)}{\partial y^2} & 0 \\ \dfrac{\partial u_{y_0}(x,y)}{\partial x} - z\dfrac{\partial^2 u_z(x,y)}{\partial x^2} & \dfrac{\partial u_{y_0}(x,y)}{\partial y} - z\dfrac{\partial^2 u_z(x,y)}{\partial y^2} & 0 \\ 0 & 0 & 0 \end{bmatrix}. \tag{10.12}$$

Based on that, the infinitesimal strain tensorial field for the plate is determined from (8.110) as

$$[\mathbf{E}] = \begin{bmatrix} \varepsilon_{xx}(x,y,z) & \gamma_{xy}(x,y,z) & 0 \\ \gamma_{xy}(x,y,z) & \varepsilon_{yy}(x,y,z) & 0 \\ 0 & 0 & 0 \end{bmatrix}. \tag{10.13}$$

The nonzero strain components are given by the sum of membrane (m) and bending (b) contributions, that is,

$$\begin{aligned} \varepsilon_{xx}(x,y,z) &= \varepsilon_{xx}^m(x,y) + \varepsilon_{xx}^b(x,y,z), \\ \gamma_{xy}(x,y,z) &= \gamma_{xy}^m(x,y) + \gamma_{xy}^b(x,y,z), \\ \varepsilon_{yy}(x,y,z) &= \varepsilon_{yy}^m(x,y) + \varepsilon_{yy}^b(x,y,z), \end{aligned} \tag{10.14}$$

where

$$\varepsilon_{xx}^m(x,y) = \frac{\partial u_{x_0}(x,y)}{\partial x}, \tag{10.15}$$

$$\varepsilon_{yy}^m(x,y) = \frac{\partial u_{y_0}(x,y)}{\partial x}, \tag{10.16}$$

$$\gamma_{xy}^m(x,y) = \frac{1}{2}\left(\frac{\partial u_{x_0}(x,y)}{\partial y} + \frac{\partial u_{y_0}(x,y)}{\partial x}\right), \tag{10.17}$$

$$\varepsilon_{xx}^b(x,y,z) = -z\frac{\partial^2 u_z(x,y)}{\partial x^2}, \tag{10.18}$$

$$\varepsilon_{yy}^b(x,y,z) = -z\frac{\partial^2 u_z(x,y)}{\partial y^2}, \tag{10.19}$$

$$\gamma_{xy}^b(x,y,z) = -2z\frac{\partial^2 u_z(x,y)}{\partial x \partial y}. \tag{10.20}$$

Example 10.2 *The strain measure components for the transversal displacement given in Example 10.1 are*

$$\begin{aligned} \varepsilon_{xx}^b(x,y,z) &= -k\left[2(x-1)(5x-1) + x^2\left(2 - \pi^2(x-1)^2\right)\sin(\pi x)\right. \\ &\quad + \left. 4\pi x(x-1)(2x-1)\cos(\pi x)\right]y^2(y-1)^2\sin(\pi y)z, \\ \varepsilon_{yy}^b(x,y,z) &= -kx^2(x-1)^2\sin(\pi x)\left[2(y-1)(5y-1) + y^2\left(2 - \pi^2(y-1)^2\right)\sin(\pi y)\right. \\ &\quad + \left. 4\pi y(y-1)(2y-1)\cos(\pi y)\right]z, \\ \gamma_{xy}^b(x,y,z) &= -8kx(x-1)(2x-1)y(y-1)(2y-1) \\ &\quad [\sin(\pi x) + \pi x(x-1)\cos(\pi x)][\sin(\pi y) + \pi y(y-1)\cos(\pi y)]z. \end{aligned}$$

These strain components on the upper face of the plate for $z = 0.5$ cm are illustrated in Figure 10.5 for $k = 10^{-3}$.

□

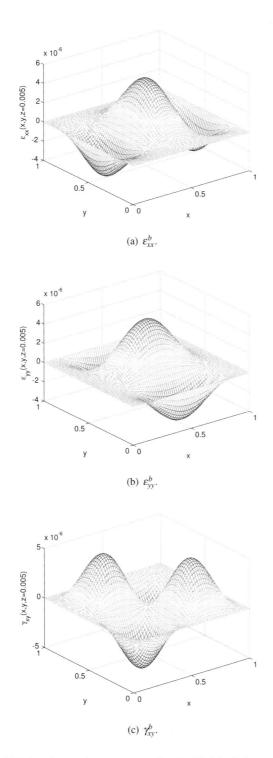

(a) ε_{xx}^b.

(b) ε_{yy}^b.

(c) γ_{xy}^b.

Figure 10.5 Example 10.2: bending strain components for the Kirchhoff plate.

10.4 RIGID ACTIONS

In order to obtain the rigid kinematic actions due to the bending effect, we make zero each of the bending strain components, that is,

$$
\begin{aligned}
\varepsilon_{xx}^{b}(x,y,z) &= -z\frac{\partial^2 u_z(x,y)}{\partial x^2} = 0, \\
\varepsilon_{yy}^{b}(x,y,z) &= -z\frac{\partial^2 u_z(x,y)}{\partial y^2} = 0, \\
\gamma_{xy}^{b}(x,y,z) &= -2z\frac{\partial^2 u_z(x,y)}{\partial x \partial y} = 0.
\end{aligned}
\tag{10.21}
$$

These equations are satisfied if $u_z(x,y) = u_z = \text{cte}$ or $\dfrac{\partial u_z(x,y)}{\partial x} = \dfrac{\partial u_z}{\partial x} = \text{cte}$ and $\dfrac{\partial u_z(x,y)}{\partial y} = \dfrac{\partial u_z}{\partial y} = \text{cte}$. The first condition $(u_z = \text{cte})$ implies the plate translation along direction z. The second and third conditions $\left(\dfrac{\partial u_z}{\partial x} = \dfrac{\partial u_z}{\partial y} = \text{cte} \right)$ represent respectively rigid rotations about the y and x axes. These rigid actions are respectively illustrated in Figures 10.6(a), 10.6(b), and 10.6(c).

Analogously, to determine the rigid kinematic components due to the membrane effect, we zero the membrane strain components, that is,

$$
\begin{aligned}
\varepsilon_{xx}^{m}(x,y) &= \frac{\partial u_{x_0}(x,y)}{\partial x} = 0, \\
\varepsilon_{yy}^{m}(x,y) &= \frac{\partial u_{y_0}(x,y)}{\partial y} = 0, \\
\gamma_{xy}^{m}(x,y) &= \frac{1}{2}\left(\frac{\partial u_{y_0}(x,y)}{\partial x} + \frac{\partial u_{x_0}(x,y)}{\partial y} \right) = 0.
\end{aligned}
\tag{10.22}
$$

The first and second equations in (10.22) are satisfied if $u_{x_0} = \text{cte}$ and $u_{y_0} = \text{cte}$, respectively. These conditions are related to rigid translations along axes x and y, as shown in Figures 10.6(d) and 10.6(e). The third expression in (10.22) is satisfied if $\dfrac{\partial u_{y_0}}{\partial x} = -\dfrac{\partial u_{x_0}}{\partial y}$. This implies in the rigid rotation about axis z as seen in Figure 10.6(f).

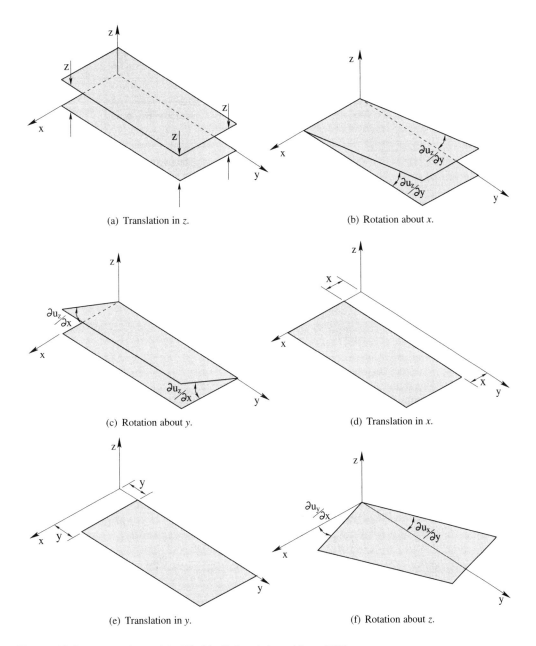

(a) Translation in z.

(b) Rotation about x.

(c) Rotation about y.

(d) Translation in x.

(e) Translation in y.

(f) Rotation about z.

Figure 10.6 Rigid actions of the Kirchhoff plate (adapted from [41]).

10.5 DETERMINATION OF INTERNAL LOADS

Substituting the nonzero strain components in (8.169), we obtain the following expression for the internal work for the volume element dV around any point P of the Kirchhoff plate:

$$dW_i = \left(\sigma_{xx} \varepsilon_{xx} + \sigma_{yy} \varepsilon_{yy} + 2\tau_{xy} \gamma_{xy} \right) dV, \tag{10.23}$$

with σ_{xx}, σ_{yy}, and τ_{xy} respectively the normal and shear stresses associated to the normal and distortion strain components. Thus, the stress tensor in each point of the plate is

$$[\mathbf{T}] = \begin{bmatrix} \sigma_{xx}(x,y,z) & \tau_{xy}(x,y,z) & 0 \\ \tau_{xy}(x,y,z) & \sigma_{yy}(x,y,z) & 0 \\ 0 & 0 & 0 \end{bmatrix}. \tag{10.24}$$

In order to obtain the total strain internal work, we integrate the internal work (10.23) over the volume V of the plate as

$$W_i = \int_V dW_i = \int_V \left[\sigma_{xx} \varepsilon_{xx} + \sigma_{yy} \varepsilon_{yy} + 2\tau_{xy} \gamma_{xy} \right] dV. \tag{10.25}$$

Introducing the expressions for the strain components ε_{xx}, ε_{yy}, and γ_{xy} given in (10.13), we obtain

$$\begin{aligned} W_i = & \int_V \left[\sigma_{xx}(x,y,z) \left(\frac{\partial u_{x_0}(x,y)}{\partial x} - z \frac{\partial^2 u_z(x,y)}{\partial x^2} \right) + \right. \\ & \sigma_{yy}(x,y,z) \left(\frac{\partial u_{y_0}(x,y)}{\partial y} - z \frac{\partial^2 u_z(x,y)}{\partial y^2} \right) + \\ & \left. \tau_{xy}(x,y,z) \left(\frac{\partial u_{y_0}(x,y)}{\partial x} + \frac{\partial u_{x_0}(x,y)}{\partial y} - 2z \frac{\partial^2 u_z(x,y)}{\partial x \partial y} \right) \right] dV. \end{aligned} \tag{10.26}$$

As the displacement components u_{x_0}, u_{y_0}, and u_z vary only with coordinates x and y, the volume integral in (10.26) can be split as the product of integrals along the plate thickness t and plane xy. Thus, we have for each term of the previous integrand

$$\begin{aligned}
-\int_V \left[\sigma_{xx}(x,y,z) \left(z \frac{\partial^2 u_z(x,y)}{\partial x^2} \right) \right] dV &= -\int_A \left(\int_{-t/2}^{t/2} z\, \sigma_{xx}(x,y,z) dz \right) \frac{\partial^2 u_z(x,y)}{\partial x^2} dA, \\
-\int_V \left[\sigma_{yy}(x,y,z) \left(z \frac{\partial^2 u_z(x,y)}{\partial y^2} \right) \right] dV &= -\int_A \left(\int_{-t/2}^{t/2} z\, \sigma_{yy}(x,y,z) dz \right) \frac{\partial^2 u_z(x,y)}{\partial y^2} dA, \\
-2\int_V \left[\tau_{xy}(x,y,z) \left(z \frac{\partial^2 u_z(x,y)}{\partial x \partial y} \right) \right] dV &= -2\int_A \left(\int_{-t/2}^{t/2} z\, \tau_{xy}(x,y,z) dz \right) \frac{\partial^2 u_z(x,y)}{\partial x \partial y} dA, \\
\int_V \left[\sigma_{xx}(x,y,z) \left(\frac{\partial u_{x_0}(x,y)}{\partial x} \right) \right] dV &= \int_A \left(\int_{-t/2}^{t/2} \sigma_{xx}(x,y,z) dz \right) \frac{\partial u_{x_0}(x,y)}{\partial x} dA, \\
\int_V \left[\sigma_{xx}(x,y,z) \left(\frac{\partial u_{y_0}(x,y)}{\partial y} \right) \right] dV &= \int_A \left(\int_{-t/2}^{t/2} \sigma_{xx}(x,y,z) dz \right) \frac{\partial u_{y_0}(x,y)}{\partial y} dA, \\
\int_V \left[\tau_{xy}(x,y,z) \left(\frac{\partial u_{y_0}(x,y)}{\partial x} + \frac{\partial u_{x_0}(x,y)}{\partial y} \right) \right] dV &= \int_A \left(\int_{-t/2}^{t/2} \tau_{xy}(x,y,z) dz \right) \\
& \left(\frac{\partial u_{y_0}(x,y)}{\partial x} + \frac{\partial u_{x_0}(x,y)}{\partial y} \right) dA.
\end{aligned} \tag{10.27}$$

The interpretation of the physical meaning of each integral along the plate thickness is now considered.

Figure 10.7 illustrates the volume element $dxdydz$ located at z units from the plate reference surface. From Figure 10.7(a), we observe that the product $\sigma_{xx}dydz$ gives force dF_x in direction x. Multiplying this force by z, we obtain the respective moment $dM_{xx} = zdF_x = z\sigma_{xx}dydz$ in direction y. The term $\sigma_{xx}dz$ represents a force by units of length in direction x. When multiplied by z, it results in the moment $dM_{xx} = z\sigma_{xx}dz$ by units of length in direction y. Integrating on the thickness, we have the bending moment by units of length relative to axis y on each point of the reference surface. Thus,

$$M_{xx}(x,y) = \int_{-t/2}^{t/2} z\sigma_{xx}(x,y,z) \, dz. \tag{10.28}$$

Despite the bending moment in direction y, we use symbol M_{xx} for compatibility with the stress component σ_{xx} that originated the bending moment. The idea of employing moments by units of length comes from the fact that the above integral is calculated only on the plate thickness t.

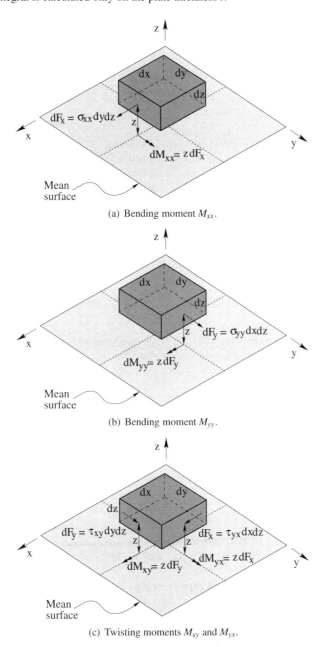

Figure 10.7 Interpretation of the integrals along the thickness of the Kirchhoff plate (adapted from [41]).

We also verify that due to the membrane effect, force $\sigma_{xx}dz$ by units of length, integrated on the plate thickness, results in the normal force by units of length in direction x. This normal force is denoted by $N_{xx}(x,y)$ and given by the following expression

$$N_{xx}(x,y) = \int_{-t/2}^{t/2} \sigma_{xx}(x,y,z)dz. \tag{10.29}$$

From Figure 10.7(b), we observe that $\sigma_{yy}dxdz$ indicates the force dF_y in direction y. When multiplying this force by z, we have the moment $dM_{yy} = zdF_y = z\sigma_{yy}dxdz$ in direction x. Thus, the compatible bending moment M_{yy} by units of length in direction x with stress component σ_{yy} that originated it is,

$$M_{yy}(x,y) = \int_{-t/2}^{t/2} z\sigma_{yy}(x,y,z)dz. \tag{10.30}$$

The force by units of length $\sigma_{yy}dz$ integrated on the thickness results in the normal force by units of length $N_{yy}(x,y)$ given by

$$N_{yy}(x,y) = \int_{-t/2}^{t/2} \sigma_{yy}(x,y,z)dz. \tag{10.31}$$

To interpret the third integral on the thickness in (10.27), consider Figure 10.7(c). The terms $dF_x = \tau_{yx}dxdz$ and $dF_y = \tau_{xy}dydz$ represent, respectively, forces in directions x and y. When multiplying these forces by z, we have respectively the twisting moments $dM_{yx} = zdF_x = z\tau_{yx}dxdz$ and $dM_{xy} = zdF_y = z\tau_{xy}dydz$ in the differential element in the y and x directions. On the other hand, $\tau_{xy}zdz$ and $\tau_{yx}zdz$ represent the twisting moments by units of length. Note that in Figure 10.7(c), the effect of these moments is to twist the differential element. Analogous to the previous cases, twisting moments M_{xy} and M_{yx} by units of length are obtained by integrating the respective differential moments by units of length on the plate thickness, that is,

$$M_{xy}(x,y) = \int_{-t/2}^{t/2} z\tau_{xy}(x,y,z)dz, \tag{10.32}$$

$$M_{yx}(x,y) = \int_{-t/2}^{t/2} z\tau_{yx}(x,y,z)dz. \tag{10.33}$$

Finally, forces $\tau_{yx}dz$ and $\tau_{xy}dz$ by units of length, when integrated on the plate thickness, result in tangent forces by units of length in the points of the reference surface, that is,

$$N_{xy}(x,y) = \int_{-t/2}^{t/2} \tau_{xy}(x,y,z)dz, \tag{10.34}$$

$$N_{yx}(x,y) = \int_{-t/2}^{t/2} \tau_{yx}(x,y,z)dz. \tag{10.35}$$

Due to the symmetry of the Cauchy stress tensor, shear stress components τ_{xy} and τ_{yx} are equal, and consequently

$$M_{xy}(x,y) = M_{yx}(x,y), \tag{10.36}$$
$$N_{xy}(x,y) = N_{yx}(x,y). \tag{10.37}$$

Figures 10.8(a), 10.8(b), and 10.8(c) illustrate respectively bending moments M_{xx} and M_{yy}, twisting moments M_{xy} and M_{yx}, and normal forces N_{xx}, N_{yy}, N_{xy}, and N_{yx} in the plate differential element with volume $dxdydz$.

Using (10.28) to (10.37), the internal work expression (10.26) can be rewritten as

$$\begin{aligned}
W_i &= -\int_A M_{xx}(x,y)\frac{\partial^2 u_z(x,y)}{\partial x^2}dA - \int_A M_{yy}(x,y)\frac{\partial^2 u_z(x,y)}{\partial y^2}dA \\
&\quad -2\int_A M_{xy}(x,y)\frac{\partial^2 u_z(x,y)}{\partial x\partial y}dA + \int_A N_{xx}(x,y)\frac{\partial u_{x_0}(x,y)}{\partial x}dA \\
&\quad +\int_A N_{yy}(x,y,z)\frac{\partial u_{x_0}(x,y)}{\partial y}dA + \int_A N_{xy}(x,y,z)\left(\frac{\partial u_{y_0}(x,y)}{\partial x} + \frac{\partial u_{x_0}(x,y)}{\partial y}\right)dA.
\end{aligned} \tag{10.38}$$

Displacement $u_z(x,y)$ in the above expression has second-order derivatives, while membrane displacements $u_{x_0}(x,y)$ and $u_{x_0}(x,y)$ have first-order derivatives. We wish to obtain a equation containing only displacement

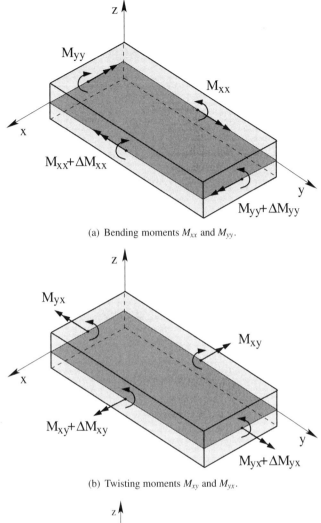

(a) Bending moments M_{xx} and M_{yy}.

(b) Twisting moments M_{xy} and M_{yx}.

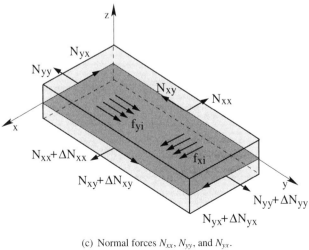

(c) Normal forces N_{xx}, N_{yy}, and N_{yx}.

Figure 10.8   Bending and twisting moments and normal forces in the Kirchhoff plate (adapted from [41]).

functions $u_z(x,y)$, $u_{x_0}(x,y)$, and $u_{y_0}(x,y)$. For this purpose, we integrate twice by parts the terms containing $u_z(x,y)$ and once the terms with $u_{x_0}(x,y)$ and $u_{y_0}(x,y)$. The first integration by parts gives

$$
\begin{aligned}
W_i &= \int_A \frac{\partial M_{xx}(x,y)}{\partial x} \frac{\partial u_z(x,y)}{\partial x} dA - \int_{\partial A} M_{xx}(x,y) \frac{\partial u_z(x,y)}{\partial x} n_x(x,y) d\partial A \\
&+ \int_A \frac{\partial M_{yy}(x,y)}{\partial y} \frac{\partial u_z(x,y)}{\partial y} dA - \int_{\partial A} M_{yy}(x,y) \frac{\partial u_z(x,y)}{\partial y} n_y(x,y) d\partial A \\
&+ \int_A \left(\frac{\partial M_{xy}(x,y)}{\partial x} \frac{\partial u_z(x,y)}{\partial y} + \frac{\partial M_{xy}(x,y)}{\partial y} \frac{\partial u_z(x,y)}{\partial x} \right) dA \\
&- \int_{\partial A} \left(M_{xy}(x,y) \frac{\partial u_z(x,y)}{\partial y} n_x(x,y) + M_{xy}(x,y) \frac{\partial u_z(x,y)}{\partial x} n_y(x,y) \right) d\partial A \\
&- \int_A \frac{\partial N_{xx}(x,y)}{\partial x} u_{x_0}(x,y) dA + \int_{\partial A} N_{xx}(x,y) u_{x_0}(x,y) n_x(x,y) d\partial A \\
&- \int_A \frac{\partial N_{yy}(x,y)}{\partial y} u_{y_0}(x,y) dA + \int_{\partial A} N_{yy}(x,y) u_{y_0}(x,y) n_y(x,y) d\partial A \\
&- \int_A \frac{\partial N_{xy}(x,y)}{\partial x} u_{y_0}(x,y) dA + \int_{\partial A} N_{xy}(x,y) u_{y_0}(x,y) n_x(x,y) d\partial A \\
&- \int_A \frac{\partial N_{xy}(x,y)}{\partial y} u_{x_0}(x,y) dA + \int_{\partial A} N_{xy}(x,y) u_{x_0}(x,y) n_y(x,y) d\partial A,
\end{aligned}
$$

where $n_x(x,y)$ and $n_y(x,y)$ are the components of the normal vector $\mathbf{n}$ on each point with coordinates (x,y) of the mean surface boundary ∂A. Integrating the terms in $u_z(x,y)$ by parts again and rearranging the result, we have

$$
\begin{aligned}
W_i &= -\int_A \left(\frac{\partial^2 M_{xx}(x,y)}{\partial x^2} + 2 \frac{\partial^2 M_{xy}(x,y)}{\partial x \partial y} + \frac{\partial^2 M_{yy}(x,y)}{\partial y^2} \right) u_z(x,y) dA \\
&+ \int_{\partial A} \left(\frac{\partial M_{xx}(x,y)}{\partial x} + \frac{\partial M_{xy}(x,y)}{\partial y} \right) n_x(x,y) u_z(x,y) d\partial A \\
&+ \int_{\partial A} \left(\frac{\partial M_{yy}(x,y)}{\partial y} + \frac{\partial M_{xy}(x,y)}{\partial x} \right) n_y(x,y) u_z(x,y) d\partial A \\
&- \int_{\partial A} \left(M_{xx}(x,y) n_x(x,y) + M_{xy}(x,y) n_y(x,y) \right) \frac{\partial u_z(x,y)}{\partial x} d\partial A \\
&- \int_{\partial A} \left(M_{xy}(x,y) n_x(x,y) + M_{yy}(x,y) n_y(x,y) \right) \frac{\partial u_z(x,y)}{\partial y} d\partial A \\
&- \int_A \left(\frac{\partial N_{xx}(x,y)}{\partial x} + \frac{\partial N_{xy}(x,y)}{\partial y} \right) u_{x_0}(x,y) dA \\
&- \int_A \left(\frac{\partial N_{yy}(x,y)}{\partial y} + \frac{\partial N_{xy}(x,y)}{\partial x} \right) u_{y_0}(x,y) dA \\
&+ \int_{\partial A} \left(N_{xx}(x,y) n_x(x,y) + N_{xy}(x,y) n_y(x,y) \right) u_{x_0}(x,y) d\partial A \\
&+ \int_{\partial A} \left(N_{xy}(x,y) n_x(x,y) + N_{yy}(x,y) n_y(x,y) \right) u_{y_0}(x,y) d\partial A.
\end{aligned}
\tag{10.39}
$$

Now we interpret the physical meaning of each area integrand of the previous expression. Suppose that the forces are given in Newtons (N) and the length units in meters (m). The dimensional analysis of term

$$
\left[\frac{\partial^2 M_{xx}}{\partial x^2} \right] = \frac{\frac{\mathrm{Nm}}{\mathrm{m}}}{\mathrm{m}^2} = \frac{\mathrm{N}}{\mathrm{m}^2},
$$

reveals that it represents a transversal distributed load, because M_{xx} is a moment in y, when divided by x^2 results in a distributed force in z. The same occurs for the other terms in the first integrand of expression (10.39). Thus, the sum of the terms in the first integrand results in an internal distributed load $q_i(x,y)$ in direction z, as illustrated in Figure 10.9. Thus,

$$q_i(x,y) = \frac{\partial^2 M_{xx}(x,y)}{\partial x^2} + 2\frac{\partial^2 M_{xy}(x,y)}{\partial x \partial y} + \frac{\partial^2 M_{yy}(x,y)}{\partial y^2}. \tag{10.40}$$

The term $\dfrac{\partial M_{xx}}{\partial x}$ represents a shear force by units of length in direction z. Observe that M_{xx} is a bending moment by units of length in direction y, which when divided by x results in a force in z, according to the following dimensional analysis:

$$\left[\frac{\partial M_{xx}}{\partial x}\right] = \frac{\frac{Nm}{m}}{m} = \frac{N}{m}.$$

Analogously, $\dfrac{\partial M_{xy}}{\partial y}$ represents a shear force in z, because M_{xy} is a twisting moment in x which, when divided by y, results in a force in z. Thus, the shear force Q_{xz} is defined by the following relation:

$$Q_{xz}(x,y) = \frac{\partial M_{xx}(x,y)}{\partial x} + \frac{\partial M_{xy}(x,y)}{\partial y}. \tag{10.41}$$

Analogously, shear force Q_{yz} is defined as

$$Q_{yz}(x,y) = \frac{\partial M_{yy}(x,y)}{\partial y} + \frac{\partial M_{xy}(x,y)}{\partial x}. \tag{10.42}$$

The shear forces Q_{xz} and Q_{yz} are illustrated in Figure 10.9. Regardless of the shear forces are in transversal direction z, the denomination Q_{xz} and Q_{yz} is only for compatibility with the notation for the bending moments M_{xx} and M_{yy}, which generated the respective forces.

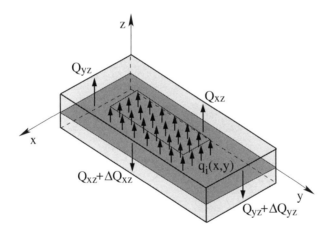

Figure 10.9 Shear forces Q_x and Q_y and distributed load q_i.

Considering now the area integrals in terms of the normal forces, we observe from the dimensional analysis of term $\dfrac{\partial N_{xx}}{\partial x}$

$$\left[\frac{\partial N_{xx}}{\partial x}\right] = \frac{\frac{N}{m}}{m} = \frac{N}{m^2},$$

that the derivatives of the normal forces define tangent distributed forces to the reference surface. Then we define the following distributed load components in directions x and y, illustrated in Figure 10.8(c). Thus,

$$f_{xi}(x,y) = \frac{\partial N_{xx}(x,y)}{\partial x} + \frac{\partial N_{xy}(x,y)}{\partial y}, \tag{10.43}$$

$$f_{yi}(x,y) = \frac{\partial N_{yy}(x,y)}{\partial y} + \frac{\partial N_{xy}(x,y)}{\partial x}. \tag{10.44}$$

Using expressions (10.40) to (10.43), the internal work (10.39) can be rewritten as

$$
\begin{aligned}
W_i \; = \; & -\int_A q_i(x,y)u_z(x,y)dA \\
& +\int_{\partial A} \left(Q_{xz}(x,y)n_x(x,y)+Q_{yz}(x,y)n_y(x,y) \right) u_z(x,y)d\partial A \\
& -\int_{\partial A} \left(M_{xx}(x,y)n_x(x,y)+M_{xy}(x,y)n_y(x,y) \right) \frac{\partial u_z(x,y)}{\partial x}d\partial A \\
& -\int_{\partial A} \left(M_{xy}(x,y)n_x(x,y)+M_{yy}(x,y)n_y(x,y) \right) \frac{\partial u_z(x,y)}{\partial y}d\partial A \\
& -\int_A f_{xi}(x,y)u_{x_0}(x,y)dA-\int_A f_{yi}(x,y)u_{y_0}(x,y)dA \\
& +\int_{\partial A} \left(N_{xx}(x,y,z)n_x+N_{xy}(x,y) \right) n_y(x,y)u_{x_0}(x,y)d\partial A \\
& +\int_{\partial A} \left(N_{xy}(x,y)n_x+N_{yy}(x,y) \right) n_y(x,y)u_{y_0}(x,y)d\partial A.
\end{aligned}
\tag{10.45}
$$

For nonrectangular plates, it is essential to indicate the boundary integrands ∂A in (10.45) along the normal (n) and tangent (t) directions on each point of the boundary, as illustrated in Figure 10.10. Taking α the angle between the directions x and n, the following relations are valid for the coordinates (x,y) and (n,t) of any point:

$$
\begin{aligned}
n &= x\cos\alpha+y\sin\alpha = xn_x+yn_y, \\
t &= -x\sin\alpha+y\cos\alpha = yn_x-xn_y.
\end{aligned}
\tag{10.46}
$$

with $n_x=\cos\alpha$ and $n_y=\sin\alpha$ the direction cosines of the unit vectors $\mathbf{e}_n$ and $\mathbf{e}_t$.

Analogously, the displacements u_{0_n} and u_{0_t} in the normal and tangent directions and u_{x_0} and u_{y_0} in directions x and y are related as (see Figure 10.10)

$$
\begin{aligned}
u_{0_n}(x,y,z) &= u_{x_0}(x,y,z)n_x+u_{y_0}(x,y,z)n_y, \\
u_{0_t}(x,y,z) &= -u_{x_0}(x,y,z)n_y+u_{y_0}(x,y,z)n_x.
\end{aligned}
\tag{10.47}
$$

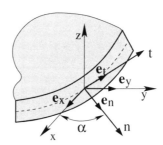

Figure 10.10 Coordinate system ntz on the boundary of a nonrectangular plate.

Given the function $u_z(n(x,y),t(x,y))$, the partial derivatives relative to x and y can be expressed in terms of the normal n and tangent t directions using the chain rule, i.e.,

$$
\frac{\partial u_z}{\partial x} = \frac{\partial u_z}{\partial n}\frac{\partial n}{\partial x}+\frac{\partial u_z}{\partial t}\frac{\partial t}{\partial x} = \frac{\partial u_z}{\partial n}n_x-\frac{\partial u_z}{\partial t}n_y,
\tag{10.48}
$$

$$
\frac{\partial u_z}{\partial y} = \frac{\partial u_z}{\partial n}\frac{\partial n}{\partial y}+\frac{\partial u_z}{\partial t}\frac{\partial t}{\partial y} = \frac{\partial u_z}{\partial t}n_x+\frac{\partial u_z}{\partial n}n_y.
\tag{10.49}
$$

The transformation of the stress tensor from the Cartesian system (x,y,z) to the circumferential system (n,t,z) is obtained using the transformation for a second-order tensor, that is,

$$
\begin{bmatrix} \sigma_{nn} & \tau_{nt} & \tau_{nz} \\ \tau_{nt} & \sigma_{tt} & \tau_{tz} \\ \tau_{nz} & \tau_{tz} & \sigma_{zz} \end{bmatrix} = \begin{bmatrix} n_x & n_y & 0 \\ -n_y & n_x & 0 \\ 0 & 0 & 1 \end{bmatrix} \begin{bmatrix} \sigma_{xx} & \tau_{xy} & \tau_{xz} \\ \tau_{xy} & \sigma_{yy} & \tau_{yz} \\ \tau_{xz} & \tau_{yz} & \sigma_{zz} \end{bmatrix} \begin{bmatrix} n_x & -n_y & 0 \\ n_y & n_x & 0 \\ 0 & 0 & 1 \end{bmatrix}.
\tag{10.50}
$$

Developing the previous product, we obtain

$$
\begin{aligned}
\sigma_{nn} &= \sigma_{xx}n_x^2 + 2\tau_{xy}n_y n_x + \sigma_{yy}n_y, \\
\tau_{nt} &= -\sigma_{xx}n_x n_y + \tau_{xy}(n_x^2 - n_y^2) + \sigma_{yy}n_y n_x, \\
\tau_{nz} &= \tau_{xz}n_x + \tau_{yz}n_y, \\
\sigma_{tt} &= \sigma_{xx}n_y^2 - 2\tau_{xy}n_y n_x + \sigma_{yy}n_x^2, \\
\tau_{tz} &= -\tau_{xy}n_y + \tau_{yz}n_x.
\end{aligned}
$$

Using these expressions, we can find the internal forces in the Kirchhoff plate in the circumferential coordinate system.

The loads due to the membrane effect are calculated as

$$
\begin{aligned}
N_{nn} &= \int_{-t/2}^{t/2} \sigma_{nn}dz = \int_{-t/2}^{t/2} \left(\sigma_{xx}n_x^2 + 2\tau_{xy}n_x n_y + \sigma_{yy}n_y^2 \right) dz \\
&= \left(\int_{-t/2}^{t/2} \sigma_{xx}dz \right) n_x^2 + 2 \left(\int_{-t/2}^{t/2} \tau_{xy}dz \right) n_x n_y + \left(\int_{-t/2}^{t/2} \sigma_{yy}dz \right) n_y \\
&= N_{xx}n_x^2 + 2N_{xy}n_x n_y + N_{yy}n_y^2,
\end{aligned}
\tag{10.51}
$$

$$
\begin{aligned}
N_{tt} &= \int_{-t/2}^{t/2} \sigma_{tt}dz = \int_{-t/2}^{t/2} \left(\sigma_{xx}n_y^2 - 2\tau_{xy}n_x n_y + \sigma_{yy}n_x^2 \right) dz \\
&= \left(\int_{-t/2}^{t/2} \sigma_{xx}dz \right) n_y^2 - 2 \left(\int_{-t/2}^{t/2} \tau_{xy}dz \right) n_x n_y + \left(\int_{-t/2}^{t/2} \sigma_{yy}dz \right) n_x^2 \\
&= N_{xx}n_y^2 - 2N_{xy}n_x n_y + N_{yy}n_x^2,
\end{aligned}
\tag{10.52}
$$

$$
\begin{aligned}
N_{nt} &= \int_{-t/2}^{t/2} \tau_{nt}dz = \left[\int_{-t/2}^{t/2} \left(\sigma_{yy} - \sigma_{xx} \right) n_x n_y + \tau_{xy} \left(n_x^2 - n_y \right) \right] dz \\
&= \left(\int_{-t/2}^{t/2} \sigma_{yy}dz - \int_{-t/2}^{t/2} \sigma_{xx}dz \right) n_x n_y + \left(\int_{-t/2}^{t/2} \tau_{xy}dz \right) \left(n_x^2 - n_y^2 \right) \\
&= \left(N_{yy} - N_{xx} \right) n_x n_y + N_{xy} \left(n_x^2 - n_y^2 \right).
\end{aligned}
\tag{10.53}
$$

Similarly, we find the shear loads in the circumferential system as

$$
\begin{aligned}
Q_n &= \int_{-t/2}^{t/2} \tau_{nz}dz = \int_{-t/2}^{t/2} \left(\tau_{xz}n_x + \tau_{yz}n_y \right) dz \\
&= -\left(\int_{-t/2}^{t/2} \tau_{xz}dz \right) n_x + \left(\int_{-t/2}^{t/2} \tau_{yz}dz \right) n_y \\
&= Q_{xz}n_x + Q_{yz}n_y,
\end{aligned}
\tag{10.54}
$$

$$
\begin{aligned}
Q_t &= \int_{-t/2}^{t/2} \tau_{tz}dz = \int_{-t/2}^{t/2} \left(-\tau_{xy}n_y + \tau_{yz}n_x \right) dz \\
&= -\left(\int_{-t/2}^{t/2} \tau_{xz}dz \right) n_y + \left(\int_{-t/2}^{t/2} \tau_{yz}dz \right) n_x \\
&= -Q_{xz}n_y + Q_{yz}n_x.
\end{aligned}
\tag{10.55}
$$

The bending and twisting moments in the circumferential system are

$$
\begin{aligned}
M_{nn} &= -\int_{-t/2}^{t/2} \sigma_{nn} z \, dz = -\int_{-t/2}^{t/2} \left(\sigma_{xx}n_x^2 + 2\tau_{xy}n_x n_y + \sigma_{yy}n_y \right) z \, dz \\
&= \left(-\int_{-t/2}^{t/2} \sigma_{xx} z \, dz \right) n_x^2 + 2 \left(-\int_{-t/2}^{t/2} \tau_{xy} z \, dz \right) n_x n_y + \left(-\int_{-t/2}^{t/2} \sigma_{yy} z \, dz \right) n_y^2 \\
&= M_{xx}n_x^2 + 2M_{xy}n_x n_y + M_{yy}n_y^2,
\end{aligned}
\tag{10.56}
$$

$$
\begin{aligned}
M_{tt} &= -\int_{-t/2}^{t/2} \sigma_{tt}\, z\, dz = -\int_{-t/2}^{t/2} \left(\sigma_{xx} n_y - 2\tau_{xy} n_x n_y + \sigma_{yy} n_x^2 \right) z\, dz \\
&= \left(-\int_{-t/2}^{t/2} \sigma_{xx}\, z\, dz \right) n_y^2 + 2\left(-\int_{-t/2}^{t/2} \tau_{xy}\, z\, dz \right) n_x n_y + \left(-\int_{-t/2}^{t/2} \sigma_{yy}\, z\, dz \right) n_x^2 \\
&= M_{xx} n_y + 2 M_{xy} n_x n_y + M_{yy} n_x^2,
\end{aligned}
\tag{10.57}
$$

$$
\begin{aligned}
M_{nt} &= -\int_{-t/2}^{t/2} \tau_{nt}\, z\, dz = -\int_{-t/2}^{t/2} \left[\left(\sigma_{yy} - \sigma_{xx} \right) n_x n_y + \tau_{xy} \left(n_x^2 - n_y \right) \right] z\, dz \\
&= \left[-\int_{-t/2}^{t/2} \sigma_{yy}\, dz - \left(-\int_{-t/2}^{t/2} \sigma_{xx}\, dz \right) \right] n_x n_y + \left(-\int_{-t/2}^{t/2} \tau_{xy}\, dz \right) \left(n_x^2 - n_x^2 \right) \\
&= \left(M_{yy} - M_{xx} \right) n_x n_y + M_{xy} \left(n_x^2 - n_y \right).
\end{aligned}
\tag{10.58}
$$

Equation (10.54) allows to express the first boundary integrand in (10.45) using the shear force Q_n, that is,

$$
\int_{\partial A} \left(Q_{xz} n_x + Q_{yz} n_y \right) u_z\, d\partial A = \int_{\partial A} Q_n u_z\, d\partial A.
\tag{10.59}
$$

Substituting (10.48) and (10.49) and equations (10.56) and (10.58) in the second boundary integrand in (10.45), we verify that

$$
\int_{\partial A} \left[\left(M_{xx} n_x + M_{xy} n_y \right) \frac{\partial u_z}{\partial x} + \left(M_{xy} n_x + M_{yy} n_y \right) \frac{\partial u_z}{\partial y} \right] d\partial A = \int_{\partial A} \left[M_{nn} \frac{\partial u_z}{\partial n} + M_{nt} \frac{\partial u_z}{\partial t} \right] d\partial A.
\tag{10.60}
$$

Replacing also (10.51) and (10.53) in the boundary integrals of the terms related to membrane effects, we have

$$
\int_{\partial A} \left[\left(N_{xx} n_x + N_{xy} n_y \right) u_{x_0} + \left(N_{xy} n_x + N_{yy} n_y \right) u_{y_0} \right] d\partial A = \int_{\partial A} \left[N_{nn} u_{0_n} + N_{nt} u_{0_t} \right] d\partial A.
\tag{10.61}
$$

Substituting (10.59), (10.60), and (10.61) in (10.45) we obtain

$$
\begin{aligned}
W_i &= -\int_A q_i(x,y) u_z(x,y)\, dA - \int_A f_{xi}(x,y)\, u_{x_0}(x,y)\, dA - \int_A f_{yi}(x,y) u_{y_0}(x,y)\, dA \\
&\quad - \int_{\partial A} \left[M_{nn}(x,y) \frac{\partial u_z(x,y)}{\partial n} + M_{nt}(x,y) \frac{\partial u_z(x,y)}{\partial t} \right] d\partial A \\
&\quad - \int_{\partial A} Q_n(x,y) u_z(x,y)\, d\partial A + \int_{\partial A} \left(N_{nn}(x,y) u_{0_n}(x,y) + N_{nt}(x,y) u_{0_t}(x,y) \right) d\partial A.
\end{aligned}
\tag{10.62}
$$

We can also integrate by parts the term M_{nt} to obtain

$$
\int_{\partial A} M_{nt}(x,y) \frac{\partial u_z(x,y)}{\partial t}\, d\partial A = \left. M_{nt} u_z \right|_1^2 - \int_{\partial A} \frac{\partial M_{nt}(x,y)}{\partial t} u_z(x,y)\, d\partial A,
$$

with 1 and 2 points on boundary ∂A with no geometric discontinuities between them, as illustrated in Figure 10.11. For N points P_i on ∂A with geometrical discontinuities, the above expression is rewritten as

$$
\int_{\partial A} M_{nt}(x,y) \frac{\partial u_z(x,y)}{\partial t}\, d\partial A = -\int_{\partial A} \frac{\partial M_{nt}(x,y)}{\partial t} u_z(x,y)\, d\partial A + \sum_{i=1}^{N} \left[\left(M_{nt}^+ - M_{nt}^- \right) u_z(x,y) \right]_{P_i}.
\tag{10.63}
$$

In this case, M_{nt}^+ and M_{nt}^- represent respectively the values to the right and left of point P_i, when circulating along ∂A in the positive tangent direction t, as illustrated in Figure 10.11.

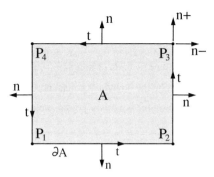

Figure 10.11 Points with geometrical discontinuities on the reference surface for a rectangular plate (adapted from [41]).

Substituting (10.63) in (10.62), expression for the internal work is given by

$$
\begin{aligned}
W_i \;=\; & -\int_A q_i(x,y)u_z(x,y)dA - \int_A f_{xi}(x,y)\,u_{x_0}(x,y)dA - \int_A f_{yi}(x,y)u_{y_0}(x,y)dA \\
& + \int_{\partial A}\left[Q_n(x,y) + \frac{\partial M_{nt}(x,y)}{\partial t}\right] u_z(x,y)d\partial A - \int_{\partial A} M_{nn}(x,y)\frac{\partial u_z(x,y)}{\partial n}d\partial A \\
& + \int_{\partial A}\left[N_{nn}(x,y)u_{0_n}(x,y) + N_{nt}(x,y)u_{0_t}(x,y)\right]d\partial A \\
& - \sum_{i=1}^{N}\left[\left(M_{nt}^{+} - M_{nt}^{-}\right)u_z(x,y)\right]_{P_i}.
\end{aligned}
\tag{10.64}
$$

Analogous to (10.41), shear force Q_n is given by

$$
Q_n(x,y) = \frac{\partial M_{nn}(x,y)}{\partial n} + \frac{\partial M_{nt}(x,y)}{\partial t}.
\tag{10.65}
$$

Note that M_{nn} represents a bending moment by units of length in the tangent direction t. Then, derivative $\dfrac{\partial M_{nn}}{\partial n}$ indicates a transversal force by units of length in direction z. Analogously for the term $\dfrac{\partial M_{nt}}{\partial t}$. Thus, we denote

$$
V_n(x,y) = Q_n(x,y) + \frac{\partial M_{nt}(x,y)}{\partial t} = \frac{\partial M_{nn}(x,y)}{\partial n} + 2\frac{\partial M_{nt}(x,y)}{\partial t}
\tag{10.66}
$$

as the internal transversal force by units of length in direction z on the boundary of the reference surface. Thus, the final expression of the internal work for the Kirchhoff plate is written as

$$
\begin{aligned}
W_i \;=\; & \int_A q_i(x,y)u_z(x,y)dA + \int_A f_{xi}(x,y)\,u_{x_0}(x,y)dA + \int_A f_{yi}(x,y)u_{y_0}(x,y)dA \\
& - \int_{\partial A} V_n(x,y)u_z(x,y)d\partial A + \int_{\partial A} M_{nn}(x,y)\frac{\partial u_z(x,y)}{\partial n}d\partial A \\
& - \int_{\partial A}\left(N_{nn}(x,y)u_{0_n}(x,y) + N_{nt}(x,y)u_{0_t}(x,y)\right)d\partial A \\
& + \sum_{i=1}^{N}\left[\left(M_{nt}^{+} - M_{nt}^{-}\right)u_z(x,y)\right]_{P_i}.
\end{aligned}
\tag{10.67}
$$

Hence, the internal loads illustrated in Figure 10.12(a) for the Kirchhoff plate model, including the membrane effect, are:

- q_i: internal transversal distributed load to the reference surface of the plate, originated by bending moments (M_{xx} and M_{yy}) and twisting moments (M_{xy} and M_{yx}) by units of length

- V_n: shear force by units of length in direction z on the boundary points

- M_{nn}: bending moment by units of length in the tangent direction t on the boundary points

- M_{nt}^+ and M_{nt}^-: internal concentrated transversal loads on the right and left of points P_i of the plate boundary with geometric discontinuities

- f_{xi} and f_{yi}: internal tangent distributed loads to the mean surface of the plate, from the normal (N_{xx} and N_{yy}) and shear (N_{xy} and N_{yx}) forces by units of length

- N_{nn}: normal force by units of length on the boundary points, normal to the lateral surface of the plate

- N_{nt}: tangent force by units of length on the boundary points, tangent to the lateral surface of the plate

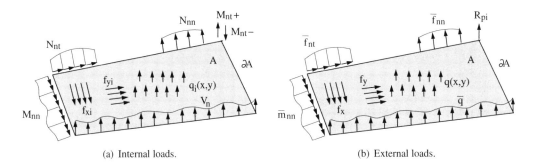

(a) Internal loads.   (b) External loads.

Figure 10.12 Internal and external loads due to the bending and membrane effects in the Kirchhoff plate (adapted from [41]).

10.6 DETERMINATION OF EXTERNAL LOADS AND EQUILIBRIUM

The external loads that can be applied to a Kirchhoff plate are those which can be balanced by the respective internal loads defined in the previous section. Thus, the external loads are the following:

- q: external transversal distributed load to the mean surface of the plate

- $\bar{q}$: transversal distributed load to the mean surface on the boundary ∂A

- $\bar{m}_{nn}$: distributed moment in the tangent direction t of boundary ∂A

- R_{P_i}: concentrated forces in the transversal direction of the plate on points P_i of the boundary with geometrical discontinuities

- f_x: external distributed load in direction x, tangent to the mean surface of the plate

- f_y: external distributed load in direction y, tangent to the mean surface of the plate

- $\bar{f}_{nn}$: distributed normal force to the lateral surface of the plate boundary

- $\bar{f}_{nt}$: distributed tangent force to the lateral surface of the plate boundary

This set of external loads is illustrated in Figure 10.12(b). The respective external work W_e for the displacement components $\left(u_{x_0}, u_{y_0}, u_z\right)$ is

$$
\begin{aligned}
W_e = & \int_A q(x,y) u_z(x,y) dA + \int_A f_x(x,y) u_{x_0}(x,y)\, dA + \int_A f_y(x,y) u_{y_0}(x,y) dA \\
& + \int_{\partial A} \bar{q}(x,y) u_z(x,y) d\partial A + \int_{\partial A} \bar{m}_{nn} \frac{\partial u_z(x,y)}{\partial n} + \int_{\partial A} \bar{f}_{nn}(x,y) u_n(x,y) d\partial A \qquad (10.68) \\
& + \int_{\partial A} \bar{f}_{nt}(x,y) u_t(x,y) d\partial A + \sum_{i=1}^{N} \left[R_{P_i} u_z(x,y) \right]_{P_i}.
\end{aligned}
$$

The PVW states that the work done by the internal and external loads must be equal for any virtual kinematic action $\delta \mathbf{u} = \left(\delta u_{x_0}, \delta u_{y_0}, \delta u_z\right)$, that is,

$$
\delta W_e = \delta W_i. \qquad (10.69)
$$

Substituting (10.67) and (10.68) in (10.69) and simplifying the common terms, we obtain

$$\int_A \left[q(x,y) + \frac{\partial^2 M_{xx}(x,y)}{\partial x^2} + 2\frac{\partial^2 M_{xy}(x,y)}{\partial x \partial y} + \frac{\partial^2 M_{yy}(x,y)}{\partial y^2} \right] \delta u_z(x,y) dA$$

$$\int_A \left[f_x(x,y) + \frac{\partial N_{xx}(x,y)}{\partial x} + \frac{\partial N_{xy}(x,y)}{\partial y} \right] \delta u_{x_0}(x,y) dA +$$

$$\int_A \left[f_y(x,y) + \frac{\partial N_{yy}(x,y)}{\partial y} + \frac{\partial N_{xy}(x,y)}{\partial x} \right] \delta u_{y_0}(x,y) dA +$$

$$\int_{\partial A} \left[\bar{q}(x,y) - V_n(x,y) \right] \delta u_z(x,y) d\partial A + \int_{\partial A} \left[\bar{m}_{nn}(x,y) + M_{nn}(x,y) \right] \frac{\partial \delta u_z(x,y)}{\partial n} +$$

$$\int_{\partial A} \left[f_{nn}(x,y) - N_{nn}(x,y) \right] \delta u_{0_n}(x,y) d\partial A + \int_{\partial A} \left[f_{nt}(x,y) - N_{nt}(x,y) \right] \delta u_{0_t}(x,y) d\partial A +$$

$$\sum_{i=1}^{N} \left[\left(R_{P_i} + M_{nt}^+ - M_{nt}^- \right) \delta u_z(x,y) \right]_{P_i} = 0.$$

As the virtual displacement action $\delta \mathbf{u} = \left(\delta u_{x_0}, \delta u_{y_0}, \delta u_z \right)$ is arbitrary, the previous expression is zero only if all terms inside brackets are simultaneously zero. This results in the following equilibrium BVP for the Kirchhoff plate:

$$\begin{cases} \dfrac{\partial^2 M_{xx}(x,y)}{\partial x^2} + \dfrac{\partial^2 M_{yy}(x,y)}{\partial y^2} + 2\dfrac{\partial^2 M_{xy}(x,y)}{\partial x \partial y} + q(x,y) = 0 & x,y \in A, \\[2mm] V_n - \bar{q} = 0 & x,y \in \partial A, \\[1mm] M_{nn}(x,y) + \bar{m}_{nn}(x,y) = 0 & x,y \in \partial A, \\[1mm] \left(M_{nt}^+ - M_{nt}^- \right) + R_{P_i} = 0 & i = 1,2,\ldots,N, \\[3mm] \dfrac{\partial N_{xx}(x,y)}{\partial x} + \dfrac{\partial N_{xy}(x,y)}{\partial y} + f_x(x,y) = 0 & x,y \in A, \\[2mm] \dfrac{\partial N_{yy}(x,y)}{\partial y} + \dfrac{\partial N_{xy}(x,y)}{\partial x} + f_y(x,y) = 0 & x,y \in A, \\[2mm] f_{nn}(x,y) - N_{nn}(x,y) = 0 & x,y \in \partial A, \\[1mm] f_{nt}(x,y) - N_{nt}(x,y) = 0 & x,y \in \partial A. \end{cases} \qquad (10.70)$$

The above expressions define the local form or boundary value problem (BVP) of the plate free of kinematic constraints. We have a second-order differential equation in terms of bending and twisting moments, as well as three boundary conditions for the bending problem. The first boundary condition refers to the balance of the shear force V_n with the distributed load $\bar{q}$. The second boundary condition considers the balance of internal M_{nn} and external $\bar{m}_{nn}$ distributed bending moments. The third condition states the balance of internal and external concentrated forces on the points with geometrical discontinuities.

In addition, there are also two first-order equations in terms of the normal forces and two boundary conditions for the membrane problem. These conditions consider the balance of internal tangent forces N_{nn} and N_{nt} with the respective external forces f_{nn} and f_{nt}.

The plate boundary loads are shown in Figure 10.13. The kinematic and force boundary conditions at the plate boundary are

1. Clamped: $u_z = 0$, $\dfrac{\partial u_z}{\partial t} = 0$, $\dfrac{\partial u_z}{\partial n} = 0$

2. Simply supported: $u_z = 0$, $M_{nt} = 0$, $M_{nn} = 0$ or $u_z = 0$, $\dfrac{\partial u_z}{\partial t} = 0$, $M_{nn} = 0$

3. Free: $Q_n = 0$, $M_{nt} = 0$, $M_{nn} = 0$

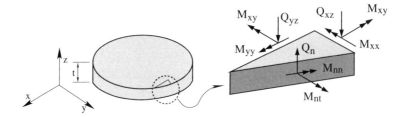

Figure 10.13   Loads on the boundary of a Kirchhoff plate.

10.7   APPLICATION OF THE CONSTITUTIVE EQUATION

We can decompose the stress components, according to the bending and membrane effects, as

$$\sigma_{xx} = \sigma_{xx}^m + \sigma_{xx}^b, \qquad \sigma_{yy} = \sigma_{yy}^m + \sigma_{yy}^b, \qquad \tau_{xy} = \tau_{xy}^m + \tau_{xy}^b.$$

The bending stress and strain components are associated with Hooke's law as

$$
\begin{aligned}
\sigma_{xx}^b(x,y,z) &= \frac{E(x,y)}{1 - v^2(x,y)} \left(\varepsilon_{xx}^b(x,y,z) + v\, \varepsilon_{yy}^b(x,y,z) \right), \\
\sigma_{xx}^b(x,y,z) &= \frac{E(x,y)}{1 - v^2(x,y)} \left(\varepsilon_{xx}^b(x,y,z) + v\, \varepsilon_{yy}^b(x,y,z) \right), \\
\tau_{xy}^b(x,y,z) &= G(x,y)\gamma_{xy}^b(x,y,z) = \frac{E(x,y)}{2(1 + v(x,y))}\gamma_{xy}^b(x,y,z) \\
&= \frac{E(x,y)\left[1 - v(x,y)\right]}{2\left(1 - v^2(x,y)\right)}\gamma_{xy}^b(x,y,z).
\end{aligned}
\tag{10.71}
$$

We suppose that Young's modulus and the Poisson ratio are constant, that is, $E(x,y) = E$ and $v(x,y) = v$. Introducing the strain components in expressions (10.71), we obtain

$$
\begin{aligned}
\sigma_{xx}^b(x,y) &= -z\frac{E}{1 - v^2}\left(\frac{\partial^2 u_z(x,y)}{\partial x^2} + v\frac{\partial^2 u_z(x,y)}{\partial y^2} \right), \\
\sigma_{yy}^b(x,y) &= -z\frac{E}{1 - v^2}\left(\frac{\partial^2 u_z(x,y)}{\partial y^2} + v\frac{\partial^2 u_z(x,y)}{\partial x^2} \right), \\
\tau_{xy}^b(x,y) &= -z\frac{E(1 - v)}{1 - v^2}\frac{\partial^2 u_z(x,y)}{\partial x \partial y}.
\end{aligned}
\tag{10.72}
$$

Substituting the previous expressions in equations (10.28), (10.30), and (10.32), the bending and twisting moments for a plate with an elastic, linear, and isotropic material are given by

$$
\begin{aligned}
M_{xx}(x,y) &= -\int_{-t/2}^{t/2} z^2 \frac{E}{1 - v^2}\left(\frac{\partial^2 u_z(x,y)}{\partial x^2} + v\frac{\partial^2 u_z(x,y)}{\partial y^2} \right) dz \\
&= -D\left(\frac{\partial^2 u_z(x,y)}{\partial x^2} + v\frac{\partial^2 u_z(x,y)}{\partial y^2} \right), \\
M_{yy}(x,y) &= -\int_{-t/2}^{t/2} z^2 \frac{E}{1 - v^2}\left(\frac{\partial^2 u_z(x,y)}{\partial y^2} + v\frac{\partial^2 u_z(x,y)}{\partial x^2} \right) dz \\
&= -D\left(\frac{\partial^2 u_z(x,y)}{\partial y^2} + v\frac{\partial^2 u_z(x,y)}{\partial x^2} \right), \\
M_{xy}(x,y) &= -\int_{-t/2}^{t/2} z^2 \frac{E}{1 + v}\frac{\partial^2 u_z(x,y)}{\partial x \partial y} dz = -D(1 - v)\frac{\partial^2 u_z(x,y)}{\partial x \partial y},
\end{aligned}
\tag{10.73}
$$

where the plate bending stiffness is

$$D = \frac{Et^3}{12\left(1 - v^2\right)}. \tag{10.74}$$

The shear forces Q_x and Q_y given in (10.41) and (10.42) are written as

$$
\begin{aligned}
Q_x(x,y) &= -D\left(\frac{\partial^3 u_z(x,y)}{\partial x^3} + v\frac{\partial^3 u_z(x,y)}{\partial x \partial y^2}\right) - D(1-v)\frac{\partial^3 u_z(x,y)}{\partial x \partial y^2} \\
&= -D\left(\frac{\partial^3 u_z(x,y)}{\partial x^3} + \frac{\partial^3 u_z(x,y)}{\partial x \partial y^2}\right),
\end{aligned} \tag{10.75}
$$

$$
\begin{aligned}
Q_y(x,y) &= -D\left(\frac{\partial^3 u_z(x,y)}{\partial y^3} + v\frac{\partial^3 u_z(x,y)}{\partial x^2 \partial y}\right) - D(1-v)\frac{\partial^3 u_z(x,y)}{\partial x^2 \partial y} \\
&= -D\left(\frac{\partial^3 u_z(x,y)}{\partial y^3} + \frac{\partial^3 u_z(x,y)}{\partial x^2 \partial y}\right).
\end{aligned} \tag{10.76}
$$

Substituting (10.73) in the bending differential equation given in (10.70), we obtain the differential equation in terms of the transversal displacement $u_z(x,y)$, that is,

$$
\frac{\partial^2}{\partial x^2}\left[D\left(\frac{\partial^2 u_z(x,y)}{\partial x^2} + v\frac{\partial^2 u_z(x,y)}{\partial y^2}\right)\right] + \frac{\partial^2}{\partial y^2}\left[D\left(\frac{\partial^2 u_z(x,y)}{\partial y^2} + v\frac{\partial^2 u_z(x,y)}{\partial x^2}\right)\right]
$$
$$
+2\frac{\partial^2}{\partial x \partial y}\left[D(1-v)\frac{\partial^2 u_z(x,y)}{\partial x \partial y}\right] - q(x,y) = 0. \tag{10.77}
$$

We simplify the above expression for D and v constants to

$$
D\left(\frac{\partial^4 u_z(x,y)}{\partial x^4} + \frac{\partial^4 u_z(x,y)}{\partial y^4} + 2\frac{\partial^4 u_z(x,y)}{\partial x^2 \partial y^2}\right) = q(x,y). \tag{10.78}
$$

From equation (10.73), we have

$$
\begin{aligned}
\left(\frac{\partial^2 u_z(x,y)}{\partial x^2} + v\frac{\partial^2 u_z(x,y)}{\partial y^2}\right) &= -\frac{M_{xx}(x,y)}{D}, \\
\left(\frac{\partial^2 u_z(x,y)}{\partial y^2} + v\frac{\partial^2 u_z(x,y)}{\partial x^2}\right) &= -\frac{M_{yy}(x,y)}{D}, \\
\frac{\partial^2 u_z(x,y)}{\partial x \partial y} &= -\frac{M_{xy}(x,y)}{D(1-v)}.
\end{aligned}
$$

Replacing the previous relations in (10.72), we obtain the following expressions for the normal and shear stress components:

$$
\begin{aligned}
\sigma_{xx}^b(x,y) &= \frac{12 M_{xx}(x,y)}{t^3} z, \\
\sigma_{yy}^b(x,y) &= \frac{12 M_{yy}(x,y)}{t^3} z, \\
\tau_{xy}^b(x,y) &= \frac{12 M_{xy}(x,y)}{t^3} z.
\end{aligned} \tag{10.79}
$$

The stress and strain components due to the membrane effect are associated by

$$
\begin{aligned}
\sigma_{xx}^m(x,y,z) &= \frac{E(x,y)}{1 - v^2(x,y)}\left(\varepsilon_{xx}^m(x,y,z) + v\,\varepsilon_{yy}^m(x,y,z)\right), \\
\sigma_{xx}^m(x,y,z) &= \frac{E(x,y)}{1 - v^2(x,y)}\left(\varepsilon_{xx}^m(x,y,z) + v\,\varepsilon_{yy}^m(x,y,z)\right), \\
\tau_{xy}^m(x,y,z) &= G(x,y)\gamma_{xy}^m(x,y,z) = \frac{E(x,y)}{2(1 + v(x,y))}\gamma_{xy}^m(x,y,z) \\
&= \frac{E(x,y)[1 - v(x,y)]}{2[1 - v^2(x,y)]}\gamma_{xy}^m(x,y,z).
\end{aligned} \tag{10.80}
$$

Introducing the membrane strain componentst in the previous equations, we have

$$
\begin{aligned}
\sigma_{xx}^m(x,y) &= \frac{E}{1-v^2}\left(\frac{\partial u_{x_0}(x,y)}{\partial x} + v\frac{\partial u_{y_0}(x,y)}{\partial y}\right), \\
\sigma_{yy}^m(x,y) &= \frac{E}{1-v^2}\left(\frac{\partial u_{y_0}(x,y)}{\partial y} + v\frac{\partial u_{x_0}(x,y)}{\partial x}\right), \\
\tau_{xy}^m(x,y) &= \frac{E(1-v)}{2(1-v^2)}\left(\frac{\partial u_{y_0}(x,y)}{\partial x} + \frac{\partial u_{x_0}(x,y)}{\partial y}\right).
\end{aligned}
\tag{10.81}
$$

Once the stress components are known, we can now relate the normal forces given in (10.29), (10.31), and (10.34) with the displacement field components. Thus, we have

$$
\begin{aligned}
N_{xx}(x,y) &= \int_{-t/2}^{t/2} \frac{E}{1-v^2}\left(\frac{\partial u_{x_0}(x,y)}{\partial x} + v\frac{\partial u_{y_0}(x,y)}{\partial y}\right)dz \\
&= T\left(\frac{\partial u_{x_0}(x,y)}{\partial x} + v\frac{\partial u_{y_0}(x,y)}{\partial y}\right), \\
N_{yy}(x,y) &= \int_{-t/2}^{t/2} \frac{E}{1-v^2}\left(\frac{\partial u_{y_0}(x,y)}{\partial y} + v\frac{\partial u_{x_0}(x,y)}{\partial x}\right)dz \\
&= T\left(\frac{\partial u_{y_0}(x,y)}{\partial y} + v\frac{\partial u_{x_0}(x,y)}{\partial x}\right), \\
N_{xy}(x,y) &= \int_{-t/2}^{t/2} \frac{E}{1+v}\left(\frac{\partial u_{y_0}(x,y)}{\partial x} + \frac{\partial u_{x_0}(x,y)}{\partial y}\right)dz \\
&= T(1-v)\left(\frac{\partial u_{y_0}(x,y)}{\partial x} + \frac{\partial u_{x_0}(x,y)}{\partial y}\right),
\end{aligned}
\tag{10.82}
$$

where the plate stiffness due to the membrane effect is

$$
T = \frac{Et}{(1-v^2)}.
$$

Substituting the above expression in the membrane differential equations given in (10.70), we obtain

$$
\frac{\partial}{\partial x}\left[T\left(\frac{\partial u_{x_0}(x,y)}{\partial x} + v\frac{\partial u_{y_0}(x,y)}{\partial y}\right)\right] + \frac{\partial}{\partial y}\left[T\frac{(1-v)}{2}\left(\frac{\partial u_{y_0}(x,y)}{\partial x} + \frac{\partial u_{x_0}(x,y)}{\partial y}\right)\right] + f_x(x,y) = 0,
$$

$$
\frac{\partial}{\partial y}\left[T\left(\frac{\partial u_{y_0}(x,y)}{\partial y} + v\frac{\partial u_{x_0}(x,y)}{\partial x}\right)\right] + \frac{\partial}{\partial x}\left[T\frac{(1-v)}{2}\left(\frac{\partial u_{y_0}(x,y)}{\partial x} + \frac{\partial u_{x_0}(x,y)}{\partial y}\right)\right] + f_y(x,y) = 0.
\tag{10.83}
$$

These are the differential equations that describe the membrane effect for the Kirchhoff plate. For constant material properties, equations (10.83) simplify to

$$
T\left[\frac{(1-v)}{2}\frac{\partial^2 u_{x_0}}{\partial y^2} + \frac{\partial^2 u_{x_0}}{\partial x^2} + \frac{\partial^2 u_{y_0}}{\partial x\partial y}\right] = -f_x,
$$

$$
T\left[\frac{(1-v)}{2}\frac{\partial^2 u_{y_0}}{\partial x^2} + \frac{\partial^2 u_{y_0}}{\partial y^2} + \frac{\partial^2 u_{x_0}}{\partial x\partial y}\right] = -f_y.
\tag{10.84}
$$

Substituting equations (10.82) in expressions (10.81), we obtain

$$
\begin{aligned}
\sigma_{xx}^m(x,y) &= \frac{N_{xx}(x,y)}{t}, \\
\sigma_{yy}^m(x,y) &= \frac{N_{yy}(x,y)}{t}, \\
\tau_{xy}^m(x,y) &= \frac{N_{xy}(x,y)}{t}.
\end{aligned}
\tag{10.85}
$$

Note that the stress components due to the bending effect have a linear variation with coordinate z, as illustrated in Figure 10.14(a). The stresses due to the membrane effect are constant as shown in Figure 10.14(b).

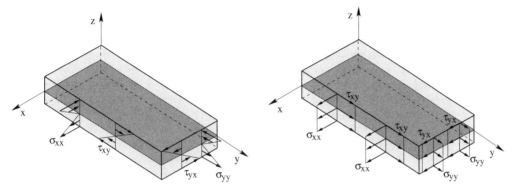

(a) Normal and shear stress distributions due to bending effect.

(b) Normal and shear stress distributions due to membrane effect.

Figure 10.14   Stress distributions in the Kirchhoff plate.

10.8   APPROXIMATED SOLUTION

The bending differential equation (strong form) for the Kirchhoff plate in terms of the transversal displacement u_z is given by equation (10.78).

In order to obtain the bending weak form, we employ the general equation (10.77) and multiply it by the test function $v(x,y)$. We assume initially that the test function belongs to the set $C^4(x,y)$ of functions with continuous partial derivatives up to the fourth order and satisfies the boundary conditions. Later, we will see that there is no need for functions v and u_z to have such regularity and satisfy all boundary conditions. Multiplying (10.77) by v and integrate in the reference surface area A we have

$$\int_A \frac{\partial^2}{\partial x^2}\left[D\left(\frac{\partial^2 u_z(x,y)}{\partial x^2} + v\frac{\partial^2 u_z(x,y)}{\partial y^2}\right)\right]v(x,y)dA +$$
$$\int_A \frac{\partial^2}{\partial y^2}\left[D\left(\frac{\partial^2 u_z(x,y)}{\partial y^2} + v\frac{\partial^2 u_z(x,y)}{\partial x^2}\right)\right]v(x,y)dA + \qquad (10.86)$$
$$\int_A 2\frac{\partial^2}{\partial x\partial y}\left[D(1-v)\frac{\partial^2 u_z(x,y)}{\partial x\partial y}\right]v(x,y)dA - \int_A q(x,y)v(x,y)dA = 0.$$

To simplify the notation, variables (x,y) will not be included in the following equations. Integrating the above expression by parts, we have

$$-\int_A \frac{\partial}{\partial x}\left[D\left(\frac{\partial^2 u_z}{\partial x^2} + v\frac{\partial^2 u_z}{\partial y^2}\right)\right]\frac{\partial v}{\partial x}dA + \int_A \frac{\partial}{\partial y}\left[D\left(\frac{\partial^2 u_z}{\partial y^2} + v\frac{\partial^2 u_z}{\partial x^2}\right)\right]\frac{\partial v}{\partial y}dA$$
$$+\int_A \frac{\partial}{\partial x}\left[D(1-v)\frac{\partial^2 u_z}{\partial x\partial y}\right]\frac{\partial v}{\partial y}dA + \int_A \frac{\partial}{\partial y}\left[D(1-v)\frac{\partial^2 u_z}{\partial x\partial y}\right]\frac{\partial v}{\partial x}dA \qquad (10.87)$$
$$+\int_{\partial A}\left[\left(\frac{\partial M_{xx}}{\partial x} + \frac{\partial M_{xy}}{\partial y}\right)n_x + \left(\frac{\partial M_{yy}}{\partial y} + \frac{\partial M_{xy}}{\partial x}\right)n_y\right]v\, d\partial A - \int_A q\, v\, dA = 0.$$

Note that the term $2\dfrac{\partial^2}{\partial x\partial y}\left[D(1-v)\dfrac{\partial^2 u_z(x,y)}{\partial x\partial y}\right]$ was first integrated in x and again in y. Definition of moments given in (10.73) were employed in the boundary integral.

Applying the definitions of shear forces (10.75) and (10.76) and expression (10.54), the boundary integral is simplified to

$$\int_{\partial A}\left[\left(\frac{\partial M_{xx}}{\partial x}+\frac{\partial M_{xy}}{\partial y}\right)n_x+\left(\frac{\partial M_y}{\partial y}+\frac{\partial M_{xy}}{\partial x}\right)n_y\right]v\,d\partial A \;=\; \int_{\partial A}\left[Q_xn_x+Q_yn_y\right]v\,d\partial A$$

$$=\; \int_{\partial A}Q_n\,v\,d\partial A.$$

Employing this relation in (10.87) and integrating again by parts we have

$$\int_A\left[D\left(\frac{\partial^2 u_z}{\partial x^2}+v\frac{\partial^2 u_z}{\partial y^2}\right)\right]\frac{\partial^2 v}{\partial x^2}dA+\int_A\left[D\left(\frac{\partial^2 u_z}{\partial y^2}+v\frac{\partial^2 u_z}{\partial x^2}\right)\right]\frac{\partial^2 v}{\partial y^2}dA+$$

$$\int_A 2\left[D(1-v)\frac{\partial^2 u_z}{\partial x\partial y}\right]\frac{\partial^2 v}{\partial x\partial y}dA+\int_{\partial A}Q_n\,v\,d\partial A \qquad (10.88)$$

$$-\int_{\partial A}\left[\left(M_{xx}\frac{\partial v}{\partial x}+M_{xy}\frac{\partial v}{\partial y}\right)n_x+\left(M_{yy}\frac{\partial v}{\partial y}+M_{xy}\frac{\partial v}{\partial x}\right)n_y\right]d\partial A-\int_A q\,v\,dA=0.$$

The second boundary integral can be simplified analogously to (10.60), that is,

$$\int_{\partial A}\left[(M_{xx}n_x+M_{xy}n_y)\frac{\partial v}{\partial x}+(M_{xy}n_x+M_{yy}n_y)\frac{\partial v}{\partial y}\right]d\partial A=\int_{\partial A}\left[M_{nn}\frac{\partial v}{\partial n}+M_{nt}\frac{\partial v}{\partial t}\right]d\partial A.$$

Substituting the above relation in (10.88) and organizing the area integrals, we obtain

$$D\int_A\left[\frac{\partial^2 u_z}{\partial x^2}\frac{\partial^2 v}{\partial x^2}+v\left(\frac{\partial^2 u_z}{\partial x^2}\frac{\partial^2 v}{\partial y^2}+\frac{\partial^2 u_z}{\partial y^2}\frac{\partial^2 v}{\partial x^2}\right)+\frac{\partial^2 u_z}{\partial y^2}\frac{\partial^2 v}{\partial y^2}+2(1-v)\frac{\partial^2 u_z}{\partial x\partial y}\frac{\partial^2 v}{\partial x\partial y}\right]dA$$

$$-\int_{\partial A}Q_n\,v\,d\partial A-\int_{\partial A}\left[M_{nn}\frac{\partial v}{\partial n}+M_{nt}\frac{\partial v}{\partial t}\right]d\partial A-\int_A q\,v\,dA=0. \qquad (10.89)$$

Integrating the term $M_{nt}\dfrac{\partial v}{\partial t}$ by parts, we get an analogous expression to (10.63), which, when substituted in equation (10.89), results in

$$D\int_A\left[\frac{\partial^2 u_z}{\partial x^2}\frac{\partial^2 v}{\partial x^2}+v\left(\frac{\partial^2 u_z}{\partial x^2}\frac{\partial^2 v}{\partial y^2}+\frac{\partial^2 u_z}{\partial y^2}\frac{\partial^2 v}{\partial x^2}\right)+\frac{\partial^2 u_z}{\partial y^2}\frac{\partial^2 v}{\partial y^2}+2(1-v)\frac{\partial^2 u_z}{\partial x\partial y}\frac{\partial^2 v}{\partial x\partial y}\right]dA$$

$$-\int_{\partial A}\left(Q_n+\frac{\partial M_{nt}}{\partial t}\right)v\,d\partial A-\int_{\partial A}M_{nn}\frac{\partial v}{\partial n}d\partial A+\sum_{i=1}^{N}\left[(M_{nt}^+-M_{nt}^-)\,v\right]_{P_i}-\int_A q\,v\,dA=0.$$

$$(10.90)$$

Finally, employing the boundary conditions given in the strong form (10.70), we have the weak form for the Kirchhoff plate

$$D\int_A\left[\frac{\partial^2 u_z}{\partial x^2}\frac{\partial^2 v}{\partial x^2}+v\left(\frac{\partial^2 u_z}{\partial x^2}\frac{\partial^2 v}{\partial y^2}+\frac{\partial^2 u_z}{\partial y^2}\frac{\partial^2 v}{\partial x^2}\right)+\frac{\partial^2 u_z}{\partial y^2}\frac{\partial^2 v}{\partial y^2}+2(1-v)\frac{\partial^2 u_z}{\partial x\partial y}\frac{\partial^2 v}{\partial x\partial y}\right]dA$$

$$=\; \int_A q\,v\,dA+\int_{\partial A}\bar{q}\,v\,d\partial A-\int_{\partial A}\bar{m}_{nn}\frac{\partial v}{\partial n}d\partial A-\sum_{i=1}^{N}\left[R_{P_i}\,v\right]_{P_i}. \qquad (10.91)$$

While the strong form (10.78) has derivatives up to the fourth order, the weak form has second-order derivatives. Functions u_z and v can be less regular in the weak form. These functions have to belong only to the sets $C^1(x,y)$ of continuous functions with continuous first-order partial derivatives or $C^1_{cp}(x,y)$ of piecewise continuous first-order derivatives. In the strong form, these functions must be continuous in $C^4(x,y)$ or piecewise continuous in $C^4_{cp}(x,y)$.

From (10.91), we observe that the natural boundary conditions, which are expressed in terms of forces and moments, are automatically satisfied by the weak form. This implies that functions v and u_z in (10.91) need only to satisfy the kinematic or essential boundary conditions. The test function v is required to satisfy only the homogeneous essential boundary conditions.

To obtain the weak form of equations in (10.83) due to the membrane effect, we follow the same procedure we used to obtain (10.91). Thus, we multiply equations in (10.83) by the test function $v(x,y)$, which belongs to the set $C^2(x,y)$, and satisfy the boundary conditions. Hence,

$$\int_A \frac{\partial}{\partial x}\left[T\left(\frac{\partial u_{x_0}(x,y)}{\partial x}+v\frac{\partial u_{y_0}(x,y)}{\partial y}\right)\right]v(x,y)dA+$$

$$\int_A \frac{\partial}{\partial y}\left[T\frac{(1-v)}{2}\left(\frac{\partial u_{y_0}(x,y)}{\partial x}+\frac{\partial u_{x_0}(x,y)}{\partial y}\right)\right]v(x,y)dA+\int_A f_x(x,y)v(x,y)dA=0,$$

$$\int_A \frac{\partial}{\partial y}\left[T\left(\frac{\partial u_{y_0}(x,y)}{\partial y}+v\frac{\partial u_{x_0}(x,y)}{\partial x}\right)\right]v(x,y)dA+$$

$$\int_A \frac{\partial}{\partial x}\left[T\frac{(1-v)}{2}\left(\frac{\partial u_{y_0}(x,y)}{\partial x}+\frac{\partial u_{x_0}(x,y)}{\partial y}\right)\right]v(x,y)dA+\int_A f_y(x,y)v(x,y)dA=0.$$

Integrating each term of the above equation by parts and simplifying, as in (10.87), we have the membrane weak form

$$T\int_A\left[\frac{\partial u_{x_0}}{\partial x}\frac{\partial v}{\partial x}+v\frac{\partial u_{y_0}}{\partial y}\frac{\partial v}{\partial x}+\frac{(1-v)}{2}\left(\frac{\partial u_{y_0}}{\partial x}\frac{\partial v}{\partial y}+\frac{\partial u_{x_0}}{\partial y}\frac{\partial v}{\partial y}\right)\right]dA$$

$$=\int_{\partial A}\left[N_{xx}n_x+N_{xy}n_y\right]v\,d\partial A-\int_A f_x\,v\,dA, \tag{10.92}$$

$$T\int_A\left[\frac{\partial u_{y_0}}{\partial y}\frac{\partial v}{\partial y}+v\frac{\partial u_{x_0}}{\partial x}\frac{\partial v}{\partial y}+\frac{(1-v)}{2}\left(\frac{\partial u_{y_0}}{\partial x}\frac{\partial v}{\partial x}+\frac{\partial u_{x_0}}{\partial y}\frac{\partial v}{\partial x}\right)\right]dA$$

$$=\int_{\partial A}\left[N_{xy}n_x+N_{yy}n_y\right]v\,d\partial A-\int_A f_y\,v\,dA. \tag{10.93}$$

While the strong form has differentiations up to the second order, the weak form has only first-order derivatives. Functions u_{x_0}, u_{y_0}, and v can be less regular in the weak form. These functions need to belong only to the set $C^0(x,y)$ of continuous functions or the set $C^0_{cp}(x,y)$ of piecewise continuous functions. In the strong form, these functions must be continuous up to the second-order derivatives in $C^2(x,y)$ or piecewise continuous in $C^2_{cp}(x,y)$.

In the same way, the natural boundary conditions are automatically satisfied by the weak form. Thus, functions u_{x_0} and u_{y_0} must satisfy only the essential boundary conditions. On the other hand, test function v must satisfy the homogeneous essential boundary conditions.

The approximated solution $u_{z_N}(x,y)$ for the transversal displacement $u_z(x,y)$ of the Kirchhoff plate is given by the following linear combination of the N global basis functions $\{\phi_i\}_{i=1}^N$:

$$u_{z_N}(x,y)=\sum_{i=1}^N a_i\phi_i(x,y). \tag{10.94}$$

The basis functions must have differentiation order compatible with the order of the weak form (10.91). It is also important that they satisfy the essential boundary conditions, once the natural boundary conditions are satisfied by the weak form. In order to obtain the coefficients a_i of the approximated solution (10.94), we use the Galerkin method.

In this case, v_N is written as

$$v_N(x,y)=\sum_{j=1}^N b_j\phi_j(x,y). \tag{10.95}$$

The weak form approximation is obtained by substituting u_z and v in (10.90) by their respective approximations u_{z_N} and v_N, namely,

$$D\int_A\left[\frac{\partial^2 u_{z_N}}{\partial x^2}\frac{\partial^2 v_N}{\partial x^2}+v\left(\frac{\partial^2 u_{z_N}}{\partial x^2}\frac{\partial^2 v_N}{\partial y^2}+\frac{\partial^2 u_{z_N}}{\partial y^2}\frac{\partial^2 v_N}{\partial x^2}\right)+\frac{\partial^2 u_{z_N}}{\partial y^2}\frac{\partial^2 v_N}{\partial y^2}+\right.$$

$$\left.2(1-v)\frac{\partial^2 u_{z_N}}{\partial x\partial y}\frac{\partial^2 v_N}{\partial x\partial y}\right]dA-\int_{\partial A}\left(Q_n+\frac{\partial M_{nt}}{\partial t}\right)v_N\,d\partial A-\int_{\partial A}M_{nn}\frac{\partial v_N}{\partial n}d\partial A+$$

$$\sum_{i=1}^N\left[\left(M_{nt}^+-M_{nt}^-\right)v_N(x,y)\right]_{P_i}-\int_A q\,v_N\,dA=0. \tag{10.96}$$

Substituting (10.94) and (10.95) in the approximated weak form (10.96), we obtain

$$
\begin{aligned}
&\sum_{i,j=1}^{N} \Bigg\{ D \int_A \frac{\partial^2 \phi_i}{\partial x^2}\frac{\partial^2 \phi_j}{\partial x^2} + v\left(\frac{\partial^2 \phi_i}{\partial x^2}\frac{\partial^2 \phi_j}{\partial y^2} + \frac{\partial^2 \phi_i}{\partial y^2}\frac{\partial^2 \phi_j}{\partial x^2} \right) + \frac{\partial^2 \phi_i}{\partial y^2}\frac{\partial^2 \phi_j}{\partial y^2} \\
&+ 2(1-v)\frac{\partial^2 \phi_i}{\partial x \partial y}\frac{\partial^2 \phi_j}{\partial x \partial y} \Bigg] a_i\, dA - \int_{\partial A}\left(Q_n + \frac{\partial M_{nt}}{\partial t} \right)\phi_j\, d\partial A \\
&- \int_{\partial A} M_{nn}\frac{\partial \phi_j}{\partial n}\, d\partial A + \sum_{k,j=1}^{N}\left[(M_{nt}^+ - M_{nt}^-)\phi_j \right]_{P_k} - \int_A q\,\phi_j\, dA \Bigg\} b_j = 0.
\end{aligned} \tag{10.97}
$$

As the coefficients b_j of the linear combination (10.95) are arbitrary, the above expression is zero if

$$
\begin{aligned}
&\sum_{i,j=1}^{N} \Bigg\{ D \int_A \left[\frac{\partial^2 \phi_i}{\partial x^2}\frac{\partial^2 \phi_j}{\partial x^2} + v\left(\frac{\partial^2 \phi_i}{\partial x^2}\frac{\partial^2 \phi_j}{\partial y^2} + \frac{\partial^2 \phi_i}{\partial y^2}\frac{\partial^2 \phi_j}{\partial x^2} \right) + \frac{\partial^2 \phi_i}{\partial y^2}\frac{\partial^2 \phi_j}{\partial y^2} \right. \\
&+ 2(1-v)\frac{\partial^2 \phi_i}{\partial x \partial y}\frac{\partial^2 \phi_j}{\partial x \partial y}\, dA \Bigg] \Bigg\} a_i = \int_{\partial A}\left(Q_n + \frac{\partial M_{nt}}{\partial t} \right)\phi_j\, d\partial A + \\
&\int_{\partial A} M_{nn}\frac{\partial \phi_j}{\partial n}\, d\partial A - \sum_{k,j=1}^{N}\left[(M_{nt}^+ - M_{nt}^-)\phi_j \right]_{P_k} + \int_A q\,\phi_j\, dA = 0.
\end{aligned} \tag{10.98}
$$

The above expression represents the system of equations

$$
[\mathbf{K}_b]\{\mathbf{a}\} = \{\mathbf{f}_b\}, \tag{10.99}
$$

with the coefficients of the global stiffness matrix $[\mathbf{K}^b]$ and load vector $\{\mathbf{f}^b\}$ given by

$$
\begin{aligned}
K_{b,ij} =\ & D \int_A \left[\frac{\partial^2 \phi_i}{\partial x^2}\frac{\partial^2 \phi_j}{\partial x^2} + v\left(\frac{\partial^2 \phi_i}{\partial x^2}\frac{\partial^2 \phi_j}{\partial y^2} + \frac{\partial^2 \phi_i}{\partial y^2}\frac{\partial^2 \phi_j}{\partial x^2} \right) \right. \\
&+ \frac{\partial^2 \phi_i}{\partial y^2}\frac{\partial^2 \phi_j}{\partial y^2} + 2(1-v)\frac{\partial^2 \phi_i}{\partial x \partial y}\frac{\partial^2 \phi_j}{\partial x \partial y} \bigg] dA, \quad i,j = 1,\ldots,N,
\end{aligned} \tag{10.100}
$$

$$
\begin{aligned}
f_{b,j} =\ & \int_{\partial A}\left(Q_n + \frac{\partial M_{nt}}{\partial t} \right)\phi_j\, d\partial A + \int_{\partial A} M_{nn}\frac{\partial \phi_j}{\partial n}\, d\partial A \\
&- \sum_{k,j=1}^{N}\left[(M_{nt}^+ - M_{nt}^-)\phi_j \right]_{P_k} + \int_A q\,\phi_j\, dA, \quad j = 1,\ldots,N.
\end{aligned} \tag{10.101}
$$

The stiffness matrix can be also denoted by the following matrix product:

$$
[\mathbf{K}_b] = \frac{t^3}{12}[\mathbf{B}_b]^T [\mathbf{D}][\mathbf{B}_b], \tag{10.102}
$$

The strain-displacement matrix $[\mathbf{B}_b]$ is given by

$$
[\mathbf{B}_{b,i}] = \begin{bmatrix} -\dfrac{\partial^2 \phi_1}{\partial x^2} & \cdots & -\dfrac{\partial^2 \phi_N}{\partial x^2} \\[2mm] -\dfrac{\partial^2 \phi_1}{\partial y^2} & \cdots & -\dfrac{\partial^2 \phi_N}{\partial y^2} \\[2mm] -\dfrac{\partial^2 \phi_1}{\partial x \partial y} & \cdots & -\dfrac{\partial^2 \phi_N}{\partial x \partial y} \end{bmatrix}. \tag{10.103}
$$

The elasticity matrix $[\mathbf{D}]$ is determined from Hooke's law (10.71). Hence,

$$
[\mathbf{D}] = \frac{E}{(1-v^2)} \begin{bmatrix} 1 & v & 0 \\ v & 1 & 0 \\ 0 & 0 & \dfrac{(1-v)}{2} \end{bmatrix}. \tag{10.104}
$$

Analogously, we can determine the weak form approximation of equations (10.92) and (10.93) for the membrane effect substituting u_{x_0} and u_{y_0} by their respective approximations $u_{x_{0_N}}$ and $u_{y_{0_N}}$, that is,

$$u_{x_{0_N}}(x,y) \quad = \quad \sum_{i=1}^{N} c_i \varphi_i(x,y), \tag{10.105}$$

$$u_{y_{0_N}}(x,y) \quad = \quad \sum_{i=1}^{N} d_i \varphi_i(x,y), \tag{10.106}$$

with $\{\varphi_i\}_{i=1}^{N}$ the set of linearly independent functions. As in the bending case, the chosen functions must have a compatible differentiation order with the ones of the weak form (10.92) and (10.93). It is also important that they satisfy the essential boundary conditions, once the natural boundary conditions are satisfied by the membrane weak form.

In this case, the approximation v_N of the test function v is expressed as

$$v_N(x,y) = \sum_{j=1}^{N} h_j \varphi_j(x,y). \tag{10.107}$$

The approximated weak form is obtained substituting u_{x_0}, u_{y_0} and v in (10.92) and (10.93) by their respective approximations u_{0_n}, $u_{y_{0_N}}$, and v_N, that is,

$$-T \int_A \left[\frac{\partial u_{x_{0_N}}}{\partial x} \frac{\partial v_N}{\partial x} + v \frac{\partial u_{y_{0_N}}}{\partial y} \frac{\partial v_N}{\partial x} + \frac{(1-v)}{2} \left(\frac{\partial u_{y_{0_N}}}{\partial x} \frac{\partial v_N}{\partial y} + \frac{\partial u_{x_{0_N}}}{\partial y} \frac{\partial v_N}{\partial y} \right) \right] dA$$
$$+ \int_{\partial A} \left[N_{xx} n_x + N_{xy} n_y \right] v_N \, d\partial A = \int_A f_x \, v_N \, dA, \tag{10.108}$$

$$-T \int_A \left[\frac{\partial u_{y_{0_N}}}{\partial y} \frac{\partial v_N}{\partial y} + v \frac{\partial u_{x_{0_N}}}{\partial x} \frac{\partial v_N}{\partial y} + \frac{(1-v)}{2} \left(\frac{\partial u_{y_{0_N}}}{\partial x} \frac{\partial v_N}{\partial x} + \frac{\partial u_{x_{0_N}}}{\partial y} \frac{\partial v_N}{\partial x} \right) \right] dA$$
$$+ \int_{\partial A} \left[N_{xy} n_x + N_{yy} n_y \right] v_N \, d\partial A = \int_A f_y \, v_N \, dA. \tag{10.109}$$

Replacing equations (10.105), (10.106), and (10.107) in the previous equation and following the same steps used for the bending effect, we find the membrane approximated weak form, that is,

$$\sum_{i=1}^{N} T \int_A \left[\frac{\partial \varphi_i}{\partial x} \frac{\partial \varphi_j}{\partial x} c_i + v \frac{\partial \varphi_i}{\partial y} \frac{\partial \varphi_j}{\partial x} d_i + \frac{(1-v)}{2} \left(\frac{\partial \varphi_i}{\partial x} \frac{\partial \varphi_j}{\partial y} d_i + \frac{\partial \varphi_i}{\partial y} \frac{\partial \varphi_j}{\partial y} c_i \right) \right] dA$$
$$= \int_{\partial A} \left[N_{xx} n_x + N_{xy} n_y \right] \varphi_j \, d\partial A - \int_A f_x \, \varphi_j \, dA, \qquad i,j = 1,\dots,N, \tag{10.110}$$

$$\sum_{i=1}^{N} T \int_A \left[\frac{\partial \varphi_i}{\partial y} \frac{\partial \varphi_j}{\partial y} d_i + v \frac{\partial \varphi_i}{\partial x} \frac{\partial \varphi_j}{\partial y} c_i + \frac{(1-v)}{2} \left(\frac{\partial \varphi_i}{\partial x} \frac{\partial \varphi_j}{\partial x} d_i + \frac{\partial \varphi_i}{\partial y} \frac{\partial \varphi_j}{\partial x} c_i \right) \right] dA$$
$$= \int_{\partial A} \left[N_{xy} n_x + N_{yy} n_y \right] \varphi_j \, d\partial A - \int_A f_y \, \varphi_j \, dA. \qquad i,j = 1,\dots,N. \tag{10.111}$$

The above expressions represent the system of equations

$$\begin{bmatrix} [K_m^{11}] & [K_m^{12}] \\ [K_m^{12}]^T & [K_m^{22}] \end{bmatrix} \begin{Bmatrix} \{c_m\} \\ \{d_m\} \end{Bmatrix} = \begin{Bmatrix} \{f_m^1\} \\ \{f_m^2\} \end{Bmatrix}. \tag{10.112}$$

The coefficients of the global stiffness matrix $[\mathbf{K}^m]$ and load vector $\{\mathbf{f}^m\}$ are written as

$$
\begin{aligned}
K^{11}_{m,ij} &= T \int_A \left[\frac{\partial \varphi_i}{\partial x} \frac{\partial \varphi_j}{\partial x} + \frac{(1-v)}{2} \frac{\partial \varphi_i}{\partial y} \frac{\partial \varphi_j}{\partial y} \right] dA \quad i,j = 1,\ldots,N, \\
K^{12}_{m,ij} &= T \int_A \left[v \frac{\partial \varphi_i}{\partial y} \frac{\partial \varphi_j}{\partial x} + \frac{(1-v)}{2} \frac{\partial \varphi_i}{\partial x} \frac{\partial \varphi_j}{\partial y} \right] dA \quad i,j = 1,\ldots,N, \\
K^{22}_{m,ij} &= T \int_A \left[\frac{(1-v)}{2} \frac{\partial \varphi_i}{\partial x} \frac{\partial \varphi_j}{\partial x} + \frac{\partial \varphi_i}{\partial y} \frac{\partial \varphi_j}{\partial y} \right] dA \quad i,j = 1,\ldots,N, \\
f^1_{m,j} &= \int_{\partial A} \left[N_{xx} n_x + N_{xy} n_y \right] \varphi_j \, d\partial A - \int_A f_x \, \varphi_j \, dA \quad j = 1,\ldots,N, \\
f^2_{m,j} &= \int_{\partial A} \left[N_{xy} n_x + N_{yy} n_y \right] \varphi_j \, d\partial A - \int_A f_y \, \varphi_j \, dA \quad j = 1,\ldots,N.
\end{aligned}
\tag{10.113}
$$

The membrane stiffness matrix can be also denoted by the product

$$
[\mathbf{K}_m] = [\mathbf{B}_m]^T [\mathbf{D}] [\mathbf{B}_m],
\tag{10.114}
$$

with the strain-displacement matrix $[\mathbf{B}_m]$ given by

$$
[\mathbf{B}_m] = \begin{bmatrix}
\dfrac{\partial \varphi_1}{\partial x} & 0 & \cdots & \dfrac{\partial \varphi_N}{\partial x} & 0 \\[2mm]
0 & \dfrac{\partial \varphi_1}{\partial y} & \cdots & 0 & \dfrac{\partial \varphi_N}{\partial y} \\[2mm]
\dfrac{\partial \varphi_1}{\partial y} & \dfrac{\partial \varphi_i}{\partial x} & \cdots & \dfrac{\partial \varphi_N}{\partial y} & \dfrac{\partial \varphi_N}{\partial x}
\end{bmatrix}.
\tag{10.115}
$$

10.8.1 PLATE FINITE ELEMENTS

The plate domain may be discretized by a finite element mesh of squares. To select the bending shape functions $\left\{ \phi_i^{(e)} \right\}_{i=1}^N$ of finite element e, we analyze equation (10.100). Note that the stiffness matrix coefficients K^b_{ij} have second-order partial derivatives of the shape functions. Thus, the minimum regularity required for the shape functions is 1, that is, they should belong to the set $C^1(\Omega)$ of continuous functions with first-order partial derivatives which are also continuous in the domain $\Omega \subset R^2$. This implies that both $u_z^{(e)}$ and the partial derivatives $\dfrac{\partial u_z^{(e)}}{\partial x}$ and $\dfrac{\partial u_z^{(e)}}{\partial y}$ are continuous between two elements.

However, the stress and strain components will be discontinuous, because they involve second-order partial derivatives. The functions obtained by the tensor product of the one-dimensional cubic Hermite polynomials have the required properties. These polynomials are presented in Chapter 5.

The determination of finite element shape functions of class $C^1(\Omega)$ for $\Omega \subset R^2$ is not a simple task. For the case of square elements, we have a recurrence form to generate the local shape functions. For this purpose, consider the Pascal triangle illustrated in Figure 10.15. The following procedure can be applied to generate a family of square elements of class C^m [23, 44]:

- Identify in the Pascal triangle the squares whose sides contain complete polynomials of odd orders. In Figure 10.15, this corresponds to squares $1, 3, 5, \cdots, m, \cdots$. Each of these squares contains $4, 16, 36, \cdots, (m+1)^2, \cdots$ independent monomials.

- Consider that the order of each polynomial contained in the square of order m corresponds to the derivative order that must be prescribed in the vertices for the four-node element. If m is the order of the derivatives, the local interpolation polynomials of this element contain all monomials of the square $2m+1$. Hence, each finite element leads to an element of class C^m.

Consider then the following examples:

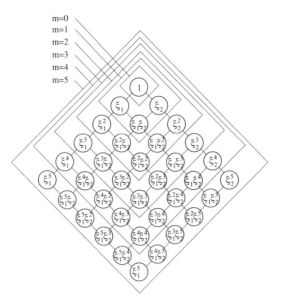

Figure 10.15 Pascal triangle for a square element.

$m = 0$: The square of zero order contains all monomials of order 0 in x and y. Hence, we only prescribe the value of the function at each finite element vertex. The shape functions correspond, in this case, to polynomials containing the monomials of the square $2m + 1 = 1$, that is, 1, x, y, xy. This element corresponds to the Lagrangian element of class $C^m = C^0$.

$m = 1$: The monomials of this square are 1, x, y, xy. Thus, for each vertex node of the four-node element, we must prescribe the values of function and derivatives relative to x, y, and xy, that is, we consider the values of u_z, $u_{z,x}$, $u_{z,y}$, and $u_{z,xy}$. The shape functions contain all monomials corresponding to the square $2m + 1 = 3$. These polynomials contain 16 coefficients uniquely determined from the imposed constraints on each node. Hence, we define the $C^m = C^1$ class element.

$m = 2$: Observing the square for $m = 2$, we should prescribe the values of u_z, $u_{z,x}$, $u_{z,y}$, $u_{z,xx}$, $u_{z,xy}$, $u_{z,yy}$, $u_{z,xxy}$, $u_{z,xyy}$, and $u_{z,xxyy}$ on each node. This four-node element corresponds to a family of class $C^m = C^2$.

Figure 10.16 illustrates several square elements. The conforming Lagrangian element of class C^0 prescribes only the value of the function on each node [see Figure 10.16(a)]. The bilinear element illustrated in Figure 10.16(b) is nonconforming, because it has three values per node and a total number of 12 shape functions. Thus, it is not possible to guarantee the continuity of the partial derivatives between two elements, because 16 coefficients are required. On the other hand, the biquadratic element of Figure 10.16(c) has nine nodes and reaches a total of 27 coefficients, guaranteeing the continuity of the approximation and partial derivatives. Finally, Figures 10.16(d) and 10.16(e) illustrate conforming and nonconforming Hermite elements. Comparing the Lagrange and Hermite conforming elements, we observe that the latter requires 16 coefficients, while the former has 27 coefficients to be calculated.

Consider then the shape functions for $m = 1$ of the four-node square element. As mentioned, there are four shape functions for each node corresponding to the values of u_z, $u_{z,x}$, $u_{z,y}$, and $u_{z,xy}$. Considering four nodes, we have a total of 16 shape functions corresponding to the 16 monomials of the Pascal triangle for order

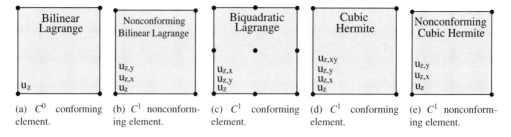

Figure 10.16   Square elements of C^0 and C^1 classes.

$2m+1 = 3$. The local representation of the displacement approximation $u_{z_N}^{(e)}(\xi_1, \xi_2)$ is written as

$$
\begin{aligned}
u_{z_N}^{(e)}(\xi_1, \xi_2) &= a_1 \phi_1^{(e)}(\xi_1, \xi_2) + a_2 \phi_2^{(e)}(\xi_1, \xi_2) + a_3 \phi_3^{(e)}(\xi_1, \xi_2) + a_4 \phi_4^{(e)}(\xi_1, \xi_2) + \\
&\quad a_5 \phi_5^{(e)}(\xi_1, \xi_2) + a_6 \phi_6^{(e)}(\xi_1, \xi_2) + a_7 \phi_7^{(e)}(\xi_1, \xi_2) + a_8 \phi_8^{(e)}(\xi_1, \xi_2) + \\
&\quad a_9 \phi_9^{(e)}(\xi_1, \xi_2) + a_{10} \phi_{10}^{(e)}(\xi_1, \xi_2) + a_{11} \phi_{11}^{(e)}(\xi_1, \xi_2) + a_{12} \phi_{12}^{(e)}(\xi_1, \xi_2) + \\
&\quad a_{13} \phi_{13}^{(e)}(\xi_1, \xi_2) + a_{14} \phi_{14}^{(e)}(\xi_1, \xi_2) + a_{15} \phi_{15}^{(e)}(\xi_1, \xi_2) + a_{16} \phi_{16}^{(e)}(\xi_1, \xi_2) \\
&= \sum_{i=1}^{16} a_i \phi_i^{(e)}(\xi_1, \xi_2).
\end{aligned}
\tag{10.116}
$$

The nodal shape functions $\phi_1^{(e)}$, $\phi_5^{(e)}$, $\phi_9^{(e)}$, and $\phi_{13}^{(e)}$ are associated to the values of u_z; $\phi_2^{(e)}$, $\phi_6^{(e)}$, $\phi_{10}^{(e)}$, and $\phi_{14}^{(e)}$ to u_{z_x}; $\phi_3^{(e)}$, $\phi_7^{(e)}$, $\phi_{11}^{(e)}$, and $\phi_{15}^{(e)}$ to u_{z_y}; $\phi_4^{(e)}$, $\phi_8^{(e)}$, $\phi_{12}^{(e)}$, and $\phi_{16}^{(e)}$ to $u_{z_{xy}}$. These two-dimensional functions, illustrated in Figure 10.17, are obtained by the tensor product of the one-dimensional cubic Hermite polynomials given in (5.123). Thus,

$$
\begin{aligned}
&\phi_1^{(e)}(\xi_1, \xi_2) = \phi_1^{(e)}(\xi_1)\phi_1^{(e)}(\xi_2), &&\phi_2^{(e)}(\xi_1, \xi_2) = \phi_2^{(e)}(\xi_1)\phi_1^{(e)}(\xi_2), \\
&\phi_5^{(e)}(\xi_1, \xi_2) = \phi_3^{(e)}(\xi_1)\phi_1^{(e)}(\xi_2), &&\phi_6^{(e)}(\xi_1, \xi_2) = \phi_4^{(e)}(\xi_1)\phi_1^{(e)}(\xi_2), \\
&\phi_9^{(e)}(\xi_1, \xi_2) = \phi_3^{(e)}(\xi_1)\phi_3^{(e)}(\xi_2), &&\phi_{10}^{(e)}(\xi_1, \xi_2) = \phi_4^{(e)}(\xi_1)\phi_3^{(e)}(\xi_2), \\
&\phi_{13}^{(e)}(\xi_1, \xi_2) = \phi_1^{(e)}(\xi_1)\phi_3^{(e)}(\xi_2), &&\phi_{14}^{(e)}(\xi_1, \xi_2) = \phi_2^{(e)}(\xi_1)\phi_3^{(e)}(\xi_2), \\
&\phi_3^{(e)}(\xi_1, \xi_2) = \phi_1^{(e)}(\xi_1)\phi_2^{(e)}(\xi_2), &&\phi_4^{(e)}(\xi_1, \xi_2) = \phi_2^{(e)}(\xi_1)\phi_2^{(e)}(\xi_2), \\
&\phi_7^{(e)}(\xi_1, \xi_2) = \phi_3^{(e)}(\xi_1)\phi_2^{(e)}(\xi_2), &&\phi_8^{(e)}(\xi_1, \xi_2) = \phi_4^{(e)}(\xi_1)\phi_2^{(e)}(\xi_2), \\
&\phi_{11}^{(e)}(\xi_1, \xi_2) = \phi_3^{(e)}(\xi_1)\phi_4^{(e)}(\xi_2), &&\phi_{12}^{(e)}(\xi_1, \xi_2) = \phi_4^{(e)}(\xi_1)\phi_4^{(e)}(\xi_2), \\
&\phi_{15}^{(e)}(\xi_1, \xi_2) = \phi_1^{(e)}(\xi_1)\phi_4^{(e)}(\xi_2), &&\phi_{16}^{(e)}(\xi_1, \xi_2) = \phi_2^{(e)}(\xi_1)\phi_4^{(e)}(\xi_2).
\end{aligned}
\tag{10.117}
$$

The bending stiffness matrix for the four-node square element employing Hermite polynomials in the local coordinate system is given by

$$
\left[\mathbf{K}_b^{(e)} \right] = \frac{t^3}{12} \int_{A^{(e)}} \left[\mathbf{B}_b^{(e)} \right]^T [\mathbf{D}] \left[\mathbf{B}_b^{(e)} \right] \det[\mathbf{J}] \, d\xi_1 d\xi_2.
$$

The strain-displacement matrix $\left[\mathbf{B}_b^{(e)} \right]$ is written as

$$
\left[\mathbf{B}_b^{(e)} \right] = \left[\; \left[\mathbf{B}_{b,1}^{(e)} \right] \quad \left[\mathbf{B}_{b,2}^{(e)} \right] \quad \left[\mathbf{B}_{b,3}^{(e)} \right] \quad \left[\mathbf{B}_{b,4}^{(e)} \right] \; \right],
$$

with $\left[\mathbf{B}_{b,i}^{(e)} \right]$ $(i = 1, \ldots, 4)$ the nodal strain-displacement matrix, relating the strain field with the nodal displace-

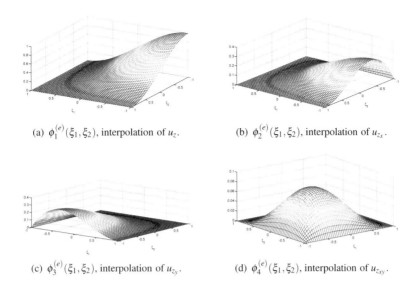

(a) $\phi_1^{(e)}(\xi_1,\xi_2)$, interpolation of u_z.

(b) $\phi_2^{(e)}(\xi_1,\xi_2)$, interpolation of u_{z_x}.

(c) $\phi_3^{(e)}(\xi_1,\xi_2)$, interpolation of u_{z_y}.

(d) $\phi_4^{(e)}(\xi_1,\xi_2)$, interpolation of $u_{z_{xy}}$.

Figure 10.17 Hermite shape functions for a square element with four nodes and four degrees of freedom per node.

ments, given by

$$\left[\mathbf{B}_{b,i}^{(e)}\right] = \begin{bmatrix} -\dfrac{\partial^2 \phi_1^{(e)}}{\partial x^2} & -\dfrac{\partial^2 \phi_2^{(e)}}{\partial x^2} & -\dfrac{\partial^2 \phi_3^{(e)}}{\partial x^2} & -\dfrac{\partial^2 \phi_4^{(e)}}{\partial x^2} \\[2mm] -\dfrac{\partial^2 \phi_1^{(e)}}{\partial y^2} & -\dfrac{\partial^2 \phi_2^{(e)}}{\partial y^2} & -\dfrac{\partial^2 \phi_3^{(e)}}{\partial y^2} & -\dfrac{\partial^2 \phi_4^{(e)}}{\partial y^2} \\[2mm] -\dfrac{\partial^2 \phi_1^{(e)}}{\partial x \partial y} & -\dfrac{\partial^2 \phi_2^{(e)}}{\partial x \partial y} & -\dfrac{\partial^2 \phi_3^{(e)}}{\partial x \partial y} & -\dfrac{\partial^2 \phi_4^{(e)}}{\partial x \partial y} \end{bmatrix}.$$

Numbers 1, 2, 3, and 4 indicate the functions of a given node of the element.

We observe that the coefficients of the above strain matrices involve the calculation of the second-order global derivatives of the local nodal functions $\phi_j^{(i)}(\xi_1,\xi_2)$, $(i,j=1,\ldots,4)$. For this purpose, we apply the following relation for plane problems [52]:

$$\left\{ \begin{array}{c} \dfrac{\partial^2}{\partial x^2} \\[2mm] \dfrac{\partial^2}{\partial y^2} \\[2mm] \dfrac{\partial^2}{\partial x \partial y} \end{array} \right\} = [\mathbf{T}_1] \left\{ \begin{array}{c} \dfrac{\partial}{\partial \xi_1} \\[2mm] \dfrac{\partial}{\partial \xi_2} \end{array} \right\} + [\mathbf{T}_2] \left\{ \begin{array}{c} \dfrac{\partial^2}{\partial \xi_1^2} \\[2mm] \dfrac{\partial^2}{\partial \xi_2^2} \\[2mm] \dfrac{\partial^2}{\partial \xi_1 \partial \xi_2} \end{array} \right\},$$

with

$$[\mathbf{T}_2] = \begin{bmatrix} j_{11}^2 & j_{12}^2 & 2 j_{11} j_{12} \\ j_{21}^2 & j_{22}^2 & 2 j_{21} j_{22} \\ j_{11} j_{21} & j_{12} j_{22} & j_{11} j_{22} + j_{12} j_{21} \end{bmatrix}$$

and

$$[\mathbf{T}_1] = -[\mathbf{T}_2][\mathbf{C}_1][\mathbf{j}],$$

with $[\mathbf{j}] = [\mathbf{J}]^{-1}$ and

$$[\mathbf{C}_1] = \begin{bmatrix} \dfrac{\partial J_{11}}{\partial \xi_1} & \dfrac{\partial J_{12}}{\partial \xi_1} \\[3mm] \dfrac{\partial J_{21}}{\partial \xi_2} & \dfrac{\partial J_{22}}{\partial \xi_2} \\[3mm] \dfrac{1}{2}\left(\dfrac{\partial J_{11}}{\partial \xi_2} + \dfrac{\partial J_{21}}{\partial \xi_1}\right) & \dfrac{1}{2}\left(\dfrac{\partial J_{12}}{\partial \xi_2} + \dfrac{\partial J_{22}}{\partial \xi_1}\right) \end{bmatrix}.$$

To select the shape functions of the membrane effect $\{\phi_i\}_{i=1}^{N}$, we must analyze expressions (10.113) and (10.114). The coefficients of the membrane stiffness matrix involve first-order partial derivatives only. Thus, we use shape functions that belong to the set $C^0(\Omega)$ of continuous functions with discontinuous first-order partial derivatives in the domain $\Omega \subset R^2$. Hence, the displacement field is continuous between two elements, but the stress and strain components are discontinuous between elements, because they involve first-order derivatives. Functions constructed with Lagrange or Jacobi polynomials presented in Chapter 8 are sufficient to attend the continuity requirements for the displacement field between the elements.

As for the bending effect, the membrane stiffness matrix can be written as

$$\left[\mathbf{K}_m^{(e)}\right] = \int_{A^{(e)}} \left[\mathbf{B}_m^{(e)}\right]^T [\mathbf{D}] \left[\mathbf{B}_m^{(e)}\right] \det[\mathbf{J}]\, d\xi_1 d\xi_2.$$

The strain-displacement matrix $\left[\mathbf{B}_m^{(e)}\right]$ is

$$\left[\mathbf{B}_m^{(e)}\right] = \left[\ \left[\mathbf{B}_{m,1}^{(e)}\right]\ \cdots\ \left[\mathbf{B}_{m,N}^{(e)}\right]\ \right],$$

with N the number of nodes/modes and $\left[\mathbf{B}_{m,i}^{(e)}\right]$ $(i = 1,\dots,N)$ the strain-displacement matrix given by

$$\left[\mathbf{B}_{m,i}^{(e)}\right] = \begin{bmatrix} \dfrac{\partial \phi_i^{(e)}}{\partial x} & 0 \\[3mm] 0 & \dfrac{\partial \phi_i^{(e)}}{\partial y} \\[3mm] \dfrac{\partial \phi_i^{(e)}}{\partial y} & \dfrac{\partial \phi_i^{(e)}}{\partial x} \end{bmatrix},$$

and $[\mathbf{D}]$ the elasticity matrix for the plane stress state given in (10.104).

The global derivatives of the shape functions are calculated as

$$\left\{ \begin{array}{c} \dfrac{\partial \phi_i^{(e)}}{\partial x} \\[3mm] \dfrac{\partial \phi_i^{(e)}}{\partial y} \end{array} \right\} = [\mathbf{j}] \left\{ \begin{array}{c} \dfrac{\partial \phi_i^{(e)}}{\partial \xi_1} \\[3mm] \dfrac{\partial \phi_i^{(e)}}{\partial \xi_2} \end{array} \right\}.$$

10.8.2 HIGH-ORDER FINITE ELEMENT

The high-order shape functions defined for the Euler-Bernoulli beam approximation in Section ?? of Chapter 5 can be used for the construction of square shape functions using tensor product.

The vertex functions are the same ones presented in the previous section based on the cubic Hermite polynomials. As there is only one 1D internal shape function, the tensor based procedure gives rise to two edge functions for each considered order from $P \geq 4$. These functions are related to the transversal displacement and one of its partial derivatives, as illustrated in Figure 10.18(a). Analogously, we add a face function for each considered order $P \geq 4$. The edge and face shape functions for $P = 4$ are illustrated in Figures 10.18(b) to 10.18(d).

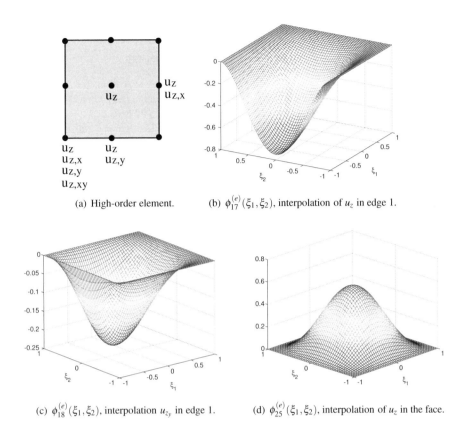

(a) High-order element.　　(b) $\phi_{17}^{(e)}(\xi_1, \xi_2)$, interpolation of u_z in edge 1.

(c) $\phi_{18}^{(e)}(\xi_1, \xi_2)$, interpolation u_{z_y} in edge 1.　　(d) $\phi_{25}^{(e)}(\xi_1, \xi_2)$, interpolation of u_z in the face.

Figure 10.18　High-order shape functions in edge 1 and face for $P = 4$.

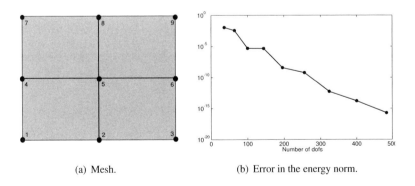

(a) Mesh.　　(b) Error in the energy norm.

Figure 10.19　Example 10.3: mesh and error in the energy norm.

Example 10.3 *Consider the transversal displacement function, similar to the one given in Example 10.1, defined in a square plate with a* 4 m *edge.*

$$u_z(x,y) = kx^2y^2(x-4)^2(y-4)^2 \sin(\pi x)\sin(\pi y),$$

with k a constant to enforce small displacements. The thickness of the plate is 1 *cm and it is clamped at the ends. Substituting this function in differential equation (10.78), we obtain the body force that must be applied to the finite element model to recover the given solution. We adopt E =* 10.92 MPa *and* v = 0.3.

We employ the four element mesh illustrated in Figure 10.19(a) for the solution of this example. The error in the energy norm in terms of the number of degrees of freedom is shown in Figure 10.19(b). There is an exponential decay of the approximation error.

□

10.9 PROBLEMS

1. Define the energy norm relative to the bending of the Kirchhoff plate.

2. Develop a MATLAB program that generates the tensor indices and plots the high-order shape functions for an edge and a face for any polynomial order.

3. Verify if the high-order edge shape functions and first-order derivatives are continuous between two elements, using a MATLAB program.

4. Develop a MATLAB program to obtain the solution and calculate the approximation error of the plate of Example 10.3.

References

1. Generalized functions and related objects. Technical report, Wolfram Mathematica, 2011.

2. O. Axelsson and V.A. Barker. *Finite Element Solution of Boundary Value Problems —- Theory and Computation*. Academic Press, Orlando, 1984.

3. I. Babuška, A. Craig, J. Mandel, and J. Pitkaranta. Efficient preconditioning for the p-version finite element method in two dimensions. *SIAM J. Numer. Anal.*, 28(3):624–661, 1991.

4. I. Babuška and M. Suri. The p and h-p versions of the finite element method, an overview. *Comp. Meth. Appl. Mech. Engr.*, 80:5–26, 1990.

5. I. Babuška, B. A. Szabó, and I. N. Katz. The p-version of the finite element method. *SIAM J. Numer. Anal.*, 18(3):515–545, 1981.

6. F.F Bargos, R.A. Augusto, and M.L. Bittencourt. A high-order finite element parallel MATLAB software. In *Proceedings of the Seventh International Conference on Engineering Computational Techonology*, Valencia, September 2010.

7. F.P. Beer, E.R. Johnston Jr, J.T. DeWolf, and D.F. Mazurek. *Mechanics of Materials*. McGraw Hill, New York, 6th edition, 2012.

8. M.A. Bhatti. *Advanced Topics in Finite Element Analysis of Structures —- With Mathematica and MAT-LAB Computations*. John Wiley & Sons, Hoboken, New Jersey, 2006.

9. M.L. Bittencourt. Fully tensorial nodal and modal shape functions for triangles and tetrahedra. *Int. J. Numer. Meth. in Eng.*, 63(2):1530–1558, 2005.

10. M.L. Bittencourt, M.G. Vazquez, and T.G. Vazquez. Construction of shape functions for the $h-$ and $p-$ versions of the fem using tensorial product. *Int. J. Numer. Meth. in Eng.*, 71(5):529–563, 2007.

11. M.L. Bittencourt and T.G. Vazquez. Tensor-based Gauss-Jacobi numerical integration for high-order mass and stiffness matrices. *Int. J. Numer. Meth. in Eng.*, 79(5):599–638, 2009.

12. J. Bonet and R.D. Wood. *Nonlinear Continuum Mechanics for Finite Element Analisys*. Cambridge University Press, Cambridge, 1997.

13. A. F. Bower. *Applied Mechanics of Solids*. CRC Press, Boca Raton, 2009.

14. J.P. Boyd. *Chebyshev and Fourier Spectral Methods*. DOVER Publications, Inc, Mineola, New York, 2000.

15. S.C. Brenner and L.R. Scott. *The Mathematical Theory of Finite Element Methods*. Springer-Verlag, New York, 1994.

16. A.F. Carbonara, D. Duarte Jr., and M.L. Bittencourt. Comparison of journal orbits under hydrodynamic lubrication regime for traditional and Newton-Euler loads in combustion engine. *Lat. Am. J. of Solids Stru.*, 6(1):13–33, 2009.

17. T.S. Chihara. *An Introduction to Orthogonal Polynomials*. Mathematics and its Applications Series. Gordon and Breach Science Publishers, New York, 1978.

18. R.D. Cook, D.S. Malkus, and M.E. Plesha. *Concepts and Applications of Finite Element Analysis*. John Wiley & Sons, USA, 3rd edition, 1991.

19. G.R. Cowper. *The Shear Coefficient in Timoshenko's Beam Theory*. American Society of Mechanical Engineers, New York, 1966.

20. E.A. de Souza Neto, D Perić, and D.R.J. Owen. *Computational Methods for Plasticity: Theory and Applications*. Wiley, New York, 2008.

21. D. A. Dunavant. High degree efficient symmetrical gaussian quadrature rules for the triangle. *Int. J. Numer. Meth. Eng.*, 21:1129–1148, 1985.

22. R.A. Feijóo, N.E. Pereira, and E. Taroco. Principios variacionales en mecanica. In *Proceedings of the 2nd Course on Theoretical and Applied Mechanics*, pages 1–200, Rio de Janeiro, July 2-27 1984. LCC/CNPq.

23. R.A. Feijóo and E. Taroco. Teoria de placas y cascaras. In *Proceedings of the 2nd Course on Theoretical and Applied Mechanics on Theory of Shells and their Applications in Engineering*, pages 117–176, Rio de Janeiro, January 3 to February 11 1983. LCC/CNPq.

24. D. Funaro. *Polynomial Approximations of Differential Equations*. Springer-Verlag, Berlin, 1992.

25. Y.C. Fung and P. Tong. *Classical and Computational Solid Mechanics*. World Scientific Publishing Company, 2001.

26. R.C. Gerardin. Dynamic modeling og the piston-conrod-crankshaft system including hydrodynamic bearings (in Portuguese). Master's thesis, University of Campinas, Campinas, Julho 2005.

27. J. M. Gere and B.J. Goodno. *Mechanics of Material*. CENGAGE Learning, Toronto, 7th edition, 2008.

28. P. Germain. *Mecanique*. Ecole Polytechnique, Paris, 1986.

29. D. Gottlieb and S.A. Orszag. *Numerical Analysis of Spectral Methods: Theory and Applications*. SIAM, Philadelfia, 1977.

30. M.E. Gurtin. *An Introduction to Continuum Mechanics*, volume 158 of *Mathematics in Science and Engineering*. Academic Press, New York, 1981.

31. J.B. Heywood. *Internal Combustion Engine Fundamentals*. McGraw Hill, New York, 1988.

32. A. Higdon, W.B. Stiles, A.W. Davis, C.R. Evces, and J.A. Weese. *Engineering Mechanics*. Prentice-Hall Inc., Englewood Cliffs, 1976.

33. T.J.R. Hughes. *The Finite Element Method: Linear Static and Dynamic Finite Element Analysis*. Prentice-Hall, New York, 1987.

34. J.R. Hutchinson. Shear coefficients for timoshenko beam theory. *J. Appl. Mech.*, pages 87–92, 2001.

35. G.E. Karniadakis and S.J. Sherwin. *Spectral/hp Element Methods for CFD, 2nd edition*. Oxford University Press, Oxford, 2nd edition, 2005.

36. W.M. Lai, D. Rubin, and E. Krempl. *Introduction to Continuum Mechanics*. Butterworth-Heinemann, Oxford, 3rd edition, 1993.

37. C. Lanczos. *The Variational Principles of Mechanics*. University of Toronto, Toronto, 1977.

38. L.E. Malvern. *Introduction to the Mechanics of a Continuous Medium*. Prentice-Hall, 1969.

39. J.T. Oden and J.N. Reddy. *Variational Methods in Theoretical Mechanics*. Springer, 2nd edition, 1983.

40. T.J. Oden and L.F. Demkowicz. *Applied Functional Analysis*. CRC Press, New York, 1996.

41. C.E.L. Pereira. Variational formulation and finite element approximation of the Kirchhoff and Reissner-Mindlin plates (in Portuguese). Master's thesis, University of Campinas, Campinas, February 2002.

42. E.P. Popov. *Engineering Mechanics of Solids*. Prentice-Hall, New York, 1st edition, 1990.

43. J. S. Przemieniecki. *Theory of Matrix Structural Analysis*. McGraw-Hill Book Company, New York, 1968.

44. J. N. Reddy. *Theory and Analysis of Elastic Plates and Shells*. CRC Press, 2006.

45. K. Rektorys. *Variational Methods in Mathematics, Science and Engineering*. Springer, Dordrecht, 2nd edition, 1980.

46. J. Shien and T. Tang. *Spectral and High-Order Methods with Applications*. Science Press, Beijing, China, 2006.

47. C.A.C. Silva. Object-oriented structural optimization and sensitivity analysis (in Portuguese). Master's thesis, University of Campinas, Campinas, 1997.

48. G. Strang and G. Fix. *An Analysis of the Finite Element Method*. Wellesley-Cambridge, New York, 2nd edition, 2008.

49. B. A. Szabó and I. Babuška. *Finite Element Analysis*. Wiley Interscience, New York, 1991.

50. S.P. Timoshenko. *History of Strength of Materials*. McGraw-Hill, New York, 1953.

51. S.P. Timoshenko and J.N. Goodier. *Theory of Elasticity*. McGraw-Hill, New York, 3rd edition, 1970.

52. Gilbert Touzot and Gouri Dahtt. *The Finite Element Displayed*. John Wiley & Sons, 1985.

53. O.C. Zienkiewicz and R.L. Taylor. *The Finite Element Method*. McGraw Hill, New York, 1989.

54. O.C. Zienkiewicz, J.Z. Zhu, and N.G. Gong. Effective and practical *h-p* version adaptive analysis procedures for the finite element method. *Int. J. Numer. Meth. in Eng.*, 28:879–891, 1989.

Index